WinkingSkull.com PLUS
Your study aid for must-know anatomy

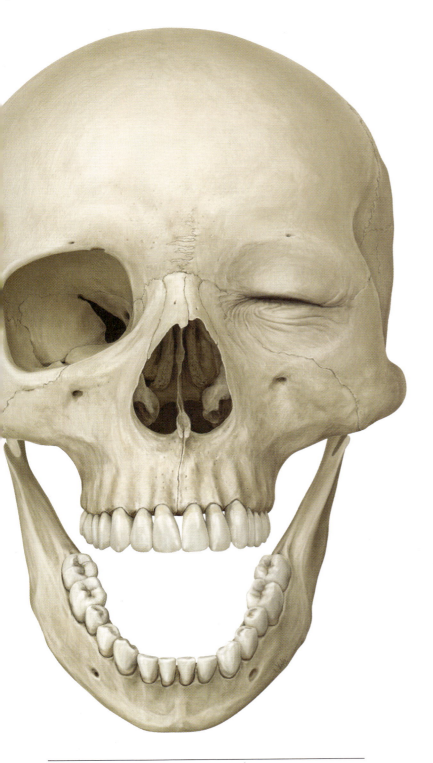

Register for WinkingSkull.com *PLUS* – master human anatomy with this unique interactive online learning tool.

Use the access code below to register for **WinkingSkull.com** *PLUS* and view over 500 full-color illustrations and radiographs from the book. After studying this invaluable image bank, you can quiz yourself on key body structures and get your score instantly to check your progress or to compare with other users' results.

WinkingSkull.com *PLUS* has everything you need for course study and exam prep:

- More than 500 full-color anatomy illustrations
- Intuitive design that simplifies navigation
- "Labels-on, labels-off" function that makes studying easy and fun
- Timed self-tests–with instant results

Simply visit WinkingSkull.com and follow these instructions to get started today.

If you do not already have a free WinkingSkull.com account, visit www.winkingskull.com, click on "Register Now," and complete the registration form. Enter the scratch-off code below.

If you already have a WinkingSkull.com account, go to the "My Account" page and click on the "Enter WinkingSkull PLUS Access Code" link. Enter the scratch-off code below.

This product cannot be returned if the access code panel is scratched off.

Some functionalities on WinkingSkull.com require support for advanced web technologies. A major browser (IE, Chrome, Firefox, Safari) within the last three major versions is suggested for use on the site.

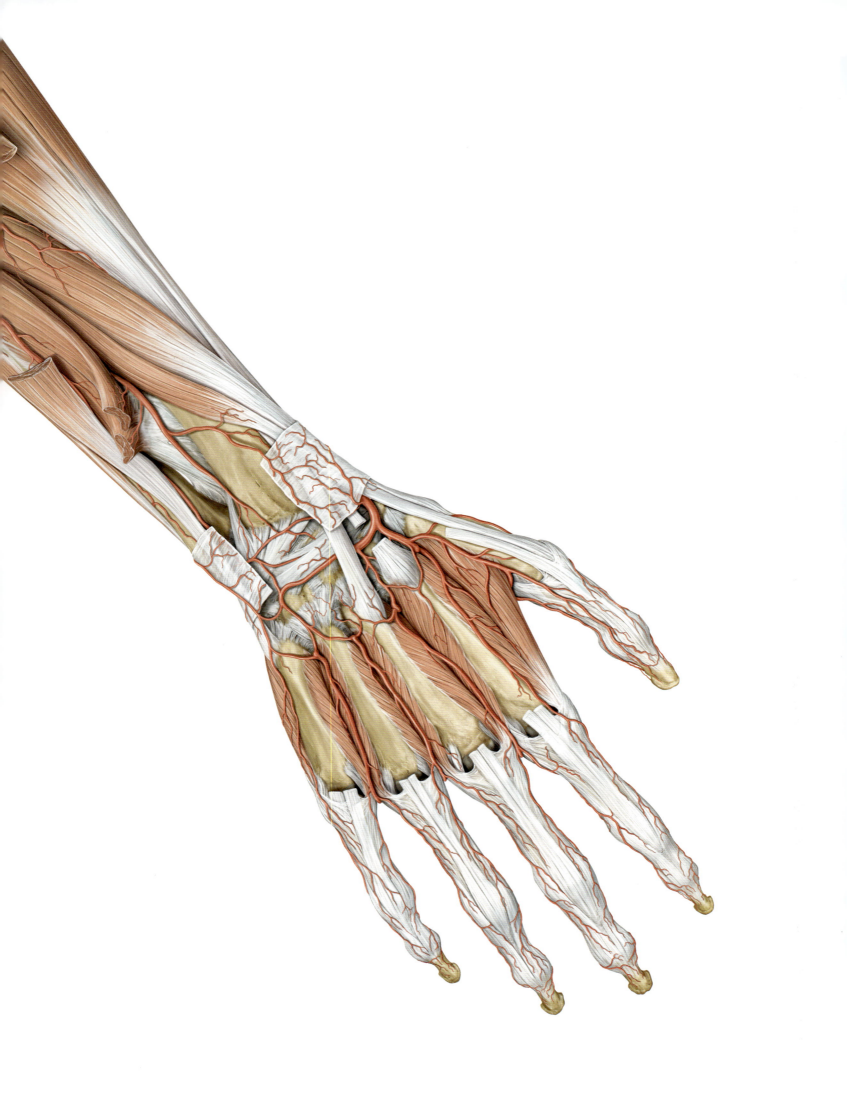

Volume 1

General Anatomy and Musculoskeletal System

THIEME Atlas of Anatomy
2nd Edition
Latin Nomenclature

Authors
Michael Schuenke, MD, PhD
Institute of Anatomy
Christian Albrecht University, Kiel

Erik Schulte, MD
Department of Anatomy and Cell Biology
Johannes Gutenberg University, Mainz

Udo Schumacher, MD
FRCPath, CBiol, FIBiol, DSc
Institute of Anatomy II: Experimental Morphology
University Medical Center, Hamburg-Eppendorf

Consulting Editor
Lawrence M. Ross, MD, PhD
Department of Neurobiology and Anatomy
The University of Texas Medical School at Houston

Consulting Editor, Latin Nomenclature
Hugo Zeberg, MD
Department of Neuroscience
Karolinska Institute, Stockholm

Illustrations by
Markus Voll
Karl Wesker

Thieme
New York • Stuttgart • Delhi • Rio de Janeiro

Editorial Director, Educational Products: Anne M. Sydor
Developmental Editor: Huvie Weinreich
Managing Editor: Judith Tomat
Director, Editorial Services: Mary Jo Casey
International Production Director: Andreas Schabert
Vice President, Editorial and E-Product Development: Vera Spillner
International Marketing Director: Fiona Henderson
International Sales Director: Louisa Turrell
Director of Sales, North America: Mike Roseman
Senior Vice President and Chief Operating Office: Sarah Vanderbilt
President: Brian D. Scanlan
Illustrations: Markus Voll and Karl Wesker
Compositor: Agnieszka & Martin Waletzko, Leonberg, Germany

Copyright ©2015 by Thieme Medical Publishers, Inc

Library of Congress Cataloging-in-Publication Data
Schünke, Michael, author.
  [Allgemeine Anatomie und Bewegungssystem. English]
  Thieme atlas of anatomy. General anatomy and musculoskeletal system / Michael Schuenke, Erik Schulte, Udo Schumacher ; illustrations by Markus Voll, Karl Wesker ; consulting editor, Lawrence M. Ross ; consulting editor, Latin nomenclature, Hugo Zeberg. -- 2nd edition, Latin nomenclature.
       p. ; cm.
  General anatomy and musculoskeletal system
  Atlas of anatomy
  Translation of the third edition of Allgemeine Anatomie und Bewegungssystem with Latin anatomical terms.
  Includes bibliographical references and index.
  Summary: "THIEME Atlas of Anatomy: General Anatomy and Musculoskeletal System, Second Edition, Latin Nomenclature is an ideal educational tool for anyone studying musculoskeletal anatomy. Each region is presented in a manner that builds understanding: starting with bones, joints, and muscles, then vasculature and nerves, and concluding with topographic illustrations. This atlas begins with a concise overview of general anatomy and moves on to the detailed anatomy of the trunk wall, upper limb, and lower limb. Key Features: Labels and anatomic terminology are in Latin nomenclature. Expanded coverage of tissue structure and development, functional testing, diagnostic imaging, and diseases of the musculoskeletal system. Exquisite full-color illustrations with clear, descriptive captions. Each two-page spread is a self-contained guide to a topic. Hundreds of clinical applications integrated into the anatomic descriptions, emphasizing the vital link between anatomic structure and function. Access to WinkingSkull.com PLUS, with over 500 images from the book for labels-on and labels-off review and timed self-tests. The THIEME Atlas of Anatomy series also features Neck and Internal Organs, and Head and Neuroanatomy. Each atlas is available in softcover"--Provided by publisher.
  ISBN 978-1-60406-923-5 (v. 1 : alk. paper) -- ISBN 978-1-62623-083-5 (ebook : v. 1)
  I. Schulte, Erik, author. II. Schumacher, Udo, author. III. Title. IV. Title: General anatomy and musculoskeletal system. V. Title: Atlas of anatomy.
  [DNLM: 1. Musculoskeletal System--anatomy & histology--Atlases. 2. Anatomy--Atlases. WE 17]
  QM25
  612.0022'3--dc23
                                                  2015010364

Printed in China by Everbest Printing Ltd   5 4 3 2 1

ISBN 978-1-60406-923-5

Also available as an e-book:
eISBN 978-1-62623-083-5

Thieme Publishers New York
333 Seventh Avenue, New York, NY 10001 USA
+1 800 782 3488, customerservice@thieme.com

Thieme Publishers Stuttgart
Rüdigerstrasse 14, 70469 Stuttgart, Germany
+49 [0]711 8931 421, customerservice@thieme.de

Thieme Publishers Delhi
A-12, Second Floor, Sector-2, Noida-201301
Uttar Pradesh, India
+91 120 45 566 00, customerservice@thieme.in

Thieme Publishers Rio, Thieme Publicações Ltda.
Edifício Rodolpho de Paoli, 25° andar
Av. Nilo Peçanha, 50 – Sala 2508,
Rio de Janeiro 20020-906 Brasil
+55 21 3172-2297 / +55 21 3172-1896

Important note: Medicine is an ever-changing science undergoing continual development. Research and clinical experience are continually expanding our knowledge, in particular our knowledge of proper treatment and drug therapy. Insofar as this book mentions any dosage or application, readers may rest assured that the authors, editors, and publishers have made every effort to ensure that such references are in accordance with **the state of knowledge at the time of production of the book**.

Nevertheless, this does not involve, imply, or express any guarantee or responsibility on the part of the publishers in respect to any dosage instructions and forms of applications stated in the book. **Every user is requested to examine carefully** the manufacturer's leaflets accompanying each drug and to check, if necessary in consultation with a physician or specialist, whether the dosage schedules mentioned therein or the contraindications stated by the manufacturers differ from the statements made in the present book. Such examination is particularly important with drugs that are either rarely used or have been newly released on the market. Every dosage schedule or every form of application used is entirely at the user's own risk and responsibility. The authors and publishers request every user to report to the publishers any discrepancies or inaccuracies noticed. If errors in this work are found after publication, errata will be posted at www.thieme.com on the product description page.

Some of the product names, patents, and registered designs referred to in this book are in fact registered trademarks or proprietary names even though specific reference to this fact is not always made in the text. Therefore, the appearance of a name without designation as proprietary is not to be construed as a representation by the publisher that it is in the public domain.

This book, including all parts thereof, is legally protected by copyright. Any use, exploitation, or commercialization outside the narrow limits set by copyright legislation, without the publisher's consent, is illegal and liable to prosecution. This applies in particular to photostat reproduction, copying, mimeographing, preparation of microfilms, and electronic data processing and storage.

Contents

Foreword by Lawrence M. Ross, Brian R. MacPherson, and Anne M. Gilroy........ XI
Preface to the Second Edition XIII
Preface to the First Edition XV
A Note on the Use of Latin Terminology XVII
Acknowledgments XIX

General Anatomy

1 Human Phylogeny and Ontogeny
1.1 Human Phylogeny 2
1.2 Human Ontogeny: Overview, Fertilization, and Earliest Developmental Stages 4
1.3 Human Ontogeny: Gastrulation, Neurulation, and Somite Formation 6
1.4 Human Ontogeny: Development of the Fetal Membranes and Placenta 8
1.5 Development of the Pharyngeal (Branchial) Arches in Humans 10
1.6 Early Embryonic Circulation and the Development of Major Blood Vessels 12
1.7 Primordial Skeleton, Prenatal Skeletal Development, and Ossification Centers 14
1.8 Bone Development and Remodeling 16
1.9 Ossification of the Limbs 18
1.10 Basic Skeletal Structure and Development and Position of the Limbs 20

2 Overview of the Human Body
2.1 The Human Body: Proportions, Surface Areas, and Body Weights 22
2.2 The Structural Design of the Human Body 24

3 Surface Anatomy of the Body, Landmarks, and Reference Lines
3.1 Terms of Location and Direction, Cardinal Planes and Axes 26
3.2 Location and Designation of Imaging Planes 28
3.3 Body Surface Anatomy 30
3.4 Body Surface Contours and Palpable Bony Prominences 32
3.5 Landmarks and Reference Lines on the Human Body 34
3.6 Body Regions (Regional Anatomy) 36

4 The Bones and Joints
4.1 The Bony Skeleton and the Structure of Tubular Bones 38
4.2 Continuous and Discontinuous Joints: Overview and False Joints (Synarthroses) 40
4.3 True Joints: Components and Intra- and Extra-articular Structures 42
4.4 True Joints: Structure of Joint Capsule and Hyaline Cartilage 44
4.5 Degenerative Joint Diseases as Exemplified by Hip Joint Arthrosis 46
4.6 Basic Principles of Joint Mechanics: Movements 48
4.7 Basic Principles of Joint Mechanics: Stability and Transmission 50
4.8 Fractures: Classification, Healing, and Treatment 52

5 The Muscles
5.1 Skeletal Muscles: Overview 54
5.2 Skeletal Muscles: Muscle Fiber Types and Pennate and Nonpennate Skeletal Muscles 56
5.3 Skeletal Muscles: Structure and Function 58
5.4 The Tendons and Mechanisms That Assist Muscle Function 60

6 The Vessels
6.1 Overview of the Human Cardiovascular System 62
6.2 The Structure of Arteries and Veins 64
6.3 The Terminal Vascular Bed 66

7 The Lymphatic System and Glands
7.1 The Human Lymphatic System 68
7.2 Exocrine and Endocrine Glands 70

8 General Neuroanatomy
8.1 Development of the Central Nervous System (CNS) 72
8.2 Neural Crest Derivatives and the Development of the Peripheral Nervous System (PNS) 74
8.3 Topography and Structure of the Nervous System 76
8.4 Cells of the Nervous System 78
8.5 Structure of a Spinal Cord Segment 80
8.6 Sensory Innervation: An Overview 82
8.7 Sensory Innervation: Principles of Dermatome and Plexus Formation 84
8.8 Sensory Innervation: Dermatomes and Cutaneous Nerve Territories 86
8.9 Motor Innervation: Organization of the Spinal Cord and Reflexes 88
8.10 Motor Innervation: Upper (First) and Lower (Second) Motor Neurons 90
8.11 Differences between the Central and Peripheral Nervous Systems 92
8.12 The Autonomic Nervous System 94
8.13 Lesions of Peripheral Nerves 96

Table of Contents

Trunk Wall

9 Bones, Ligaments, and Joints

- 9.1 The Skeleton of the Trunk ... 100
- 9.2 The Bony Spinal Column ... 102
- 9.3 Development of the Spinal Column ... 104
- 9.4 The Structure of a Vertebra ... 106
- 9.5 The Cervical Spine (Vertebrae cervicales) ... 108
- 9.6 The Thoracic Spine (Vertebrae thoracicae) ... 110
- 9.7 The Lumbar Spine (Vertebrae lumbales) ... 112
- 9.8 Os sacrum and os coccygis ... 114
- 9.9 The Intervertebral Disk (Disus intervertebralis): Structure and Function ... 116
- 9.10 The Ligaments of the Spinal Column (Columna vertebralis): Overview and Thoracolumbar Region ... 118
- 9.11 Overview of the Ligaments of the Cervical Spine ... 120
- 9.12 The Ligaments of the Upper Cervical Spine (Atlantooccipital and Atlantoaxial Joints) ... 122
- 9.13 The Intervertebral Facet Joints, Motion Segments, and Range of Motion in Different Spinal Regions ... 124
- 9.14 The Uncovertebral Joints of the Cervical Spine ... 126
- 9.15 Cross-sectional Anatomy of the Lumbar Spine ... 128
- 9.16 Degenerative Changes in the Lumbar Spine ... 130
- 9.17 The Thoracic Skeleton ... 132
- 9.18 The Sternum and Ribs (Costae) ... 134
- 9.19 The Costovertebral Joints and Thoracic Movements ... 136
- 9.20 The Bony Pelvis ... 138
- 9.21 The Pelvic Ligaments and Pelvic Measurements ... 140
- 9.22 The Sacroiliac Joint (Art. Sacroiliaca) ... 142

10 Musculature: Functional Groups

- 10.1 The Muscles of the Trunk Wall: Their Origin and Function ... 144
- 10.2 The Intrinsic Back Muscles (M. ercetor spinae): Lateral Tract ... 146
- 10.3 The Intrinsic Back Muscles (M. ercetor spinae): Medial Tract ... 148
- 10.4 The Intrinsic Back Muscles (Short Nuchal and Craniovertebral Joint Muscles) and the Prevertebral Muscles ... 150
- 10.5 The Muscles of the Abdominal Wall: Anterolateral Muscles ... 152
- 10.6 The Muscles of the Abdominal Wall: Anterior and Posterior Muscles ... 154
- 10.7 The Functions of the Abdominal Wall Muscles ... 156
- 10.8 The Muscles of the Thoracic Cage (Mm. intercostales, subcostales, scaleni, and M. transversus thoracis) ... 158
- 10.9 The Muscles of the Thoracic Cage: The Diaphragma ... 160
- 10.10 The Muscles of the Pelvic Floor and Perineum: Diaphragma pelvis and Deep and Superficial Perineal Muscles ... 162
- 10.11 Secondarily Incorporated Trunk Muscles: Spinocostal, Spinohumeral, and Thoracohumeral ... 164

11 Musculature: Topographical Anatomy

- 11.1 The Back Muscles and Fascia Thoracolumbalis ... 166
- 11.2 The Intrinsic Back Muscles: Lateral and Medial Tracts of the M. erector spinae ... 168
- 11.3 The Intrinsic Back Muscles: Short Nuchal Muscles ... 170
- 11.4 The Thoracic Wall Muscles and Fascia endothoracica ... 172
- 11.5 The Thoracoabdominal Junction: The Diaphragma ... 174
- 11.6 The Anterolateral and Anterior Abdominal Wall Muscles* ... 176
- 11.7 Structure of the Abdominal Wall and Rectus Sheath (Vagina musculi recti abdominis) ... 178
- 11.8 The Pelvic Floor Muscles: Overview of the Regio perinealis and Superficial Fasciae ... 180
- 11.9 Structure of the Pelvic Floor and Pelvic Spaces: Female versus Male ... 182
- 11.10 The Muscles of the Female Pelvic Floor and Wall ... 184
- 11.11 Pelvic Floor Muscles: M. levator ani ... 186
- 11.12 Pelvic Floor Muscles: Their Relation to Organs and Vessels in Males and Females ... 188

12 Neurovascular Systems: Forms and Relations

- 12.1 The Arteries ... 190
- 12.2 The Veins ... 192
- 12.3 The Lymphatic Vessels and Lymph Nodes ... 194
- 12.4 The Nerves ... 196

13 Neurovascular Systems: Topographical Anatomy

- 13.1 Anterior Trunk Wall: Surface Anatomy and Superficial Nerves and Vessels ... 198
- 13.2 Posterior Trunk Wall: Surface Anatomy and Superficial Nerves and Vessels ... 200
- 13.3 Posterior Trunk Wall, Posterior View ... 202
- 13.4 Posterior Trunk Wall, Anterior View ... 204
- 13.5 Anterior Trunk Wall: Overview and Location of Clinically Important Nerves and Vessels ... 206
- 13.6 Anterior Trunk Wall: Nerves, Blood Vessels, and Lymphatics in the Breast ... 208
- 13.7 Anterior Trunk Wall: The Inguinal Canal (Canalis inguinalis) ... 210
- 13.8 Anterior Abdominal Wall: Anatomy and Weak Spots ... 212
- 13.9 Inguinal and Femoral Hernias ... 214
- 13.10 Topographic Anatomy of Inguinal Hernias ... 216
- 13.11 Diagnosis and Treatment of Hernias ... 218
- 13.12 Rare External Hernias ... 220
- 13.13 Development of the External Genitalia ... 222
- 13.14 Male External Genitalia: Descensus testis and Funiculus spermaticus ... 224
- 13.15 Male External Genitalia: The Testis and Epididymis ... 226
- 13.16 Male External Genitalia: The Fasciae and Erectile Tissues of the Penis ... 228

13.17	Male External Genitalia: Nerves and Vessels of the Penis........... 230		15.5	The Posterior Muscles of the Shoulder Joint: M. deltoideus........... 304
13.18	Female External Genitalia: Overview and Episiotomy.... 232		15.6	The Posterior Muscles of the Shoulder Joint: Mm. latissimus dorsi and teres major........... 306
13.19	Female External Genitalia: Neurovascular Structures, Erectile Tissues, Erectile Muscles, and Vestibule........ 234		15.7	The Anterior Muscles of the Shoulder Joint: Mm. pectoralis major and coracobrachialis........... 308

Upper Limb

14 Bones, Ligaments, and Joints

14.1	The Upper Limb as a Whole....... 238
14.2	Integration of the Shoulder Girdle into the Skeleton of the Trunk....... 240
14.3	The Bones of the Shoulder Girdle....... 242
14.4	The Bones of the Upper Limb: The Humerus....... 244
14.5	The Bones of the Upper Limb: Torsion of the Humerus... 246
14.6	The Bones of the Upper Limb: The Radius and Ulna..... 248
14.7	The Bones of the Upper Limb: The Articular Surfaces of the Radius and Ulna....... 250
14.8	The Bones of the Upper Limb: The Hand....... 252
14.9	The Bones of the Upper Limb: The Carpal Bones....... 254
14.10	Architecture of the Radiocarpal Junction and the Metacarpus; Distal Radius and Scaphoid Fractures....... 256
14.11	The Joints of the Shoulder: Overview and Clavicular Joints....... 258
14.12	The Joints of the Shoulder: Ligaments of the Clavicular and Scapulothoracic Joints....... 260
14.13	The Joints of the Shoulder: Articular Surfaces, Articular Capsule, and Articular Cavity of the Articulatio humeri.. 262
14.14	The Joints of the Shoulder: Ligaments, Reinforcing Capsule, and Rotator Interval of the Articulatio humeri.. 264
14.15	The Joints of the Shoulder: The Subacromial Space..... 266
14.16	The Bursa subacromialis and Bursa subdeltoidea....... 268
14.17	Shoulder Arthroscopy....... 270
14.18	X-Ray and Cross-sectional Anatomy of the Shoulder.... 272
14.19	Movements of the Shoulder Girdle and Shoulder Joint... 274
14.20	The Elbow Joint (Articulatio cubiti) as a Whole......... 276
14.21	The Elbow Joint (Articulatio cubiti): Capsule and Ligaments....... 278
14.22	The Forearm: Proximal and Distal Radioulnar Joints..... 280
14.23	Movements of the Elbow and Radioulnar Joints (Articulationes radioulnares)....... 282
14.24	Overview of the Ligaments of the Hand........ 284
14.25	Intrinsic Ligaments of the Hand, Compartments of the Joint, and Ulnocarpal Complex....... 286
14.26	The Carpal Tunnel (Canalis carpi)....... 288
14.27	The Ligaments of the Fingers....... 290
14.28	The Carpometacarpal Joint of the Thumb (Articulatio carpometacarpalis pollicis)....... 292
14.29	Movements of the Hand and Finger Joints....... 294

15 Musculature: Functional Groups

15.1	Functional Muscle Groups....... 296
15.2	The Muscles of the Shoulder Girdle: Mm. trapezius, sternocleidomastoideus, and omohyoideus....... 298
15.3	The Muscles of the Shoulder Girdle: Mm. serratus anterior, subclavius, pectoralis minor, levator scapulae, and rhomboidei major and minor....... 300
15.4	The Posterior Muscles of the Shoulder Joint: The Rotator Cuff....... 302
15.5	The Posterior Muscles of the Shoulder Joint: M. deltoideus....... 304
15.6	The Posterior Muscles of the Shoulder Joint: Mm. latissimus dorsi and teres major....... 306
15.7	The Anterior Muscles of the Shoulder Joint: Mm. pectoralis major and coracobrachialis....... 308
15.8	The Anterior Muscles of the Arm: Mm. biceps brachii and brachialis....... 310
15.9	The Posterior Muscles of the Arm: Mm. triceps brachii and anconeus....... 312
15.10	The Muscles of the Anterior Forearm: The Superficial and Deep Flexors....... 314
15.11	The Muscles of the Forearm: The Radialis Muscles...... 316
15.12	The Muscles of the Posterior Forearm: The Superficial and Deep Extensors....... 318
15.13	The Intrinsic Muscles of the Hand: The Thenar and Hypothenar Muscles....... 320
15.14	The Intrinsic Muscles of the Hand: Mm. lumbricales and interossei (Metacarpal Muscles)....... 322
15.15	Overview of Muscle Functions: Shoulder Joint (Art. humeri)....... 324
15.16	Overview of Muscle Functions: Elbow Joint....... 326
15.17	Overview of Muscle Functions: Wrist....... 328

16 Musculature: Topographical Anatomy

16.1	The Posterior Muscles of the Shoulder Girdle and Shoulder Joint....... 330
16.2	The Posterior Muscles of the Shoulder Joint and Arm.... 332
16.3	The Anterior Muscles of the Shoulder Girdle and Shoulder Joint....... 334
16.4	The Muscles of the Shoulder Joint and Anterior Arm.... 336
16.5	The Muscles of the Anterior Forearm....... 338
16.6	The Muscles of the Posterior Forearm....... 340
16.7	Cross-sectional Anatomy of the Arm and Forearm....... 342
16.8	The Tendon Sheaths of the Hand....... 344
16.9	The Dorsal Digital Expansion....... 346
16.10	The Intrinsic Muscles of the Hand: Superficial Layer..... 348
16.11	The Intrinsic Muscles of the Hand: Middle Layer........ 350
16.12	The Intrinsic Muscles of the Hand: Deep Layer........ 352

17 Neurovascular Systems: Forms and Relations

17.1	The Arteries....... 354
17.2	The Veins....... 356
17.3	The Lymphatic Vessels and Lymph Nodes....... 358
17.4	The Plexus brachialis: Structure....... 360
17.5	Supraclavicular Part of the Plexus brachialis (Pars supraclavicularis)....... 362
17.6	Pars infraclavicularis of the Plexus brachialis: Overview and Short Branches....... 364
17.7	Pars infraclavicularis of the Plexus brachialis: N. musculocutaneus and N. axillaris....... 366
17.8	Pars infraclavicularis of the Plexus brachialis: N. radialis.. 368
17.9	Pars infraclavicularis of the Plexus brachialis: N. ulnaris.. 370
17.10	Pars infraclavicularis of the Plexus brachialis: N. medianus....... 372

18 Neurovascular Systems: Topographical Anatomy

18.1	Surface Anatomy and Superficial Nerves and Vessels: Anterior View....... 374
18.2	Surface Anatomy and Superficial Nerves and Vessels: Posterior View....... 376

Table of Contents

18.3	The Shoulder Region: Anterior View.	378
18.4	The Axilla (Regio axillaris): Anterior Wall	380
18.5	The Axilla (Regio axillaris): Posterior Wall	382
18.6	Conduction Anesthesia of the Plexus brachialis: Principle, Pathways, and Administering the Block	384
18.7	The Anterior Brachial Region (Regio brachialis anterior).	386
18.8	The Shoulder Region: Posterior and Superior Views	388
18.9	The Posterior Brachial Region (Regio brachialis posterior)	390
18.10	The Elbow (Regio cubitalis).	392
18.11	The Anterior Forearm Region (Regio antebrachialis anterior)	394
18.12	The Posterior Forearm Region (Regio antebrachialis posterior) and the Dorsum of the Hand (Dorsum manus).	396
18.13	The Palm of the Hand (Palma manus): Epifascial Nerves and Vessels	398
18.14	The Palm of the Hand (Palma manus): Vascular Supply	400
18.15	The Carpal Tunnel (Canalis carpi)	402
18.16	The Ulnar Tunnel and Anterior Carpal Region (Regio carpalis anterior)	404

Lower Limb

19 Bones, Ligaments, and Joints

19.1	The Lower Limb: General Aspects	408
19.2	The Anatomic and Mechanical Axes of the Lower Limb	410
19.3	The Bones of the Pelvic Girdle	412
19.4	The Femur (Os femoris): Importance of the Femoral Neck Angle	414
19.5	The Femoral Head and Deformities of the Femoral Neck	416
19.6	The Patella	418
19.7	The Tibia and Fibula	420
19.8	The Bones of the Foot from the Dorsal and Plantar Views	422
19.9	The Bones of the Foot from the Lateral and Medial Views; Accessory Tarsal Bones	424
19.10	The Hip Joint (Articulatio coxae): Articulating Bones	426
19.11	The Ligaments of the Hip Joint: Stabilization of the Femoral Head (Caput femoris)	428
19.12	The Ligaments of the Hip Joint: Nutrition of the Femoral Head (Caput femoris)	430
19.13	Cross-sectional and X-Ray Anatomy of the Hip Joint. Typical Medical Condition of the Elderly: Femoral Neck Fractures	432
19.14	Cross-sectional Anatomy of the Hip Joint (Art. coxae): Sonographic Representation of Hip Joint Effusion	434
19.15	The Movements and Biomechanics of the Hip Joint	436
19.16	The Development of the Hip Joint	438
19.17	The Knee Joint (Articulatio genus): Articulating Bones	440
19.18	The Ligaments of the Knee Joint: An Overview	442
19.19	The Knee Joint: The Cruciate and Collateral Ligaments (Ligg. cruciata et collateralia)	444
19.20	The Knee Joint: The Menisci	446
19.21	The Movements of the Knee Joint	448
19.22	The Knee Joint (Art. genus): Capsule and Joint Cavity	450
19.23	Cross-sectional Anatomy of the Knee	452
19.24	The Joints of the Foot: Overview of the Articulating Bones and Joints	454
19.25	The Joints of the Foot: Articular Surfaces	456
19.26	The Joints of the Foot: The Talocrural and Subtalar Joints	458
19.27	The Ligaments of the Foot	460
19.28	The Movements of the Foot	462
19.29	Overview of the Plantar Vault and the Transverse Arch (Arcus pedis transversus)	464
19.30	The Longitudinal Arch (Arcus pedis longitudinalis) of the Foot	466
19.31	The Sesamoid Bones (Ossa sesamoidea) and Deformities of the Toes	468
19.32	X-Ray and Cross-sectional Anatomy of the Foot	470
19.33	Human Gait	472

20 Musculature: Functional Groups

20.1	The Muscles of the Lower Limb: Classification	474
20.2	The Hip and Gluteal Muscles: The Inner Hip Muscles	476
20.3	The Hip and Gluteal Muscles: The Outer Hip Muscles	478
20.4	The Hip and Gluteal Muscles: The Adductor Group	480
20.5	The Anterior Thigh Muscles: The Extensor Group	482
20.6	The Posterior Thigh Muscles: The Flexor Group	484
20.7	The Leg Muscles: The Anterior and Lateral Compartments (Extensor and Fibularis* Group)	486
20.8	The Leg Muscles: The Posterior Compartment (Superficial Flexor Group)	488
20.9	The Leg Muscles: The Posterior Compartment (Deep Flexor Group)	490
20.10	The Short Muscles of the Foot: Dorsum, and Medial and Lateral Compartments of the Plantar Surface	492
20.11	The Short Muscles of the Foot: Central Compartment of the Plantar Surface	494
20.12	Overview of Muscle Functions: Hip Joint	496
20.13	Overview of Muscle Functions: Knee Joint	498
20.14	Overview of Muscle Functions: Ankle Joints	500

21 Musculature: Topographical Anatomy

21.1	The Muscles of the Medial and Anterior Thigh, Hip, and Gluteal Region	502
21.2	The Muscles of the Anterior Thigh, Hip, and Gluteal Region: Origins and Insertions	504
21.3	The Muscles of the Lateral and Posterior Thigh, Hip, and Gluteal Region	506
21.4	The Muscles of the Posterior Thigh, Hip, and Gluteal Region: Origins and Insertions	508
21.5	The Muscles of the Lateral and Anterior Leg: Origins and Insertions	510

21.6	The Muscles of the Posterior Leg: Origins and Insertions 512		23	**Neurovascular Systems: Topographical Anatomy**	
21.7	The Tendon Sheaths and Retinacula of the Foot 514		23.1	Surface Anatomy and Superficial Nerves and Vessels: Anterior View 544	
21.8	The Intrinsic Foot Muscles from the Plantar View: The Aponeurosis plantaris and Superficial Layer 516		23.2	Surface Anatomy and Superficial Nerves and Vessels: Posterior View 546	
21.9	The Intrinsic Foot Muscles from the Plantar View: Middle Layer 518		23.3	The Anterior Femoral Region (Regio femoralis anterior) Including the Trigonum femorale 548	
21.10	The Intrinsic Foot Muscles from the Plantar View: Deep Layer and Origins and Insertions................. 520		23.4	Arterial Supply to the Thigh 550	
21.11	Cross-sectional Anatomy of the Thigh, Leg, and Foot.... 522		23.5	The Gluteal Region (Regio glutealis): Overview of Its Vessels and Nerves..................... 552	
22	**Neurovascular Systems: Forms and Relations**		23.6	The Gluteal Region (Regio glutealis): The Foramina ischiadica and N. ischiadicus 554	
22.1	The Arteries 524		23.7	The Fossa ischioanalis 556	
22.2	The Veins.. 526		23.8	The Canalis pudendalis and Regio perinealis (Regio urogenitalis and Regio analis) 558	
22.3	The Lymphatic Vessels and Lymph Nodes 528				
22.4	The Structure of the Plexus lumbosacralis............. 530				
22.5	The Nerves of the Plexus lumbalis: Nn. iliohypogastricus, ilioinguinalis, genitofemoralis, and cutaneus femoris lateralis 532		23.9	The Posterior Thigh Region (Regio femoris posterior) and Popliteal Region (Regio genus posterior) 560	
22.6	The Nerves of the Plexus lumbalis: Nn. obturatorius and femoralis 534		23.10	The Posterior Leg Region (Regio cruris posterior) and the Tarsal Tunnel (Canalis tarsalis)................ 562	
22.7	The Nerves of the Plexus sacralis: N. gluteus superior, N. gluteus inferior, and N. cutaneus femoris posterior ... 536		23.11	The Sole of the Foot (Planta pedis) 564	
22.8	The Nerves of the Plexus sacralis: N. ischiadicus (Overview and Sensory Distribution) 538		23.12	The Anterior Leg Region and Dorsum of the Foot (Regio cruris anterior and Dorsum pedis): Cutaneous Innervation 566	
22.9	The Nerves of the Plexus sacralis: N. ischiadicus (Course and Motor Distribution).................... 540		23.13	The Arteries of the Dorsum of the Foot (Dorsum pedis)..................................... 568	
22.10	The Nerves of the Plexus sacralis: N. pudendus and N. coccygeus 542				

References 573
Index 575

Foreword

Each of the authors of the single volume *Thieme Atlas of Anatomy* was impressed with the extraordinary detail, accuracy, and beauty of the illustrations that were created for the *Thieme* three volume series of anatomy atlases. We felt these images were one of the most significant additions to anatomic education in the past 50 years. The effective pedagogical approach of this series, with two-page learning units that combined the outstanding illustrations and captions that emphasized the functional and clinical significance of structures, coupled with the numerous tables sum-marizing key information, was unique. We also felt that the overall organization of each region, with structures presented first systemically - musculoskeletal, vascular, and nervous – and then topographically, supported classroom learning and active dissection in the laboratory.

This series combines the best of a clinically oriented text and an atlas. Its detail and pedagogical presentation make it a complete support for classroom and laboratory instruction and a reference for life in all the medical, dental and allied health fields. Each of the volumes - *General Anatomy and Musculoskeletal System, Neck and Internal Organs, and Head and Neuroanatomy* - can also be used as a stand-alone text/atlas for an in-depth study of systems often involved in the allied health/medical specialty fields.

We were delighted when *Thieme* asked us to work with them to create a single-volume atlas from this groundbreaking series, and we owe a great debt to the authors and illustrators of this series in as much as their materials and vision formed the general framework for the single volume *Thieme Atlas of Anatomy*.

We thank the authors and illustrators for this very special contribution to the teaching of anatomy and recommend it for thorough mastery of anatomy and its clinically functional importance in all fields of health care-related specialties.

Lawrence M. Ross, Brian R. MacPherson, and *Anne M. Gilroy*

Preface to the Second Edition

Six years have passed since the first edition of the *Thieme Atlas of Anatomy: General Anatomy and Musculoskeletal System* was published. It has passed its first test and met the needs of students and practitioners everywhere, as evidenced by the many letters and e-mails we have received. We thank you for your praise and constructive criticism, which helps us keep improving this atlas.

Clinical knowledge presented in conjunction with anatomy is increasingly important earlier and earlier in the study of medicine. This has been further strengthened in this edition with the inclusion of about 30 new two-page spreads across the book devoted to
- osteoarthritis of the hip joint,
- compression syndromes of peripheral nerves,
- conduction anesthesia of peripheral nerves,
- shoulder arthroscopy and degenerative changes of the shoulder joint,
- functions of individual muscles and the symptoms associated with shortening or weakening of these muscles, and
- diagnostic imaging of the large joints, such as the shoulder, elbow, and wrist, and the hip, knee, and ankle.

In addition, we have added spreads on important foundational information on the common imaging planes for plain film, MRI, and CT scans, the structure of skeletal muscle fibers, the structure and chemical composition of hyaline cartilage, and the regeneration of peripheral nerves.

We have also checked, corrected, and updated all the information in this atlas.

With these improvements, this atlas is even better suited to students of medicine in what the World Health Organization (WHO) is again calling the "Decade of Bones and Joints" (first 2000 to 2010 and now 2010 to 2020) to draw attention to the continuing prominence and dramatic rise of diseases of the musculoskeletal system with the rise in average life expectancy worldwide. Today over half the chronic diseases of those over 60 involve the bones (e.g., osteoporosis) and joints (e.g., osteoarthritis), with tremendous economic consequences. One of the main reasons WHO is publicizing this is so that the world's universities appropriately prepare physicians, physical therapists, and other health care workers to address the growing global burden of these diseases due to the aging population.

This atlas emphasizes the correlations between physiologic changes in the course of life, the frequency of certain pathologic phenomena, and effective diagnostics while teaching the anatomy, better preparing students to treat patients with musculoskeletal diseases when they meet them in the clinic or in practice. When an elderly person suffers a fracture, it is not sufficient to just address the fracture. The doctor must learn why the fracture happened and address the underlying cause. Does, for example, the patient have osteoporosis, or is he or she so inflexible that any unexpected need to move leads to a fall? Interdisciplinary cooperation is needed to address these causes and provide appropriate preventive and rehabilitative care. The older we get, the more important it is for us to keep the musculoskeletal system in motion to curb degenerative disease and prevent injury.

This atlas, we hope, continues to meet your needs in the classroom and clinic, helps you attain a more nuanced understanding of the anatomy of the musculoskeletal system, and brings the fascination of anatomy in motion home to you.

Our special thanks to Prof. Dr. Cristoph Viebahn, Georg-August University, Göttingen, and Prof. Dr. Thilo Wedel, Christian-Albrechts University, Kiel, for their commitment to and constructive help on the new edition.

Michael Schuenke, Erik Schulte, Udo Schumacher,
Markus Voll, and *Karl Wesker*
Kiel, Mainz, Hamburg, Munich, and Berlin

Preface to the First Edition

When Thieme started planning this atlas, they sought the opinions of students and instructors in both the United States and Europe on what constituted an "ideal" atlas of anatomy —ideal to learn from, to master extensive amounts of information while on a busy class schedule, and, in the process, to acquire sound, up-to-date knowledge. The result of our work in response to what Thieme learned is this atlas. The *Thieme Atlas of Anatomy*, unlike most other atlases, is a comprehensive educational tool that combines illustrations with explanatory text and summary tables, introducing clinical applications throughout, and presenting anatomic concepts in a step-by-step sequence that includes system-by-system and topographical views.

Since the *Thieme Atlas of Anatomy* is based on a fresh approach to the underlying subject matter, it was necessary to create an entirely new set of illustrations for it—a task that took eight years. Our goal was to provide illustrations that would compellingly demonstrate anatomic relations and concepts, revealing the underlying simplicity of human anatomy without sacrificing detail or aesthetics.

With the *Thieme Atlas of Anatomy*, it was our intention to create an atlas that would guide students in their initial study of anatomy, stimulate their enthusiasm for this intriguing and vitally important subject, and provide a reliable reference for experienced students and professionals alike.

"If you want to attain the possible, you must attempt the impossible" (Rabindranath Tagore).

Michael Schuenke, Erik Schulte, Udo Schumacher,
Markus Voll, and *Karl Wesker*

A Note on the Use of Latin Terminology

To introduce the Latin nomenclature into an English-language textbook is a delicate task, particularly because many Latin loanwords have passed into general use. Some loanwords are so common that fluency of the text would be disturbed if they were to be translated back into Latin. These Latin loanwords have typically undergone several adaptations before becoming part of the English language. A term such as *sympathetic trunk* (lat. *truncus sympaticus*) has undergone morphological adaptation (through the loss of masculine suffix *-us*), orthographical adaptation (through the substitution of a "Germanic" *k* for a Latin *c*), and phonological adaptation (*th* and *e* instead of *t* and *i*). In addition, the word order has been reversed. The Latin term *sympaticus* is in fact borrowed from the late Greek word *sympathetikos* (from *sympathes* "having a fellow feeling, affected by like feelings"), thereby illustrating that words move between languages when cultures meet. Other anatomical terms are so colloquial (e.g. *hand*), that a Latin word (e.g. *manus*) would be inappropriate to use at all occasions. Clearly, the text would become unreadable if a strict translation of all English terms into Latin were imposed.

As a result, Latin has been used as long as it does not disrupt the flow of the text and whenever possible in figures and tables. In some cases, dual terminology has been used, with either the English or Latin word in parentheses. As much as possible, the terminology of *Terminolgia Anatomica* (1998) has been followed.

Hugo Zeberg

Acknowledgments

First we wish to thank our families. This atlas is dedicated to them.

We also thank Prof. Reinhard Gossrau, M.D., for his critical comments and suggestions. We are grateful to several colleagues who rendered valuable help in proofreading: Mrs. Gabriele Schünke, Jakob Fay, M.D., Ms. Claudia Dücker, Ms. Simin Rassouli, Ms. Heinke Teichmann, and Ms. Sylvia Zilles. We are also grateful to Dr. Julia Jürns-Kuhnke for helping with the figure labels.

We extend special thanks to Stephanie Gay and Bert Sender, who prepared the layouts. Their ability to arrange the text and illustrations on facing pages for maximum clarity has contributed greatly to the quality of the atlas.

We particularly acknowledge the efforts of those who handled this project on the publishing side: Jürgen Lüthje, M.D., Ph.D., executive editor at Thieme Medical Publishers, has "made the impossible possible." He not only reconciled the wishes of the authors and artists with the demands of reality but also managed to keep a team of five people working together for years on a project whose goal was known to us from the beginning but whose full dimensions we only came to appreciate over time. He is deserving of our most sincere and heartfelt thanks.

Sabine Bartl, developmental editor, became a touchstone for the authors in the best sense of the word. She was able to determine whether a beginning student, and thus one who is not (yet) a professional, could clearly appreciate the logic of the presentation. The authors are indebted to her.

We are grateful to Antje Bühl, who was there from the beginning as project assistant, working "behind the scenes" on numerous tasks such as repeated proofreading and helping to arrange the figure labels.

We owe a great debt of thanks to Martin Spencker, managing director of Educational Publications at Thieme, especially to his ability to make quick and unconventional decisions when dealing with problems and uncertainties. His openness to all the concerns of the authors and artists established conditions for a cooperative partnership.

Without exception, our collaboration with the entire staff at Thieme Medical Publishers was consistently pleasant and cordial. Unfortunately, we do not have room to list everyone who helped in the publication of the *Atlas*, and we must limit our acknowledgments to a few colleagues who made a particularly notable contribution: Rainer Zepf and Martin Waletzko for support in all technical matters; Susanne Tochtermann-Wenzel and Manfred Lehnert, representing all those who were involved in the production of the book; Almut Leopold for the index; Marie-Luise Kürschner and her team for creating the cover design; to Liesa Arendt, Birgit Carlsen, and Anne Döbler, representing all those who handled marketing, sales, and promotion.

The Authors

As consulting editor I was asked to review, for accuracy and appropriateness, the English translation of the *Thieme Atlas of Anatomy: General Anatomy and Musculoskeletal System*, second edition. My work involved a review and edit of the translation, conversion of nomenclature to terms in common usage in English, and some small changes in presentation to reflect accepted approaches to certain anatomic structures in North American anatomy programs. This task was eased greatly by the clear organization of the original text. In all of this, I have tried diligently to remain faithful to the intentions and insights of the authors and illustrators, whom I wish to thank for this outstanding revision.

I would also like to thank the team at Thieme Medical Publishers who worked with me. First, I wish to thank translators Terry Telger and Judith Tomat for their work with the new and altered text and Anne Vinnicombe, editorial director for Educational Products, and editorial assistants, Shira Kaye and Huvie Weinreich, for their work with the translator and for checking and correcting my work and preparing this volume with care and speed.

Finally, heartfelt thanks go to Cathrin E. Schulz, M.D., editorial consultant, for her invitation to work on the first edition of this series and her assistance and constant encouragement at that time.

Lawrence M. Ross

It has been a great honor to act as a consulting editor, with responsibility for the Latin nomenclature, for *Thieme Atlas of Anatomy: General Anatomy and Musculoskeletal System*, Second Edition. There were several people from whom I received a great deal of assistance and guidance, and must express my gratitude towards. Regarding the discussion of nomenclature, I would wish to thank my mentor Prof. Peter Århem, Ph.D., my father Lennart Zeberg, M.D., and Prof. Jonas Broman, Ph.D. In addition, I would also like to express my gratitude to Prof. Björn Meister, M.D., Ph.D., for putting forward my name for this task.

Moreover, I am deeply grateful to the staff at Thieme Medical Publishers that I have been in close contact with, in particular, the editorial director Anne Sydor, Ph.D., managing editor Judith Tomat, editorial assistant Huvie Weinreich, and marketing agent David Towle.

I would also like to acknowledge the Federative International Programme for Anatomical Terminology (FIPAT) for their work towards a standard nomenclature in the field of anatomy.

Hugo Zeberg

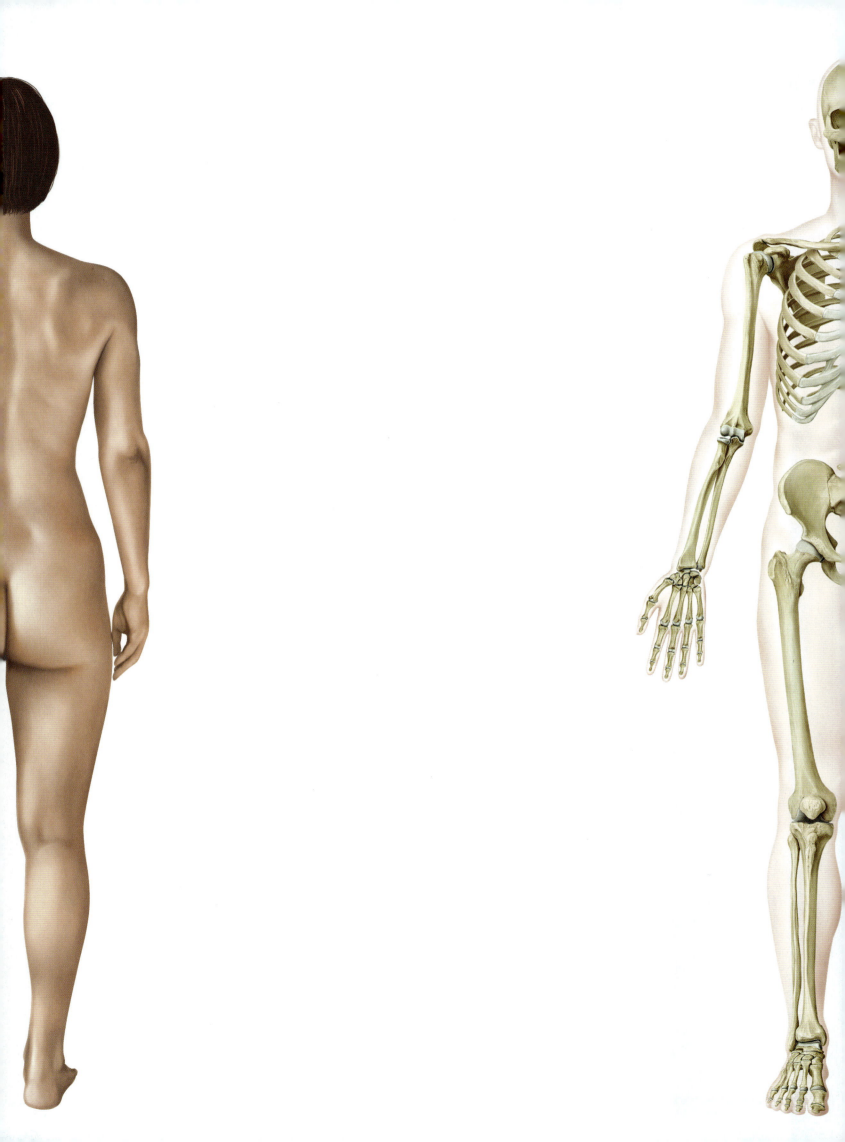

General Anatomy

1 Human Phylogeny and Ontogeny 2
2 Overview of the Human Body 22
3 Surface Anatomy of the Body, Landmarks, and Reference Lines 26
4 The Bones and Joints 38
5 The Muscles 54
6 The Vessels 62
7 The Lymphatic System and Glands 68
8 General Neuroanatomy 72

1.1 Human Phylogeny

A Brief overview of human phylogenetic development
To better understand the evolution of the human body, it is helpful to trace its phylogenetic development. Humans and their closest relatives belong to the **phylum Chordata**, which includes approximately 50,000 species. It consists of two subphyla:

- Invertebrata: the tunicates (Tunicata) and chordates without a true skull (Acraniata or Cephalochordata)
- Vertebrata: the vertebrates (animals that have a vertebral column)

Although some members of the chordate phylum differ markedly from one another in appearance, they are distinguished from all other animals by characteristic morphological structures that are present at some time during the life of the animal, if only during embryonic development (see **G**). Invertebrate chordates, such as the cephalochordates and their best-known species, the lancelet (*Branchiostoma lanceolatum*) are considered the *model of a primitive vertebrate* by virtue of their organization. They provide clues to the basic structure of the vertebrate body and thus are important in understanding the general organization of vertebrate organisms (see **D**).

All the **members of present-day vertebrate classes** (jawless fish, cartilaginous fish, bony fish, amphibians, reptiles, birds, and mammals) have a number of characteristic features in common (see **H**), including a row of vertebrae arranged in a *vertebral column (columna vertebralis)*, which gives the subphylum its name (Vertebrata). The evolution of an *amniotic egg*, i.e., the development of the embryo within a fixed shell inside a fluid-filled amniotic cavity, was a critical evolutionary breakthrough that helped the vertebrates to survive on land. This reproductive adaptation enabled the terrestrial vertebrates (reptiles, birds, and mammals) to live out their life cycles entirely on land and sever the final ties with their marine origin. When we compare the embryos of different vertebrate classes, we observe a number of morphological and functional similarities, including the formation of branchial arches (see **B**).

Mammals comprise **three major groups**: Monotremata (egg-laying mammals), Marsupialia (mammals with pouches), and Placentalia (mammals with a placenta). The placental mammals, which include humans, have a number of characteristic features (see **I**), including a tendency to invest much greater energy in the care and rearing of their young. Placental mammals complete their embryonic development inside the uterus and are connected to the mother by a placenta. Humans belong to the mammalian order of **primates**, whose earliest members were presumably small tree-dwelling mammals. Together with lemurs, monkeys, and the higher apes, human beings have features that originate from the early adaptation to an arboreal way of life. For example, primates have movable shoulder joints that enable them to climb in a hanging position while swinging from branch to branch. They have dexterous hands for grasping branches and manipulating food, and they have binocular, broadly overlapping visual fields for excellent depth perception.

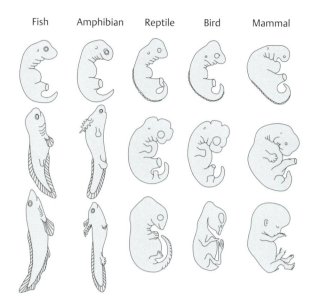

B Different stages in the early embryonic development of vertebrates
The early developmental stages (top row) of fish, amphibians, reptiles, birds, and mammals (as represented by humans) present a series of striking similarities that suggest a common evolutionary origin. One particularly noteworthy common feature is the set of branchial or pharyngeal arches in the embryonic regions that will develop into the head and neck. Although it was once thought that the developing embryo of a specific vertebrate would sequentially display features from organisms representing every previous step in its evolution ("Ontogeny recapitulates phylogeny," the "biogenetic law" of Ernst Haeckel [1834–1919]), subsequent work has shown that the vertebrates share common embryonic components that have been adapted to produce sometimes similar (fins and limbs) and sometimes radically different (gills vs. neck cartilages) adult structures.

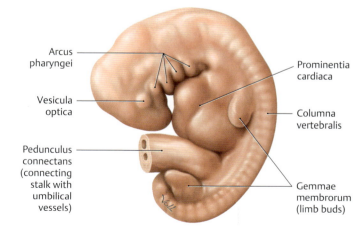

C Formation of the branchial or pharyngeal arches in a 5-week-old human embryo
Left lateral view. The branchial or pharyngeal arches (arcus pharyngei) of the vertebrate embryo have a *metameric* arrangement (similar to the somites, the primitive segments of the embryonic mesoderm); this means that they are organized into a series of segments that have the same basic structure. Among their other functions, they provide the raw material for the species-specific development of the visceral skeleton (maxilla, mandibula, auris media (middle ear), os hyoideum, and larynx), the associated facial muscles, and the pharyngeal gut (see p. 11).

General Anatomy — 1. Human Phylogeny and Ontogeny

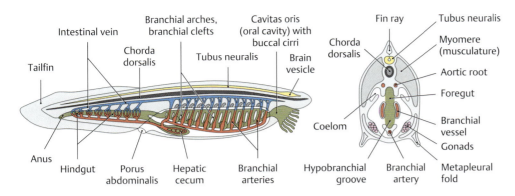

G Characteristic features of chordates

- Development of an axial skeleton (skeleton axiale, chorda dorsalis)
- Dorsal neural tube (tubus neuralis)
- Segmental arrangement of the body, particularly the muscles
- Foregut pierced by slits (branchial gut)
- Closed circulatory system
- Postanal tail

D Basic chordate anatomy, illustrated by the lancelet (Branchiostoma lanceolatum)
The vertebrates (including humans) are a subphylum of the chordates (Chordata), of which the lancelet is a typical representative. Its anatomy displays relatively simple terms of structures common to all vertebrates. The characteristic features of chordates include the development of an axial skeleton called the *chorda dorsalis*. The human body still has remnants of the chorda dorsalis, such as the nucleus pulposus of the intervertebral disks. The chorda dorsalis is present in humans only during embryonic life, however, and is not a fully developed structure. Its remnants may give rise to developmental tumors called *chordomas*. Chordates have a *tubular nervous system* lying dorsal to the chorda dorsalis. The body, particularly the muscles, is composed of multiple segments called *myomeres*. In humans, this myomeric pattern of organization is most clearly apparent in the trunk. Another distinguishing feature of chordates is the presence of a closed circulatory system.

H Characteristic features of vertebrates

- Nerve cells, sensory organs, and oral apparatus concentrated in the head (cephalization)
- Multipart brain with a hypophysis (pituitary gland)
- Replacement of the chorda dorsalis by the columna vertebralis
- Generally, two pairs of limbs
- Development of branchial arches
- Presence of neural crest cells
- Closed circulatory system with a ventral, chambered heart
- Labyrinthine organ with canales semicirculares
- Stratified epidermis
- Liver and pancreas always present
- Complex endocrine organs such as the thyroid and hypophysis
- Complex immune system
- Sexes almost always separate

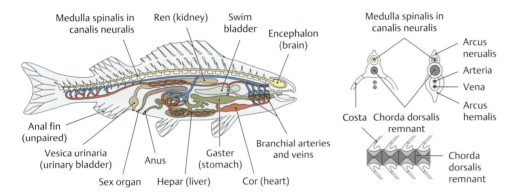

E Basic vertebrate anatomy, illustrated by the bony fish
The vertebrates are the *subphylum of chordates* from which humans evolved. With the evolution of fish, the chorda dorsalis was transformed into a vertebral column (spinal column). The segmentally arranged bony vertebrae of the spinal column encircle remnants of the chorda dorsalis and have largely taken its place. Dorsal and ventral arches arise from the vertebral bodies. The dorsal arches (vertebral or neural arches) in their entirety make up the canalis neuralis, while the ventral arches (hemal arches, arcus hemalis) form a caudal "hemal canal" that transmits the major blood vessels. The ventral arches in the trunk region are the origins of the ribs.

I Characteristic features of mammals

- Highly glandular skin covered with true hair (terminal hair)
- Females always have mammary glands for nursing offspring, which are usually born live (viviparous)
- Well-developed cerebrum
- Well-developed musculi cutanei
- Diaphragma is the major respiratory muscle and separates the thoracic and abdominal cavities
- Heterogeneous and specialized teeth
- Four-chambered heart with a (left-sided) aortic arch
- Constant body temperature (homeothermy)

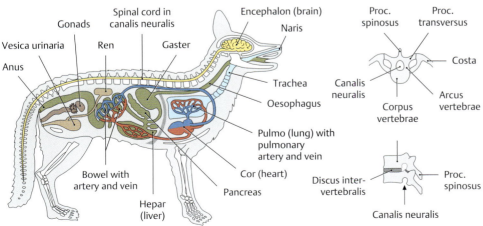

F Basic vertebrate anatomy, illustrated by the dog

1.2 Human Ontogeny: Overview, Fertilization, and Earliest Developmental Stages

Besides gross and microscopic anatomy, the developmental history of the individual organism (ontogeny) is of key importance in understanding the human body. Ontogeny is concerned with the formation of tissues (*histogenesis*), organs (*organogenesis*), and the shape of the body (*morphogenesis*).

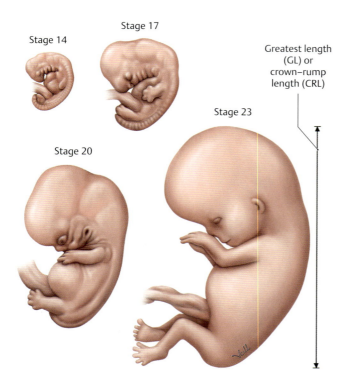

A 5- to 8-week-old human embryos
Streeter (1942) and O'Rahilly (1987) classified early human development and the embryonic period into 23 stages based on specimens from the Carnegie Collection. The Carnegie stages are defined by morphological characteristics that can be closely correlated with specific age (postovulatory days or weeks) and size (measured as the greatest length, excluding lower limb [GL], or crown–rump length [CRL], see **C**).

Stage 14: 5th week, GL 5–7 mm, future cerebral hemispheres become identifiable
Stage 17: 6th week, GL 11–14 mm, digital rays become visible.
Stage 20: 7th week, GL 18–22 mm, upper arms bent at the elbow, hands in a pronated position.
Stage 23: 8th week, GL 27–31 mm, eyelids fuse, external genitalia begin differentiation.

B Longitudinal growth and weight gain during the fetal period

Age (weeks)	Crown–rump length, CRL (cm)	Weight (g)
9–12	5–8	10–45
13–16	9–14	60–200
17–20	15–19	250–450
21–24	20–23	500–820
25–28	24–27	900–1300
29–32	28–30	1400–2100
33–36	31–34	2200–2900
37–38	35–36	3000–3400

C Timetable of antenatal human development
(The Carnegie stages are shown in parentheses.)

Weeks 1–3:	Early development
Week 1:	Tubal migration, segmentation, and blastocyst formation (stages 1–3)
Week 2:	Implantation and bilaminar embryonic disc (discus embryonicus), yolk sac (vesicula umbilicalis; stages 4–5)
Week 3:	Trilaminar embryonic disc, start of neurulation (stages 6–9)
Weeks 4–8:	**Embryonic period**
Week 4:	Folding of the embryo, neurulation concluded, axial organs, basic body shape (stages 10–13)
Weeks 5–8:	Organogenesis (formation of all essential external and internal organs, elongated limb buds) (stages 14–23)
Weeks 9–38:	**Fetal period**
Weeks 9–38:	Organ growth and functional maturation (sex-specific differentiation of the external genitalia)

Length of gestation	
• p.o. = post ovulationem	266 days = 38 weeks
• p.m. = post menstruationem	280 days = 40 weeks

Size	
• GL = greatest length, excluding lower limb	simplest, most consistent ultrasound measure
• CRL = crown–rump length	similar to GL in embryonic period, used in most descriptions of the fetal period

Fertilization

Early development (weeks 1–3)
Low rate of malformations, high rate of spontaneous abortion

Discus embryonicus (embryonic disc)

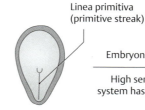

Early embryo

Linea primitiva (primitive streak)

Embryonic period (weeks 4–8)
High sensitivity; every organ system has its own sensitive phase

Embryo

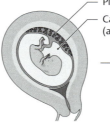

Fetus in utero

Placenta
Cavitas amniotica (amniotic cavity)

Fetal period (weeks 9–38)
Decreasing sensitivity, functional maturation

D Stages sensitive to teratogenic influences (after Sadler)

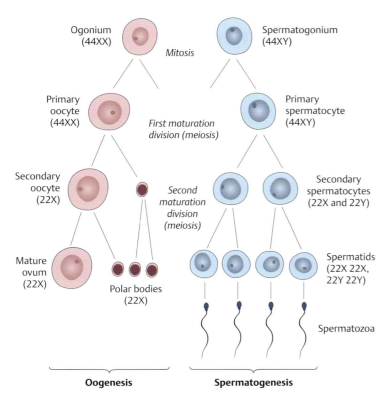

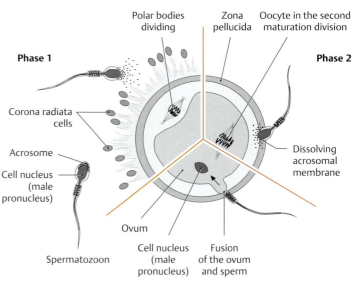

F Schematic representation of the fertilization process
(after Sadler)

In *phase 1*, the spermatozoon penetrates the corona radiata cells. In *phase 2*, the acrosome dissolves, releasing enzymes that digest the zona pellucida. In *phase 3*, the cell membranes of the ovum and sperm fuse, and the spermatozoon enters the egg.

E Formation of the ovum and sperm (after Sadler)

During the formation of the gametes (sex cells), two successive cell divisions occur (the first and second meiotic maturation divisions). This results in cells having a chromosome set that is reduced by one half (haploid). When fertilization occurs, a diploid (full) chromosome set is restored. During meiosis, extensive chromosomal rearrangement occurs, thus recombining the internal genetic information into new and different subsets.

Oogenesis: The initial oogonia first undergo a mitotic division to form primary oocytes, which still have a diploid chromosome number (44XX). Later the primary oocytes undergo a first and second maturation division by meiosis, resulting in four haploid cells (22X): one mature ovum and three polar bodies.

Spermatogenesis: Diploid spermatogonia undergo mitosis to form primary spermatocytes (44XY). These cells then divide meiotically to form four haploid spermatids, two of which have an X chromosome (22X) and two a Y chromosome (22Y). The spermatids develop into motile spermatozoa (spermatohistogenesis).

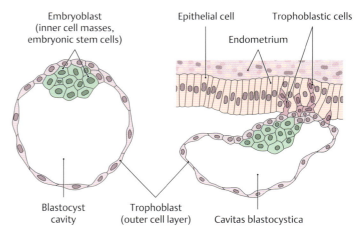

G Implantation of the blastocyst in the uterine mucosa on postovulatory day 5–6 (after Sadler)

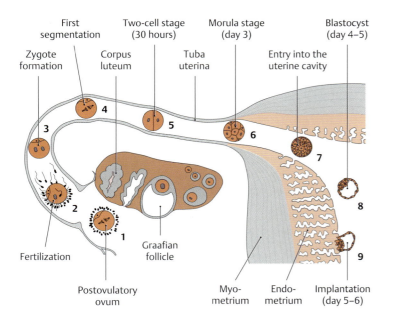

H Developmental processes during the first week of development
(after Sadler)
1. Ovum immediately after ovulation
2. Fertilized within approximately 12 hours
3. Male and female pronucleus with subsequent zygote formation
4. First segmentation
5. Two-cell stage
6. Morula stage
7. Entry into the uterine cavity
8. Blastocyst
9. Early implantation

1.3 Human Ontogeny: Gastrulation, Neurulation, and Somite Formation

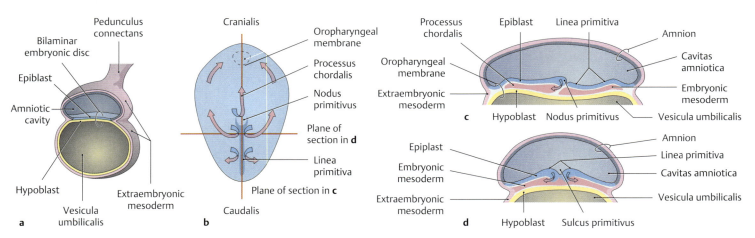

A Formation of the trilaminar human embryonic disc (gastrulation) at the start of the third postovulatory week (after Sadler)
As a result of gastrulation, the cell layers become differentiated into an *ectoderm*, *endoderm*, and *mesoderm*, from which all structures of the human body are derived (e.g., the endoderm gives rise to the central nervous system and the sensory organs). Gastrulation also establishes the primary axes of the body (ventral–dorsal, cranial–caudal, and left–right).

a Sagittal section through a conceptus at 2 postovulatory weeks. The embryonic disc is *still bilaminar* and is stretched between the amniotic cavity (cavitas amniotica) and yolk sac (vesicula umbilicalis). The extraembryonic mesoderm, whose formation commences at the posterior pole of the embryonic disc, already covers the entire conceptus, which is attached to the chorionic cavity by a connecting stalk.

b Dorsal view of an embryonic disc at the start of gastrulation. The amnion has been removed. At the start of gastrulation, the epiblast develops a primitive streak (linea primitiva), where the embryonic mesoderm is generated and migrates between epiblast and hypoblast. Shortly afterward, at the level of the primitive node (the cranial tip of the primitive streak), epiblast cells migrate cranially to form the processus chordalis and radially to form the definitive endoderm. In the process, the definitive endoderm sequentially replaces the hypoblast, while the processus chordalis only temporarily fuses with the hypoblast layer. The processus chordalis expands cranially from the primitive node to the oropharyngeal membrane.

c Sagittal section of an embryonic disc along the processus chordalis.

d Cross section of an embryonic disc at the level of the primitive groove (sulcus primitivus, arrows in **c** and **d** indicate the direction of gastrulation movements by the mesoderm).

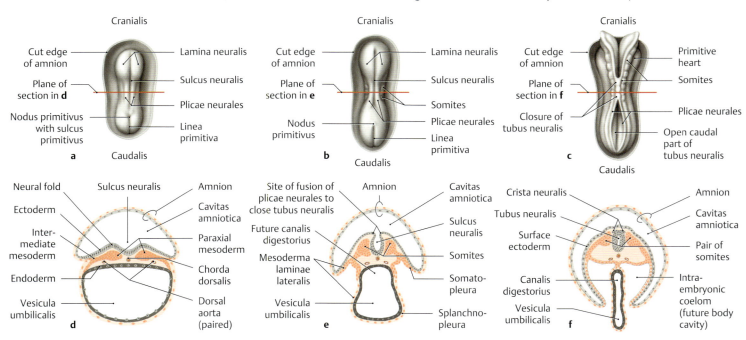

B Neurulation during early human development (after Sadler)
a–c Dorsal view after removal of the amnion.
d–f Schematic cross sections of the corresponding stages at the planes of section marked in **a–c**. Age in postovulatory days. During neurulation, the neuroectoderm differentiates from the surface ectoderm due to inductive influences from the chorda dorsalis.
a, d Embryonic disc at 19 days. The neural tube is developing in the area of the neural plate.
b, e Embryonic disc at 20 days. The first somites have formed, and the sulcus neuralis (neural groove) is beginning to close to form the tubus neuralis (neural tube), with initial folding of the embryo.
c, f Embryo at 22 days. Eight pairs of somites are seen flanking the partially closed neural tube, which has sunk below the ectoderm. At the sites where the neural folds fuse to close the neural tube, cells form a bilateral crista neuralis that detaches from the surface and migrates into the mesoderm.

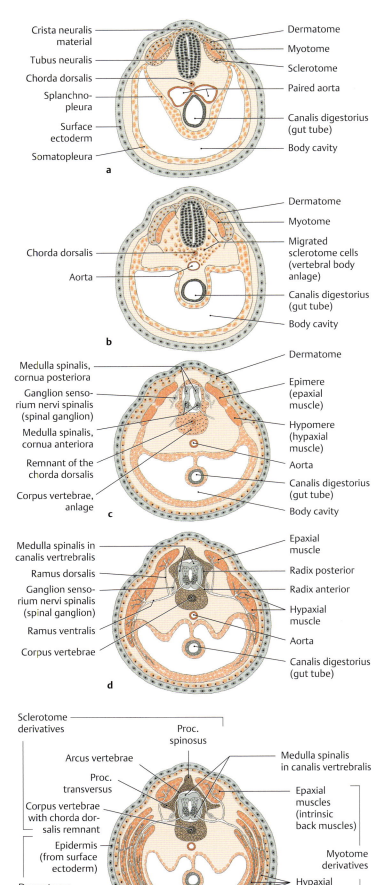

C Somite derivatives and spinal nerve formation during the embryonic period (weeks 4–8), shown in schematic cross sections (after Drews)

D Differentiation of the germ layers (after Christ and Wachtler)

Ectoderm	Tubus neuralis	Encephalon (brain), retina, medulla spinalis (spinal cord)
	Crista neuralis — Crista neuralis of the head	Sensory and parasympathetic ganglia, intramural nervous system of the bowel, parafollicular cells, smooth muscle, pigment cells, glomus caroticum (carotid body), bone, cartilage, connective tissue, dentin and cementum of the teeth, dermis and subcutaneous tissue of the head
	Crista neuralis of the trunk	Sensory and autonomic ganglia, peripheral glia, adrenal medulla, pigment cells, intramural plexuses
	Surface ectoderm — Ectodermal placodes	Anterior hypophysis, cranial sensory ganglia, olfactory epithelium, auris interna (inner ear), lens
	Surface ectoderm	Enamel organ of the teeth, epithelium of the oral cavity, salivary glands, cavitas nasi, sinus paranasales, lacrimal passages, external auditory canal, epidermis, hair, nails, cutaneous glands
Mesoderm	Axial — Chorda dorsalis, prechordal mesoderm	Extraocular muscles
	Paraxial	Columna vertebralis, costae, skeletal muscle, connective tissue, dermis and subcutis of the back and part of the head, smooth muscle, blood vessels
	Intermediate	Renes (kidneys), gonads, renal and genital excretory ducts
	Mesoderma laminae lateralis — Visceral (splanchnopleura)	Cor (heart), blood vessels, smooth muscle, bowel wall, blood, cortex glandulae suprarenalis (adrenal cortex), visceral serosa
	Mesoderma laminae lateralis — Parietal (somatopleura)	Sternum, limbs (cartilage, bones, and ligaments), dermis and subcutis of the anterolateral body wall, smooth muscle, connective tissue, parietal serosa
Endoderm		Epithelium of the bowel, respiratory tract, digestive glands, glandulae pharyngeales, eustachian tube, tympanic cavity, urinary bladder, thymus, glandulae parathyroideae, thyroid gland

(For clarity, the surrounding amnion is not shown.) The first pairs of somites appear at approximately 20 postovulatory days. All 34 or 35 of the somites ("primitive segments") have formed by day 30.

a When differentiation begins, each of these somites subdivides into a dermatome, myotome, and sclerotome (i.e., a cutaneous, muscular, and vertebral segment).

b At the end of 4 weeks, the sclerotome cells migrate toward the chorda dorsalis and form the anlage of the spinal column.

c The neural tube—the precursor of the spinal cord and brain—differentiates to form a rudimentary spinal cord with dorsal and ventral horns. Cells within the ventral horn differentiate into motor neurons that sprout axons that form the *radix anterior*. The crista neuralis has multiple derivatives, including sensory neurons that form dorsal root (spinal) ganglia, which send central processes into the spinal cord via the *radix posterior*. The myotomes become segregated into a dorsal part (epimere = epaxial muscles) and a ventral part (hypomere = hypaxial muscles).

d Each pair of radix posterior and radix anterior unites to form a spinal nerve (n. spinalis), which then divides into two main branches (ramus dorsalis and ramus ventralis). The epaxial muscles are supplied by the ramus dorsalis, the hypaxial muscles by the ramus ventralis.

e Cross section at the level of the future abdominal muscles. The epaxial muscles become the mm. erector spinae, while the hypaxial muscles develop into structures that include the lateral abdominal muscles (mm. obliqui abdominis externus and internus, m. transversus abdominis) and the anterior abdominal muscles (m. rectus abdominis).

1.4 Human Ontogeny: Development of the Fetal Membranes and Placenta

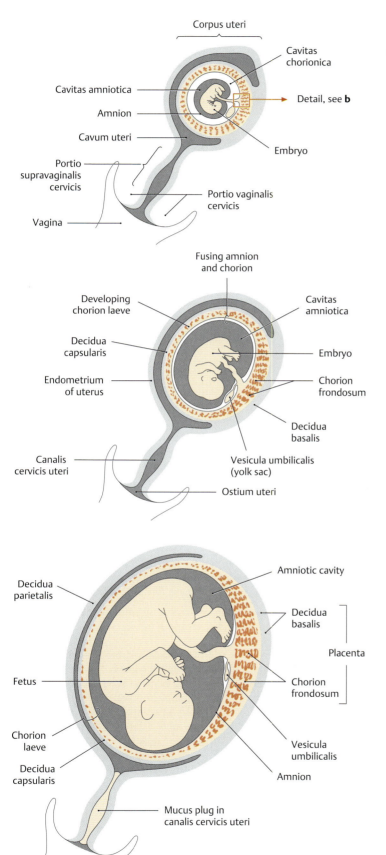

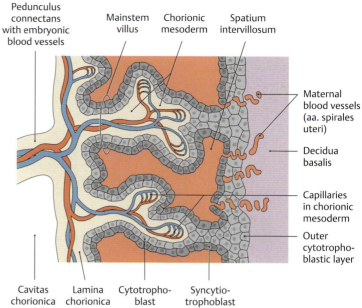

A Development of the fetal membranes and placenta (after Sadler and Drews)
a, c, and **d** Schematic sections through a pregnant uterus at different points in gestation.
b Detail from **a**.

a Embryo at 5 weeks: After the blastocyst has implanted in the uterine mucosa, the embryo initially derives its nutrition through the developing trophoblast and chorionic mesoderm. Chorionic villi are formed that surround the entire chorionic sac and embryo. They develop from primary to secondary villi and finally to tertiary villi (see close-up in **b**).

b Detail from a: The mainstem villi of the chorionic plate (lamina chorionica) are attached on the maternal side to the lamina basalis of the decidua basalis by compact columns of trophoblastic cells. Like the small villous trees that sprout and branch from them, these mainstem villi have a syncytial covering (syncytial trophoblast), which in turn rests on a continuous layer of trophoblastic cells. Inside the villi, capillaries develop in the chorionic mesoderm and communicate with the vessels in the connecting stalk. Maternal blood flows through aa. spirales uteri into the intervillous spaces (spatium intervillosum).

c Embryo at 8 weeks: While the chorionic villi continue to grow and arborize at the embryonic pole, forming the chorion frondosum, the villi outside of this zone begin to regress, forming the nonvillous chorion laeve directly below the decidua capsularis. The amniotic cavity has enlarged at the expense of the chorionic cavity, and the amnion fuses with the chorion.

d Fetus at 20 weeks: The placenta is fully formed and consists of two parts: a fetal part formed by the chorion frondosum and a maternal part, the decidua basalis.

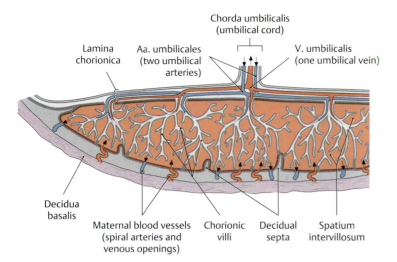

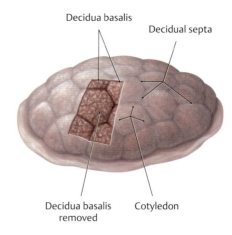

B Schematic cross section through a mature human placenta
The mature placenta is shaped like a frying pan, the maternal decidua basalis (basal plate) forming the base of the pan and the fetal chorionic plate forming the "lid." Some 40 arborizing villous trees containing fetal vessels project from the chorionic plate into the portions of the placenta that are filled with maternal blood (intervillous spaces). The maternal blood flows through approximately 80 to 100 spiral arteries into the intervillous spaces, which are divided into cotyledons by incomplete decidual septa. After the blood has bathed the villi, it is collected by irregularly distributed venous openings in the basal plate and returned to the maternal circulation.

C The postpartum placenta (after Sadler)
View of the maternal side of the delivered placenta (with a piece of the decidua basalis removed). The bulging cotyledons on the maternal surface are separated from one another by decidual septa.

E Characteristics of a mature human placenta

Size:	18–23 cm in diameter
	2–3 cm in thickness
Weight:	450–500 g
Total placental volume:	approximately 500 mL
Volume of the intervillous spaces:	approximately 150 mL
Villous surface area:	approximately 11–13 m^2
Blood circulation on the maternal side:	500–600 mL/min

Structure of the placental barrier

- Endothelium of the fetal capillaries and lamina basalis
- Fibrous villous stroma
- Syncytiotrophoblast and lamina basalis
- Continuous trophoblast cell layer (becomes discontinuous after 20 weeks of gestation)

Diffusion distance approx. 5 μm (initially approx. 50 μm)

Primary functions of the mature placenta

1. Transport of substances and exchange of metabolic products

Mother-to-fetus	Fetus-to-mother
O$_2$, water, electrolytes, carbohydrates, amino acids and lipids, hormones, antibodies, vitamins and trace elements, but also drugs, toxins, and certain viruses	CO$_2$, water, electrolytes, urea, uric acid, bilirubin, creatinine, hormones

2. Hormone production (syncytiotrophoblast)

- Human chorionic gonadotropin (HCG)
  → maintenance of the corpus luteum
- Estrogens
  → growth of the uterus and breasts (mammae)
- Progesterone
  → inhibits uterine muscle contractions

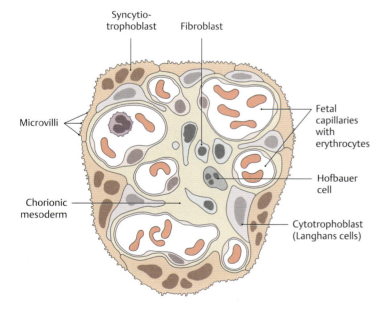

D Cross section through a terminal villus from a mature human placenta (after Kaufmann)

Clinical note: The HCG formed in the syncytiotrophoblast prevents premature breakdown of the corpus luteum and sustains the pregnancy. HCG can be detected in the maternal urine at an early stage, providing the basis for early pregnancy testing.

1.5 Development of the Pharyngeal (Branchial) Arches in Humans

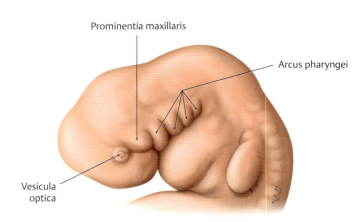

A Head and neck region of a 5-week-old human embryo, demonstrating the pharyngeal arches and clefts
Left lateral view. The pharyngeal arches are instrumental in the development of the neck and face. In *fish* and *amphibians*, the branchial arches develop into a respiratory organ (gills) for exchanging oxygen and carbon dioxide between the blood and water. *Land-dwelling vertebrates* (including humans) have *pharyngeal arches (arcus pharyngei)* rather than true branchial arches. Development of the pharyngeal arches begins in the 4th week of embryonic life as cells migrate from the crista neuralis to the future head and neck region. Within 1 week, a series of four oblique ridges (first through fourth pharyngeal arches) form that are located at the level of the cranial segment of the foregut and are separated externally by four deep grooves (pharyngeal clefts). The pharyngeal arches and grooves are prominent features of the embryo at this stage. Although the human embryo has no equivalent to the fifth and sixth branchial arches of other vertebrates, some of their components are incorporated into the human fourth pharyngeal arch.

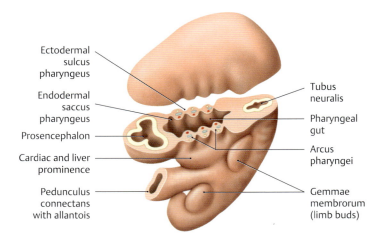

B Cross section through a human embryo at the level of the pharyngeal gut (after Drews)
Left superior oblique view. Due to the craniocaudal curvature of the embryo, the cross section passes through the pharyngeal arches and pharyngeal gut, as well as the prosencephalon and spinal cord.
The pharyngeal gut is bounded on both sides by the pharyngeal arches (see also **A**), which contain a mesodermal core. They are covered externally by ectoderm and internally by endoderm. Ectodermal pharyngeal *clefts (sulci pharyngei)* and endodermal pharyngeal *pouches (sacci pharyngei)* lie directly opposite one another. Because the embryo is curved craniocaudally, the pharyngeal gut and pharyngeal arches overlie the prominence of the rudimentary heart and liver.

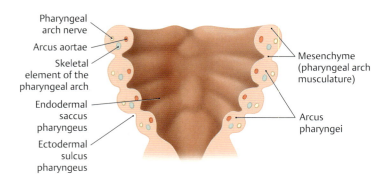

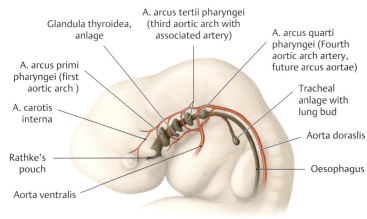

C Structure of the pharyngeal arches (after Drews)
View of the floor of the pharyngeal gut and the transversely sectioned pharyngeal arches (arcus pharyngei). The typical components of a pharyngeal arch are easily identified: the aortic arch, musculature and associated nerves, and a cartilaginous skeletal element of each pharyngeal arch. The derivatives of these structures are of key importance in the formation of the face, neck, larynx, and pharynx. Because the developmental transformation of pharyngeal arch structures is complex, it is readily disrupted, causing malformations that may involve a cluster of related derivatives. Defects in pharyngeal arch development result in branchial and lateral cervical cysts and fistulas and in a group of "first arch syndromes" involving mandibulofacial deformities.

D Location of the aortic arch and pharyngeal pouches (after Sadler)
The aortic arches (branchial arch arteries) *arise* from the paired embryonic *ventral aorta* and run between the pharyngeal pouches. They *open* dorsally into the *dorsal aorta*, which is also paired. The definitive aortic arch develops from the fourth aortic arch on the left side (the development of the aortic arch is described on p. 12). The pharyngeal pouches are paired, diverticula-like outpouchings of the endodermal pharyngeal gut. A total of four distinct pharyngeal pouches develop on each side; the fifth is often absent or rudimentary.
Note: The pouch protruding from the roof of the oral cavity is called *Rathke's pouch* (precursor of the anterior hypophysis). Note also the *lung bud* extending ventrally from the pharyngeal gut and the anlage of the *thyroid gland*.

General Anatomy — 1. Human Phylogeny and Ontogeny

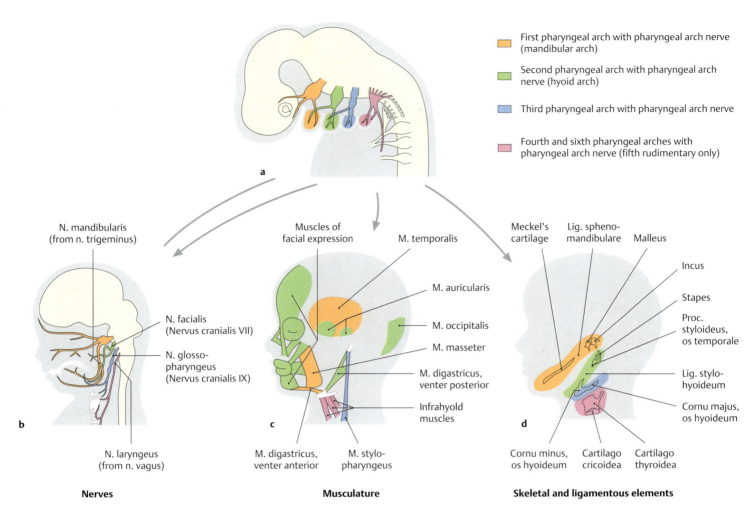

E The system of pharyngeal, or branchial, arches (after Sadler and Drews)
a Anlage (primordia) of the embryonic pharyngeal arches with the associated pharyngeal arch nerves.
b Definitive arrangement of the future nervi craniales V, VII, IX, and X.
c Muscular derivatives of the pharyngeal arches.
d Skeletal derivatives of the pharyngeal arches.

F Derivatives of the pharyngeal (branchial) arches in humans

Pharyngeal arch	Nerve	Muscles	Skeletal and ligamentous elements
First (mandibular arch)	Nervus cranialis V (n. mandibularis from n. trigeminus)	Masticatory muscles – m. temporalis – m. masseter – m. pterygoideus lateralis – m. pterygoideus medialis M. mylohyoideus M. digastricus (venter anterior) M. tensor tympani M. tensor veli palatini	Malleus and incus Portions of the mandibula Meckel's cartilage Lig. sphenomandibulare Lig. mallei anterius
Second (hyoid arch)	Nervus cranialis VII (n. facialis)	Muscles of facial expression M. stylohyoideus M. digastricus (venter posterior) M. stapedius	Stapes Proc. styloideus of the os temporale Cornu minus, os hyoideum Upper part of corpus ossis hyoidei
Third	Cranial nerve IX (n. glossopharyngeus)	M. stylopharyngeus	Cornu majus, os hyoideum Lower part of corpus ossis hyoidei
Fourth and sixth	Cranial nerve X (n. laryngeus superior and n. laryngens recurrens)	Musculi pharyngis et laryngis (pharyngeal and laryngeal muscles)	Laryngeal skeleton (cartilago thyroidea, cricoidea, arytenoidea, corniculata, and cuneiformis)

11

1.6 Early Embryonic Circulation and the Development of Major Blood Vessels

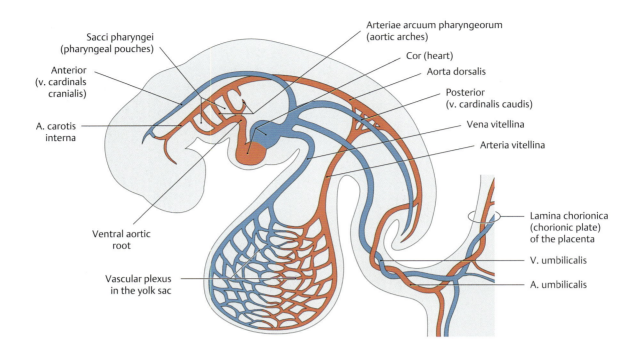

A Circulatory system of a 3- to 4-week-old human embryo
(after Drews)
Lateral view. The cardiovascular system of a 3- to 4-week-old human embryo consists of a well-functioning two-chambered heart and three distinct circulatory systems:

1. An **intraembryonic systemic circulation** (aortae ventralis and dorsalis, branchial arch and aortic arches, vv. cardinales cranialis and caudalis)
2. An **extraembryonic vitelline circulation** (aa. and vv. omphalomentericae)
3. A **placental circulation** (aa. and vv. umbilicales)

The vascular pathways still show a largely symmetrical arrangement at this stage.

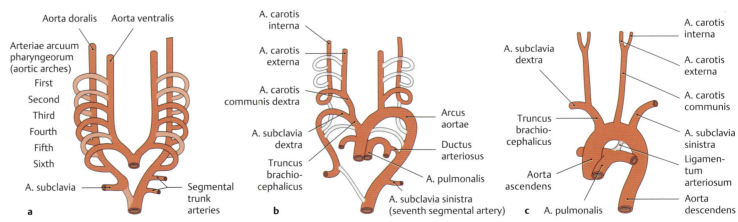

B Development of the arteries derived from the aortic arch
(after Lippert and Pabst)

a Initial stage (4-week-old embryo, ventral view). An artery develops in each of the pharyngeal arches, proceeding in the craniocaudal direction. These arteries arise from the paired ventral aortic roots, course through the mesenchyme of the pharyngeal arches, and open into an initially paired dorsal aorta. These vessels give rise to segmental trunk arteries. The six aortic arches are not all present at any one time, however. For example, while the fourth arch is forming, the first two arches are already beginning to regress. The development proceeds in such a way that the original symmetry is lost in favor of a preponderance on the left side.

b Structures that regress or persist: The first, second, and fifth aortic arches on both sides regress with continued development. The third aortic arch gives rise to an a. carotis communis on each side and the proximal portion of the a. carotis interna. The left fourth aortic arch later becomes the definitive *arcus aortae*, while the artery on the right side becomes the truncus brachiocephalicus and the a. subclavia dextra. The a. subclavia sinistra is derived from the seventh segmental artery. The trunk of the aa. pulmonalis and the ductus arteriosus are derived from the sixth aortic arch.

c Variants in the adult: Besides the typical case pictured here (77%), there are numerous variants of the truncus brachiocephalicus that occur with different frequencies. In the second most common pattern (13%), the left a. carotis communis also arises from the truncus brachiocephalicus. A right-sided aortic arch and a duplicated aortic arch each occur with a frequency of about 0.1%.

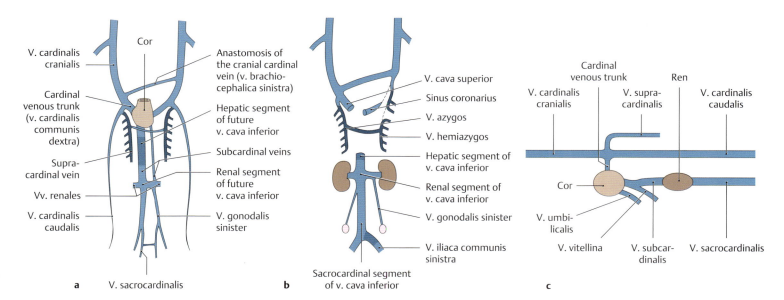

C Development of the cardinal venous system from weeks 5–7 to birth (after Sadler)
a At 5–7 weeks, ventral view, **b** at term, ventral view, **c** lateral view at 5–7 weeks, lateral view.

Up until the 4th week of development, three paired venous trunks return the blood to the heart: the vitelline, umbilical, and cardinal veins. The cardinal venous system at this stage consists of the vv. cardinales craniales, caudales, and communes. The following additional cardinal venous systems are formed between weeks 5 and 7:

- **Supracardinal veins (vv. supracardinales):** These vessels replace the posterior cardinal veins and receive blood from the intercostal veins (future azygos system: v. azygos and v. hemiazygos).
- **Subcardinal veins (vv. subcardinales):** These vessels develop to drain the kidneys—the right v. subcardinalis becoming the middle part of the v. cava inferior, and the transverse anastomosis becoming the vena renalis sinistra. The distal segment of the left subcardinal vein persists as a gonadal vein (v. testicularis or ovarica sinistra).
- **Sacrocardinal veins (vv. sacrocardinales):** These vessels develop during the formation of the lower limbs, their transverse anastomosis becoming the left v. iliaca communis.

Characteristic transverse anastomoses are formed between the individual cardinal venous systems. These connections transfer blood from the right to the left side, channeling it to the inflow tract of the heart. The transverse anastomosis between the vv. cardinales craniales, for example, forms the future v. brachiocephalica sinistra. The future v. cava superior develops from the right vv. cardinales cranialis and communis, while the left v. cardinalis communis contributes to the venous drainage of the heart (sinus coronarius).

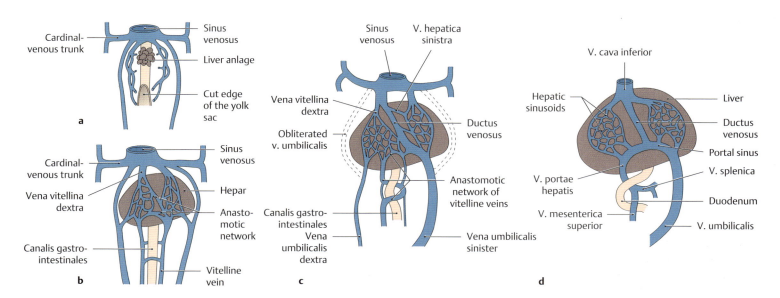

D Development of the vitelline and umbilical veins (after Sadler)
a 4th week, **b** 5th week, **c** 2nd month, **d** 3rd month. Ventral view.
Before the vitelline veins (vv. omphalomesentericae) open into the vv. umbilicales, they form a venous plexus around the duodenum, perfuse the embryonic liver anlage, and form the first hepatic sinusoids. At this stage the two umbilical veins (vv. umbilicales) still course on both sides of the liver anlage. With further development, however, they establish a connection with the hepatic sinusoids. While the vena umbilicalis dextra regresses completely during the 2nd month, the left v. umbilicalis assumes the function of transporting all blood back from the placenta to the fetus. The blood flows through a shunt (ductus venosus) into the *proximal* trunk of the vena vitellina dextra (the future post-hepatic part of the v. cava inferior) and back to the sinus venosus. The *distal* portion of the vena vitellina dextra develops into the future portal vein (v. portae hepatis), by which blood is conveyed from the unpaired abdominal organs to the liver (vv. mesentericae superior and inferior, v. splenica).

1.7 Primordial Skeleton, Prenatal Skeletal Development, and Ossification Centers

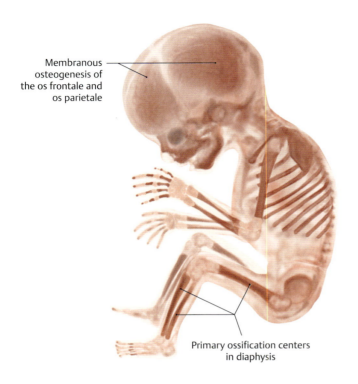

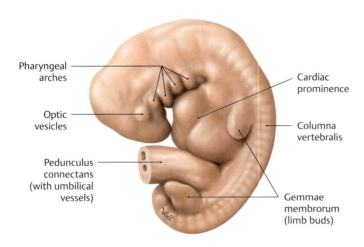

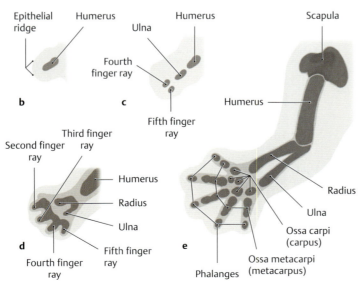

A Primordial skeleton, prenatal skeletal development, and ossification centers
(Transparent specimen of an 11-week-old fetus's with ossification centers stained with alizarin red; after Starck and Drews). The supporting tissue of the human skeleton (mainly cartilage and bone tissue) arises from embryonic connective tissue (= mesenchyme), derived from the middle germ layer (= mesoderm, see p. 7, **D**). Mesenchymal cells initially differentiate into chondroblasts (endochondral osteogenesis), which form a miniature model of the skeleton in its final form (primordial skeleton) from hyaline cartilage. It is only at a later developmental stage that the hyaline cartilage is replaced by bone tissue. Endochondral osteogenesis leads to the formation of the largest part of the human skeleton (torso, limbs, and skull base). The ossification of the primordial skeleton begins at the end of the embryonic period (8th week). It extends from the secondary ossification centers in the diaphyseal region of the long bones (perichondral ossification = bone formation directly from the mesenchyme, thus desmal osteogenesis). Soon after, with the start of endochondral ossification (= bone formation around the cartilage, thus endochondral osteogenesis), the formation of primary ossification centers in the diaphysis begins. By the 12th week, there are ossification centers in all long bones. In the epiphysis, endochondral ossification (secondary ossification centers) only begins some time after birth (exception: distal femur and proximal tibial epiphysis, which already exist at birth). At birth, many short bones, e.g., most of the tarsal bones and all of the carpal bones, consist entirely of cartilage and form an ossification center only in the ensuing months and years. Only the bones of the cranial roof, parts of the facial bones, and the collar bone (clavicula) ossify directly (membranous osteogenesis, which means mesenchyme cells develop into osteoblasts [see p. 17]).

B Development of the limbs illustrated for the arm
At the end of the 4th week of embryonic development, the limb buds (gemmae membrorum) appear as paddle-shaped outpouchings around the lateral trunk regions (**a**). These limb buds consist of mesenchymal cells (so-called limb bud blastema, see p. 145) and an ectoderm covering with an apical epithelial thickening, the epithelial ridge (**b**). The growth and arrangement of the limb buds occurs within a three-dimensional coordinate system, which has both proximal-distal and cranial-caudal axes of differentiation. The humerus of the upper arm develops first, fol- lowed by the ulna and adjacent skeletal elements of the carpal bones (ossa carpi) (**c**). The ulna and the fourth and fifth finger rays are referred to as the postaxial (caudal) section; the radius and first, second, and third finger rays, as the preaxial (cranial) section (**d**). The finger and foot rays, for example, form when the cells of the epithelial ridge divide into five segments through programmed cell death (interdigital apoptosis [**e**]). In case of a disruption of apoptosis, adjacent fingers or toes fuse together (syndactyly). The absence of entire limbs is called *amelia*; the absence of parts of the arm or leg is called *meromelia*.

Note: Achondroplasia is a genetic disorder that causes a disruption in the endochondral ossification and is one of the most frequent causes of disproportionate short stature (short limbs, short trunk, too large of a skull relative to the rest of the body). Due to a disruption in cell division and maturation of the chondrocytes in the growth plates, all bones generated by endochondral osteogenesis remain too short. Membranous osteogenesis, however, proceeds normally.

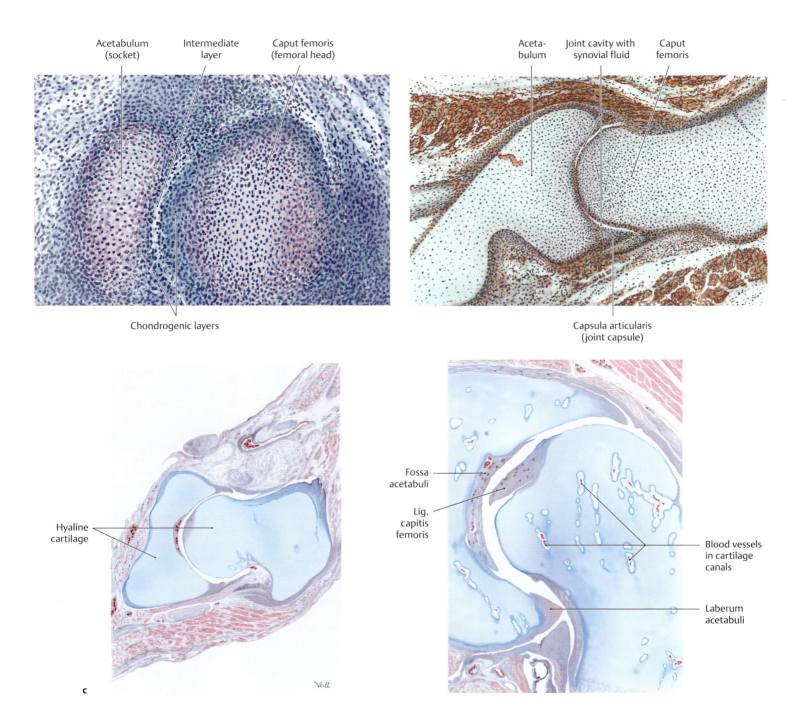

C Development of joints illustrated for the hip joint (after Uhthoff)

a In the 6th week of embryonic development, cell condensation occurs where the joints will eventually develop. A three-layered joint interzone forms. It consists of two chondrogenic layers covering the cartilage anlage (primordia) and an intermediate layer containing fewer cells.

b Around the 8th week of embryonic development, programmed apoptotic cell death leads to the formation of the joint cavity around the intermediate layer. The capsula articularis (joint capsule) arises from the peripheral region of the joint interzone and begins to produce synovial fluid.

c After the joint cavity has been formed, hyaline cartilage develops in chondrogenic layers. At the end of the 12th week of embryonic development, the development of joints is complete. The final structure of the joint is determined by functional loads (e.g., muscle force), although the joint has a genetically predetermined shape.

d The joint continues to grow (interstitial and appositional growth). With the start of the 13th week of embryonic development, nutrition through diffusion from both the perichondrium and synovia of the joint cavity is no longer sufficient, which leads to the formation of intrachondral vessels along the so-called cartilage canals. Only one area close to the joint cavity remains blood vessel-free. The vascularization of the cartilaginous epiphysis, however, stands in no relation to the development of the secondary ossification center. At the femoral head (caput femoris) of the hip joint (art. coxae), for instance, there is a delay of about 12 months between the first vascularization (3rd month of embryonic development) and the appearance of the ossification center in the proximal femoral epiphysis (6th postnatal month).

Note: There are essentially two ways in which joints can develop:
- Through segmentation (most common form), meaning cleavage of a continuous skeletal condensation (applies to almost all joints: hip joint, shoulder joint, elbow joint, etc.).
- Through appositional growth, meaning two initially separate skeletal elements grow toward one another (e.g., temporomandibular, sternoclavicular, and iliosacral joints). First, a synovial bursa forms at the junction between those elements, which then transforms into the joint cavity. In addition, intra-articular disks are typical for this type of joints (exception: iliosacral joint).

1.8 Bone Development and Remodeling

Bone development and bone remodeling are closely interrelated. During growth, for instance, bone undergoes a constant remodeling process in which immature woven bone tissue is replaced by "mature" lamellar bone. Continual remodeling also occurs in the mature skeleton, particularly in cancellous bone (called also trabecular bone, or spongiosa, see **F**). In this way, on average, approximately 10% of the entire adult skeleton is remodeled each year, meaning that the skeleton is completely renewed over about a 10-year period. This process is basically a functional adaptation of the bones to the dominant stress patterns to which they are exposed (and which vary over time). It also serves to prevent material fatigue, repair microinjuries to bone, and provide a rapidly available source of calcium.

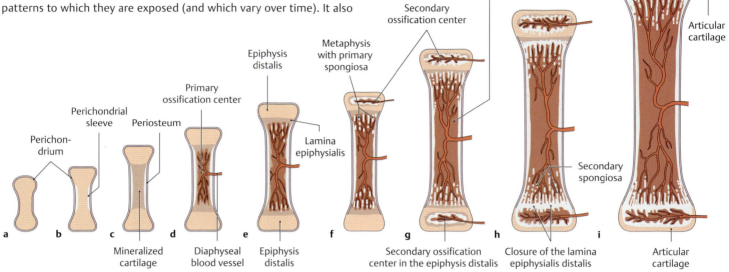

A Development of a long bone
The long bones (humerus, tibia, etc.) are mainly a product of *indirect* bone formation i.e., they form by replacing a preexisting cartilaginous model of the bone (*endochondral osteogenesis*). But portions of the long bones (the perichondrial bone collar, which allows the bone to grow in thickness) are a product of *direct* bone formation i.e., they form from the direct transformation of condensed mesenchyme (*membranous osteogenesis*, see **E**).
a Cartilaginous model of a bone in the embryonic skeleton. **b** Formation of a perichondrial bone collar (directly from mesenchyme). **c** Differentiation to hypertrophic chondrocytes and mineralization of the cartilaginous extracellular matrix. **d** Ingrowth of a diaphyseal vessel and formation of a primary ossification center. **e** Development of the proximal and distal growth centers (lamina epiphysialis). **f** Appearance of the proximal epiphyseal ossification center (secondary ossification center). **g** Formation of the distal epiphyseal ossification center. **h** Closure of the distal epiphyseal plate (lamina epiphysialis distalis). **i** Closure of the proximal epiphyseal plate (occurs at the end of skeletal growth, between about 18 and 23 years of age for most tubular bones).
Note: Osteogenesis = the formation of an individual bone; ossification = the formation of bone tissue.

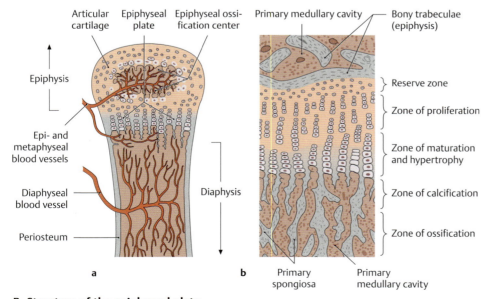

B Structure of the epiphyseal plate
a Blood supply, **b** detail from **a**: zones of the epiphyseal plate.

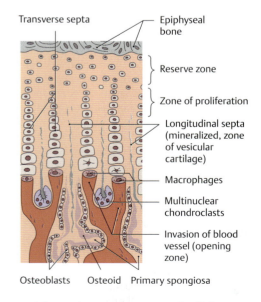

C Schematic representation of cellular processes within the epiphyseal plate

General Anatomy — 1. Human Phylogeny and Ontogeny

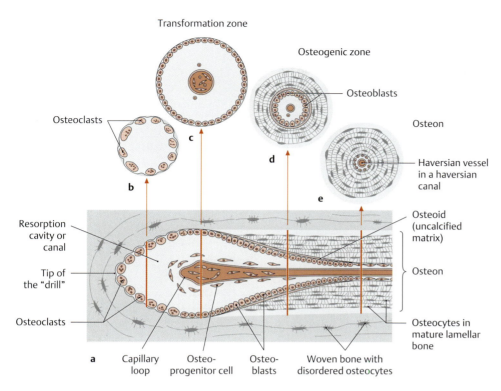

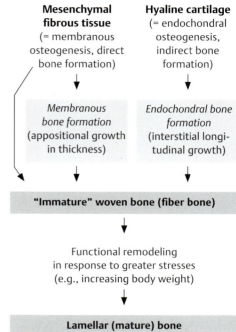

D Development of an osteon (after Hees)
The process of functional remodeling (see upper left page) begins with the invasion of blood vessels and accompanying osteoclasts ("bone-eating cells") into woven bone. They burrow through the woven bone like a drill, cutting a vascularized channel (resorption canal or cavity) that is equal in diameter to the future osteon.

- **a** Longitudinal section through a resorption canal.
- **b** Cross section at the level of the resorption canal.
- **c** Transformation zone: osteoprogenitor cells (a kind of precursor for bone-forming cells) are transformed into osteoblasts.
- **d** Osteogenic zone (osteoblasts produce bony lamellae).
- **e** Newly formed osteon.

E Types of bone development (osteogenesis)
Note: Most bones are formed by *indirect* osteogenesis. The few exceptions include the clvicula and certain bones of the calvaria. Portions of these bones develop from the mesenchyme, i.e., by *direct* osteogenesis.

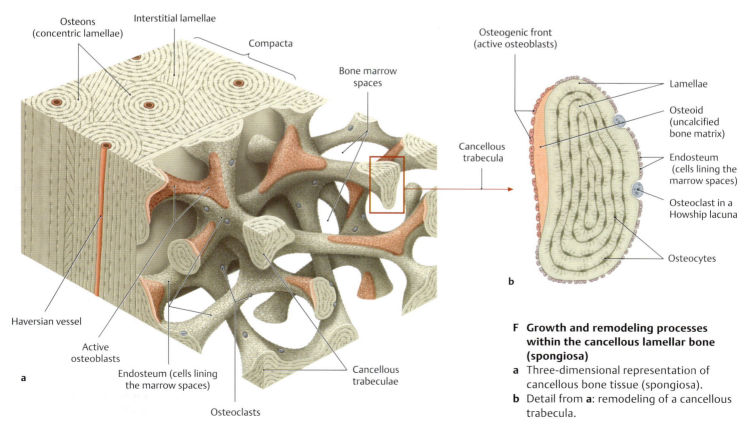

F Growth and remodeling processes within the cancellous lamellar bone (spongiosa)
- **a** Three-dimensional representation of cancellous bone tissue (spongiosa).
- **b** Detail from **a**: remodeling of a cancellous trabecula.

1.9 Ossification of the Limbs

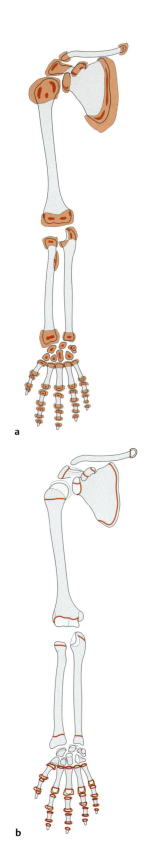

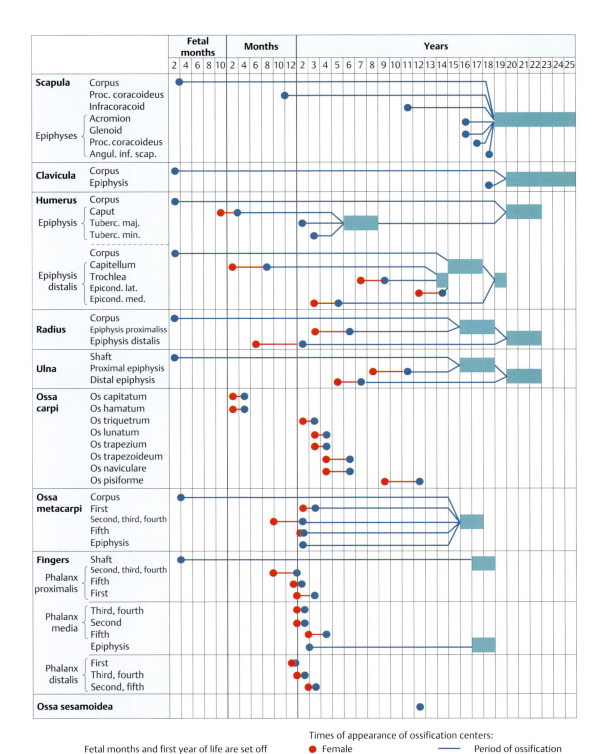

A Fetal months and first year of life are set off from the rest of the table.

Times of appearance of ossification centers:
● Female
● Male
— Period of ossification
▬ Period of synostosis

A and C Timetable of regional bone growth in the upper limb (A) and lower limb (C)
(from Niethard: *Kinderorthopädie* (Pediatric Orthopedics). Thieme, Stuttgart 1997)

The current stage of skeletal development, and thus the individual skeletal age, can be estimated from the times of appearance of the ossification centers. *Primary ossification centers*, which generally appear in the corpus region of bones *during the fetal period* (diaphyseal ossification), are distinguished from *secondary ossification centers*, which form *after birth* within the cartilaginous epiphysis and apophysis (epi-

and apophyseal ossification). Longitudinal growth ceases with the closure of the epiphyseal plate (synostosis). The tuberculum majus of the humerus, for example, begins to ossify at 2 years of age. A period of synostosis follows from 6 to 8 years of age, and after that the tuberculum majus shows only external, appositional growth. With the cessation of longitudinal growth, the ossification centers disappear and are no longer visible on X-ray films. The relationship between maturation and the appearance of secondary ossification centers is most clearly demonstrated in the ossa carpi (see also **B**). The eight ossa

B Ossification of the skeleton of the upper limb
a Location of the epiphyseal and apophyseal ossification centers.
b Location of the epiphyseal and apophyseal plates.

General Anatomy — 1. Human Phylogeny and Ontogeny

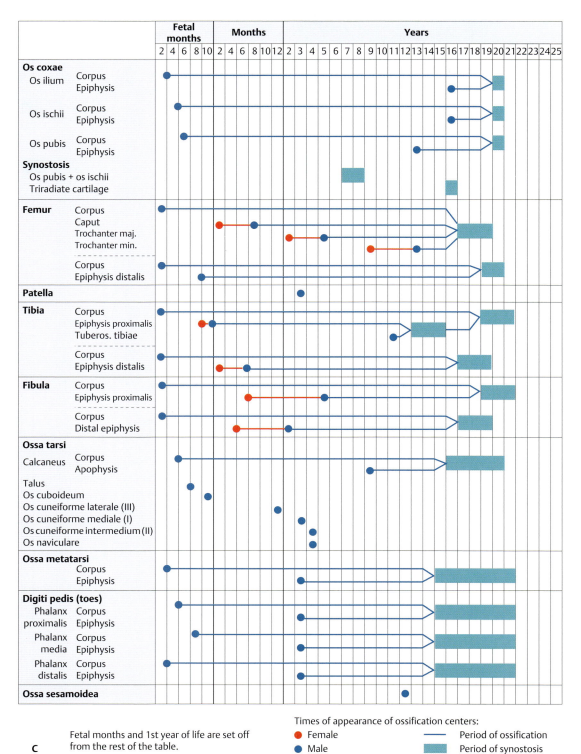

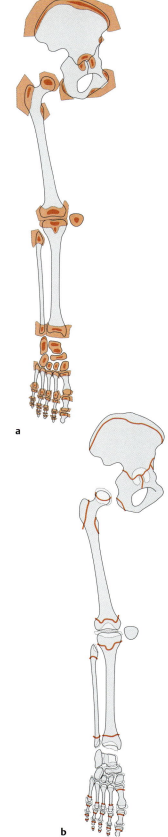

C Fetal months and 1st year of life are set off from the rest of the table.

Times of appearance of ossification centers:
- ● Female
- ● Male
- —— Period of ossification
- ▇ Period of synostosis

carpi ossify gradually over a period of approximately 9 years. The first ossification center is that of the os capitatum, which appears during the 1st year of life; the last is that of the os pisiforme, which ossifies at 9 years of age. It is standard practice to use the left or nondominant hand for radiographic examinations. The skeletal age reflects the biological maturity of the organism more than chronological age. The estimation of skeletal age, and thus of growth potential, is of key importance, for example, in the prognosis and treatment of orthopedic diseases and deformities in children. Also, given the relationship that exists between skeletal maturity and body height, the definitive adult height can usually be predicted with reasonable accuracy after 6 years of age based on the skeletal age and longitudinal measurements.

D Ossification of the skeleton of the lower limb
a Location of the epiphyseal and apophyseal ossification centers.
b Location of the epiphyseal and apophyseal plates.

1.10 Basic Skeletal Structure and Development and Position of the Limbs

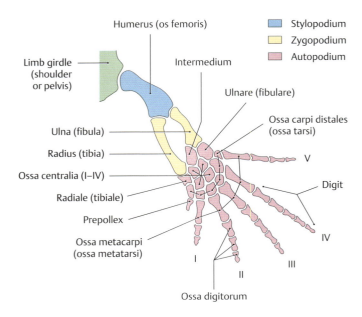

A Basic skeletal structure of a five-ray (pentadactyl) tetrapod limb (after Romer)

Both the forelimb and hindlimb of a free-ranging terrestrial vertebrate have the same basic, three-part structure consisting of a proximal, middle, and distal segment (called the *stylopodium*, *zygopodium*, and *autopodium*). The elbow or knee joint is placed between the stylopodium, which consists of a single bone (humerus or femur), and the zygopodium, which consists of two bones (the radius and ulna or the tibia and fibula). The five-ray autopodium (hand or foot) is also made up of proximal, middle, and distal units (called the *basipodium*, *metapodium*, and *acropodium*; see **C**). Additionally, there are some vertebrate classes that depart from this basic structure by showing a reduction or fusion of various bony units.

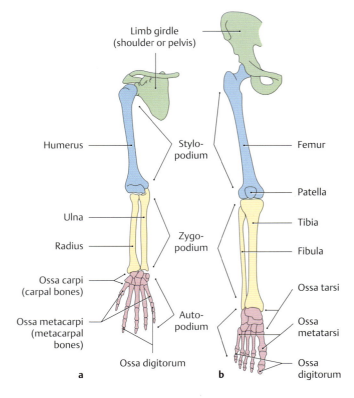

B Basic skeletal structure of the human limbs

Anterior view. **a** Right upper limb, **b** right lower limb.
The skeletal elements of the human upper and lower limbs have been colored to show how they are homologous with the tetrapod limb segments shown in **A** (stylo-, zygo-, and autopodium). Congenital malformations such as polydactyly or syndactyly (the presence of supernumerary fingers or toes or their fusion) are not uncommon.

C Bony constituents of the pentadactyl tetrapod limb

Segments	Paired forelimbs	Paired hindlimbs
Limb girdle	Shoulder girdle – Scapula and clavicula	Pelvic girdle – Os coxae
Free limbs		
Stylopodium	Arm (brachium) – Humerus	Thigh (femur) – Os femoris
Zygopodium	Forearm (antebrachium) – Radius – Ulna	Leg (crus) – Tibia – Fibula
Autopodium	Hand (manus)	Foot (pes)
– Basipodium	Carpus – Proximal row: radiale, intermedium, ulnare – Central group: ossa centralia I–IV – Distal row: ossa carpi I–V	Tarsus – Proximal row: tibiale, intermedium, fibulare – Central group: ossa centralia I–IV – Distal row: ossa tarsi I–V
– Metapodium	Metacarpus – Ossa metacarpi I–V	Metatarsus – Ossa metatarsi I–V

– Acropodium	Fingers (digiti manus)
	– Ossa digitorum manus I–V
	  (with different numbers of phalanges)
		Toes (digiti pedis)
		– Ossa digitorum pedis I–V
		  (with different numbers of phalanges)

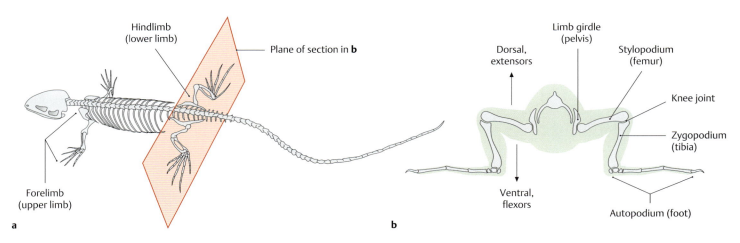

D Limb positions in a primitive terrestrial tetrapod (the lizard *Lacerta viridis*)
a Dorsal view, **b** cross section at the level of the hindlimbs.
In amphibian and reptilian tetrapods (e.g., salamanders, turtles, and lizards), the trunk is slung between the limbs and frequently touches the ground. The limbs are set almost at right angles to the body, so that the arm and thigh are nearly horizontal, and the elbow and knee point outward. The radius and ulna and the tibia and fibula are flexed at right angles at the elbow and knee. The volar surface of the hand and the plantar surface of the foot are in contact with the ground. The axes of all the joints are directed parallel to the spinal column (see **E**). *Note* that the *extensor* muscles are placed *dorsally*, while the *flexor* muscles are *ventral*. Thus, the location of the extensors and flexors relative to the bone does not change with evolution—the bone merely assumes a different alignment (see also **F**).

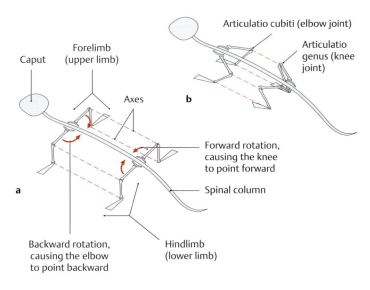

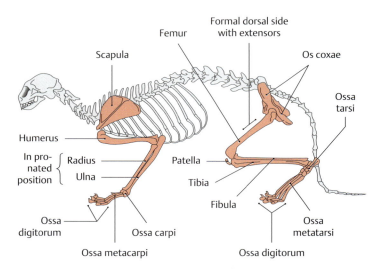

E Rotation of the limbs in mammalian evolution
a Before rotation, **b** after rotation.
An important feature of mammalian evolution involved *rotation* of the tetrapod limb. The limb was reoriented, placed parallel to the body, and moved closer to or beneath the body. This improved locomotion and supported the body more efficiently. The hindlimb rotated *forward* (with the knee pointing cephalad), while the forelimb rotated *backward* against the body (the elbow pointing caudad). As a result, both sets of limbs assumed a sagittal orientation under or alongside the trunk (see **F**).

F Skeleton of a cat (*Felis catus*)
Left lateral view. In order for the volar surfaces of the forelimbs to rest on the ground despite the backward angulation of the elbows, the forearm bones must cross to a pronated position. In the hindlimbs, there is no need for pronation of the leg bones because the thigh is rotated forward.
This arrangement of the skeletal elements in the various limb segments is essentially preserved in humans. Because the lower limb has been rotated forward, the *former dorsal side* of this limb faces *forward* in a human standing upright. As a result, the extensors of the thigh and leg (the genetically "dorsal" muscles) are on the anterior side of the limb, placed in front of the corresponding limb bones. This is one reason why the terms "anterior" and "posterior" are preferred over "dorsal" and "ventral" in the human lower limb. By contrast, the extensors and flexors of the arm and forearm have maintained their original dorsal and ventral positions, respectively.

General Anatomy — 2. Overview of the Human Body

2.1 The Human Body: Proportions, Surface Areas, and Body Weights

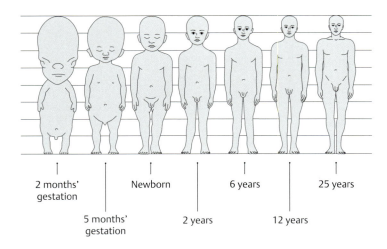

| 2 months' gestation | 5 months' gestation | Newborn | 2 years | 6 years | 12 years | 25 years |

A Change in body proportions during growth
While the head height in embryos at 2 months' gestation is equal to approximately half the total body length, it measures approximately one fourth of the body length in newborns, one sixth in a 6-year-old child, and one eighth in an adult.

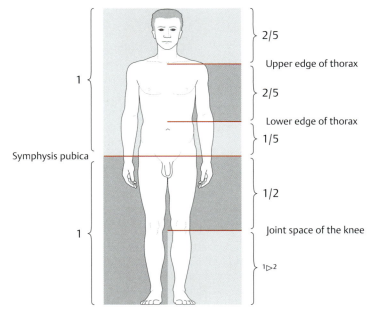

B Normal body proportions
In an adult, the midpoint of the total body height lies approximately at the level of the symphysis pubica; i.e., there is a 1:1 ratio of upper to lower body height at that level. The pelvis accounts for one fifth of the upper body height, the thorax for two fifths, and the head and neck for two fifths. The lower body height is distributed equally between the thigh and leg (plus heel) at the joint space of the knee.

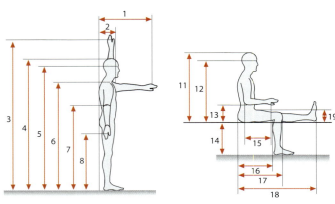

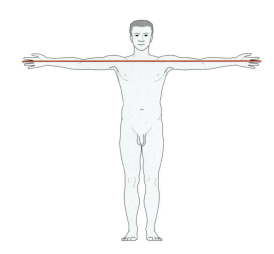

C Span of the outstretched arms
The arm span from fingertip to fingertip is slightly *greater* than the body height (approximately 103% in women and 106% in men).

Measurements (in cm) (points 9, 10, 20, and 21 not shown)	Percentiles					
	Male			Female		
	5th	50th	95th	5th	50th	95th
1 Forward reach	68.5	74.0	81.5	62.5	69.0	75.0
2 AP body thickness	26.0	28.5	38.0	24.5	29.0	34.5
3 Overhead reach (with both arms)	197.5	207.5	220.5	184.0	194.5	202.5
4 Body height	165.0	175.0	185.5	153.5	162.5	172.0
5 Ocular height	153.0	163.0	173.5	143.0	151.5	160.5
6 Shoulder height	134.5	145.0	155.0	126.0	134.5	142.5
7 Elbow-to-floor distance	102.5	110.0	117.5	96.0	102.0	108.0
8 Hand-to-floor distance	73.0	76.5	82.5	67.0	71.5	76.0
9 Shoulder width	44.0	48.0	52.5	39.5	43.5	48.5
10 Hip width, standing	34.0	36.0	38.5	34.0	36.5	40.0
11 Sitting body height (trunk height)	85.5	91.0	96.5	81.0	86.0	91.0
12 Ocular height while sitting	74.0	79.5	85.5	70.5	75.5	80.5
13 Elbow to sitting surface	21.0	24.0	28.5	18.5	23.0	27.5
14 Height of leg and foot (height of sitting surface)	41.0	45.0	49.0	37.5	41.5	45.0
15 Elbow to gripping axis	32.5	35.0	39.0	29.5	31.5	35.0
16 Sitting depth	45.0	49.5	54.0	43.5	48.5	53.0
17 Buttock–knee length	56.5	61.0	65.5	54.5	59.0	64.0
18 Buttock–leg length	96.5	104.5	114.0	92.5	99.0	105.5
19 Thigh height	13.0	15.0	18.0	12.5	14.5	17.5
20 Width above the elbow	41.5	48.0	55.5	39.5	48.5	55.5
21 Hip width, sitting	35.0	37.5	42.0	36.0	39.0	46.0

D Selected body measurements in the standing and sitting human being (unclothed, 18–65 years of age) after DIN 33402-2 (from Ergonomie—Körpermaße des Menschen—Teil 2: Werte. Berlin: Beuth; 2005)
The percentile values indicate what percentage in a population group are below the value stated for a particular body measurement (in this case, all those living in Germany, thus including migrants in the years 1999–2002). For example, the 95th percentile for body height in males 18 to 65 years of age with a body height of 185.5 cm means that 95% of this population group are shorter than 185.5 cm, and 5% are taller.

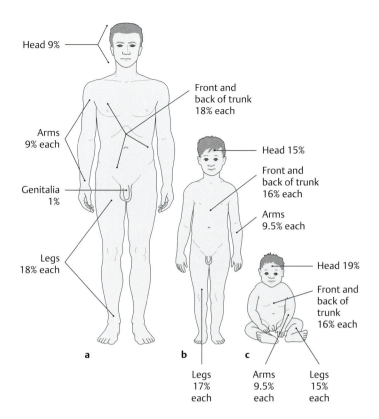

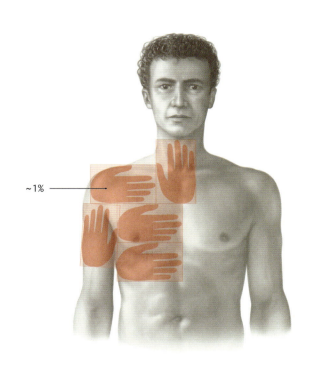

E Distribution of body surface area in adults, children, and infants

According to the "rule of nines" described by Wallace (1950), the body surface area of adults over about 15 years of age (**a**) can be divided into units that are a *multiple* of 9%: the head and each arm account for 9% each, the front and back of the trunk and each leg account for 18% (2×9) each, and the external genitalia comprise 1%. In children (**b**) and infants (**c**), the rule of nines must be adjusted for age.
Note: The rule of nines can be used in burn victims to provide a quick approximation of the area of skin that has been burned.

F Hand area rule

The percentage of the body surface affected by burns can be accurately estimated with the hand area rule, which states that the area of the patient's hand is approximately 1% of the patient's own total body surface area. The hand rule also applies to children, whose hands and total surface area are both proportionately smaller than in adults.

G Dependence of relative body surface area (skin surface area) on age, and consequences

For progressively larger solid bodies, the surface area increases as the square of the radius, but the volume increases as the cube of the body's radius. Because of this basic geometrical relationship, smaller animals generally have a larger relative surface area than larger animals. A higher ratio of surface area to volume causes smaller animals to radiate relatively more body heat. As a result, small animals, such as mice and children, tend to have a higher metabolic rate than larger animals, such as elephants and human adults.

Age	Body weight (kg)	Body surface area (cm²)	Body surface area over body weight (cm²/kg)
Newborn	3.4	2100	617.6
6 months	7.5	3500	466.7
1 year	9.3	4100	440.9
4 years	15.5	6500	419.4
10 years	30.5	10,500	344.3
Adult	70.0	18,100	258.6

H Body mass index

In anthropometry, the body mass index (BMI) has become the international standard for evaluating body weight because it correlates relatively well with total body fat. BMI is defined as the body weight in kilograms divided by the square of the height in meters:

$$BMI = \frac{kg}{m^2}$$

2.2 The Structural Design of the Human Body

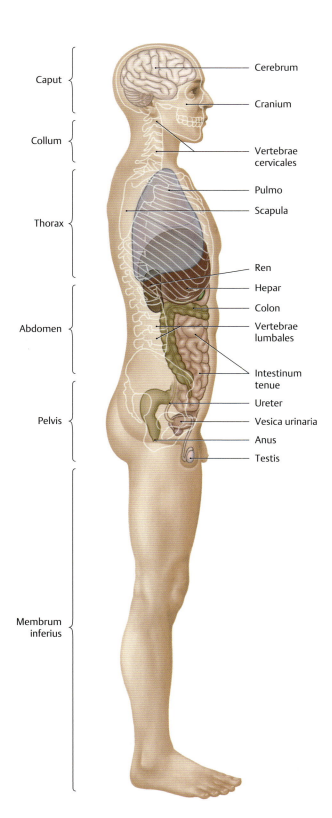

A Location of the internal organs
Lateral view.

B Regional subdivisions of the body

Caput (head)

Collum (neck)

Truncus (trunk)
- Thorax (chest)
- Abdomen
- Pelvis

Membrum superius (upper limb)
- Cingulum membri superioris (shoulder girdle)
- Pars libera membri superioris (free upper limb)

Membrum inferius (lower limb)
- Cingulum membri inferioris (pelvic girdle)
- Pars libera membri inferioris (free lower limb)

C Functional subdivisions of the body by organ system

Locomotor system (musculoskeletal system)
- Skeleton and skeletal connections (passive part)
- Striated skeletal musculature (active part)

Viscera
- Cardiovascular system
- Hemolymphatic system
- Endocrine system
- Respiratory system
- Digestive system
- Urinary system
- Male and female reproductive system

Nervous system
- Central and peripheral nervous system
- Sensory organs

The skin and its appendages

D Topography of body cavities, serous cavities, and connective-tissue spaces

Organs and organ systems are embedded either in serous cavities or in connective-tissue spaces of varying size. A serous cavity is a fully enclosed potential space that is lined by a shiny membrane (serosa) and contains a small amount of fluid. The serosa consists of two layers that are *usually* apposed (both layers are not *necessarily* in direct contact, as in the abdominal cavity): a lamina visceralis that directly invests the organ, and a lamina parietalis that lines the wall of the serous cavity.

Enclosed serous cavities
- Cavitas thoracis (chest cavity):
  – Cavitas pleuralis
  – Cavitas pericardialis

- Cavitas abdominis (abdominal cavity)
  – Cavitas peritonealis (abdominis)
- Cavitas pelvis (pelvic cavity)
  – Cavitas peritonealis (pelvis)

Connective-tissue spaces
- Space between the middle and deep layers of fascia cerrivicalis
- Mediastinum
- Spatium extraperitoneale
  – the spatium retroperitoneale (retroperitoneum) and
  – the spatium subperitoneale
- Bursa and synovial cavities

Note: The cavitates abdominis and pelvis are connected.

General Anatomy — *2. Overview of the Human Body*

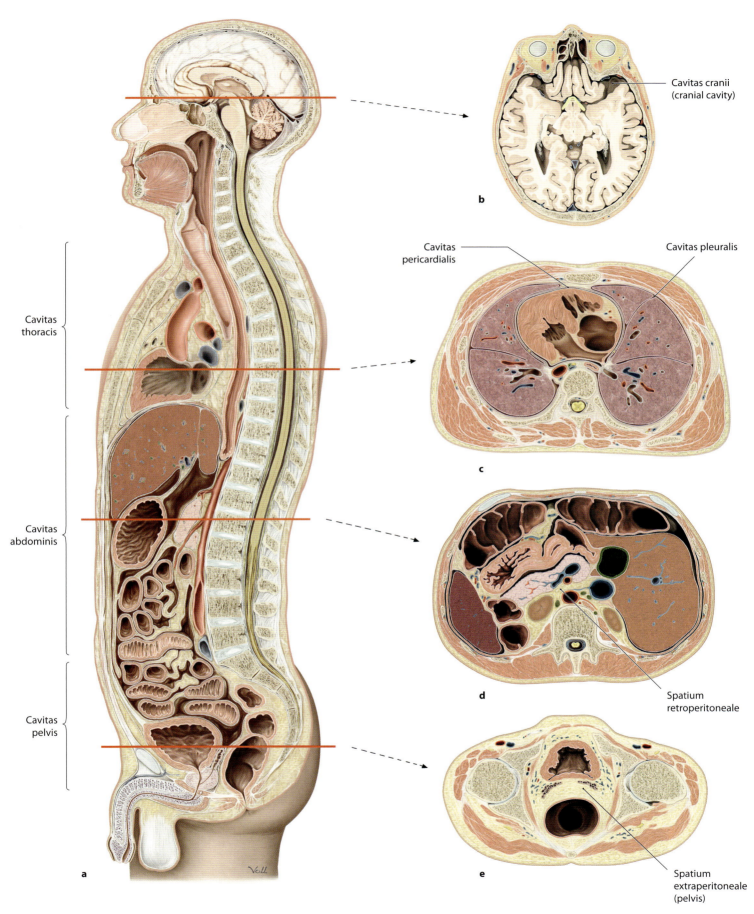

E Selected cross sections through the human body Superior view
a Midsagittal section.
b Cross section at the level of the head.
c Cross section through the thorax.
d Cross section through the abdomen.
e Cross section through the pelvis minor (see also Terms of Location and Direction, Cardinal Planes and Axes, p. 26).

3.1 Terms of Location and Direction, Cardinal Planes and Axes

A General terms of location and direction

Upper body (head, neck, and trunk)	
Cranialis	Pertaining to or located toward the head
Cephalad	Directed toward the head
Caudalis	Pertaining to or located toward the tail
Caudad	Directed toward the tail
Anterior	Pertaining to or located toward the front Synonym: Ventral (used for all animals)
Posterior	Pertaining to or located toward the back Synonym: Dorsal (used for all animals)
Superior	Upper or above
Inferior	Lower or below
Medius	Located in the middle
Flexor	Pertaining to a flexor muscle or surface
Extensor	Pertaining to an extensor muscle or surface
Axialis	Pertaining to the axis of a structure
Transversalis	Situated at right angles to the long axis of a structure
Longitudinalis	Parallel to the long axis of a structure
Horizontalis	Parallel to the plane of the horizon
Verticalis	Perpendicular to the plane of the horizon
Medialis	Toward the median plane
Lateralis	Away from the medial plane (toward the side)
Medianus	Situated in the median plane or midline
Centralis	Situated at the center or interior of the body
Peripheralis	Situated away from the center
Superficialis	Situated near the surface
Profundus	Situated deep beneath the surface
Externus	Outer or lateral
Internus	Inner or medial
Apicalis	Pertaining to the tip or apex
Basalis	Pertaining to the bottom or base
Occipitalis	Pertaining to the back of the head
Temporalis	Pertaining to the lateral region of the head (the temple)
Sagittalis	Situated parallel to the sagittal suture (sutura sagittalis)
Coronalis	Situated parallel to the coronal suture (pertaining to the crown of the head) (sutura coronalis)
Rostralis	Situated toward the nose or brow
Frontalis	Pertaining to the forehead (frons)
Basilaris	Pertaining to the skull base

Limbs	
Proximalis	Close to or toward the trunk
Distalis	Away from the trunk (toward the end of the limb)
Radialis	Pertaining to the radius or the lateral side of the forearm
Ulnaris	Pertaining to the ulna or the medial side of the forearm
Tibialis	Pertaining to the tibia or the medial side of the leg
Fibularis	Pertaining to the fibula or the lateral side of the leg
Palmaris (volaris)	Pertaining to the palm of the hand
Plantaris	Pertaining to the sole of the foot
Dorsalis	Pertaining to the back of the hand or top of the foot

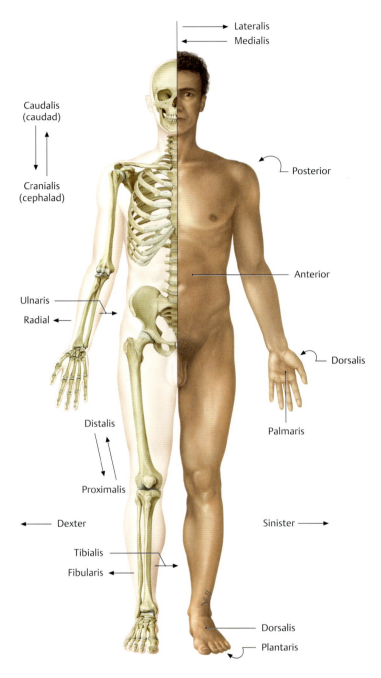

B The anatomic body position
The gaze is directed forward, the hands are supinated. The *right* half of the body is shown in light shading to demonstrate the skeleton.
Note that the designations "left" and "right" always refer to the patient.

C Abbreviations for anatomic structures

A. arteria (aa. = arteriae)
V. vena (vv. = venae)
M. musculus (mm. = musculi)
Lig. ligamentum (ligg. = ligamenta)
Nl. nodus lymphoideus (nll. = nodi lymphoidei)

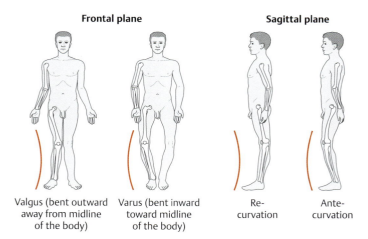

Valgus (bent outward away from midline of the body) | Varus (bent inward toward midline of the body) | Recurvation | Antecurvation

E Axis deviation of upper and lower limbs
Joint deformities at the limbs can lead to axis deviation in the two articulating bones in both frontal and sagittal planes. In accordance with international guidelines, axis deviation in the frontal plane is called varus or valgus deformity and recurvation or antecurvation in the sagittal plane. A varus deformity in the knee joint exists if the axis deviation is concave to the vertical body axis or if the distal bone (the tibia) points to the median line (e.g., genu varum = bow-leggedness). In a valgus deformity of the knee joint, the axis deviation is convex to the vertical body axis, or the tibia points away from the median line (e.g., genu valgum = knock knee).

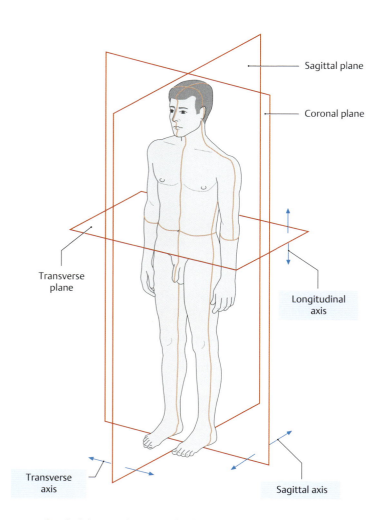

D Cardinal planes and axes in the human body (neutral position, left anterolateral view)
Although any number of planes and axes can be drawn through the human body, it is standard practice to designate *three cardinal planes and axes*. They are perpendicular to one another and are based on the three spatial coordinates.

The cardinal body planes:
- **Sagittal plane:** Any *vertical* plane that is parallel to the sutura sagittalis of the skull, passing through the body from front to back. The *midsagittal plane* (= median plane) divides the body into equal left and right halves.
- **Frontal (coronal plane):** Any plane that is *parallel to the forehead (frons)* or to the sutura coronalis of the skull. In the standing position, it passes vertically through the body from side to side.
- **Transverse plane (axial plane):** Any *horizontal*, cross-sectional plane that divides the body into upper and lower portions. It is perpendicular to the longitudinal body axis.

The cardinal body axes:
- **Vertical or longitudinal axis:** In the standing position, this axis runs through the body *craniocaudally* and is perpendicular to the ground. It lies at the intersection of the coronal and sagittal planes.
- **Sagittal axis:** This axis runs *anteroposteriorly* from the front to back surface of the body (or from back to front) and lies at the intersection of the sagittal and transverse planes.
- **Transverse or horizontal axis:** This axis runs from side to side and lies at the intersection of the coronal and transverse planes.

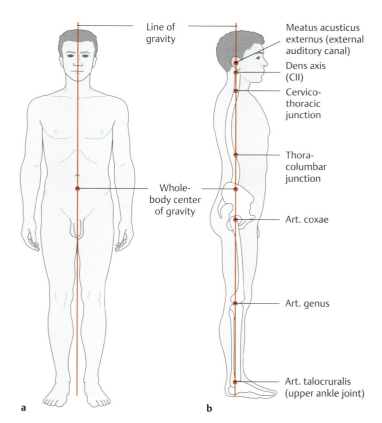

F The whole-body center of gravity and the line of gravity
a Anterior view. The line of gravity is directed vertically along the midsagittal plane, passing through the whole-body center of gravity below the promontorium ossis sacri at the level of the second sacral vertebra.
b Lateral view. The line of gravity passes through the external auditory canal, the dens of the axis (second cervical vertebra), the anatomic and functional junctions within the spinal column, the whole-body center of gravity, and through the hip, knee, and ankle joints (after Kummer).

3.2 Location and Designation of Imaging Planes

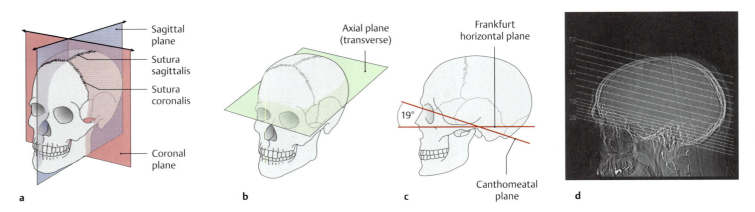

A Radiographic planes, illustrated for the skull

a Just as there are three standard anatomic sectional planes (frontal, sagittal, and transverse planes, see p. 27), certain standard sectional planes have been defined for the most important imaging techniques (computed tomography [CT] and magnetic resonance imaging [MRI]).

Radiographic plane		Anatomic plane
Coronal plane	=	Frontal plane (along or parallel to coronal suture)
Sagittal plane	=	Sagittal plane (along or parallel to sagittal suture)
Axial plane	=	Transverse plane (at the head, parallel to the Frankfurt horizontal plane, or at the trunk and limbs, perpendicular to the corresponding longitudinal axes)

b The Frankfurt horizontal plane (transverse plane) extends from the upper margin of the external auditory canal to the lower margin of the orbital cavity.

c u. d As a general rule, CT scans of the skull or the brain are slanted at a 19-degree angle from anterior to posterior (along the so-called canthomeatal plane). This plane extends from the upper margin of the external auditory canal to the canthus (corner of the eye). This protects the contents of the orbit from heavy X-ray exposure.

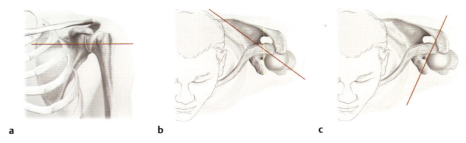

B Standard planes in MRI shoulder examination

The patient lies on his or her back with the arm of the shoulder joint that is to be examined positioned parallel to the body in arm outward rotation or neutral (null) position.

a Position of the axial (transverse) plane of the left shoulder, anterior view.
b Position of the angled coronal plane (parallel to the m. supraspinatus and vertical to the cavitas glenoidalis), superior view.
c Position of the angled sagittal plane (parallel to the cavitas glenoidalis), superior view.

Thanks to the imaging of the body in multiple sectional planes (or multiplane imaging) and the particularly good differentiation of the periarticular soft tissues, MRI is very well suited for joint exams. Depending on the sectional plane selected and the relaxation time (T1- and T2-weighted scans), the tissues can be differentiated according to their anatomic position and their fat and water percentages. *Note:* Axial (transverse) planes (see **D**) are always viewed from below (distal).

C Optimal MRI planes to show important anatomic structures of shoulder joint

MRI plane	Important visible anatomic structures
Axial (transverse) plane	• M. supraspinatus • Labrum glenoidale • Articular capsule (joint capsule) • Ligg. glenohumeralia • Tendon of biceps femoris, long head
Angled coronal plane	• M. supraspinatus • M. infraspinatus • Bursa subacromialis • Art. acromioclavicularis • Upper and lower labrum glenoidale
Angled sagittal plane	• Rotator cuff muscles • Lig. coracoacromiale • Acromion • Glenohumeral ligament • Subacromial space

General Anatomy — 3. Surface Anatomy of the Body, Landmarks, and Reference Lines

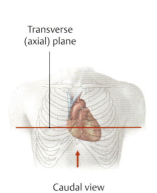

Transverse (axial) plane

Caudal view

a

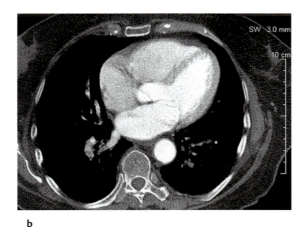

b

c

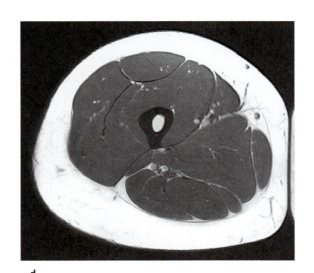

d

D Examining axial images
a and b trunk; c–e lower limb.

a Anterior view of thorax showing the location of the transverse (axial) plane at the level of T8 in **b**.
b Axial CT of thorax, inferior view, in the so-called soft-tissue window. The soft tissue (the heart) can be evaluated in high resolution; the lungs, however, because they are filled with air, are overexposed (black) and the bones are underexposed (white).
c Anterior view of right upper thigh, showing the location, and position of the transverse (axial) plane in **d**.
d Axial, T1-weighted MRI of right thigh, distal view.
e Cross section of the right upper tight in **d** with structures identified.

Axial (transverse) CT and MRI images of the trunk are viewed from below and of the limbs are viewed distally: the patient lies on his or her back. Thus, axial images always show the posteriorly positioned spinal cord (medulla spinalis) pointing downward, the anteriorly positioned skeleton of the thorax pointing upward. Correspondingly, the front of the upper thigh points upward and the back, downward; structures on the right side are displayed on the left side, and structures on the left side are displayed on the right side.
Note: Coronal (frontal) cross sections are viewed as if the patient stands facing of the examiner.

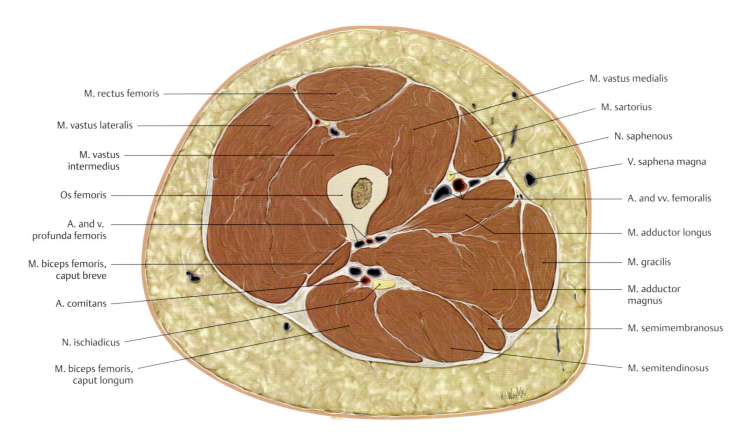

29

General Anatomy — 3. Surface Anatomy of the Body, Landmarks, and Reference Lines

3.3 Body Surface Anatomy

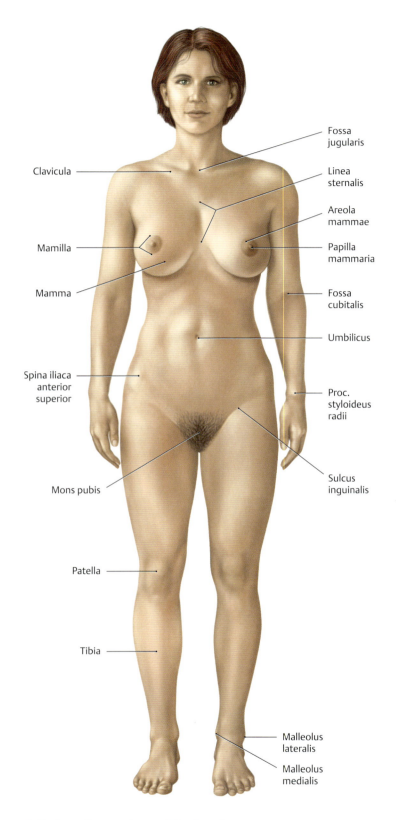

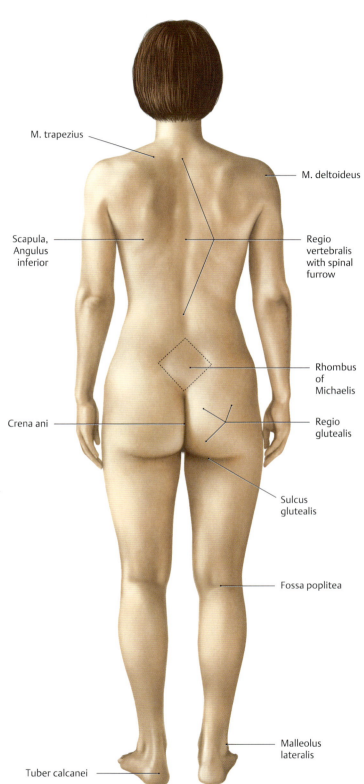

A Body surface anatomy of the female
Anterior view. Body surface anatomy deals with the surface anatomy of the living subject. It plays an important role in classic methods of examination (inspection, palpation, percussion, auscultation, function testing), and so it has particular significance in clinical examination courses. To avoid repetition, identical structures such as the olecranon have not been labeled on both the female and male bodies.

B Body surface anatomy of the female
Posterior view.

General Anatomy — 3. Surface Anatomy of the Body, Landmarks, and Reference Lines

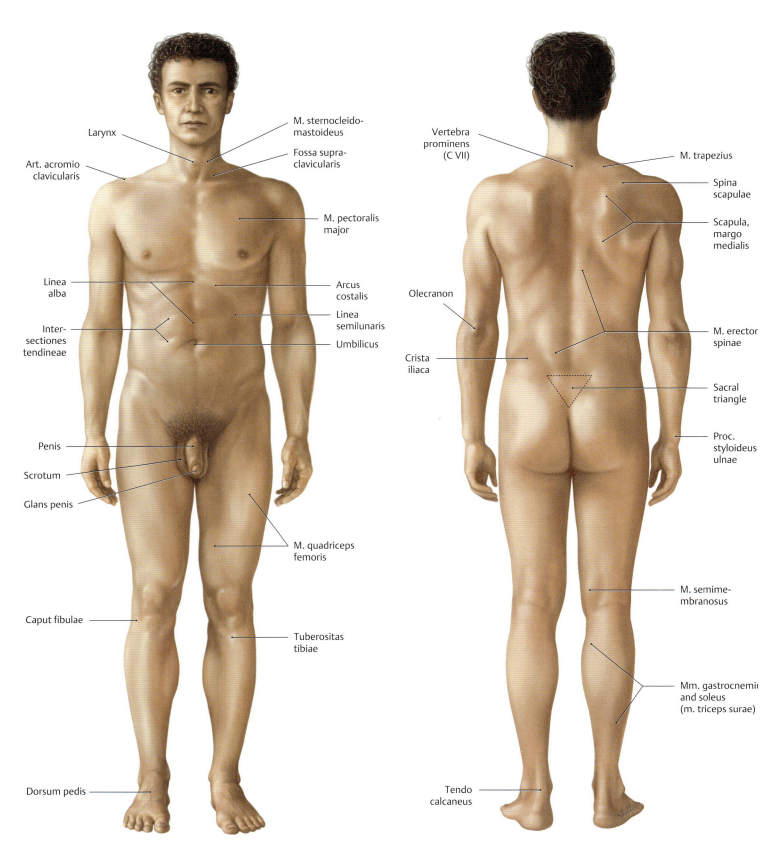

C Body surface anatomy of the male
Anterior view.

D Body surface anatomy of the male
Posterior view.

3.4 Body Surface Contours and Palpable Bony Prominences

Palpable bony prominences are important landmarks for anatomic orientation in the skeleton, as it is not always possible to palpate articulating skeletal structures (e.g., the hip joint). In these cases, the examiner must rely on palpable bony prominences as an indirect guide to the location of the inaccessible structure.

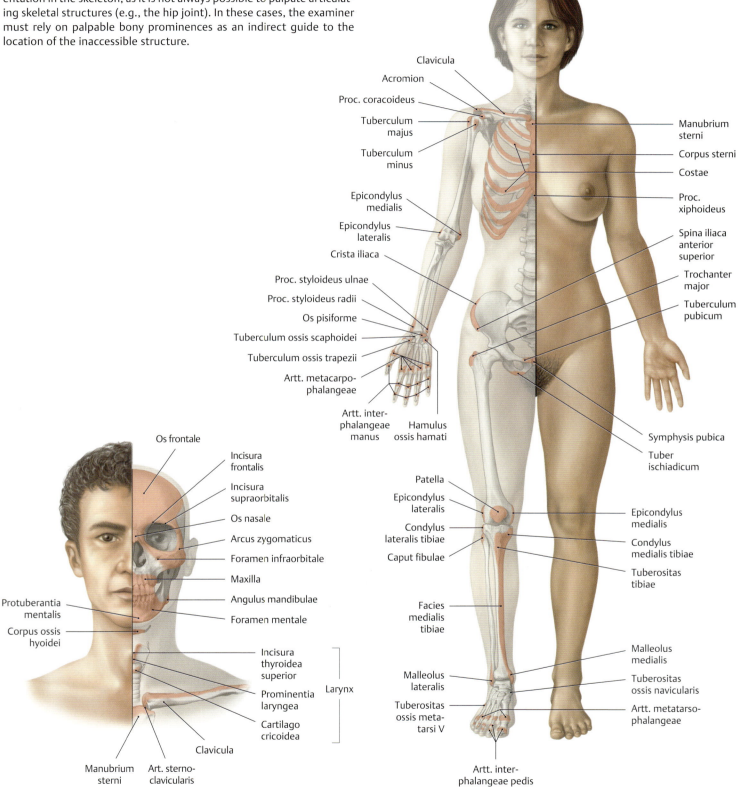

A Surface contours and palpable bony prominences of the face and neck
Anterior view.

B Surface contours and palpable bony prominences of the trunk and upper and lower limbs in the female
Anterior view.

General Anatomy — 3. Surface Anatomy of the Body, Landmarks, and Reference Lines

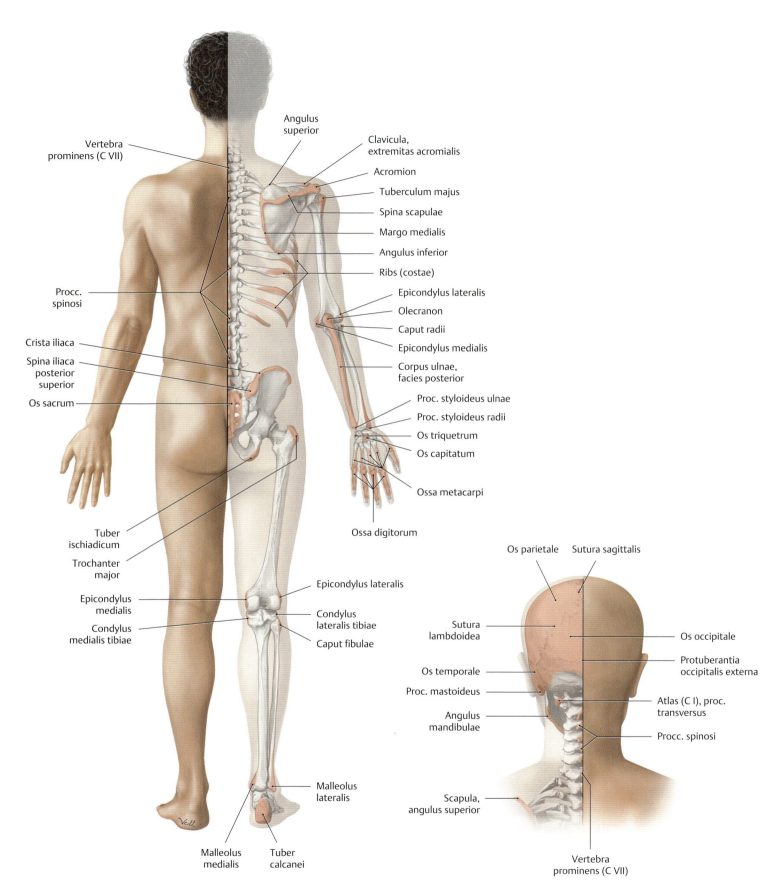

C Surface contours and palpable bony prominences of the trunk and upper and lower limbs in the male
Posterior view.

D Surface contours and palpable bony prominences of the head and neck
Posterior view.

3.5 Landmarks and Reference Lines on the Human Body

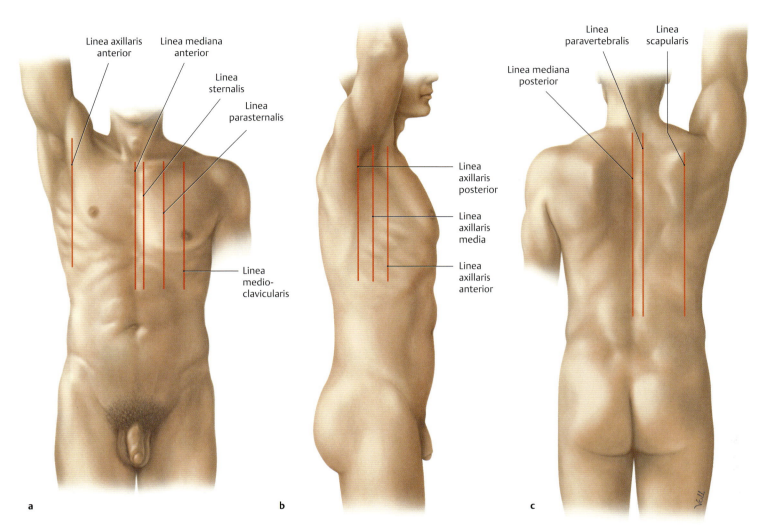

A Vertical reference lines on the trunk
a Anterior view, b right lateral view, c posterior view.

Linea mediana anterior	Anterior trunk midline passing through the center of the sternum
Linea sternalis	Line along the sternal margin
Linea parasternalis	Line midway between the linea sternalis and linea medioclavicularis
Linea medioclavicularis	Line through the midpoint of the clavicula (often identical to the linea mamillaris)
Linea axillaris anterior	Line at the level of the anterior axillary fold (m. pectoralis major)
Linea axillaris media	Line midway between the anterior and posterior axillary lines
Linea axillaris posterior	Line at the level of the posterior axillary fold (m. latissimus dorsi)
Linea mediana posterior	Posterior trunk midline at the level of the spinous processes
Linea paravertebralis	Line at the level of the transverse processes
Linea scapularis	Line through the angulus inferior of the scapula

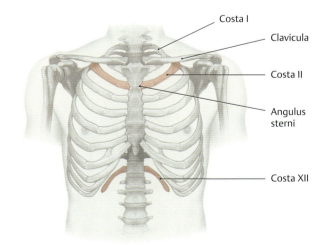

B "Rib counting" for anatomic orientation in the thorax
The first rib is covered by the clavicula. The first palpable rib is the second rib; therefore, the count begins at that level. The second rib attaches to the sternum at the level of the a ngulus sterni. At the lower end of the rib cage, it is best to start at the twelfth rib, which is palpable only in its posterior portion.

General Anatomy — 3. Surface Anatomy of the Body, Landmarks, and Reference Lines

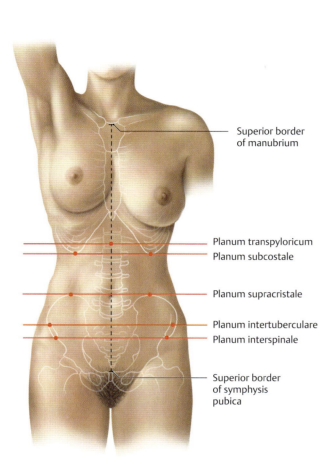

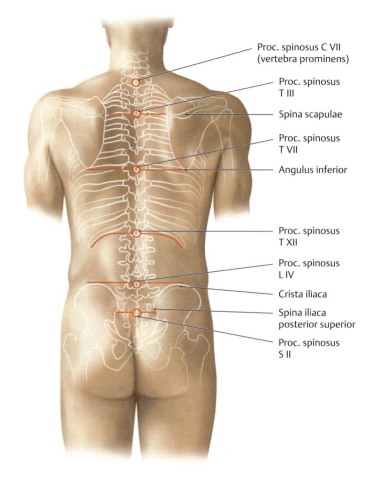

C Standard transverse planes through the cavitas abdominis
(see also p. 199)
Anterior view.

Planum transpyloricum	Transverse plane midway between the superior borders of the symphysis and manubrium
Planum subcostale	Plane at the lowest level of the costal margin (inferior margin of the tenth cartilago costalis)
Planum supracristale	Plane passing through the summits of the cristae iliacae
Planum intertuberculare	Plane at the level of the tubercula iliaca (the tuberculum iliacum lies approximately 5 cm posterolateral to the spina iliaca anterior superior)
Planum interspinale	Plane at the level of the spinae iliacae anteriores superiores

D Processi spinosi that provide useful posterior landmarks
Posterior view.

Proc. spinosus C VII	Vertebra prominens (the projecting spinous process of C VII is clearly visible and palpable)
Proc. spinosus T III	At the level of the line connecting the two spinae scapulae
Proc. spinosus T VII	At the level of the line connecting the anguli inferiores of both scapulae
Proc. spinosus T XII	Just below the twelfth rib
Proc. spinosus L IV	At the level of the line connecting the summits of the cristae iliacae
Proc. spinosus S II	At the level of the line connecting the spinae iliacae posteriores superiores (recognized by small skin depressions directly over the spinae iliacae)

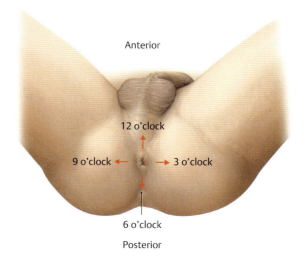

E Lithotomy position (supine with the legs, hips, and knees flexed and the thighs abducted)
The position of choice for proctological examinations. Clock-face notation is used for anatomic orientation (e.g., to describe the location of a lesion):

- Top   = toward os pubis   = 12 o'clock
- Bottom = toward os sacrum = 6 o'clock
- Right   = 3 o'clock
- Left   = 9 o'clock

35

3.6 Body Regions (Regional Anatomy)

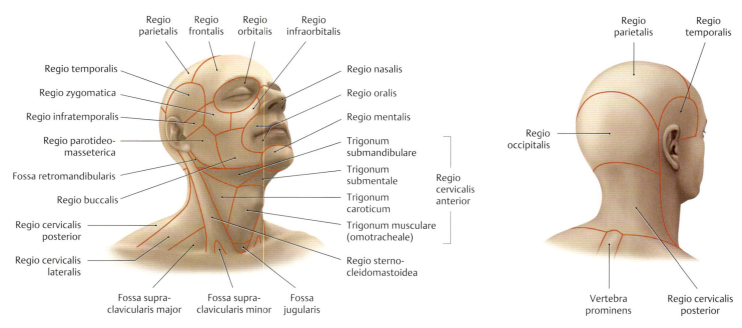

A Regions of the head and neck
Right anterolateral view.

B Regions of the head and neck
Right posterolateral view.

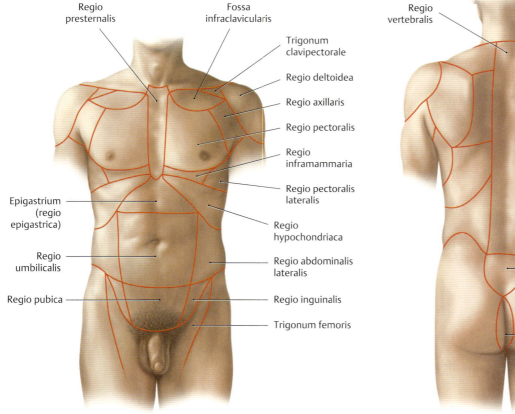

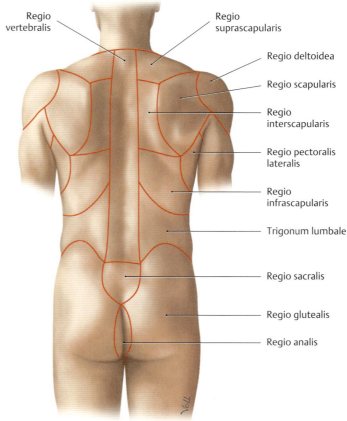

C Regions of the thorax and abdomen
Anterior view.

D Regions of the back and buttocks
Posterior view.

General Anatomy — 3. Surface Anatomy of the Body, Landmarks, and Reference Lines

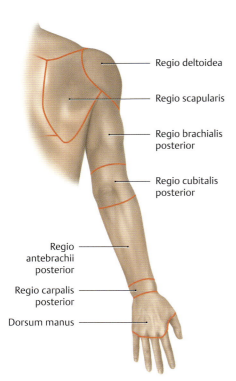

E Regions of the upper limb
Posterior view.

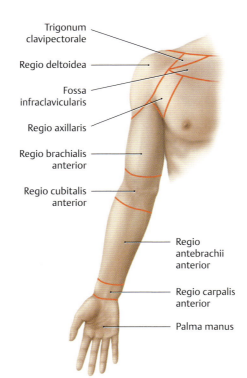

F Regions of the upper limb
Anterior view.

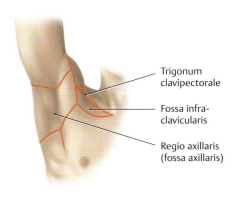

G Regions about the axilla
Anterior view.

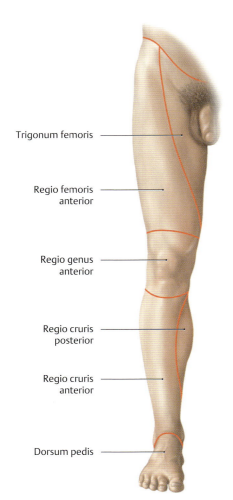

H Regions of the lower limb
Anterior view.

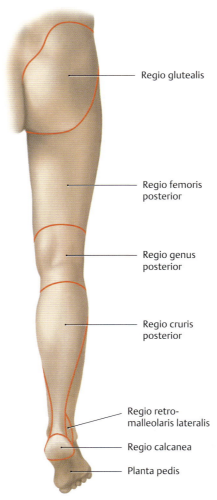

I Regions of the lower limb
Posterior view.

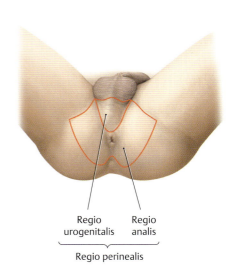

J Perineal region (lithotomy position)

37

4.1 The Bony Skeleton and the Structure of Tubular Bones

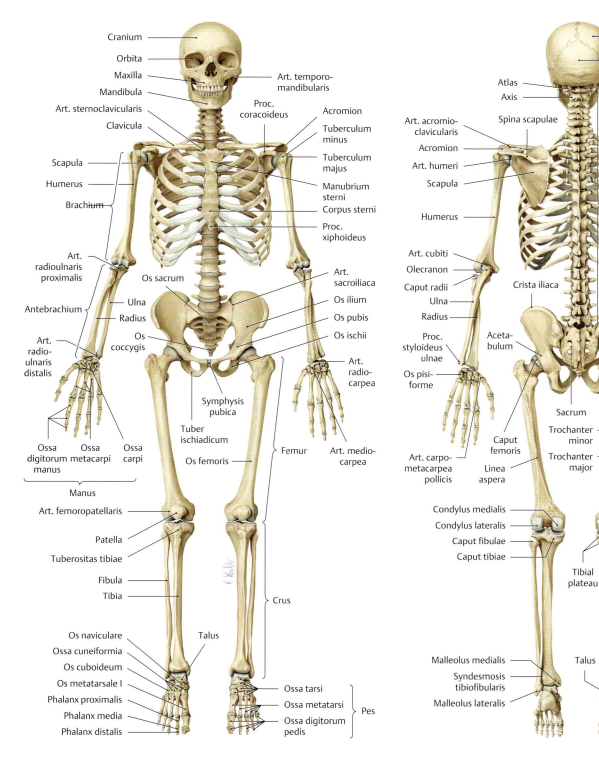

A Human skeleton anterior view
The left forearm is pronated, and both feet are in plantar flexion.

B Human the posterior view
The left forearm is pronated, and both feet are in plantar flexion.

C Types of bone

- **Ossa longa,** e.g., tubular bones of the limbs
- **Ossa brevia,** e.g., ossa carpi and ossa tarsi
- **Ossa plana,** e.g., scapula, os ilium, and bones of the calvaria
- **Ossa irregularia,** e.g., vertebrae; anomalous, supernumerary bones not consistently present, as in the skull base

- **Ossa pneumatica** (containing air-filled spaces), e.g., bones of the facial skeleton and sinus paranasales
- **Ossa sesamoidea** (bones incorporated in tendons), e.g., the patella
- **Ossa accessoria** (anomalous, supernumerary bones), as in the calvaria and foot (generally result from the failure of fusion of certain adjacent ossification centers)

General Anatomy —— 4. The Bones and Joints

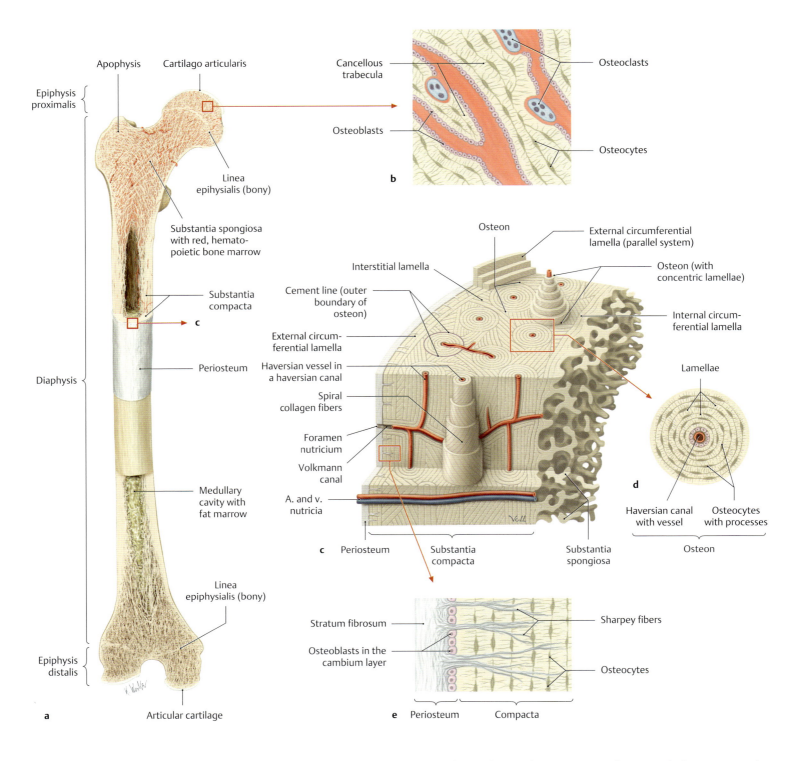

D Structure of a typical tubular bone, illustrated for the femur

a Femur (os femoris) with coronal sections through the proximal and distal parts.

b Detail from **a**: The sectioned areas display the lamellar architecture ("lamellar bone") of the cancellous trabeculae. The lamellae are arranged in contiguous plates, similar to plywood. Since the cancellous trabeculae do not have an actual vascular supply and are nourished by diffusion from the adjacent medullary cavity, the trabeculae attain a thickness of only about 200–300 μm.

c Detail from **a**: Three-dimensional representation of substantia compacta, whose structural units consist of vascularized osteons approximately 1 cm long and 250–350 μm in diameter. The haversian canals, which tend to run longitudinally in the bone, are connected to one another by short transverse and oblique Volkmann canals and also to the vessels of the periosteum and medullary cavity.

d Detail from **c**, demonstrating the microstructure of an osteon. The haversian canal at the center is surrounded by approximately 5 to 20 concentric lamellar systems composed of osteocytes and extracellular matrix. The osteocytes are interconnected by numerous fine cytoplasmic processes.

e Detail from **c**, showing the structure of the periosteum.

4.2 Continuous and Discontinuous Joints: Overview and False Joints (Synarthroses)

A Different types of joints

False joints	True joints

(Synarthroses; continuously connected by connective tissue, cartilage, or bone; low to moderate degree of mobility; see right side)
 	
• **Syndesmoses** (fibrous joints, e.g., newborn fontanelles)
• **Synchondroses** (cartilaginous joints, e.g. spinal disks; if the intervening tissue is mostly fibrocartilaginous, the joint is called a symphysis, e.g., symphysis pubica)
• **Synostoses** (sites of bony fusion, e.g., os sacrum)
	(Separated by a joint space; mobility is variable, depending on the attached ligaments = discontinuous connection)
	
	• **Diarthroses** (joints with varying degrees of freedom, see p. 42); classified according to
	  – Shape and arrangement of articular surfaces (spheroidal, ellipsoidal, and saddle joints)
	  – Number of joint axes
	  – Number of degrees of freedom
	• **Amphiarthroses** (stiff joints, e.g., art. sacroiliaca)

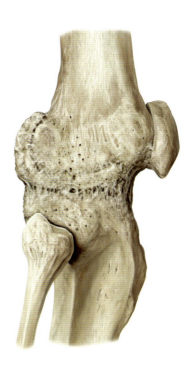

C Ankylosis of the knee joint (art. genus)
Illustration based on a specimen from the anatomic collection of the University in Kiel. Following the complete loss of cartilage, bone gradually fills the joint cavity, resulting in complete stiffness of the joint.

B Short glossary of joint diseases, imaging, and treatment

- **Pseudarthrosis:** So-called false joint (instability due to missing wedge of bone) after abnormal fracture healing (see p. 53).
- **Ankylosis:** Abnormal bony fixation of a true joint (see **C**).
- **Arthrodesis** (called spondylodesis in the spinal cord): Surgically induced joint ossification by performing osteosynthesisis procedure (see p. 53 bottom **D**).
  – Main indicators: infectious arthritis, post-traumatic joint destruction, degenerative joint diseases, and paralytic instability.
  – Principle: loss of mobility in favor of absence of pain and stability.
  – Important for good joint motion postsurgery: stabilization in mode of function, which means in the position that is crucial for the most important joint functions, such as grasping for the arm and standing and walking for the leg.
- **Arthrotomy:** Surgical opening of a joint.
- **Arthrography:** Examination of the joint cavity using x-rays with the aid of a contrast agent (has increasingly lost relevance after rise in the use of MRI).
- **Arthroscopy:** Endoscopic examination of the joint, often associated with ensuing endoscopic therapy, e.g., arthroscopic reconstruction of injured ligament and capsule structures, removal of loose bodies, and treatment of joint cartilage defects (e.g., osteochondrosis dissecans).
- **Synovectomy:** Removal of the synovial membrane, e.g., in chronic polyarthritis.
- **Joint puncture** (intra-articular injection): Puncture of a joint space to remove fluid (e.g., in an articular effusion) or application of substances (medication).
- **Joint endoprothesis:** Joint replacement surgery through total or partial endoprosthesis in advanced joint disease (arthritis) (see p. 46).

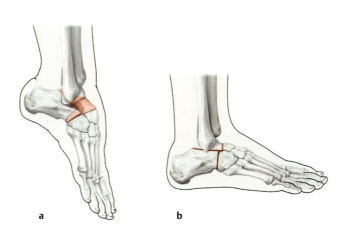

a   b

D T-arthrodesis after Lambrinudi for treatment of a club foot deformity (after Niethard and Pfeil)
Right foot before (**a**) and after (**b**) arthrodesis treatment, lateral view. The surgical goal is to repair the club foot position caused by stiffness (arthrodesis) to achieve a plantigrade foot (lower leg and foot form a 90-degree angle) while at the same time maintaining mobility in the upper ankle joint. Since three joints (art. subtalaris, art. talonavicularis, and art. calcaneocuboidea, see p. 454) are fused or stiffened, the procedure is called T-arthrodesis for "triple arthrodesis." It involves resecting the talar head and interlocking it with the navicular bone by performing osteosynthesis.
Note: In case of instability and ankle joint misalignment, arthrodesis is the safest treatment method.

General Anatomy — **4. The Bones and Joints**

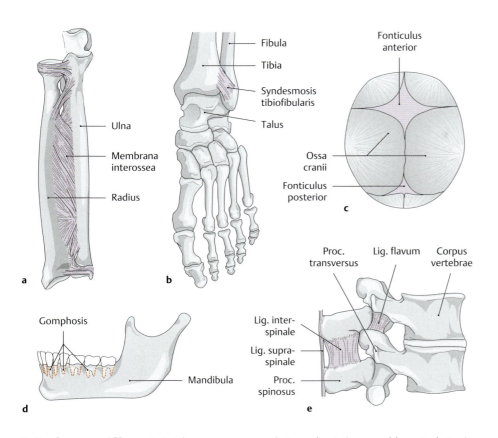

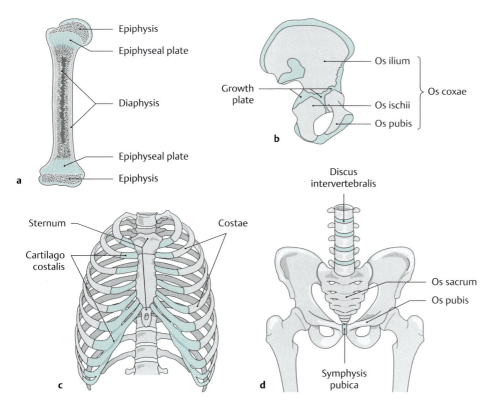

E Syndesmoses (fibrous joints)
a Membrana interossea.
b Syndesmosis tibiofibularis.
c Fontanelles.
d Gomphosis (immovable articulation).
e Lig. flavum, Lig. interspinale and Lig. supraspinale.

F Synchondroses (cartilaginous joints)
a Epiphyseal plates prior to closure.
b Os coxae before closure of the growth plates.
c Cartilago costalis.
d Symphysis pubica and Discus intervertebrales (Symphysis intervertebralis).

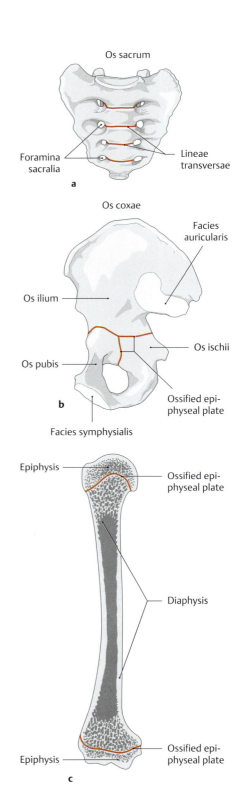

G Synostoses (sites of bony fusion)
a Os sacrum (fused sacral vertebrae).
b Os coxae (fusion of the os ilium, os ischii, and os pubis).
c Closed and ossified epiphyseal plates.

41

4.3 True Joints: Components and Intra- and Extra-articular Structures

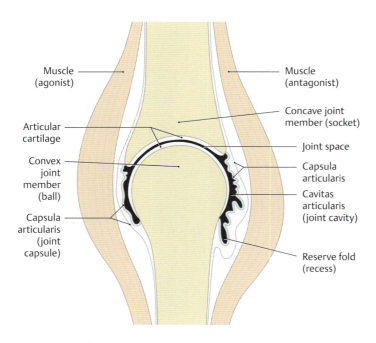

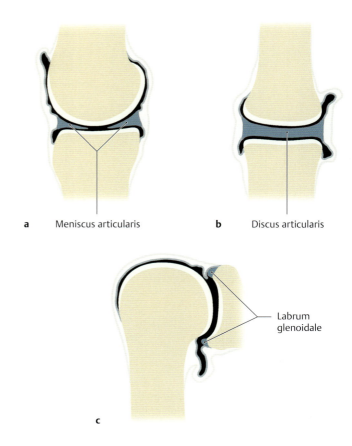

a Meniscus articularis
b Discus articularis
c Labrum glenoidale

A Structure of a true joint
In true joints, the articulating joint surfaces are separated by cavities measuring no more than a few millimeters in width (diarthroses), unlike false joints, in which the joint surfaces are connected (synarthroses). True joints are also called synovial joints, as the joint capsule produces synovial fluid. In addition to the cavity, true joints have the following characteristics:

- Variously shaped articular surfaces (facies articulares) covered by hyaline cartilage
- Joint cavity (cavitas articularis) with articular recesses, the width of which vary
- All-around closed joint capsule increased by variously shaped plicae alares, plicae synoviales, and villi synoviales
- Highly viscous synovial fluid produced by joint capsule
- Intra-articular components to improve joint congruency and increase force-absorbing surface, e.g., menisci and disci articulares and labra articularia
- Various forms of ligaments, with ligg. intracapsularia and extracapsularia acting as primary joint stabilizers
- Joint crossed by muscles that move it in opposite directions (agonist/antagonist)
- Synovial bursa, often located near the joint and thus articulating with the joint cavity
- Not all of the above are present in all true joints.

B Intra-articular structures and their importance to the functional integrity of joints
Some joints contain intra-articular structures that ensure stability and function. For example, they increase the force-transmitting surface by balancing out incongruencies between the different parts of the joint, thus considerably reducing the compressive force on the cartilage. By definition, intra-articular structures are located in the articular cavity and float within synovial fluid, which means they are in direct contact with the synovial fluid through which they are primarily nourished (menisci, disks, articular labra).

a **Menisci articulares:** Crescent-shaped structures, wedge-shaped in cross section, which are formed regularly in the knee only. They consist of collagen fibers, connective tissue, and fibrous cartilage. While their peripheral components are supplied by blood vessels of the joint capsule with which they are conjoined, the interiorly located fibrocartilaginous components are nourished by the synovia.

b **Disci articulares** partially consist of connective tissue and fibrous, cartilaginous, disk-like structures, which divide the joint into two separate chambers. Articular disks are regularly found in the art. temporomandibularis (jaw), art. sternoclavicularis and art. radiocarpea (proximal wrist joint) (see p. 287).

c **Labra articularia** are wedge-shaped structures at the outer margins of the osseous hip and shoulder joint sockets (labrum glenoidale and labrum acetabulare, see pp. 266 and 430). They consist mainly of fibrocartilage, and their outer layer is composed of connective tissue, the fibers of which are joined to the capsules. The articular labra enlarge the ar-ticulating surfaces of the shoulder and hip joint.

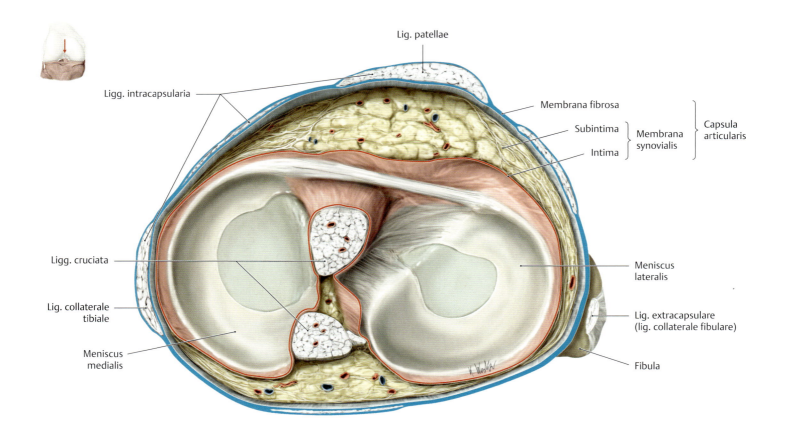

C The joint ligaments: extra-articular structures that are usually located intracapsular (inside the joint capsule)

Ligg. intracapsularia can run along both in the subintima (ligg. cruciata) and in the membrana fibrosa (e.g., ligg. iliofemorale, ischiofemorale, and pubofemorale of the hip joint). They strengthen the capsule and thus serve mainly mechanical functions. For example, they are responsible for stabilizing and guiding the joint (guiding ligaments) or, to varying degrees, limiting or restricting motion (restraining ligaments). Occasionally, the joint ligaments extend without any direct contact to the actual joint capsule (ligg. extracapsularia), such as the lig. collaterale fibulare in the knee joint. Often, ligament structures, which at first seem to run in the cavitas articularis (e.g., the ligg. cruciata in the knee joint, see p. 444, and the lig. capitis femoris in the hip joint), are incorrectly called intra-articular ligaments. However, unlike the meniscus and the labrum articulare, they are always covered by a thin synovial intima and thus run subintimally. Strictly speaking, they are located inside the capsule, which means they are intracapsular, and thus outside the actual joint capsule, which means they are extra-articular (see p. 44).

Note: Ligaments must have the "right" length in order to be able to guide and stabilize the joint. If they are overstretched or tear (as a result of chronic overuse or injuries, e.g., as sprains or dislocation), the joint becomes unstable. If ligaments shorten because the joint is not moved or held permanently in an unfavorable position (e.g., knee joint in bent position, proximal phalanges in stretched position), the joint becomes more or less immobile (joint contracture).

D Composition and function of synovial fluid

Synovial fluid is a clear, slightly yellowish, viscous fluid (product of secretion of the synovial cysts; pH value 7.4–7.7). Even in large joints, such as the knee joint, its volume is only 3–5 mL (less than 1 mL in small joints).

Compositon	Function
• Hyaluronate (or hyaluronic acid), lubricin (mucin-like glycoprotein), and phospholipids • Blood plasma components (mainly blood proteins and glucose) • Defense cells 60–150/μL (mainly macrophages and lymphocytes)	① Nourishing the hyaline cartilage, which is devoid of blood vessels, through diffusion and convection ② Lubricating the cartilaginous articular surfaces, thus enabling frictionless motion ③ Functioning as a buffer by evenly distributing pressure across the articular surfaces

4.4 True Joints: Structure of Joint Capsule and Hyaline Cartilage

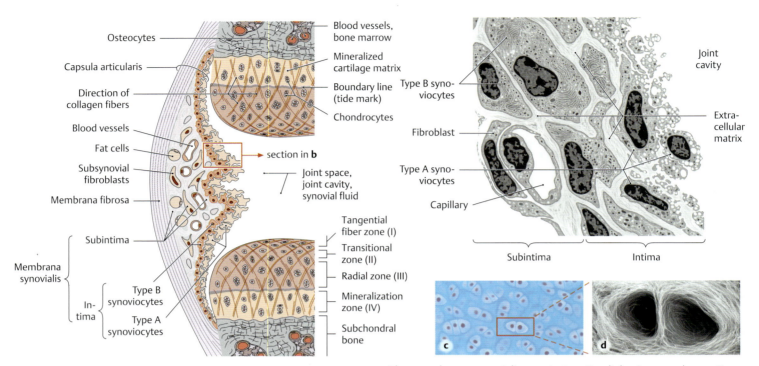

A Structure of joint capsule and hyaline cartilage
The joint cavity of true joints is fully enclosed by a joint capsule (capsula articualris), the two layers of which differ morphologically from one another and serve different functions:

Joint capsule (capsula articularis)

- **Membrane fibrosa – WALL** (firm, collagenous connective tissue; fibers extend to the periosteum). Thickness varies both within one joint and in different joints; in many joints reinforced by ligaments (ligg. intracapsularia, see p. 43)
- **Membrana synovialis – WALLPAPER** (attached to the margin of the joint capsule)

Synovial intima – lining cells (1 to 3 layers of epithelial-like [not epithelial cells because no basal membrane] covering cells); **production and resorption of synovial fluid.**

- **Type A synoviocytes** (boundary layer to the joint capsule): macrophage-like cells with vacuoles, Golgi apparatus, mitochondria, lysosomes; resorption of used synovial fluid and phagocytosis of bacteria and cell debris
- **Type B synoviocytes** (underneath the type A synoviocytes): fibroblast-like cells with rough endoplasmic reticulum, secretory granules; mainly production of synovial fluid

Synovial subintima (connective tissue with blood and lymph vessels, pain receptors [free nerve endings], and mechanoreceptors [Pacini and Ruffini bodies = proprioceptors], which react to changes in lengths or tension in the capsula articularis = **helps guide joint movement sequences**), structure differs based on topography

- **Areolar membrana synovialis** (loose, richly vascularized connective tissue)
- **Adipose membrana synovialis** (mainly fat cells in connective tissue)
- **Fibrous membrana synovialis** (vessel-poor and collagen-rich connective tissue)

Note: The **synovial membrane (membrana synovialis)** can regenerate into old age, even after complete removal (= synovectomy; necessary, e.g., in cases of chronic arthritis due to rheumatism).

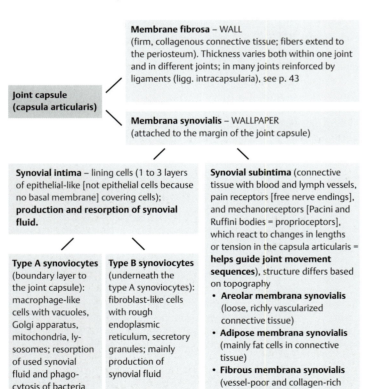

The membrana synovialis reacts to stimuli by increased secretion, which appears as articular effusion and swelling in the entire joint region. Depending on the type of stimuli (mechanical, allergic, infectious), the effusion is either clear and thin or turbid and pus-filled. A true bloody effusion (hemarthrosis) develops in the wake of injuries (e.g., cruciate ligament rupture). The ensuing pain is the result of overstretching the joint capsule and the release of inflammatory mediators (e.g., prosta-glandins, histamines, bradykinin, and cytokines). With the exception of the jaw and art. sternoclavicularis, which are covered by fibrous cartilage, all articular surfaces are covered by **hyaline cartilage**. It is malleable and reacts to pressure (elastic compression). Depending on the amount of pressure, its thickness varies—ranging from 1–2 mm in the phalangeal joints to 5–7 mm in the femoropatellar joints. Hyaline cartilage consists of extracellular matrix (ECM) and chondrocytes (approx- imately 5% volume fraction).

Note: While the synovial membrane can regenerate through the adjoining connective tissue, the hyaline cartilage (except in the primordial skeleton) does not have any surrounding connective tissue (perichondrium) and thus can hardly regenerate. In addition, it is devoid of vessels. The cartilage cells (chondrocytes) must exclusively be nourished by synovial fluid (diffusion and convection, see **D**, on the right).
Hyaline cartilage can be divided into zones. They are determined by the arcade-like course the collagen fibrils follow (**a**). From zone IV (deepest zone, where the collagen fibrils are connected to the bone underneath the cartilage), the collagen fibrils extend almost vertically (radially) to the articular surface (zone III), then turn to follow a course parallel to the surface (zone II), and again after a short tangential course extend downward (zone I).

The collagen fibrils are visible only under the transmission electron microscope. In a light microscope image, they are "masked" as they break the light just as their surroundings do. After enzymatically liberating chondrocytes and proteoglycans using hyaluronidase, looking through the transmission electron microscope creates the impression of empty "cartilage holes" that are surrounded by a thick mesh of collagenous fibrils (for composition of collagen fibrils, see **d**). Proteoglycans, a component of the extracellular matrix, however, are clearly visible under the light microscope because they carry many negative charges, which can be visualized by using dyes such as hematoxylin (see **c**).

General Anatomy — 4. The Bones and Joints

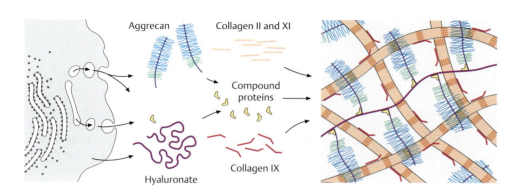

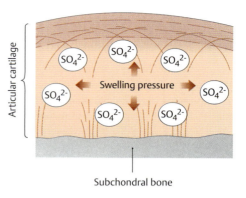

B Structure and function of extracellular matrix (ECM)
The extracellular matrix (all macromolecules that are secreted by cartilage cells and immobilized when interacting with other molecules in the extracellular space) primarily consists of

- Collagen fibers or fibrils (mainly collagen type II, IX, and XI)
- Proteoglycans (mainly aggrecan) with glycosaminoglycans (e.g., hyaluronate, chondroitin, and keratan sulfate), as well as core and compound proteins
- Glycoproteins (e.g., adhesion proteins such as chondronectin)
- Interstitial fluid (water) and electrolytes (mainly cations, Ca^+, K^+, Na^+)

The collagen fibrils (diameter of 15–130 nm) consist of collagen molecules, each of which are formed by three polypeptide chains (α-chains) and have the shape of a right-handed screw (triple helix). Individual triple helices (tropocollagens) are cross-linked by covalent connections, which give them their extraordinary tensile strength.

D Biphasic viscoelastic model of the joint cartilage (after Mow and colleagues)
Firmness and elasticity during compression of the cartilage are crucial for the proper functioning of the joints. They are both ensured by the swelling pressure (the principle of the elastic spring). It arises from the interaction of proteoglycans and collagen fibrils in the extracellular matrix (ECM), where they collaborate to carry out particular functions. The extraordinary tensile strength of the collagen fibrils allows for mechanical stability and structural integrity of the joint cartilage. The polyanionic proteoglycans ensure its physiochemical properties (e.g., reversible water-binding capacity). In free, watery fluid, proteoglycans, which are negatively charged, would normally greatly expand by repelling each other. They are prevented from doing so only by the tensile strength of the collagen fibrils, so they can take up only one fifth of the space they actually require. There is only a limited possibility for further compressing the proteoglycans. If cartilaginous tissue is subjected to mechanical pressure (normal pressure), interstitial fluid and cations escape into the joint cavity, and the ECM increasingly thickens. As a result, it is harder for fluid to drain from the cavity, the hydrostatic pressure in the tissue rises (viscid components of the viscoelastic model), until the hydrostatic pressure from inside and mechanical pressure from outside are in balance, and no further compression occurs. With the thickening of the ECM, the negatively charged glycosaminoglycans approach each other, leading to an increased repulsion force as a result of which negative pressure builds up. When the pressure is relieved, the tissue swells up to its original size due to proteoglycans repelling each other, and interstitial fluid and free cations migrate anew into the tissue (elastic component of the viscoelastic model) until the collagen network again puts a stop to it. The flow of interstitial fluid is also called *convection*. It is a crucial precondition for the nourishment of the joint cartilage, which does not contain blood vessels.

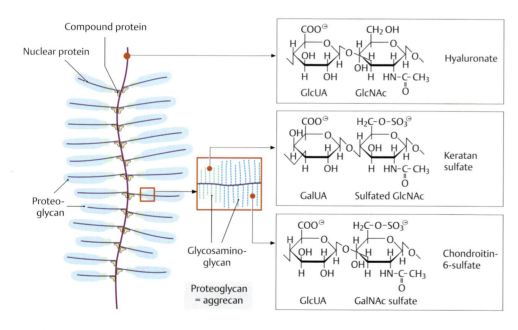

C Aggrecan, the most important proteglycan of the hyaline cartilage (after Koolman and Roehm)
Proteoglycans such as aggrecan are very large molecular complexes that consist of more than 100 glycosaminoglycan chains. They look like a bottle brush, with the individual glycosaminoglycan molecules as "bristles," affixed to a central hyaluronate molecule through so-called compound proteins. Disaccharide units, which mostly consist of a uric acid (e.g., glucuronic acid, GlcUA) and a type of amino sugar (e.g., N-acetylgalactosamine, GalNAc) and usually carry two negative charges (one each on the carboxyl and on the sulfate groups), form the basic structure of the glycosaminoglycan chains. For reasons of electroneutrality, the proteoglycans bind cations, which thanks to osmosis can reversibly bind a lot of water. Due to their strong water-binding capabilities, proteoglycan molecules take up huge volumes of water and thus create large, space-filling structures. GalUA, galacturonic acid; GlcNAc, N-acetylglucosamine.

4.5 Degenerative Joint Diseases as Exemplified by Hip Joint Arthrosis

A Etiology and pathogenesis of osteoarthritis
For many years now, degenerative joint diseases—mainly osteoarthritis—have been the top cause of disability worldwide. Due to the high cost of treatment and the large number of days employees are absent at work, osteoarthritis represents a major economic problem. Elderly people are particularly affected. The seeming increase in cases of this disease over the past few decades is due to people reaching ever higher ages, which allows them to actually experience their osteoarthritis (higher age and degenerative joint disease don't necessarily go hand in hand).

The main cause of osteoarthritis is an imbalance between the load and the capacity of the articular cartilage. Secondary osteoarthritis (in contrast to primary osteoarthritis, the cause of which is not pathologic) develops almost exclusively due to suboptimal biomechanics, e.g., in cases of incongruencies of the articulating surfaces. Over- or incorrectly loaded joints play a major role, e.g., caused by post-traumatic deformities, joint dysplasia, axis deviation, and instabilities. In particular, an increasing number of sports injuries lead to osteoarthritis. Excess weight, lack of exercise, and bad nutrition further contribute to the development of osteoarthritis. Early detection and correction of axis deviations or osteotomies to correct joint incongruencies are of particular significance for the prevention of osteoarthritis. If osteoarthritis has already developed, and conservative therapies have been exhausted, in addition to performing arthrodesis (joint stiffening, see p. 40), there is a possibility to replace the joint (endoprosthesis, see **C**).

One of the main causes of the disease is the gradual attrition, meaning the wear and tear and destruction, of the articular cartilage (see **D**). If the hyaline cartilage is too heavily loaded over the long term (see etiology), it disrupts the regular flow of interstitial fluid into the articular cartilage (see p. 40). This affects not only the elasticity but also the nourishment of the bradytrophic cartilage (no blood vessels), which, due to its limited metabolism and a missing perichondrium, is barely able to regenerate (see p. 44). As a result of increased mechanical overload of articular cartilage areas, both cartilage matrix and chondrocytes are damaged (see **D**). This damage is irreversible; complete restoration or repair is not possible. There are, however, endogenic compensatory mechanisms, which are supposed to make up for the loss of cartilage in the early stages (I and II, see **D**) and the increased pressure in the subchondral bone in the late stages (III and IV). The body's attempts to repair itself explain the shift between pain-free periods and periods marked by heavy pain. Heavily loaded joints such as hip or knee joints or the intervertebral joints of the spinal column are per se affected by osteoarthritis (see **B**).

* Synonyms: osteoarthrosis—in contrast to arthritis, in the process of which the cartilage disappears only secondarily as a result of inflammation of the articular cartilage.

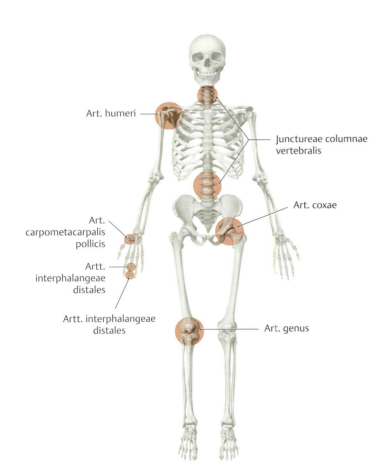

B Common locations for degenerative joint diseases

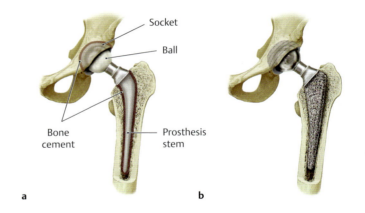

C Total endoprosthesis (TEP) of the hip joint
a Cemented TEP, b Noncemented TEP.
Endoprosthetic joint replacement has become a possibility for almost all joints. The best results, particularly in terms of long-term success, have been achieved in the hip joint (replacement of head and socket). The implants—usually made of metal, synthetic material, or ceramic —are either attached to the joint with a special cement or implanted "cement free." The cemented TEP is immediately stressable postsurgery; the noncemented TEP requires a little more time (several weeks) to firmly connect with the surrounding bone tissue. However, a noncemented prosthesis is easier to replace should it be necessary if it becomes loose, since there is no cement to remove. Which prosthesis to implant depends on several factors, such as age, primary bone stability, and possible preexisting bone diseases (osteoporosis). Complications of artificial joint replacement are foremost loosening of the joint and infection. Both cemented and noncemented prostheses currently have a 90% probability of lasting significantly longer than 10 years.

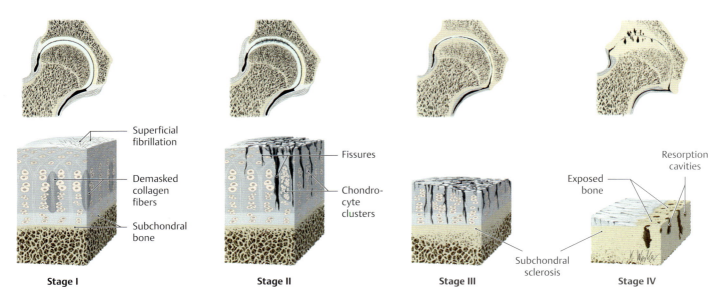

D Stages, clinical symptoms, and compensatory mechanisms of osteoarthritis

Stage I	Stage II	Stage III	Stage IV
• Articular surface is roughed up and begins to fray (surface fibrillations) • In the deeper lying cartilaginous layers, collagen fibers are exposed (demasked) by decreased proteoglycan synthesis	• Development of fissures that extend deep into the cartilage • Loss of cartilage matrix • In the fissure area, individual cartilage cells divide again (chondrocyte clusters develop)	• Increasing density of the subchondral bone as a result of increasing cartilage attrition • First osteophytes appear at the bone bridge • Inflammatory response at the articular capsule (reactive synovitis, usually with effusion) as a result of increased accumulation of deposits caused by cartilage wear	• The bone is entirely exposed; the bone surfaces rub on each other • Collapse of the subchondral bone roof (bone necrosis) • Development of large resorption cavities in the articulating joint surfaces (see E) • Increasingly strong growth of osteophytes

Compensatory mechanisms in the early stages (I to III)	Compensatory mechanisms in the advanced stage (IV)	Goal of the compensatory mechanisms
• Limited proliferation (cell division) of the previously post-mitotic cartilage cells • Formation of so-called chondrocyte clusters, or brood capsules, with newly developed extracellular matrix	• Activation of mesenchymal progenitor cells in opened subchondral cavities within the spinal cord (subchondral cysts) • Followed by formation of a new articular surface through metaplastic fibrocartilage	Compensate for cartilage loss
Reactive thickening of the subchondral bone with increasing bone density (subchondral sclerosis)	Enlargement of the articular force–transmitting surface through formation of marginal osteophytes (so-called marginal exostosis)	Equalize pressure in the subchondral bone

Clinical symptoms: After an asymptomatic early stage of the disease (so-called latent arthrosis) that can last for years, patients initially report discrete pain, which gets worse with increased joint load and is accompanied by the first signs of referred muscle pain. In a more advanced stage of the disease, in addition to stress-related pain, patients experience pain when in neutral position (so-called activated osteoarthritis). Its cause lies in the inflammatory response of the articular capsule to the increased accumulation of deposits caused by cartilage wear. The reactive synovitis is almost always accompanied by effusions. As the disease progresses, symptoms intensify, with simultaneous increased joint instability. The final stage is characterized by constant pain, heavily limited load-carrying capacity, strong muscle contraction, drastic malalignment, and stiffening of the joint.

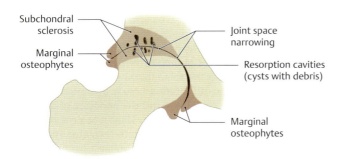

E Synopsis of the pathologic changes observed in X-ray examination of advanced hip joint arthrosis

Schematic representation, anteroposterior view of left hip joint. The destruction and degeneration of the hyaline articular cartilage leads to typical signs in the conventional X-ray, including narrowing of the cartilage gap in the joint parts that are most heavily loaded as a result of increasing loss of cartilage, the subchondral thickening (subchondral sclerosis) and lightening (subchondral cyst formation) of the bone, and the presence of osteophytes (from reactive bone changes) both in loaded and unloaded areas.

4.6 Basic Principles of Joint Mechanics: Movements

A Joints as functional units
In order to understand joints as functional units, it is best to draw on physical laws, for although joints primarily enable motion, at the same time they guarantee stability of the musculoskeletal system and assist in transmitting force between the articulating bones.

- Movements are changes of location in space (translational and rotational motion), which have both temporal (uniform or nonuniform motion in time) and spatial aspects (variable capacity for motion along the three axes) (see **B–D**).
- In order to balance the body and stand erect against gravity, almost every joint, according to their location relative to the body's center of gravity, has to be stabilized mainly through ligaments, muscles, and tendons (see p. 60).
- The forces acting on a joint (muscle and ligament force, body weight, or gravity) create joint pressure and cause movement to occur (via the torsional moment) if they are applied outside the center of rotation, e.g., via a lever arm (see p. 51).

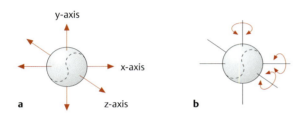

B Degrees of freedom, illustrated for the possible movements of a tennis ball in space
a Three degrees of freedom in *translation* (one each along the x-, y-, and z-axis).
b Three degrees of freedom in *rotation* (one each around the x-, y-, and z-axis).

With its similarity to a tennis ball, the spheroidal joint (see **Fc** and **Fd**) has the greatest freedom of movement.

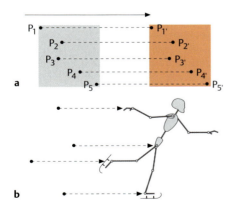

C Translation
Translation means that a body is sliding on a straight or curved path *without rotating*. As a result, all points on the moving body travel an *equal distance* in the same direction.
a All the points move the same distance in parallel lines.
b The gliding ice skater illustrates movement in translation.

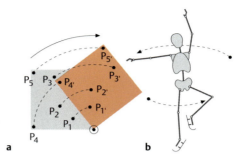

D Rotation
When a body is rotating, different points on the body move in concentric circles and travel *different distances*.
a All the points move different distances in circular arcs. **b** The spinning ice skater illustrates movement in rotation.

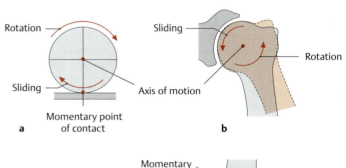

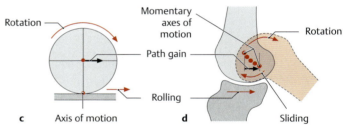

E Sliding and rolling in rotational motions
When rotating, articular surfaces can slide onto each other or roll off each other. In most cases, however, it is a combination of both (rollslide).
Note: The more congruent the articular surfaces (roughly the same radius of curvature, e.g., shoulder joint; **a, b**), the better they slide onto each other; the less congruent the articular surfaces, the more they roll off each other (e.g. knee joint; **c, d**). Rolling and sliding differ in relation to the path gain of the axis of rotation.

- **a, b Rotation of articular surfaces without path gain of the axes of motion (= sliding)**, meaning one point of the articular surface successively touches different points on the other articular surface (**a**), as happens with abduction (= rotational motion) in the shoulder joint (**b**).
- **c, d Rotation of articular surfaces with path gain of the axes of motion (= rolling)**, mean-ing the surface of the rotating body rolls off the articular surface of the other body, and so each point of the articular surface touches one point of the other surface, with the distance covered being the same for both joints. An example is flexion in the knee joint (**d**): rotation here is a combination of rolling and sliding, because the femur and the condyles of the tibia have markedly different radii of curvature. Thus, when rolling, the axis of motion moves back on a particular path (evolute), indicating only where the movement of flexion occurs at that instant. Such an axis is also called an instantaneous axis.

F Correlation between form and movement capabilities of a joint
The movements a joint can perform are most notably determined by the form, meaning the geometry, of the articulating surfaces. In general, there are two basic movements:

- Translational movement (**a**, **b**) and
- Rotational movement.

Whereas translational movements in the joints are usually limited to one or two axes of motion (e.g., femoropatellar joint in the knee [**a**], vertebral joints [**b**] in the spinal column), rotational movements that occur, for instance, in spheroidal joints (**c**, **d**) have up to three axes of motion. In the femoropatellar joint, for example, when bending the knee, the patella slides up or down in the femoral groove, meaning the joint has one degree of freedom in translation with, accordingly, two main movements. In vertebral joints with their mostly flat surfaces, translational movements usually occur along one straight line, albeit in different directions (often two degrees of freedom in translation with four main movements). A spheroidal joint, however, has three degrees of freedom in rotation relative to the number of axes of motion and thus six main movements (abduction/adduction, flexion/extension, internal/external rotation). In joints with rotational movements, depending on the form of the articulating surfaces, the number of axes of motion and thus the number of degrees of freedom can be limited, e.g., ellipsoidal or saddle joint (two axes of motion, thus four degrees of freedom, **e**, **f**), hinge or pivot joint (one axis of motion, thus two degrees of freedom, **g**, **h**).

a Femoropatellar joint (art. femoropatellaris)

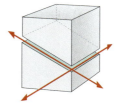

b Vertebral joint (junctureae columnae vertebralis)

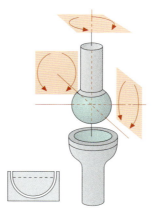

c Spheroidal joint with deep socket
A joint with three mutually perpendicular axes of motion, resulting in six primary movements (i.e., hip joint).

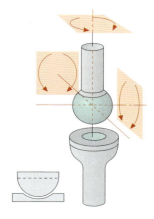

d Spheroidal joint with shallow socket
A joint with three mutually perpendicular axes of motion, resulting in six primary movements (i.e., shoulder joint).

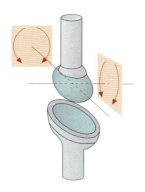

e Ellipsoidal joint
A biaxial joint with four primary movements (i.e., the radiocarpal joint).

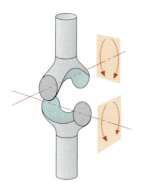

f Saddle joint
A biaxial joint with four primary movements (i.e., the carpometacarpal joint of the thumb).

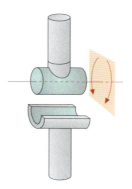

g Hinge joint
A joint with one axis of motion and two primary movements (i.e., parts of the elbow joint).

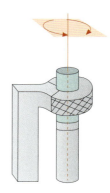

h Pivot joint
A uniaxial joint with two primary movements (i.e., the proximal radioulnar joint).

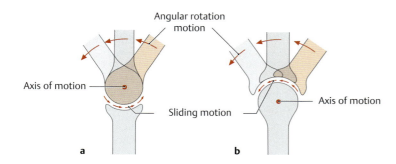

G Convex-concave rule of joint movement
The convex-concave rule describes how the sliding motion in a joint depends on the angular rotation motion.

a If the convex joint partner moves in an angular direction around the axis of motion, its rotational motion inside the joint is in the opposite direction of the angular motion.
b If the concave joint partner moves in an angular direction around the axis of motion, its rotational motion inside the joint is in the same direction as the angular motion.

4.7 Basic Principles of Joint Mechanics: Stability and Transmission

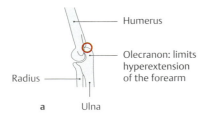
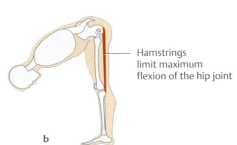
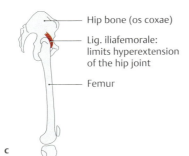
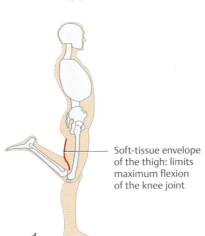

A Constraints on joint motion
The range of joint motion depends not only on the shape of the bony joint members (see D, left) but also on the muscles, ligaments, and soft-tissue envelope that surround the joint. Accordingly, all of these factors determine the total range of joint motion:

a Bony constraint.
b Muscular constraint.
c Ligamentous constraint.
d Soft-tissue constraint.

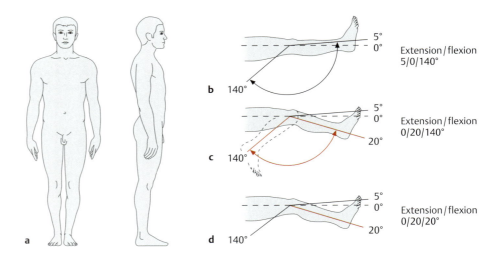

B The neutral zero method
The neutral zero method is a standardized method for measuring the range of joint motion.

a Zero-degree starting position, viewed from the anterior and lateral aspects.
b Range of motion of a normal knee joint.
c Limitation of motion caused by a flexion contracture.
d Ankylosis of the knee joint in 20° of flexion.

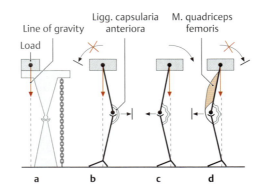

C Active and passive joint stabilization as demonstrated in the knee joint
The knee joint is actively stabilized by the quadriceps femoris and passively stabilized by the anterior m. capsular ligaments (ligg. poplitea obligua and arcuata).

a Mechanical model: A load (body weight) causes a torsional moment in the joint when the line of gravity is in front of the middle of the joint. To prevent the upper joint partner from tilting forward, a chain acts as a counterforce.
b Passive stabilization through the ligg. poplitea obligua and arcuata: The knee joint needs to shift behind the line of gravity posterior to the anterior capsular ligaments (ligg. capsularia) so that the ligaments can stabilize the knee joint.
c If the line of gravity extends behind the knee joint, the ligaments are useless; the body tilts backward, and the knee joint gives way.
d Active stabilization of the knee joint by the m. quadriceps femoris extending on its front side.

D The most important groups of muscles and ligaments responsible for the maintenance of erect posture

Muscles	Ligaments
• The lower leg muscles, mainly m. triceps surae and m. tibialis anterior, stabilize the upper ankle joint in the sagittal plane • The upper leg muscles (m. quadriceps femoris) stabilize the knee joint • The musculi glutaei stabilize the hip joint in the sagittal plane • The smaller gluteal muscles (mm. glutei medius and minimus) stabilize the hip joint in the frontal plane • The intrinsic back muscles stabilize the spinal column	• The anterior capsular ligaments (ligg. capsularia) stabilize the knee joint • The lig. iliofemorale stabilizes the hip joint • The ligg. collateralia stabilize the knee joint and the upper ankle joint in the frontal plane

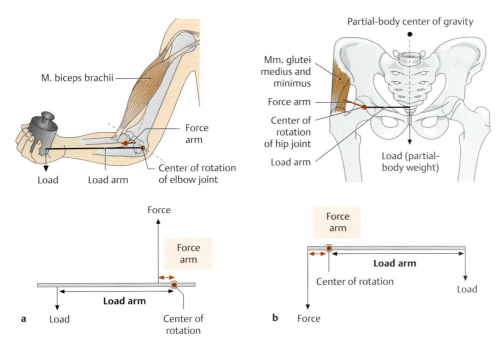

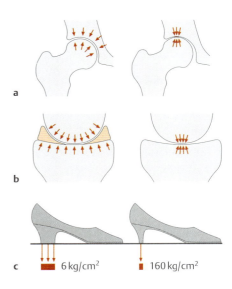

E One- and two-arm levers
a One-arm lever (elbow joint), b two-arm lever (hip joint).
Joint mechanics is based on the principles of the lever. The amount of force that a muscle can transmit to a joint depends on the length of the associated lever arm. This depends on the perpendicular distance from the muscle and its tendon to the center of rotation (= *force arm*) and is opposed by the force of the *load arm*. In the case of the elbow joint in **a**, the load arm is the distance from the joint axis (center of rotation) to the load. The *magnitude of each of the three active forces* is determined by multiplying the force by the force arm and the load by the load arm. This product is called the *torque* (= moment of rotational force) because the active forces produce a rotational movement of the associated lever. If the product of the load times the load arm equals the product of the force times the force arm, both torques are identical, and the joint is at rest. The lever in **a** is classified as a *one-arm lever* because the muscular force and load act *on the same side*—in this case, to the left of the center of joint rotation. The lever in **b** is a *two-arm lever* because the muscular force acts to the left of the center of joint rotation, while the force of the body weight acts to the right of the joint center.

F Dependence of joint load on the size of the force-transmitting surface of the joint load
The larger the force-transmitting surface, the more evenly the force is distributed and the lower the joint pressure, or joint load, per unit area.

a Hip joint with normal and limited covering of the caput femuris.
b Knee joint with and without meniscus.
c Example of a square stacked heel and a stiletto heel: if body weight is the same, but the force-transmitting surface is less, the load on the bottom of the foot is many times higher.

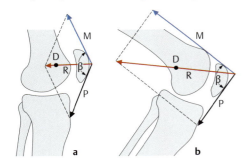

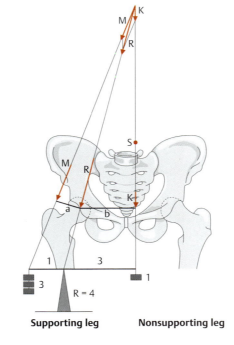

G Dependence of magnitude of the resultant force vector on the net joint force
a The resultant force vector (red) of the femoropatellar joint runs through the center of rotation of the joint (D) and is the summation of individual muscle force vectors (M) (m. quadriceps femoris) and the ligament force vector (P) (lig. patellae). It creates the joint pressure and is a measure of the forces acting on the joint.
b Due to the increased intensity of contraction of the m. quadriceps femoris when bending the knee, the resultant force in the bend position is significantly higher than in the stretching position, which means that with increased bending, the joint load in the femoropatellar joint increases.

H Load on the right hip joint in the support phase
Anterior view. In a one-legged stance or while walking in the support phase, the partial-body center of gravity (S) shifts, which results in a turning moment of the load, which tends to tilt the body above the joint toward the supporting leg. In order to remain stable, the turning moment has to be balanced by a counterforce (e.g., a muscle or ligament force). In the hip joint, it is provided by the muscle force (M) of the hip abductors (glutei medius and minimus). However, unlike the force of the partial-body weight (K) at the hip joint, it acts at a lever arm that is three times shorter, which means the ratio between the lever arm of the muscle force (**a**) to the lever arm of the partial-body weight is 1:3. The muscle force needed to stabilize the hip therefore is approximately three times that of the partial-body weight when standing on one leg. From this it follows that based on the different length of the lever arms, the compressive force that the hip joint has to withstand when walking, (resultant force R) has to be four times higher than the partial-body weight (after Pauwels). That means that the hip joint has to be able to permanently withstand extreme loads and is thus very often affected by osteoarthritis.

4.8 Fractures: Classification, Healing, and Treatment

A Fracture classification

Fractures can be classified according to different characteristics; those mentioned below are practical and have proven successful in daily use:

- According to **formation and injury mechanism:**
  - Traumatic fracture caused by direct force (transverse or comminuted fracture, see **Ba** and **Bg**, e.g., as a result of being hit while walking) or indirect force (lever arm and rotational motion directed away from the force effect, e.g., torsion fracture to the lower leg as a result of a ski crash, see **Bf**).
  - Nontraumatic fracture ("spontaneous fracture") as a result of preexisting localized or generalized bone disease, such as tumors, osteomyelitis, or osteoporosis (= pathologic fracture, can occur without any external forceful impact) or as a result of microtrauma, which is caused by repeated moderate application of force at the same spot over and over again eventually leading to a fracture (= fatigue fracture, e.g., metatarsal fracture such as a so-called march fracture).
- According to **fracture morphology** (e.g., degree of break in continuity [fissure or fracture]), course of fracture lines, number of fragments, see **B** and **C**).
- According to **severity of soft-tissue damage:**
  - Closed fracture or
  - Open fracture: generally, it can be assumed to give direct access to germs; however, what is important in the end is the severity of soft-tissue damage, meaning to what extent muscles, vessels, or nerves have also been damaged.

Furthermore, with regards to long bones, when considering the location of the fracture (epiphyseal, metaphyseal, and diaphyseal), a distinction is made between proximal fractures, fractures of the long bones, and distal fractures with or without joint involvement. In addition to pain, clinical symptoms of a bone fracture include obvious deformation, abnormal mobility, and palpable and distinguishable crepitus (creaking, crackling as a result of crepitation).

Note: Fractures suffered during childhood show distinct characteristics and have to be disinguished from those suffered in adulthood. The most important anatomic difference involves the growth plates and the periosteum, which is thicker and more robust in children. If, for example, a fracture causes damage in the proliferative zone in the growth plate, cellular demise and callous bridges can lead to a premature closure of the epiphyseal plate. In so-called greenstick fractures, the cortex fractures on one side while maintaining the thick periosteal tube (intraperiosteal fracture).

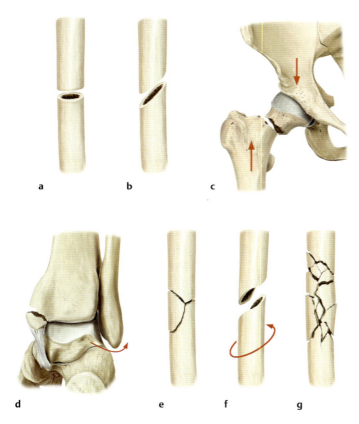 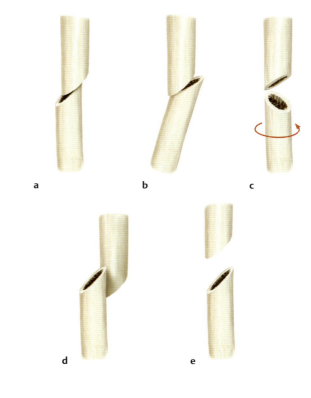

B The most important types of fractures
a Transverse fracture; **b** oblique fracture; **c** shearing fracture (e.g., fracture of the femoral neck); **d** avulsion fracture (e.g., avulsion fragment in the intra-articular ligament of the malleolus medialis); **e** bending fracture with bending wedge as third fragment; **f** torsion, or spiral, fracture; **g** comminuted fracture with more than six fragments.

C Typical fracture dislocations
Fractures can be primarily (by acting force) or secondarily (by a disrupted muscle balance, e.g., muscular traction at the fragments) more or less distinctly displaced or dislocated.
a Sideways dislocation (dislocatio ad latus); **b** axial dislocation (dislocatio ad axim); **c** angular malalignment; **d, e** dislocation with overlapping and separated fragments.

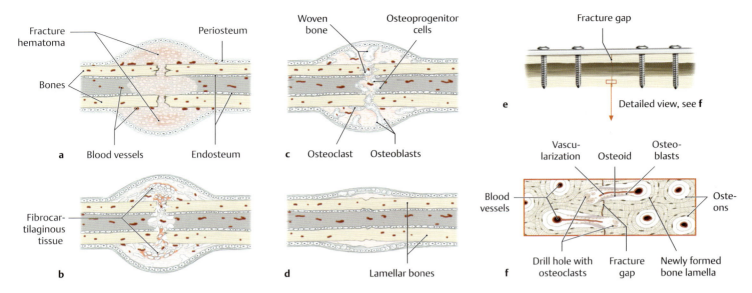

D Fracture healing

Preconditions for optimal fracture healing are repositioning and retaining the bone fragments of the fracture. In principle, this can be achieved both conservatively (e.g., with a cast) and surgically (see **E**). In addition, fracture healing requires cells with osteogenic potential, sufficient vascularization, and absolute mechanical rest in the fracture area. During the healing process, active movement therapy is crucial in order to prevent muscle atrophy, due to osteoporosis inactivity, stiffening of the joints, and circulatory disturbances. Endochondral and membranous ossification are part of the healing process. Thus, these processes, which occur in the course of the normal growth and the development of the skeleton, are repeated.

- **Indirect (natural or secondary) fracture healing (a–d),** which has the following physiologic stages:
  - Development of a fracture hematoma (inflammatory phase, **a**) and organization of the fracture hematoma with the help of loose, vessel-rich connective tissue (granulation stage)
  - Development of a fibrocartilaginous cartilage while involving cells of the periosteum and the endosteum (so-called fibrocartilaginous callus or fixation callus, **b**)
  - Mineralization of the cartilage (callus hardening within 6 to 8 weeks after fracture occurrence; so-called osseous callus or fracture callus; at first thick and bulky)
  - Loss of cartilage substance and new bone formation (woven bone) with the help of osteoprogenitor cells (**c**)
  - Bone restructuring in the lamellar bone, which takes months (**d**)

- **Direct (or primary) fracture healing (e, f).** In this case, no chondral fracture callus develops, but ideally (though only rarely) direct angiogenic ossification occurs, meaning the sprouting osteons grow directly from one fragment into another (**e, f**). However, this requires surgical osteosynthesis and that the fracture gap is no more than 0.5-mm wide (contact healing). Conservative stabilization using synthetic dressing will not result in direct fracture healing.

Note: Pseudarthrosis (failure of bone formation after 6 months) is among the most common complications of fracture healing. It arises due to both excessive mechanical load on the regenerating bone and absence of callus formation as a result of poor circulation.

E Basic principles of osteosynthesis

Surgical osteosynsthesis creates a temporary mechanical bone union using screws, plates, wires, or nails. The procedure neither replaces nor expedites bone healing. However, there are a number of **advantages** compared to a conservative cast treatment:

- Exact repositioning of bone fragments in joint fractures
- Immediate mobilization (prevents thrombosis, emboli, decubitus, and fracture disease, e.g., edema and dystrophy)
- Timely stabilization exercises (mobilization of joints without load) and partial load-bearing capacity to allow for healing

Disadvantages include the risks associated with anesthesia and surgery, as well as the possibility of infection. There are five different osteosynthetic procedures (which can be combined as needed):

- Screw osteosynthesis (e.g., compression with corticalis-traction screw, **a**)
- Plate osteosynthesis (e.g., compression plate, **b**)
- Intramedullary nail, rod, or fixation (e.g., in diaphyseal fracture, **c**)
- Tension-band wiring (e.g., using a wire loop, **d**)
- External fixation (e.g., three-dimensional frame construction, **e**)

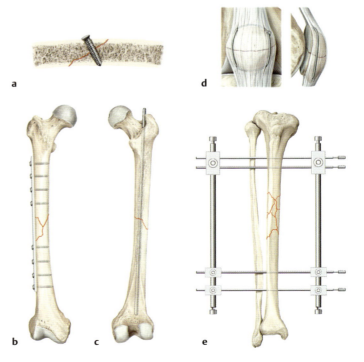

General Anatomy — 5. The Muscles

5.1 Skeletal Muscles: Overview

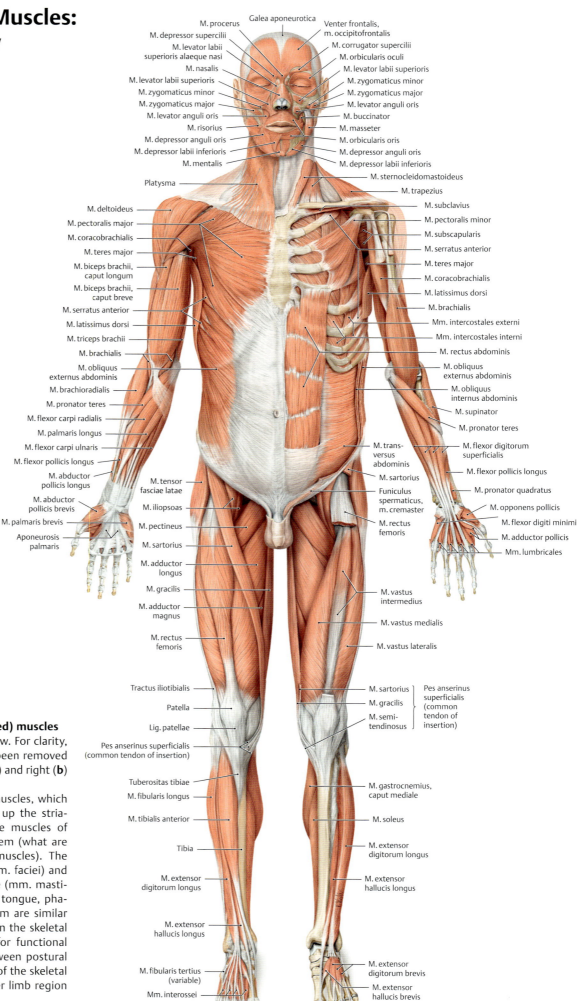

A Overview of skeletal (striated) muscles
a Anterior view, **b** posterior view. For clarity, some superficial muscles have been removed or windowed on both the left (**a**) and right (**b**) side of the body.
Approximately 220 individual muscles, which differ in shape and size, make up the striated muscles. Overall, these are muscles of the active musculoskeletal system (what are often referred to as skeletal muscles). The muscles of facial expression (mm. faciei) and masticatory muscles of the face (mm. masticatorii) and the muscles of the tongue, pharynx, larynx, eye, and tympanum are similar in form to these muscles. Within the skeletal muscles (extrafusal muscles), for functional reasons, one differentiates between postural and phasic muscles. Two thirds of the skeletal muscles are located in the lower limb region

General Anatomy — 5. The Muscles

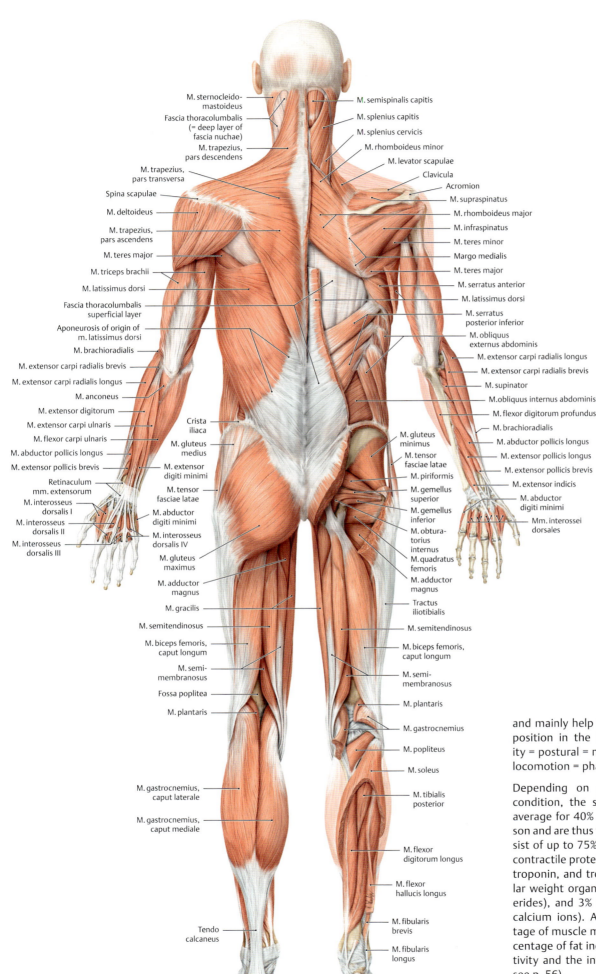

and mainly help to hold the body in an erect position in the field of gravity (= antigravity = postural = muscles) and in movement (= locomotion = phasic muscles).

Depending on gender, age, and physical condition, the skeletal muscles account on average for 40% of the total weight of a person and are thus the heaviest organ. They consist of up to 75% water, 20% proteins (mainly contractile proteins of muscles: myosin, actin, troponin, and tropomyosin), 2% low molecular weight organic components (e.g., triglycerides), and 3% inorganic components (e.g., calcium ions). As people age, their percentage of muscle mass decreases, and their percentage of fat increases (due to physical inactivity and the influence of hormonal factors, see p. 56).

b

5.2 Skeletal Muscles: Muscle Fiber Types and Pennate and Nonpennate Skeletal Muscles

A Muscle fiber types for postural muscles and muscles of movement: a comparison

All striated skeleton muscles basically consist of two different muscle fiber types (type I and type II fibers), which differ in their metabolic, physiological, histochemical, and biochemical characteristics (see **B**). The type II muscle fibers can be further divided into type IIA and type IIB fibers based on weight of isoforms in their myosin chains. Since the type I and type II fibers respond to a single action potential from the innervated axon with a single twitch in an all-or-none response, they are also known as twitch fibers. In contrast, tonic fibers, which are only found in muscle spindles and external eye muscles, gradually depolarize, leading to gradual fiber shortening.

The twitch duration of type I and II fibers differs. Type I fibers twitch more slowly (slow-twitch, or ST, fibers) than type II fibers (fast-twitch, or FT, fibers). The basic distribution pattern of type I and type II fibers in individual skeletal muscles is apparently genetically determined. In each case, the muscle fiber type of a motor unit is determined by the innervating neuron, so that all fibers of a motor unit belong to the same fiber type. Muscles with small motor units (< 100) have mostly fast-twitch (type II) fibers; muscles with large motor units (several thousand muscle fibers) have mostly slow-twitch (type I) fibers (see p. 59). The fiber type and its twitch duration are linked to a specific form of capacity and thus a particular type of twitch duration. Due to their high oxidative metabolism, type I fibers have endurance capacity and are thus mainly found in the postural muscles. Due to their high glycolytic metabolism, type II fibers can perform short, forceful contractions and are thus mainly found in phasic or elastic muscles. However, as skeletal muscles are highly adaptable, the genetically programmed distribution of type I and type II fibers can be influenced through neuromuscular activity (exercise). Accordingly, muscles that predominantly consist of type I fibers are found in endurance athletes, such as long-distance runners, cyclists, rowers, and cross-country skiers. Muscles that predominantly consist of type II fibers are found in explosive athletes, such as sprinters, high or long jumpers, and weight lifters (Pette and Saron, 2001). This phenotypic plasticity of the muscle is based on qualitative and quantitative changes in gene expression and mainly involves contractile and regulatory proteins of the myofibril apparatus, as well as the enzymes in metabolism (Pette, 1999).

	Postural muscles (red muscles)	Muscles of movement (white muscles)
Characteristics:	• Phylogenetically older • Predominantly slow-twitch fibers (type I fibers, approximately 100 ms) • Function best in endurance • Fatigue slowly • Large motor units • Rich in myoglobin • Abundant mitochondria • Energy derived from oxidative (aerobic) metabolism • Little glycogen (PAS-negative) • Relatively highly vascularized • Prone to shortening (increased resting tonus) and require regular stretching	• Phylogenetically more recent • Predominantly fast-twitch fibers (type II fibers, approximately 30 ms) • Brief periods of intense activity • Fatigue more rapidly • Small motor units • Scant myoglobin • Few mitochondria • Energy derived mainly from anaerobic glycolysis • Abundant glycogen (PAS-positive) • Much smaller capillary supply • Prone to atrophy and require regular strengthening
Examples:	Mm. intercostales, mm. masticatorii, m. trapezius (Pars descendens), mm. ischiocrurales, m. iliopsoas, mm. adductores, m. rectus femoris, m. soleus, m. erector spinae (mainly the cervical and lumbar part)	M. biceps brachii, mm. vastus lateralis and medialis, m. tibialis anterior, m. serratus anterior, m. gluteus maximus, m. gastrocnemius

Abbreviation: PAS, periodic acid–Schiff.

Note: The loss of muscle mass (sarcopenia) in old age and the associated loss of muscular force and endurance are the most important causes for impaired physical mobility and, as a result, impaired ability to live independently. Sarcopenia is prevalent among the old and is a heavy financial burden on the health care system. Between the ages of 25 and 75, a person loses almost 40% of muscle mass (mainly type II muscle fibers). Especially dramatic is the loss of muscle mass after the age of 50 (15% per decade). Taking into account that muscle weakness is the most common factor in the risk of falling, exercises that target a specific muscle group (especially for elasticity) can efficiently counteract the loss of strength and in combination with balance exercises significantly reduce the risk of falling.

B Histochemical differences between type I and type II skeletal muscle fibers

Microscopic cross section of a skeletal muscle (m. tibialis anterior, rat, 200X magnification, 8-μm-thick frozen section). Succinated dehydrogenase (SDH), a mitochondrial enzyme of muscle metabolism, catalyzes the transformation of succinate into fumarate (reduction of fumarate leads to a brownish black reaction product, histochemical proof for the presence of SDH). The activity of mitochondrial SDH makes the varying percentage of mitochondria in the individual muscle fibers visible: heavily colored type I fibers with numerous mitochondria between the myofibrils and underneath the sarcolemma, and pale-colored type II fibers with few mitochondria.

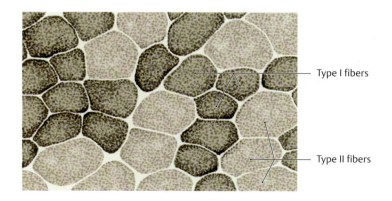

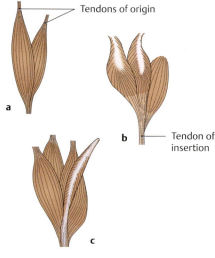
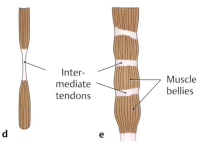
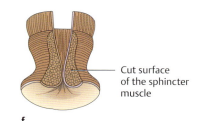
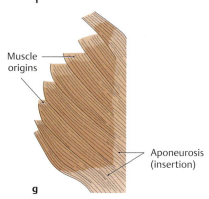

C Morphological forms of muscles
a Two heads = bicipital
  (e.g., m. biceps brachii).
b Three heads = tricipital
  (e.g., m. triceps surae).
c Four heads = quadricipital
  (e.g., m. quadriceps femoris).
d Two bellies = digastric
  (e.g., m. digastricus).
e Multiple bellies = multigastric
  (e.g., m. rectus abdominis).
f Radial (e.g., m. sphincter ani externus).
g Flat (e.g., m. obliquus externus abdominis).

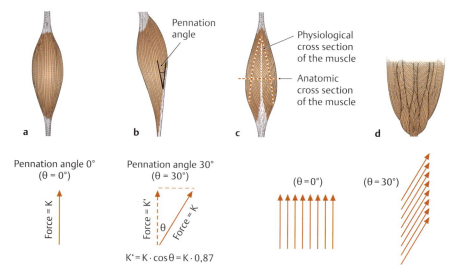

D Pennate and nonpennate skeletal muscles
Muscle fibers may have a parallel arrangement or may show varying degrees of convergence, or pennation, at their attachment.
a Nonpennate muscle (with parallel fibers)(m. fusiformis); **b** unipennate muscle (m. unipennatus); **c** bipennate muscle (m. bipennatus); **d** multipennate muscle (m. multipennatus); **e** effect of pennation angle on force (at a 30° pennation angle, the tendon force F' amounts to 87% of the force generated in the muscle fibers); **f** effect of fiber arrangement on the number of fibers in a given cross section of muscle.
Whereas almost all fibers of skeletal muscles are of the same thickness (average diameter approximately 60 μm, see p. 58), their length varies considerably (ranging from a few millimeters to up to 20 cm), as does their ratio of fiber length to muscle length (between 0.2 and 0.6, meaning the fiber length amounts to between 20 and 60% of the length of the muscle) and their pennation angle.
All three factors play a significant role in lifting force and lifting height and thus in muscle "power output" (power = force x velocity or lifting force x lifting height):

- The longer the fibers, the greater the maximum shortening and thus the maximum lifting height of the muscle.
- The longer the fibers relative to the length of the muscle, the smaller the physiological cross-sectional area of the muscle and thus the lifting force of the muscle.
- The greater the pennation angle, the greater the physiological cross-sectional area and thus the lifting force.

Note: The physiological cross section is perpendicular to the cross sections of the muscle fibers, whereas the anatomic cross section is measured at the thickest part of the muscle.

E Nonpennate and pennate muscles: a comparison

Nonpennate (parallel-fibered) muscle (m. fusiformis)	Pennate muscle (m. pennatus)
• Fibers run roughly along the longitudinal direction of the tendon (= muscle line of action) and are thus able to transmit almost all of their force to the tendon, see **f** • Maximum fiber shortening (= lifting height) and extent of actual fiber shortening during muscle activity are almost identical • Anatomic cross section (perpendicular to the longitudinal axis of the muscle, at its thickest part) and physiological cross section (perpendicular to the longitudinal axis of the fibers, see **c**) are roughly the same size	• The fibers form a (pennation) angle (up to 30°) with the longitudinal axis of the tendon and thus can transmit only a portion of their force to the tendon • Due to pennation angle, maximum fiber shortening (= lifting height) is greater than actual fiber shortening during muscle activity (= path gain) • Physiological cross section is larger than anatomic cross section, meaning (due to the angle) more muscle fibers can insert into the tendon in a given cross section as compared to a parallel-fibered muscle, thus increasing the lifting height, which depends on the size of the physiological cross section

Note: In nonpennate (parallel-fibered) muscles, force production is greater (direct force transmission from muscle to tendon, as there is no pennation angle); in pennate muscles, force development is greater (there are more fibers in a given cross section, since fibers are arranged obliquely to a tendon, resulting in greater lifting height, see above). The lesser force production of pennate muscles is compensated for by the greater force development. A huge advantage of pennate muscles compared to nonpennate muscles is their smaller size for similar size power output. If there were only nonpennate muscles, the body, in a lot of places, would not have space for a sufficient number of strong muscles.

5.3 Skeletal Muscles: Structure and Function

A Structure of a skeletal muscle
a Cross section of a skeletal muscle; b detail from a (cross section); c detail from a (longitudinal section); d structure of a muscle fiber (= muscle cell); e structure of a myofibril.
In striated skeletal muscle, muscle fibers and connective tissue are tightly interwoven and collaborate to carry out particular functions. The connective tissue is arranged in sheaths, which perform the following functions:

- Endomysium: most interior connective tissue sheaths (important for the tensile strength of the muscles); surrounds and condenses 200 to 250 muscle fibers (= muscle cells) into primary bundles. Each primary bundle, or muscle cell, contains the end branches of the motor axons to the motor end plates, as well as numerous (300–400 per mm²) heavily gyrose capillaries, see c (important for the supply of the muscles).
- Perimysium (important for the transmission of muscle tensile force to tendons): surrounds and condenses numerous primary bundles into secondary bundles (see a), which are several millimeters thick and thus are clearly visible to the naked eye (meat fibers); in contrast, the primary bundles (see above) have an average medial cross-sectional area of 1 mm².
- Epimysium: loose connective tissue layer directly beneath the muscle fascia (see b); establishes the connection to the muscle.

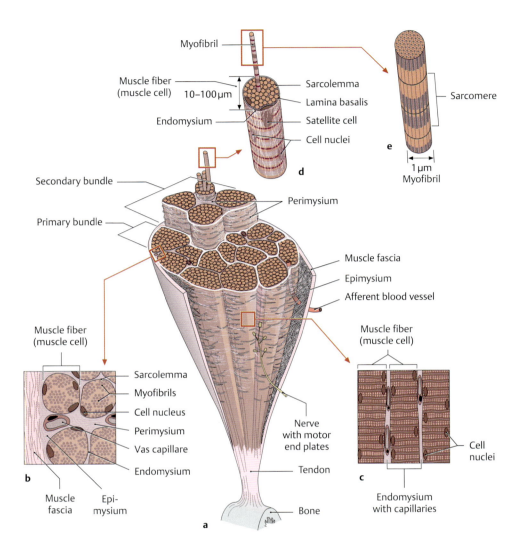

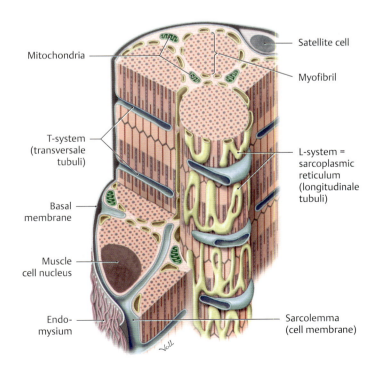

B Structure of a skeletal muscle fiber
Skeletal muscle fibers are especially large cells, with average diameters of roughly 60 μm (10–100 μm) and lengths up to 20 cm. The dominant structures in the cytoplasm are myofibrils, mitochondria, and the L- and T-systems. The L-, or longitudinal, system is a cavity system (sarcoplasmic reticulum, longitudinal tubules), which is arranged lengthwise to the myofibrils and offers a pool for calcium ions. The T-, or transverse, system is created when a cell membrane periodically invaginates deeply into the interior of the muscle cell and as a result develops transverse tubules. It enlarges the membrane surface of a muscle fiber by 5 to 10 times, and the extracellular space can extend over the entire muscle fiber cross section. It guarantees the fast spread of an action potential deep into the muscle fiber.
Characteristic for the skeletal muscle cell are the numerous cell nuclei (approximately 50 nuclei/mm length), which are located directly underneath the cell membrane (sarcolemma). This very large number of cell nuclei arises during embryonic development, when the progenitor cells (myoblasts), arranged in a chainlike pattern, fuse with each other. Scattered between the sarcolemma and basal membrane are satellite cells (approximately 800/mm³ muscle tissue), which as dormant myoblasts (stem cells) represent some sort of cell pool.

General Anatomy — 5. The Muscles

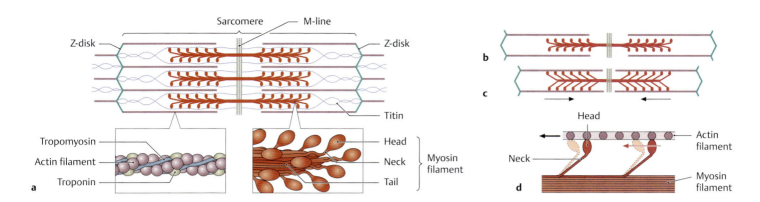

C Structure of a sarcomere

a Schematic representation of a sarcomere (= structural unit of a striated muscle cell); **b** myosin heads at rest; **c** myosin heads during contraction; **d** interaction between myosin heads and actin.

The arrangement of myofibrils in skeletal muscle fibers (see **B**) shows a cross-striation under the light microscope. It arises as the result of an alternating arrangement of thin actin (diameter: 7 nm) and thick myosin filaments (diameter: 15 nm). While actin filaments and their accompanying proteins (tropomyosin and troponin) of two adjacent sarcomeres are connected with each other at the Z-disks, the myosin filaments in the M-line among others are cross-linked by the myosin protein and in addition are kept in place by the elastic protein titin.

With the bipolar structure of the myosin filaments, a distinction is made between three parts: the head, neck, and tail. During muscle contraction, the myosin heads migrate along the actin filaments (so-called filament sliding) toward the Z-disks. Thereby, when at rest, each sarcomere shortens maximally to 70% of their optimal length of 2.2 μm; however, the individual filaments maintain their original length. This sliding mechanism is based on a quick succession of reaction cycles, during which cross-bridges between myosin heads and actin filaments are developed and detached again. It is associated with a tilting and rowing movement of the myosin head toward the middle of the sarcomere, with the myosin heads snapping off in an articulated manner (red arrow), and the actin filaments shoving on (black arrow). A single tilting movement shifts the actin filaments by about 10 to 20 mm, meaning the sarcomere shortens by about 1% of its original length. Greater changes in length are enabled by numerous repetitions of the interaction between actin and myosin. The basis for these cyclical processes is the fission of adenosine triphosphate (ATP) through the myosin head ATPase and the increase in cytosolic Ca+ concentration (electromechanical coupling; for more details, see books about physiology).

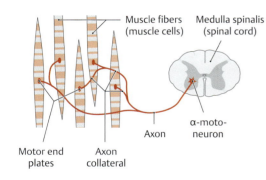

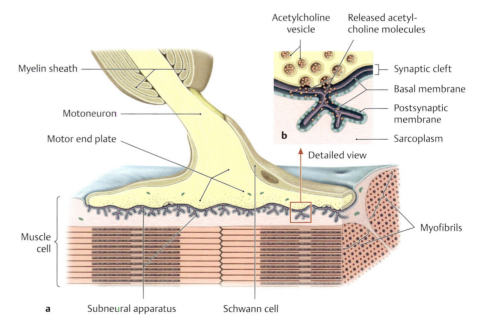

D Motor Unit

A motor unit is the sum of all muscle fibers that are innervated by a motor nerve fiber (the axon of a motor nerve cell of the spinal cord = α-motoneuron). There are small motor units (< 100 fibers) with mostly fast-twitching muscle fibers and large motor units (several thousand muscle fibers) with mostly slow-twitching muscle fibers. The fewer the muscle fibers combined in a motor unit, the more precise and finely tuned the movements of the muscle (e.g., finger muscles and external eye muscles). However, in muscles whose main function is to maintain posture or to produce movement (e.g., gluteal and back muscles), several thousand muscle fibers are innervated by one motor nerve cell.

E Motor end plate

a Schematic representation of a motor end plate; **b** components of the synaptic contact (detailed view of **a**).

The motor axon divides at its end into several branches (one branch per muscle fiber), loses its myelin sheath, and develops on each of the branches a motor end plate (neuromuscular synapse), where the synaptic transmission of action potentials of the motoneuron occurs. Acetylcholine serves as a neurotransmitter and is stored in synaptic vesicles in the axoplasm of the end plate. As a result of folds in the sarcolemma (see **B**), the postsynaptic membrane with its acetylcholine receptors is greatly enlarged (subneural folding apparatus). In the entire 100-nm-wide system, there is one lamina basalis, to which the acetylcholinesterase (an enzyme responsible for breaking down acetylcholine) is attached. The functional sequence at the motor end plate is generally identical to that at other synapses.

5.4 The Tendons and Mechanisms That Assist Muscle Function

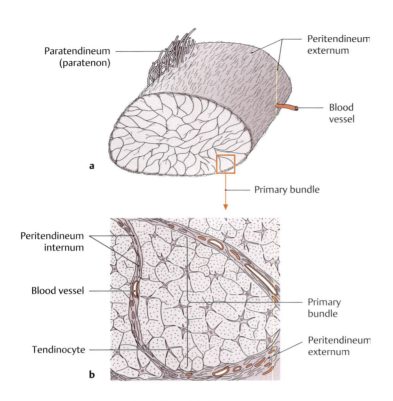

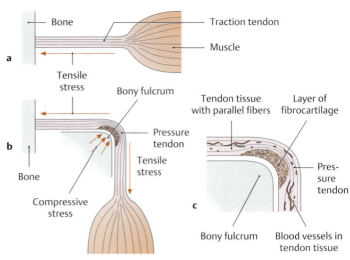

A Structure of a tendon (after Kristic)
a The tendon is connected to its surroundings by the loose, richly vascularized paratendineum (paratenon).
b Detail from **a**: The individual primary bundles are surrounded by peritendineum internum and are grouped into the actual tendon by the peritendineum externum. The function of a tendon is to transmit force from the muscle to the bone.

B Pressure tendons and traction tendons
a Traction tendons are subject to tensile stresses and consist of strong connective tissue with parallel fibers.
b Pressure tendons are strained by pressure and change their direction by running around the bone (unlike traction tendons). They consist of fibrocartilage on the side in contact with the bone, which acts as a fulcrum.
c Detail from **b**: The fibrocartilage layer in the compressed area, unlike the strong connective tissue in a traction tendon, is not vascularized.

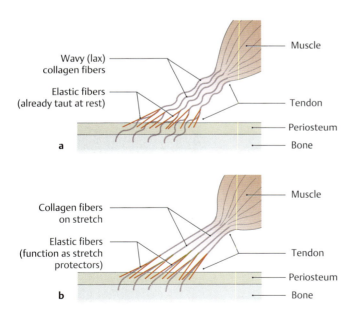

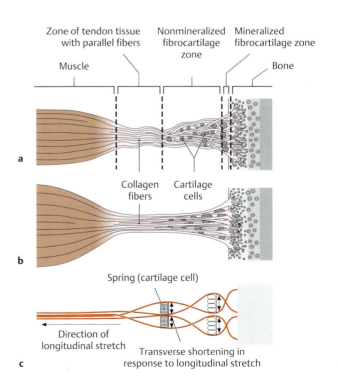

C Structure and function of a periosteal diaphyseal tendon insertion
a Tendon in the lax condition (muscle relaxed).
b Tendon on stretch (muscle contracted).

D Structure and function of a chondral apophyseal tendon insertion
a Tendon in the lax condition (muscle relaxed).
b Tendon on stretch (muscle contracted).
c The principle of stretch protection: cartilage cells in the nonmineralized fibrocartilage zone act like taut springs to resist transverse shortening.

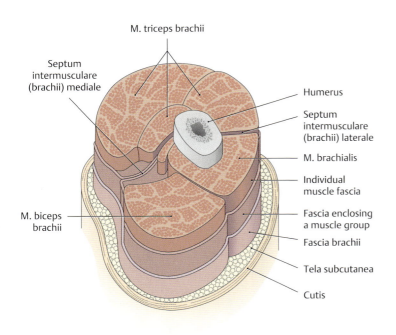

E Muscle fasciae
Cross section through the middle third of the right arm, proximal view. Muscle fasciae (fibrous sheaths enclosing muscle) are composed of tough collagenous connective tissue. They help to maintain the shape and position of muscles and permit adjacent muscles or muscle groups to glide past each other with relatively little friction (less friction means less loss of force).

F Structure of a tendon sheath (synovial sheath)
Tendon sheaths serve to protect and facilitate the gliding of tendons that run directly on bone. The wall structure of the sheath, consisting of an outer fibrous membrane and an inner synovial membrane, resembles that of a capsula articularis. The inner layer of the membrana synovialis is firmly attached to the tendon, while its outer layer is attached to the membrana fibrosa of the tendon sheath. The space between the two layers is filled with synovial fluid (synovia). The mesotendineum (sometimes referred to as *vincula brevia* and *longa* in different locations) transmits blood vessels to the tendon.

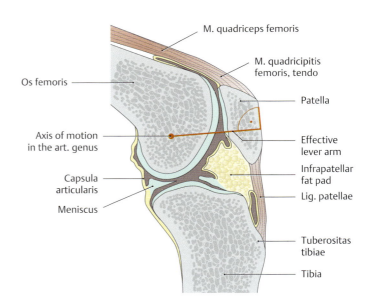

G Bursae synoviales in the shoulder region
Right shoulder anterior view, with some of the muscles removed. Bursae are pouchlike structures of varying size, usually flattened, that contain synovial fluid. Their wall structure is similar to that of a capsula articularis. The bursae may become inflamed (*bursitis*), causing severe pain.

H Functional significance of sesamoid bones (ossa sesamoidea)
Sagittal section through a knee joint (art. genus). Sesamoid bones are bones that are embedded in tendons and protect the tendons from excessive friction. Their occurrence is variable, so everyone does not have an equal number of sesamoid bones. Their main functional role is to *lengthen the effective lever arm of a musc*le, increasing its mechanical efficiency. The diagram illustrates this principle for the patella, which is the largest sesamoid bone in the body. The patella significantly lengthens the effective lever arm, represented by a perpendicular line from the joint axis to the tendon of insertion of the m. quadriceps femoris.

General Anatomy — 6. The Vessels

6.1 Overview of the Human Cardiovascular System

A Schematic representation of the circulatory system

Special circulatory organs are needed to transport and distribute the blood, ensuring that it is made accessible to all the cells in the body. These organs consist of the heart and vascular system (blood vessels and lymphatics). The **system of blood vessels** consists of arteries, capillaries, and veins. The *arteries (arteriae)* carry the blood from the heart and distribute it throughout the body. The *veins (venae)* return the blood to the heart. The exchange of gases, nutrients, and waste products takes place in the *capillary (vasa capillaria)* region. All blood vessels leading away from the heart are called arteries, and all vessels leading toward the heart are called veins, regardless of their oxygen content (the umbilical vein, for example, carries oxygen-rich blood). The blood flow in this closed vascular system is maintained by the **pumping action of the heart**. The **lymphatic system** runs parallel to the venous system. It originates with blind-ended vessels in the capillary region, collects the *extracellular* fluid that is deposited there, and returns it to the venous blood through *lymphatic vessels (vasa lymphatica)*. Lymph nodes (nodi lymphoidei) are interposed along these pathways to filter the lymph. Functionally, the circulatory system is divided into two main circuits:

- **The pulmonary circulation:** *Deoxygenated venous blood* from the upper and lower body regions is returned through the vv. cavae superior and inferior to the *right* atrium. It then enters the *right* ventricle, which pumps it through the aa. pulmonales to the lungs.
- **The systemic circulation:** *Oxygen-enriched blood* from the lungs returns through the vv. pulmonales to the *left* atrium. From the left atrium it enters the *left* ventricle, which pumps the blood through the aorta into the systemic circulation. A special part of the systemic circuit is the **portal circulation,** which includes two successive capillary beds. *Before* venous blood returns to the v. cava inferior from the capillary beds of the unpaired abdominal organs (stomach, bowel, pancreas, and spleen), it is carried by the v. portae hepatis to the capillary bed of the liver. This ensures that nutrient-rich blood from the digestive organs undergoes numerous filtering and metabolic processes in the liver before it is returned to the v. cava inferior via the vv. hepaticae.

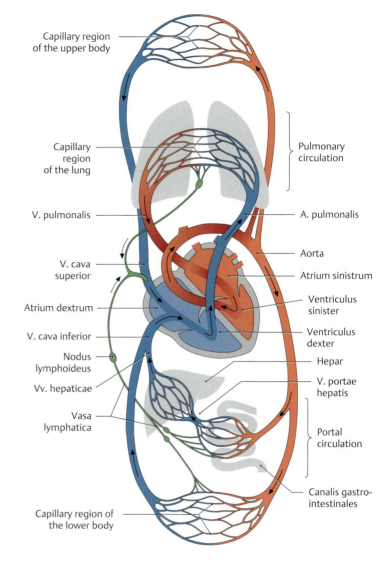

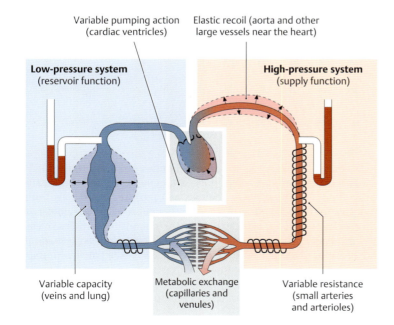

B Basic functional diagram of the circulatory system
(no distinction is made between the systemic and pulmonary systems in the diagram; after Klinke, Silbernagl)

Blood is transported through the circulatory system along a *pressure gradient* created by the different pressure levels in the arterial and venous systems. While the average blood pressure in the *arterial high-pressure system* is approximately 100 mmHg (13.3 N), the pressure in the *venous low-pressure system* generally does not exceed 20 mmHg (2.6 N). The two systems meet in the capillary region of the terminal vascular bed, where metabolic exchange takes place. When the heart expels blood during systole, the *arteries surrounding the heart (elastic-type arteries)* can temporarily expand to accommodate the ejected blood volume. During the diastole that follows, the vessel lumen undergoes an *elastic recoil* that transforms the intermittently ejected blood volumes into continuous flow. Arteries distant from the heart (*muscular-type arteries*) can actively expand (*vasodilation*) and contract (*vasoconstriction*), providing a very effective means of controlling vascular resistance and regulating local blood flow. The veins are also called *capacitance vessels* because of the high volume of blood contained within the veins. They can accommodate 80% of the total blood volume and thus serve an important reservoir function.

General Anatomy — 6. The Vessels

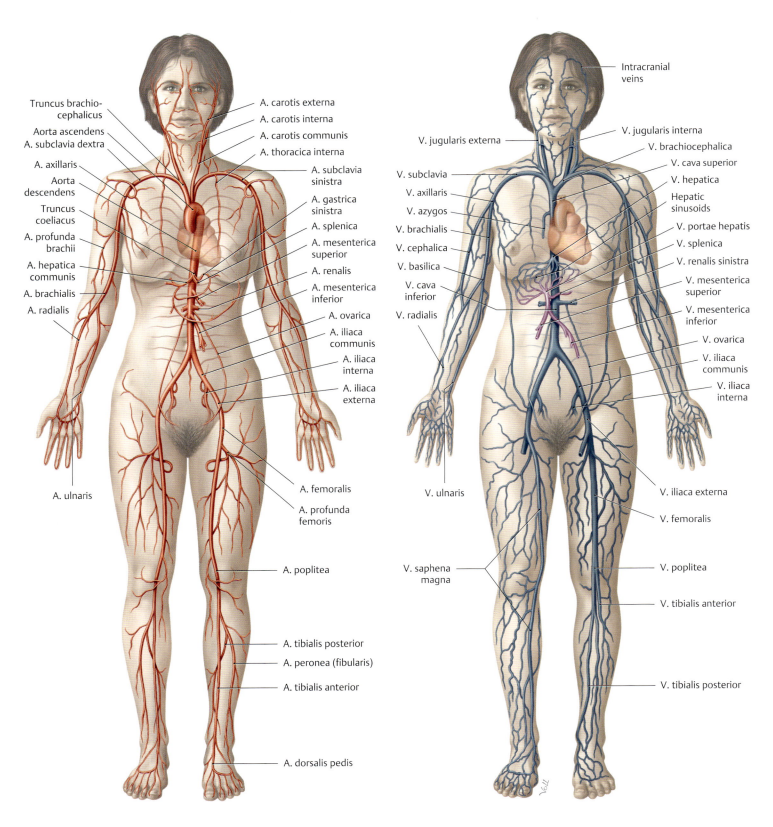

C Anterior view. Overview of the principal arteries in the systemic circulation

D Overview of the principal veins in the systemic circulation
Anterior view. The venous system is comprised of superficial veins, deep veins, and perforator veins, which interconnect the superficial and deep venous systems.
Note the portal circulation (v. portae hepatis), which carries nutrient-rich blood (shown here in purple) from the digestive organs directly to the liver (compare to the left side of **A**).

6.2 The Structure of Arteries and Veins

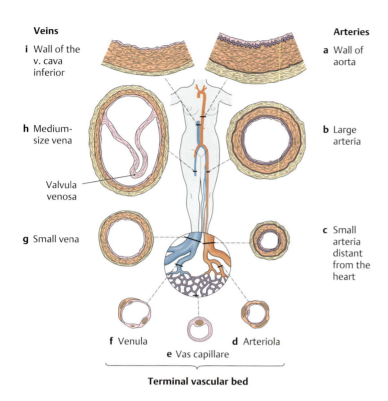

A Structure of the blood vessels in different regions of the systemic circulation
Consistent with changing demands, the vessels in different regions of the systemic circulation (high- and low-pressure systems, microcirculation) show significant local structural differences despite a basic similarity in the arrangement of their wall layers. Whereas a relatively high internal pressure prevails throughout the *arterial system*, and the arterial vessels have correspondingly thick walls, the *veins* have a considerably lower intravascular pressure, resulting in thinner walls and larger luminal diameters than in the arteries. In the terminal vascular bed, on the other hand, the vessel wall layers are reduced to permit the exchange of gases, fluids, and other substances.
a–c Arteries, **d–f** terminal vascular bed, **g–i** veins. **a** Close-up view of the aortic wall (elastic-type artery). **b, c** Large and small arteries distant from the heart (muscular-type arteries). **d** Arteriola. **e** Vas capillare. **f** Venula. **g, h** Small and medium-size veins (some with venous valves). **i** Close-up view of the wall of the v. cava inferior.

B Organization of the blood vessel system

Arteries (high-pressure system = supply function)
- Elastic-type arteries
- Muscular-type arteries

Terminal vascular bed (microcirculation = exchange function)
- Arterioles (arteriolae)
- Capillaries (vasa capillaria)
- Venules (venulae)

Veins (low-pressure system = reservoir function)
- Small and medium-size veins (with valves)
- Large venous trunks

	Arteries		Terminal vascular bed		Veins	
	Aorta	Small arteria	Arteriola	Venula	Vena	Vena cava
Wall thickness (w)	2.5 mm	1 mm	20 µm	5 µm	0.5 mm	1.5 mm
Luminal radius (ri)	12.5 mm	2 mm	20 µm	20 µm	2.5 mm	15 mm

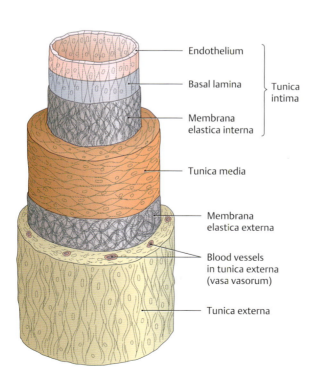

C Wall structure of a blood vessel, illustrated for a muscular-type artery
The wall of a blood vessel basically consists of *three layers*: the tunica *intima*, *media*, and *externa*. The three-layered structure is clearly apparent in the walls of arteries and is less conspicuous in veins (see **D**).

- The **tunica intima** consists of a layer of spindle-shaped endothelial cells that are aligned along the vessel axis and rest upon a basement membrane and a thin layer of subendothelial connective tissue. In muscular-type arteries, the tunica intima is consistently separated from the media by membrana elastica interna.
- The **tunica media** consists of an approximately circular arrangement of smooth muscle cells, elastic and collagenous fibers, and proteoglycans. Muscular-type arteries may have an membrana elastica externa that separates the media from the tunica externa.
- The **tunica externa**, like the tunica intima, is composed of longitudinally aligned elements, mostly connective tissue. The tunica externa of veins may additionally contain smooth muscle. The tunica externa transmits autonomic nerves to the muscle of the vessel wall, and, especially in larger vessels, it also transmits the vasa vasorum, which supply blood to the outer third of the vessel wall.

Specific functions can be assigned to all three layers. The intima is concerned with the exchange of gases, fluids, and other substances through the vessel wall; the media regulates blood flow, and the tunica externa integrates the blood vessel into its surroundings.

General Anatomy — 6. The Vessels

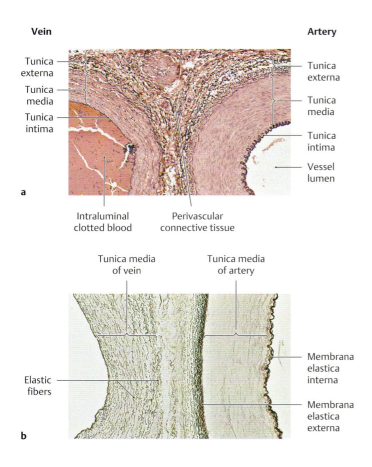

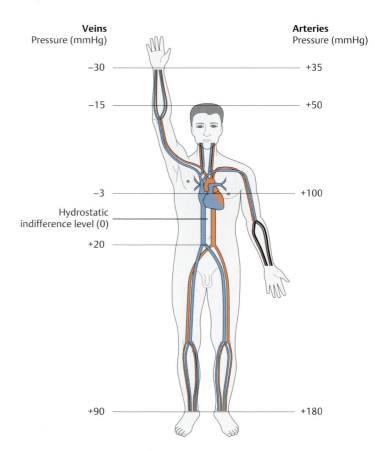

D Differences in the wall structure of arteries and veins
Wall sections from a muscular-type artery and an accompanying vein. Comparison of tissue cross sections treated with different stains. **a** H & E-resorcin-fuchsin stain of the a. and v. tibialis posterior. **b** Resorcin-fuchsin stain of the a. and v. femoralis.
Note the characteristic structural differences in the tunica media. While the arterial tunica media consists of densely packed layers of smooth muscle cells, the venous tunica media contains a far greater amount of connective tissue elements (collagenous and elastic fibers), giving it a much looser structure. The veins also lack a conspicuous layered structure and a membrana elastica interna (from Lüllmann-Rauch: *Histologie*, 2nd ed. Stuttgart: Thieme; 2006).

E Arterial and venous pressure changes in the standing position
Changing from a recumbent to a standing position radically alters the pressure relationships in the circulatory system. The hydrostatic effects of this change cause the pressure to rise sharply in the lower parts of the body, while the pressures in the upper body decrease (the pressures remain unchanged at the "hydrostatic indifference level" just below the diaphragm). Along with the hydrostatic pressure changes, approximately 500 mL of blood volume is shifted into the lower limb veins. This rise of venous pressure greatly increases the transmural pressure in the lower limb veins, while the pressure in the head and neck veins may fall so low that the veins collapse. This explains why analogous veins in the lower and upper body regions have wall layers of different thickness; i.e., the veins on the dorsum of the foot are much more muscular than on the dorsum of the hand. The wall of the v. cava inferior, however, is paper-thin due to the low venous pressures at that level.

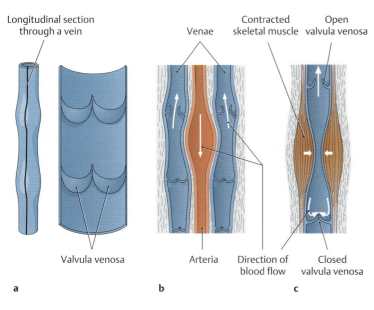

F Venous return to the heart
The following factors promote the return of venous blood to the heart: **a** opening and closing of the venous valves, **b** arteriovenous coupling (the pulse wave in the artery is transmitted to the accompanying vein), and **c** the muscle pump.
Venous return is also aided by the "suction effect" of the heart, i.e., the negative pressure produced when the valve plane moves toward the cardiac apex during systole. A lack of muscular movement due to prolonged standing or sitting, for example, can cause the damming back of venous blood, leading to a raised intravascular pressure and incompetence of the venous valves. This can result in edema, varicose veins, and circulatory impairment.

6.3 The Terminal Vascular Bed

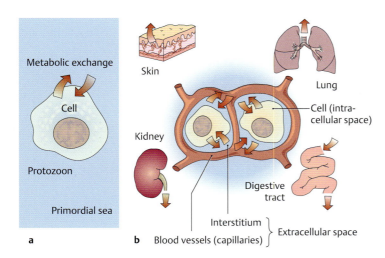

A The milieu in which a cell lives (after Silbernagl, Despopoulos)
a Protozoan: The first single-cell organisms lived in an environment, the primordial sea, that provided a milieu of constant composition. The internal and external milieus were the same, and so neither of them changed during metabolism.
b Human: The cells of a multicellular organism are bathed by extracellular fluid, whose volume is substantially smaller than the intracellular volume. The extra- and intracellular fluids also have a different composition. In this situation the internal milieu would change very quickly if the intercellular space (interstitium) were not linked via the bloodstream to organs such as the lungs, kidneys, and digestive tract, which absorb nutrients and excrete metabolic products. Nutrients absorbed from the bowel are distributed to the cells of the various organs (interstitium of the capillary beds) via the bloodstream. The blood also transports the metabolic products of the cells to the organs that are responsible for their excretion (e.g., the lungs and kidneys).

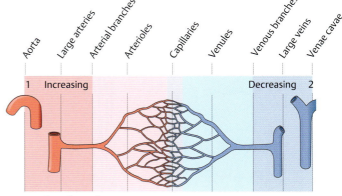
Number of vessels

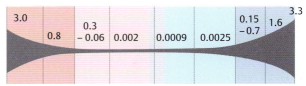
Diameter of the individual vessel (cm)

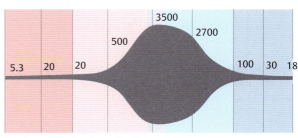
Total cross-sectional area (cm²)

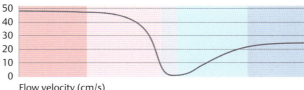
Flow velocity (cm/s)

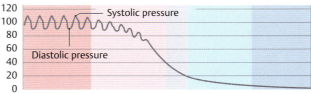
Intravascular pressure (mmHg)

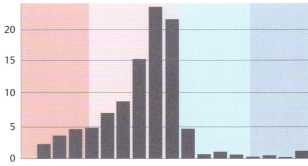
Percentage of total resistance

B Characteristics of different vascular regions (after Silbernagl, Despopoulos)
The terminal vascular bed is the site of the microcirculation and therefore is the site where gases, fluids, and other substances are exchanged. It consists of

- an *afferent arterial limb* (precapillary arterioles),
- the *capillary bed* itself, and
- an *efferent venous limb* (postcapillary venules).

The smallest vessels, the capillaries, consist only of an endothelial layer and a basal lamina to which pericytes may be externally attached (contrast with the more complex structure of large vessels, p. 46). Owing to the extensive branching of the vessels in the capillary bed, the total vascular cross section is greatly increased (approximately 800 times), while the flow velocity is correspondingly reduced (from 50 cm/s in the aorta to 0.05 cm/s in the capillaries). With an average capillary length of 0.5 mm, a time of approximately 1 second is available for metabolic exchange. The increased vascular resistance in the arterioles and capillaries caused by contact of the blood with the large endothelial surface area (increased friction) lowers the blood pressure and eliminates pressure spikes. Thus, the capillaries provide ideal conditions for exchange processes to occur between the blood and the interstitial fluid that bathes the body's cells.

General Anatomy — 6. The Vessels

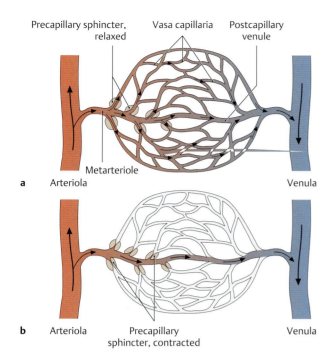

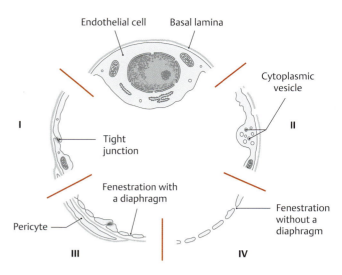

C Blood flow in the capillary bed
a Sphincter relaxed, b sphincter contracted.
Precapillary sphincters, with their circular array of muscle cells, are located at the junction of the metarterioles and capillaries and regulate blood flow within the capillary network. When the sphincters contract, the branching capillaries are closed, and the capillary bed is unperfused except for the metarterioles (e.g., only about 25 to 35% of all capillaries are perfused under resting conditions). The arterioles and venules may also be interconnected by shunts called *arteriovenous anastomoses*.

D Different forms of capillary endothelial cells (scheme of ultrastructural features)
Capillaries range from 5 to 15 µm in diameter and consist of endothelial cells, basal lamina, and external pericytes. Pericytes have various properties and functions, including a role in vascular development and angiogenesis. The individual endothelial cells are connected to one another by adhesion contacts, tight junctions, and gap junctions, largely preventing any metabolic exchange between individual endothelial cells. The endothelia of different capillaries have varying degrees of permeability, and several types of endothelium are distinguished on that basis:

I Closed endothelium without fenestrations and with a continuous basal lamina (e.g., nervous system)
II Closed endothelium with pinocytotic activity (e.g., cardiac and skeletal muscle)
III Endothelial cells fenestrated by a diaphragm (e.g., gastrointestinal tract)
IV Endothelial cells with intercellular gaps (large fenestrations) and without a continuous basal lamina (e.g., liver)

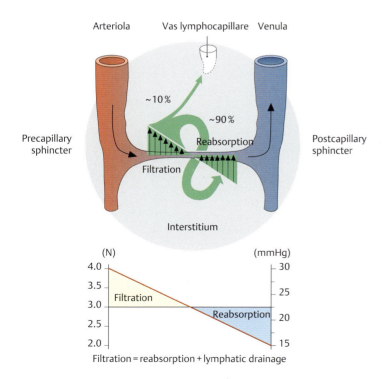

E Mechanism of fluid exchange in a capillary (after Silbernagl, Despopoulos)
Fluid exchange between capillaries and the surrounding tissue (interstitium) is regulated by a changing pressure gradient between the blood pressure in the capillaries (hydrostatic pressure) and the intravascular colloid osmotic pressure. The driving force behind the fluid exchange is the hydrostatic blood pressure. The pressure at the arterial end of the capillary is 35 mmHg (= 4.6 N), which is 10 mmHg *higher* than the colloid osmotic pressure of approximately 25 mmHg (= 3.3 N). This positive pressure differential makes it possible for fluid as well as dissolved particles to filter out of the capillaries and into the surrounding tissue. These relationships are reversed at the venous end of the capillary—there, the hydrostatic blood pressure falls to approximately 15 mmHg (2.0 N), while the colloid osmotic pressure remains essentially unchanged at about 25 mmHg. As a result of this, the hydrostatic pressure on the *venous side of the capillary* is 10 mmHg *lower* than the colloid osmotic pressure (15–25 = –10 mmHg), causing fluid with its solute particles to flow back into the vessel (*reabsorption*).
Of the 20 L of fluid that leave the capillaries each day, only about 18 L (90%) are reabsorbed. Approximately 2 L (10%) of the filtered volume is removed by lymphatic vessels in the form of lymph. If this fluid exchange does not occur in the manner described, edema may develop (i.e., a persistent accumulation of fluids in the interstitium). The reason for this may be an *elevated hydrostatic pressure* (due to blood pooling on the venous side of the capillaries) or a *decreased colloid osmotic pressure* (due to a decrease in plasma proteins). In both cases there is an imbalance of fluid exchange, allowing fluid to accumulate in the tissue, known as edema.

7.1 The Human Lymphatic System

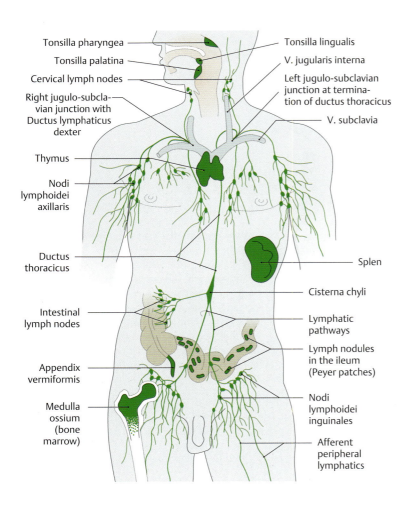

A The human lymphatic system
This system includes the lymphatic vessels and the lymphatic organs (immune organs, see **B**). The **lymphatic vascular system** runs parallel to the venous system and performs several functions:

- Its primary function is to clear the interstitial spaces of tissue fluid and substances that cannot be reabsorbed in the venous capillary bed. The composition of the lymph varies in different regions and is similar to that of the surrounding interstitial fluid.
- It carries away food lipids (chylomicrons) that are absorbed in the bowel.
- It returns lymphocytes from the lymphatic organs to the blood.

The lymphatic vascular system consists of

- *lymphatic capillaries*, which begin peripherally as blind-ended vessels;
- the *lymphatic vessels* and interposed *lymph nodes*; and
- the *major lymphatic trunks* (Ductus thoracicus and Ductus lymphaticus dexter).

The lymphatic capillaries collect fluid from the interstitium and transport it via the lymphatic vessels and lymph nodes to the major lymphatic trunks. The fluid reenters the venous system from these trunks at the junctions of the vena subclavia and vena jugularis interna. The lymph drained from three body quadrants enters the left jugulo-subclavian venous junction, while only lymph from the right upper quadrant enters the right jugulo-subclavian venous junction.
The **lymphatic organs** are part of the specific immune system and, as such, are situated at likely portals of entry for infectious microorganisms. The spleen is the only immune organ that is directly integrated into the bloodstream.

B Primary and secondary lymphatic organs
The functions of the lymphatic organs include mounting a specific immune response. A distinction is drawn between primary and secondary lymphatic organs. The primary lymphatic organs are concerned with the production, maturation, and selection of immune cells. The secondary lymphatic organs are subsequently populated by the immunocompetent lymphocytes and are sites for various processes, such as antigen presentation, lymphocyte proliferation, and antibody formation.

- **Primary lymphatic organs:**
  - Thymus (selection of T-lymphocytes)
  - Medulla ossium (bone marrow, selection of B-lymphocytes)
- **Secondary lymphatic organs:**
  - Splen
  - Nodi lymphoidei
  - Mucosa-associated lymphatic tissue (MALT) and the pharyngeal lymphatic (Waldeyer's) ring—the tonsilla pharyngea, palatina, and lingualis.
  - Bronchus-associated lymphatic tissue (BALT)
  - Gut-associated lymphatic tissue (GALT), such as Peyer patches and the appendix vermiformis

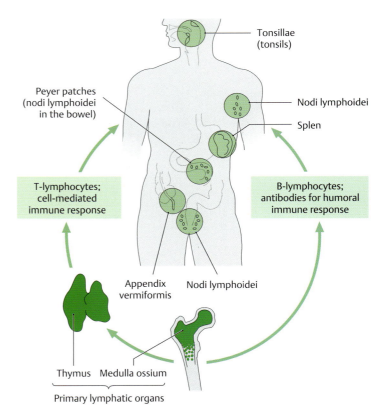

C Organization of the lymphatic vascular system (after Kubik)

Three compartments can be distinguished in the lymphatic vascular system based on **topographical** and **functional** criteria:

1. A superficial system → drains the skin (cutis) and subcutaneous tissue (subcutis).
2. A deep system → drains lymph from the muscles, joints, tendon sheaths, and nerves.
3. An organ-specific system → drains the organs and shows organ-specific differences.

A system of **perforator vessels** interconnects the superficial and deep systems, conveying lymphatic fluid toward the surface from deeper tissues.

The lymphatic vascular system can be subdivided into *four different regions* based on the **histologic structure of the vessel walls**:

1. Lymphatic capillaries
2. Precollectors
3. Collectors
4. Lymphatic trunks

Lymphatic capillaries and precollectors are also known as *initial lymphatics*.

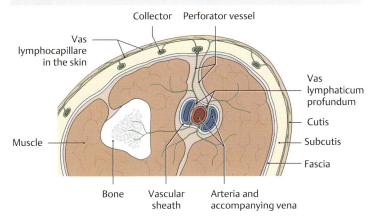

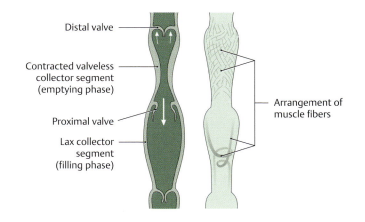

D Organization and structure of the different lymphatic regions
(after Kubik)
a Lymphatics in the skin and muscles.
b Detail from **a**, showing the structure and function of a collector segment.

Both the superficial and deep lymphatics originate with the extremely thin-walled **lymphatic capillaries**, which are approximately 50 μm in diameter. Their endothelium is bounded by an incomplete basal lamina, and they are attached by collagenous "anchoring filaments" to elastic fibers and collagen fibers in their surroundings. The network of lymphatic capillaries opens into larger **precollectors** approximately 100 μm in diameter. Unlike the lymphatic capillaries, these vessels contain valve cusps, and their walls are reinforced by a layer of connective tissue. They open into **collectors**, which also contain valves and have a transverse diameter of 150 to 600 μm. Like the larger lymphatic vessels and the lymphatic trunks, the collectors have a venous-type wall structure divided indistinctly into an intima (endothelium and basement membrane), a smooth-muscle media, and a fibrous adventitia. *Lymph transport* is effected by a series of rhythmic contractile waves (10–12/min) that are generated in the smooth-muscle, valveless collector segments. The *direction* of lymph flow is controlled by closing the distal valves and opening the proximal valves of the precollectors and collectors.

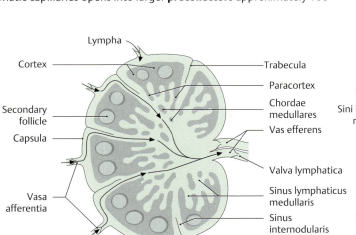

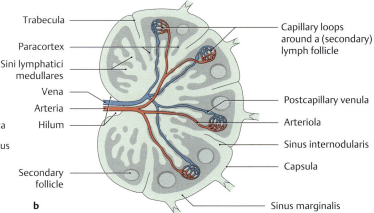

E Structure of a lymph node
a Lymph circulation, b blood supply to the lymph node.

Lymph nodes are small filtering stations located in the course of lymphatic vessels and are components of the specific immune response (they contain T- and B-lymphocytes). *Regional lymph nodes* are distinguished from the *collecting lymph nodes* that receive lymph from multiple regional nodes. The lymph enters the lymph node through multiple afferent vessels (vasa afferentia). As the fluid passes along the various lymph sinuses to the efferent vessels (vasa efferentia), it comes into contact with the lymph node tissue over a broad surface area. From outside to inside, a lymph node consists of the cortex, paracortex, and medulla. The numerous secondary follicles in the *cortex* form the *B-lymphocyte region*, and the lymphocyte-rich areas between and below the secondary follicles are the *T-lymphocyte regions (paracortex)*. Lymphocytes leave the bloodstream in the high-endothelial postcapillary venules of the T-lymphocyte region; then, after differentiating, they leave the lymph node with the draining lymph via efferent lymph vessels (vasa efferentia), which often become the afferent vessel (vasa afferentia) of another lymph node of a lymph node group.

7.2 Exocrine and Endocrine Glands

A Development and classification of glands
Glands (Glandulae) are epithelial aggregations of highly specialized single cells (goblet cells, multicellular intraepithelial glands) or of larger cell groups that have migrated to deeper levels. Their function is to synthesize and release secretions. Glands fall into two main categories:

- **Exocrine glands** (**glandulae exocrinae**, e.g., salivary glands, sweat glands): These glands release their secretion *externally* to the skin or mucosa, either directly or through excretory ducts.
- **Endocrine glands (glandulae endocrinae):** Their secretions (in this case hormonal messengers) are released *internally*, i.e., into the bloodstream, lymphatics, or intercellular spaces. Endocrine glands do not have excretory ducts (see **F** for mechanisms of hormone release). Once released into the bloodstream, the hormones are distributed throughout the body and are transported to their target cells, where they bind to specific receptors and exert their effect.

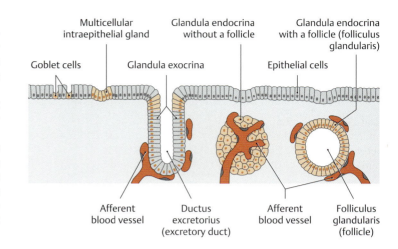

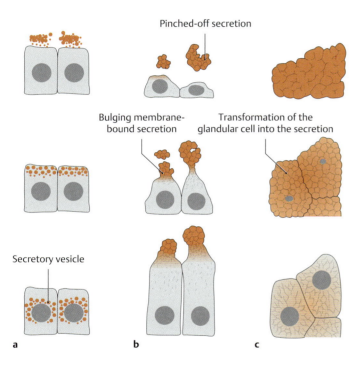

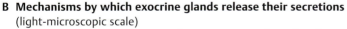

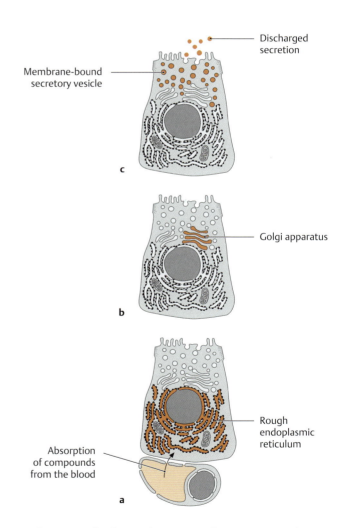

B Mechanisms by which exocrine glands release their secretions
(light-microscopic scale)

a Exocytosis: In this mechanism, the secretion is released *without an enclosing membrane* (merocrine or eccrine secretion). The membrane-bound vesicles containing the secretion fuse with the apical cell membrane and discharge their contents to the outside with no loss of membranous material (the secretory mechanism of most glands, see also **C**).

b Apocytosis: The membrane-bound vesicles form a bulge in the apical cell membrane and are finally *pinched off* by it (apocrine secretion). The pinched-off secretory products are enclosed within a membrane. This mechanism is necessary in the secretion of fats. The membrane encloses the fats and keeps them emulsified (e.g., scent glands, mammary glands).

c Holocytosis: In this mechanism the *entire glandular cell disintegrates and becomes the secretory product* (holocrine secretion). As a result, the glandular cells must be constantly replaced by a basal regenerative cell layer (e.g., sebaceous glands in the skin).

C Production and release of secretions by exocytosis (electron-microscopic scale)

After the glandular cell has absorbed essential compounds from the blood and synthesized necessary materials such as secretory proteins in the rough endoplasmic reticulum (**a**), the secretions are transported by the Golgi apparatus (**b**) to the apical part of the cell, where they are discharged by exocytosis (**c**).

D Principal sites where hormones and hormonelike substances are formed

Hormones are vitally important chemical messengers that enable cells to communicate with one another. Usually, very small amounts of these messengers act on metabolic processes in their target cells. Different hormones can be classified on the basis of their

- site of formation,
- site of action,
- mechanism of action, or
- chemical structure.

Examples are steroid hormones (e.g., testosterone, aldosterone), amino acid derivatives (e.g., epinephrine, norepinephrine, dopamine, serotonin), peptide hormones (e.g., insulin, glucagon), and fatty acid derivatives (e.g., prostaglandins).

Principal sites of formation	Hormones and hormonelike substances
Classic endocrine hormonal glands	
Glandula pituitaria (lobus anterior and lobus posterior)	ACTH (adrenocorticotropic hormone, corticotropin)
	TSH (thyroid-stimulating hormone, thyrotropin)
	FSH (follicle-stimulating hormone, follitropin)
	LH (luteinizing hormone, lutropin)
	STH (somatotropic hormone, somatotropin)
	MSH (melanocyte-stimulating hormone, melanotropin)
	PRL (prolactin)
	ADH (antidiuretic hormone or vasopressin)
	Oxytocin (formed in the hypothalamus and secreted by the glandula pituitaria, lobus posterior)
Corpus pineale	Melatonin
Glandula thyroidea	Thyroxine (T_4) and triiodothyronine (T_3)
C cells of the glandula thyroidea	Calcitonin
Glandulae parathyroideae	Parathyroid hormone
Glandulae suprarenales	Mineralocorticoids and glucocorticoids
	Androgens
	Epinephrine and norepinephrine
Pancreatic islet cells (Langerhans cells)	Insulin, glucagon, somatostatin, and pancreatic polypeptide
Ovarium	Estrogens and progestins
Testis	Androgens (mainly testosterone)
Placenta	Chorionic gonadotropin, progesterone
Hormone-producing tissues and single cells	
Central and autonomic nervous system	Neuronal transmitters
Parts of the diencephalon (e.g., the hypothalamus)	Releasing and inhibitory hormones (liberins and statins)
System of gastrointestinal cells in the GI tract	Gastrin, cholecystokinin, secretin
Cardiac atria	Atrial natriuretic peptide
Kidney	Erythropoietin, renin
Liver	Angiotensinogen, somatomedins
Immune organs	Thymus hormones, cytokines, lymphokines
Tissue hormones	Eicosanoids, prostaglandins, histamine, bradykinin

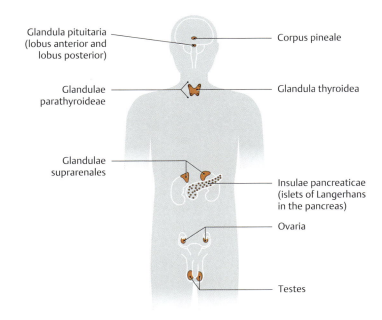

E Overview of the human endocrine glands

The diffuse or disseminated endocrine cell system (individual endocrine cells dispersed among the cells of the surface epithelium) in the gastrointestinal tract is not shown.

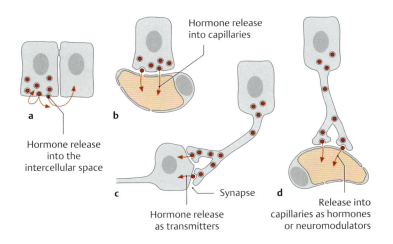

F Types of hormone-mediated information transmission

The endocrine system is closely linked to the autonomic nervous system and immune system in terms of its biological tasks. It functions as a kind of wireless communication system that coordinates the functions of target tissues and target organs, which may be located at distant sites.

a **Paracrine and autocrine secretion:** The hormones are not released into the bloodstream but into the intercellular space. Hence they act only in close proximity to their site of synthesis.

b **Endocrine secretion:** The hormones are synthesized and released into the bloodstream (fenestrated capillaries).

c **Neurocrine secretion:** Hormones of the neurocrine system (neurotransmitters) act in the form of synaptic transmitter substances and are concerned with local information transmission.

d **Neurosecretion:** Hormones or neuromodulators (neurohormones) are produced in specialized nerve cells and released to blood vessels in neurohemal regions (e.g., the glandula pituitaria). This enables them to act on distant organs.

8.1 Development of the Central Nervous System (CNS)

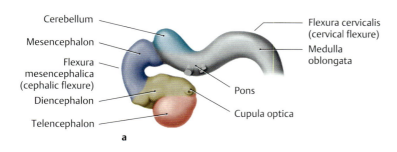

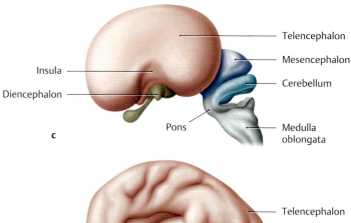

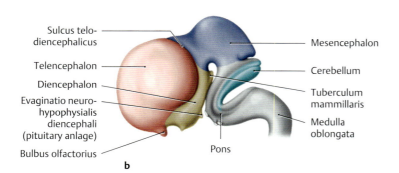

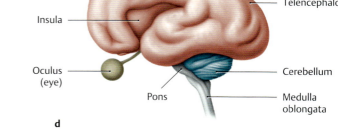

A Development of the brain

a Embryo with a greatest length (GL, see p. 4) of 10 mm, *at the beginning of the 2nd month of development*. Even at this stage we can see the differentiation of the neural tube into segments that will generate various brain regions (see **C**):

- Medulla oblongata (gray)
- Pons (gray)
- Cerebellum (light blue)
- Midbrain (mesencephalon, dark blue)
- Interbrain (diencephalon, yellow)
- Forebrain (telencephalon, red)

Note: The telencephalon grows over all the other brain structures as development proceeds.

b Embryo with a GL of 27 mm, *near the end of the 2nd month of development* (end of the embryonic period). The olfactory bulb (bulbus olfactorius) is developing from the telencephalon, part of the pituitary anlage (*neurohypophysis*) from the diencephalon.

c Fetus with a GL of 53 mm, in approximately the *3rd month of development*. By this time the telencephalon has begun to overgrow the other brain areas. The insula is still on the surface but will subsequently be covered by the cerebral hemispheres (compare with **d**).

d Fetus with a crown–rump length (CRL, see p. 4) of 27 cm (270 mm), in approximately the *7th month of development*. The brain has begun to develop conspicuous gyri and sulci.

B Brain vesicles and their derivatives

The cranial end of the neural tube expands to form three primary brain vesicles: the forebrain (prosencephalon), the midbrain (mesencephalon), and the hindbrain (rhombencephalon). The forebrain (telencephalon) and interbrain (diencephalon) develop from the prosencephalon. The mesencephalon gives rise to the colliculus superior and inferior and related structures. The rhombencephalon differentiates into the pons, cerebellum, and medulla oblongata. The pons and cerebellum are also known collectively as the metencephalon. Some important structures of the adult brain are listed at far right to illustrate the derivatives of the brain vesicles. They can be traced back in the diagram to their developmental precursors.

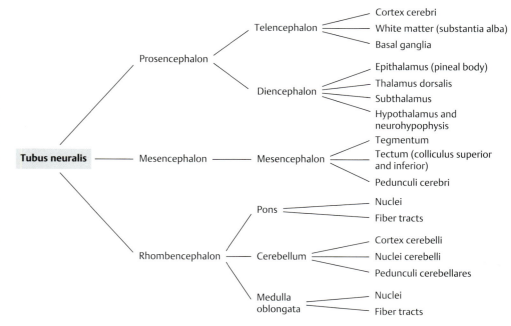

General Anatomy — 8. General Neuroanatomy

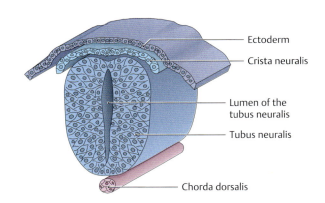

C Development of the nervous system: cross section through the neural tube, neural crest, and dorsal ectoderm

During development, the neural groove folds away from the overlying dorsal ectoderm and closes to form the *neural tube (tubus neuralis)*. Cells migrate from the lateral portions of the neural groove to form the *neural crest* on each side. The *central* nervous system (systema nervosum centrale; brain and spinal cord) develops from the neural *tube*, while the *peripheral* nervous system develops from derivatives of the neural *crest* (see p. 74).

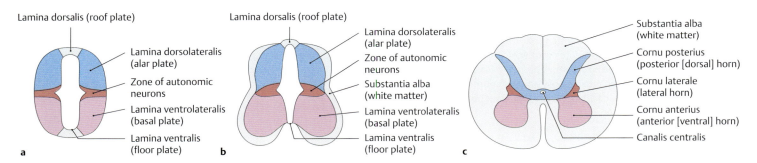

D Differentiation of the neural tube in the spinal cord region during development

Cross section, cranial view.
a Early neural tube, **b** intermediate stage, **c** adult spinal cord.
The neurons that form in the *basal plate* are *efferent* (motor) neurons, while those that form in the *alar plate* are *afferent* (sensory) neurons.

The area between them—the future thoracic, lumbar, and sacral cord—is another zone that gives rise to preganglionic autonomic neurons. The roof plate and floor plate do not form neurons. A knowledge of how these neuron populations are distributed is helpful in understanding the structure of the hindbrain (rhombencephalon, see **E**).

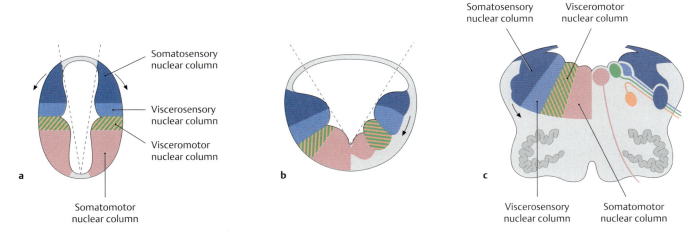

E Embryonic migratory movements of neuron populations and their effect on the location of the cranial nerve nuclei

Cross section, cranial view. (Visual aid: If we compare the spinal cord to a book, it would be closed in **a** and open in **b** and **c**.)

a In the **initial stage,** the motor neurons are ventral (anterior), and the sensory neurons are dorsal (posterior). The arrows indicate the directions of migration.
b In the **early embryonic stage,** the neurons of the alar plate migrate laterally and ventrally (anteriorly).

c In the adult brain (medulla oblongata and pons, derivatives of the rhombencephalon), we can distinguish **four nuclear columns** (after His and Herrick) that contain functionally distinct cranial nerve nuclei (from medial to lateral):

1. Somatomotor column (lilac)
2. Visceromotor column (orange and green stripes)
3. Viscerosensory column (light blue)
4. Somatosensory column (dark blue)

8.2 Neural Crest Derivatives and the Development of the Peripheral Nervous System (PNS)

A Development of the neural crest cells
At 3 weeks' development, the chorda dorsalis induces surface ectoderm in the medial embryonic disc to thicken and form the neural plate (lamina neuralis; neuroectoderm). The neural plate differentiates to form the primordia of the nervous system. Neural folds (plicae neurales) are raised on each side of the neural plate, and a median groove develops between them—the neural groove (sulcus neuralis). This groove subsequently deepens and closes to form the neural tube (tubus neuralis), which sinks below the ectoderm. Portions of the folds that do not contribute to neural tube formation differentiate to form the neural crest (crista neuralis). While neural crest cells in the future head region start to migrate even before the neural tube is closed, crest migration in the trunk is delayed until tube closure. Cells destined to form the neural crest detach from the ectoderm at the fusing margins of the neural tube and undergo an *epithelio-mesenchymal transition*, diving into underlying mesoderm, where they begin a long and tortuous migration. As the neural tube differentiates into the central nervous system (systema nervosum centrale), migrating neural crest cells settle in different locations and develop into sensory and autonomic ganglia, endocrine glands, melanocytes, cartilage, and other structures (see **B** and **C**) (after Wolpert).

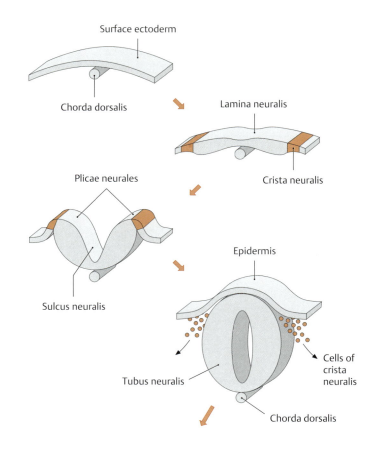

B Main migratory pathways and derivatives of the neural crest
(after Christ and Wachtler)
Neural crest cells originating in different regions have different migratory paths and fates. Those from cranial levels contribute to cartilage and bone in the head and neck, and to cranial parasympathetic ganglia (see **C**). Those from thoracolumbar levels do not become skeletal cells, but generate peripheral neurons, endocrine cells, melanocytes, and Schwann cells. This diagram shows the migration of the neural crest cells *in the trunk of the early embryo* (4 weeks' development, see p. 7). They follow three **main migratory pathways**:

① Dorsolateral pathway (melanoblasts, which differentiate into melanocytes)
② Ventrolateral pathway (ganglioblasts, which differentiate into sensory nerve cells in the dorsal [posterior] root [spinal] ganglia)
③ Ventral pathway (cells differentiate into neurons and associated cells of the paravertebral sympathetic ganglia [ganglia sympathica], into chromaffin cells of the adrenal medulla [glandula suprarenalis], and into autonomic plexuses [e.g. plexus myentericus] in the gastrointestinal tract)

Thus, the neural crest can develop into a variety of seemingly unrelated *nonneuronal* cells as well as peripheral ganglion cells. These unusual characteristics of the neural crest—its pluripotential capacity and wide-ranging migration—have consequences when its differentiation or migration is defective. Disruption of neural crest development may deprive organs of their autonomic innervation (Hirschsprung disease). Tumors derived from neural crest cells tend to be highly malignant and difficult to treat. (See **D**.)

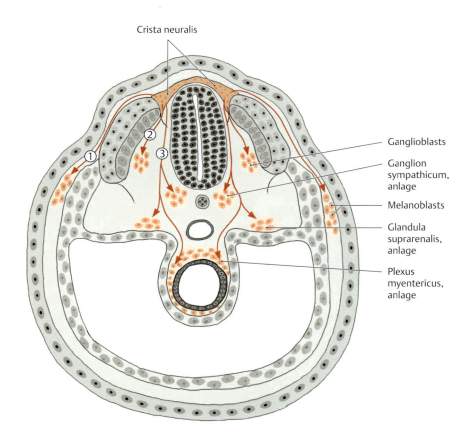

General Anatomy — 8. General Neuroanatomy

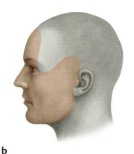

a b

C Neural crest derivatives in the head and neck region
Besides the equivalents of the structures named in **B** (e.g., melanocytes), other structures in the head and neck region that originate from the cranial neural crest are skeletal and cartilaginous muscles and muscles of facial expression.

a Cranial neural crest derivatives in the adult skeleton: ossa faciei (facial bones) os hyoideum, portions of the thyroid cartilage.
b Most of the facial skin is derived from the neural crest.

D Diseases of neural crest derivatives (selected examples)

Crista neuralis	Disease
Parasympathetic visceral ganglia	Neuroblastoma (malignant childhood tumor)
Enteric nervous system	Hirschsprung disease (aganglionic colon)
Glial cells (Schwann cells, satellite cells)	Neurofibromatosis (Recklinghausen disease)
Melanocytes	Malignant melanoma, albinism
Adrenal medulla	Pheochromocytoma (adrenal gland tumor)
Endocrine cells of the lung and heart	Carcinoids (malignant tumors with endocrine activity)
Parafollicular cells (C cells) of the glandula thyroidea	Medullary thyroid carcinoma

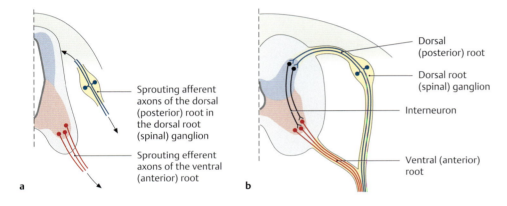

E Development of a peripheral nerve
Afferent (blue) and efferent (red) axons sprout *separately* from the neuron somata during early development (**a**). Primary afferent (sensory) neurons develop in the dorsal (posterior) root (spinal) ganglia, and primary motor neurons develop from the basal plate of the spinal cord (**b**). *Interneurons* (black), which connect sensory ganglia and motor neurons, develop later.

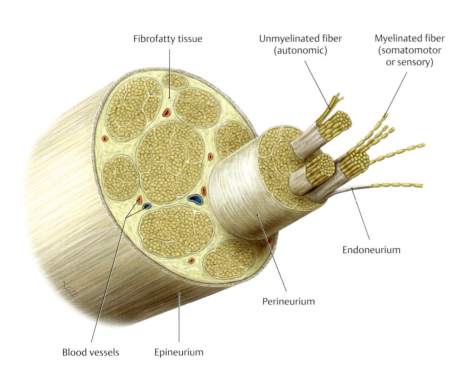

F Structure of a peripheral nerve
A peripheral nerve consists entirely of axons (also called neurites) and sheath tissue (Schwann cells, fibroblasts, blood vessels). The axons transmit information either from the periphery to the CNS (*afferents*) or in the opposite direction from the CNS to the periphery (*efferents*). Axons may be myelinated or unmyelinated. The latter have a much slower conduction velocity and are usually fibers of the autonomic nervous system (see p. 95). Among the investing layers of the nerve, the *perineurium* ensheaths the nerve fascicles and provides an important tissue barrier (see p. 93).

8.3 Topography and Structure of the Nervous System

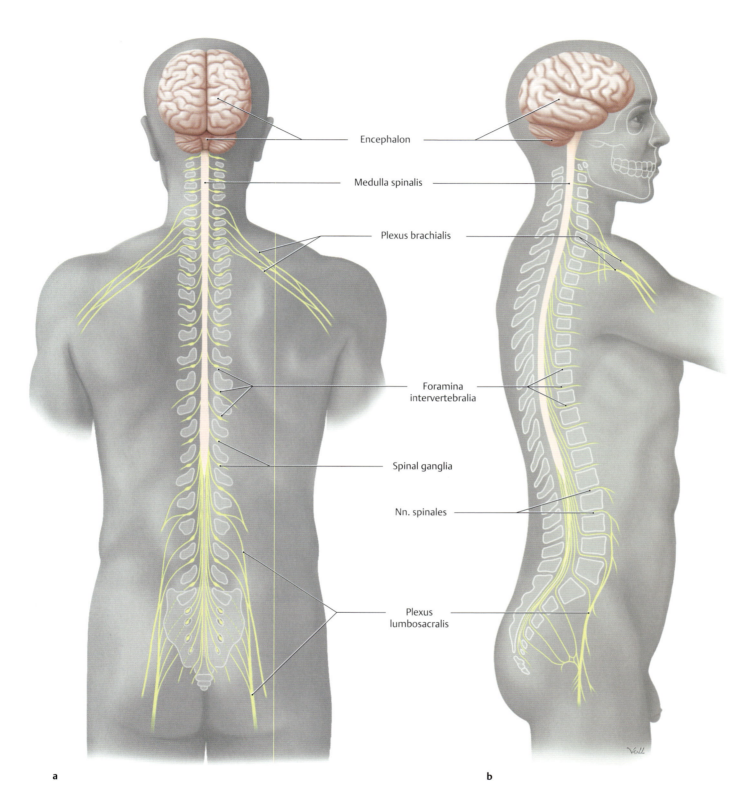

A Topography of the nervous system
a Posterior view, **b** right lateral view.
The *central nervous system* (CNS), consisting of the brain (encephalon) and spinal cord (medulla spinalis), is shown in pink. The *peripheral nervous system* (PNS), consisting of nerves and ganglia, is shown in yellow. The nerves arising from the medulla spinalis leave their bony canal through the *foramina intervertebralia* and are distributed to their target organs. The *spinal nerves* are formed in the foramina by the union of their dorsal (posterior) roots and ventral (anterior) roots (see p. 81). The small *spinal ganglion* in the foramina intervertebralia appears as a slight swelling of the dorsal root (visible only in the posterior view; its function is described on p. 81).
In the limbs, the ventral rami of the spinal nerves come together to form plexuses. These plexuses then give rise to the peripheral nerves that supply the limbs.

General Anatomy — 8. General Neuroanatomy

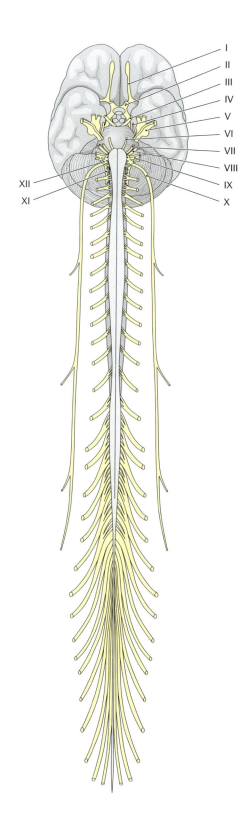

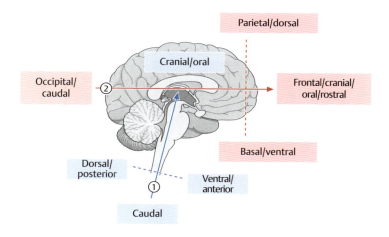

C Terms of location and direction in the CNS
Midsagittal section, right lateral view.
Note two important axes:

① The almost vertical brainstem axis (corresponds approximately to the body axis).
② The horizontal axis through the diencephalon and telencephalon.
Keep these reference axes in mind when using directional terms in the CNS.

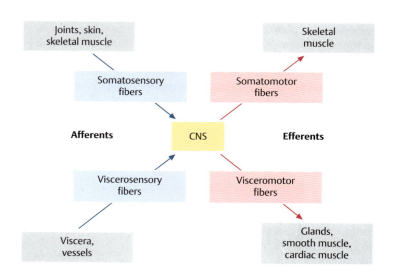

B Spinal nerves and cranial nerves
Anterior view. *Thirty-one pairs of spinal nerves* arise from the spinal cord in the peripheral nervous system, compared to *12 pairs of cranial nerves* that arise from the brain. The cranial nerve pairs are traditionally designated by roman numerals.
Note: The first two cranial nerves, the olfactory (I) and optic (II) nerves, are not peripheral nerves in the strict sense, but outpouchings from the brain, which means they are pathways of the central nervous system. They are enclosed within the meninges and contain cells that are exclusively found in the central nervous system, oligondendrocytes and microglial cells.

D Schematic representation of information flow in the nervous system
The information encoded in nerve fibers is transmitted either *to the CNS* (brain and spinal cord) or *from the CNS* to the periphery (PNS, including the peripheral parts of the autonomic nervous system, see p. 72). Fibers that carry information to the CNS are called afferent fibers or *afferents* for short; fibers that carry signals away from the CNS are called efferent fibers or *efferents*.

77

8.4 Cells of the Nervous System

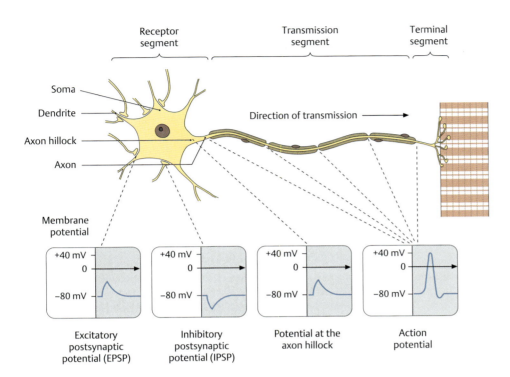

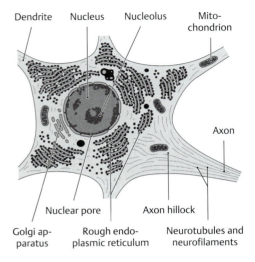

B Electron microscopy of the neuron
Neurons are rich in *rough endoplasmic reticulum* (protein synthesis, active metabolism). This endoplasmic reticulum (known also as *Nissl substance*) is easily demonstrated by light microscopy using cationic dyes, which bind to the phosphodiester backbone of the ribosomal RNAs. The distribution pattern of the Nissl substance is used in neuropathology to evaluate the functional integrity of neurons. Neurotubules and neurofilaments are referred to collectively as *neurofibrils* in light microscopy, as they are too fine to be identified as separate structures under a light microscope. Neurofibrils can be demonstrated in light microscopy by impregnating the nerve tissue with silver salts. This is of interest in neuropathology because the clumping of neurofibrils is an important histologic feature of Alzheimer disease.

A The nerve cell (neuron)
The neuron is the smallest functional unit of the nervous system. Neurons communicate with other nerve cells through synapses. The *synapses that end at nerve cells* usually do so at dendrites (as seen here). The transmitter substance that is released at the synapses to act on the dendrite membrane may have an *excitatory* or *inhibitory* action, meaning that the transmitter either increases or decreases the local action potential at the nerve cell membrane. All of the excitatory and inhibitory potentials of a nerve cell are integrated in the axon hillock. If the excitatory potentials predominate, the stimulus exceeds the excitation threshold of the neuron, causing the axon to fire (transmit an impulse) according to the all-or-nothing rule.

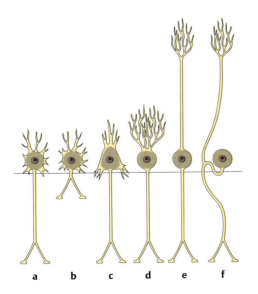

C Basic forms of the neuron and its functionally adapted variants
The horizontal line marks the region of the axon hillock, which represents the initial segment of the axon. (The structure of a peripheral nerve, consisting only of axons and sheath tissue, is shown on p. 75.)

a Multipolar neuron (multiple dendrites) with a *long* axon (= long transmission path). Examples are projection neurons such as α-motoneurons in the spinal cord.
b Multipolar neuron with a short axon (= short transmission path). Examples are interneurons like those in the gray matter (substantia grisea) of the brain and spinal cord.
c Pyramidal cell: Dendrites are present only at the apex and base of the *tridentate* cell body, and the axon is long. Examples are efferent neurons of the cerebral motor cortex.
d Purkinje cell: An elaborately branched dendritic tree arises from a circumscribed site on the cell body. The Purkinje cell receives many synaptic contacts from afferents to the cerebellum and is also the efferent cell of the cortex cerebelli.
e Bipolar neuron: The dendrite branches in the periphery. Examples are bipolar cells of the retina.
f Pseudounipolar neuron: The dendrite and axon are not separated by a cell body. An example is the primary afferent (= first sensory) neuron in the spinal ganglion (see p. 89).

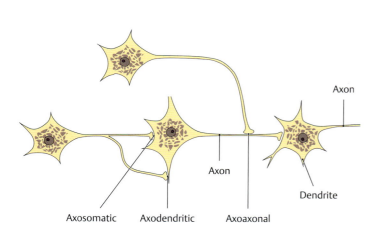

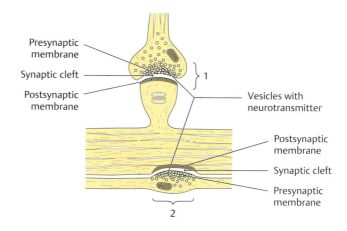

D Synaptic patterns in a small group of neurons
Axons can terminate at various sites on the target neuron and form synapses there. The synaptic patterns are described as axodendritic, axosomatic, or axoaxonal. Axodendritic synapses are the most common (see also **A**).

E Electron microscopy of synapses in the CNS
Synapses are the functional connection between two neurons. They consist of a presynaptic membrane, a synaptic cleft, and a postsynaptic membrane. In a *spine synapse* (1), the presynaptic knob (bouton) is in contact with a specialized protuberance (spine) of the target neuron. The side-by-side synapse of an axon with the flat surface of a target neuron is called a parallel contact or *bouton en passage* (2). The vesicles in the presynaptic expansion contain the neurotransmitters that are released into the synaptic cleft by exocytosis when the axon fires. From there the neurotransmitters diffuse to the postsynaptic membrane, where their receptors are located. A variety of drugs and toxins act upon synaptic transmission (antidepressants, muscle relaxants, toxic gases, botulinum toxin).

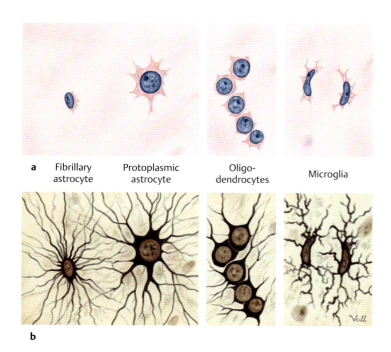

F Cells of the neuroglia in the CNS
Neuroglial cells surround the neurons, providing them with structural and functional support (see **G**). Various staining methods are available in light microscopy for selectively demonstrating different portions of the neuroglial cells:

a Cell nuclei demonstrated by a basic stain.
b Cell bodies demonstrated by silver impregnation.

G Summary: cells of the CNS and PNS and their functional importance

Type of cell	Function
Neurons (CNS and PNS)	1. Impulse formation 2. Impulse conduction 3. Information processing
Glial cells	
Astrocytes (CNS only)	1. Maintain a constant internal milieu in the CNS 2. Contribute to the structure of the blood–brain barrier (see p. 93) 3. Phagocytize dead synapses 4. Form scar tissue in the CNS (e.g., in multiple sclerosis or following a stroke)
Microglial cells (CNS only)	Phagocytosis ("macrophages of the brain")
Oligodendrocytes (CNS only)	Myelin sheath formation in the CNS
Schwann cells (PNS only)	Myelin sheath formation in the PNS
Satellite cells (PNS only)	Modified Schwann cells; surround the cell body of neurons in PNS ganglia

8.5 Structure of a Spinal Cord Segment

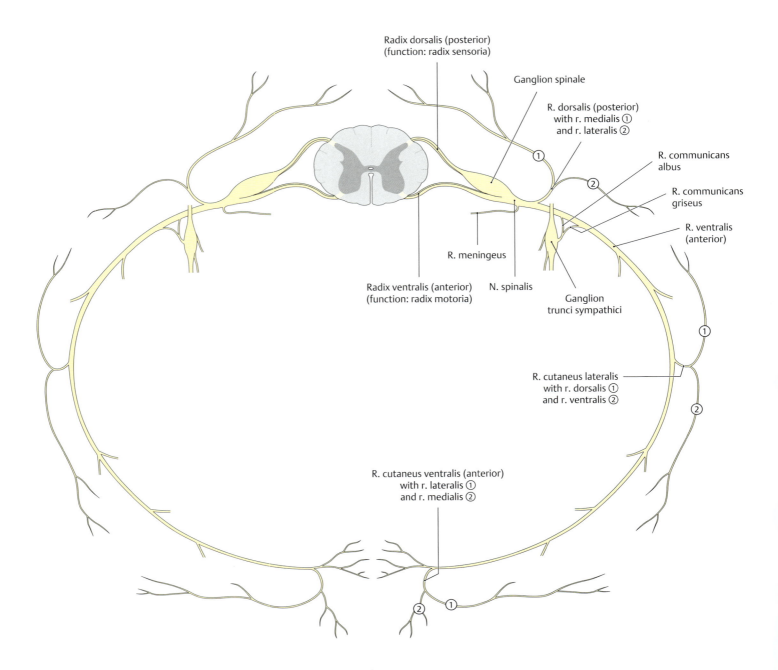

A Structure of a spinal cord segment with its spinal nerve
Superior view. The spinal cord is made up of 31 consecutive segments arranged one above the other (see **B**). A *radix ventralis (anterior root)* and a *radix dorsalis (posterior root)* emerge from the sides of each segment. The *anterior* root consists of *efferent* (motor) fibers, while the *posterior* root consists of *afferent* (sensory) fibers. Both roots from one segment unite in the foramen intervertebrale to form the *nervus spinalis*. The afferent (sensory) fibers and efferent (motor) fibers intermingle at this junction, so that the branches into which the spinal nerve divides (see below) contain motor *and* sensory elements (except for the r. meningeus, which is purely sensory). This division into branches (rami) occurs shortly after the radix ventralis and dorsalis unite to form the spinal nerve. Consequently, the spinal nerve itself is only about 1 cm long.

The principal branches (rami) of the n. spinalis have the following functions:

- The r. ventralis innervates the anterior and lateral body wall and the limbs.
- The r. dorsalis innervates the skin of the back and the intrinsic back muscles.
- The r. meningeus reenters the spinal canal, providing sensory innervation to the spinal membranes and other structures.
- The r. communicans albus carries white (= myelinated) fibers *to* the ganglion truncus sympathicus.
- The r. communicans griseus carries gray (= unmyelinated) fibers *from* the ganglion trunci sympathici back to the n. spinalis (the functional significance of this is described on p. 95). The anterior and posterior rami of the nn. spinalis subdivide into further branches.

General Anatomy — 8. General Neuroanatomy

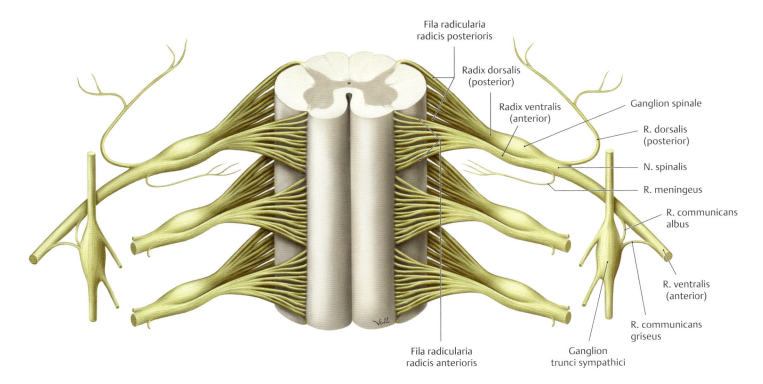

B Spinal cord segments
Anterior view. The spinal cord is composed of many segments, each of which gives rise to one pair of spinal nerves (for clarity, only the upper segment of the spinal cord is shown). Several rootlets (fila radicularia) unite to form the anterior and the posterior roots of a spinal nerve, which then divides into five branches (see **A**). The segmental structure of the spinal cord is visible externally only at the rootlets; the spinal cord itself does not show any external segmentation.

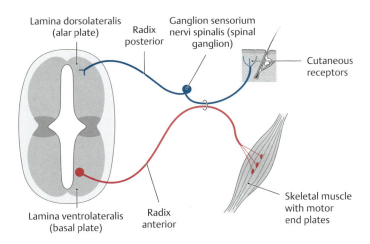

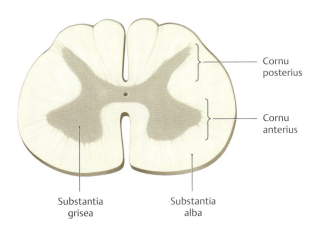

C Embryological origins of the topographical and functional anatomy of a spinal cord segment
The afferent fibers (e.g., from cutaneous receptors) pass through the *posterior root (radix posterior)* into the posterior horn of the spinal cord, which is derived embryologically from the alar plate. The efferent fibers arise from neurons located in the anterior horn of the spinal cord, which is a derivative of the basal plate. They leave the spinal cord by the *anterior root (radix anterior)* and are distributed to their target organ, such as a skeletal muscle.

D Cross section of a spinal cord segment
Superior view. A cross section of the spinal cord shows the centrally located gray matter (substantia grisea) arranged in a butterfly-shaped pattern and the white matter (substantia alba) surrounding it. Cell bodies of neurons are located in the gray matter; axons move up and down in the white matter. The gray matter is divided into anterior and posterior horns (cornu anterius et posterius). The cell bodies of efferent neurons are located in the anterior horns, and those of afferent neurons in the posterior horns. Thus, the spinal cord is functionally arranged from anterior (motor) to posterior (sensory) (horizontal arrangement). For the relationship between spinal cord segments and dermatomes (vertical arrangement of the spinal cord), see p. 82.

General Anatomy — 8. General Neuroanatomy

8.6 Sensory Innervation: An Overview

A Location and designation of spinal cord segments in relation to the vertebral canal

Right lateral view. The spinal cord (medulla spinalis) consists of 31 segments that are arranged from top to bottom:

- 8 cervical segments,
- 12 thoracic segments,
- 5 lumbar segments,
- 5 sacral segments, and
- 1 coccygeal segment (no clinical significance)

The growth of the spinal cord lags behind that of the spinal column, so in an adult it extends only to the first lumbar vertebra, as shown here (compare p. 128). The spinal cord segments are equivalent to the dermatomes on the body surface (see **C**). In other words, a dermatome is an area on the body surface, the sensory receptors of which (e.g., for pressure, temperature, pain, vibration) pull with their afferents into a spinal cord segment, thus enabling a 1:1 relationship between body surface and spinal cord segment. In clinical terms, it means that in the case of sensory disorder (dermatome malfunction), it is possible to determine at which level of the spinal cord the malfunction is located. A myotome (see p. 7) is the equivalent of a dermatome, as regards motor function.

Note: There are seven cervical vertebrae (C I–C VII) but eight pairs of cervical nerves (C I–C VIII). The highest pair of cervical nerves exit the vertebral canal superior to the first cervical vertebra. The remaining pairs of cervical nerves, like all the other spinal nerve pairs, exit inferior to the cervical vertebral body. The pair of nn. coccygei (gray) has no clinical importance.

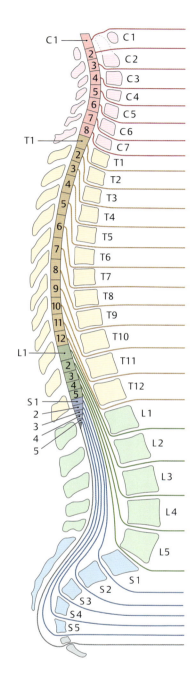

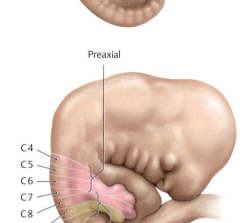

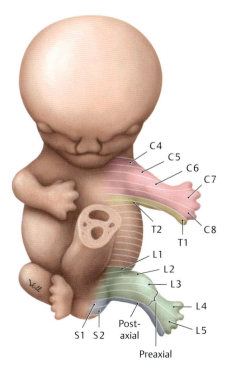

B Dermatomes and limb buds (after Sadler)

From the somites, which give rise to limbs, cells migrate, and limb buds form. These migrating cells drag their original segmental innervation along behind them. Thus, dermatomes reflect the segmental arrangement, which is based on the somites (see p. 7).

- **a** In a 5-week-old embryo, the dermatomes are still segmentally arranged.
- **b** In a 6-week-old embryo, migrations have already taken place: the preaxial segments are positioned cranially, the postaxial segments caudally. The segments lying in between have shifted in a distal direction toward the hand.
- **c** In a 7-week-old embryo, the proportions are similar to those postbirth.

This dermatome diagram, similar to the diagram in **C**, is simplified, because the cells, which form the dermatomes, migrate (for details, see p. 14).

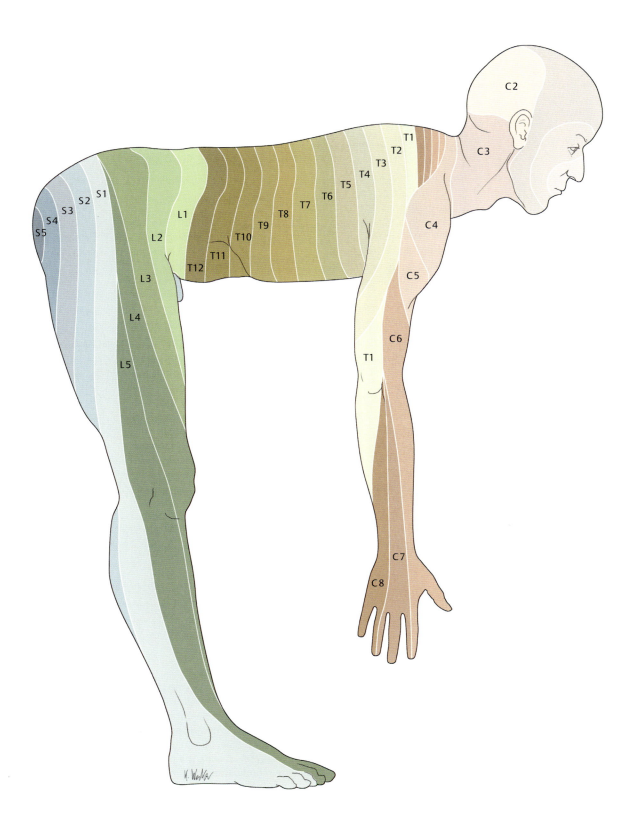

C Simplified scheme for learning the dermatomes (after Mumenthaler)

The distribution of dermatomes on the human limbs results from the sprouting of the limb buds that occurs during embryonic development. When the limbs are positioned at right angles to the body as in a quadruped, it is easier to appreciate the pattern of innervation. Compare the location of the dermatomes to those of the spinal cord sections in **A**. The more complex pattern of dermatomes resulting from cell migration is shown on p. 86.

Note: The C1 segment contains only motor fibers; thus, there is not a C1 dermatome.

8.7 Sensory Innervation: Principles of Dermatome and Plexus Formation

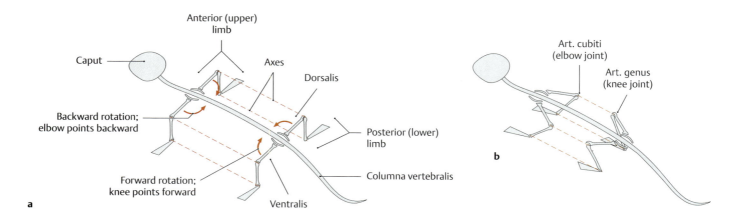

A Phylogenetic development of dermatomes
In lower tetrapods, the trunk is located between the limbs (**a**). In mammals, the limbs are also rotated (**b**). The posterior (= lower) limbs are rotated forward, and the original posterior position of the limb muscles shifts anteriorly when picturing the mammal walking erect (for details, see p. 20). This is not the case for the anterior (= upper) limbs, which are rotated backward. That is why the cells that form dermatomes in the lower limb region migrate more than similar cells in the upper limb region. That leads to a screwlike distribution of dermatomes in the lower limbs, unlike that in the upper limbs (see p. 86).

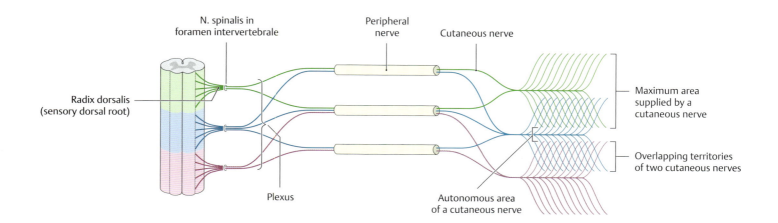

B Course of sensory fibers from the posterior root to the dermatome
Sensory fibers pass from the dorsal root to the foramen intervertebrale, where they unite with motor fibers to form the spinal nerve. The sensory fibers are then distributed to the ventral and dorsal rami of the spinal nerve. The simple segmental arrangement of the sensory territories that occurs on the trunk is not found in the limbs (see **D**). This results from the migratory movements of various muscular and cutaneous primordia, or precursors, during the development of the limbs. Because these primordia carry their segmental innervation with them, the sensory fibers from different segments become intermingled in the limbs (plexus formation, see **C**). After fibers have been distributed in the plexus, the fibers in peripheral nerves travel to their destinations, their terminal portions often consisting entirely of sensory cutaneous nerves. The area of skin that is innervated by one spinal cord segment is called a *dermatome*. The dermatomes of adjacent spinal cord segments are often located so close together that their territories broadly overlap. This explains why the clinically detectable area of sensory loss caused by a segmental nerve lesion may be consid- erably smaller than the dermatome itself. The area that receives all of its sensory supply from one cutaneous nerve is called the autonomous area of that nerve. When a segmental nerve lesion occurs, the cutaneous nerves from the two adjacent spinal cord segments are still available to supply at least the peripheral part of the affected skin area. This helps in understanding the difference between radicular (= segmental) and peripheral sensory innervation. When a nerve root becomes damaged (e.g., due to a herniated intervertebral disk), the resulting sensory loss will follow a radicular innervation pattern (see p. 86). But when a peripheral nerve is damaged (e.g., due to a limb injury), the sensory loss will conform to a peripheral innervation pattern (see p. 87).

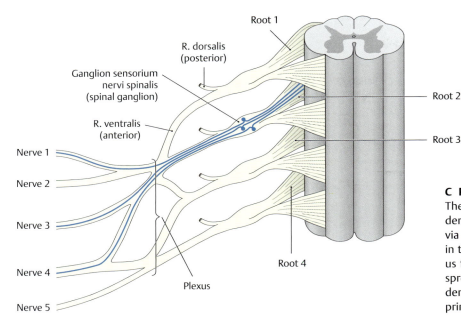

C Principles of plexus formation
The axons, which form the afferents from a dermatome, extend from that dermatome via several peripheral nerves to a single root in the spinal cord. As part of embryonic plexus formation, afferent axons of that one root spread via several peripheral nerves to the dermatome, in response to signals from the primordia of the dermatome.

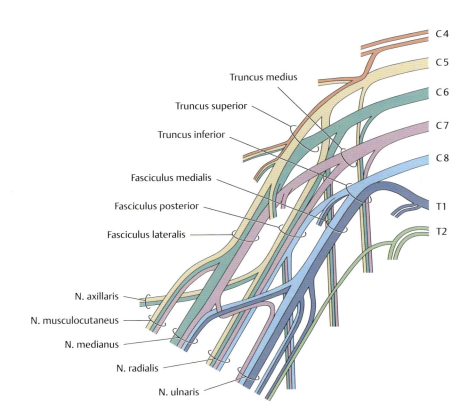

D Plexus formation as illustrated by the plexus brachialis
During embryonic development, migration and, associated with it, mixing of dermatome and myotome primordia occur with the primordia dragging their innervation along behind them. Dermatomes and myotomes have an identical developmental pattern. In the following, we will discuss only dermatomes (for more on myotomes, see p. 88). During their development, the primordia of dermatomes transmit signals to the budding sensory axons (see p. 75). This leads to every dermatome receiving its axons from its spinal cord segment. (To exhibit features more clearly, every spinal cord segment is individually colored). In order for the axons to leave their spinal cord segment and reach "their" dermatome, they have to be paired with the various peripheral nerves. The place where the shifting and mixing of axons (= fibers) occurs is called a plexus. With the plexus brachialis, a distinction is made between different parts of the plexus. Prior to plexus formation, axons from one spinal cord segment combine to form roots. The axons from the roots of C5 and C6 merge to form the truncus superior, the axons from the root of C7 form the truncus medius, and those of C8 and T1 form the truncus inferior. The anterior branches of the truncus superior and medius form the fasciculus lateralis and fasciculus medialis, and the posterior branches of all three trunks form the fasciculus posterior. Finally, from the fasciculi arise the large nerves of the arm and shoulder, where the axons travel on their way to the dermatomes.

8.8 Sensory Innervation: Dermatomes and Cutaneous Nerve Territories

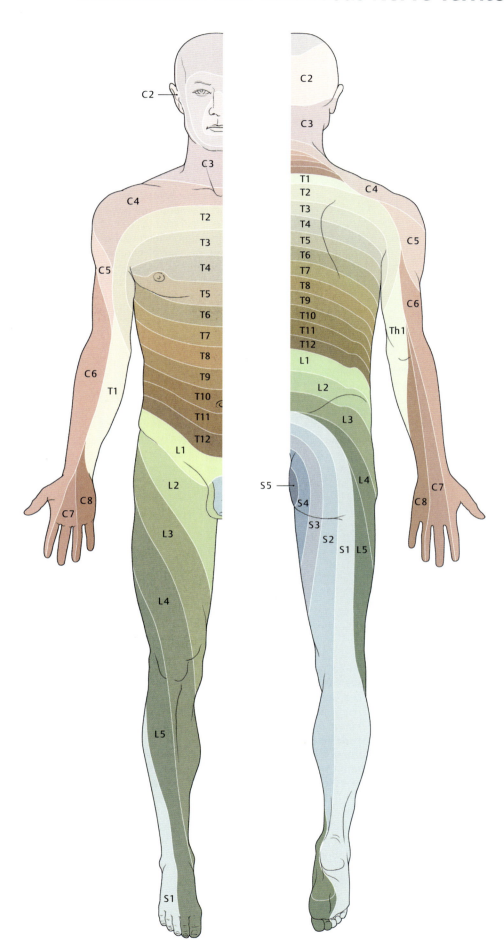

A Pattern of radicular (segmental) sensory innervation (dermatomes)

The skin area supplied by a posterior spinal nerve root (radix dorsalis) is called a *dermatome*. Because the C 1 segment consists entirely of motor fibers, it lacks a corresponding sensory field. A knowledge of radicular innervation is very important clinically. For example, when a herniated intervertebral disk is impinging on a sensory root, it will cause sensory losses in the affected dermatome. The area of sensory loss can then be used to locate the level of the lesion: Which intervertebral disk is affected? In a patient with shingles (herpes zoster inflammation of a spinal ganglion), the dermatome supplied by that ganglion will be affected (after Mumenthaler).

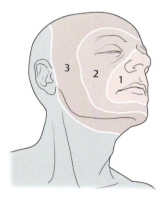

B Pattern of nuclear sensory innervation in the head region

The head receives its sensory innervation from the n. trigeminus (nervus cranialis V). A lesion of the sensory nucleus of the n. trigeminu within the brain (= central lesion) causes an onion-skin pattern of sensory alteration along the concentric *Sölder lines* that encircle the mouth and nostrils. The pattern of these lines corresponds to the distribution of the neurons in the sensory nucleus of the n. trigeminus (somatotopic organization, i.e., certain groups of neurons in the CNS are linked to certain territories in the periphery). Territory 1 is supplied by the cranial nuclear column, territory 2 by the middle column, and territory 3 by the caudal column. This pattern of sensory loss is like that associated with radicular neuropathy affecting a peripheral nerve.

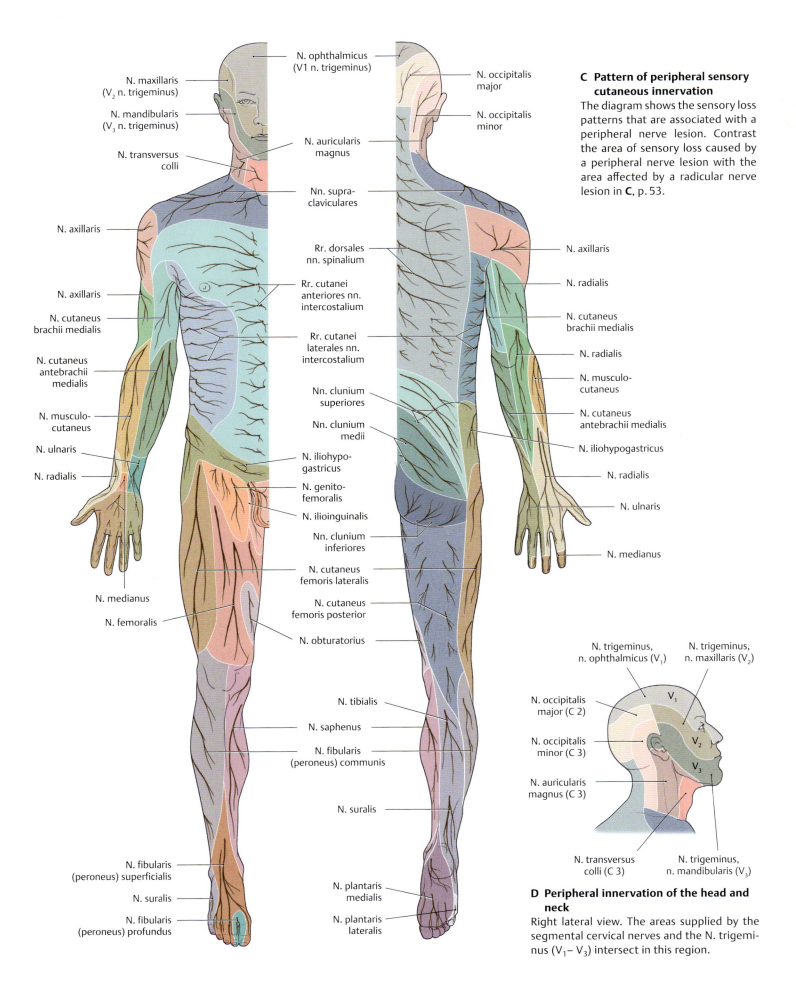

C Pattern of peripheral sensory cutaneous innervation
The diagram shows the sensory loss patterns that are associated with a peripheral nerve lesion. Contrast the area of sensory loss caused by a peripheral nerve lesion with the area affected by a radicular nerve lesion in **C**, p. 53.

D Peripheral innervation of the head and neck
Right lateral view. The areas supplied by the segmental cervical nerves and the N. trigeminus (V_1–V_3) intersect in this region.

8.9 Motor Innervation: Organization of the Spinal Cord and Reflexes

A Organization of anterior horns (cornua anteriora) in the spinal cord
Whereas the dermatomes are areas of sensory innervation (= afferents), myotomes are areas of motor innervation of the skeletal muscles (= efferents). Just as in the posterior horn, in the anterior horn, too, the spinal cord segments are stacked on top of each other. With regard to innervation, a distinction is generally made between two types of skeletal muscles:

- Those with monosegmental innervation
- Those with polysegmental innervation

In monosegmentally innervated muscles (green muscle), the cell bodies of the motoneurons innervating these muscles are located at the level of a single spinal cord segment. In polysegmentally innervated muscles (blue and orange muscle), the cell bodies of the motoneurons reside inside a column that extends over several spinal cord segments. Muscles that are exclusively or predominantly innervated by one spinal cord segment are called segment-indicating muscles. They can be clinically evaluated by testing a corresponding reflex.

B Principles of plexus formation
In the plexus region, efferent axons from various spinal cord roots that innervate a polysegmentally innervated muscle combine into one peripheral nerve, which then extends to the muscle.

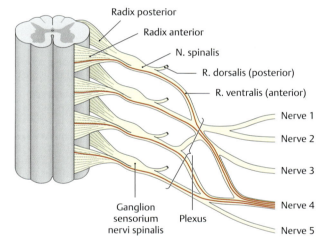

C Reflexes
The gray matter in the spinal cord, among other things, helps to coordinate muscle action on the spinal cord level (without involvement of the cortex), that is, reflexes. Broadly speaking, a distinction is made between a monosynaptic proprioceptive reflex (shown on the left) and a polysynaptic exteroceptive reflex (shown on the right).
Monosynaptic proprioceptive reflex: Receptors in a muscle transmit information about the condition of the muscle (length, muscle tension) via neurons in the spinal ganglia (ganglia sensoria nervorum spinalium), the synapses of which end on the motor neuron of the muscle in the spinal cord, thereby influencing its actvity.
Polysynaptic exteroceptive reflex: Receptors in the skin, rather than a muscle, transmit information to an interneuron; the interneuron synapses with the motoneuron of the muscle in the spinal cord. Since there is more than one neuron involved in the transmission of sensory information, the reflex is referred to as polysynaptic.

General Anatomy — 8. General Neuroanatomy

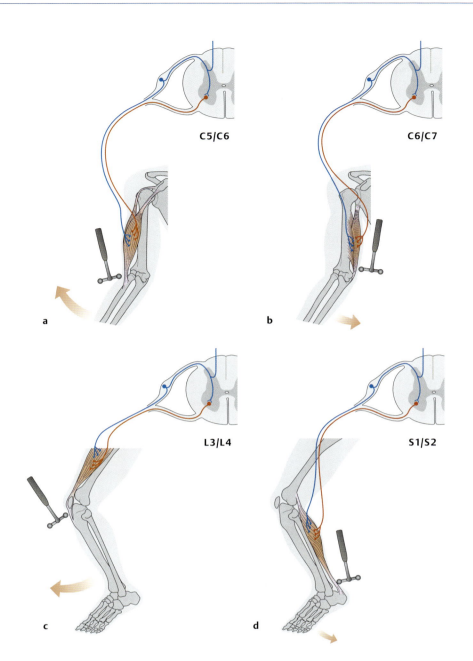

D Clinically important reflexes
a Biceps reflex; b triceps reflex; c patellar tendon reflex (quadriceps reflex); d calcaneal (Achilles') tendon reflex.

For each reflex, the muscles, reflex trigger points, involved nerves (afferents in blue, efferents in red), and corresponding spinal cord segments are shown. The most important proprioceptive reflexes should be tested during every clinical examination. The reflex is triggered by a brief blow with the reflex hammer, in most cases hitting the tendon of the muscle. As a result, the muscle stretches. If the muscle then contracts in reaction to the expansion, the reflex arc is intact. Although the reflex involves one muscle and the nerve supplying it, several spinal cord segments are involved in the innervation (multisegmental muscles, see **A**). As part of reflex testing, it is important to compare each reflex to its contralateral counterpart in order to detect one-sided augmentation, impairment, or pathologic deviations.

E Topographical and functional organization of a spinal cord segment

The **afferent fibers** from the skin, muscles, and joints (somatosensory, blue) and from the viscera (viscerosensory, green) pass through the posterior root into the spinal cord and terminate in the posterior horn (cornu posterius). Both fibers arise from pseudounipolar cells in the posterior root (spinal) ganglion (ganglion sensorium nervi spinalis).

The **efferent fibers** for the skeletal musculature (somatomotor, red) and for the viscera (visceromotor, brown) pass through the anterior root to their respective target organs, i.e., the skeletal muscles, the vascular smooth muscles, and the smooth muscles of the internal organs. The fibers differ in their origin. Fibers for the skeletal muscles originate in the anterior horn (cornu anterius) of the spinal cord, and fibers for the viscera originate in the lateral horn (cornu laterale). The axons of somatic motoneurons in the anterior horn synapse directly on skeletal muscle fibers (p. 90). Visceral motoneurons in the lateral horn, however, provide indirect innervation to most of their target organs. Their axons form synapses with autonomic neurons in larger discrete sympathetic ganglia or in scattered clus- ters embedded in the visceral organs (p. 95).

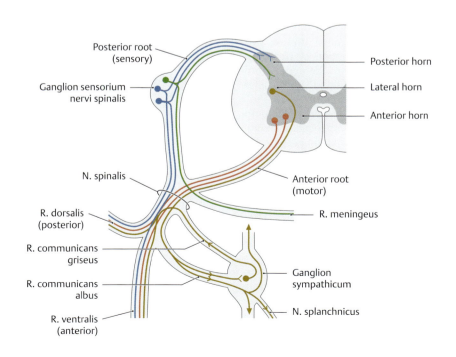

8.10 Motor Innervation: Upper (First) and Lower (Second) Motor Neurons

A Simplified scheme of motor innervation
The simplest common path of motor innervation originates in the primary cerebral motor cortex. Large neurons (dark red) contribute axons to bundles collected into massive *corticospinal* tracts. Many axons pass uninterrupted to the spinal cord, where they synapse upon motor neurons in the anterior horn (cornu anterius; orange). In the brainstem (truncus encephali) the bundles of corticospinal axons are designated *pyramids* or *pyramidal tracts (tractus corticospinalis)*. In the medulla oblongata, most corticospinal axons cross the midline in the *pyramidal decussation* and continue on the opposite side. Axons from motor neurons in the anterior horn of the spinal cord exit in *anterior roots* (radix anterior; p. 62) and reach their targets, skeletal muscles, via peripheral nerves, where they synapse directly at *neuromuscular junctions*.

The typical spinal motor neuron receives inputs from multiple sources. The simplest local circuit involves afferent (sensory) synapses from fibers from posterior root (spinal) ganglion cells (blue). Although most afferent information is relayed to spinal motor neurons through multiple intermediate *interneurons*, some sensory input—particularly from tendon stretch receptors—is transmitted directly via a single synapse, as depicted. This chain of connections is called a *reflex arc*. One such arc produces the knee-jerk or patellar tendon reflex.

This pattern of motor connections is responsible for two fundamental principles in the diagnosis of neurologic disorders. Because of their positions in the pathway, the cerebral cortical cells are referred to as *upper motor neurons*, those in the spinal cord as *lower motor neurons*. Damage to lower motor neurons or their axons leads to denervation of muscles, with *flaccid paralysis* and eventual muscle atrophy. In contrast, interruption of corticospinal axons or destruction of the upper motor neurons themselves, resulting, for instance, from a cerebral infarction (stroke), leads to a loss of voluntary control over the lower motor neurons. With such an upper motor neuron lesion, the lower motor neuron is controlled solely by local spinal circuits. Local reflexes remain or are enhanced, but muscles show sustained contractions and increased tone, with *spastic paralysis*.

The second fundamental correlation between the anatomy of the motor pathway and neurologic diagnosis involves the crossing of corticospinal axons in the pyramidal decussation. Upper motor lesions that occur above (cranial to) this decussation will produce spastic paralysis on the *contralateral* side of the body; lesions in corticospinal axons below (caudal to) this point will cause spastic paralysis on the same side as (*ipsilateral* to) the lesion.

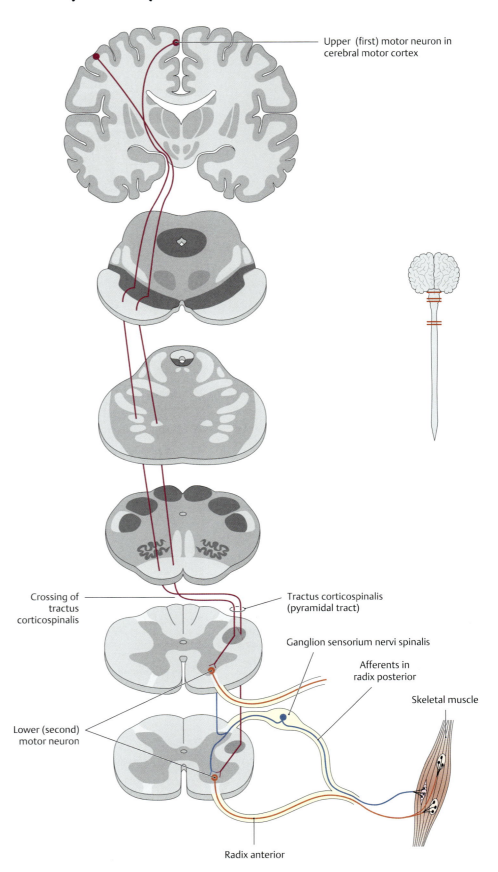

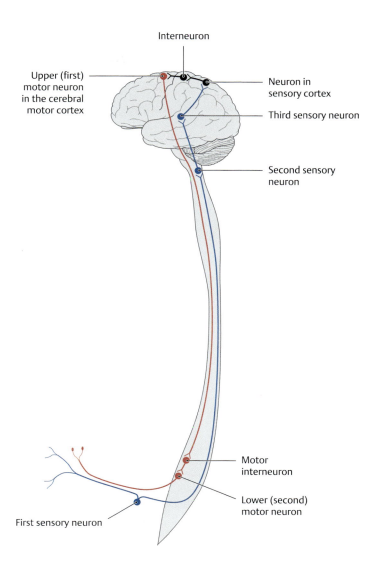

B Neural circuit for sensory and motor innervation
Left lateral view. Information processing in the CNS is actually more complex than shown in **A** because motor innervation is influenced by afferents of a kind not illustrated here (= sensorimotor circuit). Although a great many neurons are active at each step in the circuit, the diagram shows only one at each level. Sensory information, encoded as electrical impulses, enters the spinal cord through a primary afferent neuron (first sensory neuron) whose cell body is in the spinal ganglion (afferent neurons shown in blue). The information is relayed synaptically to the second and third afferent (sensory) neurons, which transfer it to the sensory cortex (black). Associative neurons (interneurons, black) transmit the information to the upper motor neuron in the cerebral motor cortex, where the information is modified by many associated inputs. The cortical motor neuron (upper motor neuron) redirects the information to the spinal cord, where it reaches the spinal (lower) motor neuron, either directly or via interneurons and additional synaptic relays. Finally, the lower motor neuron in the spinal cord transmits the impulses to the voluntary skeletal muscle. Corticospinal projections are most direct and numerous for lower motor neurons that innervate muscles, like those in the human hand, which are often under fine control and coordinated cognitive direction. Other motor pathways, not depicted here, are more important for controlling posture and balance. Because the axons in some of these other pathways are not in the corticospinal and pyramidal tracts, they are sometimes referred to as *extrapyramidal*.

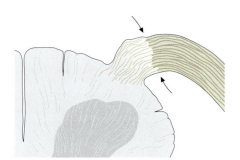

C Obersteiner–Redlich zone in a posterior root
The Obersteiner–Redlich zone marks the morphological junction between the CNS and PNS (arrows). Oligodendrocytes in the CNS form a myelin sheath around the axons only as far as this zone, which is located just past the emergence of the posterior root from the spinal cord. The myelin sheaths in the PNS are formed by Schwann cells (see p. 92 for details). The myelin is so thin at these sites that the fibers appear almost unmyelinated. This creates a site of predilection for immunologic diseases such as the immune reactions that occur in the late stages of syphilis.

8.11 Differences between the Central and Peripheral Nervous Systems

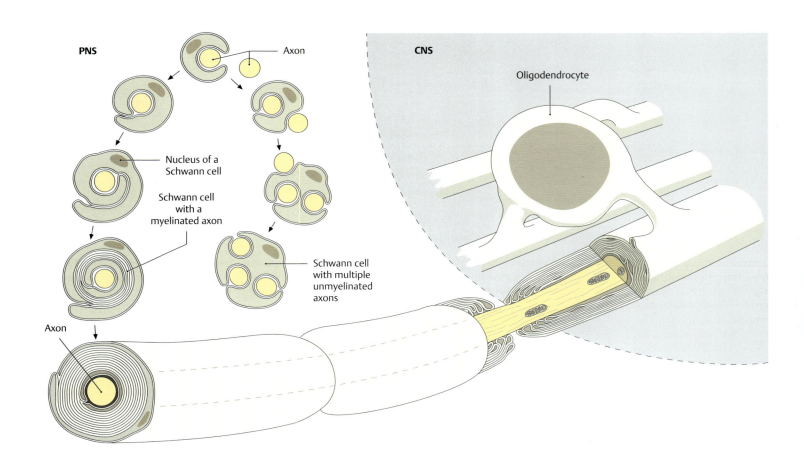

A Myelination differences in the PNS and CNS
The purpose of myelination is to provide the axons with electrical insulation, which significantly increases the nerve conduction velocity. The very lipid-rich membranes of myelinating cells are wrapped around the axons to produce this insulation. Schwann cells (left) myelinate the axons in the *PNS*, while oligodendrocytes (right) form the myelin in the *CNS*.
Note: In the *CNS*, *one* oligodendrocyte *always* wraps around *multiple* axons. In the *PNS*, *one* Schwann cell *may* ensheath *multiple* axons if the peripheral nerve is *unmyelinated*. If the peripheral nerve is *myelinated*, *one* Schwann cell *always* wraps around *one* axon.

Owing to this improved insulation, the nerve conduction velocity is higher in myelinated nerves than in unmyelinated nerves. Myelinated fibers occur in areas where fast reaction speeds are needed (muscular contractions), while unmyelinated fibers occur in areas that do not require rapid information transfer, as in the transmission of visceral pain. Because of the different cell types, myelin has a different composition in the CNS and PNS. This difference in myelination is also important clinically. An example is multiple sclerosis, in which the oligodendrocytes are damaged, but the Schwann cells are not, so that the central myelin sheaths are disrupted, while the peripheral myelin sheaths remain intact.

B Myelination

Principles
• Myelinated axons are generally thicker than unmyelinated axons.
• Myelinated axons transmit impulses faster than unmyelinated axons.
• Unmyelinated axons reside only in the PNS.

CNS
• Oligodendrocytes myelinate axons, with the oligodendrocyte enveloping from several up to 50 axons.

PNS
• With myelinated axons, Schwann cells wrap many layers around a single axon.
• Somatomotor and somatosensory axons are usually myelinated (exception: small pain fibers).
• Axons of the autonomic nervous system are often unmyelinated (exception: e.g., white rami communicantes of preganglionic fibers).
• With unmyelinated axons, Schwann cells wrap one layer around several axons.
• The transition from Schwann cell to oligodendrocyte occurs in the Obersteiner–Redlich zone (see **C**, p. 91).

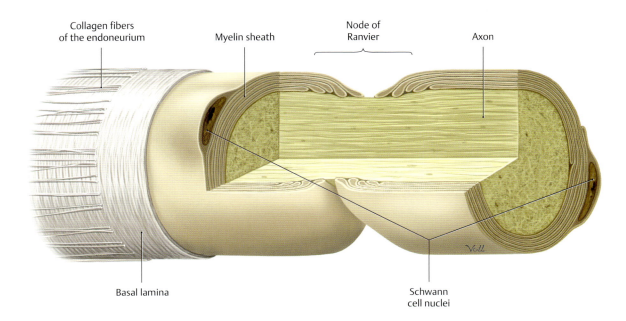

C Structure of a node of Ranvier in the PNS
In the PNS, the *node of Ranvier* is the site where two Schwann cells come together. That site is marked by a small gap in the myelin sheath, which forms the morphological basis for *saltatory nerve conduction*, allowing impulses to be transmitted at a higher velocity.

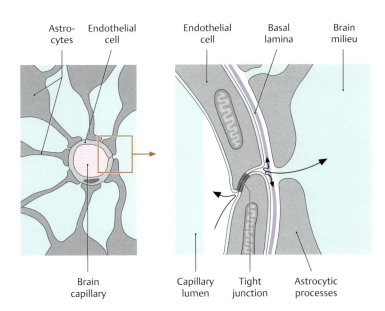
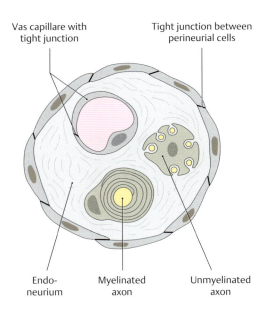

D Structure of the blood–brain barrier in the CNS
Besides the type of myelination, there is also a difference in tissue barriers between the CNS and PNS. The CNS is isolated from surrounding tissues by the blood–brain barrier. The components of the blood–brain barrier include (1) most importantly, a continuous capillary endothelial cell layer, sealed by tight junctions; (2) a continuous basal lamina surrounding the endothelial cells; and (3) enveloping astrocytic processes surrounding the brain capillary. This barrier serves to exclude macromolecules, as well as many small molecules that are not actively transported by the endothelial cells, thus protecting the delicate environment of the CNS. The barrier is vulnerable, however, to lipid-soluble molecules that can traverse the endothelial cell membranes. The perineurial sheath creates a similar barrier in the PNS (see **E**).

E Structure of the perineurial sheath in the PNS
The perineurial sheath, like the blood–brain barrier, is formed by tight junctions between the epithelium-like fibroblasts (perineurial cells; the perineurium is described on p. 75). It isolates the milieu of the axon from that of the surrounding endoneural space (endoneurium), thereby preventing harmful substances from invading the axon. This tissue barrier must be surmounted by drugs that are designed to act on the axon, such as local anesthetic agents.

8.12 The Autonomic Nervous System

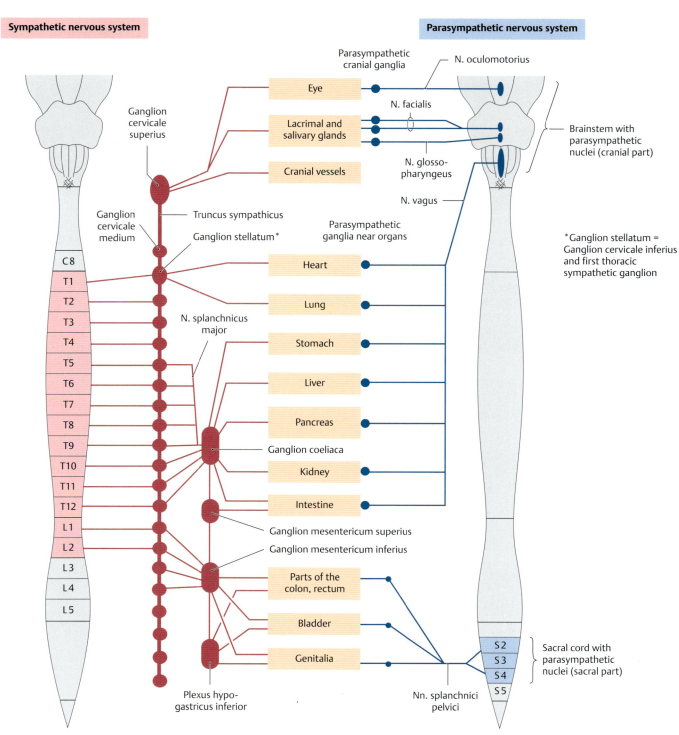

A Structure of the autonomic nervous system
The system of motor innervation of skeletal muscle is complemented by the *autonomic nervous system*, with two divisions: sympathetic (red) and parasympathetic (blue). Both divisions have a two-neuron path between the CNS and their targets: a *preganglionic* CNS neuron and a ganglion cell in the PNS that is close to the target. The preganglionic neurons of the sympathetic system are in the lateral horns of the cervical, thoracic, and lumbar spinal cord. Their axons exit the CNS via the anterior roots and synapse in sympathetic ganglia in bilateral *paravertebral* chains (*trunci sympathici*), or as single midline *prevertebral ganglia* (see **E**). Axons from these ganglion cells course in unmyelinated bundles on blood vessels or in peripheral nerves to their targets. The preganglionic neurons of the parasympathetic system are located in the brainstem and sacral spinal cord. Their axons exit the CNS via cranial and pelvic splanchnic nerves to synapse with parasympathetic ganglion cells. In the head, these cells are in discrete ganglia associated with the cranial nerves. In other locations the parasympathetic ganglion cells are in tiny clusters embedded in their target tissues. The sympathetic and parasympathetic systems regulate blood flow, secretions, and organ function; the two divisions often act in antagonistic ways on the same target (see **B**). Although this basic dichotomy of visceral motor activity was identified early, by Langley (1905) and others, it has been shown more recently that autonomic control of various organs, particularly in the gastrointestinal and urogenital tracts, is highly sophisticated, dependent upon feedback from local visceral afferents that relay pain and stretch information, etc., through complex local circuits.

B Synopsis of the sympathetic and parasympathetic nervous systems
1. The sympathetic nervous system can be considered the excitatory part of the autonomic nervous system that prepares the body for a "fight or flight" response.
2. The parasympathetic nervous system is the part of the autonomic nervous system that coordinates the "rest and digest" responses of the body.
3. Although there are separate control centers for the two divisions in the brainstem and spinal cord, they have close anatomic and functional ties in the periphery.
4. The principal transmitter at the target organ is *acetylcholine* in the parasympathetic nervous system and *norepinephrine* in the sympathetic nervous organ.
5. Stimulation of the sympathetic and parasympathetic nervous systems produces the following different effects on specific organs:

Organ	Sympathetic nervous system	Parasympathetic nervous system
Eye	Pupillary dilation	Pupillary constriction and increased curvature of the lens
Salivary glands	Decreased salivation (scant, viscous)	Increased salivation (copious, watery)
Heart	Elevation of the heart rate	Slowing of the heart rate
Lungs	Decreased bronchial secretions and bronchial dilation	Increased bronchial secretions and bronchial constriction
Gastrointestinal tract	Decreased secretions and motor activity	Increased secretions and motor activity
Pancreas	Decreased secretion from the endocrine part of the gland	Increased secretion
Male sex organs	Ejaculation	Erection
Skin	Vasoconstriction, sweat secretion, piloerection	No effect

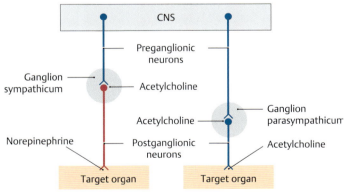

C Circuit diagram of the autonomic nervous system
The synapse of the central, preganglionic neuron uses *acetylcholine* as a transmitter in both the sympathetic *and* parasympathetic nervous systems (cholinergic neuron, shown in blue). In the sympathetic nervous system, the transmitter changes to *norepinephrine* at the synapse of the postganglionic neuron with the target organ (adrenergic neuron, shown in red), while the parasympathetic system continues to use acetylcholine at that level.
Note: Various types of receptors for acetylcholine (= neurotransmitter sensors) are located in the membrane of the target cells. As a result, acetylcholine can produce a range of effects depending on the receptor type.

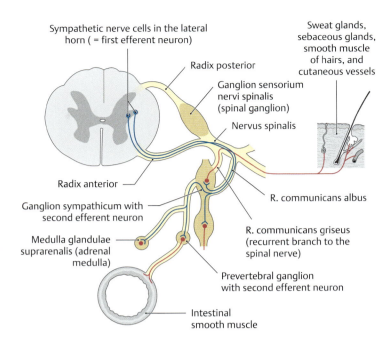

D Distribution of sympathetic fibers in the periphery
Skin (blood vessels, cutaneous glands, and smooth muscle cells attached to hair follicles) and blood vessels of skeletal muscles are only innervated sympathetically. For these, synapses between the first and second neurons occur in the ganglia of the sympathetic trunk. For the sympathetic fibers to the organs, the synapses occur in the prevertebral ganglia or in the organ itself (see **E**).
Note: Skin and skeletal muscles are not parasympathetically innervated.

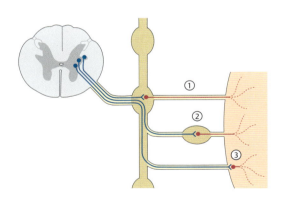

E Location of synapses in the sympathetic nervous system
Acetylcholine is the neurotransmitter released by both the first and second neurons in the parasympathetic system, but in the sympathetic system the second neuron releases norepinephrine. Synapses of the first (preganglionic cholinergic) neuron (blue) with the second neuron (red) may be located

① In the paravertebral ganglia on either side of the vertebral column, which are linked to one another as the sympathetic trunk. The synapses for the blood vessels of the skin and skeletal muscles mainly reside in these ganglia (see **D**).
② In prevertebral ganglia located in front of the spinal column. The synapses for visceral fibers of the abdomen and pelvis reside here.
③ In the target organ, or effector, itself, i.e., the adrenal medulla.

8.13 Lesions of Peripheral Nerves

A Damage to a peripheral nerve
A peripheral nerve is a cluster of afferent (pain, temperature, pressure, vibration, sense of position) and efferent (somatomotor and visceromotor) axons. Pressure (left) or cuts (right) can damage the occasionally long course of the nerve.
Damage through pressure: It only takes slight pressure for the peripheral nerves to respond sensitively, with the afferent system being the first to be affected. Already a mild stimulus leads to paresthesia (e.g., tingling when the n. ulnaris is stimulated due to a slight tap on the medial elbow); longer and heavier pressure leads to loss of sensation in the area innervated by the receptive field. Finally, the efferent system is damaged, causing motor loss. There are numerous anatomic bottlenecks (e.g., tunnel or slit-shaped [osteo] fibrous canals), which can occasionally lead to compression syndromes. They are first perceived as pain and later as loss of function in the affected muscles. Some of the important syndromes are listed under **B**.
Damage through cutting: As a result of a cut (a mixed nerve has been completely severed), the afferent and efferent systems fail at the same time. The muscles innervated by this nerve can no longer be moved, resulting in flaccid paralysis (for more about the difference between flaccid and spastic paralysis, see p. 90). In addition, the loss of afferents leads to a loss of sensation, and the loss of sympathetic fibers leads to vegetative symptoms (increased sweat secretion and increased cutaneous perfusion) in the skin area innervated by the nerve.

B Selected nerve compression (bottleneck) syndromes
This provides an overview of the syndromes or symptoms that can be caused by the compression of individual nerves (for more details about the syndromes, see textbooks of neurology).

Affected nerve	Anatomic region	Syndrome/symptoms
Shoulder girdle and upper limb		
Plexus brachialis	Upper thoracic aperture	Thoracic outlet syndrome (e.g., scalene syndrome, cervical rib syndrome, costoclavicular syndrome)
N. suprascapularis	Incisura scapulae	Incisura scapulae syndrome
N. axillaris	Lateral axillary hiatus	Lateral axillary hiatus syndrome
N. ulnaris	Sulcus ulnaris Flexor carpi ulnaris tendon of origin, r. palmaris n. ulnaris (palmar branches)	Ulnar nerve entrapment (sulcus ulnaris syndrome) Cubital tunnel syndrome Ulnar tunnel syndrome (Guyon–Logen syndrome)
N. radialis	Sulcus n. radialis	Park bench paralysis
– R. profundus	M. supinator (arcade of Frohse)	Supinator syndrome (distal radial compression syndrome)
– R. superficialis	Distal radial forearm	Wartenberg disease
N. medianus	Passage through m. pronator teres Canalis carpi	Pronator teres syndrome Carpal tunnel syndrome
Pelvic girdle and lower limb		
N. ischiadicus	Gluteal region	Piriformis syndrome
N. femoralis	Inguinal region	Pain/weakness of the front upper leg
N. obturatorius	Anterior pubic bone area	Howship–Romberg syndrome
N. saphenus	Canalis adductorius	Dysesthesia at the interior lower leg
N. cutaneus femoris lateralis	Lateral inguinal/upper leg region	Meralgia paraesthetica (inguinal ligament syndrome)
N. fibularis communis	Caput/collum fibulae	Fibular tunnel syndrome
N. fibularis profundus	Anterior upper ankle joint	Anterior tarsal tunnel syndrome
N. tibialis/Nn. plantares	Inner malleolus region	Posterior tarsal tunnel syndrome
Nn. digitales plantares	Toe ball area 3/4	Interdigital Morton neuralgia

General Anatomy — 8. General Neuroanatomy

C Damage to a peripheral nerve exemplified by loss of sensitivity
Functionally speaking, in the innervated area of an afferent nerve, the nerve is only a part of a conduit cable for stimuli, which ends in the sensory area of the cortex cerebri, where conscious perception of the respective sensory stimulus occurs (see p. 91). The brain always locates damage to an afferent nerve in the area innervated by that cutaneous nerve, regardless of whether the damage occurs close to the spinal cord or not. In case of sensitivity loss, the damage is not necessarily located where sensation is lost.

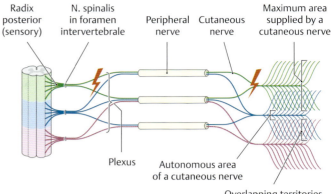

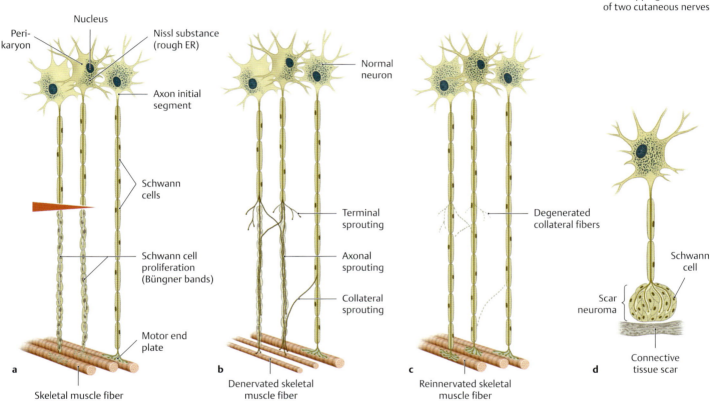

D Regeneration of a peripheral nerve after having been cut
In the peripheral nervous system (PNS), a nerve can usually regenerate after its axon has been severed.

a A few days after the axon has been severed, it first degenerates distally to the lesion (so-called Waller degeneration). Its myelin sheath disintegrates, and the axon dissipates, as it is no longer nourished through the perikaryon (the so-called trophic function of the perikaryon is no longer required). In the affected perikarya, the cell nucleus shifts to the margin, and the Nissl substance partially dissolves (tigrolysis). At the same time, the Schwann cells divide, forming a cluster of cells from which the so-called Büngner bands arise.

b Weeks later, the severed axons begin to sprout again proximally (axonal sprouting) and extend by 1 mm/day into the areas they originally innervated. From adjacent axons, which have not been affected by the lesions, axon collaterals can also extend into the affected area. The process of axonal sprouting can be tested with the help of the Hoffmann–Tinel sign. When percussing over the course of the nerve, patients experience a tingling around the area where the axons sprout.

c Months later, the regeneration is complete. The axons have reached the skeletal muscles again and innervate them. Axons that have not made it to their relevant skeletal muscles and axons that have sprouted from the axon collaterals perish. As the Schwann cells divide distally along the lesion, the same segment of axons is now being myelinated by more Schwann cells than previously. This leads to an increased number of nodes of Ranvier, thus slowing down the saltatory conduction. This process can be measured by determining the nerve conduction velocity. The position of the nuclei and the texture of the Nissl substance in the perikarya have normalized.

After the peripheral nerve has been severed, its ends are sutured back together end-to-end. The goal of the nerve suture is to maintain the continuity of the connective tissue tube with its Schwann cell tubes as the anatomic landmark for the sprouting axons. In case of an injury to a large segment of the nerve, a nerve transplant may be performed. This transplant bridges the gap; it, too, only provides a conduit for the spouting axons. Such a conduit is crucial for sprouting axons because axons cannot penetrate into the scar tissue. If, because of such a scar, axons cannot find their way back to their muscle, it can lead to amputation, or scar, neuroma (see **d**), which is a conglomerate of Schwann cells and axonal sprouts.

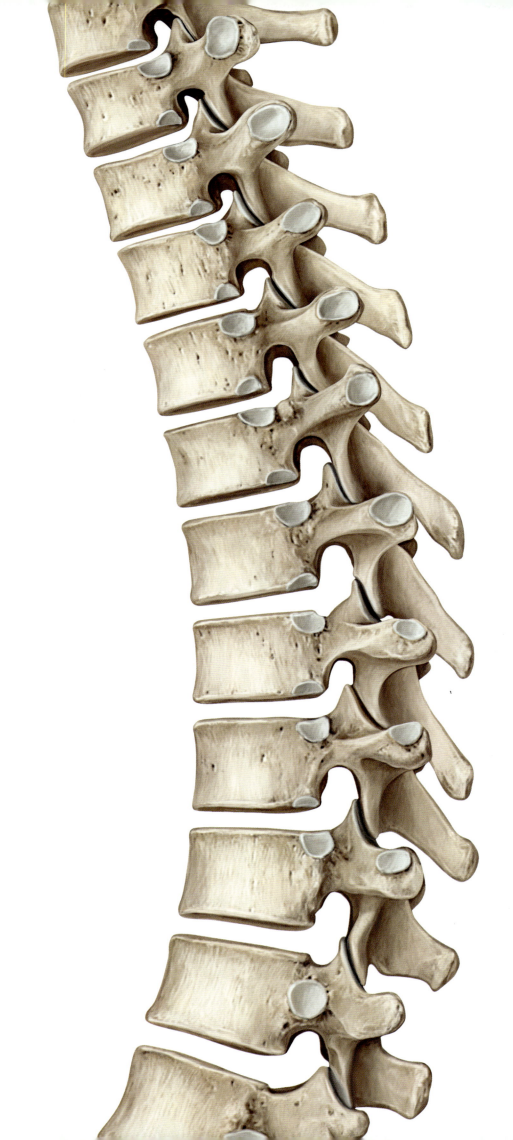

Trunk Wall

9　Bones, Ligaments, and Joints 100

10　Musculature: Functional Groups 144

11　Musculature: Topographical Anatomy 166

12　Neurovascular Systems: Forms and Relations 190

13　Neurovascular Systems: Topographical Anatomy ... 198

Trunk Wall — 9. Bones, Ligaments, and Joints

9.1 The Skeleton of the Trunk

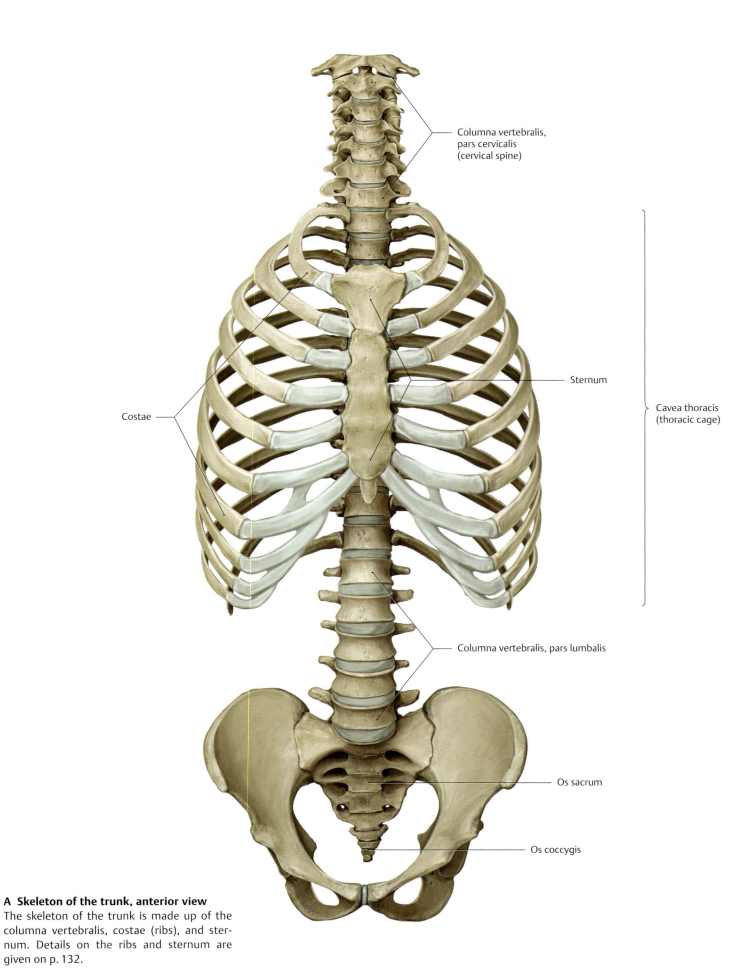

A Skeleton of the trunk, anterior view
The skeleton of the trunk is made up of the columna vertebralis, costae (ribs), and sternum. Details on the ribs and sternum are given on p. 132.

Trunk Wall —— *9. Bones, Ligaments, and Joints*

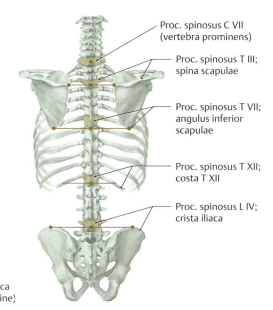

C The spinous processes as anatomic landmarks

Posterior view. The spinous processes of the vertebrae (procc. spinosi vertebrales) appear as variable prominences beneath the skin and provide important landmarks during the physical examination. With few exceptions, they are easily palpated.

- The proc. spinosus of the seventh cervical vertebra, located at the junction of the cervical and thoracic spine. Usually it is the most prominent of the procc. spinosi, causing the seventh cervical vertebra to be known also as the *vertebra prominens*.
- The proc. spinosus of the third thoracic vertebra, located on a horizontal line connecting the spinae scapulae
- The proc. spinosus of the seventh thoracic vertebra, located at the level of the inferior angles of the scapulae (anguli inferiores scapulae)
- The proc. spinosus of the twelfth thoracic vertebra, located slightly below the attachment of the last rib
- The proc. spinosus of the fourth lumbar vertebra, located on a horizontal line connecting the highest points of the cristae iliacae

Note: The proc. spinosus of the thoracic vertebrae are angled downward (see p. 110), so the proc. spinosus of the fifth thoracic vertebra, for example, is at the level of the sixth thoracic corpus vertebrae.

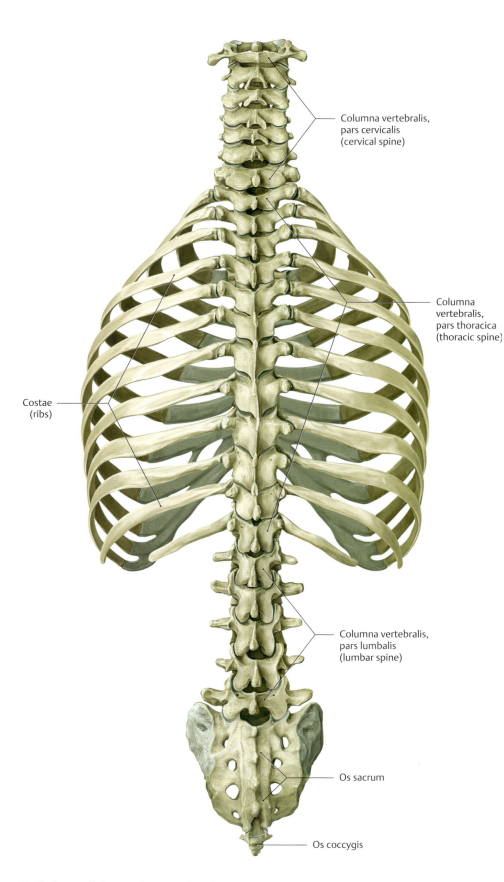

B Skeleton of the trunk, posterior view

101

9.2 The Bony Spinal Column

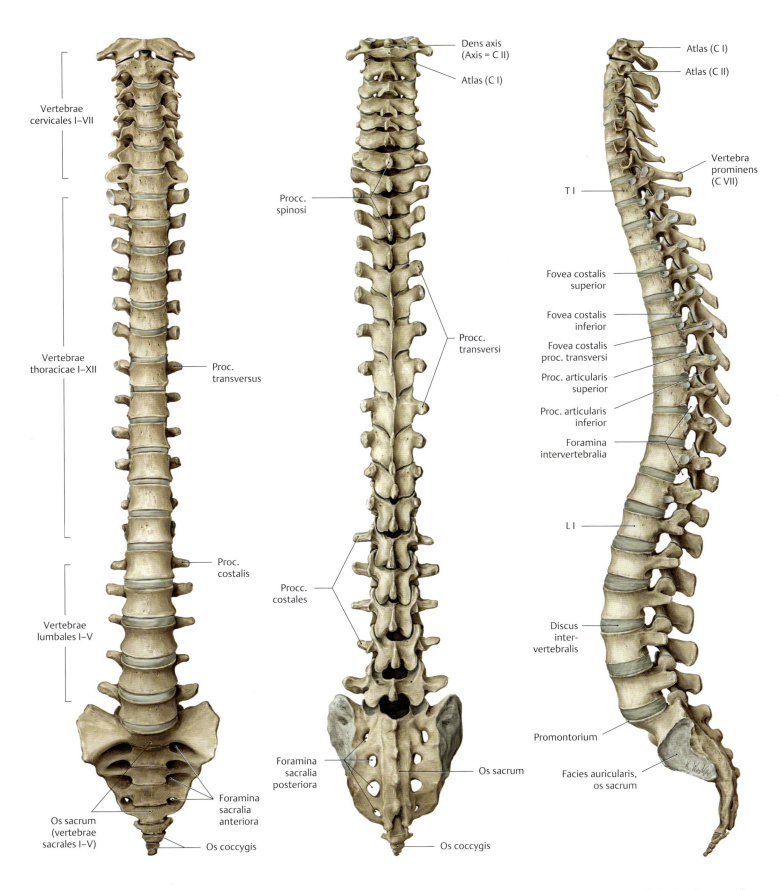

A The bony spinal column (columna vertebralis)
a Anterior view, b posterior view, c left lateral view.

Note: Phylogenetically, the procc. transversi of the lumbar vertebrae are rudimentary ribs (costae). They therefore are known as procc. costales (see also p. 106).

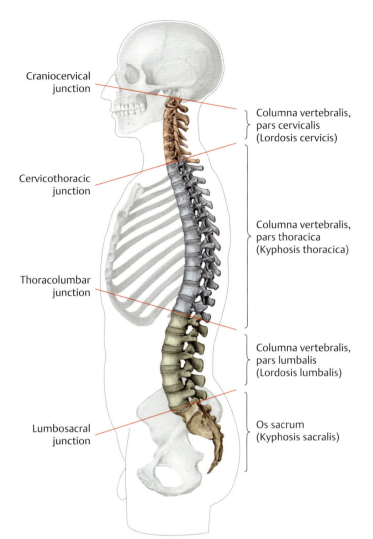
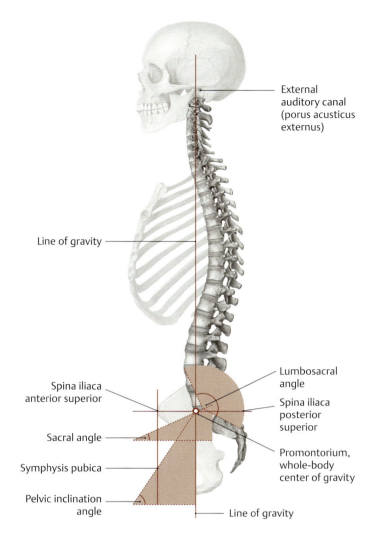

B Regions and curvatures of the spinal column
Left lateral view. The spinal column of an adult is divided into four regions and presents four characteristic curvatures in the sagittal plane. These curves are the result of human adaptation to upright bipedal locomotion, acting as springs to cushion axial loads. The following regions and curvatures are distinguished in the craniocaudal direction:

- Cervical spine – cervical lordosis (lordosis cervicis)
- Thoracic spine – thoracic kyphosis (kyphosis thoracica)
- Lumbar spine – lumbar lordosis (lordosis lumbalis)
- Sacral spine – sacral kyphosis (kyphosis sacralis)

The cervical, thoracic, and lumbar regions of the spinal column are also known collectively as the *presacral spine*. The transitional areas between the different regions are of clinical importance because they are potential sites for spinal disorders (e.g., herniated disks). Occasionally, the vertebrae in these transitional areas have an atypical morphology that identifies them as *transitional vertebrae*. This is particularly common at the lumbosacral junction, where *sacralization* or *lumbarization* may be seen, depending on the appearance of the atypical vertebra. With lumbarization, the first sacral vertebra is not fused to the os sacrum and constitutes an extra lumbar vertebra. With sacralization, there are only four lumbar vertebrae, the fifth being "sacralized" by fusion to the os sacrum. These *assimilation disorders* are often unilateral (hemilumbarization, hemisacralization).

C Integration of the spinal column into the pelvic girdle
Skeleton of the trunk with the skull and pelvic girdle, left lateral view. Normally, the spinal column is curved and integrated into the pelvic girdle in such a way that characteristic angles are formed between certain imaginary lines and axes. These angles and lines are useful in the radiographic evaluation of positional abnormalities and deformities of the spine and trunk.
Lumbosacral angle: angle formed by the axes of the L V and S I vertebrae, averaging 143°. It results from the fact that the os sacrum is a fixed component of the pelvic ring (see p. 138) and thus contributes little to straightening the columna vertebralis. The result is a characteristic sharp angle at the junction of the presacral part of the spinal column with the os sacrum.
Sacral angle: angle between the horizontal plane and the superior surface of the os sacrum, averaging approximately 30°
Pelvic inclination angle (Inclinatio pelvis): angle formed by the apertura pelvis superior (connecting the promontorium ossis sacri to the upper border of the symphysis pubica) with the horizontal. It measures approximately 60° in upright stance. The pelvic inclination angle increases or decreases as the pelvis is tilted forward or backward (see p. 157). With an ideal pelvic position in upright stance, the spina iliaca anterior superior and spina iliaca posterior superior are at the same horizontal level, and the spina iliaca anterior superior is directly above the symphysis pubica. By knowing this, the examiner can easily evaluate the position of the pelvis by using palpable bony landmarks.
Line of gravity: the line of gravity passes through landmarks that include the external auditory canal, the dens of the axis (C II), the functional-anatomical transition points in the spinal column (between lordosis and kyphosis), and the whole-body center of gravity just anterior to the promontorium ossis sacri

9.3 Development of the Spinal Column

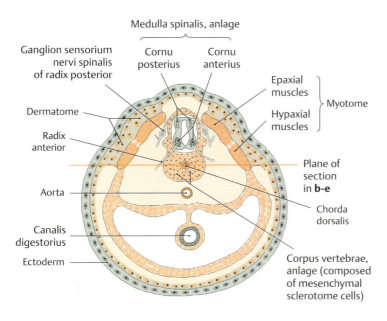

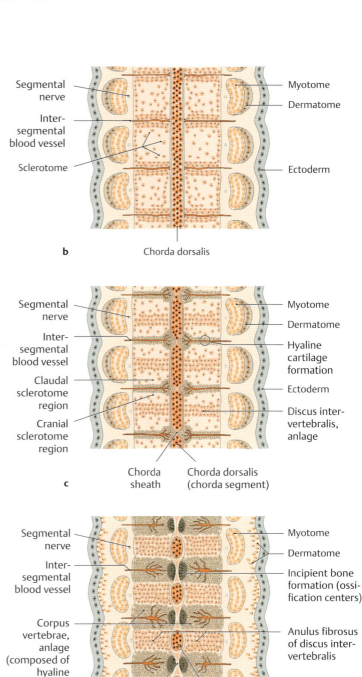

A Development of the spinal column (columna vertebralis), weeks 4–10
a Schematic cross section, b–e schematic coronal sections (the plane of section in b–e is indicated in a).

- **a, b** The former somites have differentiated into the myotome, dermatome, and sclerotome. The sclerotome cells separate from the other cells at 4 weeks, migrate toward the chorda dorsalis, and form a cluster of mesenchymal cells around the chorda (anlage of the future spinal column).
- **c** Adjacent cranial and caudal sclerotome segments above and below the intersegmental vessels join together and begin to chondrify in the sixth week, displacing the chorda material superiorly and inferiorly (segments of the chorda dorsalis).
- **d** The intervertebral disks with their nucleus pulposus and anulus fibrosus develop between the rudimentary vertebral bodies. Ossification begins at the center of the corpus vertebrae in the eighth week of development.
- **e** By fusion of the caudal and cranial sclerotome segments, the segmentally arranged myotomes interconnect the processes of two adjacent vertebral anlages, bridging the gap across the disci intervertebrales. This is how the *motion segments* are formed (see p. 124). The segmental spinal nerve courses at the level of the future foramen intervertebrale, and the intersegmental vessels become the *nutrient vessels* (vasa nutricia) of the vertebral bodies (week 10).

Clinical aspects: If the neural tube or the dorsal portions of the vertebral arches fail to close normally during embryonic development, the result is *spina bifida*—a cleft spine. In this anomaly, the spinal column is open posteriorly, and the procc. spinosi are absent (the various forms and manifestations are described in textbooks of embryology). Usually there is a bilateral defect in the vertebral arches (generally affecting the LIV and LV region), known as *spondylolysis*. This defect may be congenital or acquired (e.g., due to trauma). Acquired cases are common in sports that pose a risk of vertebral arch fractures (javelin throwing, gymnastics, high jumping). If the associated discus intervertebralis is also damaged, the corpus vertebrae will begin to slip forward (*spondylolisthesis*). In cases of *congenital spondylolysis* (which are associated with varying degrees of spondylolisthesis), the slippage progresses slowly during growth, and the condition tends to stabilize after 20 years of age.

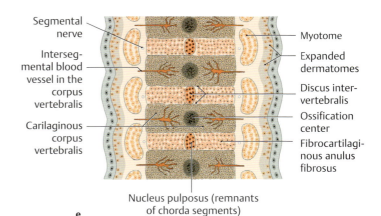

Trunk Wall — 9. Bones, Ligaments, and Joints

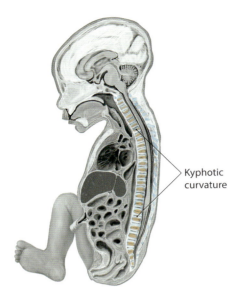

B Neonatal kyphosis
Midsagittal section through a newborn, left lateral view. Owing to the curved intrauterine position of the fetus, the newborn has a "kyphotic" spinal curvature with no lordotic straightening of the cervical and lumbar spine (after Rohen, Yokochi, Lütjen-Drecoll).

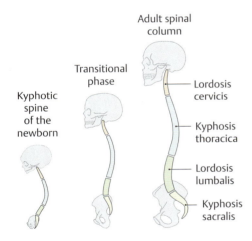

C Straightening of the spine during normal development (after Debrunner)
The characteristic curvatures of the adult spine are only partially present in the newborn (compare with **B**) and appear only during the course of postnatal development. First cervical lordosis (lordosis cervicis) develops to balance the head in response to the growing strength of the posterior neck muscles. Lumbar lordosis (lordosis lumbalis) develops later as the child learns to sit, stand, and walk. The degree of lordosis increases until the legs can be fully extended at the hips, and it finally becomes stable during puberty. A similar transformation of the spinal column is observed in the phylogenetic transition from quadrupedal to bipedal locomotion.

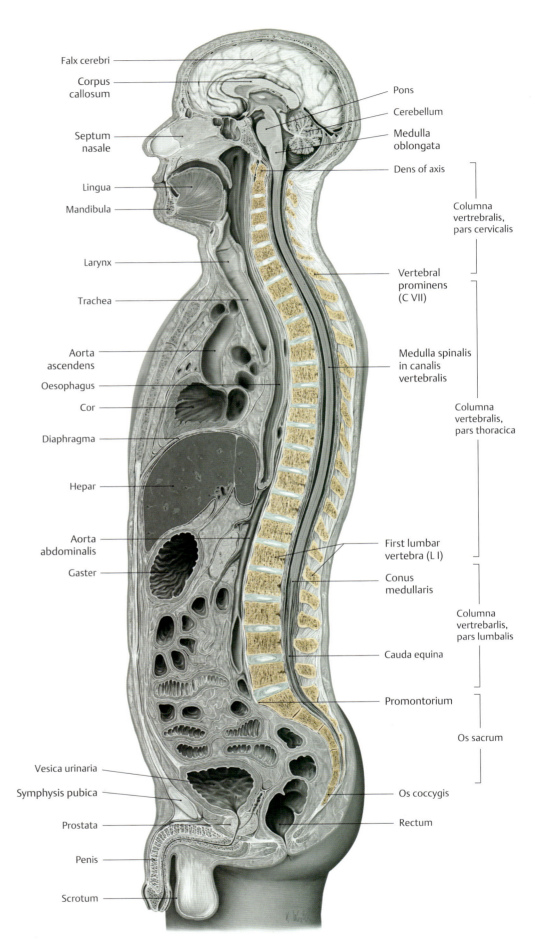

D Physiological curvatures of the adult spine

Midsagittal section through an adult male, left lateral view.

105

9.4 The Structure of a Vertebra

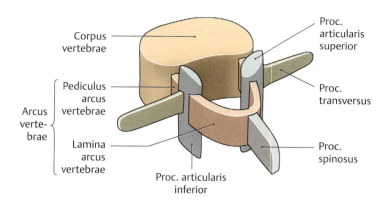

A Structural elements of a vertebra
Left posterosuperior view. All vertebrae except the atlas and axis (see p. 109) consist of the same basic structural elements:

- A vertebral body (corpus vertebrae)
- A vertebral arch (arcus vertebrae)
- A spinous process (proc. spinosus)
- Two transverse processes (procc. transversi; called procc. costales in the lumbar vertebrae)
- Four articular processes (procc. articulares)

The processes give attachment to muscles and ligaments, and the bodies of the thoracic vertebrae have costovertebral joints. The vertebral bodies and arches enclose the foramen vertebrale, and all of the vertebral foramina together constitute the vertebral (spinal) canal (canalis vertebralis).

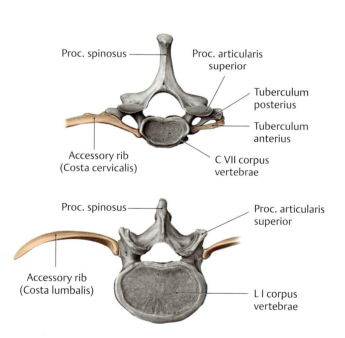

B Accessory ribs
Superior view. **a** Costa cervicalis, **b** Costa lumbalis.
The presence of anomalous cervical ribs (costa cervicalis) can narrow the interscalene space, causing compression of the plexus brachialis and a. subclavia (scalenus syndrome or cervical rib syndrome, see also p. 362). An accessory lumbar rib (costa lumbalis), on the other hand, has no adverse clinical effects.

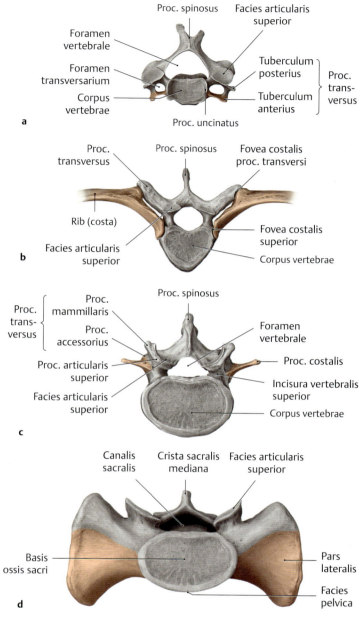

C Costal elements in different regions of the spinal column
Superior view. The shape and configuration of the vertebrae are closely related to the development of the ribs and their rudiments (indicated here by color shading).

a Cervical vertebrae: Here the rudimentary rib forms a process called the tuberculum anterius. It unites with the tuberculum posterius to form the foramen transversarium.
b Thoracic vertebrae: Because these vertebrae give attachment to the ribs, their bodies and transverse processes bear corresponding cartilage-covered articular surfaces (fovea costalis proc. transversi, also fovea costalis superior and inferior).
c Lumbar vertebrae: The costal elements in the lumbar spine take the form of "transverse processes," which are much larger than in the cervical spine. Because of their size, and the fact they are rudimentary ribs they are also known as procc. costales.
d Sacrum: Here the rudimentary rib forms the anterior portion of the partes laterales of the sacral vertebra. It is fused to the transverse processes.

Trunk Wall — 9. Bones, Ligaments, and Joints

D Typical vertebrae from different regions of the spinal column
Superior and left lateral view.

a, b First cervical vertebra (atlas).
c, d Second cervical vertebra (axis).
e, f Fourth cervical vertebra.
g, h Sixth thoracic vertebra.
i, j Fourth lumbar vertebra.
k, l Sacrum.

The vertebrae in different regions of the spinal column (columna vertebralis) differ not only in their size but also in their special features. While the corpora vetebrales gradually become larger from superior to inferior to accommodate the increasing stresses imposed by the gravity and body weight, the *foramina* vertebralia gradually become smaller to match the decreasing diameter of the spinal cord (medulla spinalis). The arrangement of the arcus vertebrae and adjacent processes also varies at different levels in the spine (for details see pp. 109, 111, and 113).

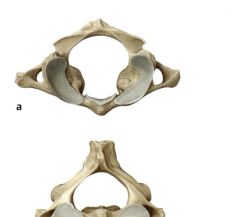
a

b

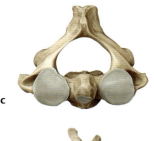
c

d

e

f

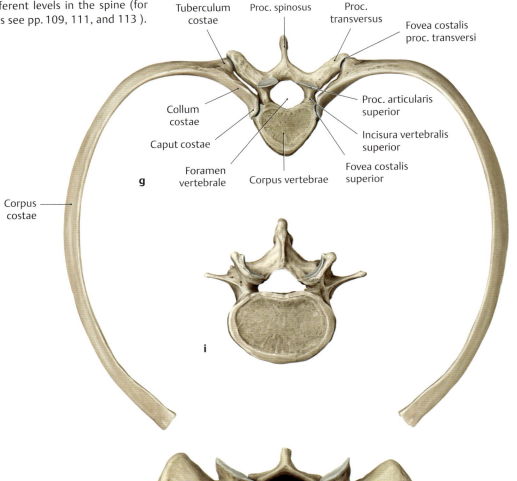

Tuberculum costae, Proc. spinosus, Proc. transversus, Fovea costalis proc. transversi, Collum costae, Proc. articularis superior, Caput costae, Incisura vertebralis superior, Foramen vertebrale, Corpus vertebrae, Fovea costalis superior, Corpus costae

g

h

j

i

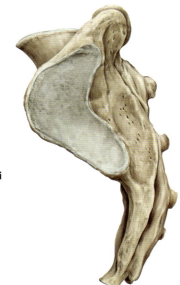
i

k

107

9.5 The Cervical Spine (Vertebrae cervicales)

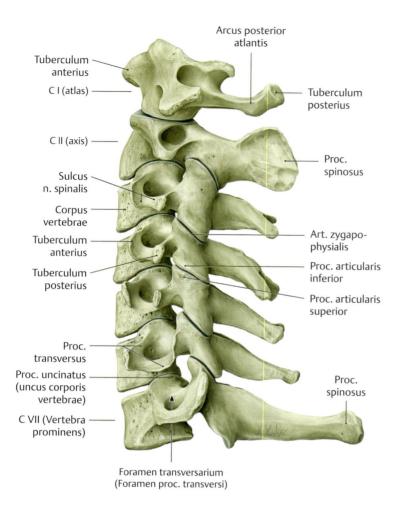

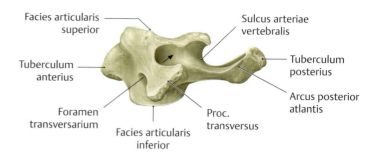

a First cervical vertebra (C I, atlas)

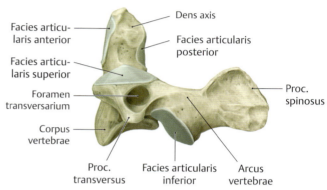

b Second cervical vertebra (C II, axis)

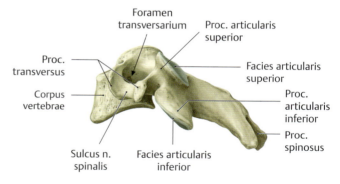

c Fourth cervical vertebra (C IV)

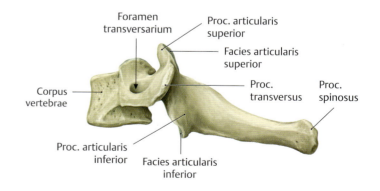

d Seventh cervical vertebra (C VII, vertebra prominens)

B Cervical vertebrae, left lateral view

A Cervical spine, left lateral view
Of the cervical vertebrae, which number seven in all, the **first and second cervical vertebrae** (C I and C II, atlas and axis) differ most conspicuously from the common vertebral morphology. They are specialized for bearing the weight of the head and allowing the head to move in all directions, similar to a ball-and-socket joint. Each of the **remaining five cervical vertebrae** (C III–C VII) has a relatively small body, which presents a more or less square shape when viewed from above, and a large, triangular foramen vertebrale (see **Cc**). The superior and inferior surfaces of the vertebral bodies are saddle-shaped, the superior surfaces bearing lateral procc. uncinati that do not appear until about the tenth year of life (see p. 126). The proc. transversus consists of an anterior and a posterior bar, which terminate laterally in two small tubercles (tuberculi anterius and posterius). These bars enclose the foramen transversarium (foramen proc. transversi), through which the arteria vertebralis ascends from the C VI to the C I level. The superior surface of the proc. transversus of the first three cervical vertebrae bears a broad, deep notch (sulcus n. spinalis), in which lies the emerging n. spinalis at that level. The procc. articulares superior and inferior are broad and flat. Their articular surfaces are flat and are inclined approximately 45° from the horizontal plane. The procc. spinosi of the third through sixth cervical vertebrae are short and bifid. The proc. spinosus of the seventh cervical vertebra is longer and thicker than the others and is the first of the procc. spinosi that is distinctly palpable through the skin (vertebra prominens).

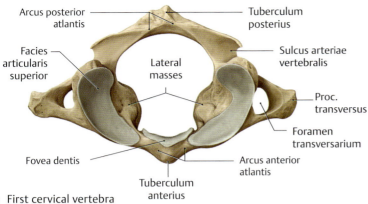

a First cervical vertebra (C I, atlas)

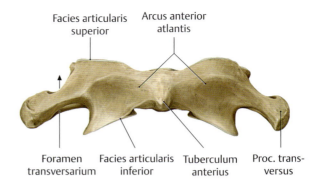

a First cervical vertebra (C I, atlas)

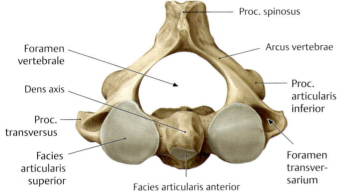

b Second cervical vertebra (C II, axis)

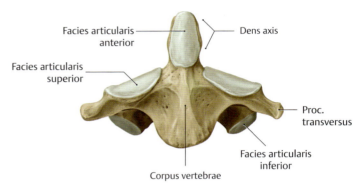

b Second cervical vertebra (C II, axis)

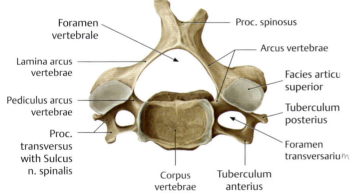

c Fourth cervical vertebra (C IV)

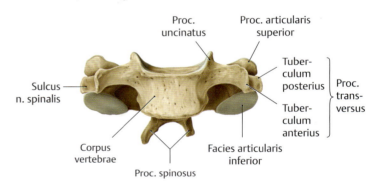

c Fourth cervical vertebra (C IV)

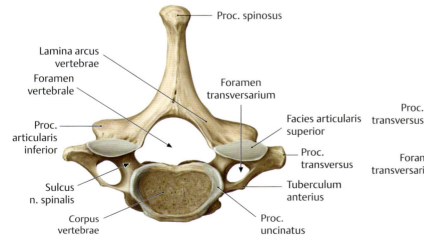

d Seventh cervical vertebra (C VII, vertebra prominens)

C Cervical vertebrae, superior view

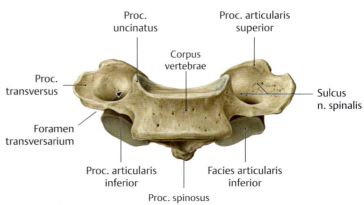

d Seventh cervical vertebra (C VII, vertebra prominens)

D Cervical vertebrae, anterior view

9.6 The Thoracic Spine (Vertebrae thoracicae)

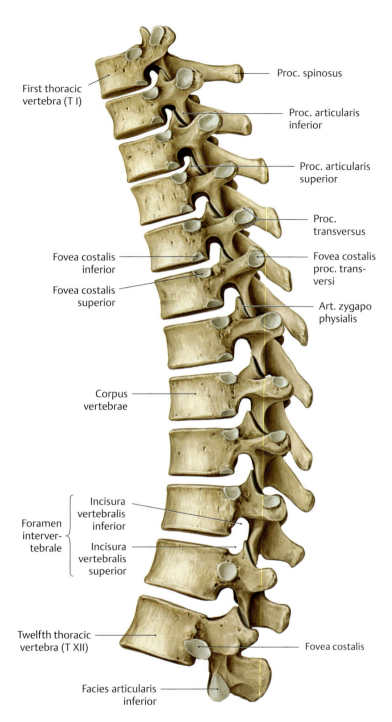

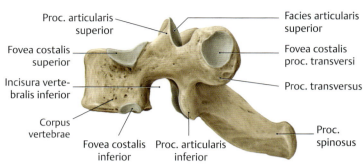

a Second thoracic vertebra (T II)

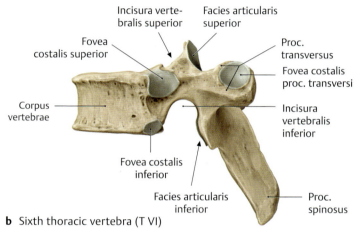

b Sixth thoracic vertebra (T VI)

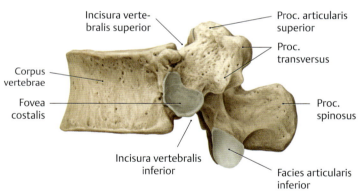

c Twelfth thoracic vertebra (T XII)

A Thoracic spine, left lateral view
The thoracic corpora vertebrae gradually become taller and broader from superior to inferior, the lower vertebral bodies assuming a transverse oval shape like that of the lumbar vertebrae. The foramen vertebrale is roughly circular and is smaller than in the cervical and lumbar vertebrae. The end plates are rounded and triangular. The procc. spinosi are long and angled sharply inferiorly, creating an overlapping arrangement that interlinks the thoracic vertebrae. The facets of the proc. articularis inferior are directed anteriorly, while the facets of the proc. articularis superior face posterior so that they can articulate with the procc. articulares inferiores to form the artt. zygophysiales or facet joints (p.124). Another feature of the thoracic vertebrae is that their procc. transversi are angled backward to allow for articulations with the ribs.

B Thoracic vertebrae, left lateral view
The costal facets are cartilage-covered surfaces that articulate with the corresponding ribs (see p. 137). The bodies of the second through ninth thoracic vertebrae (T I–T IX) bear two articular facets on each side — a fovea costalis superior and a fovea costalis inferior — such that two adjacent vertebrae combine to articulate with the head of a rib (caput costae). Thus, a given numbered rib articulates with its own numbered vertebra and the vertebra above. Exceptions to this scheme are ribs I and X to XII, which articulate only with the vertebral body of the same number. The body of the tenth thoracic vertebra has only superior articular facets, and the tenth rib usually articulates only with the tenth thoracic vertebrae. As mentioned, the transverse processes of the thoracic vertebrae (except for T XI and T XII) also bear articular facets for the ribs (fovea costalis proc. transversi).

Trunk Wall ——— 9. *Bones, Ligaments, and Joints*

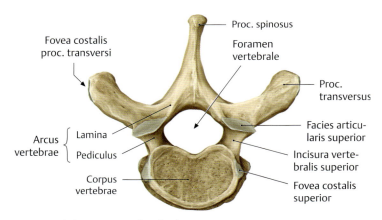

a Second thoracic vertebra (T II)

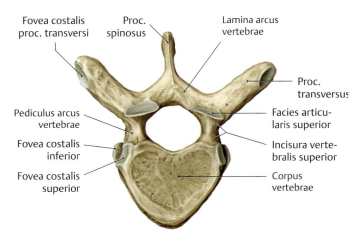

b Sixth thoracic vertebra (T VI)

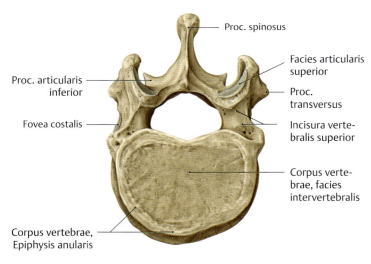

c Twelfth thoracic vertebra (T XII)

C Thoracic vertebrae, superior view
The lamina arcus vertebrae and pediculus arcus vertebrae make up the arcus vertebrae.

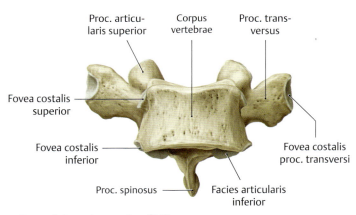

a Second thoracic vertebra (T II)

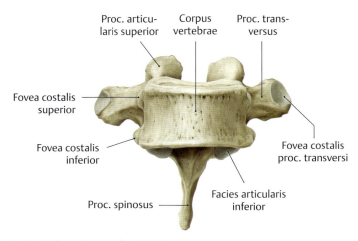

b Sixth thoracic vertebra (T VI)

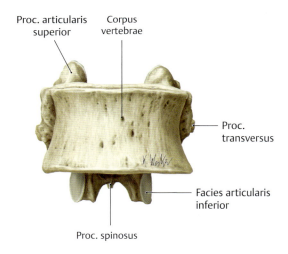

c Twelfth thoracic vertebra (T XII)

D Thoracic vertebrae, anterior view

111

9.7 The Lumbar Spine (Vertebrae lumbales)

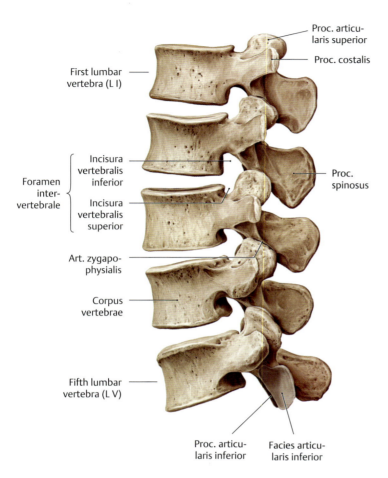

A Lumbar spine, left lateral view
The bodies of the lumbar vertebrae are large and have a transverse oval shape when viewed from above (see **C**). The massive vertebral arches enclose an almost triangular foramen vertebrale, and they unite posteriorly to form a thick proc. spinosus that is flattened on each side. The "transverse processes" of the lumbar vertebrae correspond phylogenetically to rudimentary ribs (see p. 106). Thus, they are more accurately termed Procc. costales and are not homologous with the procc. transversi of the other vertebrae. The thick proc. costalis is fused with the actual proc. transversus, which is a small, pointed eminence located at the root of each proc. costalis (proc. accessorius, see **Cb**). The relatively massive procc. articulares superior and inferior bear slightly inclined articular facets that have a vertical, almost sagittal orientation. The articular facets of the procc. articulares superiores are slightly concave and face medially, while those of the procc. articulares inferiores are slightly convex and face laterally. The procc. mamillares on the lateral surfaces of the procc. articulares superiores serve as origins and insertions for the intrinsic back muscles (see **Bb** and **Ca**).

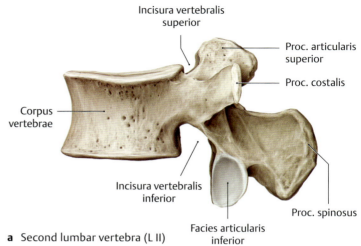

a Second lumbar vertebra (L II)

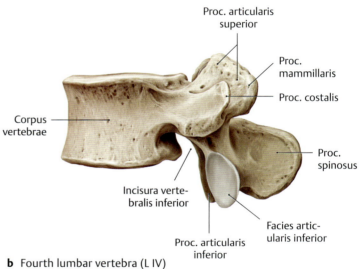

b Fourth lumbar vertebra (L IV)

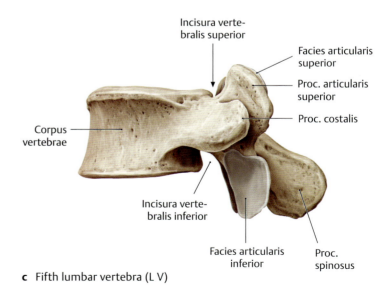

c Fifth lumbar vertebra (L V)

B Lumbar vertebrae, left lateral view

Trunk Wall — **9. Bones, Ligaments, and Joints**

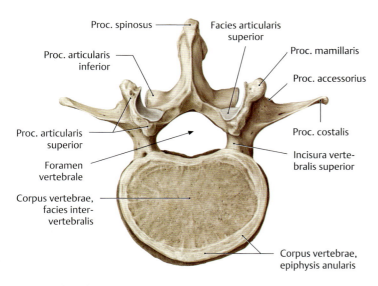

a Second lumbar vertebra (L II)

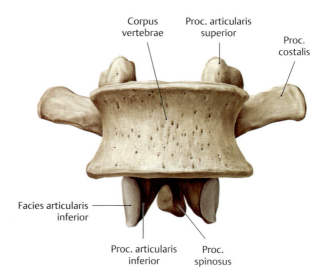

a Second lumbar vertebra (L II)

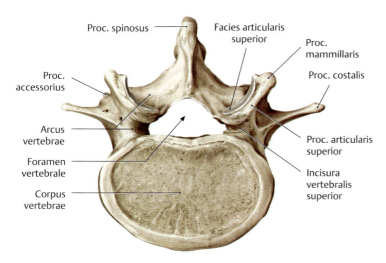

b Fourth lumbar vertebra (L IV)

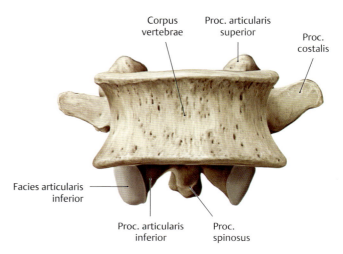

b Fourth lumbar vertebra (L IV)

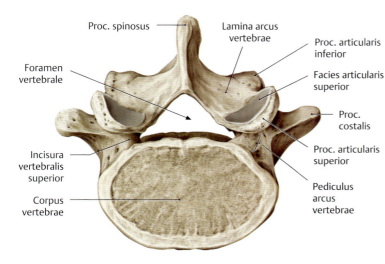

c Fifth lumbar vertebra (L V)

C Lumbar vertebrae, superior view

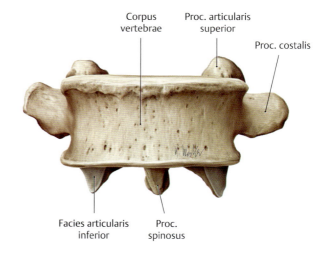

c Fifth lumbar vertebra (L V)

D Lumbar vertebrae, anterior view

113

9.8 Os sacrum and os coccygis

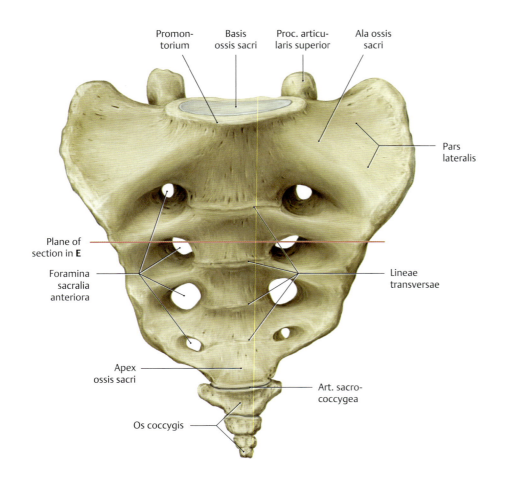

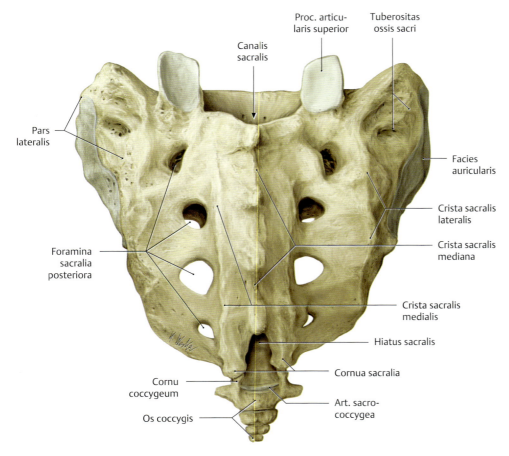

A Os sacrum and os coccygis, anterior (pelvic) view
The os sacrum at birth is composed of five separate sacral vertebrae. Postnatally, they fuse to form a single bone that is flattened anteroposteriorly and has a triangular shape when viewed from the front. The *base of the sacrum* (basis ossis sacri), located at the superior end of the bone, articulates with the body of the fifth lumbar vertebra by a wedge-shaped intervertebral disk. The *apex of the sacrum* (apex ossis sacri) is at the inferior end of the bone and articulates with the os coccygis. The anterior or *pelvic surface* (facies pelvica) of the os sacrum is concave in both the sagittal and transverse planes (see **C**). Between the foramina sacralia anteriora are four transverse ridges (lineae transversae), which mark the sites of fusion between the five sacral vertebrae. The os coccygis consists of three or four rudimentary vertebrae. Only the first coccygeal vertebra exhibits some of the typical structural elements of a fully formed vertebra. A cartilaginous disk usually connects the base of the os coccygis to the apex ossis sacri (art. sacrococcygea). This joint allows passive forward and backward motion of the os coccygis, increasing the anteroposterior diameter (conjugata recta) of the apertura pelvis inferior during childbirth (see p. 139).

B Os sacrum and os coccygis, posterior view
The fused proc. spinosus form a jagged bony ridge on the convex dorsal surface (facies dorsalis) of the os sacrum, the *crista sacralis mediana*. This ridge is flanked on each side by the paired *cristae sacrales mediales*, formed by fusion of the articular processes. The medial crests are continuous below with the rudimentary inferior articular processes (processi articulares inferiores) of the fifth sacral vertebra (cornua sacralia); they are continuous above with the two procc. articulares superiores that face posteriorly. Between the cornua sacralia is an aperture, the hiatus sacralis, which is formed by the incomplete vertebral arch (arcus vertebrae) of the fifth sacral vertebrae and provides access to the canalis sacralis (e.g., for anesthesia). Lateral to the foramina sacralia posteriora are another pair of longitudinal ridges, the *cristae sacrales laterales*, formed by the fused transverse processes. The bony union of the transverse processes with the rudimentary ribs forms the thick *lateral parts* of the sacral wing that flank the body of the sacrum. Each lateral part bears an ear-shaped ("auricular") surface (facies auriculares) that articulates with the ilium (see **C**). In this view, the two small coccygeal cornua representing the superior articular processes and the two rudimentary transverse processes of the os coccygis are visible.

C Os sacrum, left lateral view

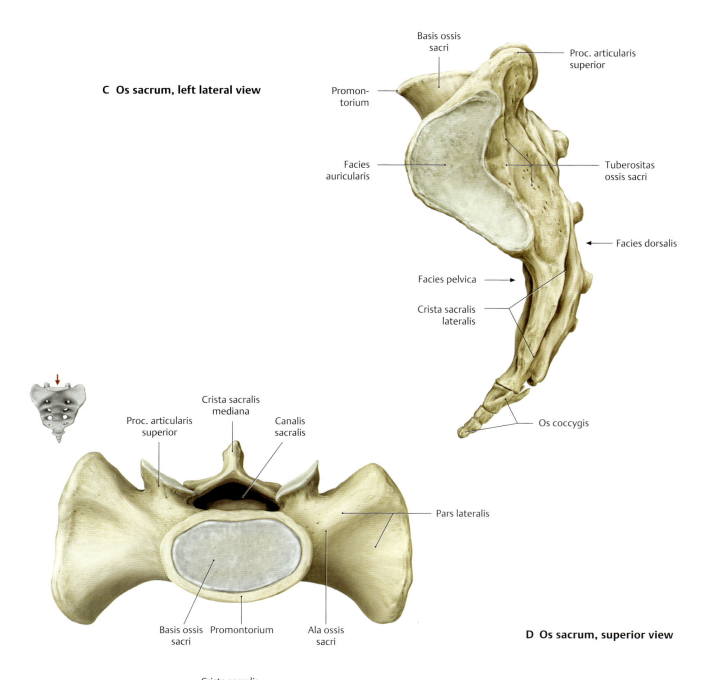

D Os sacrum, superior view

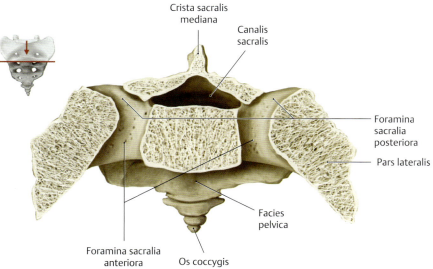

E Transverse section through the os sacrum

Superior view (the level of the section is shown in **A**). Fusion of the upper four sacral vertebrae creates four T-shaped bony canals at the level of the foramina intervertebralia on each side, providing sites of emergence for the first through fourth nervi sacrales. The corresponding rr. ventales and dorsales of the spinal nerves exit the bony canals through the foramina sacralia anteriora and posteriora (see p. 237).

9.9 The Intervertebral Disk (Discus intervertebralis): Structure and Function

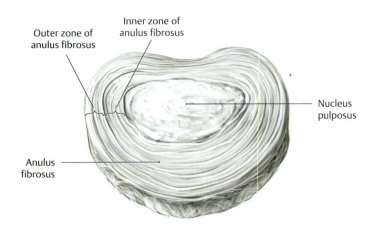

A Structure of an intervertebral disk (discus intervertebralis)
Isolated lumbar discus intervertebralis, anterosuperior view.
The intervertebral disk consists of an external fibrous ring, the *anulus fibrosus*, and a gelatinous core, the *nucleus pulposus*. The anulus fibrosus consists of an outer zone and an inner zone. The outer zone is a fibrous sheath that possesses high tensile strength and is made up of concentric laminae of type I collagen fibers. Its fiber systems crisscross due to their varying obliquity and interconnect the marginal ridges of two adjacent vertebrae (see **B**), to which they are attached. At the junction with the inner zone of the anulus fibrosus, the tough fibrous tissue of the outer zone blends with a fibrocartilaginous tissue whose type II collagen fibers are attached to the hyaline cartilage end plates of the corpora vertebrae (see **Da** and **Ea**).

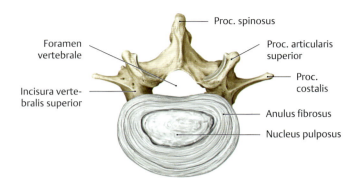
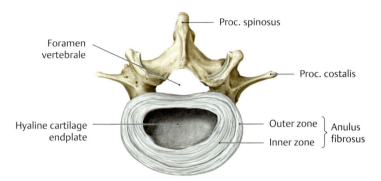
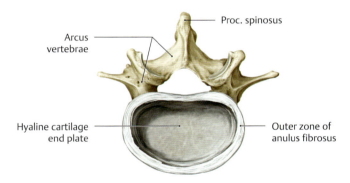
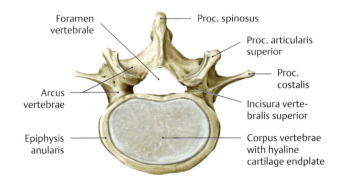

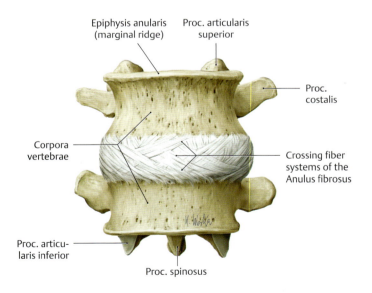

B Outer zone of the anulus fibrosus
Discus intervertebrales between the third and fourth lumbar vertebrae, anterior view.
The connective-tissue fiber bundles in the outer zone of the anulus fibrosus cross one another at various angles, interconnecting the bony marginal ridges of two adjacent corpora vertebrae.

C Main structural components of a discus intervertebralis
Fourth lumbar vertebra with its associated upper disk, superior view.

a Intervertebral disk with the anulus fibrosus and nucleus pulposus.
b Anulus fibrosus (with the nucleus pulposus removed).
c Outer zone of the anulus fibrosus (with the inner zone removed).
d Hyaline cartilage end plate within the bony epiphysis anularis (entire discus intervertebralis removed).

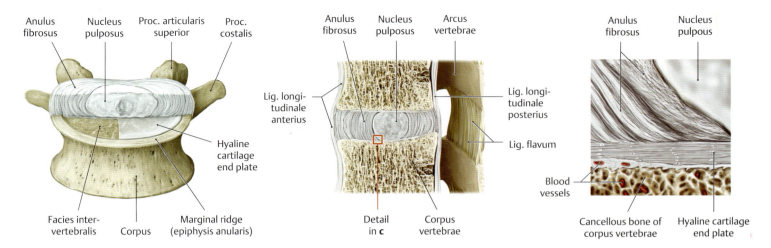

D Position of the discus intervertebralis in the motion segment
a Hyaline cartilage end plate, anterosuperior view (the anterior half of the disk and right half of the end plate have been removed).
b Sagittal section through a motion segment (see p.124), left lateral view.
c Detail from b.

Except for the outer zone of the anulus fibrosus, the entire disk is in contact superiorly and inferiorly with the hyaline cartilage layer of the adjacent vertebral end plates. The subchondral, bony portion of the end plate consists of compact bone (facies intervertebralis) permeated by myriad pores (see c) through which the vessels in the bone marrow spaces of the corpora vertebrae can supply nutrients to the disk tissue.

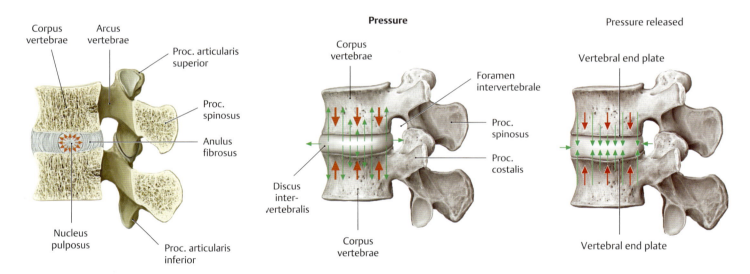

E Load-dependent fluid shifts in the discus intervertebralis
a The nucleus pulposus functions as a "water cushion" to absorb transient axial loads on the discus intervertebralis. Mechanically, the disk represents a hydrostatic system that is resilient under pressure. It is composed of a tension-resistant sheath (the anulus fibrosus) and a watery, incompressible core, the nucleus pulposus. This core consists of 80 to 85% water, which it can reversibly bind in its paucicellular, gelatinous, mucoviscous tissue (owing to its high content of glycosaminoglycans). The nucleus pulposus is under a very high hydrostatic pressure, particularly when acted upon by gravity and other forces. This pressure can be absorbed by the adjacent cartilaginous end plates and also by the anulus fibrosus (which transforms compressive forces into tensile forces). In this way the nucleus pulposus functions as a "water cushion" or hydraulic press between two adjacent corpora vertebrae. Combined with the anulus fibrosus, it acts as an effective shock absorber that can distribute pressures uniformly over the adjacent vertebral end plates.

b Fluid outflow from the discus intervertebralis (green arrows) in response to a sustained pressure load (thick red arrows). While transient loads are cushioned by the shock-absorber function of the nucleus pulposus and anulus fibrosus (see a), sustained loads cause a gradual but permanent outflow of fluid from the disk. The turgor and height of the disk are reduced, while the end plates and eventually the bony vertebral elements move closer together (see p.131 for further details on disk degeneration).

c Fluid uptake by the discus intervertebralis (green arrows) when pressure is released (thin red arrows). The process described in b is reversed when the pressure on the disk is released, and the height of the disk increases. This increase is caused by fluid uptake from the subchondral blood vessels in the bone marrow spaces, which are instrumental in disk nutrition (see Dc). As a result of pressure-dependent fluid shifts (convection) in the disci intervertebrales, the overall body height decreases temporarily by approximately 1% (1.5–2.0 cm) relative to the initial body height during the course of the day.

9.10 The Ligaments of the Spinal Column (Columna vertebralis): Overview and Thoracolumbar Region

A The ligaments of the columna vertebralis at the level of the thoracolumbar junction (T11–L3)
Left lateral view. The two uppermost thoracic vertebrae have been sectioned in the midsagittal plane.

B The ligaments of the columna vertebralis
The ligaments of the spinal column bind the vertebrae securely to one another and enable the spine to withstand high mechanical loads and shearing stresses. The ligaments are subdivided into vertebral *body* ligaments and vertebral *arch* ligaments.

Vertebral body ligaments
• Lig. longitudinale anterius
• Lig. longitudinale posterius

Vertebral arch ligaments
• Ligg. flava
• Ligg. interspinalia
• Lig. supraspinale
• Lig. nuchae*
• Ligg. intertransversaria

* The sagittally oriented lig. nuchae runs between the protuberantia occipitalis externa and the vertebra prominens (C VII); it corresponds to a lig. supraspinale that is broadened superiorly (see p. 121).

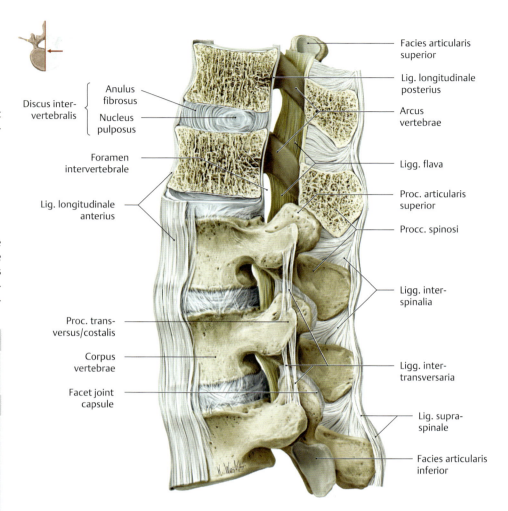

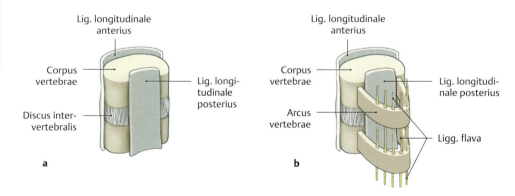

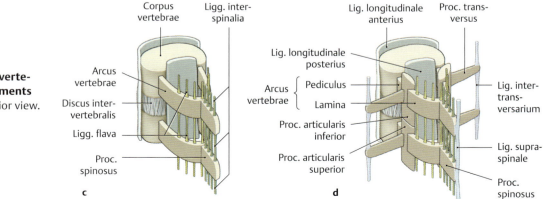

C Schematic representation of the vertebral body and vertebral arch ligaments
Viewed obliquely from the left posterior view.
a Vertebral body ligaments.
b–d Vertebral arch ligaments.

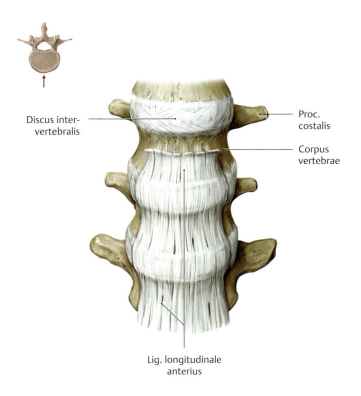
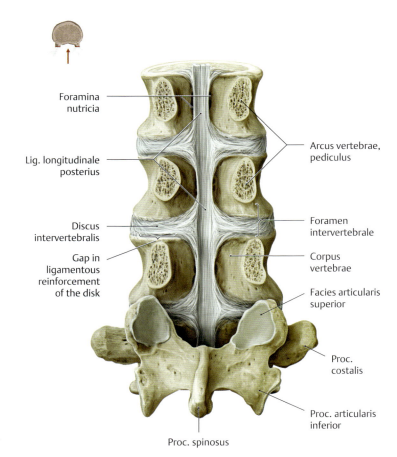

D The ligaments surrounding the lumbar spine
a Lig. longitudinale anterius, anterior view.
b Lig. longitudinale posterius, posterior view after removal of the vertebral arches at the pediculi arcus vertebrae.
c Ligg. flava and ligg. intertransversaria, anterior view (after removal of the L II–L IV corpora vertebralis). (The other vertebral arch ligaments are not visible in this view.)

The **lig. longitudinale anterius** runs broadly on the anterior side of the corpora vertebrae, extending from the skull base to the os sacrum. Its deep fibers bind adjacent corpora vertebrae together, while its superficial fibers span multiple segments. Its collagenous fibers are attached firmly to the corpora vertebrae but have only a loose attachment to the disci intervertebrales. The thinner **lig. longitudinale posterius** descends from the clivus along the posterior surface of the corpora vertebrae, passing into the canalis sacralis. The portion over the corpora vertebrae is narrow and is attached to their superior and inferior margins. The ligament broadens at the levels of the disci intervertebrales, to which it is firmly attached by tapered lateral extensions. Despite the attachment of the ligament to the anulus fibrosus of the intervertebral disks (not seen clearly here, being hidden by the lig. logitudinale posterius), a large portion of the discus intervertebralis has no ligamentous reinforcement, especially laterally (predisposing these disks to lateral disk herniation, see p.131). Both longitudinal ligaments contribute to maintaining the normal curvature of the columna vertebralis. The **ligg. flava** consist mainly of elastic fibers, which give these ligaments their characteristic yellow color. They are thick, powerful ligaments that connect the laminae of adjacent vertebral arches and reinforce the wall of the canalis vertebralis posterior to the foramina intervertebralia (see **A**). When the columna vertebralis is erect, the ligg. flava are under tension and help the back muscles to stabilize the spine in the sagittal plane. They also act as reins to limit forward flexion of the columna vertebralis, thereby helping to maintain the position of the flexed spine. The tips of the transverse processes are connected on each side by the **ligg. intertransversaria**, which serve to limit rocking movements of one vertebra upon another.

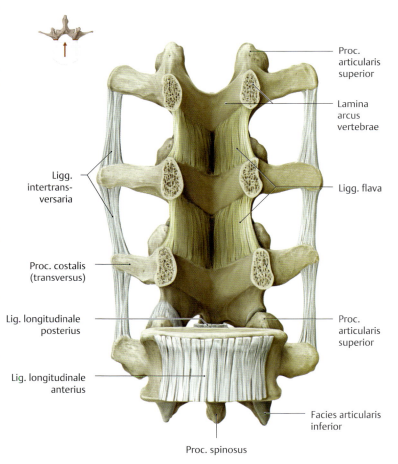

9.11 Overview of the Ligaments of the Cervical Spine

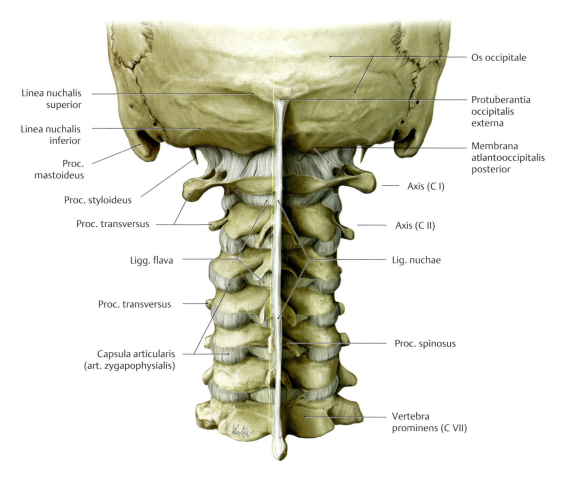

A The ligaments of the cervical spine
a Posterior view.
b Anterior view after removal of the anterior skull base (see p. 122 for the ligaments of the upper cervical spine, especially the craniovertebral joints).

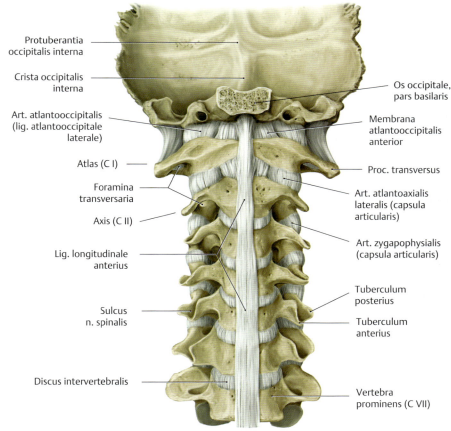

B The craniovertebral joints
The craniovertebral joints are the articulations between the atlas (C I) and os occipitale (Art. atlantooccipitalis) and between the atlas and axis (C II, artt. atlantoaxiales). While these joints, which number six in all, are anatomically distinct, they are mechanically interlinked and comprise a functional unit (see p. 125).

Atlantooccipital joints
(Art. atlantooccipitalis)

Paired joints where the oval, slightly concave fascies articulares superiores of the atlas articulate with the convex condyli occipitales.

Atlantoaxial joints (Artt. atlantoaxiales)

- Art. atlantoaxialis lateralis = paired articulation between the facies articulares inferiores of the atlas and the facies articulares superiores of the axis
- Art. atlantoaxialis mediana = unpaired articulation (comprising an anterior and posterior compartment) between the dens axis (fovea dentis atlantis), and the cartilage-covered anterior surface of the lig. transversum atlantis (see p. 123)

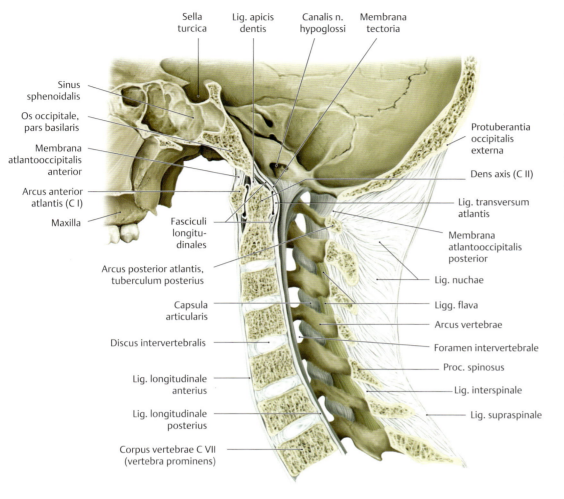

C The ligaments of the cervical spine: lig. nuchae
Midsagittal section, left lateral view. The lig. nuchae is the broadened, sagittally oriented part of the lig. supraspinale that extends from the vertebra prominens (C VII) to the protuberantia occipitalis externa (see **A**; see also p. 122 for the ligaments of the atlantooccipital and atlantoaxial joints).

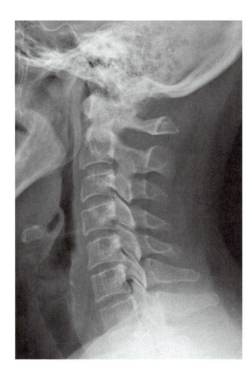
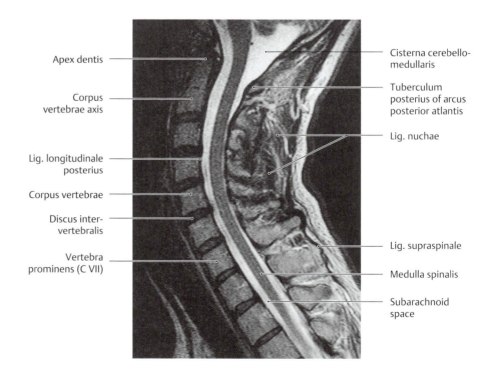

D Plain lateral radiograph of the cervical spine

E Magnetic resonance image of the cervical spine
Midsagittal section, left lateral view T2-weighted TSE sequence (from Vahlensieck M, Reiser M. MRT des Bewegungsapparates. 2nd ed. Stuttgart: Thieme; 2001).

9.12 The Ligaments of the Upper Cervical Spine (Atlantooccipital and Atlantoaxial Joints)

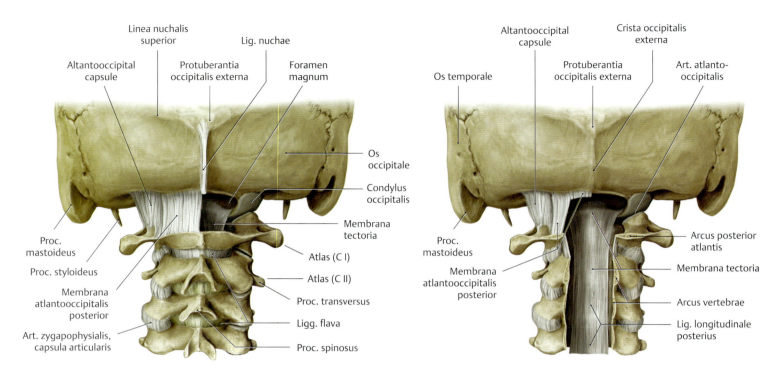

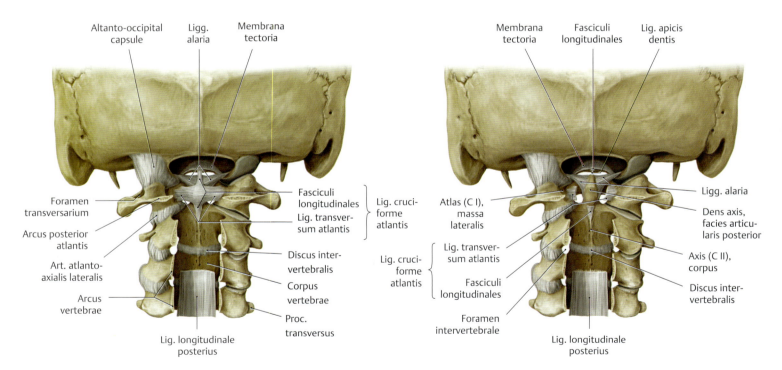

A The ligaments of the craniovertebral joints
Skull and upper cervical spine, posterior view.

a The membrana atlantooccipitalis posterior—the "ligamentum flavum" between the atlas and os occipitale (see p. 120)—stretches from the arcus posterior atlantis of the atlas to the posterior rim of the foramen magnum. This membrane has been removed on the right side.

b With the canalis vertebralis opened and the spinal cord removed, the membrana tectoria, a broadened expansion of the lig. longitudinale posterius, is seen to form the anterior boundary of the canalis vertebralis at the level of the craniovertebral joints.

c With the membrana tectoria removed, the lig. cruciforme atlantis can be seen. The lig. transversum atlantis forms the thick horizontal bar of the cross, and the fasciculi longitudinales form the thinner vertical bar.

d The lig. transversum atlantis and fasciculi longitudinales have been partially removed to demonstrate the paired ligg. alaria, which extend from the lateral surfaces of the dens to the corresponding inner surfaces of the condyli occipitalis, and the unpaired lig. apices dentis, which passes from the tip of the dens to the anterior rim of the foramen magnum.

Trunk Wall — *9. Bones, Ligaments, and Joints*

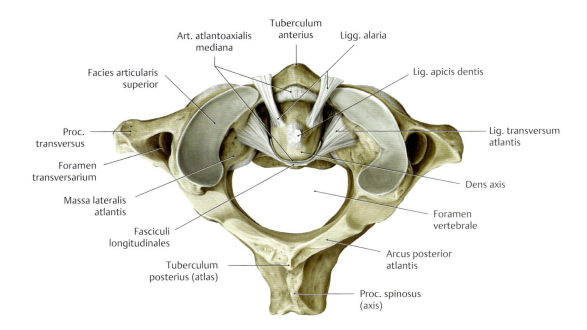

B The ligaments of the art. atlantoaxialis mediana
Atlas and axis, superior view. (The fovea dentis, while part of the art. atlantoaxialis mediana, is hidden by the capsula articularis.)

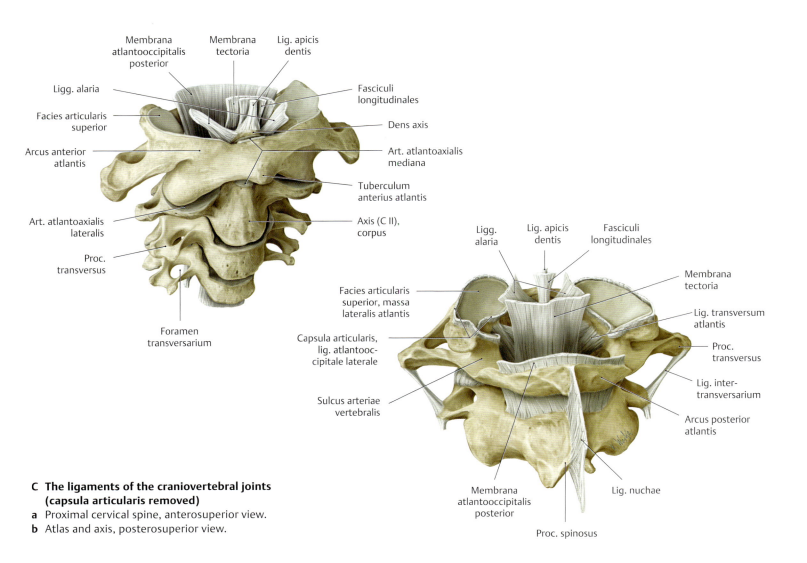

C The ligaments of the craniovertebral joints (capsula articularis removed)
a Proximal cervical spine, anterosuperior view.
b Atlas and axis, posterosuperior view.

123

9.13 The Intervertebral Facet Joints, Motion Segments, and Range of Motion in Different Spinal Regions

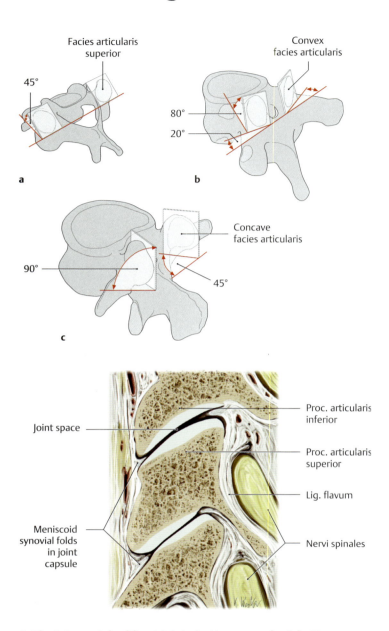

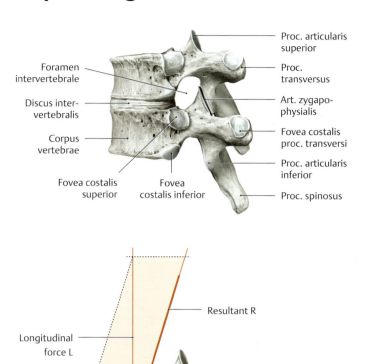

A The intervertebral facet joints (artt. zygapophysiales))
The diagrams show the position of the articular surfaces of the intervertebral facet joints in different regions of the columna vertebralis, shown from the left posterosuperior view: **a** cervical spine, **b** thoracic spine, **c** lumbar spine. Panel **d** is a sagittal section through the facet joints at the level of the third, fourth, and fifth cervical vertebrae, shown from the lateral view (drawing based on a specimen from the Anatomical Collection at Kiel University).

The paired facet joints are true synovial joints formed by the procc. articulares of the vertebral arches (see p. 106). Their articular surfaces, called **facets,** show varying degrees of inclination from the horizontal (and vertical) in different spinal regions and are therefore specialized for certain directions and ranges of motion (the movements that can occur in different spinal regions are shown in **D**). The **capsula articularis** of the facet joint is inserted into the margins of the articular facets and is often firmly attached to the lig. flavum (see **d**). The capsula articularis tends to be broad and lax in the cervical spine but is narrower in the thoracic and lumbar regions. Almost all the intervertebral joints contain *meniscoid synovial folds,* which form a crescent-shaped projection into the joint space. The folds consist of some loose but mostly firm connective tissue that has a rich vascular supply. The function of these folds is to fill the spaces around the articular surfaces (**d**).

B Structure and loading of a motion segment, illustrated for two thoracic vertebrae
Lateral view. "Motion segment" is the term applied to the articular and muscular connection between two adjacent vertebrae (**a**). It consists of the discus intervertebralis, the paired intervertebral facet joints (artt. zygapophysiales), and the associated ligaments and muscles (not depicted here). For clinical purposes, the motion segment is also considered to include the contents of the foramina intervertebralia (nerves and blood vessels, see pp. 190 and 196) and the contents of the canalis vertebralis. In all, the columna vertebralis contains 25 of these motion segments, which constitute distinct functional and morphological units. However, because of the interdependence of these units, abnormalities in a circumscribed portion of the columna vertebralis tend to affect the other spinal segments as well. Each of these motion segments is subjected to certain loads that can be represented as applied forces (**b**): a forward-directed *shear force* and a downward-directed *longitudinal force,* which combine to give the resultant force R. The longitudinal force acts on the corpus vertebrae and discus intervertebralis, while the shear force is absorbed mainly by the ligaments and facet joints (counterforce S'). The shear force can be subdivided into a normal force (Sn) and a tangential force (St). Because the shearing forces are not perpendicular to the articular surfaces of the facet joints, these surfaces are loaded by the axially directed *normal force* (Sn), which is weaker than the original shear force. Displacement of the vertebrae by the upward-directed *tangential force* (St) is prevented by the ligaments and intrinsic back muscles (after Kummer).

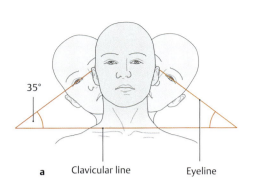
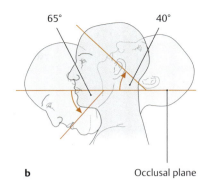
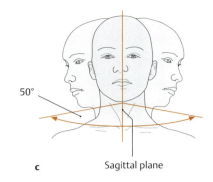

C Total range of motion of the cervical spine
a Lateral flexion, b flexion/extension, c rotation.

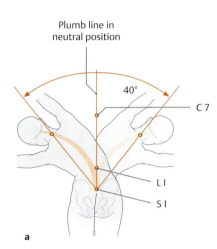
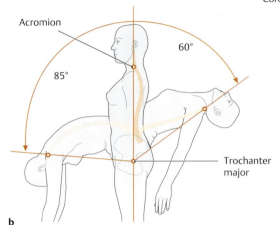
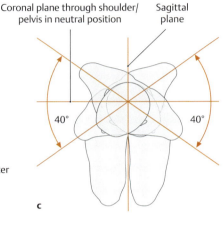

D Total range of motion of the thoracic and lumbar spine
a Lateral flexion, b flexion/extension, c rotation.
The clinical examination, particularly functional testing, is of key importance in examinations of the spine. Because the total range of spinal motion is comprised of the movements of a total of 25 motion segments, generally the examiner can detect only movement disorders that affect individual regions. For example, clinical testing can show clear evidence of ankylosis affecting particular spinal segments. The examiner makes use of standard reference lines (e.g., the clavicular line or occlusal plane) in determining whether the range of spinal motion is normal or decreased.

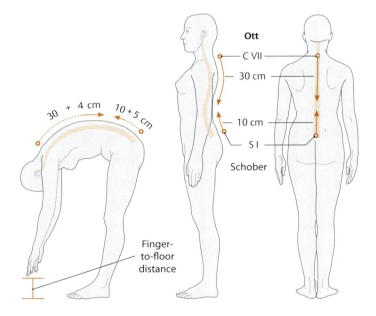

F Measurement of the range of thoracic and lumbar spinal flexion by the method of Schober and Ott
In the method of Schober and Ott, the patient stands erect while the examiner marks the S I processus spinosus and a second point 10 cm higher. When the patient bends as far forward as possible, the distance between the two skin markings will increase to approximately 15 (10 + 5) cm (range of motion of the lumbar spine). The thoracic range of motion is determined by measuring 30 cm down from the processus spinosus of the C VII vertebra (vertebra prominens) and marking that point on the skin. When the patient bends forward, the distance between the markings may increase by up to 4 cm. An alternative method is to measure the smallest finger-to-floor distance (FFD) with the knees extended.

E Average ranges of motion in different spinal regions (degrees)

	Cervical spine			Thoracic spine	Lumbar spine	Cervical + thoracic + lumbar
	A-o joint	A-a joint	Entire cervical spine			
Flexion	20	—	65	35	50	150
Extension	10	—	40	25	35	100
Lateral flexion*	5	—	35	20	20	75
Rotation*	—	35	50	35	5	90

A-o joint = Atlantooccipital joint
A-a joint = Atlantoaxial joint   *To each side

9.14 The Uncovertebral Joints of the Cervical Spine

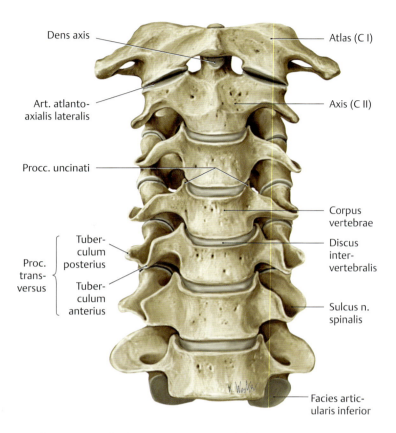

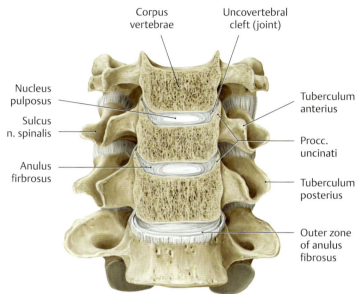

A The uncovertebral joints in a young adult
Cervical spine of an 18-year-old man, anterior view.

a The upper end plates of the C III through C VII corpora vertebrae have lateral projections (unci corporis vertebrae or procc. uncinati) that develop during childhood. Starting at about 10 years of age, the procc. uncinati as they develop gradually come into contact with the oblique, crescent-shaped margin on the undersurface of the next higher vertebral body. This results in the formation of lateral clefts (uncovertebral clefts or joints, see **b**) in the outer portions of the disci intervertebrales.

b C IV through C VII vertebrae. The bodies of the C IV–C VI vertebrae have been sectioned in the coronal plane to demonstrate more clearly the uncovertebral joints or clefts. These clefts are bounded laterally by a connective tissue structure, a kind of joint capsule, which causes them to resemble true joint spaces. These clefts or fissures in the discus intervertebralis were first described by the anatomist Hubert von Luschka in 1858, who called them *hemiarthroses laterales*. He interpreted them as primary mechanisms designed to enhance the flexibility of the cervical spine and confer a functional advantage (drawings based on specimens from the Anatomical Collection at Kiel University).

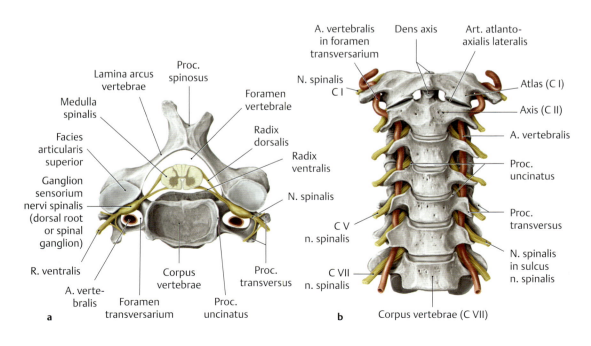

B Topographical relationship of the spinal nerve and a. vertebralis to the proc. uncinatus

a Fourth cervical vertebra with spinal cord (medulla spinalis), spinal roots (radix nervi), spinal nerves (n. spinalis), and aa. vertebrales, superior view.

b Cervical spine with both arteriae vertebrales and the emerging nervi spinales, anterior view.

Note the course of the a. vertebralis through the foramina transversaria and the course of the n. spinalis at the level of the foramina intervertebralia. Given their close proximity, both the artery and nerve may be compressed by osteophytes (bony outgrowths) caused by uncovertebral arthrosis (see **D**).

Trunk Wall —— **9. Bones, Ligaments, and Joints**

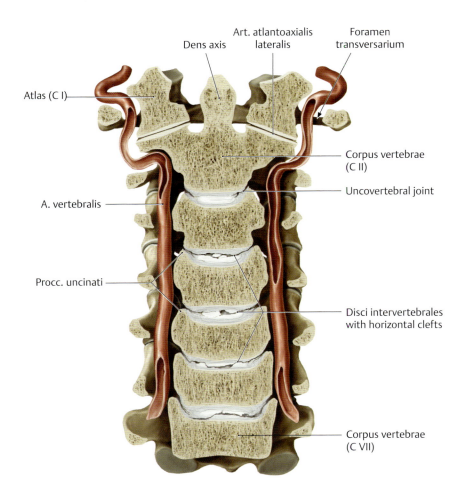

C Degenerative changes in the cervical spine (uncovertebral arthrosis)
Coronal section through the cervical spine of a 35-year-old man, anterior view. *Note* the course of the a. vertebralis on both sides of the corpora vertebrae.
The development of the uncovertebral joints at approximately 10 years of age initiates a process of cleft formation in the disci intervertebrales. This process spreads toward the center of the disk with aging, eventually resulting in the formation of complete transverse clefts that subdivide the disci intervertebrales into two slabs of roughly equal thickness. The result is a progressive degenerative process marked by flattening of the disks and consequent instability of the motion segments (drawing based on specimens from the Anatomical Collection at Kiel University).

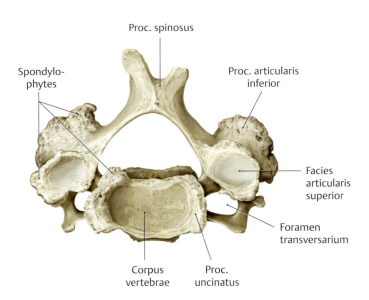
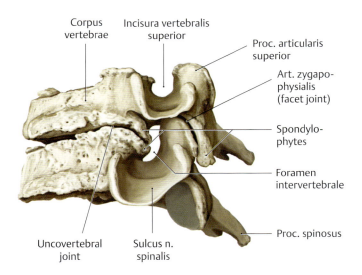

D Advanced uncovertebral arthrosis of the cervical spine
a Fourth cervical vertebra, superior view.
b Fourth and fifth cervical vertebrae, lateral view (drawings based on specimens from the Anatomical Collection at Kiel University).

The uncovertebral joints undergo degenerative changes comparable to those seen in other joints, including the formation of osteophytes (called spondylophytes when they occur on corpora vertebrae). These sites of new bone formation serve to distribute the imposed forces over a larger area, thereby reducing the pressure on the joint. With progressive destabilization of the corresponding motion segment, the facet joints undergo osteoarthritic changes leading to osteophyte formation. Osteophytes of the uncovertebral joints have major clinical importance because of their relation to the foramen intervertebrale and a. vertebralis (uncovertebral arthrosis). They cause a progressive narrowing of the foramen intervertebrale, with increasing compression of the n. spinalis and often of the a. vertebralis as well (see **C**). Meanwhile, the spinal canal itself may become significantly narrowed (spinal stenosis) by the same process.

9.15 Cross-sectional Anatomy of the Lumbar Spine

A Midsagittal section through the lower part of the spinal column
Left lateral view.
Note: The caudal end of the medulla spinalis, the conus medullaris, terminates in most people at the level of the first lumbar vertebra but in a few at the level of the second lumbar vertebra. The medulla spinalis and spinal canal are approximately the same length until the 12th week of prenatal development, so that each spinal cord segment and its pair of nervi spinales lie adjacent to the vertebra of the same number. With further growth, however, the columna vertebralis lengthens more rapidly than the medulla spinalis, resulting in an increasing cephalad displacement of the conus medullaris. At birth the conus medullaris has already reached the level of the third lumbar vertebra, and it continues its gradual upward shift until about the 10th year of life. Because of these disparate growth rates, the spinal roots run obliquely downward from their segment of origin in the cord to reach their corresponding foramen intervertebrale. The spinal roots that descend from the lower end of the cord are collectively termed the *cauda equina* ("horse's tail"). Because the membranes that enclose the medulla spinalis (the meninges) extend into the canalis sacralis (cisterna lumbalis), a needle can be safely introduced into the subarachnoid space below the conus medullaris to sample liquor cerebrospinalis without injuring the cord *(lumbar puncture)*. This site is also used for *lumbar spinal anesthesia* to block both the afferent nerve roots (for analgesia) and the efferent nerve roots (for muscular paralysis) that supply the pelvic region and lower limbs.

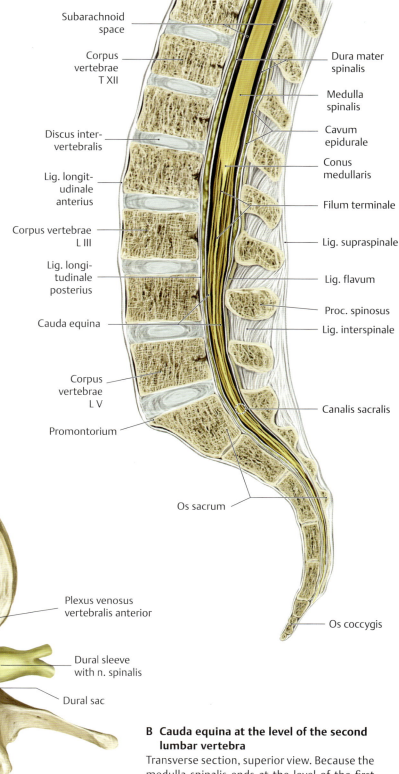

B Cauda equina at the level of the second lumbar vertebra
Transverse section, superior view. Because the medulla spinalis ends at the level of the first lumbar vertebra in most people, at the level of the second lumbar vertebra, the dural sac, which ends at the level of the second sacral vertebrae, consists only of the cauda equina and the filum terminale (see **A**). Here the epidural space (spatium epidurale) is enlarged and filled with extended venous plexus and fatty tissue.

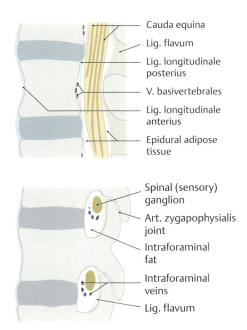
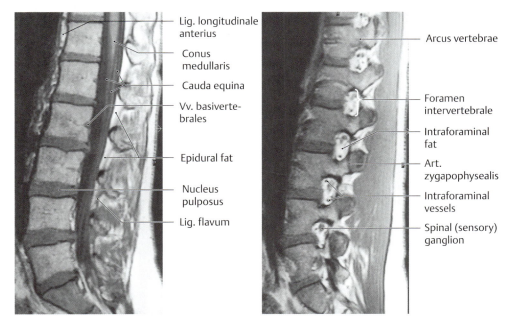

C MRI of the normal lumbar spine: Sagittal and transverse sections (photos from Vahlensieck M, Reiser M. MRT des Bewegungsapparates. 3rd ed. Stuttgart: Thieme; 2006)

a, b structures visible in the median and parasagittal section.

c, d T1-weighted SE sequence of the lumbar spine in median and parasagittal section.

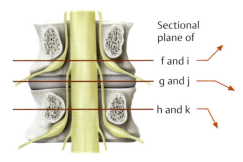

e motion segment L4–L5 after removal of arcus vertebrae, dorsal view; position of the transverse planes:

f supradisk level.
g disk level L4–L5.
h infradisk level.

In **f** and **g**, the foramen intervertebrale is shown; in **h**, the pediculus arcus vertebrae is shown.

i–k T1-weighted SE sequences of the motion segment L4–L5 in the corresponding axial (transverse) planes.

Note the position of the spinal roots in relation to the dural sac. The spinal roots extend in so-called dural bags, which reach into the foramen intervertebrale, so the roots and the spinal ganglion are completely bathed in fluid (see Fig. **B**).

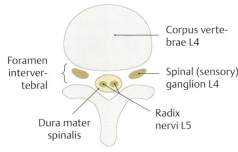
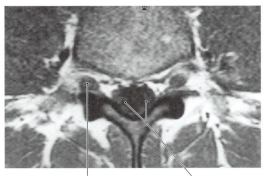
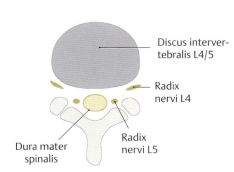
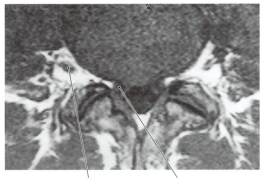
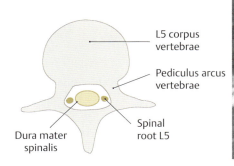
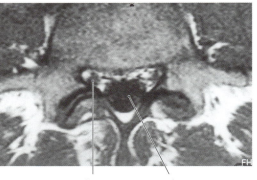

9.16 Degenerative Changes in the Lumbar Spine

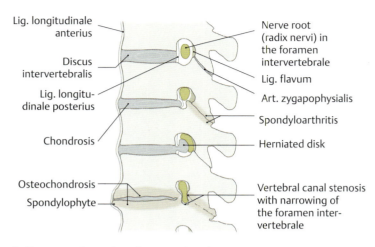

A Degenerative spine changes at a glance
Degenerative spine changes are among the most common disorders. They are age-dependent and increase dramatically after the age of 30. The intervertebral disk, the adjacent osseous end plates, the vertebrae, and the ligaments of the participating motion segment are most often affected.

Degeneration of the intervertebral disks and disk herniation: As a result of fluid depletion in the nucleus pulposus (decreased water-binding capacity — a typical symptom of old age), the intervertebral disk space increasingly narrows, and the affected motion segment becomes increasingly unstable. The resulting locally increased mechanical loading of the disk causes fraying and cracking in the anulus fibrosus (chondrosis) and eventually leads to a herniated disk (protrusion or prolapse, see **D**).

Lumbar spinal stenosis: The body tries to withstand disk degeneration through reactive bone changes, similar to what happens with osteoarthritis in the limb joints (see p. 46). At the corpora vertebrae, marginal spikes (osteophytes = spondylophytes) form, which are supposed to enlarge the force-transmitting surface, thereby relieving the relevant vertebral joint and stabilizing the motion segment, causing end plate sclerosis (osteochondrosis). Similar processes occur at the small vertebral joints (spondylarthrosis). The increasing narrowing of the canalis vertebralis and the foramina intervertebrale leads to degenerative lumbar spinal stenosis (see **B**). Later, the marginal spikes grow toward each other until they finally touch and as a result bridge and ossify the motion segment ("formation of bone bridges," see **Cc**). This increasingly impairs spine movement, often decreasing discomfort. A herniated disk is no longer possible, which is why herniated disks do not occur in old age.

Vertebral body fractures and deformities: In old age, the vertebral body is often the weakest link. Advanced osteoporosis or osteolytic bone metastasis accompanied by loss of stability in the bone can lead to vertebral body fractures and deformities. The symptoms include initially localized pain, which can spread if, at the same time, nerve roots are compressed.

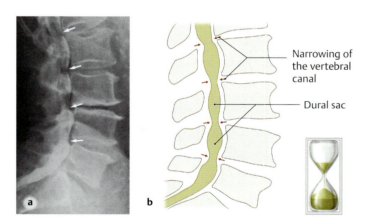

B Degenerative lumbar spinal stenosis
Symptoms of this disorder include the concentric narrowing of the canalis vertebralis at the level of the relevant motion segment. It develops based on a congenital narrowing and/or acquired degenerative changes, such as dorsal spondylophytes and arthrotically transformed vertebral joints (spondylarthrosis). Although rare, ligamentous hypertrophy, too, particularly of the lig. longitudinale posterius and the lig. flava, can contribute to narrowing of the canalis vertebralis. Typical symptoms include load-dependent neurogenic pain in the lumbar region and legs. It is caused by walking for long periods of time but also by standing, and subsides only by supporting the body (e.g., by propping oneself up) and by kyphosis of the lumbar spine (sitting with the upper body bent forward). Pain, sensory disturbance, and paralysis lead to drastically impaired walking distance (claudicatio spinalis). The diagnosis is made using MRI and/or with the help of lumbar myelography with the spine positioned in lateral projection (**a**).

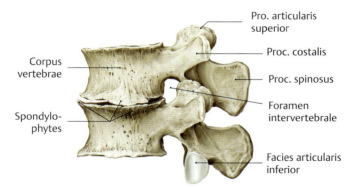

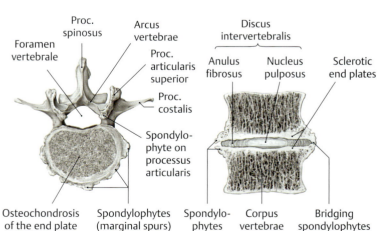

C Lumbar disk herniation
a Posterolateral herniation, superior view
b Posterior herniation, superior view
c Posterolateral herniation, posterior view (the arci vertebrae have been removed to demonstrate the lumbar dural sac and corresponding nerve roots)

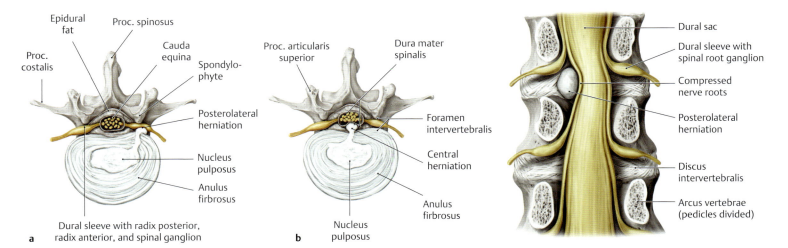

D Lumbar disk herniation
a Posterolateral herniation, superior view.
b Posterior herniation, superior view.
c Posterolateral herniation, posterior view (the arci vertebrae have been removed at the level of the pediculus arcus vertebrae to demonstrate the lumbar dural sac and corresponding nerve roots).

In old age, not only does the discus intervertebralis get thinner (due to loss of water in the nucleus pulposus, see **A**), it also shifts more easily, as a result of a decrease in resistance in the discus intervertebralis sleeve, the anulus fibrosus, which begins to tear and fray. The tissue of the jelly-like nucleus pulposus shifts initially toward the weak spots of the anulus fibrosus (intervertebral disk protrusion). If it tears as a result of continuous strain, it protrudes (disk prolapse) and compresses the content of the foramen intervertebrale. Sequestration occurs when a part of the intervertebral disk that has shifted forward no longer shows any connection to the rest of the discus intervertebralis. A posterolateral herniation generally compresses the subjacent lying radix spinalis (**c**), causing pain and paralysis that affect inferiorly adjacent dermatomes and corresponding muscles (see **E**).

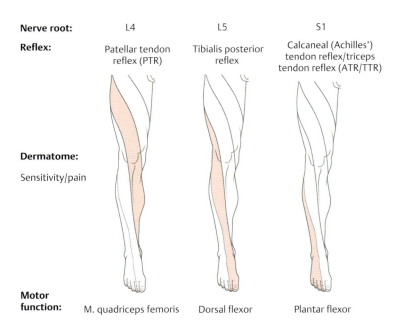

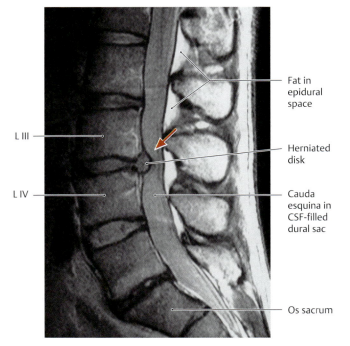

E Schematic representation of neurologic disorders in lumbar root compression syndromes as a result of herniated disks (LIV, LV, and SI)

In over 90% of all cases, herniated lumbar disks affect the LV and SI roots. Patients complain about pain and sensitivity loss in the affected dermatomes, as well as weakness or paralysis in the muscles innervated by these nerve roots. Patients experience the symptoms of root compression syndrome in the following order with increasing severity: pain, sensitivity loss, and paralysis. Often the cause is a weakening or abnormal proprioceptive reflex associated with the affected root.

F Posterior disk herniation in the lumbar spine
Midsagittal T2-weighted magnetic resonance image of the lumbar spine, left lateral view (from Vahlensieck M, Reiser M. MRT des Bewegungsaparates. 2nd ed. Stuttgart: Thieme; 2001). The image shows a conspicuous herniated disk at the L III–L IV level (red arrow), which protrudes posteriorly (transligamentous herniation). The dural sac at that level is deeply indented (see **Db**).

9.17 The Thoracic Skeleton

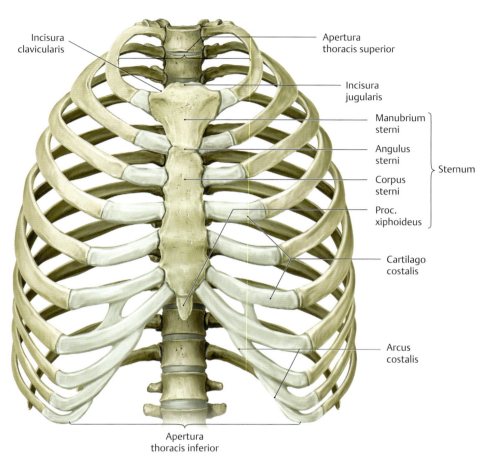
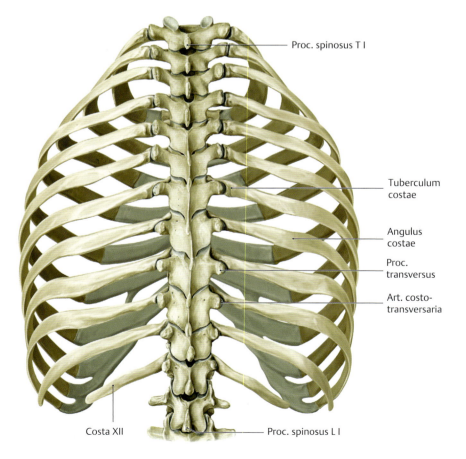

A Thoracic skeleton
a Anterior view, b posterior view.
The thoracic skeleton (thoracic cage, thorax) consists of the columna vertebralis, the 12 pairs of ribs, and the sternum. These structures are movably interconnected by ligaments, true joints, and synchondroses. Tension is imparted to the thorax by the mm. intercostales. The thoracic cage (cavea thoracis) encloses the cavitas thoracis and has an apertura thoracis superior (the thoracic inlet) and an apertura thoracis inferior (the thoracic outlet). Its shape is subject to marked variations relating to age and gender. The ribs of an infant have very little obliquity and are approximately horizontal. With increasing age, the ribs incline downward, while the thoracic cage becomes flattened anteroposteriorly. This is associated with a relative decrease in the size of the thoracic outlet. Generally, the female thorax is narrower and shorter than in the male. From a functional standpoint, the thoracic skeleton and its muscular wall structures form a rugged, stable enclosure that permits the respiratory excursions that are necessary for normal breathing. This is apparent in patients with severe chest injuries such as multiple rib fractures due to blunt trauma, where instability of the chest wall leads to paradoxical respiration: the affected side of the rib cage moves inward during inspiration and outward during expiration. This results in a *pendelluft* effect (from the German *Pendel*, pendulum, and *Luft*, air): air streams back and forth between the lungs, leading to increased dead space ventilation, decreased alveolar gas exchange, and respiratory failure. Patients thus affected generally require intubation.

Trunk Wall — 9. Bones, Ligaments, and Joints

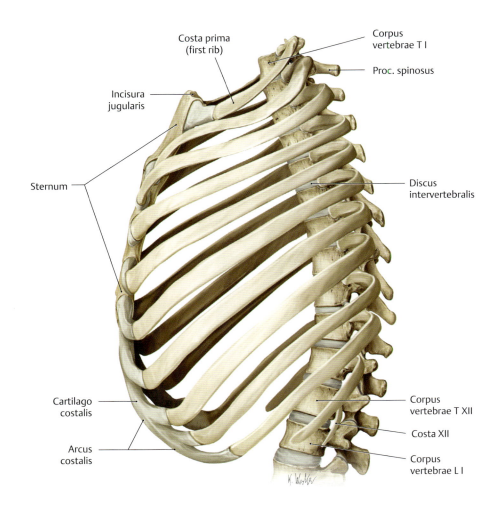

B Thoracic skeleton, lateral view

C True, false, and floating ribs
Lateral view. Each of the 12 pairs of ribs is bilaterally symmetrical, but the shapes of the ribs vary at different levels. The first seven pairs of ribs, called the *true ribs* (costa verae), are normally connected anteriorly to the sternum. Of the remaining five, called the *false ribs* (costa spuriae), the eighth, ninth, and tenth ribs are joined by their cartilago costalis to the cartilage of the rib directly above, contributing to the structure of the costal margin (arch) (arcus costalis; see **Aa**). The last two pairs of "false" ribs, called the *floating ribs* (costa fluctuantes), usually terminate freely between the muscles of the lateral abdominal wall.

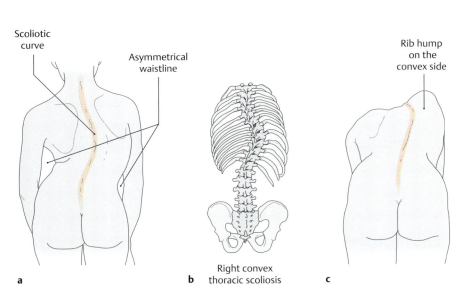

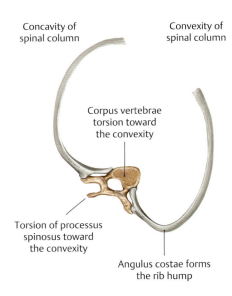

D Lateral curvature of the spinal column (scoliosis)
a, b Posterior view. Scoliosis commonly presents as a *right convex* curve of the spinal column at the level of the T VIII–T IX vertebrae (**b**). It is manifested by a typical postural deformity in upright stance (**a**).

c, d When a patient with right convex scoliosis bends forward, a typical rib hump appears on the convex side of the curve (**c**). This happens because the adjacent ribs also occupy an abnormal position due to torsion of the corpora vertebrae (**d**, superior view).

133

9.18 The Sternum and Ribs (Costae)

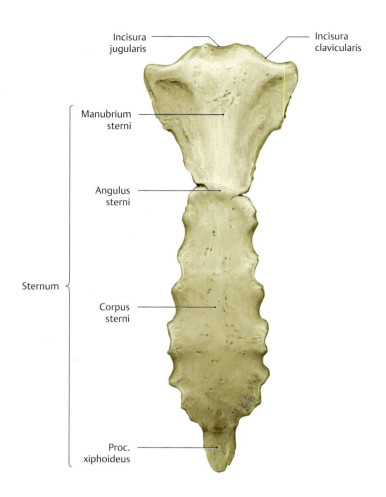
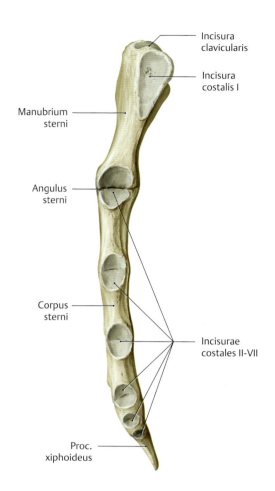

A The sternum
a Anterior view.
b Lateral view.

The sternum is a flattened bone, slightly convex anteriorly, with multiple indentations along its lateral borders (incisurae costales). The adult sternum consists of three bony parts:
- The manubrium sterni
- The corpus sterni
- The proc. xiphoideus

The manubrium, corpus, and proc. xiphoideus in adolescents and young adults are connected to one another by cartilaginous plates (synchondrosis manubriosternalis and synchondrosis xiphosternalis), which gradually ossify with increasing age. The fully ossified, mature adult form is depicted here. The depression in the superior border of the manubrium (incisura jugularis) is clearly palpable through the skin and marks the inferior margin of the fossa jugularis. On each side of the incisura jugularis is a depression for articulation with the clavicula on that side (incisura clavicularis), and just below that is a shallow concavity (incisura costalis I) for the synchondrosis with the first rib. At the junction of the manubrium sterni and corpus sterni is an articular facet (facies articularis) for the second rib (incisura costalis II). At that site the manubrium is usually angled slightly backward in relation to the body of the sternum (angulus sterni). The lateral borders of the corpus sterni bear additional costal notches (incisurae costales III-VII) that articulate with the third through seventh costal cartilages, the notches for the sixth and seventh costal cartilages being placed very close together. The sometimes bifid and perforated proc. xiphoideus itself is devoid of costal attachments and is highly variable in its form. Frequently, the proc. xiphoideus is still cartilaginous in adults.

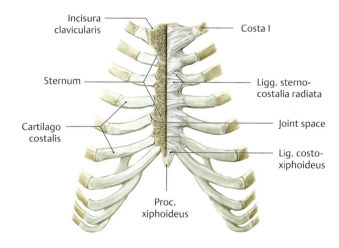

B The sternocostal joints (artt. sternocostales)
Anterior view (the right half of the sternum has been sectioned frontally to show the sternocostal joints). The connections between the costal cartilages for the first through seventh ribs and the incisurae costales sterni consist partly of synchondroses and partly of true joints. Generally, a joint space is found only at the second through fifth ribs, while the first, sixth, and seventh ribs are attached to the sternum by synchondroses. In the true joints as well as the synchondroses, ligaments (ligg. sternocostalia radiata) radiate from the perichondrium of the cartilago costalis to the anterior surface of the sternum, where they blend with the periosteum to form a dense fibrous membrane (the membrana sterni). The second rib is typically the most superior palpable rib.

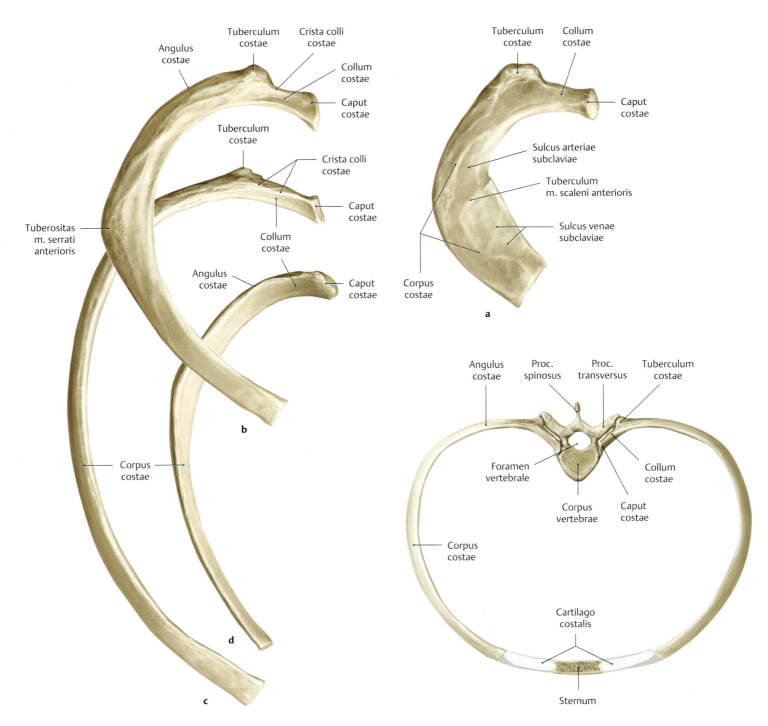

C Variable size and shape of the ribs
a First rib, b second rib, c fifth rib, d eleventh rib (all are right ribs, viewed from above).
The collum costae extends from the caput costae to the tuberculum costae. Except on the first rib, the neck bears a sharp superior ridge (crista colli costae). Lateral to the costal tubercle is the shaft (corpus) of the rib, which curves forward to form the costal angle. The shafts of the second through twelfth ribs in particular show irregular curvatures (on the flat and on the edges) and are also twisted about their long axis. Because of this torsion, the external surfaces of the ribs face slightly downward at their vertebral end and slightly upward at their anterior end. Normally, the first and twelfth ribs are the shortest, while the seventh rib is the longest. The cartilago costalis increases in length from the first to the seventh rib and shortens again past the eighth rib. Every rib except the first, eleventh, and twelfth has a groove along its inferior border (sulcus costae) that affords some protection for the intercostal vessels and nerves (see pp. 173 and 205).

D Segments of the rib and structure of a thoracic segment
Sixth pair of ribs, superior view. Each rib consists of a bony part (os costale) and a cartilaginous part (cartilago costalis). The bony part consists of the following segments, starting from its vertebral end:

- Head (caput costae)
- Neck (collum costae)
- Costal tubercle (tuberculum costae)
- Shaft (corpus costae), which includes the costal angle (angulus costae)

9.19 The Costovertebral Joints and Thoracic Movements

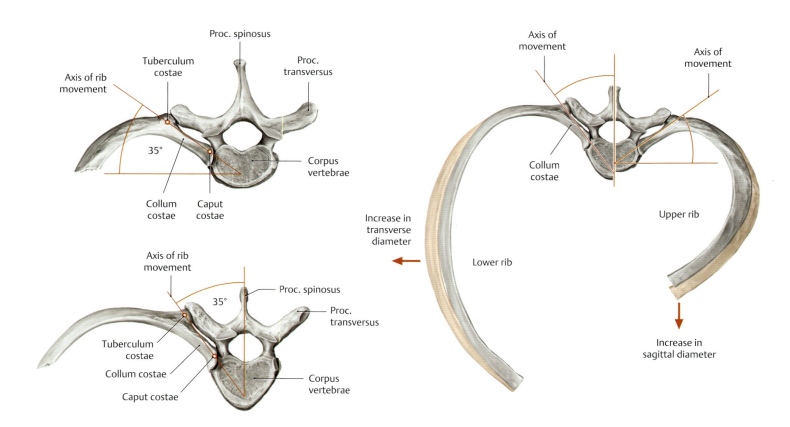

A Axes for movements of the costovertebral joints and ribs
(after Kapandji)
Superior view.

a Axis of upper rib movements.
b Axis of lower rib movements.
c Direction of rib movements (see **C** for the costovertebral joints).

The axes of rib movements are directed parallel to the necks of the ribs (collum costae). The axes for the upper ribs are closer to the coronal plane (**a**), while those for the lower ribs are closer to the sagittal plane (**b**). For this reason, a rib excursion in the upper part of the rib cage mainly increases the sagittal thoracic diameter, while a lower rib excursion increases the transverse diameter (see **B**).

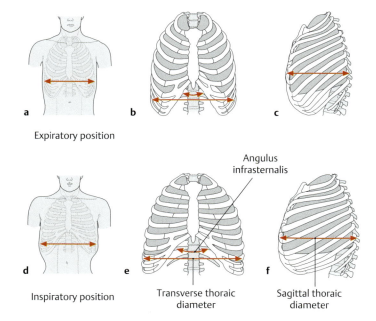

B Movements of the rib cage during costal or chest breathing (sternocostal breathing)

Respiration (ventilation) is effected through changes in the thoracic volume. The increase in thoracic volume that is necessary for inspiration can be accomplished in two ways:

1. By lowering the diaphragma (costodiaphragmatic breathing or abdominal breathing, see p.160)
2. By elevating the ribs (sternocostal breathing, also known as costal or chest breathing)

While breathing at rest is almost entirely abdominal, respiration during physical effort is augmented by chest breathing with the mm. intercostales and the auxiliary muscles of respiration. The drawings illustrate the thoracic volume changes that occur during *chest or costal breathing*, in which the thoracic volume decreases and increases in both the coronal and sagittal planes. Panels **a–c** show the *decrease* in the transverse and sagittal chest diameters at end-expiration, and **d–f** show the *increase* at end-inspiration.

Trunk Wall — 9. Bones, Ligaments, and Joints

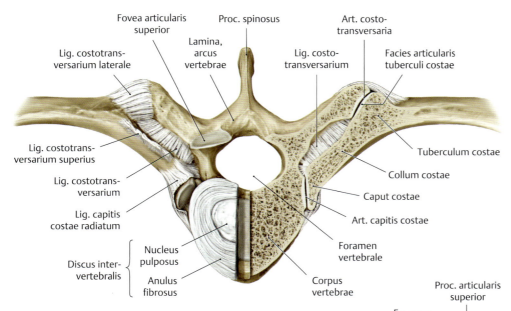

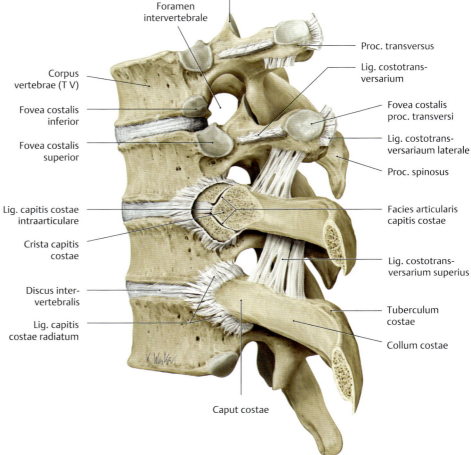

C Ligaments of the costovertebral joints
The costovertebral joints are the joints by which the ribs articulate with the vertebrae. They consist of two types: the joints of the heads of the ribs (art. capitis costae) and the costotransverse joints (art. costotransversaria). Though morphologically distinct, these different joint types are functionally interrelated.

a Articulation of the eighth rib with the eighth thoracic vertebra, superior view (the joint of the caput costae and costotransverse joint on the left side have been transversely sectioned).

b The fifth through eighth thoracic vertebrae and associated ribs (seventh and eighth ribs), left lateral view (the joint of the head of the seventh rib has been transversely sectioned).

Joint of the head of the rib (art. capitis costae): This joint consists of two articular surfaces:

1. An articular facet on the head of the rib (facies articularis capitis costae)
2. A costal facet on the vertebral body (fovea costalis)

The articular facets on the head of the second through tenth ribs (defined by the crista capitis costae) articulate with the fossa formed by the foveae costales superior and inferior on two adjacent corpora vertebrae and the discus intervertebralis between them. The lig. capitis costae intraarticulare, which extends from the crista capitis costae to the discus intervertebralis, separates the joint cavity of the second through tenth rib heads into two compartments. By contrast, the heads of the first, eleventh, and twelfth ribs each articulate with only one thoracic corpus vertebrae (see A, p. 110). In all the joints of the heads of the ribs, the capsula articularis is reinforced by the lig. capitis costae radiatum.

Costotransverse joint (art. costotransversaria): In the costotransverse joints of the first through tenth ribs, the facies articularis tuberculi costae articulates with the fovea costalis proc. transversi of the corresponding thoracic vertebra. The eleventh and twelfth ribs do not have a art. costotransversaria because their proc. transversus do not have articular facets (see A, p. 110). Three ligaments stabilize the art. costotransversaria and also strengthen the joint capsule:

1. The lig. costotransversarium laterale (from the tip of the proc. transversus to the tuberculum costae)
2. The lig. costotransversarium (between the collum costae and the proc. transversus)
3. The lig. costotransversarium superius (between the collum costae and the proc. transversus of the vertebra above it)

9.20 The Bony Pelvis

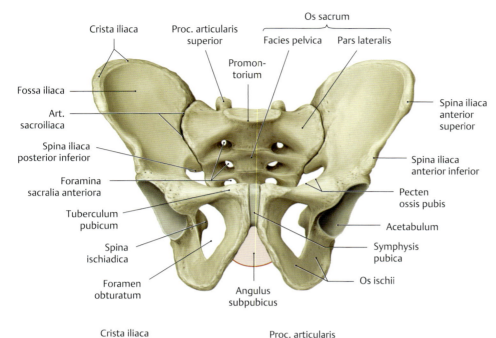

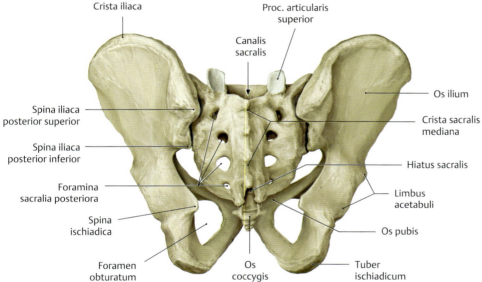

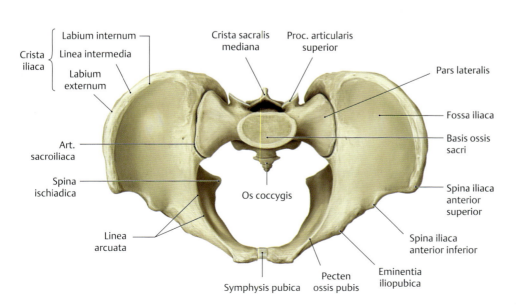

A The male pelvis
a Anterior view.
b Posterior view.
c Superior view.

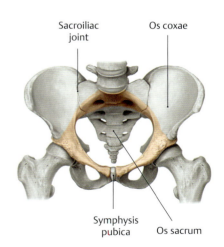

B The pelvic girdle and pelvic ring
Anterosuperior view. The pelvic girdle (cingulum pelvicum) consists of the two hip bones (ossa coxae). The sacroiliac joints and the cartilaginous pubic symphysis (symphysis pubica) unite the bony parts of the pelvic girdle with the os sacrum to form a stable ring called the pelvic ring (indicated by color shading). It allows very little mobility, because stability throughout the pelvic ring is an important prerequisite for transmitting the trunk load to the lower limbs.

Trunk Wall — 9. Bones, Ligaments, and Joints

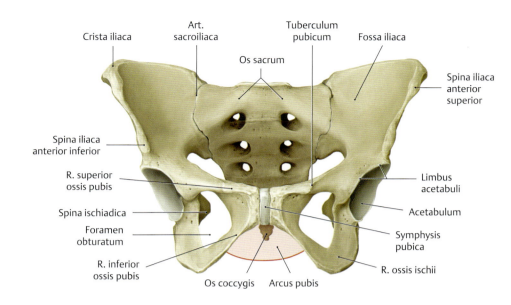

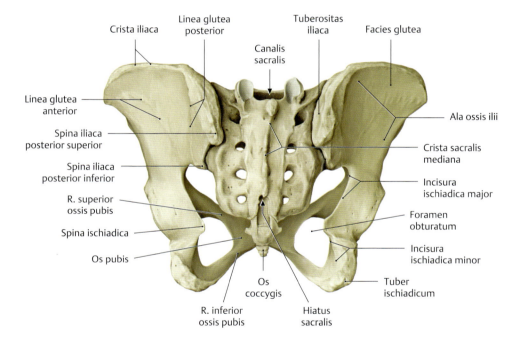

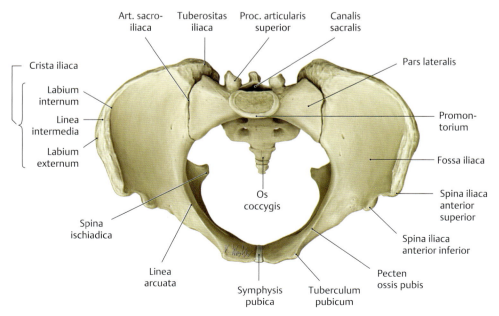

C The female pelvis
a Anterior view.
b Posterior view.
c Superior view.

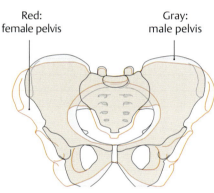

D Gender-specific features of the pelvis
Anterosuperior view. A male and female pelvis have been superimposed to illustrate the gender-specific differences. Comparison reveals that the female pelvis is larger and broader than the male pelvis, while the latter is taller, narrower, and more massive than the female pelvis. The inlet of the female pelvis is larger and has an almost oval shape, while the sacral promontorium shows a greater projection in the male pelvis (**Cc**). Sex differences are also noted in the angle of the arcus pubicus formed by the rami inferiores ossis pubis, which is acute in men (70°) but significantly larger in women (almost 90–100°). Accordingly, this angle is called the angulus subpubicus in men and the arcus pubicus in women (see **D**, p. 141). The os sacrum also exhibits differences between the sexes. In women the os sacrum is angled at the level of the third and fourth vertebrae (see p. 114), while in men it presents a uniform curvature.

9.21 The Pelvic Ligaments and Pelvic Measurements

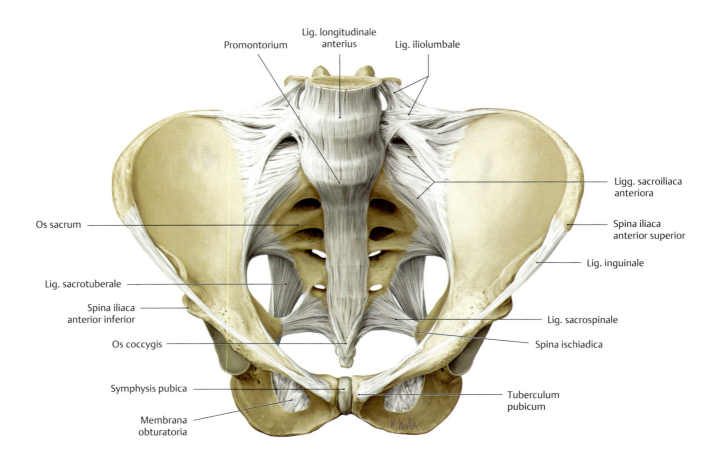

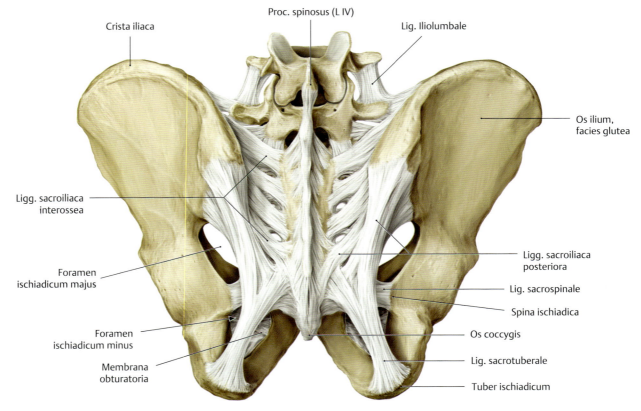

A Ligaments of the male pelvis
a Anterosuperior view.
b Posterior view.

Trunk Wall — 9. Bones, Ligaments, and Joints

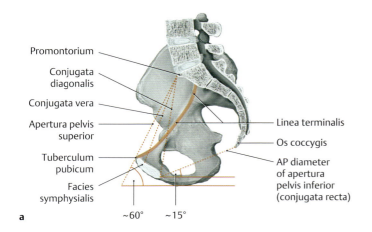

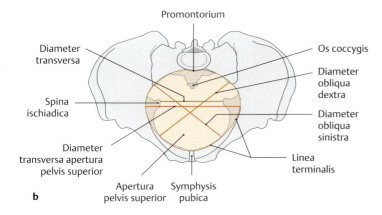

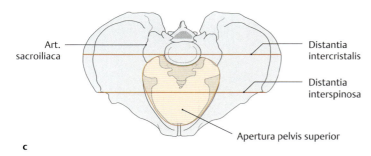

B Internal and external pelvic measurements
a Right half of a female pelvis, medial view.
b Female pelvis, superior view.
c Male pelvis, superior view.

The linea terminalis is marked in red in **a**. The apertura pelvis superior is color-shaded in **b** and **c**.

C Internal and external pelvic measurements, linea terminalis, and apertura pelvis superior

The internal and external pelvic measurements provide direct or indirect information on the size and shape of the bony boundaries of the lesser pelvis (pelvis minor). Because the lesser pelvis functions as the birth canal, the internal and external pelvic dimensions have special practical significance in obstetrics, determining whether the cavitas pelvis is broad enough to allow for a vaginal delivery. A particularly important measurement is the conjugata vera of the pelvic inlet (the obstetric conjugate), which is the smallest anteroposterior dimension of the lesser pelvis. With *pelvimetry*, a method of measuring pelvic dimensions, potential obstructions to labor can be identified prior to the delivery. Generally, the measurements are performed by transvaginal sonography. Some pelvic dimensions, such as the conjugata diagonalis, can be accurately determined by bimanual examination.

Internal pelvic measurements in women (see Ba and Bb)

- Diameter conjugata (conjugata vera) = 11 cm (distance from the promontorium to the posterior border of the symphysis)
- Diameter diagonalis (conjugata diagonalis) = 12.5–13 cm (distance from the sacral promontorium to the lower border of the symphysis)
- Diameter sagittalis of the apertura pelvis inferior (conjugata recta) = 9 (+2) cm (distance from the lower border of the symphysis to the tip of the os coccygis)
- Diameter transversa of the apertura pelvis superior = 13 cm (greatest distance between the lineae terminales)
- Diameter transversa = 11 cm (distance between the spinae ischiadicae)
- Diameter obliqua dextra (I) and sinistra (II) = 12 cm (distance from the art. sacroiliaca at the level of the linea terminalis to the eminentia iliopectinea on the opposite side)

External pelvic measurements in men (see Bc)

- Distantia interspinosa = 25–26 cm (distance between the spinae iliacae anteriores superiores)
- Distantia intercristalis = 28–29 cm (greatest distance between the left and right crista iliaca in the coronal plane)
- Conjugata externa = 20–21 cm (distance from the upper border of the symphysis pubica to the proc. spinosus of the L V vertebra)

Linea terminalis (see Ba)

Boundary line between the greater and lesser pelvis (pelvis major and pelvis minor), consists of the symphysis pubica, crista ossis pubis, pecten ossis pubis, linea arcuata ossis ilium and promontorium.

Apertura pelvis superior (see Bb and Bc)

Plane through the pelvic inlet at the level of the linea terminalis, below which is the lesser pelvis

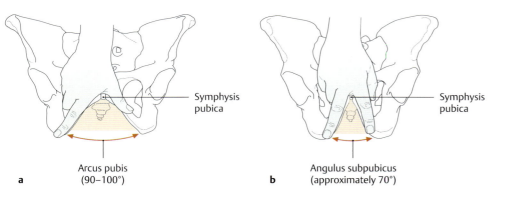

D Subpubic angle
Anterior view.

a Female pelvis: Arcus pubis.
b Male pelvis: Angulus subpubicus.

141

9.22 The Sacroiliac Joint (Art. Sacroiliaca)

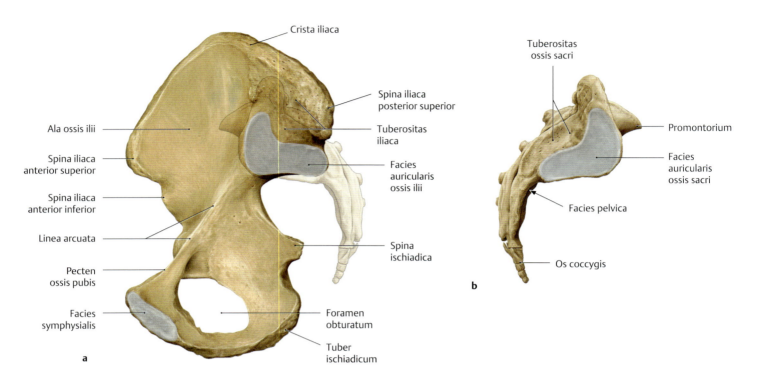

A Articular surfaces of the sacroiliac joint
a Facies auricularis ossis ilii, right hip bone (os coxae), medial view (the os sacrum is transparent in this view).
b Facies auricularis ossis sacri, right lateral view.

The two ear-shaped articular surfaces of the os ilium and os sacrum (auricular surfaces) are brought together at the sacroiliac joint (art. sacroiliaca). The facies auricularis of the os sacrum is slightly notched at its center, and there is a reciprocal ridge on the auricular surface of the os ilium. The shape and size of both auricular surfaces show considerable individual variation—more so than in other joints. Their cartilaginous covering is generally irregular, the articular cartilage on the sacral side being approximately twice as thick as on the iliac side.

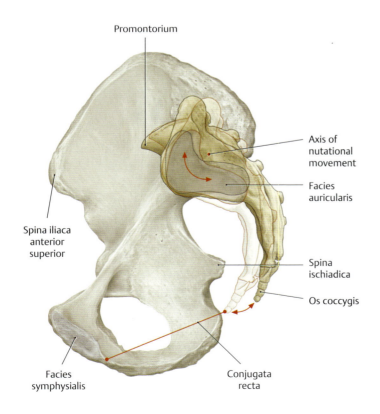

B Nutation in the sacroiliac joint
Right half of pelvis, medial view. Movements in the sacroiliac joints alter the width of the pelvic ring and thus have practical importance in obstetrics. The amplitude of the movements is greatly limited by tight ligaments and varies considerably in different individuals and between the sexes. Basically, very slight rotational and translational movements can be distinguished in the joints. Nutation, as shown here, is a rotational or "tilting" movement of the os sacrum about an axis located at the attachments of the ligg. sacroiliaca interossea. With *anterior rotation of the os sacrum*, the promontorium moves forward and downward, while the os coccygis moves upward and backward, thereby increasing the conjugata recta of the apertura pelvis inferior. With *posterior rotation of the os sacrum*, the AP diameter of the apertura pelvis superior increases, while the AP diameter (conjugata recta) of the apertura pelvis inferior is decreased.

Trunk Wall — 9. Bones, Ligaments, and Joints

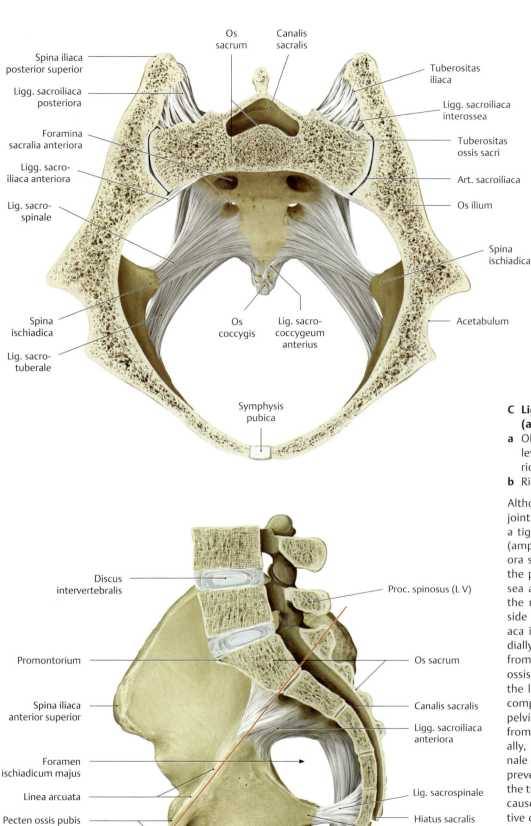

C Ligaments of the sacroiliac joint (articulatio sacroiliaca)
a Oblique section through the pelvis at the level of the apertura pelvis superior, superior view (plane of section indicated in **b**).
b Right half of pelvis, medial view.

Although the articulatio sacroiliaca is a true joint, its movements are greatly restricted by a tight joint capsule and powerful ligaments (amphiarthrosis). The ligg. sacroiliaca anteriora stabilize the joint on the anterior side of the pelvis, while the ligg. sacroiliaca interossea and posteriora and ligg. iliolumbalia are the main stabilizing elements on the dorsal side (see p.140). The powerful ligg. sacroiliaca interossea are deep bands that run medially just behind the articulatio sacroiliaca from the tuberositas iliaca to the tuberositas ossis sacri. They are completely covered by the ligg. sacroiliaca posteriora. The ligament complex helps to anchor the os sacrum in the pelvic ring during upright stance and keep it from sliding into the cavitas pelvis. Additionally, the lig. sacrotuberale and lig. sacrospinale (see **b**) stabilize both sacroiliac joints and prevent posterior tilting of the pelvis about the transverse axis. *Sacroiliac joint pain* may be caused by chronic inflammatory or degenerative changes (e.g., ankylosing spondylitis, osteoarthritis) or by trauma (e.g., sport-related injuries). Hypermobility of the articulatio sacroiliaca may also develop as a result of general ligamentous weakness or a pregnancy- or hormone-related laxity of the ligaments.

Sacroiliac dysfunction may include joint locking (a sudden force like that occurring in a straight-legged jump may cause a "wedging" of the os sacrum with joint locking). This stretches the joint capsule and can cause excruciating pain during most body movements.

10.1 The Muscles of the Trunk Wall: Their Origin and Function

Overview of the Trunk Wall Muscles
The muscles of the trunk wall in the strict sense consist of the intrinsic back muscles and the muscles of the chest and abdominal wall. In the broad sense, they include the muscles of the pelvic floor (which form the inferior boundary of the cavitas abdominopelvica) and the diaphragma (which separates the cavitas thoracis and cavitas abdominis). Besides the muscles of the trunk wall, the back and thorax also contain muscles of the shoulder girdle and upper limb that migrated to the trunk during the course of development (*nonintrinsic back and thoracic muscles*). Examples are the *thoracohumeral muscles* anteriorly, the *spinohumeral* muscles laterally and posteriorly, and the *spinocostal* muscles. Other muscles that have migrated to the trunk, such as m. trapezius, are derived from the mesenchyme of the branchial arches (branchial musculature). They are innervated by nervi craniales (m. trapezius by the n. accessorius) and were incorporated secondarily into the trunk wall (see p. 298 ff).

A Trunk wall muscles in the strict sense

Intrinsic back muscles

Lateral tract
- Sacrospinal system
  - M. iliocostalis
  - M. longissimus
- Spinotransverse system
  - M. splenius
- Intertransverse system
  - Mm. intertransversarii
  - Mm. levatores costarum

Medial tract
- Spinal system
  - Mm. interspinales
  - M. spinalis
- Transversospinal system
  - Mm. rotatores breves and longi
  - M. multifidus
  - M. semispinalis

Short nuchal and craniovertebral joint muscles (Mm. capitis and suboccipitales)
- M. rectus capitis posterior major
- M. rectus capitis posterior minor
- M. obliquus capitis superior
- M. obliquus capitis inferior

Prevertebral neck muscles (belong topographically to the group of deep neck muscles but act mainly on the cervical spine)

- M. longus capitis
- M. longus colli
- M. rectus capitis lateralis
- M. rectus capitis anterior

Muscles of the thoracic cage and wall

- Mm. intercostales
- M. transversus thoracis
- Mm. subcostales
- Mm. scaleni (belong topographically to the group of deep neck muscles but are functionally related to thoracic breathing)

Muscles of the abdominal wall

Anterolateral abdominal muscles
- M. obliquus externus abdominis
- M. obliquus internus abdominis
- M. transversus abdominis

Anterior (straight or strap) abdominal muscles
- M. rectus abdominis
- M. pyramidalis

Posterior (deep) abdominal muscles
- M. quadratus lumborum
- M. psoas major (belongs functionally to the hip muscles, see p. 476)

B Trunk wall muscles in the broad sense

Muscles of the pelvic floor and perineum

Diaphragma pelvis
- M. levator ani
  - M. puborectalis
  - M. pubococcygeus
  - M. iliococcygeus
- M. coccygeus

Deep perineal muscles
- M. transversus perinei profundus (in male)
- M. urethrae externus
- M. compressor urethrae (in female)
- M. sphincter urethrae (in female)

Superficial perineal muscles
- M. bulbospongiosus
- M. ischiocavernosus
- M. transversus perinei superficialis
- M. sphincter ani externus

Diaphragma

- Pars costalis
- Pars lumbalis
- Pars sternalis

C Muscles that migrated secondarily to the trunk wall
(described in the Upper Limb unit, p. 238)

Spinocostal muscles (see also p. 164)

- M. serratus posterior superior
- M. serratus posterior inferior

Spinohumeral muscles between the trunk and shoulder girdle

- M. rhomboideus major and minor
- M. levator scapulae
- M. serratus anterior
- M. subclavius
- M. pectoralis minor
- M. trapezius

Spinohumeral muscles between the trunk and arm

- M. latissimus dorsi

Thoracohumeral muscles

- M. pectoralis major

* The mm. suboccipitales in the strict sense are the short or deep nuchal muscles that are counted among the intrinsic back muscles (because they are innervated by a r. dorsalis). The mm. recti capitis anterior and lateralis are *not* classified as intrinsic back muscles because they are innervated by rr. ventrales, even though they are also suboccipital in their location.

The Trunk Wall — 10. Musculature: Functional Groups

Origin of the Trunk Wall Muscles
The striated muscles of the trunk wall (including the muscles of the diaphragma and pelvic floor), like the limb muscles, develop embryologically from the myotomes of the somites (see p. 6) and are therefore called the *somatic muscles*. In all, approximately 42 to 44 pairs of segmental somites are formed in the paraxial mesoderm between the 20th and 30th day of development. Five occipital, 7 cervical, 12 thoracic, 5 lumbar, 5 sacral, and 8 to 10 coccygeal somites are formed in a craniocaudal sequence (see **D**). Some of these somites regress with further development, particularly the first occipital somites and most of the coccygeal somites, so that the number of original somites is greater than the number of subsequent vertebral segments. The boundary between the head and neck runs through the fifth pair of occipital somites. At the end of the sixth week of development, the somite myotomes migrate in the dorsoventral direction and become separated into a dorsal part (epimere or epaxial muscles) and a ventral part (hypomere or hypaxial muscles) (**E**). While the epaxial muscles develop into intrinsic (local) back muscles and retain their original location, the hypomere develops into the anterolateral muscles of the chest and abdominal wall and the limb musculature (**F**). As the myotomes become segregated, the nervi spinales undergo a corresponding division into a r. dorsalis for the epaxial muscles and a r. ventralis for the hypaxial muscles (see **Ea**). The original segmental (metameric) arrangement of the trunk muscles mostly disappears with further development. It persists only in the deep layers of the intrinsic back muscles (e.g., the mm. rotatores, interspinales, and intertransversarii) and the thoracic muscles (e.g., the mm. intercostales interni and externi), while the superficial portions of the myotomes fuse together to form long, continuous muscles ("polymerization") in which only the neurovascular supply still exhibits the original segmental arrangement (see **F**).

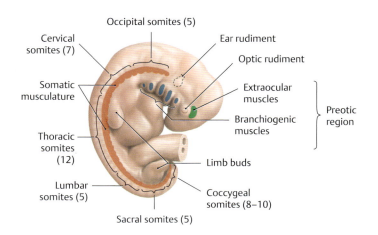

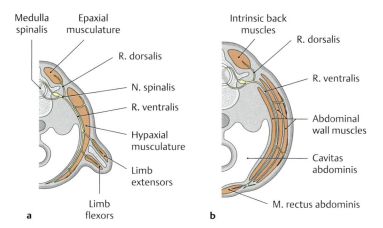

D Somites in a 5-week-old human embryo
Right lateral view. The somites formed from the paraxial mesoderm are classified as *preotic* (shown in blue and green) or *postotic* (shown in red), meaning that they are located cranial or caudal to the ear rudiment. The somatic muscles develop from postotic somites. Segmentation into different somites is not observed in the preotic region, which contains the rudiments for the branchiogenic pharyngeal arch muscles and the extraocular muscles. The structures in this region are innervated by nervi craniales (after Boyd, Hamilton, and Mossmann, quoted in Starck).

E Transverse sections through a 6-week-old human embryo
a Transverse section at the level of a limb bud.
b Transverse section through the abdominal wall.

Muscle precursor cells with replicative capacity located at the level of the limb buds migrate from the myotomes into the limb buds, and the myotome that remains in these regions develops into intrinsic back muscles. With further growth of the limb bud, the muscle tissue differentiates into a dorsal rudiment (blastema) for the extensor muscles and a ventral rudiment for the flexor muscles of the upper and lower limbs (see p. 20). As in the hypomere, the limb muscles are innervated by r. anterior nervi spinalis (plexus brachialis and plexus lumbosacralis, see pp. 360 and 530).
Note the different innervation of the epaxial muscles (r. dorsalis) and hypaxial muscles (r. ventralis).

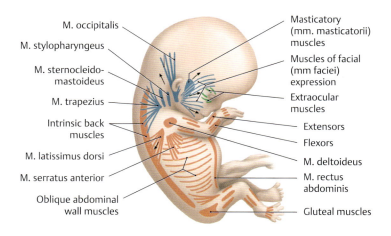

F Diagram of the principal muscle groups in an 8-week-old human embryo
Right lateral view. Red = somatic muscles, blue = branchiogenic (branchial arch) muscles, green = extraocular muscles (after Boyd, Hamilton, and Mossmann, quoted in Starck).

10.2 The Intrinsic Back Muscles (M. ercetor spinae): Lateral Tract

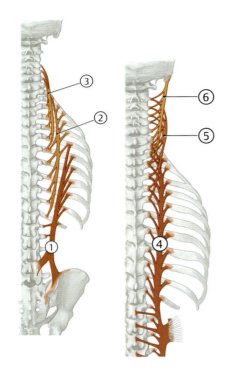

A Lateral tract of the intrinsic back muscles (m. erector spinae): schematic of the sacrospinal system
a M. iliocostalis.
b M. longissimus.

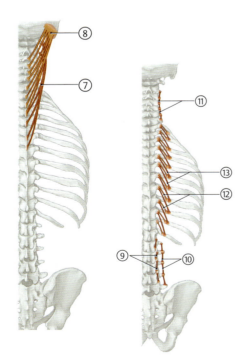

B Lateral tract of the intrinsic back muscles (m. erector spinae): schematic of the spinotransverse and intertransverse systems
a M. splenius.
b Mm. intertransversarii and levatores costarum.

M. iliocostalis (*see p. 147, opposite, lower right)

Origin:
① M. iliocostalis lumborum: os sacrum, crista iliaca, fascia thoracolumbalis
② M. iliocostalis thoracis: 7th–12th ribs
③ M. iliocostalis cervicis: 3rd–7th ribs

Insertion:
- M. iliocostalis lumborum: 6th–12th ribs, deep layer of fascia thoracolumbalis, transverse processes of upper lumbar vertebrae
- M. iliocostalis thoracis: 1st–6th ribs
- M. iliocostalis cervicis: transverse processes of C IV–C VI vertebrae

Action: Entire muscle: bilateral contraction extends the spine, unilateral contraction bends the spine laterally to the same side

Innervation: Lateral branches of rr. dorsales of nervi spinales C8–L1

M. longissimus

Origin:
④ M. longissimus thoracis: os sacrum, crista iliaca (common tendon of origin with m. iliocastalis), spinous processes of lumbar vertebrae, transverse processes of lower thoracic vertebrae
⑤ M. longissimus cervicis: transverse processes of T I–T VI vertebrae
⑥ M. longissimus capitis: transverse processes of T I–T III vertebrae and transverse and articular processes of C IV–C VII vertebrae

Insertion:
- M. longissimus thoracis: 2nd–12th ribs, costal processes of lumbar vertebrae, transverse processes of thoracic vertebrae
- M. longissimus cervicis: transverse processes of C II–C V vertebrae
- M. longissimus capitis: proc. mastoideus

Action:
- Entire muscle: bilateral contraction extends the spine, unilateral contraction bends the spine laterally to the same side
- M. longissimus capitis: bilateral contraction extends the head, unilateral contraction flexes and rotates the head to the same side

Innervation: Lateral branches of rr. dorsales of nervi spinales C1–L5

M. splenius

Origin:
⑦ M. splenius cervicis: spinous processes of the T III–T VI vertebrae
⑧ M. splenius capitis: spinous processes of the C III–T III vertebrae

Insertion:
- M. splenius cervicis: transverse processes of C I and C II
- M. splenius capitis: linea nuchalis superior, proc. mastoideus

Action: Entire muscle: bilateral contraction extends the cervical spine and head, unilateral contraction flexes and rotates the head to the same side

Innervation: Lateral branches of rr. dorsales of nervi spinales C1–C6

M. intertransversarii

Origin and insertion:
⑨ Mm. intertransversarii mediales lumborum: course between adjacent procc. mamillares of all lumbar vertebrae
⑩ Mm. intertransversarii laterales lumborum: course between adjacent procc. costales of all lumbar vertebrae
⑪ Mm. intertransversarii posteriores cervicis: course between adjacent tubercula posteriora of the C II–C VII vertebrae
- Mm. intertransversarii anteriores cervicis: course between adjacent tubercula anteriora of the C II–C VII vertebrae

Action:
- Bilateral contraction stabilizes and extends the cervical and lumbar spine
- Unilateral contraction bends the cervical and lumbar spine laterally to the same side

Innervation: Rr. dorsales of the nervi spinales except for mm. intertransversarii anteriores cervicis (rr. ventrales of the nervi spinales)

Mm. levatores costarum

Origin:
⑫ Mm. levatores costarum breves: transverse processes of the C VII and TI–T XI vertebrae

Insertion:
⑬ Mm. levatores costarum longi: transverse processes of the C VII and TI–T XI vertebrae

Action:
- Mm. levatores costarum breves: angulus costae of the next lower rib
- Mm. levatores costarum longi: angulus costae of second lower rib
- Bilateral contraction extends the thoracic spine
- Unilateral contraction bends the thoracic spine to the same side, rotates it to the opposite side

Innervation: Rr. dorsales of the nervi spinales

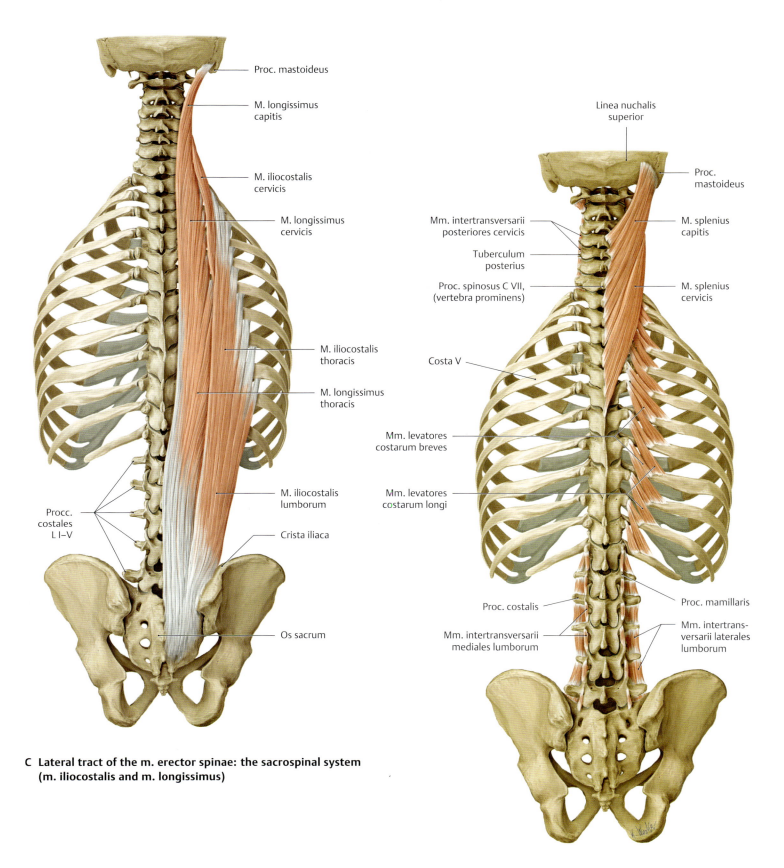

C Lateral tract of the m. erector spinae: the sacrospinal system (m. iliocostalis and m. longissimus)

* The schematic muscle diagrams with accompanying tables on p. 146 are intended to give a systematic overview of the muscles and their actions. The drawings on the right-hand page are intended to display the muscles as they would appear in a dissection. Not all the structures listed in the tables on p. 146 are labeled in the drawings above, as they are not all visible in the views shown.

D Lateral tract of m. erector spinae: the spinotransverse system (m. splenius) and intertransverse system (mm. intertransversarii and mm. levatores costarum))

10.3 The Intrinsic Back Muscles (M. ercetor spinae): Medial Tract

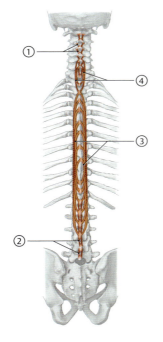

A Medial tract of the m. erector spinae: schematic of the spinal system
Mm. interspinales and m. spinalis.

Mm. interspinales

Origin and insertion:	① Mm. interspinales cervicis: course between the spinous processes of the cervical vertebrae
	② Mm. interspinales lumborum: course between the spinous processes of the lumbar vertebrae
Action:	Extends the cervical and lumbar spine
Innervation:	Rr. dorsales of the nervi spinales

M. spinalis

Origin:	③ M. spinalis thoracis: lateral surface of the spinous processes of the T X–T XII and L I–L III vertebrae
	④ M. spinalis cervicis: spinous processes of the C V–C VII and T I–T II vertebrae
Insertion:	• M. spinalis thoracis: lateral surface of the spinous processes of the T II–T VIII vertebrae
	• M. spinalis cervicis: spinous processes of the C II–C V vertebrae
Action:	• Bilateral contraction extends the cervical and thoracic spine
	• Unilateral contraction bends the cervical and thoracic spine to the same side
Innervation:	Rr. dorsales of the nervi spinales

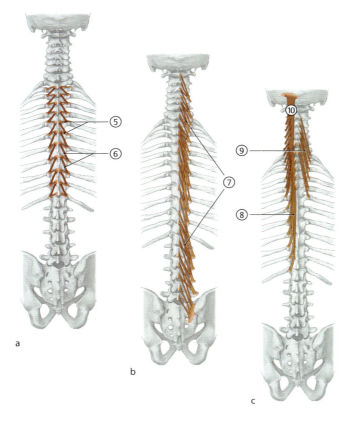

B Medial tract of the m. erector spinae: schematic of the transversospinal system
a Mm. rotatores breves and longi.
b M. multifidus.
c M. semispinalis.

Mm. rotatores breves and longi

Origin and insertion:	⑤ Mm. rotatores breves: from proc. transversus of thoracic vertebrae to proc. spinosus of next higher vertebra
	⑥ Mm. rotatores longi: from proc transversus of thoracic vertebrae to proc. spinosus of the vertebra two levels higher
Action:	• Bilateral contraction extends the thoracic spine
	• Unilateral contraction rotates it to the opposite side
Innervation:	Rr. dorsales of the nervi spinales

⑦ M. multifidus (most fully developed in lumbar spine)

Origin:	Os sacrum, os ilium, proc. mamillares of L–LV, transverse and articular processes of T I–T IV and, C IV–C VII
Insertion:	Superomedially to proc. spinosus of vertebra two to four levels higher
Action:	• Bilateral contraction extends the spine
	• Unilateral contraction flexes to the same side and rotates to the opposite side
Innervation:	Rr. dorsales of the nervi spinales

M. semispinalis

Origin:	⑧ M. semispinalis thoracis: transverse processes of the T VI–T XII vertebrae
	⑨ M. semispinalis cervicis: procc. transversi of the T I–T VI vertebrae
	⑩ M. semispinalis capitis: procc. transversi of the C III–T VI vertebrae
Insertion:	• M. semispinalis thoracis: proc. spinosi of the C VI –T IV vertebrae
	• M. semispinalis cervicis: proc. spinosi of the C II–C V vertebrae
	• M. semispinalis capitis: os occipitale between the linea nuchalis superior and linea nuchalis inferior
Action:	• Bilateral contraction extends the thoracic spine, cervical spine, and head (stabilizes the craniovertebral joints)
	• Unilateral contraction bends the head, cervical spine, and thoracic spine to the same side and rotates them to the opposite side
Innervation:	Rr. dorsales of the nervi spinales

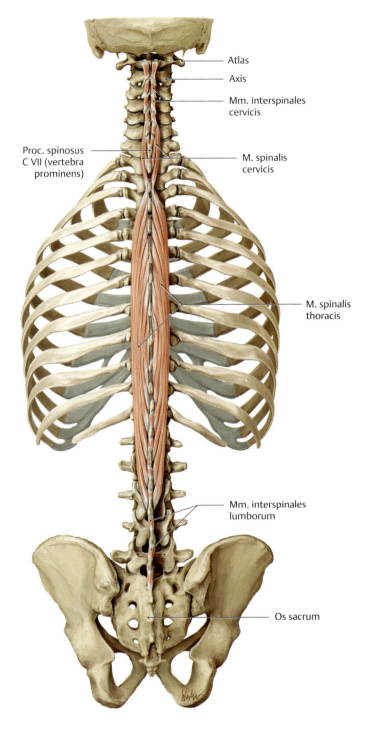

C Medial tract of the m. erector spinae: the spinal system (mm. interspinales and m. spinalis)

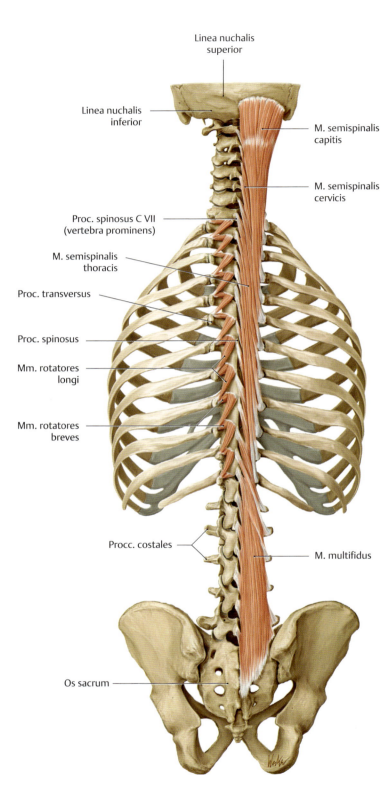

D Medial tract of the m. erector spinae: the transversospinal system (mm. rotatores breves and longi, m. multifidus and m. semispinalis)

10.4 The Intrinsic Back Muscles (Short Nuchal and Craniovertebral Joint Muscles) and the Prevertebral Muscles

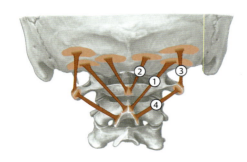

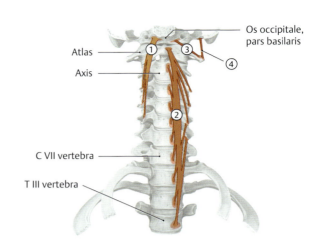

A Schematic of the short nuchal and craniovertebral joint muscles (mm. suboccipitales): mm. recti capitis posterior major and minor and mm. obliquii capitis superior and inferior
Posterior view.

B Schematic of the prevertebral neck muscles (mm. colli and cervicis): mm. longi capitis and colli and mm. recti capitis anterior and lateralis
Anterior view.

① M. rectus capitis posterior major

- **Origin:** Proc. spinosus of the axis
- **Insertion:** Middle third of the linea nuchalis inferior
- **Action:**
  - Bilateral: extends the head
  - Unilateral: rotates the head to the same side
- **Innervation:** R. dorsalis of C1 (N. suboccipitalis)

② M. rectus capitis posterior minor

- **Origin:** Tuberculum posterior of the atlas
- **Insertion:** Inner third of the linea nuchalis inferior
- **Action:**
  - Bilateral: extends the head
  - Unilateral: rotates the head to the same side
- **Innervation:** R. dorsalis of C1 (N. suboccipitalis)

③ M. obliquus capitis superior

- **Origin:** Proc. transversus of the atlas
- **Insertion:** Above the insertion of the m. rectus capitis posterior major
- **Action:**
  - Bilateral: extends the head
  - Unilateral: tilts the head to the same side and rotates it to the opposite side
- **Innervation:** R. dorsalis of C1 (N. suboccipitalis)

④ M. obliquus capitis inferior

- **Origin:** Proc. spinosus of the axis
- **Insertion:** Proc. transversus of the atlas
- **Action:**
  - Bilateral: extends the head
  - Unilateral: rotates the head to the same side
- **Innervation:** R. dorsalis of C1 (N. suboccipitalis)

Note: The prevertebral muscles are not included among the intrinsic back muscles as they are innervated by the anterior rami of the nervi spinales.

① M. longus capitis

- **Origin:** Tubercula anteriora of the proc. transversus of the C III–C VI vertebrae
- **Insertion:** Pars basilaris of the os occipitale
- **Action:**
  - Unilateral: tilts and slightly rotates the head to the same side
  - Bilateral: flexes the head
- **Innervation:** Direct branches from the plexus cervicalis (C1–C4)

② M. longus colli (cervicis)

- **Origin:**
  - Pars recta (medial part): anterior sides of the C V–C VII and T I–T III vertebral bodies
  - Pars obliqua superior: tubercula anteriora of the processus transversus of the C III–C V vertebrae
  - Pars obliqua inferior: anterior sides of the T I–T III vertebral bodies
- **Insertion:**
  - Pars recta: anterior sides of the C II–C IV vertebrae
  - Pars obliqua superior: tuberculum anterius of the atlas
  - Pars obliqua inferior: tubercula anteriora of the processus transversus of the C V and C VI vertebrae
- **Action:**
  - Unilateral: tilts and rotates the cervical spine to the same side
  - Bilateral: flexes the cervical spine
- **Innervation:** Direct branches from the plexus cervicalis (C2–C6)

③ M. rectus capitis anterior

- **Origin:** Massa lateralis of the atlas
- **Insertion:** Pars basilaris of the Os occipitale
- **Action:**
  - Unilateral: lateral flexion of art. atlantooccipitalis
  - Bilateral: flexion at the atlantooccipital joint
- **Innervation:** R. ventralis of the C1 nerve

④ M. rectus capitis lateralis

- **Origin:** Proc. transversus of the atlas
- **Insertion:** Pars basilaris of the os occipitale (lateral to the condyli occipitales)
- **Action:**
  - Unilateral: lateral flexion of art. atlantooccipitalis
  - Bilateral: flexion of art. atlantooccipitalis
- **Innervation:** R. ventralis of the C1 nerve

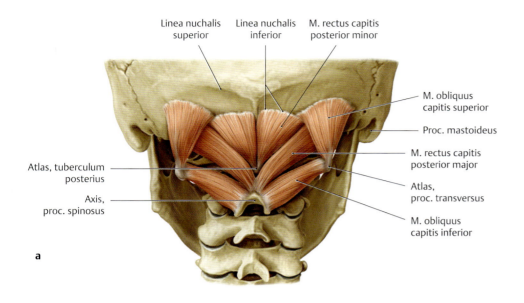

C The short nuchal and craniovertebral joint muscles: mm. recti capitis posterior and mm. obliquii capitis
a Posterior view, b lateral view.
In a strict sense, the short nuchal muscles consist only of the muscles innervated by the R. dorsalis of the first nervus spinalis (n. suboccipitalis). They include representatives of the lateral tract (m. obliquus capitis inferior) and the medial tract (m. obliquus capitis superior and mm. recti capitis posterior major and minor). The anterior group of short nuchal muscles (mm. recti capitis lateralis and anterior) is innervated by the rr. ventrales, placing them among the prevertebral neck muscles (see **D**).

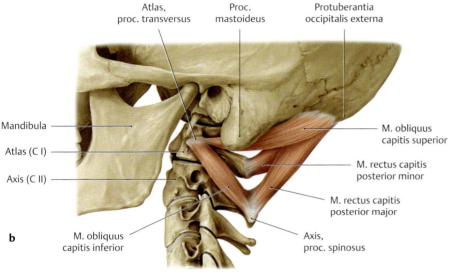

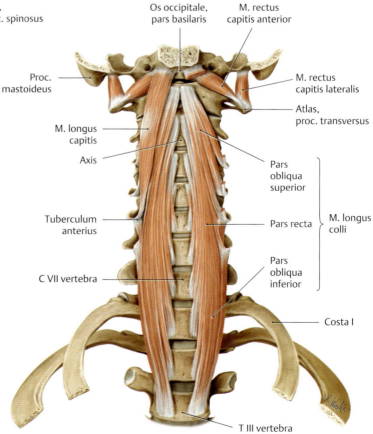

D The prevertebral neck muscles: m. longus capitis, m. longus colli, and mm. recti capitis anterior and lateralis
Anterior view after removal of the cervical viscera. M. longus capitis has been partially removed on the left side.

10.5 The Muscles of the Abdominal Wall: Anterolateral Muscles

A Schematic of m. obliquus externus abdominis

M. obliquus externus abdominis

Origin:	Outer surface of the 5th–12th ribs
Insertion:	• Labium externum of the crista iliaca
	• Anterior layer of the rectus sheath, linea alba
Action:	• Unilateral: bends the trunk to the same side, rotates the trunk to the opposite side
	• Bilateral: flexes the trunk, stabilizes the pelvis, active in expiration, compresses the abdomen
Innervation:	Nn. intercostales (T 5–T 11), n. subcostalis (T 12), n. iliohypogastricus

B Schematic of m. obliquus internus abdominis

M. obliquus internus abdominis

Origin:	Deep layer of the fascia thoracolumbalis, linea intermedia of the crista iliaca, spina iliaca anterior superior, lateral half of the lig. inguinale
Insertion:	• Lower borders of the 10th–12th ribs
	• Anterior and posterior layers of the rectus sheath (vagina musculi recti abdominis), linea alba
Action:	• Junction with the m. cremaster
	• Unilateral: bends the trunk to the same side, rotates the trunk to the same side
	• Bilateral: flexes the trunk, stabilizes the pelvis, active in expiration, compresses the abdomen
Innervation:	• Nn. intercostales (C 8–T 11), n. subcostalis (T 12), n. iliohypogastricus,
	• N. ilioinguinalis
	• M. cremaster (r. genitalis of n. genitofemoralis)

C Schematic of m. transversus abdominis

M. transversus abdominis

Origin:	• Inner surfaces of the 7th–12th costal cartilages
	• Deep layer of the fascia thoracolumbalis
	• Labium internum of the crista iliaca, spina iliaca anterior superior
	• Lateral part of the lig. inguinale
Insertion:	• Posterior layer of the rectus sheath, linea alba
Action:	• Unilateral: rotates the trunk to the same side
	• Bilateral: active in expiration, compresses the abdomen
Innervation:	Nn. intercostales (T 5–T 11), n. subcostalis (T 12), and nn. iliohypogastricus, ilioinguinalis, and genitofemoralis

The Trunk Wall ——— *10. Musculature: Functional Groups*

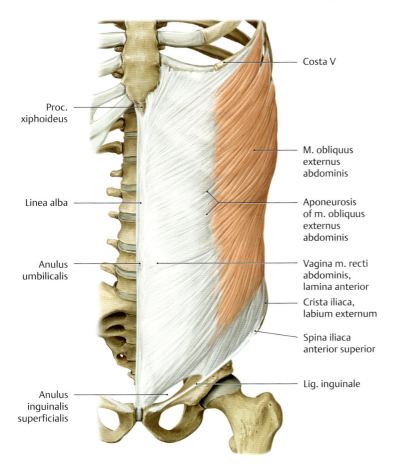

D M. obliquus externus abdominis
Left side, anterior view.

E M. obliquus internus abdominis
Left side, anterior view.

F M. transversus abdominis
Left side, anterior view.
(For structure of the rectus sheath, see p.179.)

153

10.6 The Muscles of the Abdominal Wall: Anterior and Posterior Muscles

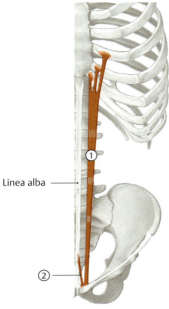

① **M. rectus abdominis**

Origin:	Cartilages of the fifth through seventh ribs, proc. xiphoideus of the sternum
Insertion:	Pubis (between the tuberculum pubicum and symphysis pubica)
Action:	Flexes the lumbar spine, stabilizes the pelvis, active in expiration, compresses the abdomen
Innervation:	Nn. intercostales (T V –T XI), n. subcostalis (T XII)

② **M. pyramidalis**

Origin:	Pubis (anteriorly at the insertion of the m. rectus abdominis)
Insertion:	Linea alba (runs within the rectus sheath)
Action:	Tenses the Linea alba
Innervation:	N. subcostalis (T XII)

A Schematic of the anterior (straight, or strap) muscles of the abdominal wall: mm. rectus abdominis and pyramidalis

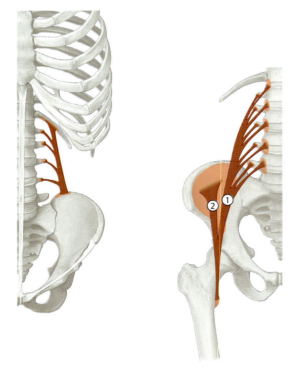

M. quadratus lumborum

Origin:	Crista iliaca
Insertion:	Twelfth rib, costal processes of the L I – L IV vertebrae
Action:	• Unilateral: bends the trunk to the same side • Bilateral: bearing down and expiration
Innervation:	N. subcostalis (n. intercostalis XII)

M. iliopsoas (① m. psoas major and ② m. iliacus)*

Origin:	• M. psoas major (superficial layer): lateral surfaces of the T XII corpus vertebrae, the L I–L IV vertebral bodies, and the associated disci intervertebrales • M. psoas major (deep layer): procc. costarii of the L I–L V vertebrae • M. iliacus: fossa iliaca
Insertion:	Insert jointly as the M. iliopsoas on the trochanter minor of the os femoris
Action:	• Articulatio coxae: flexion and external rotation • Lumbar spine: unilateral contraction (punctum fixum on the femur) bends the trunk laterally, bilateral contraction raises the trunk from the supine position
Innervation:	Direct branches from the nervi spinales (L1–L4) and n. femoralis (L2–L4).

* Of these two muscles, only the M. psoas major belongs topographically to the posterior abdominal muscles. It is classified functionally as a hip muscle (see p. 476).

B Schematic of the posterior (deep) muscles of the abdominal wall: mm. quadratus lumborum and psoas major

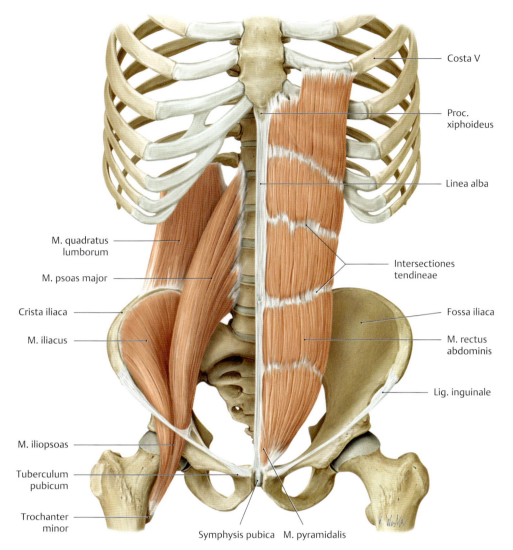

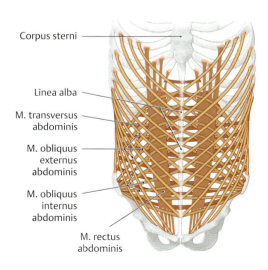
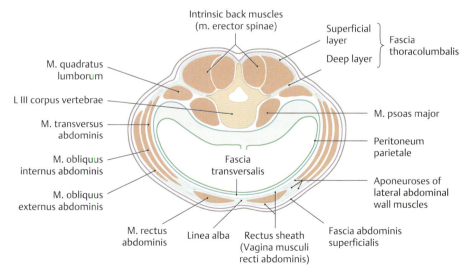

C Anterior (mm. rectus abdominis and pyramidalis) and posterior muscles of the abdominal wall (mm. quadratus lumborum and iliopsoas)
Anterior view. The anterior abdominal wall muscles are shown on the left side and the posterior muscles on the right side.

D Arrangement of the muscles of the abdominal wall and rectus sheath
a Anterior view, b transverse section at the level of L III.
The m. rectus abdominis and lateral oblique abdominal wall muscles and their aponeuroses comprise a functional unit. Fusion of the aponeuroses of the oblique muscles creates a sheath enclosing the rectus muscles that in turn meets the aponeurosis from the opposite side to form the linea alba in the abdominal midline. For the upper three fourths of the length of the rectus muscles, the aponeurosis of the m. obliquus internus splits, so portions pass both anterior and posterior to the rectus muscles. Therefore, in the upper region of the rectus, an anterior rectus sheath (aponeurosis m. obliqui externi abdominis and a portion of the aponeurosis m. obliqui interni abdominis) and a posterior rectus sheath (the rest of the aponeurosis m. obliqui interni abdominis and aponeurosis m. transversi abdominis) are formed (see p. 179). Over the lower one fourth of the rectus muscles, the aponeuroses of all three abdominal muscles pass anterior to the rectus muscles. No posterior sheath is present. The inferior edge of the posterior rectus sheath is called the linea arcuata.

10.7 The Functions of the Abdominal Wall Muscles

Functions of the Abdominal Wall Muscles

The different abdominal wall muscles perform numerous functions, which very often are carried out in concert with other muscle groups (e.g., the back and gluteal muscles and the diaphragma). The principal actions of the abdominal wall muscles are as follows:

- Maintenance of abdominal tone: tensing the abdominal wall and compressing the abdominal viscera (abdominal press)
- Stabilizing the vertebrae and reducing stresses on the spinal column
- Moving the trunk and pelvis
- Assisting in respiration

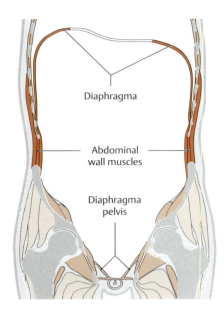

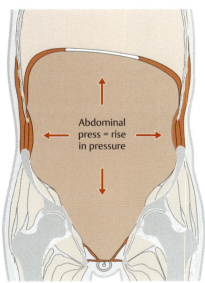

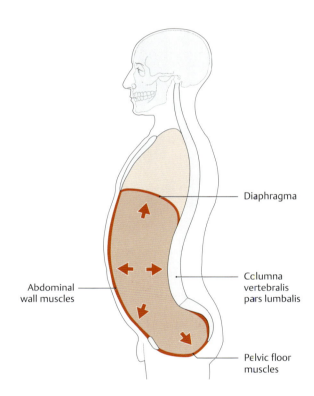

A Abdominal press = raising the intra-abdominal pressure by tensing the abdominal wall and pelvic floor muscles and the diaphragma

Schematic coronal section through the cavitas abdominis, anterior view.

a The walls of the abdominal and pelvic cavity (cavitas abdominis et pelvis) are formed by bony structures (spinal column, thoracic cage, pelvis) and also by muscles (diaphragma, muscles of the abdominal wall and pelvic floor).

b When the muscles about the abdomen contract (abdominal press), they reduce the volume of the cavitas abdominis, thereby raising the intra-abdominal pressure and actively compressing the abdominal viscera. This action is important, for example, in expelling stool from the rectum (defecation), expelling urine from the bladder (micturition), and emptying the gastric contents (vomiting). The abdominal press is also an essential part of maternal pushing during the expulsive phase of labor.

B Abdominal press = stabilizing the spinal column by raising the intra-abdominal pressure

Schematic midsagittal section through the trunk, left lateral view. Simultaneous contraction of the diaphragma and the muscles of the abdominal wall and pelvic floor raises the pressure in the cavitas abdominis (abdominal press). The hydrostatic effect of this maneuver stabilizes the trunk, reduces stresses on the spinal column (especially at the lumbar level), and stiffens the trunk wall like the wall of an inflated ball. This action is performed automatically during the lifting of heavy loads. The "inflatable space" of the trunk can be employed in this way to lighten the pressure load on the intervertebral disks by up to 50% in the upper lumbar spine and by approximately 30% in the lower lumbar spine. Meanwhile, the forces exerted by the intrinsic back muscles are reduced by more than 50%. This explains the importance of well-conditioned abdominal muscles in preventing and treating diseases of the spinal column.

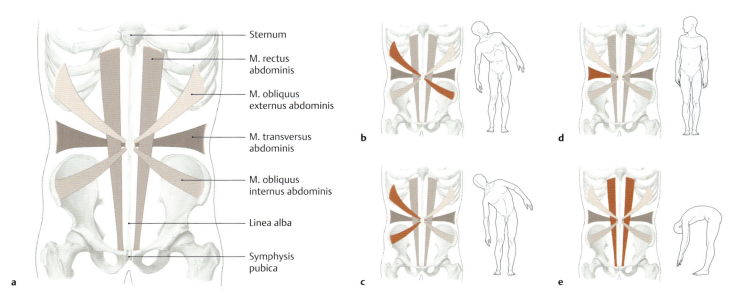

C Trunk movements involving the straight and oblique abdominal wall muscles

a Course and arrangement of the straight and oblique abdominal wall muscles.
b Lateral flexion to the right side with simultaneous rotation of the trunk to the left side by contraction of the m. obliquus externus abdominis on the right side and m. obliquus internus abdominis on the left side.
c Lateral flexion to the right side by contraction of the right m. obliquii externus and internus (aided by the right quadratus lumborum).
d Rotating the trunk to the right side is effected by the right m. transversus abdominis.
e Flexion of the trunk is effected by bilateral contraction of the abdominis mm. recti.

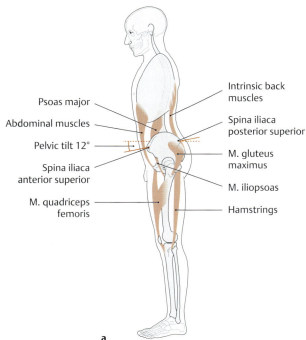  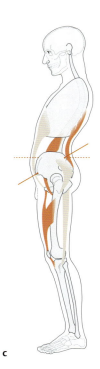

D Effect of the abdominal wall muscles on pelvic movements: active and passive posture

a Normal active posture, **b** active rigid posture, **c** passive slumped posture.

An imbalance between the intrinsic back muscles and abdominal muscles is particularly evident in the curvature of the lower spine and in the degree of pelvic tilt. In a normal active posture, the pelvis is tilted forward by approximately 12° (**a**). When a rigid posture is assumed ("stomach in, chest out"), the pelvis is held in a more upright position so that the spina iliaca anterior superior and the spina iliaca posterior superior are at the same level (**b**). The most active muscles are the abdominal wall muscles, the gluteal muscles, and the hamstrings. When the abdominal muscles are lax and are not well conditioned, the result is a passive slumped posture (**c**) with an excessive degree of anterior pelvic tilt. Also, the lordotic curvature of the lumbar spine is accentuated due to progressive shortening of the intrinsic back muscles. This posture is reinforced by the tendency of m. iliopsoas (m. psoas major and m. iliacus) to become shortened.

10.8 The Muscles of the Thoracic Cage (Mm. intercostales, subcostales, scaleni, and M. transversus thoracis)

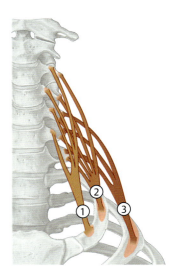

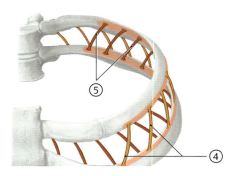

A Schematic of the thoracic wall muscles:
Anterior view.
(**a**) mm. scaleni and (**b**) mm. intercostales

Mm. scaleni

Origin:	① M. scalenus anterior: tubercula anteriora of the proc. transversus of the C III–C VI vertebrae
	② M. scalenus medius: tubercula posteriora of the proc. transversus of the C III–C VII vertebrae
	③ M. scalenus posterior: tubercula posteriora of the proc. transversus of the C V–C VII vertebrae
Insertion:	• M. scalenus anterior: tuberculum musculi scaleni anterioris of first rib
	• M. scalenus medius: first rib (posterior to the sulcus arteriae subclaviae)
	• M. scalenus posterior: outer surface of the second rib
Action:	• With the ribs mobile: raises the upper ribs (in inspiration)
	• With the ribs fixed: bends the cervical spine toward the same side (with unilateral contraction) and flexes the neck (with bilateral contraction)
Innervation:	• M. scalenus anterior: anterior rami of C4–C6
	• M. scalenus medius: anterior rami of C3–C8
	• M. scalenus posterior: anterior rami of C6–C8

Mm. intercostales

Origin and insertion:	④ Mm. intercostales externi (tuberculum costae to chondro-osseous junction): arise at the lower margin of a rib and insert on the upper margin of the next lower rib (course obliquely forward and downward)
	⑤ Mm. intercostales interni (angulus costae to sternum): arise at the lower margin of a rib and insert on the upper margin of the next lower rib (course obliquely backward and downward)
Action:	• Mm. intercostales intimi: division of the mm. intercostales interni (same course and action)
	• Mm. intercostales externi: elevate (raise) the ribs (in inspiration), support the intercostal spaces (spatium intercostale), stabilize the chest wall
	• Mm. intercostales interni and intimi: depress (lower) the ribs (in expiration), support the intercostal spaces (spatium intercostale), stabilize the chest wall
Innervation:	First through eleventh nn. intercostales I–XI

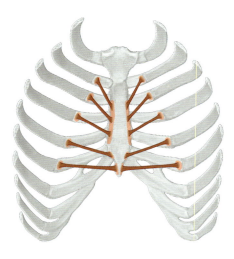

B Schematic of the thoracic wall muscles: m. transversus thoracis
Posterior view.

Mm. transversus thoracis

Origin:	Inner surface of the corpus sterni and the proc. xiphoideus sterni
Insertion:	Inner surface of the cartilago costalis of the second through sixth ribs
Action:	Lowers the ribs (in expiration)
Innervation:	Nn. intercostales II–VII

The Trunk Wall — 10. Musculature: Functional Groups

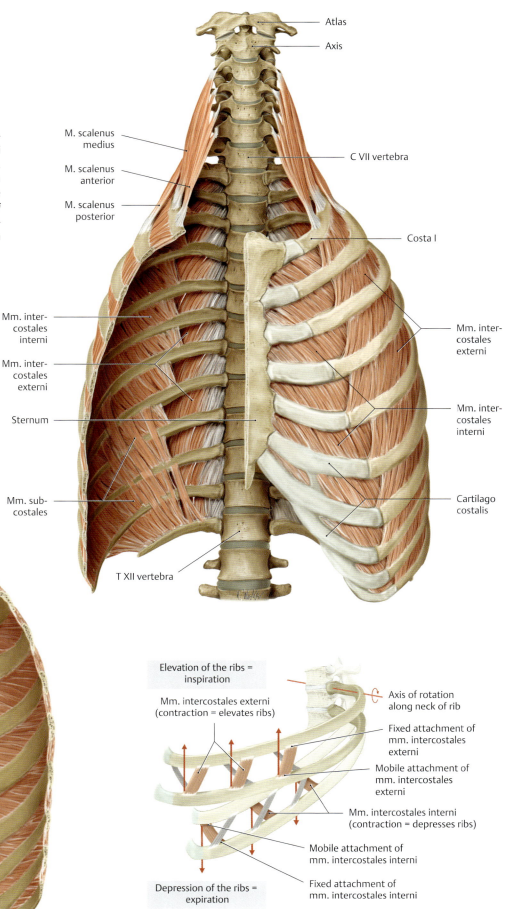

C Mm. scaleni anterior, medius, and posterior and the mm. intercostales interni and externi
Anterior view with anterior thoracic cage partially removed. Topographically, mm. scaleni are included among the deep neck muscles, but functionally they play an important role in thoracic breathing. Mm. subcostales have the same orientation as mm. intercostales *interni* but skip over one or two ribs to form continuous sheets, especially at the angle of the sixth through eleventh ribs.

D M. transversus thoracis
Posterior surface of the anterior thoracic cage removed in **C** (right half of m. transversus thoracis).

E Action of the mm. intercostales interni and externi
Note the position of the axis of motion along the neck of the rib.

159

10.9 The Muscles of the Thoracic Cage: The Diaphragma

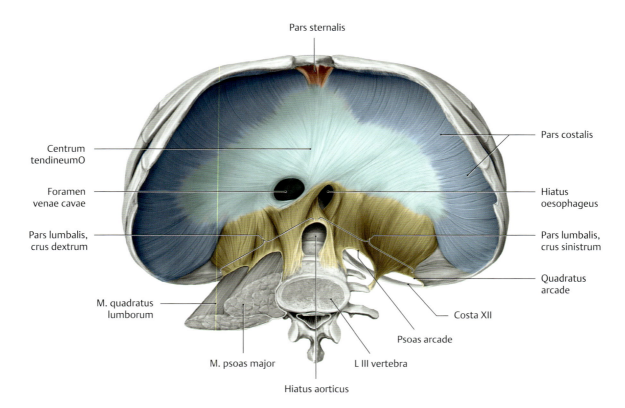

A Overview of the diaphragma
Superior view.

Origin:	• Pars costalis: lower margin of costal arch (inner surface of seventh through twelfth ribs)
	• Pars lumbalis (crus dextrum and crus sinistrum):
	  – Medial parts: L I–L III vertebral bodies, second and third intervertebral disks, lig. longitudinale anterius
	  – Lateral parts: first tendinous arch of the aorta abdominalis (lig. arcuatum mediale) at L I associated with its anterior surface; second tendinous arch of the psoas arcade (lig. arcuatum mediale) from L II corpus vertebrae to associated processus costalis; third tendinous arch of the quadratus lumborum arcade (lig. arcuatum laterale) from the processus costalis of L II to the tip of the twelfth rib
	• Pars sternalis: posterior surface of the proc. xiphoideus sterni

Insertion:	Centrum tendineum
Action:	Principal muscle of respiration (diaphragmatic and thoracic breathing), aids in compressing the abdominal viscera (abdominal press)
Innervation:	N. phrenicus from the plexus cervicalis (C3–C5)

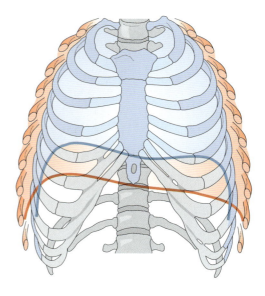

B Position of the diaphragma and ribs at full inspiration and expiration
Thoracic cage (cavea thoracis), anterior view.
Note the different positions of the diaphragma at full inspiration (red) and full expiration (blue). During a physical examination, the posterior lung boundaries can be identified by percussion (tapping the body surface). The respiratory movement of the diaphragma from end-expiration to end-inspiration should be determined; it is approximately 4 to 6 cm (see p. 174f).

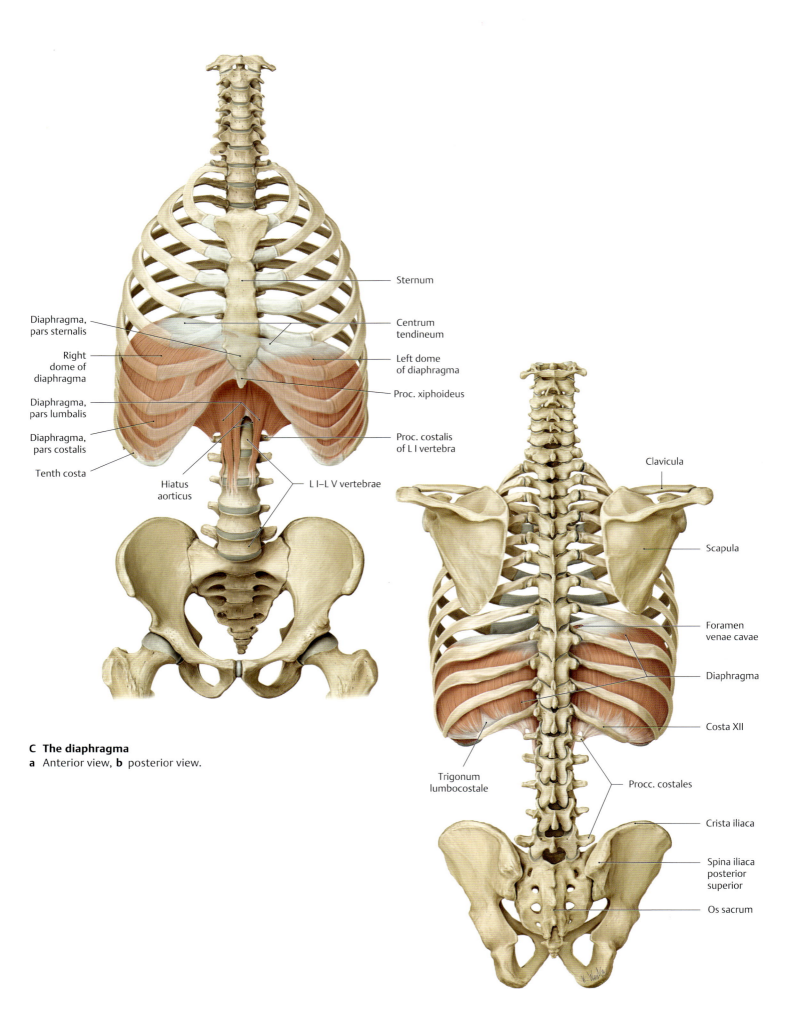

C The diaphragma
a Anterior view, b posterior view.

10.10 The Muscles of the Pelvic Floor and Perineum: Diaphragma pelvis and Deep and Superficial Perineal Muscles

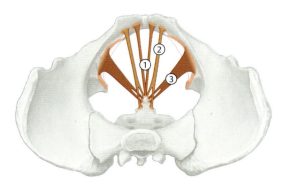

A Schematic of the diaphragma pelvis: m. levator ani (Mm. puborectalis, pubococcygeus, and iliococcygeus) and m. coccygeus (not shown)
Superior view.

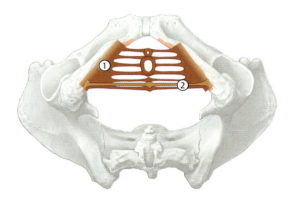

B Schematic of the mm. transversi perinei profundus and superficialis
Inferior view.

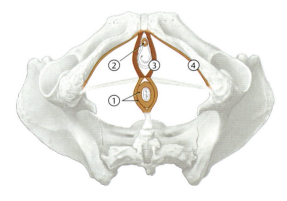

C Schematic of the mm. sphincteres ani externus, sphincter urethrae, bulbospongiosus, and ischiocavernosus
Inferior view.

③ M. bulbospongiosus
Runs anteriorly from the centrum tendineum to the clitoris in females or to the raphe penis in males
Action: Narrows the vaginal introitus in females, surrounds the corpus spongiosum in males
Innervation: N. pudendus (S2–S4)

M. levator ani

① M. puborectalis
Origin: R. superior ossis pubis on both sides of the symphysis pubica
Insertion: Loop-shaped around the anorectal junction, interwoven with the pars profunda of m. sphincter ani externus
Innervation: N. pudendus and direct branches of plexus sacralis (S2–S4)

② M. pubococcygeus
Origin: Os. pubis (lateral at origin of the m. puborectalis)
Insertion: Lig. anococcygeum, os coccygis
Innervation: N. pudendus (S2–S4)

③ M. iliococcygeus
Origin: Tendinous arch of the internal obturator fascia (arcus tendineus m. levatoris ani)
Insertion: Levator (iliococcygeal) raphe, lig. anococcygeum, os coccygis
Innervation: N. pudendus (S2–S4)
Function of the pelvic diaphragm: Holds the pelvic organs in place

M. coccygeus (not shown)
Origin: Inferior end of os sacrum
Insertion: Spina ischiadica
Innervation: Direct branches of plexus sacralis (S4, S5)
Function: Holds pelvic organs in place, flexes the os coccygis

① M. transversus perinei profundus (in males; smooth muscle, in females–see note p. 163)
Origin: R. inferior ossis pubis, r. ossis ischii
Insertion: Wall of vagina or prostata, urethra
Innervation: N. pudendus (S2–S4)

② M. transversus perinei superficialis
Origin: R. ossis ischii
Insertion: Centrum tendineum
Innervation: N. pudendus (S2–S4)
Function of the transverse perineal muscles: Holds the pelvic organs in place, closes the urethra

① M. sphincter ani externus
M. sphincter ani externus is a circular muscle that closes the orifice around the canalis analis, extends from the centrum tendineum to the lig. anococcygeum, and is divided pars subcutanea, pars superficialis, and pars profunda.
Action: Closes the anus
Innervation: N. pudendus (S2–S4)

② M. sphincter urethrae externus
Division of the m. transversus perinei profundus (encircles the urethra)
Action: Closes the urethra
Innervation: N. pudendus (S2–S4)

④ M. ischiocavernosus
Origin: R. ossis ischii
Insertion: Crus penis/crus clitoridis
Action: Compresses the crura clitoridis/penis; helps to promote and maintain erection
Innervation: N. pudendus (S2–S4)

The Trunk Wall — 10. Musculature: Functional Groups

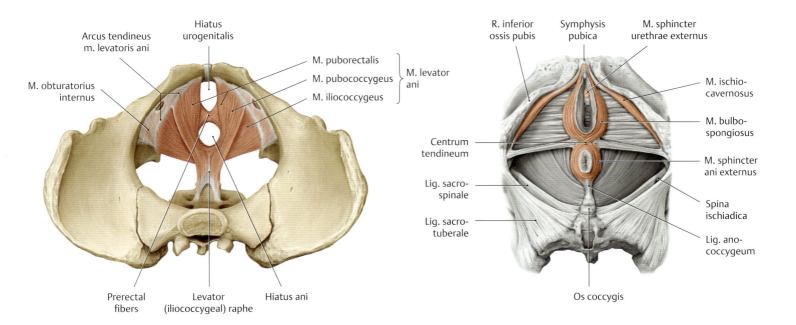

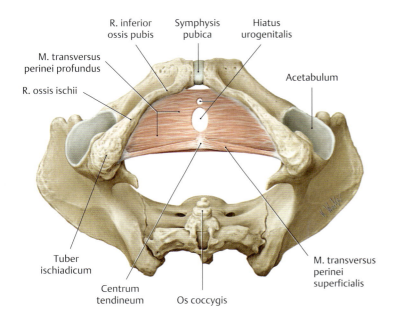

D The muscles of the female pelvic floor
a Diaphragma pelvis, superior view.
b Mm. transversi perinei, inferior view.
c Sphincter and erectile muscles, inferior view.

Note: The existence of the m. transversus perinei profundus in the female has been called into question. This is because this muscle becomes heavily permeated by connective tissue with aging and especially after vaginal deliveries. In older women, the spatium profundum perinei (see p. 183) basically contains connective tissue that completely occupies the hiatus urogenitalis at the openings of the urethra and vagina (see p. 188). We have noted this replacement by smooth muscle in each illustration of the female pelvis.

Important Functions of the Pelvic Floor Muscles
The pelvic floor performs a dual function:

- It supports the abdominal and pelvic organs by closing the cavitas abdominopelvica inferiorly, bearing the bulk of the visceral load.
- It controls the openings of the rectum and urogenital passages (sphincter functions), which mechanically weaken the pelvic floor by piercing it.

To accomplish these inherently conflicting functions (sealing off the cavitas pelvis while maintaining several apertures), the pelvic floor is lined by overlapping sheets of funnel-shaped muscle and connective tissue. This complex structure, however, also makes the pelvic floor highly susceptible to damage, especially in women.
Repetitive, extreme fluctuations in intra-abdominal pressure and other stresses, particularly at the end of pregnancy, can weaken the connective tissue apparatus and damage the pelvic floor muscles. Stretching and other injuries of the pelvic floor or the nerves supplying it, as during labor and delivery (in multiparous women), can eventually lead to pelvic floor insufficiency and its various clinical sequelae:

- Descent of the pelvic floor and thus the pelvic organs (e.g., *descensus uteri*)
- In extreme cases, prolapse of the uterus with eversion of the vagina (*prolaps uteri*)

Visceral descent is generally associated with urinary or fecal incontinence in response to coughing or other acts (*stress incontinence*). Mild degrees of descent often respond well to regular pelvic floor exercises, but more serious cases may require surgical treatment with *pelvic floor repair* (surgical exposure and approximation of the two levator crura = puborectalis muscle) (e.g., approximating the levator muscles by tightening mm. puborectales) or surgically attaching the pelvic organs (e.g., colpopexy or rectopexy) to the pelvic wall or sacrum.

10.11 Secondarily Incorporated Trunk Muscles: Spinocostal, Spinohumeral, and Thoracohumeral

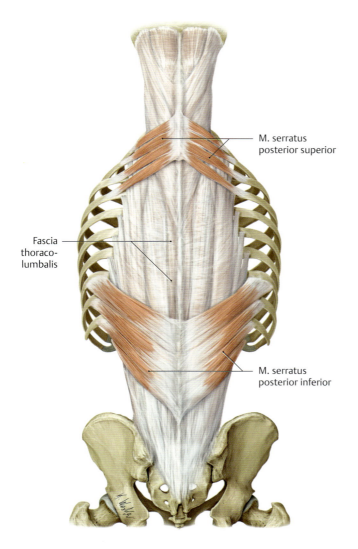

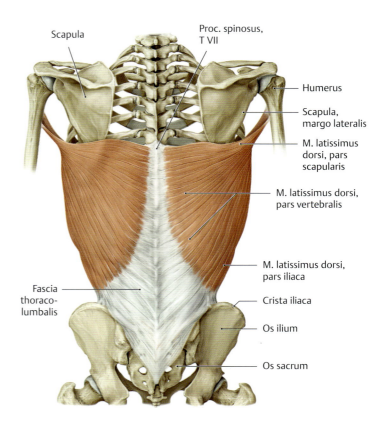

B M. latissimus dorsi
Posterior view.

A Overview of spinocostal muscles
Due to their position above the fascia thoracolumbalis and their innervation by the anterior rami of the nervi spinales, the two posterior serratus muscles are counted among the secondarily incorporated trunk muscles. These muscles, which are mostly very thin and often markedly segmented, act as auxiliary inspiratory muscles of respiration (see below).

M. serratus posterior superior	
Origin:	Lig. nuchae and procc. spinosi of the vertebrae C VII–T III
Insertion:	Superior borders of costae II–V, lateral to angulus costae
Function:	Lifts the ribs, thus supporting inspiration
Innervation:	Nn. intercostales (T2–T5)

Serratus posterior inferior	
Origin:	Procc. spinosi of vertebrae T XI–L II and fascia thoracolumbalis
Insertion:	Inferior borders of costae IX–XII, near angulus costae
Function:	Also aids inspiration; by drawing the lower ribs backward and downward, it assists in preventing further narrowing of the thoracic aperture, thus providing a fixed stable base of the diaphragma
Innervation:	Nn. spinales (T9–T12), rr. ventrales

C List of secondarily incorporated trunk muscles
(with the exception of mm. serrati, which are on p. 298 in the Upper Limb unit). All of the intrinsic back muscles are covered by large muscles, which in part extend all the way to the pelvis. In the course of phylogenesis, they have extended their origin to the trunk (incorporated trunk muscles, compare p. 145). These muscles initially migrated from the back as hypaxial muscles (innervation: rr. ventrales) toward the upper limb (thus shoulder girdle muscles) in order to later relocate again to the back. This probably happened as part of the development of tetrapods, when the pectoral girdle muscles grew and had to carry more weight. A distinction is made between:

Spinocostal or trunk rib muscles (between spinal column and ribs)
• M. serratus posterior superior • M. serratus posterior inferior

Spinohumeral or muscles of the trunk and shoulder girdle or trunk and arm (between spinal column and shoulder girdle or upper arm)	
• Muscles of the trunk and shoulder girdle – Mm. rhomboidei major u. minor – M. levator scapulae – M. serratus anterior – M. subclavius – M. pectoralis minor – M. trapezius	• Muscles of the trunk and arm – M. latissimus dorsi • Muscles of the thorax and arm – M. pectoralis major

Note: All incorporated muscles are innervated by rr. ventrales of nn. spinales.

The Trunk Wall — 10. Musculature: Functional Groups

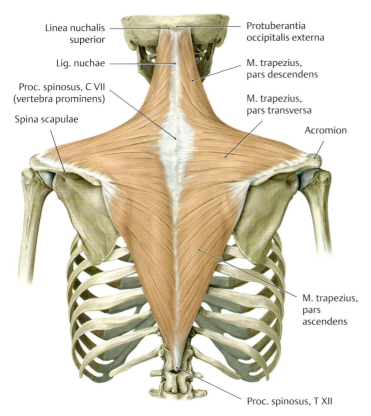

D M. trapezius
Posterior view.

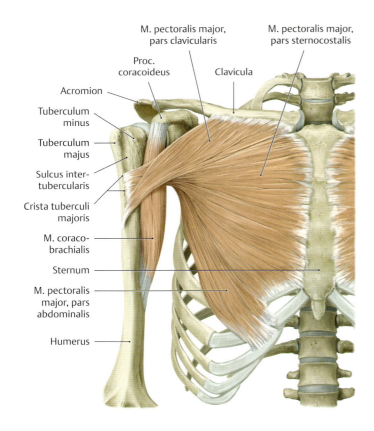

E Mm. pectoralis major and coracobrachialis
Right side, anterior view.

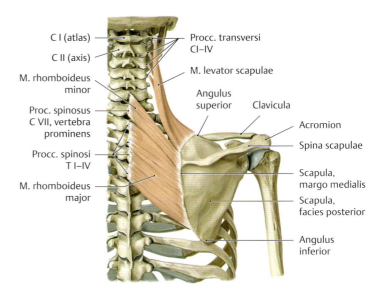

F M. levator scapulae, m. rhomboideus major, and m. rhomboideus minor
Right side, posterior view.

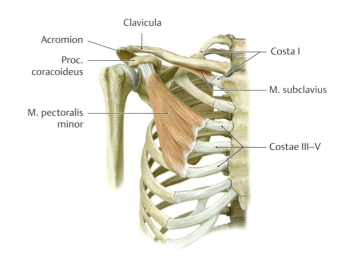

G M. pectoralis minor and m. subclavius
Right side, anterior view.

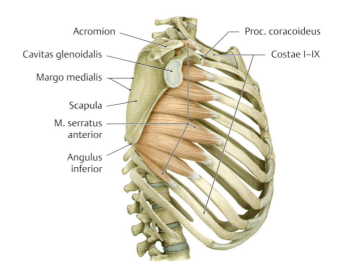

H M. serratus anterior
Right side, lateral view.

165

11.1 The Back Muscles and Fascia Thoracolumbalis

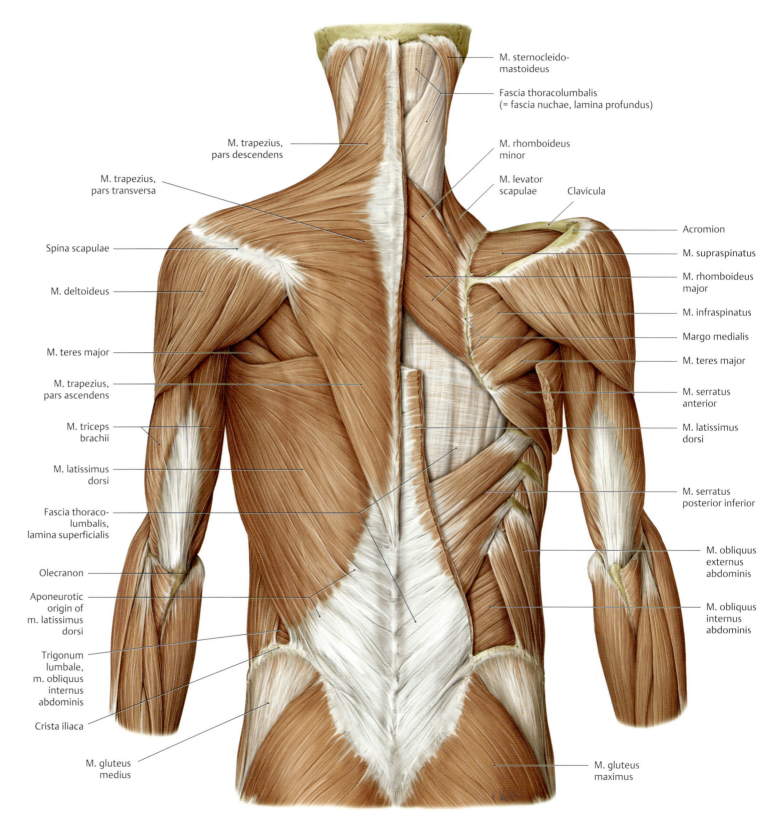

A The fascia thoracolumbalis as a partition between the intrinsic and nonintrinsic back muscles
Posterior view. The m. trapezius has been completely removed and the m. latissimus dorsi has been partially removed on the right side to reveal the fascia thoracolumbalis. The superficial layer (lamina superficialis) of the fascia thoracolumbalis separates the intrinsic back muscles from the nonintrinsic muscles that have migrated to the back.

Note: The lamina superficialis of the fascia thoracolumbalis is located close to the columna vertebrale just beneath the skin, where it is heavily reinforced and serves as the origin for the extensive tendon of the m. latissimus dorsi. Lateral to the origin, the lamina superficialis is pushed beneath both the tendon and the muscle, thereby losing direct contact to the subcutaneous tissue (tela subcutanea). Along its entire width, medial to lateral, the lamina superficialis covers the intrinsic back muscles.

The Trunk Wall —— 11. Musculature: Topographical Anatomy

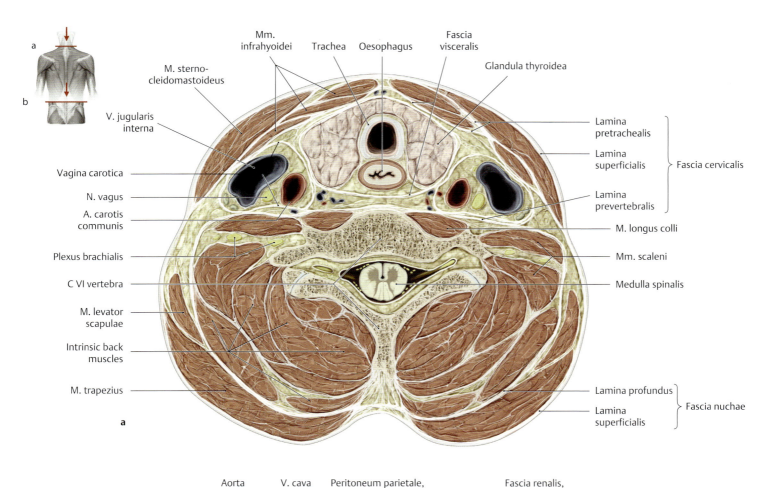

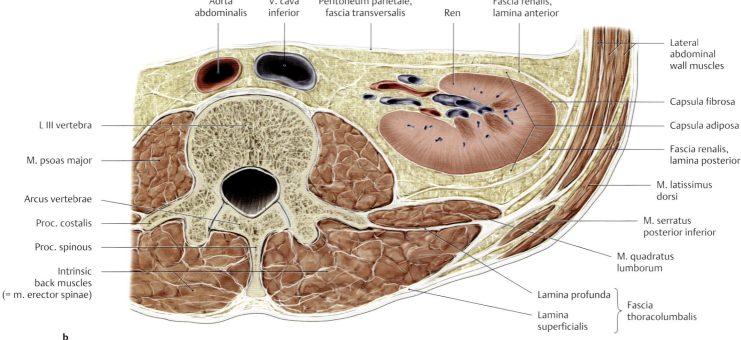

B Fascia thoracolumbalis
a Transverse section through the neck at the level of the C VI vertebra, superior view.
b Transverse section through the posterior trunk wall at the level of the L III vertebra (cauda equina removed), superior view.

The fascia thoracolumbalis forms the lateral portion of an osseofibrous canal that encloses all of the *intrinsic back muscles*. In addition to the fascia thoracolumbalis, this canal is formed by the arcus vertebrae and the proc. spinosus and proc. costalis of the associated vertebrae. The fascia thoracolumbalis consists of a superficial and a deep layer (laminae superficilis et profunda), especially in the lumbar region (regio lumbalis); both layers unite at the lateral margin of the intrinsic back muscles. At the back of the neck, the lamina superficialis of the fascia thoracolumbalis blends with the fascia nuchae (lamina profunda), becoming continuous with the lamina prevertebralis of the fascia cervicalis.

11.2 The Intrinsic Back Muscles: Lateral and Medial Tracts of the M. erector spinae

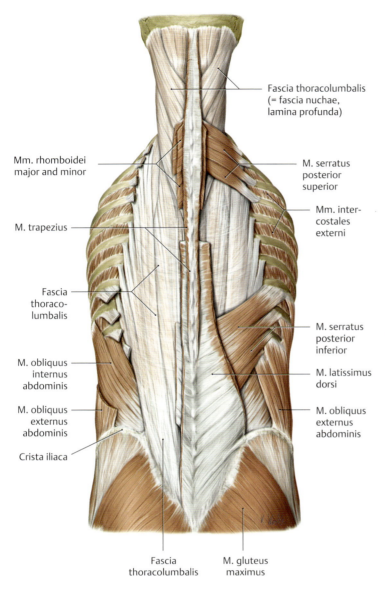

A Course of the fascia thoracolumbalis
Posterior view. To reveal the fascia thoracolumbalis, both shoulder girdles and the extrinsic back muscles have been removed (except for the mm. serrati posterior superior and inferior and the aponeurotic origin of the m. latissimus on the right side).

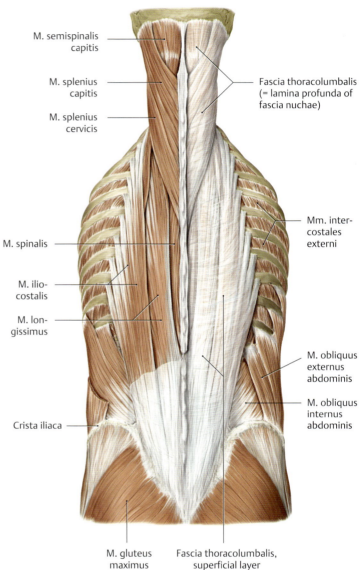

B Lateral tract of the m. erector spinae
Posterior view. Portions of the lamina superficialis of the fascia thoracolumbalis have been removed on the left side of the back to expose the lateral tract muscles (mm. iliocostalis, longissimus, splenii cervicis and capitis). The mm. levatores costarum and intertransversarii, also part of the lateral tract, are covered here by the mm. iliocostalis and longissimus (see **C** and **D**).
Note that the fascia thoracolumbalis on the back of the neck is continuous with the lamina profunda of the fascia nuchae.

The Trunk Wall —— **11. Musculature: Topographical Anatomy**

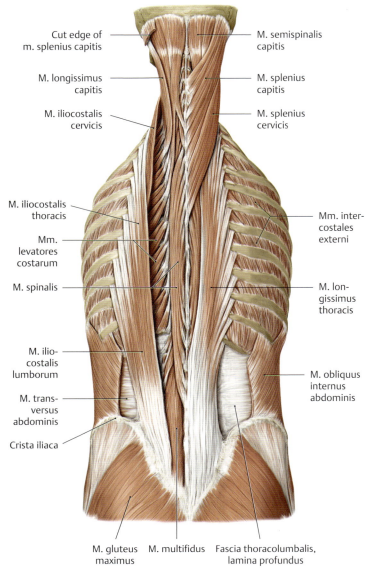

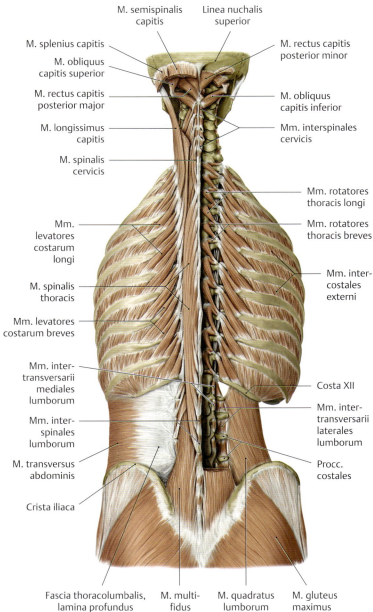

C Medial tract of the m. erector spinae (portions of the lateral tract left in place)

Posterior view. The m. longissimus (except for m. longissimus capitis) and mm. splenii cervicis and capitis have been removed on the left side of the back, and the m. iliocostalis have been removed on the right side (see **D** for the mm. rotatores).

Note the lamina profundus of the fascia thoracolumbalis from which both the m. obliquus internus abdominis and the m. transversus abdominis originate (see **D**).

D Medial tract of the m. erector spinae (with the entire lateral tract removed)

Posterior view. The entire lateral tract (except for the mm. intertransversarii and levatores costarum)) has been removed, along with portions of the medial tract, to demonstrate the various individual muscles of the medial tract.

Note the origin of the m. transversus abdominis from the lamina profundus of the fascia thoracolumbalis in the lumbar region (left side). On the right side, the deep fascial layer tand m. multifidus have been removed to display the mm. intertransversarii (lateral tract) and the m. quadratus lumborum (posterior [deep] abdominal muscle).

11.3 The Intrinsic Back Muscles: Short Nuchal Muscles

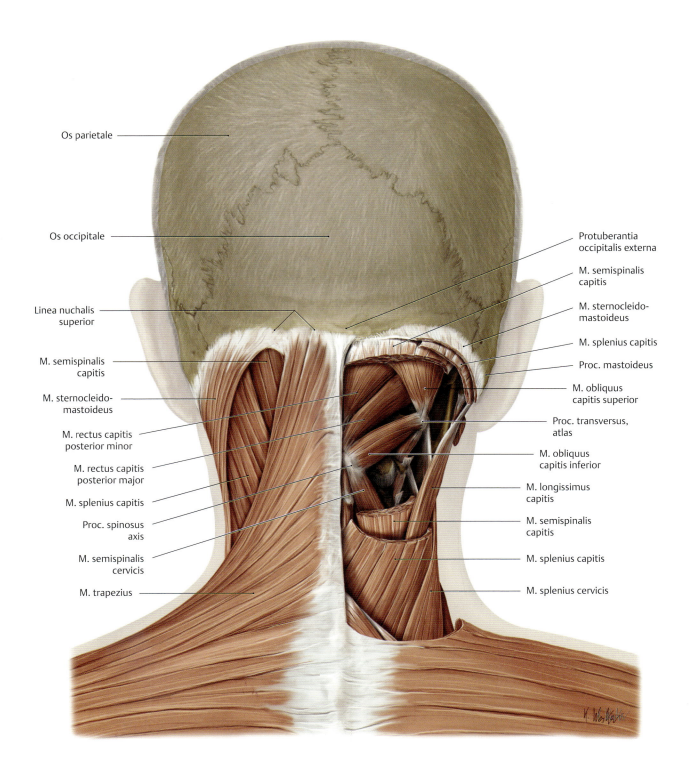

A Location of the short nuchal muscles (mm. suboccipitales)
Nuchal region, posterior view. The mm. suboccipitales in the strict sense are the short (deep) nuchal muscles that belong to the intrinsic back muscles (mm. recti capitis posteriores major and minor and obliquii capitis superior and inferior). They meet the criterion of being innervated by a r. dorsalis—in this case the r. dorsalis C I, the n. suboccipitalis. The *mm. recti capitis anterior* and *lateralis* are *not* classified as intrinsic back muscles, despite their suboccipital location, because they are innervated by rr. ventrales. The short nuchal muscles lie within the fascia thoracolumbalis deep at the back of the neck and course between the os occiput and the first two cervical vertebrae. They act mainly on the craniovertebral joints (see p. 120) and support differentiated head movements (e.g., for fine adjustments of head position). The following muscles have been partially removed to reveal their location in the right nuchal region: mm. trapezius, sternocleidomastoideus, splenius capitis, and semispinalis capitis. An important landmark in the deep nuchal region is the proc. spinosus axis.

The Trunk Wall — 11. Musculature: Topographical Anatomy

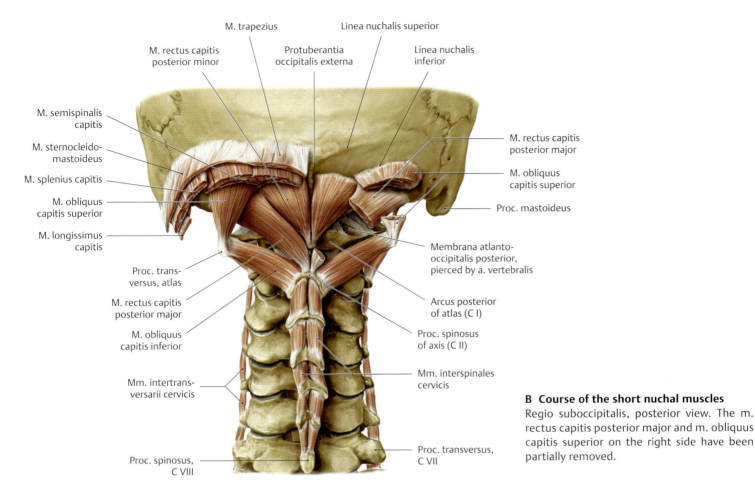

B Course of the short nuchal muscles
Regio suboccipitalis, posterior view. The m. rectus capitis posterior major and m. obliquus capitis superior on the right side have been partially removed.

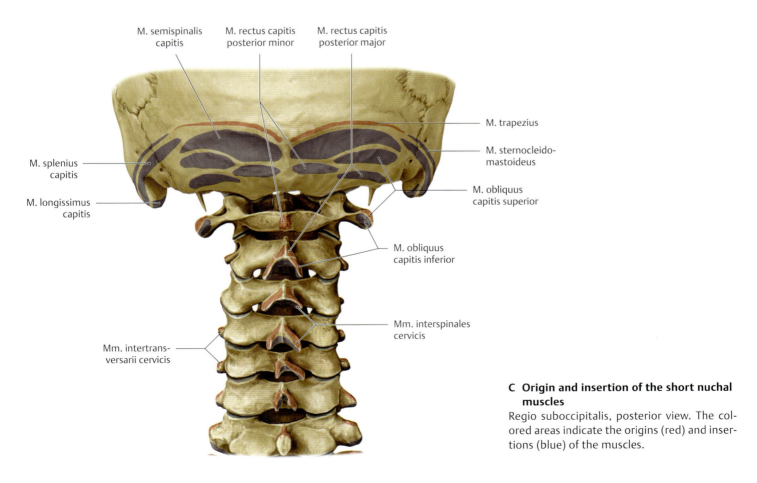

C Origin and insertion of the short nuchal muscles
Regio suboccipitalis, posterior view. The colored areas indicate the origins (red) and insertions (blue) of the muscles.

11.4 The Thoracic Wall Muscles and Fascia endothoracica

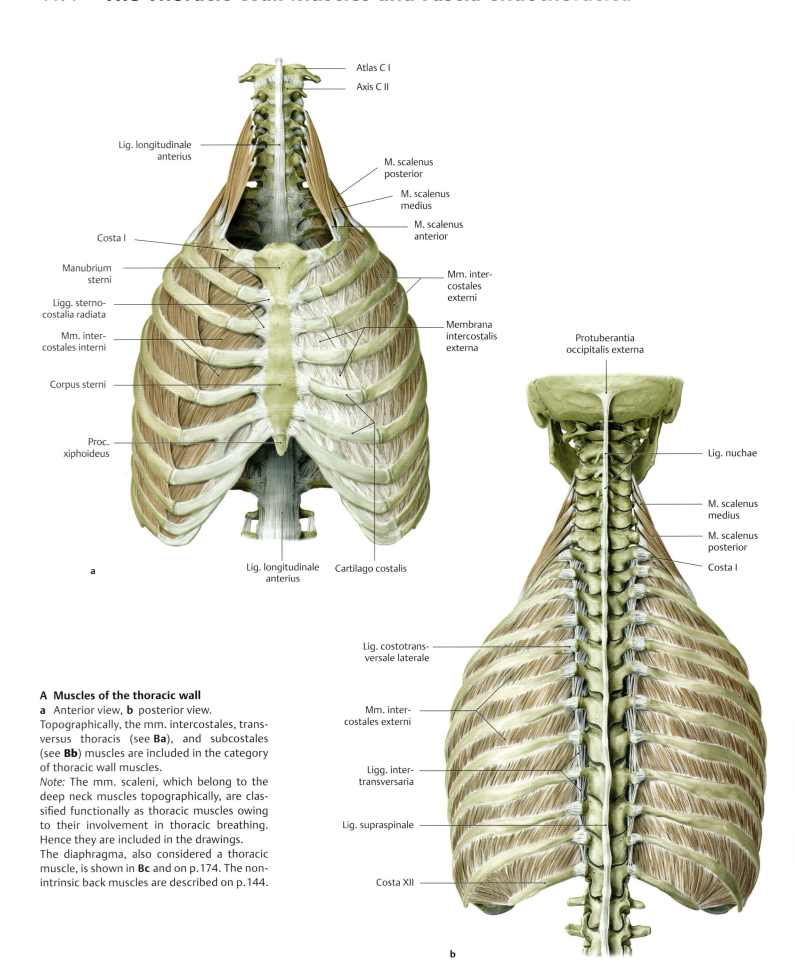

A Muscles of the thoracic wall
a Anterior view, b posterior view.
Topographically, the mm. intercostales, transversus thoracis (see **Ba**), and subcostales (see **Bb**) muscles are included in the category of thoracic wall muscles.
Note: The mm. scaleni, which belong to the deep neck muscles topographically, are classified functionally as thoracic muscles owing to their involvement in thoracic breathing. Hence they are included in the drawings.
The diaphragma, also considered a thoracic muscle, is shown in **Bc** and on p. 174. The nonintrinsic back muscles are described on p. 144.

The Trunk Wall — 11. Musculature: Topographical Anatomy

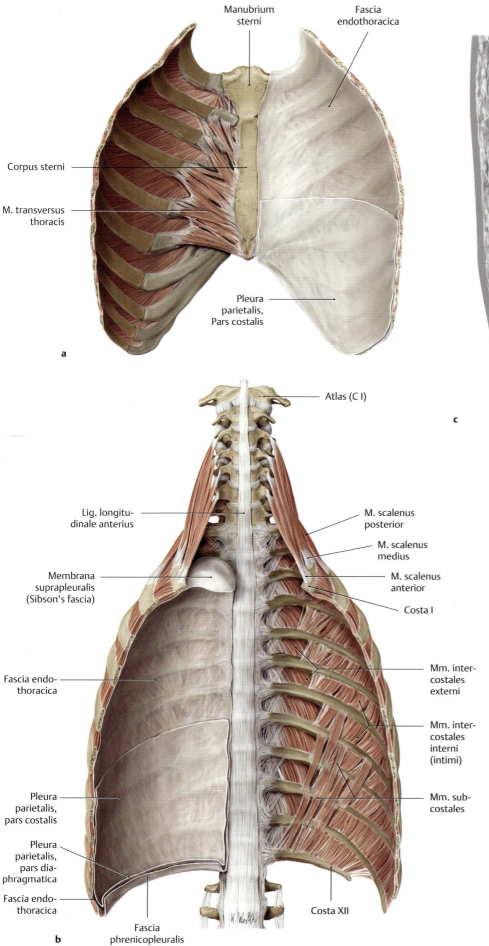

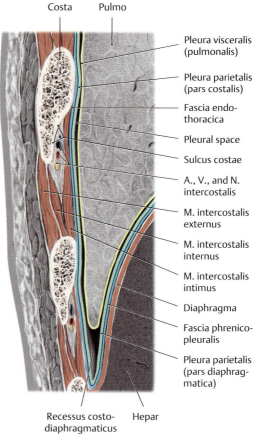

B Fascia endothoracica
a Posterior surface of the anterior thoracic wall segment removed in **b**.
b Posterior thoracic wall, anterior view (fascia endothoracica removed on the left side).
c Coronal section through the lateral thoracic wall and recessus costodiaphragmaticus.

The cavitas thoracis is lined by a fascia-like layer of connective tissue, the *fascia endothoracica*. It lies between the deep muscles of the thoracic wall and the pars costalis of the pleura parietalis, to which it is firmly attached, and is analogous to the fascia transversalis of the cavitas abdominis (**a**). The fascia endothoracica is thickened over the pleural apex to form the *membrana suprapleuralis (Sibson's fascia)*. The *fascia phrenicopleuralis* is the portion of the fascia endothoracica that connects the pars diaphragmatica of the pleura parietalis to the upper surface of the diaphragma (**b**). The recessus costodiaphragmaticus (**c**) between the chest wall and diaphragma is a potential space that enlarges on inspiration (during lowering of the diaphragma) to accommodate the expanding lung. The pleural space is the potential space located between the pars costalis of the pleura parietalis and the pleura visceralis/pulmonalis, which directly invests the lung tissue.

11.5 The Thoracoabdominal Junction: The Diaphragma

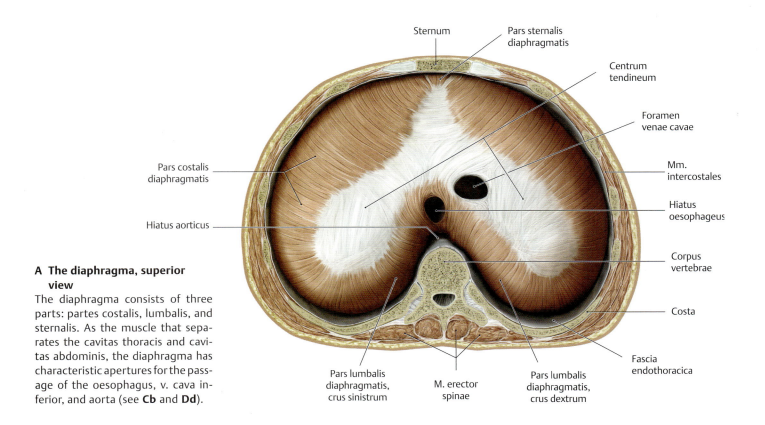

A The diaphragma, superior view
The diaphragma consists of three parts: partes costalis, lumbalis, and sternalis. As the muscle that separates the cavitas thoracis and cavitas abdominis, the diaphragma has characteristic apertures for the passage of the oesophagus, v. cava inferior, and aorta (see **Cb** and **Dd**).

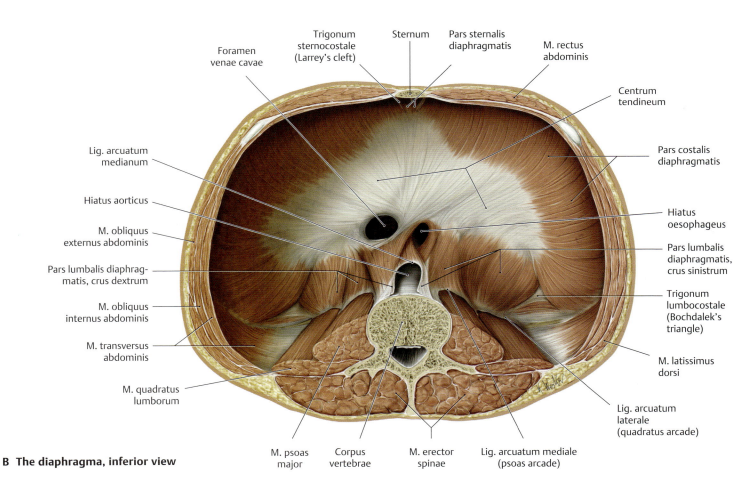

B The diaphragma, inferior view

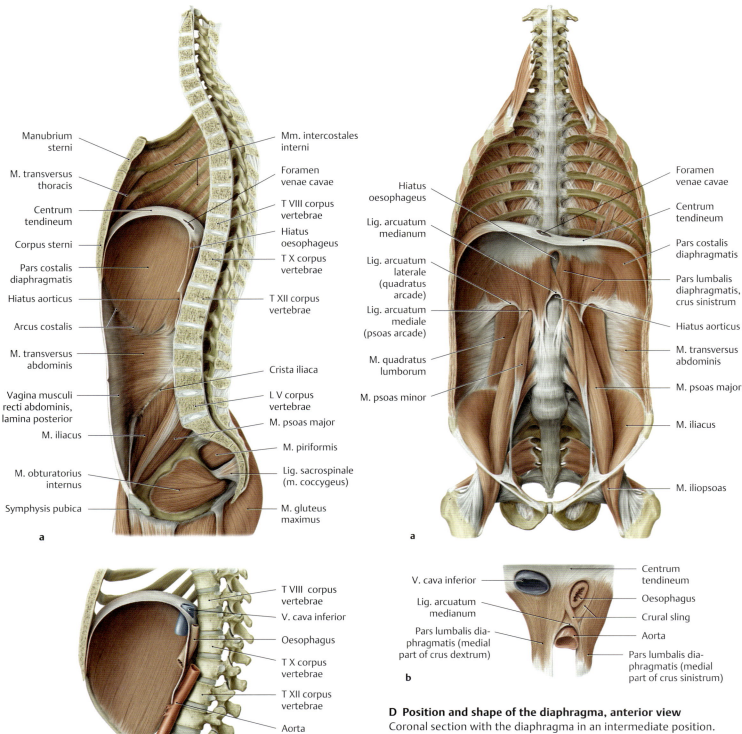

C Position and shape of the diaphragma, viewed from the left side

Midsagittal section demonstrating the right half of the body. The diaphragma is in an intermediate position at end-expiration.

a The apertures in the diaphragma are located at vertical positions corresponding to the following landmarks in the lower thoracic spine: foramen venae cavae = T VIII corpus vertebrae, hiatus oesophageus = T X corpus vertebrae, hiatus aorticus = T XII corpus vertebrae.
b The diaphragmatic apertures and the structures that they transmit (see p. 205).

D Position and shape of the diaphragma, anterior view
Coronal section with the diaphragma in an intermediate position.

a The apertures in the diaphragma are located in the region of the centrum tendineum (v. cava inferior) and in the pars lumbalis diaphragmatis (hiatus oesophageus and hiatus aorticus).
b Enlarged view of the diaphragmatic apertures, with vessels transected. The foramen venae cavae is located to the right of the midline, the hiatus oesophageus and hiatus aorticus to the left.

In a diaphragmatic hernia (diaphragmatic rupture), abdominal viscera prolapse into the chest cavity through a congenital or acquired area of weakness in the diaphragma. By far the most common herniation site is the hiatus oesophageus, accounting for 90% of cases. Typically, the distal end of the oesophagus and the gastric cardia (gastric inlet) "slide" upward through the hiatus oesophageus into the chest (axial hiatal hernia or sliding hernia; approximately 85% of all hiatal hernias). Typical symptoms are acid reflux, heartburn, and a feeling of retrosternal pressure after meals. More severe cases may present with nausea, vomiting, and functional cardiac complaints.

11.6 The Anterolateral and Anterior Abdominal Wall Muscles*

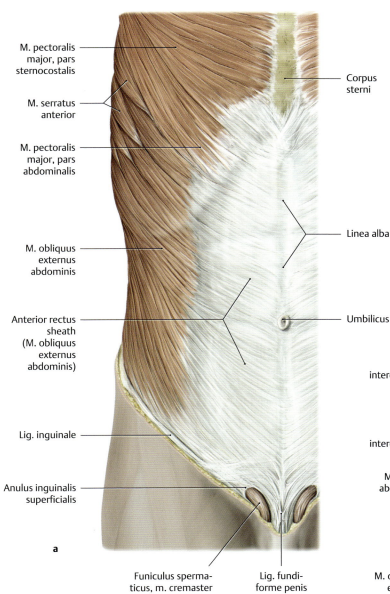

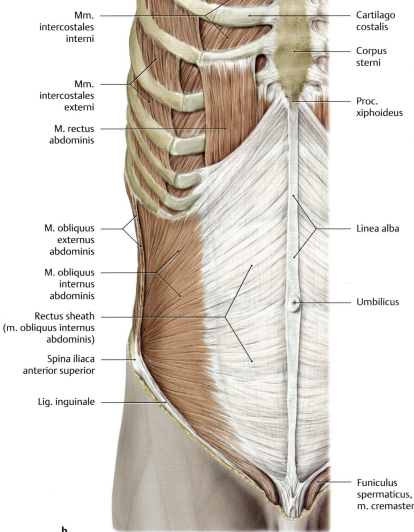

* The posterior (deep) abdominal wall muscles, most notably the m. psoas major, are actually hip muscles in a functional sense because they act predominantly on the art. coxae. For this reason they are described in the Lower Limb unit (see p. 476).

A Anterolateral abdominal wall muscles in the male

Right side, anterior view. The oblique muscles of the abdominal wall consist of the mm. obliquii externus and internus abdominis and the m. transversus abdominis.

a The aponeurosis of the m. obliquus externus abdominis borders the superficial inguinal ring (anulus inguinalis superficialis), its inferior margin forming the inguinal ligament (lig. inguinale).

b The mm. obliquus externus abdominis, pectoralis major, and serratus anterior have been removed. The inferior border of the m. obliquus internus abdominis forms the roof of the canalis inguinalis (see p. 211) and is continued onto the funiculus spermaticus in the male as the m. cremaster and fascia cremasterica.

The Trunk Wall — 11. Musculature: Topographical Anatomy

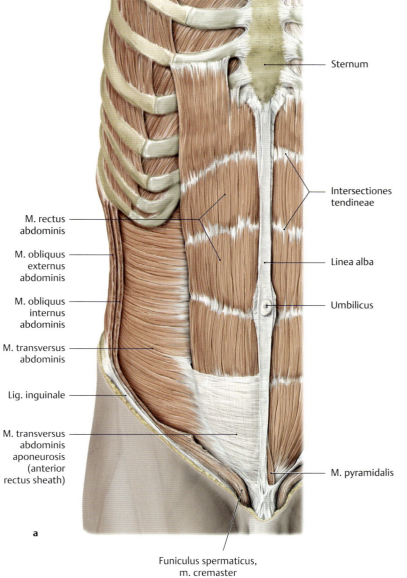

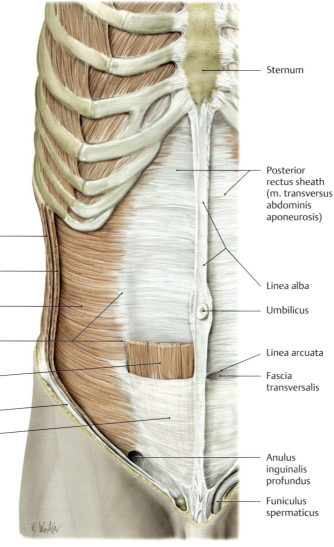

B Anterior (straight) abdominal wall muscles in the male

Right side, anterior view. The straight, or strap, muscles of the abdominal wall include the mm. rectus abdominis and pyramidalis.

a The m. obliquus internus abdominis has been removed.
b In addition, the upper portion of the m. rectus abdominis has been removed.

Note: Below the linea arcuata, the aponeurosis of the m. transversus abdominis and the aponeurosis of the m. obliquus internus abdominis lie anterior to the m. rectus abdominis (see also p. 179).

11.7 Structure of the Abdominal Wall and Rectus Sheath (Vagina musculi recti abdominis)

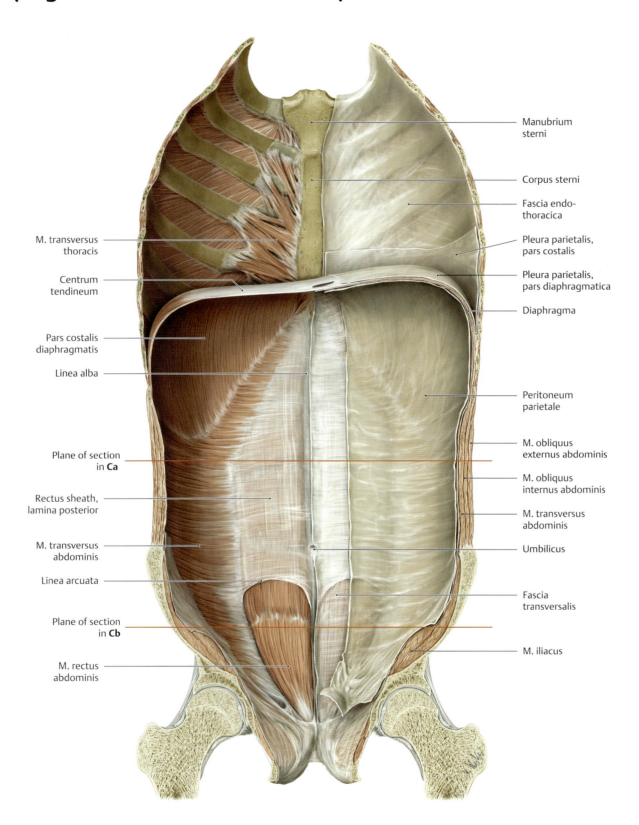

A Overview of the abdominal wall and rectus sheath
Posterior view with the viscera removed. To show how the diaphragma separates the cavitas thoracis from the cavitas abdominis, the fascia transversalis and peritoneum parietale have been removed from the left abdominal wall, while the fascia endothoracica and pleura parietalis have been removed from the left chest wall. The rectus sheath (enclosing the vagina m. recti abdominis) plays a special role in the abdominal wall, because its structure changes below the linea arcuata to accommodate the increasing pressure of the viscera against the body wall (see **C**). The vagina musculi recti abdominis is formed by the aponeuroses of the lateral abdominal muscles (of which only the m. transversus is visible here; the others are hidden) and is divided into a lamina anterior and a lamina posterior.

The Trunk Wall —— 11. Musculature: Topographical Anatomy

B Structure of the abdominal wall
Cross section through the abdominal wall above the umbilicus, superior view. The following layers are distinguished in the lateral abdominal wall, from inside to outside:
- Peritoneum parietale
- Fascia transversalis
- M. transversus abdominis
- M. obliquus internus abdominis
- M. obliquus externus abdominis
- Fascia abdominis superficialis
- Fascia abdominis superficialis, fatty layer
- Skin (cutis)

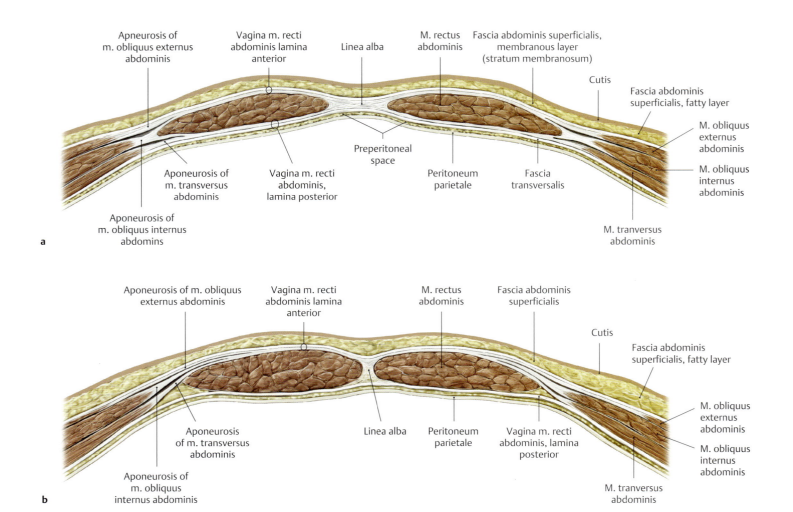

C Structure of the rectus sheath (vagina musculi recti abdominis)
Cross sections through the rectus sheath superior (**a**) and inferior (**b**) to the linea arcuata, superior view. The aponeuroses of the lateral abdominal wall muscles ensheath the anterior strap muscles to form the rectus sheath (vagina musculi recti abdominis). This creates a muscular compartment consisting of an anterior layer and a posterior layer (lamina anterior and lamina posterior). While the aponeuroses of the three lateral abdominal muscles contribute equally to the anterior and posterior layers of the sheath above the umbilicus, the two layers blend together approximately 3 to 5 cm below the umbilicus (at the level of the linea arcuata) to form a single (and consequently more stable) sheet that passes in front of the m. rectus abdominis. Below the linea arcuata, the lamina posterior of the agina musculi recti abdominis is absent. The preperitoneal space, a fine layer of loose connective tissue and fat (see also p. 216), is located between the fascia transversalis and the peritoneum parietale.

11.8 The Pelvic Floor Muscles: Overview of the Regio perinealis and Superficial Fasciae

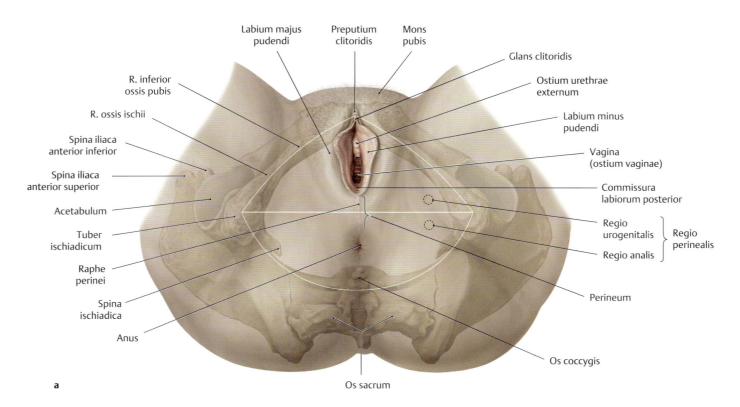

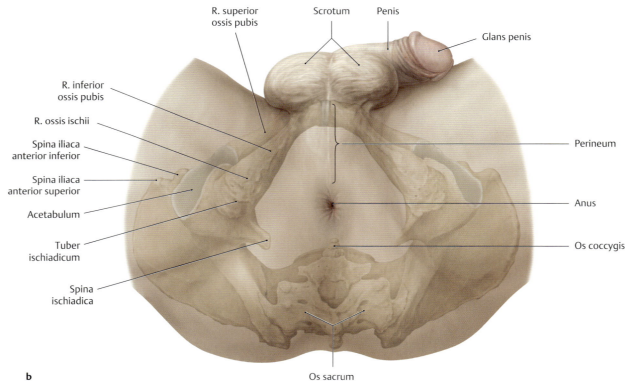

A Regio perinealis in the female (a) and male (b)
Lithotomy position, inferior view. The perineal region (regio perinealis) in both sexes consists of the *regio urogenitalis* anteriorly and the adjacent *regio analis* posteriorly. The two regions are separated by a line that runs between the tubera ischiadica. The perineum denotes the soft-tissue area between the thighs and the buttocks. In the female, obstetricians refer to the perineum as the region from the anterior margin of the commissura labiorum posterior of the vulva. It is considerably longer in the male, extending from the anal margin to the root of the scrotum. The perineum is a region containing fibrous and fatty tissue, and the *corpus perineale* (see **Ba**) is a fibromuscular mass, also referred to as the *centrum perinei of the corpus perineale*. A more detailed structural description of the perineum is given on p. 184.

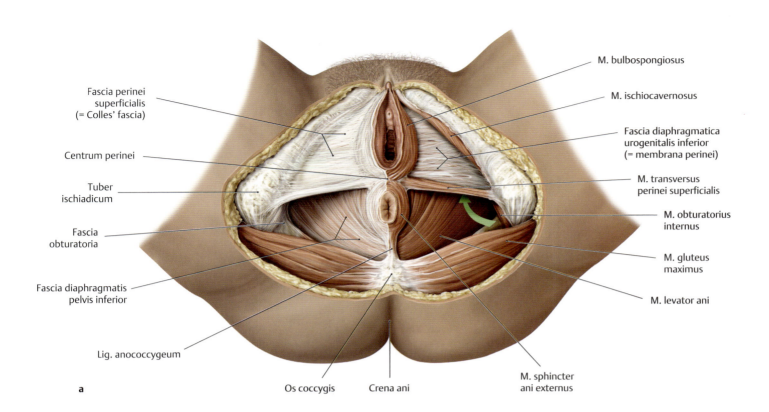

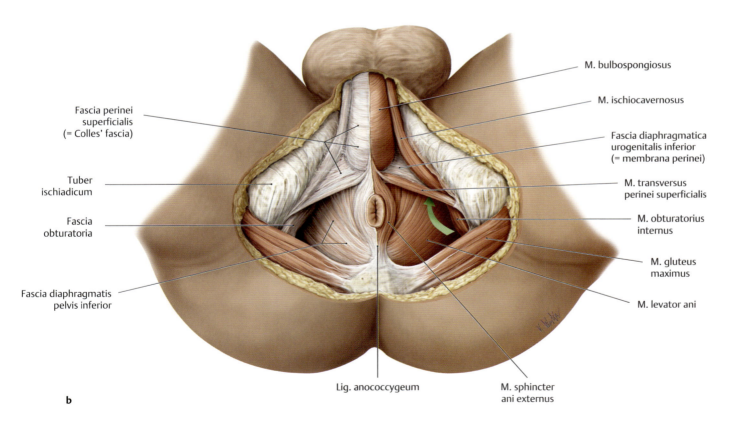

B Superficial fasciae of the female (a) and male pelvic floor (b)
Lithotomy position, inferior view. The fascia perinei superficialis (regio urogenitalis) and the fascia diaphragmatis pelvis inferior (regio analis) are intact on the right side but have been removed on the left side. In this way the spatium superficiale perinei has been opened in the regio urogenitalis on the left side of each dissection, and the m. levator ani has been exposed in the regio analis. The spatium superficiale perinei is bounded posteriorly by the m. transversus perinei superficialis and superiorly by the membrana perinei. The m. bulbospongiosus is located in the medial part of this superficial perineal space; the m. ischiocavernosus is located laterally. The green arrow in each figure points to the anterior recess of the left fossa ischioanalis (see also p. 556).

11.9 Structure of the Pelvic Floor and Pelvic Spaces: Female versus Male

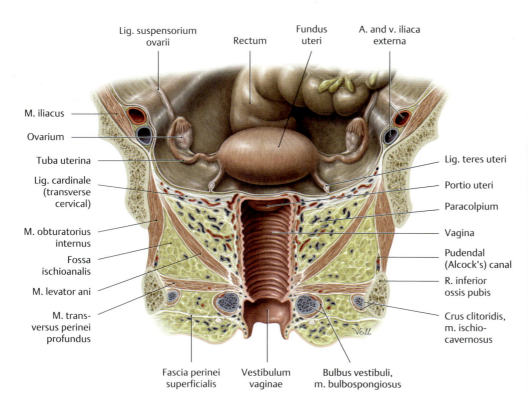

A Coronal section through the female pelvis
Anterior view.

Plane of section in **A** and **B**

C Subdivisions of the pelvis and structure of the pelvic floor (in both sexes)

Subdivisions of the pelvis

The pelvis is the portion of the cavitas abdominis located in the lesser pelvis. It is surrounded by the skeleton of the lesser pelvis, which meets the greater pelvis at the *linea terminalis* (see p. 141). The pelvic cavity (cavitas pelvis) lies above the pelvic floor (supralevator) and is subdivided into upper and lower levels by the peritoneum:

- **Upper level:** cavitas peritonealis of the lesser pelvis
- **Lower level:** subperitoneal space

The perineum lies below the pelvic floor (infralevator) and is separated into

- Deep perineal space with ischioanal fossa (spatium profundum perinei with fossa ischioanalis)
- Superficial perineal space (spatium superficiale perinei)
- Subcutaneous perineal space

See **B** and **E**.

Subjacent to the pelvis are three more spaces that are considered to be separate from the pelvis itself: the *spatium profundum perinei*, the *spatium superficialie perinei*, and the *subcutaneous perineal space* (see **B** and **F**).

Structure of the pelvic floor

The three muscular and connective-tissue sheets that contribute to the structure of the pelvic floor are arranged in three levels:

- **Upper level:** diaphragma pelvis
- **Middle level:** deep urogenital muscles
- **Lower level:** sphincters and erectile muscles of the urogenital and intestinal tract

The funnel-shaped diaphragma pelvis is formed by the mm. levator ani and coccygeus and the superior and inferior fasciae of the diaphragma pelvis. The deep urogenital muscles (see p. 184) lie in the **deep perineal pouch (spatium profundum perinei)**. They include the deep tranverse perineal muscle in the male, a sheet of smooth muscle in the female, and the inferior portion of the m. sphincter urethrae externus. The membrana perinei separates the deep from the **superficial perineal pouch (spatium superficiale perinei)**, the muscles of which are the mm. transversus perinei superficialis, bulbospongiosus, ischiocavernosus, and the m. sphincter ani externus more posteriorly.

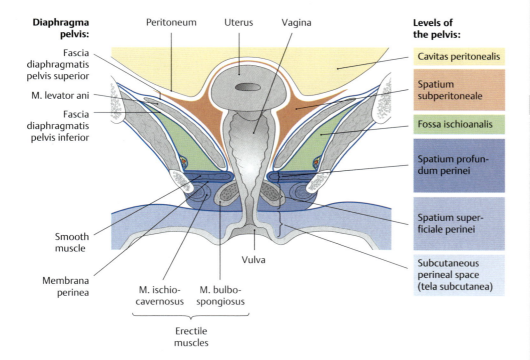

B Pelvic spaces, fasciae, and arrangement of the pelvic floor muscles in the female pelvis
Coronal section at the level of the vagina (see small diagram above for the exact location of the plane). The levels of the pelvis are shown in different colors.

The Trunk Wall — **11. Musculature: Topographical Anatomy**

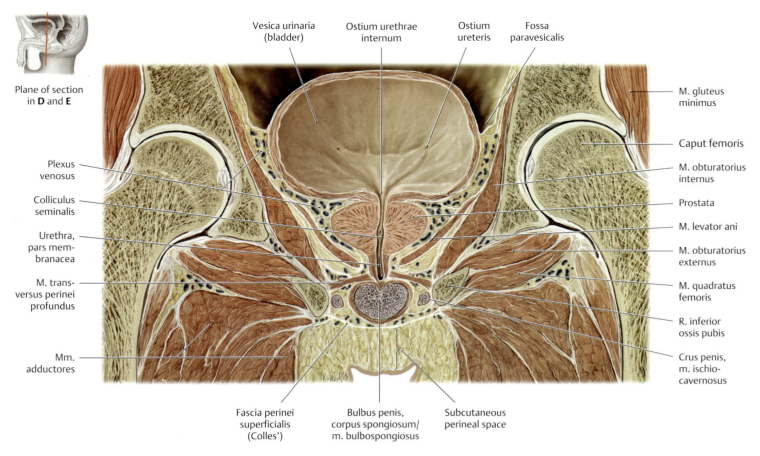

D Coronal section through the male pelvis
Anterior view.

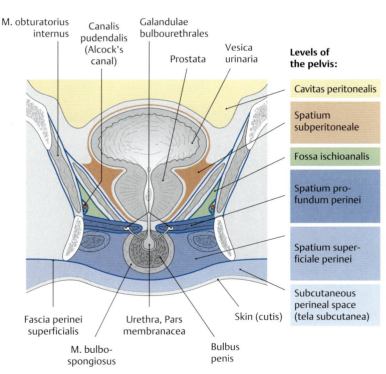

E Pelvic spaces, fasciae, and arrangement of the pelvic floor muscles in the male pelvis
Coronal section at the level of the prostate (see small diagram above for the exact location of the plane). Different colors indicate the various levels of the pelvis and the perineal spaces.

F Boundaries and contents of the spatium profundum perinei and spatium superficiale perinei (Colles' space) (in both sexes)

Spatium profundum perinei (deep perineal space)

- Boundaries:
  - Membrana perinei
  - Fascia diaphragmatis urogenitalis inferior
  - Inferior portion of inner fascia of m. obturatorius internus
- Contents:
  - M. transversus perinei profundus
  - Inferior portion of m. sphincter urethrae externus
  - Compressor urethra muscles (in female)
  - Anterior extension of ischioanal fat pad
  - Intermediate (membranous) portion of urethra (in male)
  - Bulbourethral glands (in male)
  - Branches of related neurovascular structures

Spatium profundum superficiale (superficial perineal space)

- Boundaries:
  - Fascia perinei superficialis (Colles')
  - Membrana perinei
  - Ischiopubic rami
- Contents:
  - M. transversus perinei superficialis
  - Mm. bulbospongiosi
  - Mm. ischiocavernosi
  - Radix penis and proximal portion of pars spongiosa urethrae masculinae (in males)
  - Clitoris, bulbus vestibuli, and glandulae vestibulares majores (in females)
  - Branches of related neurovascular structures

The **subcutaneous perineal space** is located between the fascia perinei superficialis and skin and contains mostly fatty tissue.

183

11.10 The Muscles of the Female Pelvic Floor and Wall

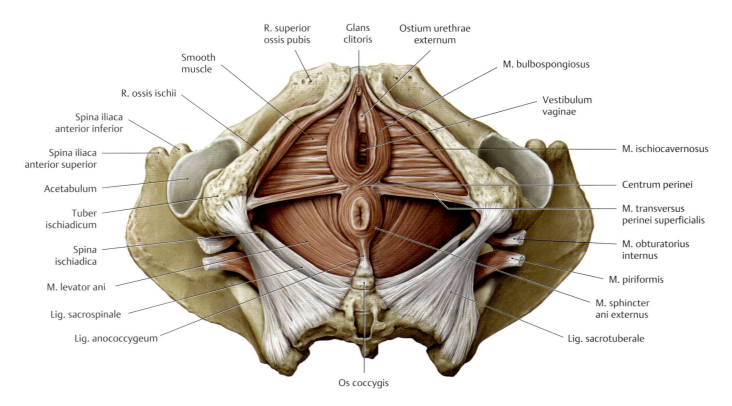

A Muscles of the pelvic floor with the fasciae removed
Female pelvis, inferior view. The muscular layers are progressively removed from **B** through **D** to demonstrate the underlying muscles from a consistent perspective. The m. levator ani is described more fully on pp. 186–187.

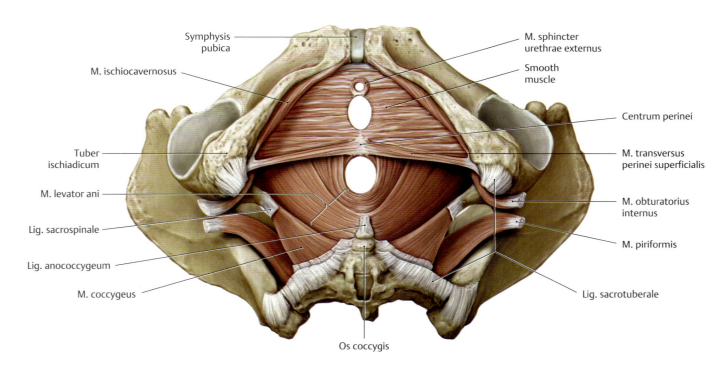

B Muscles of the pelvic floor with the sphincters removed
Female pelvis, inferior view. The sphincter muscles have been removed from the urogenital and intestinal tracts (= mm. bulbospongiosi and m. sphincter ani externus), leaving the m. sphincter urethrae externus intact.

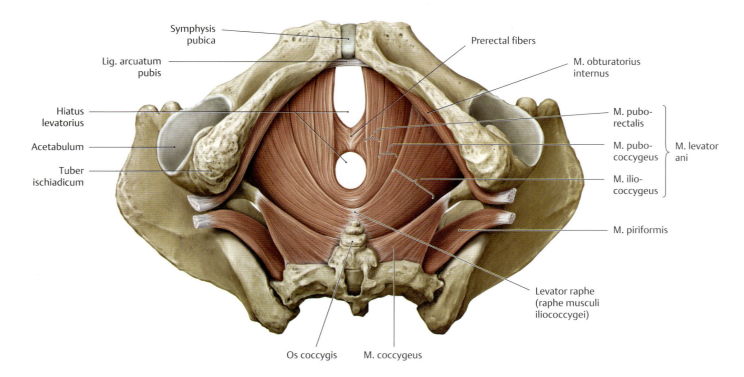

C Muscles of the pelvic floor with the urogenital muscles removed
Female pelvis, inferior view. The urogenital muscles (= mm. transversi perinei profundus and superficialis and mm. ischiocavernosi) have been removed.

Note the opening of the hiatus levatorius, which is bounded by the two crura of the m. puborectalis (the levator crura). *Note* also the prerectal fibers that have split off from the m. puborectalis. The prerectal fibers are interwoven with connective-tissue fibers and smooth muscle to form the fibromuscular framework of the perineum (see also p. 180).

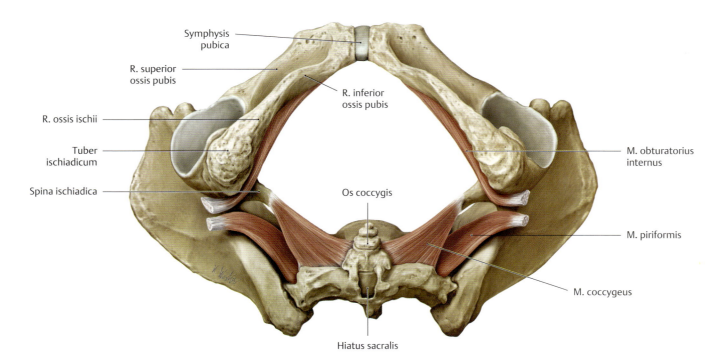

D Muscles of the pelvic wall (parietal muscles of the lesser pelvis)
Female pelvis, inferior view. All of the pelvic floor muscles have been removed, leaving the parietal muscles intact (mm. obturatorius internus, coccygeus, and piriformis). Along with the skeleton of the lesser pelvis, these muscles contribute structurally to the pelvic wall and assist in closing the posterior apertura pelvis inferior. The m. obturatorius internus and its fascia provide a tendon of origin for the m. iliococcygeus, which is part of the m. levator ani complex (arcus tendineus m. levatoris ani, see p. 187).

185

11.11 Pelvic Floor Muscles: M. levator ani

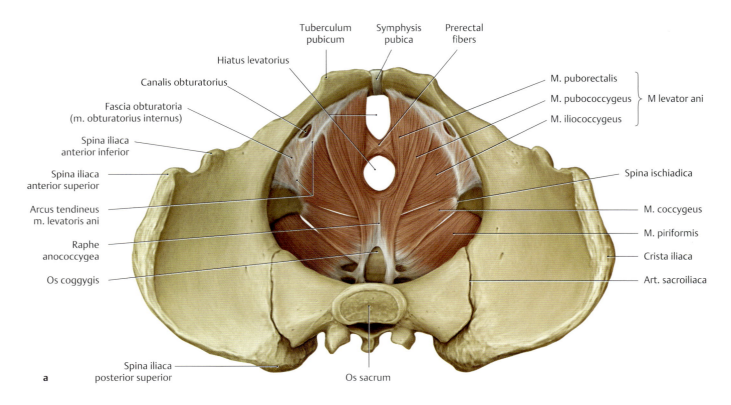

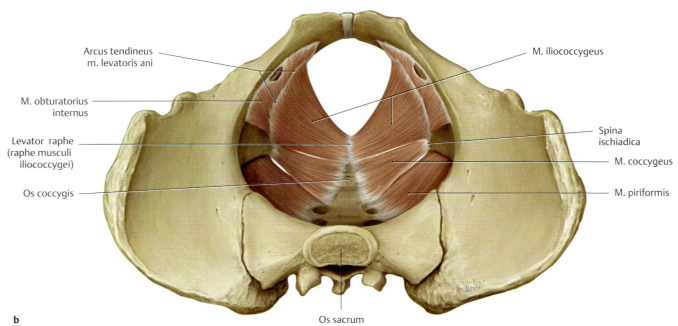

A Parts of the m. levator ani and the parietal muscles of the pelvic wall
Female pelvis, superior view.

a The m. levator ani consists of three parts: the mm. puborectalis, pubococcygeus, and iliococcygeus. It arises from the anterior and lateral pelvic wall on a line running from the center of the symphysis pubica to the spina ischiadica (= arcus tendineus m. levatoris ani). As a contributor to continence, the m. puborectalis assists the m. sphincter ani externus (not visible here) in keeping the anus closed. It arises from the ramus superior ossis pubis on both sides of the symphysis pubica and runs in the shape of a loop around the rectum, where the fibers are connected to the deep part of the m. sphincter ani externus. It has the shape of an arched gateway, with its two crura (the levator crura) forming the boundaries of the hiatus levatorius. In contraction of the puborectal sling, the rectal wall is pulled forward toward the pubic bone, which increases the so-called anorectal angle (angle between rectum and canalis analis) and tightens the canalis analis by squeezing it.

b The mm. puborectalis and pubococcygeus have been removed. The m. coccygeus (muscle fibers on the lig sacrospinale) and the m. piriformis complete the apertura pelvis inferior posteriorly on both sides of the sacrum.

The Trunk Wall —— 11. Musculature: Topographical Anatomy

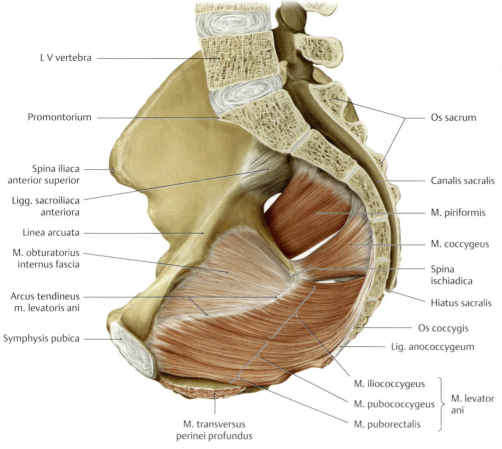

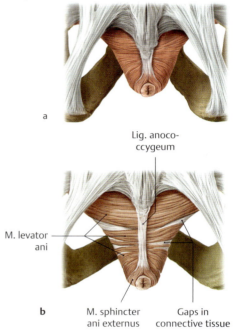

B Arcus tendineus m. levatoris ani
Right hemipelvis, medial view. The arcus tendineus m. levatoris ani is a thickening of the m. obturatorius internus fascia that serves as the origin of the m. iliococcygeus.

D Gender-related differences in structure of the m. levator ani (**a** male, **b** female)
Posterior view.
Note the gaps in the connective tissue between the muscular parts of the m. levator ani in female (**b**).

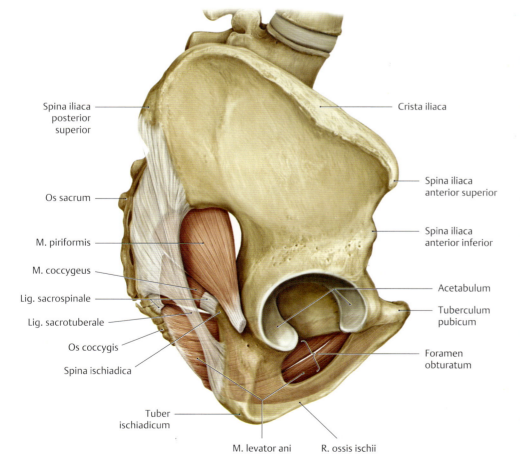

C Funnel shape of the m. levator ani
Pelvis viewed from the right side. A portion of the os ischii is shown as transparent. A contraction of the funnel-shaped m. levator ani leads to lifting of the anus (levatore = to rise) and to forward motion caused by extension of the m. puborectalis. At its neutral position, the levator funnel stands at a steep angle but flattens when contracted. Because the muscle has a cone or funnel shape, during defecation, only the peripheral parts of the m. levator ani contract; the four areas close to the sphincter, including the m. puborectalis, go limp so that the anus descends, and the stool is pushed through the opened canalis analis.

The Trunk Wall — 11. Musculature: Topographical Anatomy

11.12 Pelvic Floor Muscles: Their Relation to Organs and Vessels in Males and Females

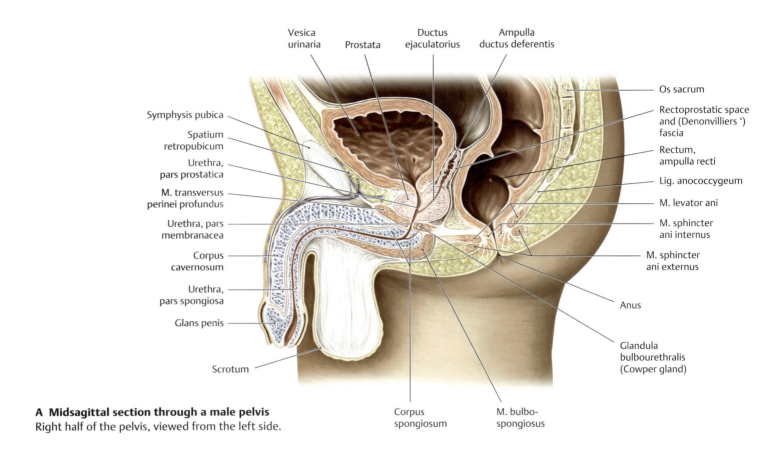

A Midsagittal section through a male pelvis
Right half of the pelvis, viewed from the left side.

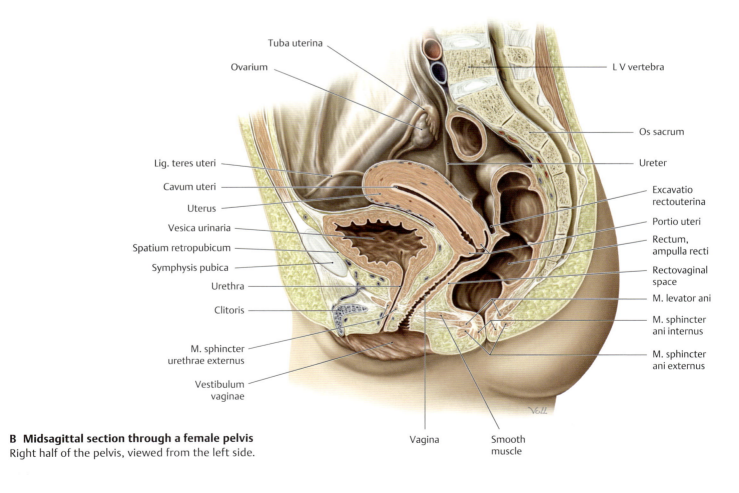

B Midsagittal section through a female pelvis
Right half of the pelvis, viewed from the left side.

The Trunk Wall —— 11. Musculature: Topographical Anatomy

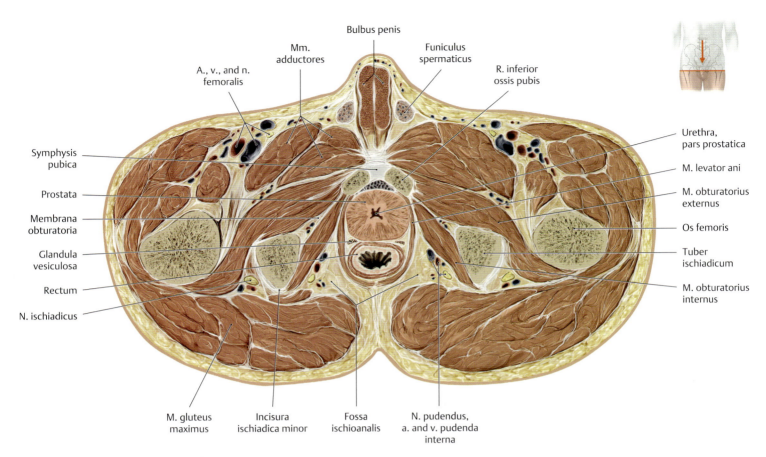

C Cross section through a male pelvis
Superior view.

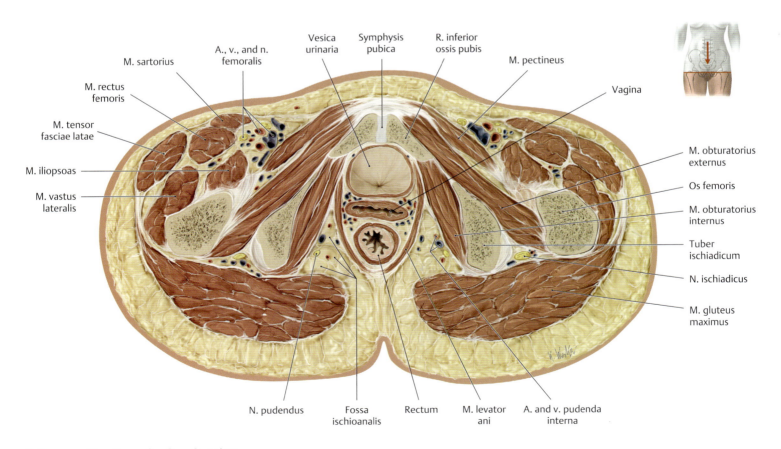

D Cross section through a female pelvis
Superior view.

12.1 The Arteries

A Overview of the arteries of the trunk wall

The arrangement of the neurovascular structures in the trunk reflects the segmental anatomy of the trunk wall, particularly in the thoracic region. Accordingly, each of the intercostal spaces is traversed by arteria, vena and nervus intercostalis.

The chest wall is supplied principally by the aa. intercostales posteriores, which arise from the aorta, and by rr. intercostales anteriores arising from the a. thoracica interna:

- Aa. intercostales posteriores I and II, which are given off by the a. intercostalis suprema (= branch of the truncus costocervicalis, see **D a**)
- Aa. intercostales posteriores III–XI (each giving off a r. dorsalis, collateralis, and a r. cutanus lateralis, see **D b**)
- A. musculophrenica (one of the two terminal branches of the a. thoracica interna), which runs behind the costal arch, see **B**
- A. subcostalis (a. intercostalis XII), see **B**
- Aa. intercostales anteriores, which arise from the a. thoracica interna, see **B**

Many other "regional" arteries supply the anterior, lateral, and posterior trunk wall.

Anterior trunk wall
- Rr. perforantes (from the a. thoracica interna, e.g., the rr. mammarii mediales that supply the mamma), see **D b**
- A. epigastrica superior (continuation of the a. thoracica interna, see **B** and **C**)
- A. epigastrica inferior (from the a. iliaca externa, see **B** and **C**)
- A. epigastrica superficialis, see **B**
- A. circumflexa ilium superficialis, see **B**
- A. circumflexa ilium profunda, see **B**

Posterior trunk wall
- Rr. dorsales (from the a. intercostalis posterior) each with a r. cutaneus medialis and lateralis, and a r. spinalis, see **D c**
- Aa. lumbales I–IV (each with a r. dorsalis and spinalis), see **B**
- A. sacralis mediana, see **B**

Lateral trunk wall
- A. thoracica superior, see **B**
- A. thoracoacromialis, see **B**
- A. thoracica lateralis, see **B**
- Rr. cutanei laterales (from the a. intercostalis) which distribute branches mainly to the mamma (rr. mammarii laterales, see **D b**)
- A. iliolumbalis (from the a. iliaca interna) which gives off a r. iliacus, lumbalis, and spinalis, see **B**

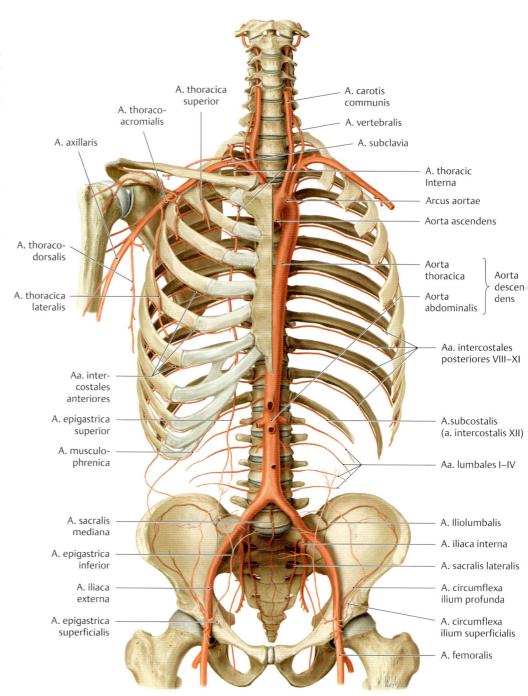

B Arteries of the trunk wall
Anterior view. The anterior portions of the ribs have been removed on the left side.

Trunk Wall — 12. Neurovascular Systems: Forms and Relations

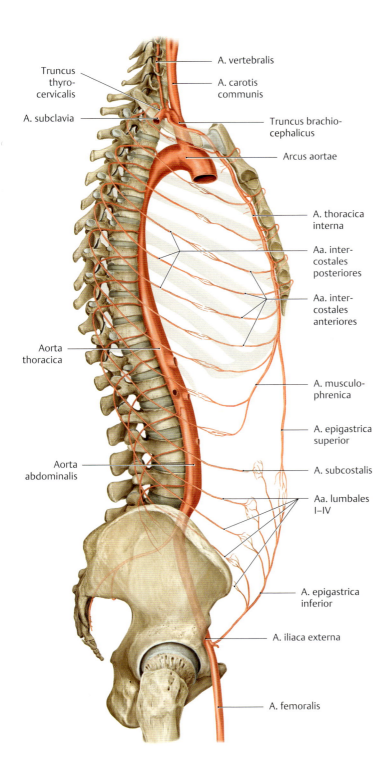

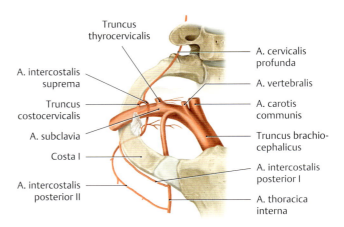

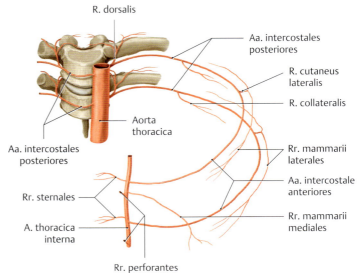

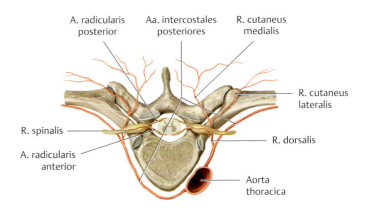

C Arteries of the trunk wall
Right lateral view.

D Course and branches of the intercostal arteries

a Anterior view of the a. intercostalis suprema, which gives off the first two aa. intercostales.
Note: The aa. intercostales posteriores I and II are not branches of the aorta thoracica but arise from the a. intercostalis suprema (branch of the truncus costocervicalis), which are branches of the a. subclavia.

b Anterior view of the aa. intercostales posteriores, which are segmental branches of the aorta thoracica.
Note: The aa. intercostales *anteriores* arise from the a. subclavia (via the a. thoracica interna), while the aa. intercostales *posteriores* arise directly from the aorta thoracica.

c Branches of the aa. intercostales posteriores, viewed from the superior view.

191

12.2 The Veins

A Overview of the veins of the trunk wall
The veins of the trunk wall drain into both the vena caval and azygos systems (see **B**). Within the vena caval system, we can distinguish between the tributary regions of the v. cava inferior and the v. cava superior. Connections between the superior and inferior vena cavae are called *cavocaval anastomoses* (collateral channels).

Tributaries of the v. cava superior
- V. intercostalis suprema (v. brachiocephalica) (see **B**)
- Vv. intercostales anteriores (v. thoracica interna, v. subclavia) (see **D**)
- V. epigastrica superior (v. thoracica interna, v. subclavia)
- V. thoracica lateralis (v. axillaris) (see **C**)
- V. thoracoepigastrica (v. axillaris) (see **C**)

Tributaries of the v. cava inferior (see **B**)
- Vv. intercostales posteriores
- V. subcostalis   } V. lumbalis ascendens
- Vv. lumbales I–IV

- V. iliolumbalis   } V. iliaca communis
- V. sacralis mediana

- V. circumflexa ilium profunda   } V. iliaca externa
- V. epigastrica inferior

- V. sacralis lateralis
- V. obturatoria (see p. 212)   } V. iliaca interna
- V. pudenda interna (see p. 230)

- Vv. pudendae externae
- V. circumflexa ilium superficialis   } V. femoralis
- V. epigastrica superficialis

Tributaries of the v. azygos (see **B**)
- Vv. intercostales superiores
- Vv. intercostales posteriores
- V. hemiazygos
- V. hemiazygos accessoria
- Vv. columnae vertebrales, see **Ea**

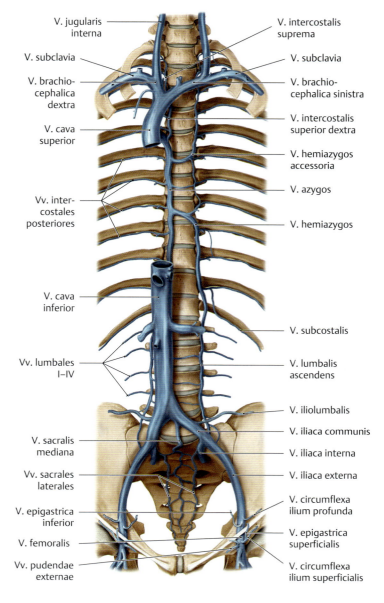

B Major veins in the trunk
Anterior view.

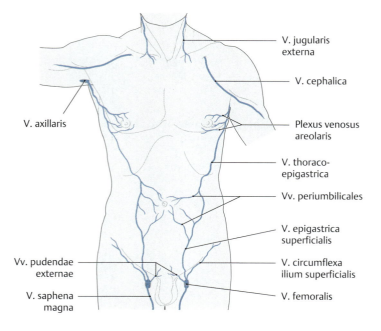

C Superficial veins of the anterior trunk wall
Anterior view. Normally, these veins are not palpable, but they are of key importance in the development of portocaval anastomoses, in which the former v. umbilicalis connects the v. portae hepatis to the vv. cavae superior and inferior. In cases where portal hypertension develops as a result of liver disease (hepatic cirrhosis due to alcohol abuse), the portal venous blood must partially bypass the liver, flowing through the vv. paraumbilicales (see p. 210) to the superficial truncal veins in the umbilical region (vv. periumbilicales) and finally to the heart. Since the superficial veins (vv. superficiales) must carry considerably more blood in this situation, they undergo varicose dilation, becoming visible and palpable on the abdomen. This is also called a "Medusa's head" (*caput medusae*) because the serpentine, dilated veins around the umbilicus resemble the snake-haired head of the Gorgon.

Trunk Wall — 12. Neurovascular Systems: Forms and Relations

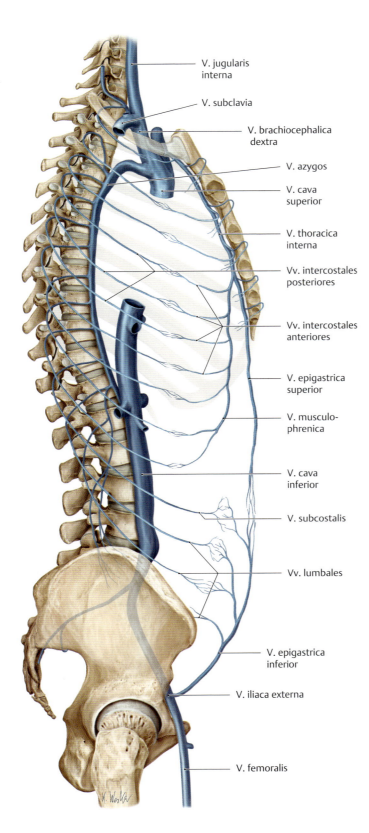

D Veins of the trunk wall
Right lateral view.

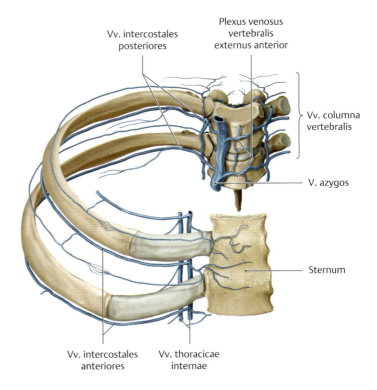

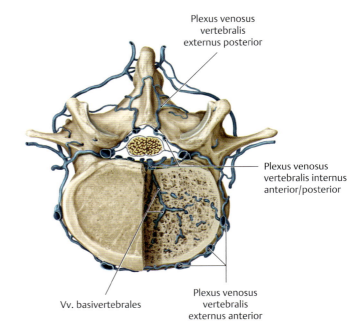

E Intercostal veins and venous plexuses of the vertebral canal
a Columna vertebralis and costa segment, anterosuperior view.
b Lumbar vertebra, superior view.

193

12.3 The Lymphatic Vessels and Lymph Nodes

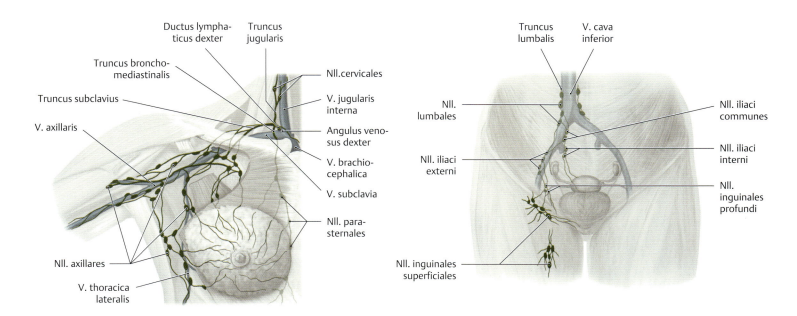

A Regional lymph nodes and their associated lymphatics
Anterior view.

a Axillary, parasternal, and cervical lymph nodes (right thoracic and axillary region with the arm abducted). The lymph node levels are described on p. 209.
b Lymph nodes of the inguinal region and lesser pelvis.

B Left and right venous angles
Anterior view. The approximately 1-cm-long *ductus lymphaticus dexter* collects lymph from the right upper quadrant of the body (see **Ca**) and empties into the **right venous angle** (angulus venosus dexter) at the junction of the right v. jugularis interna with the right v. subclavia. Its major tributaries are

- the truncus jugularis dexter (right half of the head and neck),
- the truncus subclavius dexter (right upper limb, right side of the chest and back wall), and
- the truncus bronchomediastinalis dexter (organs of the right thoracic cavity).

The *ductus thoracicus* is approximately 40 cm long and transports lymph from the entire lower half of the body and left upper quadrant. It empties into the **left venous angle** (angulus venosus sinister) between the left v. jugularis interna and left v. subclavia. Its main tributaries are

- the truncus jugularis sinister (left half of the head and neck),
- the truncus subclavius sinister (left upper limb, left side of the chest and back wall),
- the truncus bronchomediastinalis sinister (organs of the left thoracic cavity),
- the trunci intestinales (abdominal organs), and
- the truncus lumbalis dexter and sinister (right and left lower limb; pelvic viscera; right and left pelvic, abdominal, and back wall).

The vasa lymphatica intercostalia transport lymph from the upper left and right intercostal spaces to the corresponding ipsilateral lymphatic duct. The lower intercostal spaces (spatii intercostales) from both sides drain to the cisterna chyli.

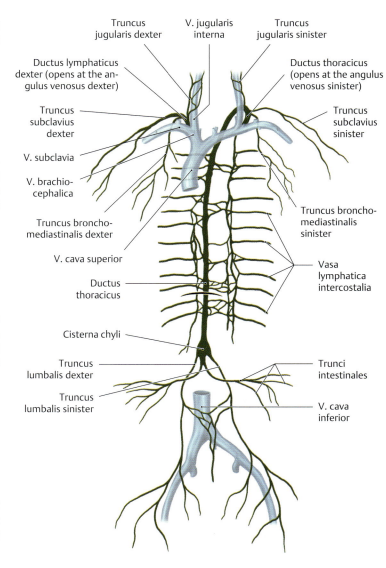

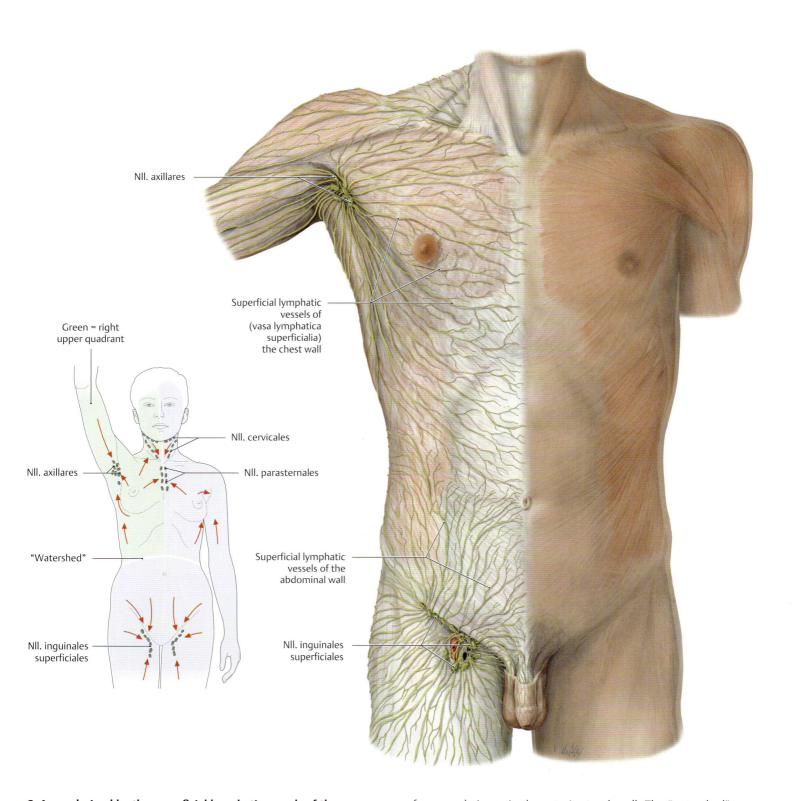

C Areas drained by the superficial lymphatic vessels of the anterior trunk wall

Anterior view.

a Lymphatic pathways and regional lymph nodes (nodi lymphoidei regionales) of the anterior trunk wall (arrows indicate the direction of lymph flow).
b Superficial network of lymphatic vessels in the right anterior trunk wall (vasa lymphatica superficialia).

Lymph from the skin of the trunk wall is collected mainly by the the nll. axillares and inguinales superficiales, following the general pattern of venous drainage in the anterior trunk wall. The "watershed" zone between the two drainage regions is defined by a curved line above the umbilicus and below the costal arch. Lymph from the regional axillary and inguinal lymph nodes (nodi lymphoidei inguinales) is finally collected by two lymphatic trunks, each of which drains into the jugulosubclavian angulus venosus on the corresponding side of the body (right or left angulus venosus, see **B**). Lymphatic fluid from the right upper quadrant (green) is returned to the venous system by the *ductus lymphaticus dexter*, while lymph from the other three body quadrants (violet) is returned to the veins by the *ductus thoracicus*.

12.4 The Nerves

A Anterior and posterior branches (rami) of the spinal nerves
The trunk wall receives most of its sensory nerve supply (nn. sensorii) from the anterior and posterior rami of the T 1–T 12 spinal cord segments (nn. intercostales and rr. dorsales nn. spinalium) (see also p. 200).

Spinal cord segment (medulla spinalis)	Anterior branches (rr. ventrales)	Posterior branches (rr. dorsales)
C1	} Plexus cervicalis	N. suboccipitalis
C2		N. occipitalis major
C3		N. occipitalis tertius (see p. 200)
C4		
C5	} Plexus brachialis	
C6		
C7		
C8		
T1		
T2	} Nn. intercostales and n. subcostalis	} Rr. dorsales nn. spinalium*
T3		
T4		
T5		
T6		
T7		
T8		
T9		
T10		
T11		
T12		
L1	} Plexus lumbalis	
L2		
L3		
L4		
L5		
S1	} Plexus sacralis	
S2		
S3		
S4		
S5	} Plexus coccygeus (see p. 542)	
Co1		
Co2		

* The rr. dorsales n. spinalium of the L 1–L 3 nervi spinales are also known as the nn. clunium superiores, those from S 1–S 3 as the nn. clunium medii (see **C**).
Note: The nn. clunium *inferiores* are anterior rami from the plexus sacralis; see also p. 536.

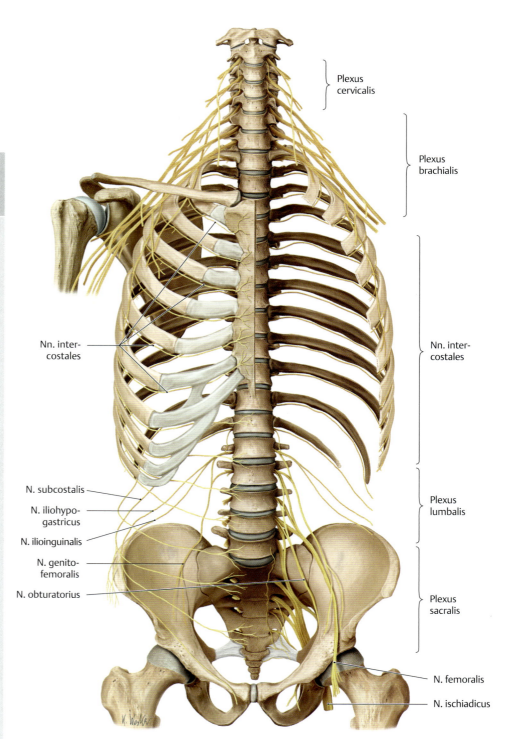

B Nerves of the trunk wall
Anterior view. The anterior part of the left half of the thoracic cage (cavea thoracis) has been removed. The trunk wall receives most of its motor and sensory innervation from the 12 thoracic nervi spinales. Of all the nervi spinales, these 12 pairs most clearly reflect the original segmental (metameric) organization of the body. The rr. ventrales run foward in the intercostal spaces (spatium intercostale) to become the nn. intercostales, while the rr. dorsales are distributed to the intrinsic back muscles and to the skin of the back. Portions of the trunk wall are also supplied by nerves from the plexus cervicalis (nn. supraclaviculares), plexus brachialis (e.g., n. thora-cicus longus), and plexus lumbalis (e.g., n. ilioinguinalis).

Trunk Wall — *12. Neurovascular Systems: Forms and Relations*

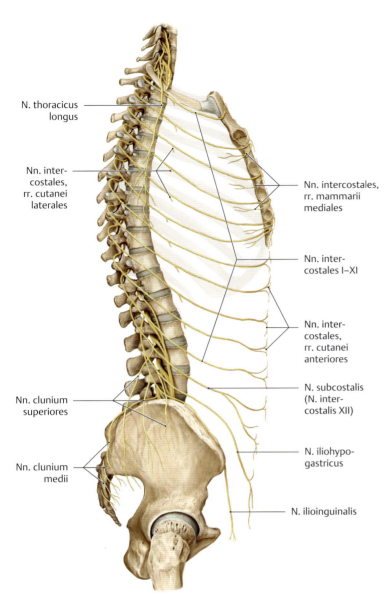

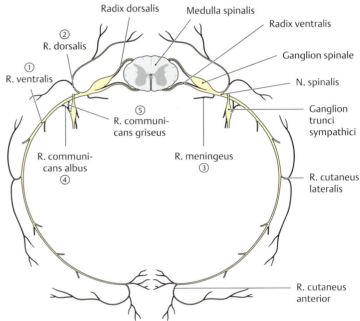

D Course of the intercostal nerves
Right side, anterior view.

C Course of the nerves on the lateral trunk wall
Right lateral view.
Note the segmental arrangement of the nn. intercostales (compare with the segmental arrangement of the arteriae and veinae, pp. 190 and 192).

E Branches of a spinal nerve
Formed by the union of the posterior (sensory) and anterior (motor) roots (radix spinalis), the approximately 1-cm-long spinal nerve (n. spinalis) courses through the foramen intervertebrale and divides into five branches after exiting the canalis vertebralis (see **F**).

F Spinal nerve branches and the territories they supply

Spinal nerve branch	Motor or visceromotor territory	Sensory territory
① R. ventralis (anterior)	All somatic muscles except for the intrinsic back muscles	Skin of the lateral and anterior trunk wall and of the upper and lower limbs
② R. dorsalis (posterior)	Intrinsic back muscles	Posterior skin of the head and neck, skin of the back and buttock
③ R. meningeus		Spinal meninges, ligaments of the spinal column, capsules of the facet joints
④ R. communicans albus	Carries preganglionic fibers from the nervus spinalis to the truncus sympathicus ("white" because the preganglionic fibers are myelinated)	
⑤ R. communicans griseus*	Carries postganglionic fibers from the truncus sympatheticus back to the nervus spinalis ("gray" because the fibers are unmyelinated)	

*Strictly speaking, the r. communicans griseus is not a branch of the n. spinalis but a branch passing from the truncus sympathicus to the n. spinalis.

13.1 Anterior Trunk Wall: Surface Anatomy and Superficial Nerves and Vessels

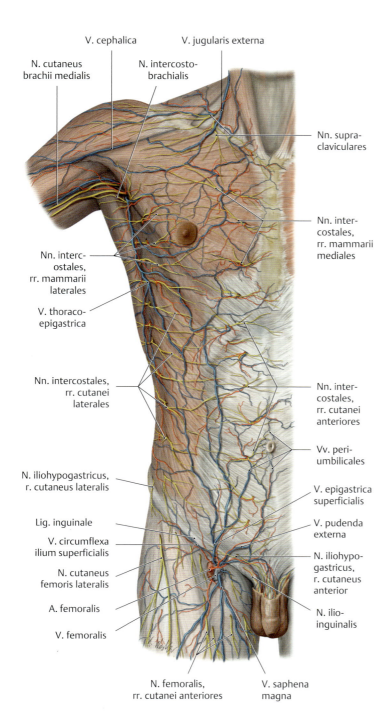

A Superficial cutaneous vessels and nerves of the anterior trunk wall
Anterior view.
Superficial vessels: Most of the *arterial supply* to the anterior trunk wall comes from two sources: the a. thoracica interna and the a. epigastrica superficialis. The superficial *veins* drain chiefly into the v. axillaris (via the v. thoracoepigastrica) and into the v. femoralis (via the vv. epigastrica superficialis and circumflexa ilium superficialis). The vv. peri- and paraumbilicales provide the main communication between the superficial veins of the trunk wall and the portal veins (portocaval anastomoses).
Superficial nerves: The *sensory supply* to the anterior trunk wall has a largely segmental arrangement (provided, for example, by rr. cutanei laterales and anteriores from the intercostal nerves). The plexus cervicalis (nn. supraclaviculares)) is additionally involved in the thoracic region, as is the plexus lumbalis (e.g., nn. iliohypogastricus, ilioinguinalis) in the lower abdominal region.

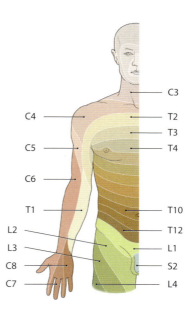

B Segmental (radicular) cutaneous innervation of the anterior trunk wall (dermatomes)
Right half of the trunk and adjacent upper limb, anterior view. Every sensory nerve root (radix dorsalis) innervates a specific skin area with its fibers. These "dermatomes" (see p. 86), then, correspond to associated spinal cord segments. The dermatomes are arranged in bandlike patterns that encircle the chest wall and upper abdomen. Below the umbilicus, the dermatomes become angled slightly downward toward the median plane. A "segmental gap" exists between the C4 and T2 dermatomes because the phylogenetic outgrowth of the human upper limb has removed the sensory fibers of C5–C8 and T1 from the trunk wall (after Mumenthaler).

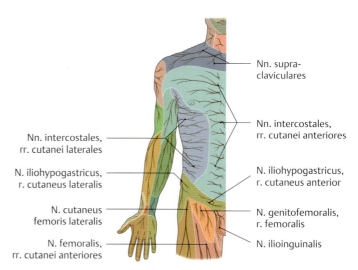

C Peripheral sensory cutaneous innervation of the anterior trunk wall
Right half of the trunk and adjacent upper limb, anterior view. The color-coded map of the peripheral cutaneous nerve territories (rr. cutanei) follows the branching pattern of the cutaneous nerves in the subcutaneous connective tissue. Besides the cutaneous branches of the intercostal nerves (rr. cutanei anteriores and laterales), it is chiefly the nn. supraclaviculares and the nn. iliohypogastricus and ilioinguinalisthat supply the skin of the anterior trunk wall (after Mumenthaler).

Trunk Wall — 13. Neurovascular Systems: Topographical Anatomy

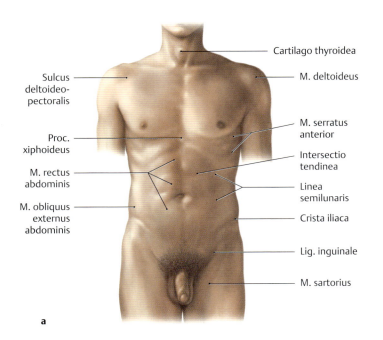
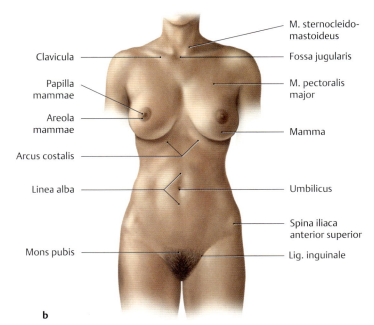

D Surface anatomy of the anterior trunk wall
a Male, b female.

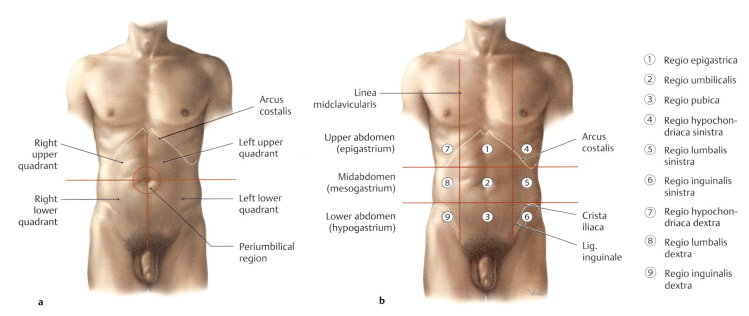

E Criteria for dividing the abdomen into regions
a The abdomen is divided into four quadrants by two perpendicular lines that intersect at the umbilicus.
b Coordinate system composed of two vertical and two horizontal lines. They divide the abdomen into nine regions, each located in either the upper, middle, or lower abdomen. The two vertical lines represent the left and right midclavicular lines. The two horizontal lines pass through the lowest point of the tenth ribs or the summit of the two cristae iliacae (see p. 35).

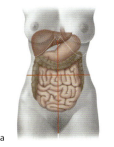

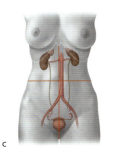

F Projection of the abdominal organs onto the four quadrants of the anterior abdominal wall
a Organs of the anterior layer, b organs of the middle layer,
c organs of the posterior layer.
The organs of the *anterior* layer abut the anterior abdominal wall. The organs of the *middle* layer are located in the posterior part of the cavitas abdominis (some are partially retroperitoneal), and those of the *posterior* layer are located outside or behind the actual abdominal cavity (i.e., they are retroperitoneal).

199

13.2 Posterior Trunk Wall: Surface Anatomy and Superficial Nerves and Vessels

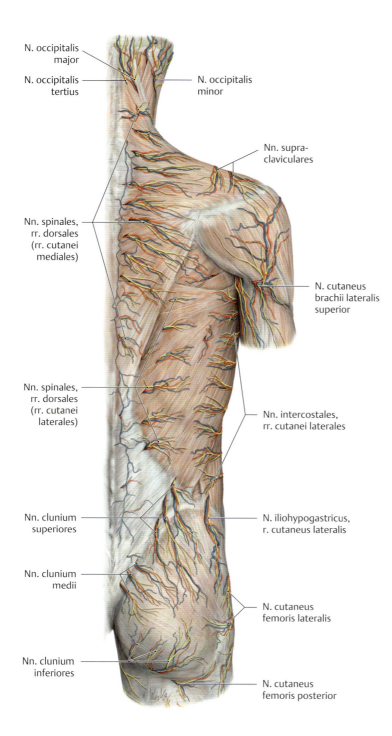

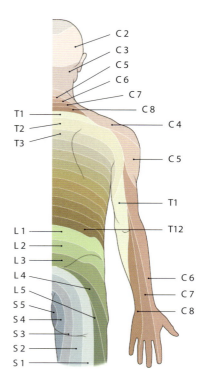

B Segmental (radicular) cutaneous innervation of the posterior trunk wall (dermatomes)
Right half of the trunk and adjacent upper limb, posterior view (after Mumenthaler).

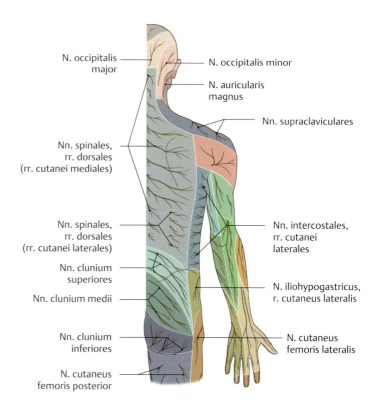

A Superficial cutaneous vessels and nerves of the posterior trunk wall

Posterior view. Except for the lower buttocks and lateral portions of the trunk wall, the posterior trunk wall derives its sensory innervation from rr. dorsales of the nn. spinales and from rr. cutanei laterales of the nn. intercostales. This is a predominantly segmental innervation pattern, analogous to that described in the anterior trunk wall. Both the rr. cutanei medialis and lateralis pass with the cutaneous vessels through the intrinsic back muscles to the skin of the back. The skin of the buttocks is supplied by lateral branches from the three cranial lumbar and sacral nerves (nn. clunium superiores and nn. clunium medii).

Note: The lower part of the buttock is supplied by the nn. clunium inferiores, which are branches of the plexus sacralis; thus, they are derived from the *rr. ventrales* of the nn. spinales.

C Peripheral sensory cutaneous innervation of the posterior trunk wall
Right half of the trunk and adjacent upper limb, posterior view (after Mumenthaler).

Trunk Wall — 13. Neurovascular Systems: Topographical Anatomy

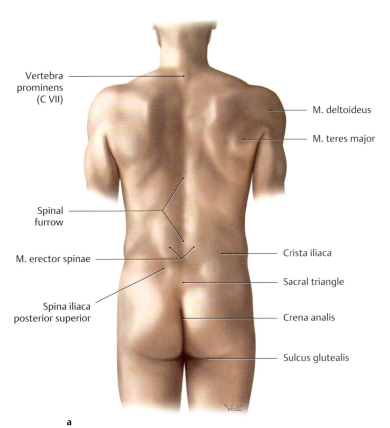

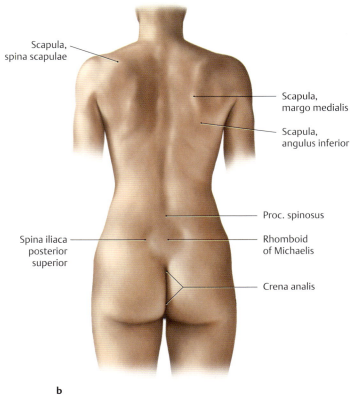

a

b

D Surface anatomy of the posterior trunk wall
a Male, b female.
In both sexes a *spinal furrow* runs vertically in the posterior midline of the trunk below the C VII proc. spinosus. It is formed by the fixation of the subcutaneous tissue to the corresponding processi spinosi. At the sacral level in males, the furrow widens to form the *sacral triangle* (bounded by the right and left spina iliaca posterior superior and the upper part of the crena analis). The corresponding diamond-shaped area in females is called the *rhomboid of Michaelis* (see **F**).

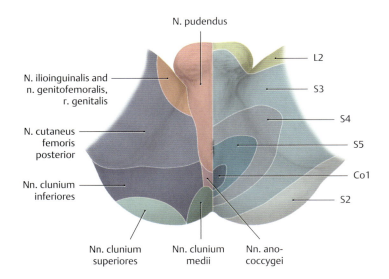

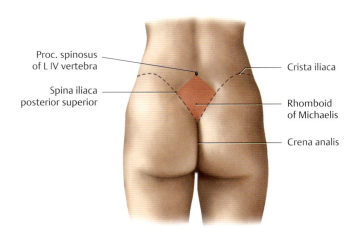

E Segmental and peripheral cutaneous innervation of the male perineal region
Lithotomy position. The segments or dermatomes have been mapped on the left side of the body, and the areas supplied by the peripheral cutaneous nerves are shown on the right side (after Mumenthaler).

F Anatomic boundaries of the Michaelis rhomboid
Female gluteal region, posterior view. In women the sacral triangle is expanded to form a diamond-shaped figure with the following boundaries: the left and right spina iliaca posterior superior, the proc. spinosus of the LIV vertebra, and the upper part of the crena analis. With a normal female pelvis, the vertical and horizontal dimensions of the rhomboid are approximately equal. The shape of the Michaelis rhomboid (named for the German gynecologist G. A. Michaelis, 1798–1848) reflects the width of the female pelvis, providing an indirect indicator of the size of the birth canal.

13.3 Posterior Trunk Wall, Posterior View

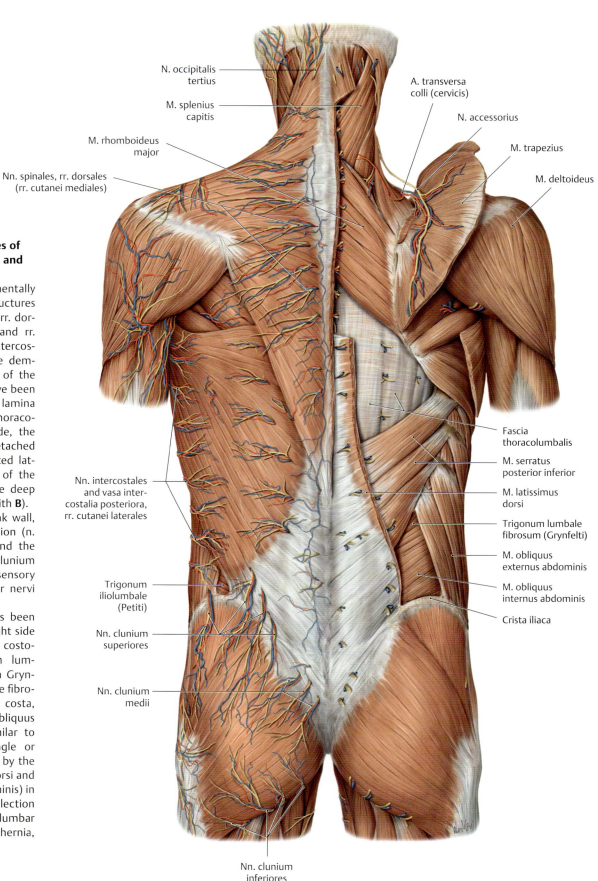

A Neurovascular structures of the posterior trunk wall and nuchal region

Posterior view. The segmentally arranged neurovascular structures of the posterior trunk wall (rr. dorsales of the nn. spinales and rr. dorsales of the posterior intercostal and lumbar vessels) are demonstrated on the left side of the trunk (all muscle fasciae have been removed except for the lamina superficialis of the fascia thoracolumbalis). On the right side, the m. trapezius has been detached from its origins and reflected laterally to show the course of the a. transversa cervicis in the deep scapular region (compare with **B**).

Note: On the posterior trunk wall, only the lateral nuchal region (n. occipitalis minor, see **C**) and the lower gluteal region (nn. clunium inferiores) receive their sensory supply from ramus anterior nervi spinalis.

The m. latissimus dorsi has been partially removed on the right side to demonstrate the upper costolumbar triangle (trigonum lumbale fibrosum or "trigonum Grynfelti"). The trigonum lumbale fibrosum (boundaries: twelfth costa, m. erector spinae, and m. obliquus internus abdominis) is similar to the lower iliolumbar triangle or "trigonum Petiti" (bounded by the crista iliaca, m. latissimus dorsi and m. obliquus externus abdominis) in that it creates a site of predilection for rare, usually acquired lumbar hernias (Grynfeltt or Petit hernia, see also p. 221).

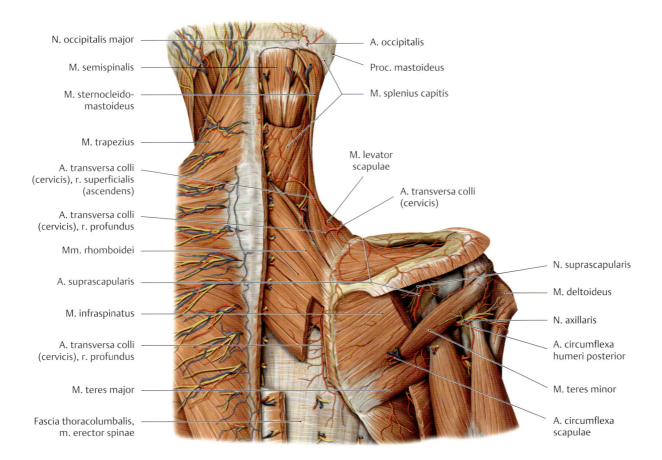

B Arteries of the deep scapular region
Right scapular region, posterior view. The mm. trapezius, splenius capitis, deltoideus, infraspinatus, and rhomboideus major and minor have been completely or partially removed on the right side. The deep scapular region is supplied by the aa. transversa colli (cervicis), cervicalis profunda (see **C**), suprascapularis, circumflexa scapulae, and circumflexa humeri posterior. All of these vessels arise directly or indirectly—via the truncus thyrocervicalis—from the a. subclavia (neither is visible here). The aa. suprascapularis, circumflexa scapulae, dorsalis scapulae, and circumflexa humeri posterior form the *"scapular arcade"* (see p. 389). Medial to the processus mastoideus, the a. occipitalis appears below the tendon of insertion of the m. sternocleidomastoideus and runs upward with the sensory n. occipitalis major to the skin of the occiput. The n. occipitalis major pierces both the m. trapezius and m. semispinalis capitis in the area of their firm tendinous attachments. It may become compressed at these sites, leading to *occipital neuralgia*.

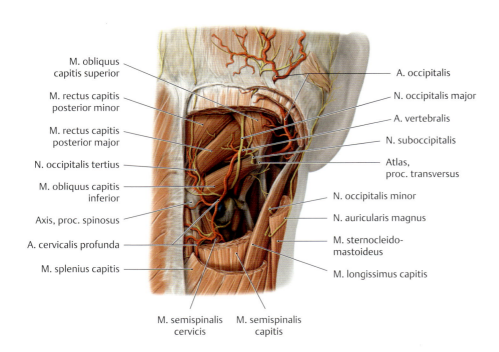

C Suboccipital triangle (trigonum arteriae vertebralis or trigonum suboccipitale)
Posterior view. The mm. trapezius, sternocleidomastoideus, splenius capitis, and semispinalis capitis have been removed to display the suboccipital region on the right side. The suboccipital triangle is bounded by the mm. suboccipitales (mm. recti capitis posteriores major, and the mm. obliquii capitis superior and inferior). In the deep portion of the triangle, the a. vertebralis runs through sulcus a. vertebralis of the atlas. The n. suboccipitalis (C 1), which is purely motor, emerges above the arcus posterior atlantis to supply the short muscles of the head. The n. occipitalis major (C 2) and, at a lower level, the n. occipitalis tertius (C 3) wind posteriorly as they pass the lower margin of the m. obliquus capitis inferior. The a. cervicalis profunda, a branch of the truncus costocervicalis, runs between the mm. semispinales capitis and cervicis.

13.4 Posterior Trunk Wall, Anterior View

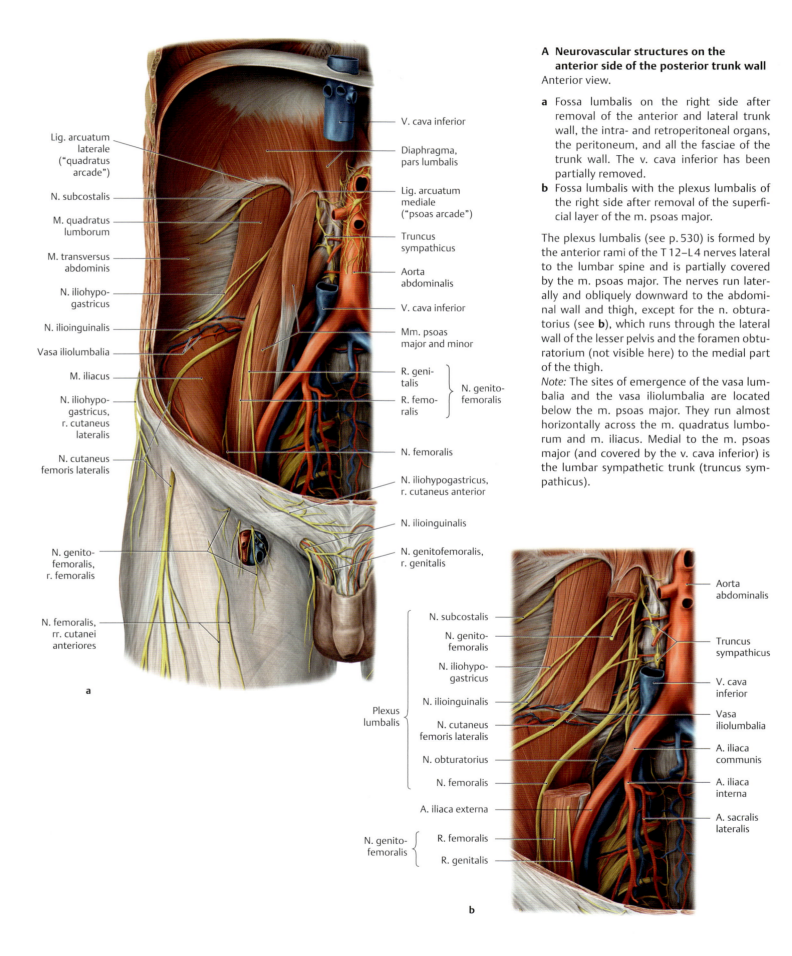

A Neurovascular structures on the anterior side of the posterior trunk wall
Anterior view.

a Fossa lumbalis on the right side after removal of the anterior and lateral trunk wall, the intra- and retroperitoneal organs, the peritoneum, and all the fasciae of the trunk wall. The v. cava inferior has been partially removed.

b Fossa lumbalis with the plexus lumbalis of the right side after removal of the superficial layer of the m. psoas major.

The plexus lumbalis (see p. 530) is formed by the anterior rami of the T 12–L 4 nerves lateral to the lumbar spine and is partially covered by the m. psoas major. The nerves run laterally and obliquely downward to the abdominal wall and thigh, except for the n. obturatorius (see **b**), which runs through the lateral wall of the lesser pelvis and the foramen obturatorium (not visible here) to the medial part of the thigh.

Note: The sites of emergence of the vasa lumbalia and the vasa iliolumbalia are located below the m. psoas major. They run almost horizontally across the m. quadratus lumborum and m. iliacus. Medial to the m. psoas major (and covered by the v. cava inferior) is the lumbar sympathetic trunk (truncus sympathicus).

Trunk Wall — 13. Neurovascular Systems: Topographical Anatomy

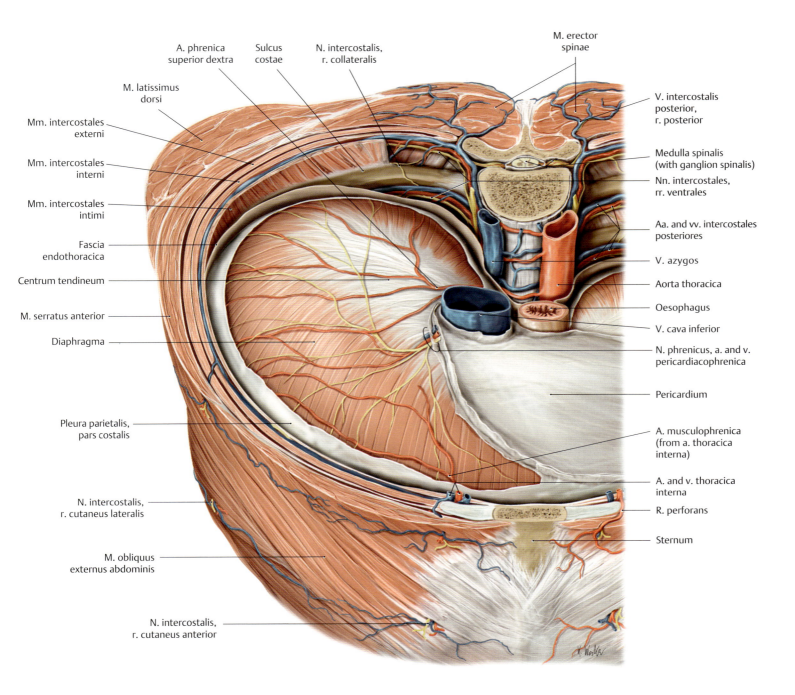

B Neurovascular structures of the posterior trunk wall at the thoracic level

Transverse section through the thorax after removal of the thoracic organs, pleura parietalis, and part of the fascia endothoracica, anterosuperior view. The chest wall receives its arterial blood supply from the aa. intercostales posteriores and is drained by the vv. intercostales, which empty into the azygos system. The intercostal vessels run with the intercostal nerves along the inferior border of the associated costa, lodged in the sulcus costae.

205

13.5 Anterior Trunk Wall: Overview and Location of Clinically Important Nerves and Vessels

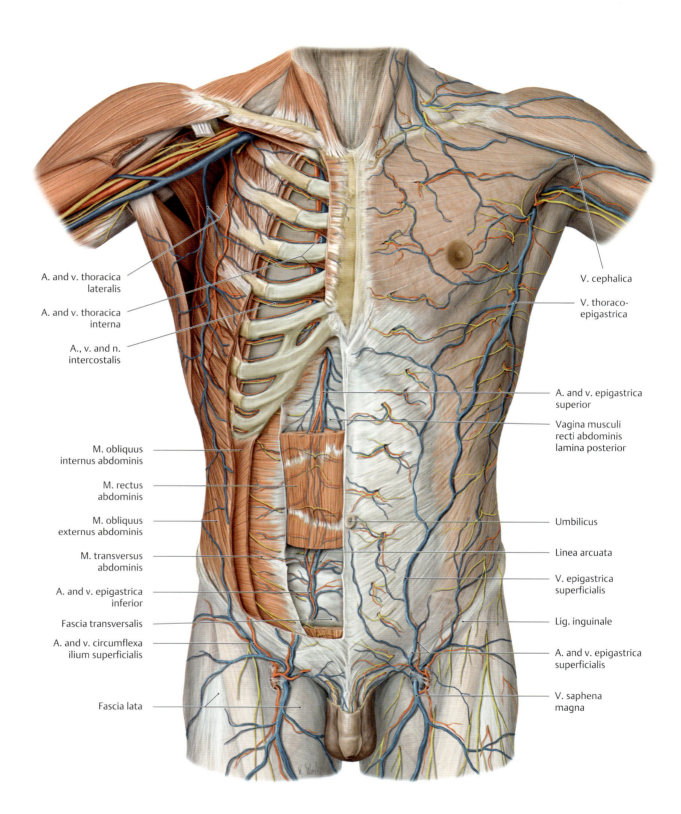

A Neurovascular structures on the anterior side of the anterior trunk wall

Anterior view. The superficial (subcutaneous) neurovascular structures are demonstrated on the left side of the trunk and the deep neurovascular structures on the right side. For this purpose the mm. pectorales major and minor have been completely removed on the right side, and the mm. obliquii externus and internus abdominis have been partially removed. Portions of the right m. rectus abdominis have been removed or rendered transparent to demonstrate the vasa epigastrica inferiora. Finally, the intercostal spaces have been exposed to display the course of the vasa intercostalia and nn. intercostales.

Trunk Wall — 13. Neurovascular Systems: Topographical Anatomy

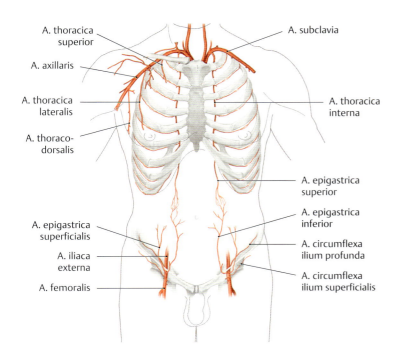

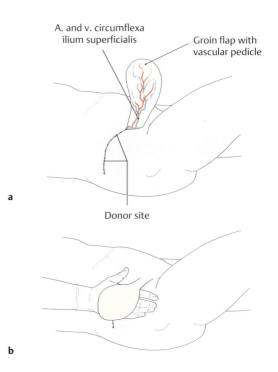

B The arterial supply of the anterior trunk wall
Anterior view. The anterior trunk wall receives its blood supply from two main sources: the a. thoracica interna, which arises from the a. subclavia, and the a. epigastrica inferior, which arises from the a. iliaca externa. It is also supplied by smaller vessels arising from the a. axillaris (a. thoracica superior, a. thoracodorsalis, and a. thoracica lateralis) and from the a. femoralis (a. epigastrica superficialis and a. circumflexa ilium superficialis).

C Importance of the a. circumflexa ilium superficialis in harvesting groin skin flaps for plastic surgery
a Dissection of a groin skin flap based on the a. circumflexa ilium superficialis.
b The groin skin flap transferred to the dorsum of the right hand (after Weber).

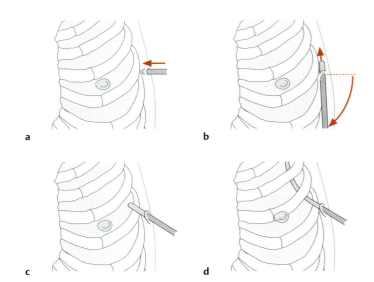

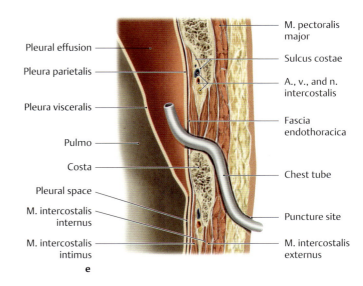

D Placement of a chest tube to drain a pleural effusion
A chest tube may be inserted to drain an abnormal fluid collection from the pleural space, such as a pleural effusion due to bronchial carcinoma. The best site for placing the chest tube can be determined by percussion or ultrasound examination. Generally, one optimum puncture site in the sitting patient is at the level of the seventh or eighth spatium intercostale on the linea axillaris posterior (see **e** and p. 34).
a–d Steps in the placement of a chest tube (after Henne-Bruns, Dürig, and Kremer). Anterior view.
a A skin incision is made under local anesthesia, and the drainage tube is introduced perpendicular to the chest wall.
b On reaching the ribs, the tube is angled 90° and advanced cephalad in the subcutaneous plane, parallel to the chest wall.
c On reaching the next higher spatium intercostale, the tube is passed through the mm. intercostales above the costa.
d The tube is then advanced into the cavitas pleuralis.
e Preserving the a., v., and n. intercostales during the insertion of a chest tube.
Longitudinal section through the chest wall at the level of the linea axillaris posterior, after placement of the chest tube. The drain should always be introduced at the *upper margin of a rib* to avoid injuring the a., v., and n. intercostales. For additional puncture sites, see textbooks on surgery.

207

13.6 Anterior Trunk Wall: Nerves, Blood Vessels, and Lymphatics in the Breast

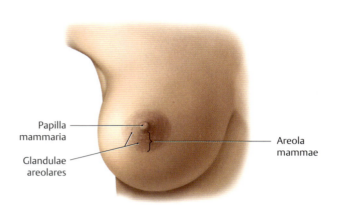

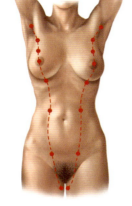

A Shape and appearance of the breast
Right breast, anterior view. The breast (mamma) is shaped like a cone that is more rounded in its lower half than in the upper quadrants. It consists of the glandular tissue (glandula mammaria) and a fibrous stroma that contains fatty tissue. The excretory ducts of the glandular tissue open on the cone-shaped nipple (papilla mammaria), which lies at the center of the more heavily pigmented areola mammae. Numerous small protuberances mark the openings of apocrine sweat glands and free sebaceous glands (glandulae areolares).

B The mammary ridges
The rudiments of the glandulae mammariae form in both sexes along the mammary ridges, appearing as an epidermal ridge extending from the axilla to the inguinal region on each side. Although rarely the mammary ridges may persist in humans to form accessory nipples (polythelia), normally all the rudiments disappear except for the thoracic pair. By the end of fetal development, ducti lactiferi have sprouted into the subcutis from the two remaining epithelial buds. After menarche, breast development in females is marked by growth of the fibrous stroma and proliferation of the glandular tree in response to stimulation by sex hormones.

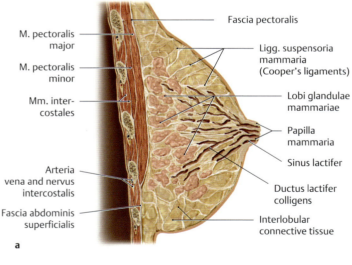

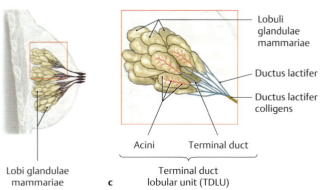

C Gross and microscopic anatomy of the breast
a Sagittal section. The base of the adult breast extends from the second to the sixth costa along the linea midclavicularis and directly overlies the mm. pectoralis major, serratus anterior, and obliquus externus abdominis. It is loosely attached to the fascia pectoralis and adjacent fascial planes (fasciae axillaris and abdominis superficialis) by connective tissue. The breast is additionally supported, especially in its upper portion, by permeating bundles of connective tissue (the ligg. suspensoria mammaria, or *Cooper's ligaments*). The glandular tissue is composed of 10 to 20 individual lobes (lobi glandulae mammariae), each of which has its major ductus lactifer colligens, which opens on the nipple (papilla mammaria) by way of a dilated segment, the sinus lactifer (structure of the lobus shown in **b**). The glands and ducti lactiferi are surrounded by firm, fibrofatty tissue that has a rich blood supply.
b Sagittal section of the duct system and portions of a lobus. A mammary lobe resembles a tree composed of branching ductus lactiferi, which terminate in smaller lobuli (approximately 0.5 mm in diameter). In the *nonlactating breast* (as shown here), these lobuli contain rudimentary acini that are arranged in clusters of small epithelial buds without a visible lumen.
c The terminal duct lobular unit (TDLU). One Lobulus and its terminal duct make up the basic secretory unit of the breast. Each lobulus is composed of acini that empty into a terminal ductule. The associated intralobular connective tissue (mantle tissue) contains stem cells that give rise to the tremendous cell growth (proliferation of the duct system and differentiation of the acini) that occurs during the transformation to the *lactating breast*. The TDLU is of key importance in pathohistology because it is the site where most malignant breast tumors originate (after Lüllmann).

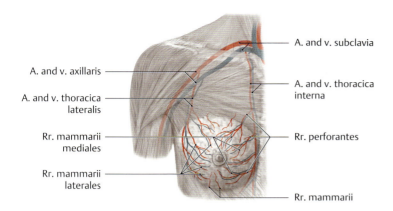
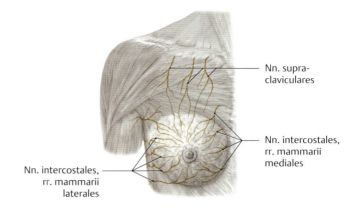

D Blood supply to the breast
The breast derives its blood supply from rr. perforantes of the a. thoracica interna (= rr. mammarii mediales from the second through fourth intercostal spaces), branches of the a. thoracica lateralis (rr. mammarii laterales), and direct branches from the second through fifth intercostal arteries (rr. mammarii). The breast is drained by the vv. thoracicae interna and lateralis.

E Nerve supply to the breast
The sensory innervation of the breast has a segmental arrangement and is supplied by branches of the second through sixth intercostal nerves (rr. mammarii laterales and mediales). Branches of the plexus cervicalis (nn. supraclaviculares) also supply the upper portion of the breast.

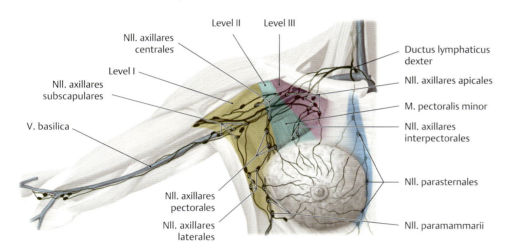
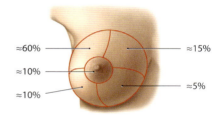

G Distribution of malignant tumors by quadrant in the breast
The numbers indicate the average percentage location of malignant breast tumors.

F Lymphatic drainage of the breast
The lymphatic vessels of the breast can be divided into a superficial, subcutaneous, and deep system. The deep system begins with lymphatic capillaries at the acinar level (see **Cb** and **c**) and is particularly important as a route for tumor metastasis. The main regional filtering stations are the axillary and parasternal lymph nodes, the approximately 30 to 60 axillary lymph nodes receiving most of the lymphatic drainage. They are the first nodes to be affected by metastasis (see **G**) and therefore have major oncological significance. The *axillary lymph nodes (nodi lymphoidei axillares)* are subdivided into levels (see p. 358):

- **Level I: lower axillary group** (lateral to the m. pectoralis minor):
  - Nll. axillares pectorales
  - Nll. axillares subscapulares
  - Nll. axillares laterales
  - Nll. paramammarii
- **Level II: middle axillary group** (at the level of the m. pectoralis minor):
  - Nll. axillares interpectorales
  - Nll. axillares centrales
- **Level III: upper infraclavicular group** (medial to the m. pectoralis minor):
  - Nll. axillares apicales

The *parasternal lymph nodes (nodi lymphoidei parasternales)*, which are distributed along the Vasa thoracica interna, chiefly drain the medial portion of the mamma. From there, tumor cells may spread across the midline to the opposite side. The survival rate in breast cancer patients correlates most strongly with the number of involved lymph nodes at the various axillary nodal levels. The parasternal lymph nodes are rarely important in this regard. According to Henne-Bruns, Dürig, and Kremer, the 5-year survival rate is approximately 65% with metastatic involvement of level I, 31% with involvement of level II, but approaches 0% in patients with level III involvement. This explains the key prognostic importance of a *sentinel lymphadenectomy* (removal of the sentinel lymph node). This technique is based on the assumption that every point in the integument drains via specific lymphatic pathways to a particular lymph node, rarely draining to more than one. Accordingly, the lymph node that is the first to receive lymph from the primary tumor will be the first node to contain tumor cells that have spread from the primary tumor by lymphogenous metastasis. The specific lymphatic drainage path, and thus the sentinel node, can be identified by scintigraphic mapping with radio-labeled colloids (^{99m}Tc sulfur microcolloid), which has superseded the older technique of patent blue dye injection. The first lymph node to be visualized is the sentinel node. That node is selectively removed and histologically examined for the presence of tumor cells. If the sentinel node does not contain tumor cells, generally the rest of the axillary nodes will also be negative. This method is 98% accurate in predicting the level of axillary nodal involvement prior to surgery.

13.7 Anterior Trunk Wall: The Inguinal Canal (Canalis inguinalis)

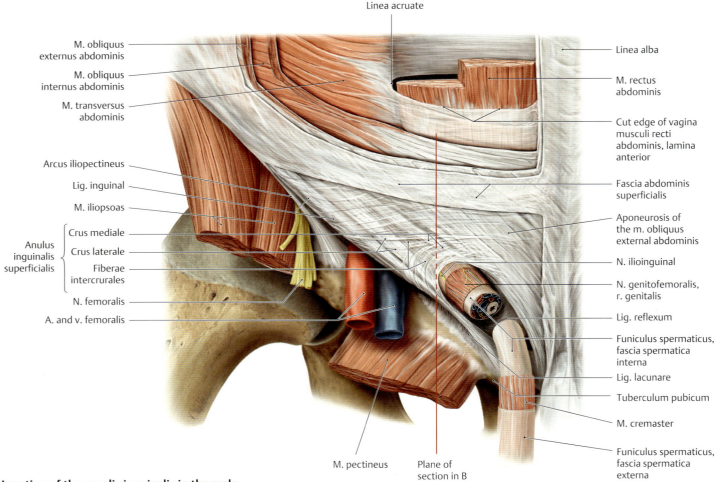

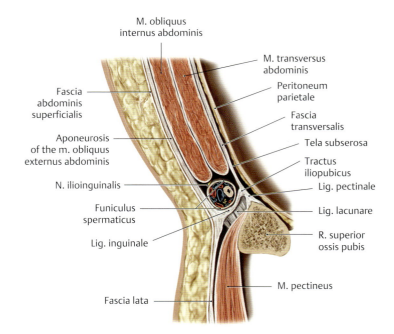

A Location of the canalis inguinalis in the male
Right inguinal region, anterior view. Approximately 4 to 6 cm long, the canalis inguinalis passes obliquely forward, downward, and medially above the inguinal ligament (lig. inguinale) to pierce the anterior abdominal wall. It begins internally at the deep inguinal ring (anulus inguinalis profundus; **D** and **E**) in the fossa inguinalis lateralis (see p. 212) and opens externally at the superficial inguinal ring, (anulus inguinalis superficialis), lateral to the tuberculum pubicum. With the fascia abdominis superficialis removed, this "external opening" of the canal can be identified as a slitlike orifice in the aponeurosis of the m. obliquus externus abdominis (external oblique aponeurosis). It is bounded by the *crus mediale* superomedially and by the *crus laterale* inferolaterally. Both crura are interconnected by the *fibrae intercrurales*. The anulus inguinalis superficialis is completed internally by arched fibers from the lig. inguinale (lig. reflexum), forming a deep groove. The canalis inguinalis in the male provides a pathway for the descent of the testis during fetal life (see p. 224). Its contents in the male (after testicular descent) include the funiculus spermaticus, and its contents in the female include the ligamentum teres uteri.

B Sagittal section through the canalis inguinalia in the male
Medial view. *Note* the structures that form the walls of the canalis inguinalis above and below the funiculus spermaticus and on the anterior and posterior sides (compare with **C**). The openings and wall structures of the canalis inguinalis bear an important relationship to the pathophysiology of hernias (after Schumpelick).

C Openings and wall structures of the inguinal canal (canalis inguinalis)

The canalis inguinalis resembles a flattened tube with an internal and external opening (see below), a floor, a roof, and anterior and posterior walls. A lumen is present only after its contents have been removed (the funiculus spermaticus in males, the lig. teres uteri/a. lig. teretis uteri in females, the n. ilioinguinalis and vasa lymphatica in both sexes). The canalis inguinalis remains patent for life, especially in males, and thus forms a path for potential herniation through the abdominal wall (see p. 213).

Openings of the canalis inguinalis (see A)

Superficial inguinal ring (anulus inguinalis superficialis)	Opening in the aponeurosis of m. obliquus externus abdominis bounded by the crus mediale, crus laterale, fibrae intercrurales, and lig. reflexum
Deep inguinal ring (anulus inguinalis profundus)	Opening between the lig. interfoveolare, lig. inguinale, and plica umbilicalis lateralis; formed by an outpouching of the fascia transversalis (becomes the fascia spermatica interna) (see p. 212)

Wall structures of the canalis inguinalis (see B)

Floor	Lig. inguinale (densely interwoven fibers of the lower aponeurosis of m. obliquus externus abdominisand adjacent fascia lata of the thigh)
Roof	M. transversus abdominis and m. obliquus internus abdominis
Anterior wall	Aponeurosis of m. obliquus externus abdominis
Posterior wall	Fascia transversalis and peritoneum (partially thickened by the lig. interfoveolare)

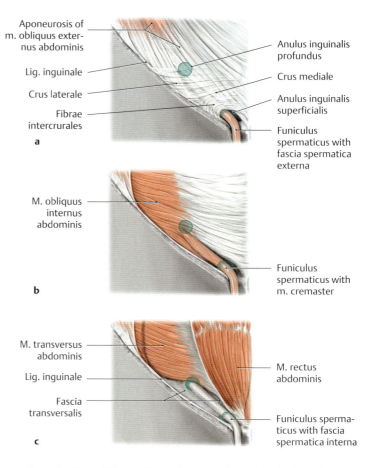

D Contribution of the oblique abdominal muscles to the structure of the male canalis inguinalis
Right inguinal region, anterior view.
a–c Progressive removal of the abdominal wall muscles.

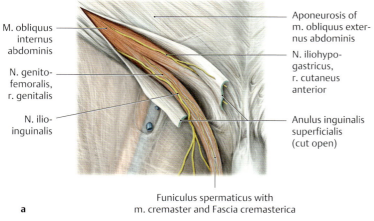

E The canalis inguinalis, progressively opened to expose the spermatic cord (funiculus spermaticus)
Right inguinal region, anterior view.

a Division of the aponeurosis of m. obliquus externus abdominis reveals the m. obliquus internus abdominis, some of whose fibers are continued onto the funiculus spermaticus as the m. cremaster. The r. genitalis of the n. genitofemoralis runs with it below the fascia cremasterica (see p. 532). The n. ilioinguinalis runs through the canalis inguinalis on the funiculus spermaticus. Its sensory fibers pass through the anulus inguinalis superficialis to the skin over the symphysis pubica and are distributed to the lateral portion of the scrotum or labia majora and medial thigh.

b With the m. obliquus internus abdominis divided and the m. cremaster split, the full course of the funiculus spermaticus through the canalis inguinalis can be displayed. The funiculus spermaticus appears at the anulus inguinalis profundus where the fascia transversalis is invaginated into the canalis inguinalis (and encloses the funiculus spermaticus on its way to the testis as the fascia spermatica interna). It runs below the m. transversus abdominis along the posterior wall of the canalis inguinalis (fascia transversalis and peritoneum). The wall at the midportion of the canal is formed by the lig. interfoveolare and is reinforced medially by the lig. reflexum. Medial to the lig. interfoveolare, deep to which run the epigastric vessels (Hesselbach triangle), and superior to the lig. inguinale is the fossa inguinalis medialis, a weak spot in the abdominal wall that is a common site for direct inguinal hernias (after Schumpelick; see also p. 213).

13.8 Anterior Abdominal Wall: Anatomy and Weak Spots

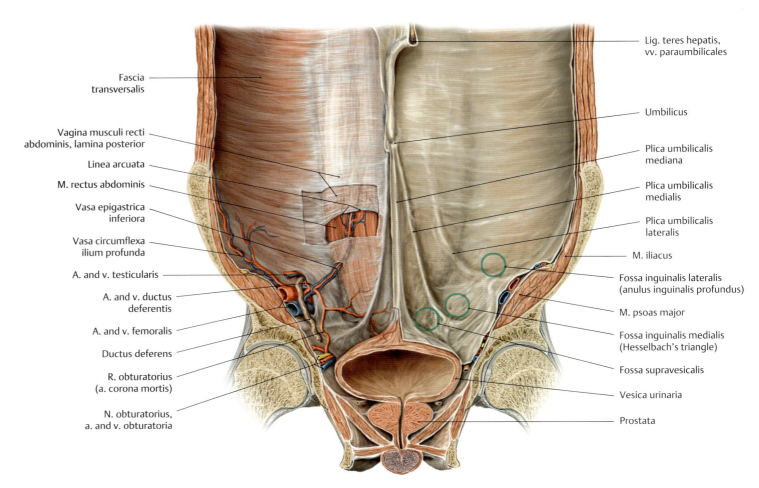

A Internal surface anatomy of the anterior abdominal wall in the male
Coronal section through the abdominal and cavitas pelvis at the level of the articulationes coxae, posterior view. All of the abdominal and pelvic organs have been removed except for the vesica urinaria and prostata. Por-tions of the peritoneum and fascia transversalis have also been removed on the left side. The internal surface anatomy of the lower abdominal wall is marked by five peritoneal folds (plicae), which extend toward the umbilicus:

- An unpaired *plica umbilicalis mediana* on the midline (contains the oblit-erated urachus)
- Paired left and right *plicae umbilicales mediales* (contain the left and right obliterated a. umbilicalis)
- Paired left and right *plicae umbilicales laterales* (contain the left and right vasa epigastrica inferiora)

Located between the peritoneal folds on each side are three more or less distinct fossae, which are sites of potential herniation through the anterior abdominal wall:
- The *fossa supravesicalis*, located between the plica umbilicalis mediana and plica umbilicalis medialis above the apex of the bladder
- The *fossa inguinalis medialis* (Hesselbach's triangle), located between the plica umbilicalis medialis and plica umbilicalis lateralis
- The *fossa inguinalis lateralis*, located lateral to the plica umbilicalis lateralis (site of the anulus inguinalis profundus)

B Internal and external openings for abdominal hernias
Above the lig. inguinale, the plicae umbilicales mediana, medialis, and lateralis (see **A**) form three sites of weakness on each side of the abdom- inal wall where indirect and direct inguinal hernias and suprapubic hernias typically occur. Another weak spot is located *below the lig. inguinale* and medial to the v. femoralis in the anulus femoralis. At that site the anulus femoralis is covered only by loose, compliant connective tissue, the septum femorale, which is permeated by numerous lymphatic vessels (vasa lymphatica). The sharp-edged lig. lacunare forms the medial border of the anulus femoralis and can contribute to the incarceration of a femoral hernia (see p. 215).

Hernia	Internal opening*	External opening*
	Above the lig. inguinale:	
Hernia supravesicalis	Fossa supravesicalis	Anulus inguinalis superficialis
Hernia inguinalis directa	Fossa inguinalis medialis (Hesselbach's triangle)	Anulus inguinalis superficialis
Hernia inguinalis indirecta	Fossa inguinalis lateralis (anulus inguinalis profundus)	Anulus inguinalis superficialis
	Below the lig. inguinale:	
Hernia femoralis	Anulus femoralis/ Femoral ring	Hiatus saphenus (fossa ovalis)

*The internal opening is where the hernia starts and the external opening is where it is observed clinically.

Trunk Wall — 13. Neurovascular Systems: Topographical Anatomy

C Definition, occurrence, and structure of an abdominal hernia
Between the thorax and bony pelvis is an extended skeletal gap that is covered by multiple abdominal wall layers composed of broad muscles, fasciae, aponeuroses, and peritoneum. The muscular foundation is deficient at certain locations, and the abdominal wall at those sites is formed entirely by connective-tissue structures. These are the areas of greatest weakness (Loci minoris resistentiae) in the abdominal wall. Occa-sionally, these weak spots cannot withstand a rise of intra-abdominal pressure, and openings are created for the herniation of abdominal viscera. The term "hernia" (L. *hernia* = "rupture") refers to the protrusion of peritoneum parietale through an anatomic opening (e.g., inguinal or femoral hernia) or a secondary defect (e.g., umbilical hernia). An *external hernia* is one that protrudes from the cavitas abdominis and is visible on the body surface, while an *internal hernia* protrudes into a peritoneal pouch that is contained within the abdomen. Hernias are also classified by the time of their occurrence as *congenital* (e.g., umbilical hernia, indirect inguinal hernia through a patent processus vaginalis) or *acquired* (e.g., direct inguinal hernia, femoral hernia). The following components of a hernia are important in terms of surgical treatment:

Hernial opening: the orifice or defect through which the viscus herniates.
Hernial sac: the pouch, generally lined by peritoneum parietale, that contains the herniated viscus. Its size is highly variable, depending on the extent of the hernia.
Hernial contents: may be almost any intra-abdominal viscus but usually consist of omentum majus or loops of small bowel.
Coverings: the tissue layers surrounding the hernial sac. The composition of the coverings depends on the location and mechanism of the hernia.

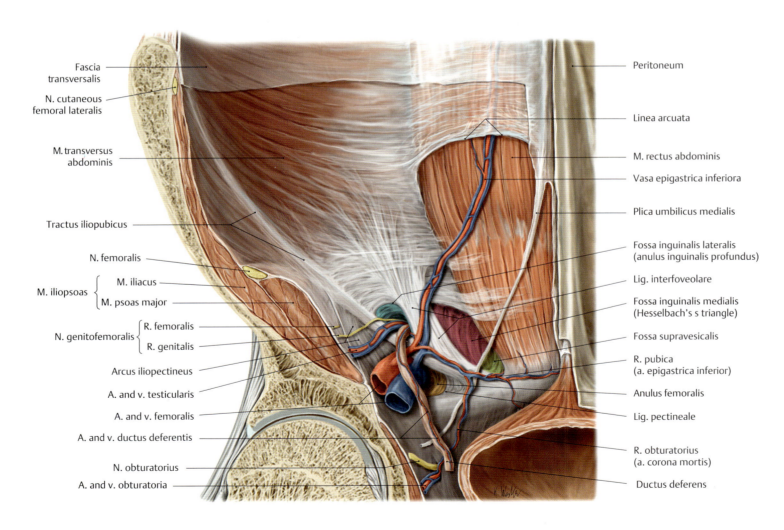

D Internal hernial openings in the male inguinal and femoral region
Detail from **A**, posterior view. The peritoneum and fascia transversalis have been partially removed to reveal the hernial openings more clearly. The internal openings (see **A** and **B**) for indirect and direct inguinal hernias, femoral hernias, and suprapubic (= supravesical) hernias are indicated by color shading.

213

13.9 Inguinal and Femoral Hernias

A Hernias of the groin region: inguinal and femoral hernias*

Hernia	Openings and course
• **Direct (medial) inguinal hernia** (hernia inguinalis directa) Always acquired	• *Internal opening:* fossa inguinalis medialis (Hesselbach's triangle), i.e., above the lig. inguinale and medial to the a. and v. epigastricae inferiores • *Course:* sac is perpendicular to the abdominal wall • *External opening:* anulus inguinalis superficialis
• **Indirect (lateral) inguinal hernia** (hernia inguinalis indirecta) Congenital (patent proc. vaginalis peritonei) or acquired	• *Internal opening:* anulus inguinalis profundus, i.e., above the lig. inguinale and lateral to the a. and v. epigastricae inferiores • *Course:* sac passes through the canalis inguinalis • *External opening:* anulus inguinalis superficialis
• **Femoral hernia** (Hernia femoralis) Always acquired	• *Internal opening:* lacuna vasorum (anulus femoralis and septum femorale), i.e., below the lig. inguinale and medial to the vasa femoralis • *Course:* sac passes through the canalis femoralis below the fascia lata • *External opening:* hiatus saphenus (fossa ovalis)

*80% of all hernias are inguinal hernias (90% of which occur in males), and approximately 10% are femoral hernias (more common in females) (see also p. 216). Inguinal hernias are among the most common structural defects in humans, accounting for some 20% of all surgical operations.

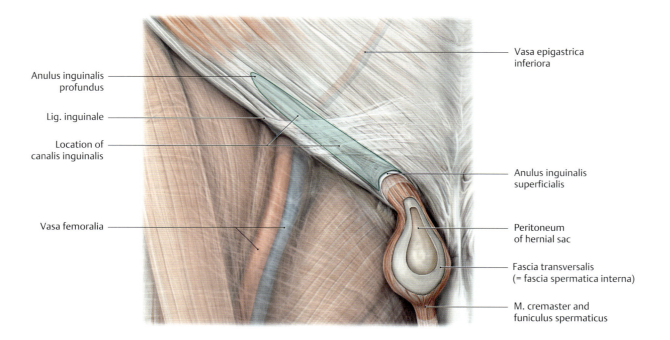

B Congenital or acquired indirect inguinal hernia (hernia inguinalis indirecta)
Right male inguinal region with the skin and superficial body fascia removed, anterior view. The fascia lata of the thigh is shown transparent, and the funiculus spermaticus is windowed. Regardless of the location of the *internal* opening, both *indirect (lateral)* and *direct (medial)* inguinal hernias (see **C**) emerge from the anulus inguinalis superficialis above the lig. inguinale. Indirect inguinal hernias may be congenital (due to a patent proc. vaginalis peritonei) or acquired and follow a tract parallel to the abdominal wall. The hernial sac enters the expanded anulus inguinalis profundus (internal opening), which is lateral to the epigastric vessels, and passes medially and obliquely along the canalis inguinalis. Invested by the fascia enclosing the funiculus spermaticus, it finally emerges at the anulus inguinalis superficialis and descends into the scrotum. The coverings of all indirect inguinal hernias include the peritoneum and the fascia transversalis (fascia spermatica interna).

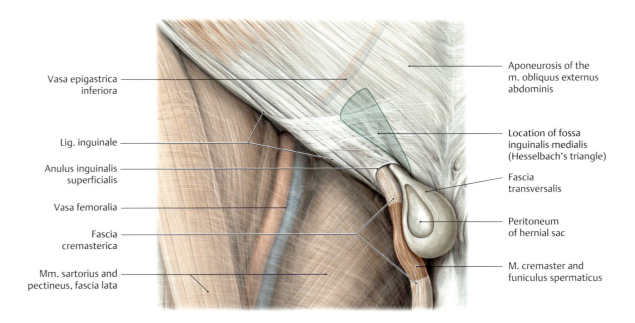

C Acquired direct inguinal hernia (hernia inguinalis directa)
Right inguinal region in the male with the skin and superficial fascia removed, anterior view. The fascia lata of the thigh is lightly shaded. Direct inguinal hernias are always acquired and follow a tract that is perpendicular to the abdominal wall. The hernia will leave the cavitas abdominis medial to the vasa epigastrica inferiora within the fossa inguinalis medialis (see p. 217). Such hernias may pass through the anulus inguinalis superficialis, and in such instances the hernial sac is medial to the funiculus spermaticus.

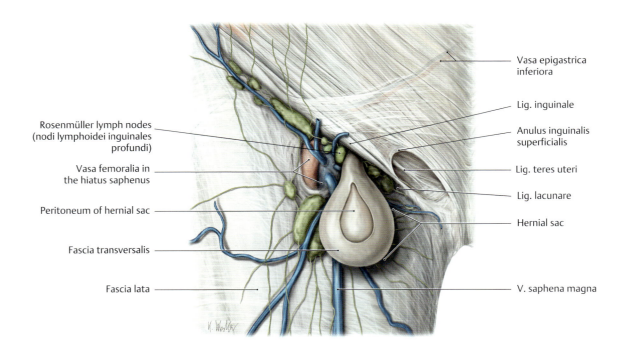

D Acquired femoral hernia (hernia femoralis)
Right female inguinal region with the skin and superficial fascia removed, anterior view. Femoral hernias are always acquired and are more common in women (broader pelvis and larger anulus femoralis). They always pass inferior to the lig. inguinale and medial to the v. femoralis, extending through the anulus femoralis to enter the canalis femoralis (not visible here). The funnel-shaped canal begins at the anulus femoralis (the internal opening of the hernia, not visible here), ends approximately 2 cm inferiorly at the hiatus saphenus (fossa ovalis), and lies anterior to the fascia pectinea. The sharp-edged lig. lacunare forms the medial boundary of the anulus femoralis (risk of incarceration). Typically, the canalis femoralis is occupied by loose fatty and connective tissue and deep inguinal lymph nodes (nodi lymphoidei inguinales profundi). Femoral hernias may emerge at the hiatus saphenus (external opening), which is covered by the thin lamina cribrosa, and thus become subcutaneous.

13.10 Topographic Anatomy of Inguinal Hernias

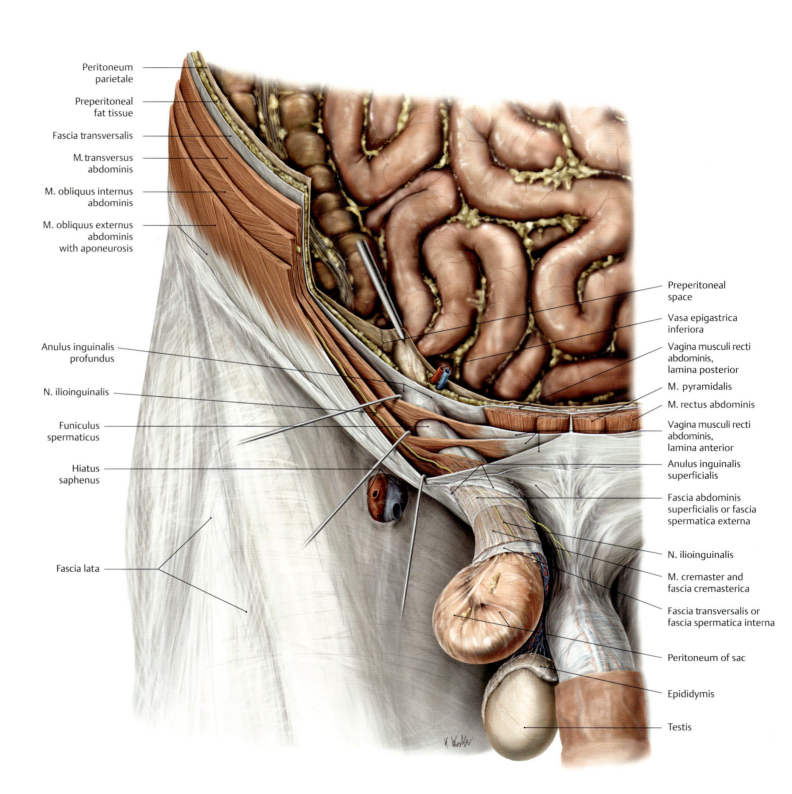

A Acquired indirect inguinal hernia (hernia inguinalis indirecta) in a male

Right side anterior view, skin and abdominal walls largely removed. In the area of the canalis inguinalis that runs from upper lateral to lower medial, the individual abdominal wall muscles have been spread apart to reveal the relationship of structures. The hernial sac of the indirect inguinal hernia (a hernia that runs through the canalis inguinalis) containing an intestinal loop is visible. Because the hernial sac is within the funiculus spermaticus, the hernia is covered by peritoneum parietale, fascia transversalis, and the m. cremaster and fascia cremasterica.

B Schematic representation of the canalis inguinalis, the abdominal wall layers, and their continuation into the covering of the testes

Anterior view; several section planes have been combined in this schematic. The two arrows point to the fossa inguinalis lateralis (= anulus inguinalis profundus) and the fossa inguinalis medialis (Hesselbach's triangle). The vasa epigastrica inferiora extend between them in the plica umbilicalis lateralis. The fossa inguinalis lateralis is the internal hernial orifice for indirect inguinal hernias (see **Da**); the fossa inguinalis medialis, for direct inguinal hernias (see **Db**). Both types of inguinal hernias use the anulus inguinalis superficialis as the external hernial orifice.

Note: The tunica vaginalis testis with its lamina visceralis and lamina parietalis is the remnant of the obliterated processus vaginalis peritonei (**b**) (compare p. 224).

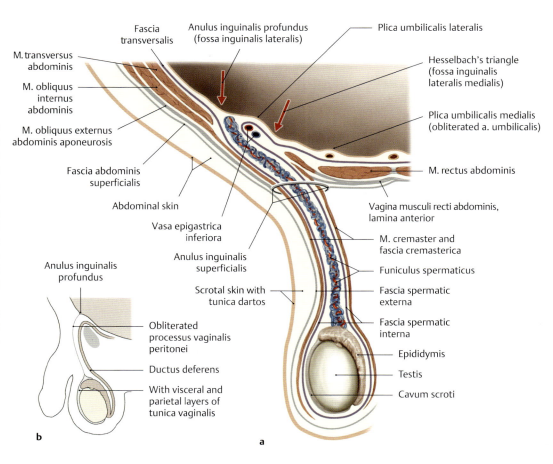

C Comparison of the layers of the abdominal wall and the corresponding coverings of the spermatic cord and testes

The coverings of the funiculus spermaticus and testes are derivatives of muscles and fasciae of the abdominal wall and envelop the funiculus spermaticus and testes in a sac (scrotum), whose superficial layer is formed from the abdominal skin.

Abdominal wall layers	Coverings of the spermatic cord and testes
• Abdominal skin and membranous layer of fascia abdomins superficialis	→ Scrotal skin with tunica dartos
• Aponeurosis m. obliqui externi abdominis	→ Fascia spermatica externa
• M. obliquus internus abdominis	→ M. cremaster with fascia cremasterica
• Fascia transversalis	→ Fascia spermatica interna
• Peritoneum	→ Tunica vaginalis testis, lamina parietalis and lamina visceralis

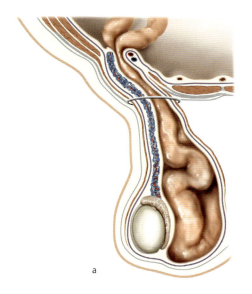 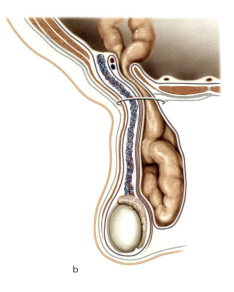

D Comparison of indirect and direct inguinal hernias

a Indirect inguinal hernia (congenital or acquired): The inner hernial orifice (fossa inguinalis lateralis = anulus inguinalis profundus) is lateral to the vasa epigastrica inferiora. The hernial contents (e.g., intestinal loops), which are enclosed by the peritoneum parietale, either push through the canalis inguinalis to the scrotum (acquired hernia) or move through the still-open proc. vaginalis peritonei (see **Bb**) to the scrotum (congenital hernia). The hernial sac is made of the same elements in both cases: peritoneum parietale, fascia transversalis and m. cremaster.

b Direct inguinal hernia (acquired): The inner hernial orifice (fossa inguinalis medialis = Hesselbach's triangle) is medial to the vasa epigastrica inferiora. The hernial contents push directly through the abdominal wall and via the anulus inguinalis superficialis to the scrotum. In comparison to indirect inguinal hernias, the hernial sac consists only of peritoneum parietale and fascia transversalis.

13.11 Diagnosis and Treatment of Hernias

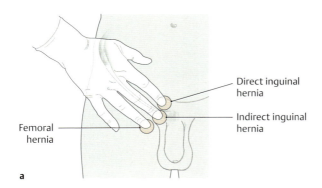

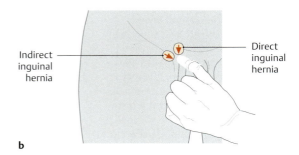

A Techniques for the examination of inguinal and femoral hernias
Hernias of the groin region, like most hernias, are typically precipitated by a rise of intra-abdominal pressure (e.g., due to coughing, sneezing, or straining) and present as a palpable bulge or swelling in the inguinal region. Usually this swelling regresses spontaneously when the patient lies down; thus, the clinical examination is performed with the patient in a standing position. Spontaneous pain is generally absent with an uncomplicated hernia, and a foreign body sensation is more common. Persistent pain accompanied by a feeling of pressure at the hernia site, nausea, and vomiting are signs of incarceration (see **C**). In patients with a bulge in the inguinal region or a scrotal mass (see p. 227), the differential diagnosis should include hydrocele testis , varicocele, ectopic testis, lymphoma, and other tumors of the testis or epididymis. Since inguinal and femoral hernias are of the external type and are easily accessible to inspection and palpation, the diagnosis is generally made clinically.

a **Palpation from the iliac spine (the three-finger rule):** The "three-finger rule" makes it easier to appreciate the topographical anatomy of inguinal and femoral hernias and differentiate among direct and indirect inguinal hernias and femoral hernias. When the examiner places the thenar on the spina iliaca anterior superior the index finger points to a direct hernia, the middle finger to an indirect hernia, and the ring finger to a femoral hernia. Thus, when a hernia is felt below the index finger, for example, this means that the patient has a direct inguinal hernia.
Caution: The hernial sac protrudes through the anulus inguinalis superficialis in both direct and indirect hernias, so there is no difference between them based on inspection or palpation.
b **Palpation from the scrotum:** This technique is particularly useful for palpating smaller inguinal hernias in the standing patient. By invaginating the scrotum and groin skin, the examiner palpates along the funiculus spermaticus to the anulus inguinalis superficialis and touches the posterior wall of the canalis inguinalis with the pad of the finger. When the patient coughs, an experienced examiner can distinguish a direct hernia, which strikes the pad of the finger, from an indirect hernia, which strikes the distal tip of the finger.

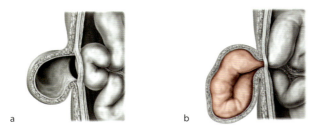

B Complete reduction of a hernia and an incarcerated hernia (strangulation)

a When the herniated viscus can move freely within the hernial sac and at the hernial opening, generally the hernia can be reduced spontaneously (e.g., by lying down) or by manual manipulation. As a result, there is no risk of acute incarceration.

b Incarceration is the most serious complication of a hernia. Strangulation at the neck of the hernia restricts blood flow to the herniated bowel, with effects ranging from ischemia to necrosis. The patient may develop symptoms of a functional or mechanical bowel obstruction with a life-threatening interruption of intestinal transit due to narrowing or obstruction of the bowel lumen. Given the risk of bowel perforation and peritonitis in these cases, immediate surgery is indicated (see **C**).

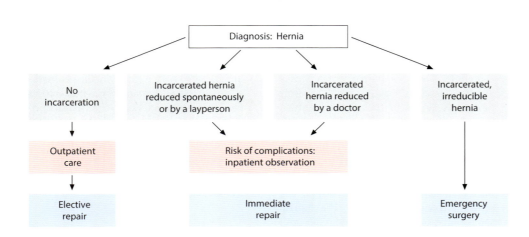

C Hernia symptoms and timing of the repair (after Henne-Bruns, Dürig, and Kremer)
As a rule, hernias do not respond definitively to conservative treatment (e.g., trusses or binders), and a permanent reduction is achieved only by surgical closure of the hernial opening (see **E**). The timing of the repair is based on the clinical presentation, i.e., the presence of a reducible, irreducible, or incarcerated hernia.

Trunk Wall — 13. Neurovascular Systems: Topographical Anatomy

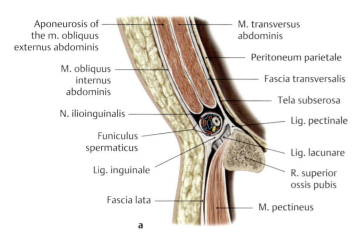
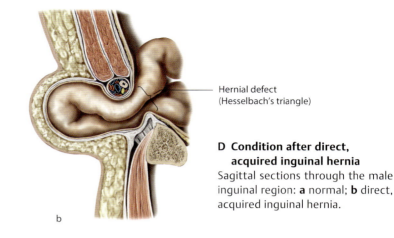

D Condition after direct, acquired inguinal hernia
Sagittal sections through the male inguinal region: **a** normal; **b** direct, acquired inguinal hernia.

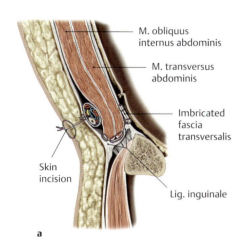
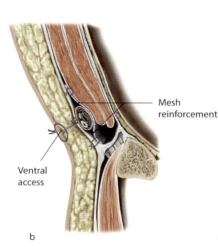
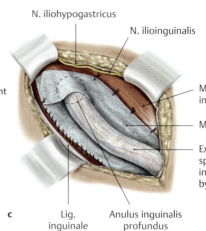

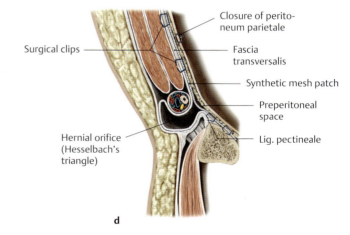

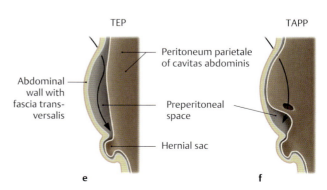

E Inguinal hernia repair
Sagittal section through the male inguinal region.

Various surgical methods are available for hernia repair. They differ mainly in the technique used to reinforce the posterior wall of the canalis inguinalis. In all methods, the steps are as follows:
- Expose the hernial sac.
- Reduce the hernial contents.
- Close the hernial orifice and thus restore the stability of the abdominal walls.

In younger patients, in which there is a strong relationship between fascia and muscle, reconstruction usually is performed without the use of foreign material to reinforce the abdominal wall (e.g., the Shouldice technique). In older patients or in the case of a recurring hernia, the abdominal wall is usually reinforced with a mesh patch (e.g., the Lichtenstein technique) during surgery.

a Shouldice technique: The posterior wall of the canalis inguinalis is reinforced by overlapping the fascia transversalis and suturing the mm. obliquus internus abdominis and transversus abdominis to the lig. inguinale in two layers.

b, c Lichtenstein technique: With this tension-free repair, the abdominal wall is reinforced by placing a synthetic mesh patch posterior to the mm. obliquus internus abdominis and transversus abdominis.

d–f Mesh reinforcement through totally extraperitoneal (TEP) or transabdominal preperitoneal (TAPP) repair. The abdominal wall is reinforced by placing a mesh patch through a peritoneal incision anterior to the mm. obliquus internus abdominis and transversus abdominis in the preperitoneal space between the fascia transversalis and peritoneum parietale (**d**). Whereas in TEP the cavitas peritonealis is not entered (extraperitoneal, **e**), in TAPP the peritoneum parietale has to be entered from inside the cavitas peritonealis (transabdominal, **f**).

219

13.12 Rare External Hernias

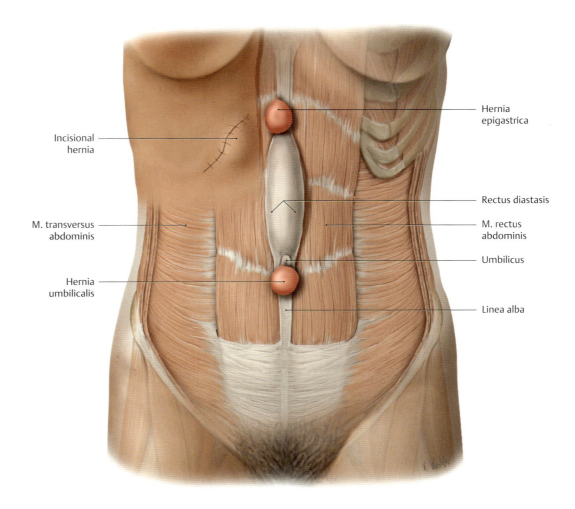

A Location of hernias of the anterior abdominal wall

B Hernias of the anterior abdominal wall*

Hernia	Location, occurrence, and typical features
• Hernia umbilicalis (umbilical hernia)	• Umbilical region, passing through the anulus umbilicalis: – Congenital umbilical hernia: incomplete regression of the normal fetal umbilical hernia due to scarring of the papilla umbilicalis (hernial sac: amnion and peritoneum) – Acquired umbilical hernia: common after multiple pregnancies, also in association with obesity, hepatic cirrhosis, or ascites (secondary widening of the anulus umbilicalis)
• Omphalocele	• Congenital persistence (1:6000) of an abdominal wall defect with an incomplete reduction of abdominal viscera during fetal life; unlike an umbilical hernia, the omphalocele is not covered by skin or subcutaneous tissue but only by peritoneum, mucous connective tissue (Wharton jelly), and amniotic epithelium (so the contents are easily recognized)
• Hernia epigastrica (epigastric hernia)	• The hernial openings are gaps in the linea alba excluding the umbilicus (on a continuum with rectus diastasis, see below)
• Rectus diastasis	• The mm. recti separate at the linea alba when the abdominal muscles are tightened, creating a site for potential herniation (the hernia reduces on relaxation, and complaints are rare)
• Incisional hernia	• Occurs at a previous incision site (usually in the upper abdominal midline)

*Umbilical and epigastric hernias comprise approximately 10% of all hernias.

Trunk Wall — 13. Neurovascular Systems: Topographical Anatomy

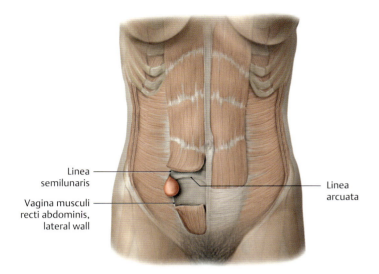

D **Spieghelianhernia**

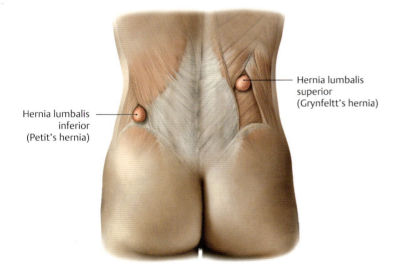

E **Hernialumbalis**

C **Other rare hernias occurring elsewhere on the trunk***

Hernia	Location
• **Spieghelian hernia**	• Anterior abdominal wall between the linea semilunaris and lateral rectus sheath, usually at the level of the linea arcuata
• **Hernia lumbalis**	• Between the twelfth costa and crista iliaca: – Hernia lumbalis superior (superior costolumbar triangle, Grynfeltt's triangle): between the twelfth costa and m. iliocostalis – Hernia lumbalis inferior (inferior iliolumbar triangle, Petit's triangle): between the crista iliaca, m. latissimus dorsi, and m. obliquus externus abdominis
• **Hernia obturatoria**	• Through the foramen obturatum and then between the mm. pectineus, adductor longus, and obturatorius externus
• **Hernia ischiadica**	• Through the foramen ischiadicum majus: – Hernia suprapiriformis (above the m. piriformis) – Hernia infrapiriformis (below the m. piriformis) – Hernia spinotuberosa (in front of the lig. sacrotuberale)
• **Hernia perinealis**	• Through the pelvic floor: – Hernia perinealis anterior (in front of the m. transversus perinei profundus) – Hernia perinealis posterior (behind the m. transversus perinei profundus) – Hernia ischiorectalis (through the m. levator ani into the fossa ischioanalis)

* Less than 1% of all hernias, generally acquired (after Schumpelick).

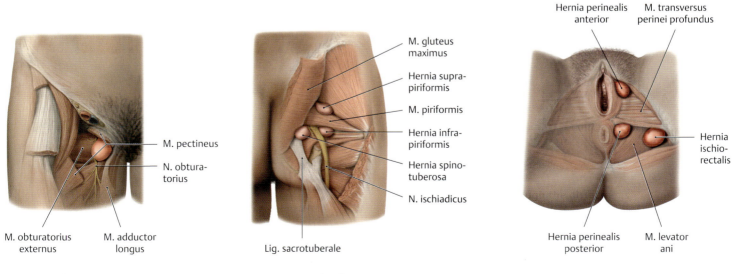

F **Hernia obturatoria**   G **Hernia ischiadica**   H **Hernia perinealis**

221

13.13 Development of the External Genitalia

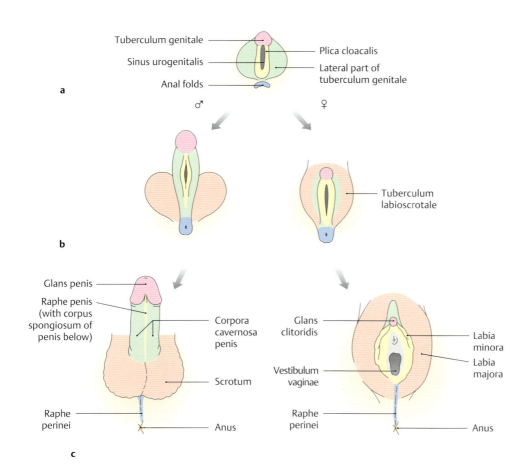

B Derivatives of the undifferentiated embryonic genital anlage (primordia) during development of the external genital organs (after Starck)*

Undifferentiated anlage	Male	Female
Tuberculum genitale	Glans penis, corpora cavernosa of penis	Clitoris, glans clitoridis
Plica cloacalis	Raphe penis, corpus spongiosum of penis	Labia minora
Tuberculum labioscrotale	Scrotum	Labia majora
Sinus urogenitalis	Pars spongiosa urethrae	Vestibulum vaginae
Plica analis	Raphe perinei	Raphe perinei

*Details on the development of the gonads and genital tracts can be found in textbooks of embryology.

A Development of the external genitalia
a Rudimentary, undifferentiated external genitalia in a 6-week-old embryo.
b Differentiation of the external genitalia along male or female lines in a 10-week-old fetus.
c Differentiated external genitalia in the newborn.

The external genital organs develop from an undifferentiated mesodermal primordium in the *cloaca* and, like the gonads, pass through an initial **indifferent stage**. The anorectal area and sinus urogenitalis (cloaca) are not yet separated from each other and are closed externally by a common cloacal membrane. The following elevations develop around the cloacal membrane due to intensive mesodermal cell divisions:

- Anterior: the genital tubercles
- Lateral: the urogenital folds
- Posterior: the anal folds
- Lateral to the urogenital folds: the genital swellings (labioscrotal swellings)

Later, between the sixth and seventh weeks of development, the septum urorectale divides the cloaca into an anterior part (sinus urogenitalis) and a posterior part (anus and rectum). The cloacal membrane disappears, and the ostium urogenitale forms anteriorly. The early perineum forms at the level of the septum urorectale (by fusion of the paired anal folds to the raphe perinei). **Differentiation of the genital organs** begins approximately in the eighth to ninth week of fetal development. Sexual differentiation is clearly evident by the 13th week and is fully developed by the 16th week.

- In the *male fetus*, the genital tubercles enlarge *under the influence of testosterone* to form the phallus and future penis. The sinus urogenitalis closes completely by fusion of the genital folds and forms the pars spongiosa of the urethra. The genital (scrotal) swellings unite to form the scrotum.
- In the *female fetus (absence of testosterone)*, the genital tubercles give rise to the clitoris. The sinus urogenitalis persists as the vestibulum vaginae, and the two genital folds form the labia minora. The genital swellings enlarge to form the labia majora.

Male sex organs develop only in the presence of the factors listed below:
- A functionally competent *SRY* (sex-determining region of the Y) gene on the Y chromosome (otherwise, ovaries and a female phenotype will develop). The SRY gene ensures that anti-müllerian hormone and Leydig cells are produced (see below).
- Among its other functions, the *anti-müllerian hormone* induces regression of the müllerian ducts. It is formed in the somatic cells of the chorda testicularis (future Sertoli cells) starting in the eighth week of fetal life.
- *Leydig cells* begin to form in the fetal testes by the ninth week and produce large amounts of androgens (testosterone) until birth. They stimulate differentiation of the *wolffian duct* into seminiferous tubules and the development of the male external genitalia.

Alteration or interruption of this process of differentiation at any stage can lead to incomplete midline fusion, which leaves persistent clefts (hypospadias, epispadias, see **C**), or to more extensive external genital anomalies (see **E**) (after Starck).

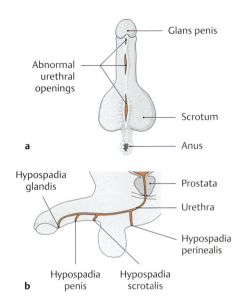

C Hypospadias: a urethral anomaly in boys
a Cleft anomalies affecting the underside of the penis and scrotum.
b Possible sites of emergence of the urethra in hypospadias (penis viewed from the lateral view).

If the genital folds do not fuse completely during sexual differentiation (see **A**), the result is a cleft anomaly of the urethra, which may open on the underside of the penis *(hypospadias)* or on its dorsal surface *(epispadias)*. Hypospadias is much more common, with an incidence of 1:3000 compared to 1:100,000 for epispadias. It is most common to find an abnormal urethral orifice in the glans region of the penis (hypospadia glandis). Additionally, the shaft of the penis is usually shortened and angled downward by the presence of ventral fibrous bands. Surgical correction is generally performed between the sixth month and second year of life (after Sökeland, Schulze, and Rübben).

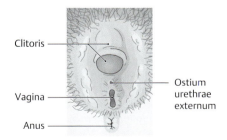

D External genitalia of a woman with adrenogenital syndrome
Anterior view. The external genitalia show definite signs of masculinization. The clitoris is markedly enlarged. The labia majora and minora are partially fused, and the sinus urogenitalis forms an undersized vestibulum vaginae (see **E**, female pseudohermaphroditism).

E Various forms of intersexuality*

Condition	Features
• **Hermaphroditismus verus****	• Very rare form of hermaphroditism (approximately 70% of cases have a female karyotype: 46,XX). The gonads contain both testicular and ovarian tissue *(ovotestis)*, but with a preponderance of ovarian tissue. Hence the external genitalia tend to have a female appearance with a markedly enlarged clitoris. A uterus is frequently present. Most hermaphrodites are raised as girls.
• **Pseudohermaphroditismus**	• The pseudohermaphrodite has a definite chromosomal sex (female: 46,XX or male: 46,XY) but a phenotype of the *opposite* gender. The condition is termed male pseudohermaphroditism when a testis is present and female pseudohermaphroditism when an ovarium is present.
– *Pseudohermaphroditismus masculinus* → Chromosomal sex: male (46,XY) → Phenotype: female	• **Etiology and pathogenesis** The female phenotype results from a lack of fetal androgen exposure: 1. Disturbance of testosterone synthesis 2. Disturbance of testosterone conversion 3. Androgen receptor defect 4. Testicular dysgenesis • **Example: testicular feminization** (1:20,000 live births): – 46,XY chromosome complement – Individual has a female phenotype (estrogen synthesis present) but lacks pubic and axillary hair ("hairless woman"). The upper vagina and uterus are also absent. – *Cause:* androgen receptor defect or a disturbance of androgen metabolism (5α-reductase-2 defect) – *Result:* absence of spermatogenesis – *Treatment:* removal of the testes, which are usually in the inguinal region (risk of malignant transformation) and estrogen replacement for life
– *Pseudohermaphroditismus femininus* → Chromosomal sex: female (46,XX) → Phenotype: male	• **Etiology and pathogenesis** The male phenotype results from fetal androgen exposure: 1. Congenital enzyme defect 2. Diaplacental androgen exposure • **Example: congenital adrenogenital syndrome** (1:5000 live births): – 46,XX chromosome complement – Female internal genital organs with masculinized *external* genitalia (enlarged clitoris, partial fusion of the labia majora, small sinus urogenitalis, see **D**) – *Cause:* adrenocortical hyperplasia with impaired steroid synthesis based on a genetic enzyme defect (most commonly a 21-hydroxylase deficiency). The low hormone level causes increased adrenocorticotropic hormone secretion, leading to the overproduction of androgens. – *Treatment:* hydrocortisone therapy for life, which may be combined with a mineralocorticoid.

*Intersexuality refers to a condition marked by contradictions in the development of general external sex characteristics, the gonads, and the chromosomal sex.
**Named after Hermaphroditos, the androgynous son of Hermes and Aphrodite from Greek mythology.

Trunk Wall — 13. Neurovascular Systems: Topographical Anatomy

13.14 Male External Genitalia: Descensus testis and Funiculus spermaticus

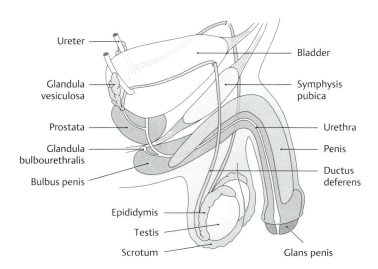

A Overview of the male genital organs
The internal and external male genitalia are distinguished by their *origins*. The internal reproductive organs originate from the two urogenital ridges located above the pelvic floor (except for the prostata and glandulae bulbourethrales (Cowper's glands), which develop from urethral epithelium and thus are derivatives of the sinus urogenitalis). By contrast, the external genital organs develop around the sinus urogenitalis from a genital anlage (primordium) located below the pelvic floor (see p. 222).

Male internal genital organs	Male external genital organs

• Testes
• Epididymis
• Ductus deferens (vas deferens)
• Accessory sex glands
  – Prostata
  – Seminal vesicles (glandulae vesiculosae)
  – Glandulae bulbourethrales (Cowper's glands)
	• Penis
	• Scrotum
	• Coverings of the testis

Topographically, however, the testis, epididymis, and a portion of the ductus deferens are classified among the external genital organs because they migrate from the cavitas abdominis into the scrotum during fetal development (descensus testis).

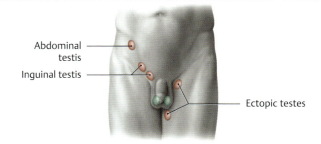

C Anomalous positions of the testes
Abnormalities in the descensus testis occur in approximately 3% of all newborns. A testis may be retained in the cavitas abdominis or in the canalis inguinalis (cryptorchidism or retained testis). A deficiency of androgen production is the presumed cause. An ectopic testis is one that strays from the normal tract and occupies an abnormal position. The principal results are infertility due to the higher ambient temperature and an increased risk of malignant transformation.

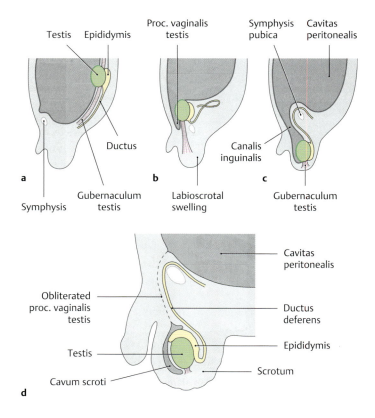

B Descensus testis
Lateral view.
a Second month, **b** third month, **c** at birth, **d** after obliteration of the proc. vaginalis testis of the peritoneum.
Near the end of the second month of development, the gonads and the rest of the mesonephros lie in a common peritoneal fold (urogenital fold), from which the "gonadal ligaments" are derived after regression of the mesonephros. The lower gonadal ligament, called the gubernaculum testis, is important for the descent of the testis. It passes below the genital ducts, pierces the abdominal wall in the area of the canalis inguinalis, and ends in the labioscrotal swelling, an outpouching of the anterior abdominal wall. Traction from this gonadal ligament (a consequence of body growth, which is more rapid than the growth of the genital organs) causes the testis and epididymis to slide downward along the posterior trunk wall external to the peritoneum **(transabdominal descent)**. By the start of the third month, the testis has already reached the entrance of the future canalis inguinalis. The proc. vaginalis peritonei, a funnel-shaped outpouching of peritoneum, forms anterior to the gubernaculum testis and is continued into the scrotal swelling with the other layers of the abdominal wall. It gives rise to the coverings of the funiculus spermaticus and testis after the testis has completed its descent. A second phase, which is completed shortly before birth **(transinguinal descent)**, culminates in passage of the testis through the canalis inguinalis into the scrotum. After descensus testis is completed (by birth), the proc. vaginalis peritonei is obliterated except for a small space that partially surrounds the testis as the cavum scroti (tunica vaginalis testis with a lamina visceralis and a lamina parietalis, see p. 226). Failure of this process of obliteration results in a persistent communication between the cavitas abdominis and testicular cavity (congenital indirect inguinal hernia, see p. 214) (after Starck).

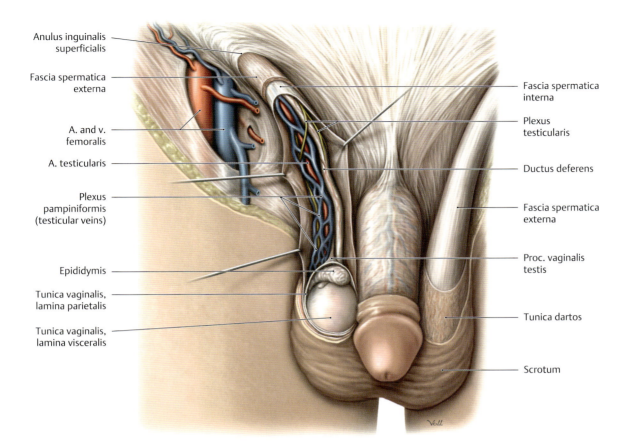

D The penis, scrotum, and funiculus spermaticus
Anterior view. The skin has been partially removed from over the scrotum and funiculus spermaticus. The tunica dartos and fascia spermatica externa are exposed on the left side, and the funiculuc spermaticus on the right side has been opened in layers. The skin of the scrotum differs in many respects from the skin of the abdominal wall. It is more pigmented, markedly thinner, more mobile, and devoid of subcutaneous fat. It also contains a network of myofibroblasts (tunica dartos), whose contraction causes a wrinkling of the skin. This reduces the surface area of the scrotum and, when accompanied by vasoconstriction of the cutaneous vessels, reduces heat loss from the testes. This mechanism regulates the temperature to optimize spermatogenesis.

E Contents of the spermatic cord (funiculus spermaticus)
Cross section through the funiculus spermaticus. The neurovascular structures that supply the testes converge at the level of the deep inguinal ring (anulus inguinalis profundus), forming a bundle the thickness of the small finger that is held together by loose connective tissue and by the coverings of the funiculus spermaticus and testes. The coverings consist of the following structures:

- M. cremaster with fascia cremasterica
- R. genitalis of the n. genitofemoralis
- A. and v. cremasterica
- Ductus deferens
- A. and v. ductus deferentis
- A. testicularis
- Vv. testiculares (plexus pampiniformis)
- Autonomic nerve fibers (plexus testicularis)
- Lymphatic vessels

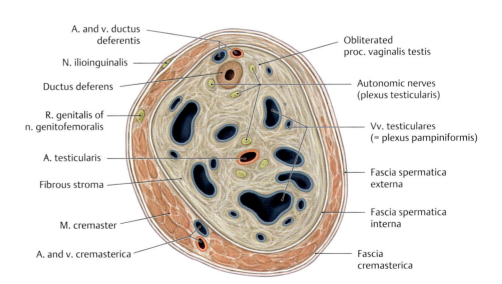

- The obliterated processus vaginalis testis

Special features: The large veins of the plexus pampiniformis are exceptionally thick-walled, with a three-layered structure, and hence are easily mistaken for arteries. In life the ductus deferens, with its compact muscular wall, is palpable through the skin as a firm cord the thickness of a knitting needle. The ease of surgical access to this site is utilized in *vasectomy*, or ligation of the ductus deferens, to interrupt the transport of sperm (sterilization).

13.15 Male External Genitalia: The Testis and Epididymis

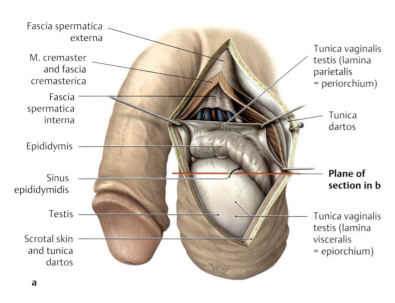

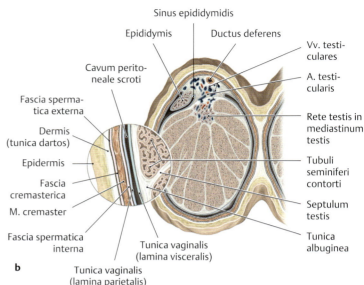

A Tunica vaginalis testis and cavum peritoneale scroti (cavum serosum testis)
a Opened tunica vaginalis of the left testis, lateral view.
b Cross section through the testis, epididymis, and scrotum, superior view.

The tunica vaginalis testis (unobliterated end of the proc. vaginalis testis, see p. 224) forms a serous coat surrounding the testis and epididymis. Its lamina visceralis is fused to the tunica albuginea of the testis.

At the mediastinum testis (a suspensory ligament where nerves and vessels enter and leave the testis), the tunica vaginalis testis is reflected to form the lamina parietalis, which is covered externally by the fascia spermatica interna. Between the two layers is a slitlike mesothelium-lined space (cavum peritoneale scroti) that contains a scant amount of fluid and is partially continuous between the testis and epididymis (sinus epididymidis). An abnormal fluid collection in the cavum serosum scroti is called a *hydrocele testis* (see **Fb**) (after Rauber and Kopsch).

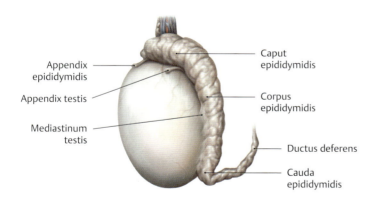

B Surface anatomy of the testis and epididymis
Left testis and epididymis, lateral view. The combined weight of the testis and epididymis at sexual maturity is approximately 20 to 30 g. The *testis* has an ovoid shape (approximately 5 cm long and 3 cm wide) and an average volume of approximately 18 mL (12–20 mL). The testicular tissue is enclosed in a tough fibrous capsule (tunica albuginea) and has a rubbery consistency. The *epididymis* consists of a head (caput epididymidis), which is attached to the upper pole of the testis, and a body and tail (corpus and cauda epididymidis) that curve down along the mediastinum on the posterior side of the testis. The tail of the epididymis becomes continuous with the ductus deferens at the lower pole of the testis.

C Structure of the testis and epididymis
Section through the testis (epididymis intact), lateral view. Fibrous septula testis extend radially from the tunica albuginea of the testis toward the mediastinum testis, subdividing the testicular tissue into approximately 370 wedge-shaped lobules (lobuli testis). Each lobule contains one or more tubuli seminiferi contorti, in whose epithelium the spermatocytes are formed (spermatogenesis, see p. 5) and which open into the rete testis. From there, approximately 10 to 15 ductuli efferentes pass to the head of the epididymis, where the ductus epididymidis begins. This single duct is continuous distally with the ductus deferens, which passes through the canalis inguinalia in the funiculus spermaticus to enter the cavitas abdominis and opens into the pars prostatica of the urethra via a short intervening segment, the ductus ejaculatorius (see p. 229).

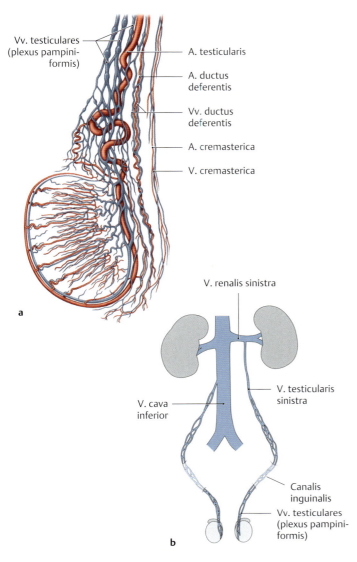

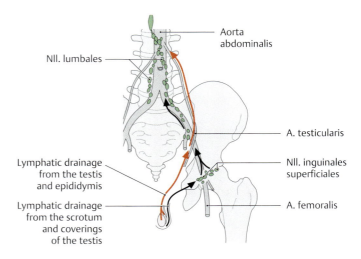

E Lymphatic drainage and regional lymph nodes of the testis, the epididymis, the coverings of the testis, and the scrotum
The lymphatic vessels of the testis and epididymis drain to the nll. lumbales, accompanied by the vasa testicularia. The lymphatic vessels from the scrotum and the coverings of the testis drain to the nll. inguinales superficiales (see p. 528).
Note: Advanced testicular tumors tend to metastasize to retroperitoneal lymph nodes because they serve as the primary lymphatic conduit from the testis and epididymis.

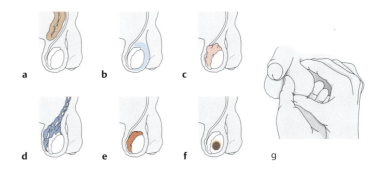

D Blood supply of the testis
a Arterial supply: The testis, epididymis, and their coverings are supplied by three different arteries, which anastomose with one another (after Hundeiker and Keller, quoted in Rauber and Kopsch):

- A. testicularis: arises directly from the aorta
- A. ductus deferentis: arises from the a.iliaca interna
- A. cremasterica: arises from the a. epigastrica inferior

The vessels supplying the scrotum arise from the a. pudenda interna (see p. 556).

b Different venous drainage patterns of the right and left testes: Venous blood from the testis and epididymis flows into the vv. testiculares in the area of the mediastinum testis. These veins form an elongated venous network, especially distally, called the plexus pampiniformis. It surrounds the branches of the a. testicularis and ascends with it through the canalis inguinalis into the retroperitoneum. There the *right* v. testicularis empties into the v. cava inferior, while the *left* v. testicularis opens into the v. renalis sinistra. This asymmetry of venous drainage has major clinical relevance: the left v. testicularis enters the v. renalis sinistra at a right angle. This creates a physiologically significant constriction that can obstruct venous outflow from the left v. testicularis and thus from the plexus pampiniformis (varicocele, see **Fd**). In this case, the plexus pampiniformis can no longer perform its "thermostat" function (cooling venous blood returning from the a. testicularis), resulting in a local heat buildup that may compromise the fertility of the left testis.

F Abnormal findings on clinical examination of the external genitalia
a–f Diseases that may present with scrotal swelling: **a** inguinal hernia, **b** hydrocele testis (serous fluid collection in the cavum serosum scroti), **c** spermatocele (retention cyst in the epididymis), **d** varicocele (painful, varicose dilation of the plexus pampiniformis), **e** epididymitis (painful bacterial inflammation of the epididymis), **f** testicular tumor (painless, usually unilateral induration of the testis).
g Bimanual examination of the testis and epididymis: Clinical examination of the external genitalia should include palpation of the testis and epididymis (bimanual examination). Based on the disease features noted above, the following questions should be addressed during the clinical examination:

- Is the mass confined to the scrotum?
- Is there transient enlargement of the mass when the patient coughs?
- Is the mass translucent when examined by *transillumination* (illumination with a flashlight)?
- Is the mass painless or tender to pressure?

Note: A painless induration of the testis, especially in young men, should always raise suspicion of a testicular tumor (after Sökeland, Schulze, and Rübben).

13.16 Male External Genitalia: The Fasciae and Erectile Tissues of the Penis

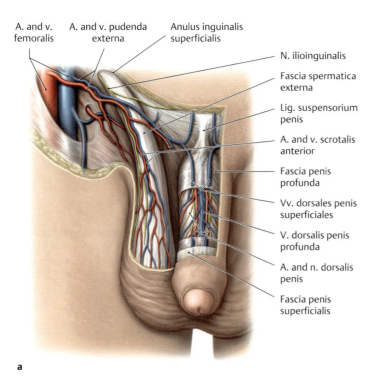

A Arrangement of the penile fasciae
a Anterior view of the penis (skin and fasciae partially removed).
b Right lateral view of the penis (skin and fasciae partially removed).
c Cross section through the shaft of the penis.

The penis is covered by thin, mobile skin that is devoid of fatty tissue. The skin over the glans penis is duplicated to form the preputium penis (foreskin), which is attached to the undersurface of the glans by the median fold of the frenulum preputii (see **b**). The erectile tissues of the penis are surrounded by a common, strong envelope of collagenous fibers, the *tunica albuginea*. The two layers of the fasciae penis superficialis and profunda also surround the corpus spongiosum penis and corpora cavernosa penis. The erectile tissues, their fibrous sheaths, and the way in which the vessels are incorporated into these fibrous sheaths are of key interest in understanding the function of the penis (see p. 231).

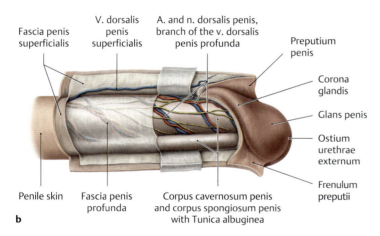

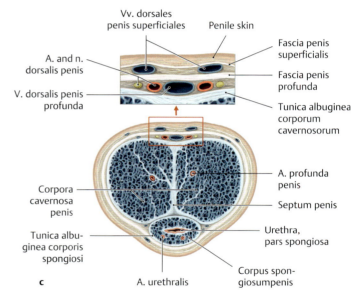

B Constriction of the preputium penis (phimosis)
a Constriction of the preputium penis in a 3-year-old boy.
b Appearance following circumcision.

The epithelium of the inner layer of the preputium penis is adherent to the surface epithelium of the glans penis in newborns and infants. Because of this, the distal junction of the outer and inner layers of the preputium penis is normally constricted, a condition characterized as **physiological phimosis**. During the first 2 years of life, the epithelial attachments become separated due to enlargement of the glans and the secretion of smegma (cellular debris sloughed from the stratified keratinized epithelium). If the preputium penis still cannot slide over the glans by 3 years of age due to a functional stenosis (e.g., persistent epithelial attachments due to an absence of smegma secretion), the phimosis should be surgically corrected by circumcision. This procedure may be conservative or radical, resecting all the preputium penis (as shown here), depending on the severity of the phimosis. Immediate surgical intervention (before 3 years of age) is necessary for **paraphimosis**—an emergency situation in which the glans is strangulated by the narrowed preputium penis (painful, livid swelling of the glans due to decreased blood flow, with risk of necrosis) (after Sökeland, Schulze, and Rübben).

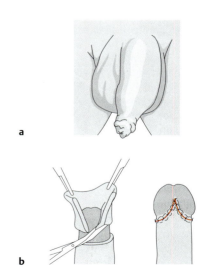

Trunk Wall — 13. Neurovascular Systems: Topographical Anatomy

C The erectile tissues and erectile muscles of the penis

a Inferior view. The corpus spongiosum is partially mobilized, and the skin and fasciae have been removed. The mm. ischiocavernosus and bulbospongiosus have been removed on the left side, along with the fascia diaphragmatis urogenitalis inferior.

b Cross section through the root of the penis. The root (radix) of the penis is firmly attached to the membrana perinei and pelvic skeleton. It is distinguished from the freely mobile shaft (corpus penis) of the penis, with its dorsal and urethral surfaces (dorsum penis and Facies urethralis), and from the glans penis, which bears the ostium urethrae externum. The penis contains two types of erectile tissue:

- The paired corpora cavernosa penis
- The unpaired Corpus spongiosum penis

At radix penis, each of the corpora cavernosa penis tapers to form a crus penis. Between the two crura lies the thickened end of the corpus spongiosum (bulbus penis). The glans penis forms the distal end of the corpus spongiosum. Its posterior margin is broadened to form the corona glandis, which is turned over the ends of the corpora cavernosa. The erectile tissues receive their blood supply from branches of the a. pudenda interna, which branches in the spatium profundum perinei (see pp. 183 and 230).

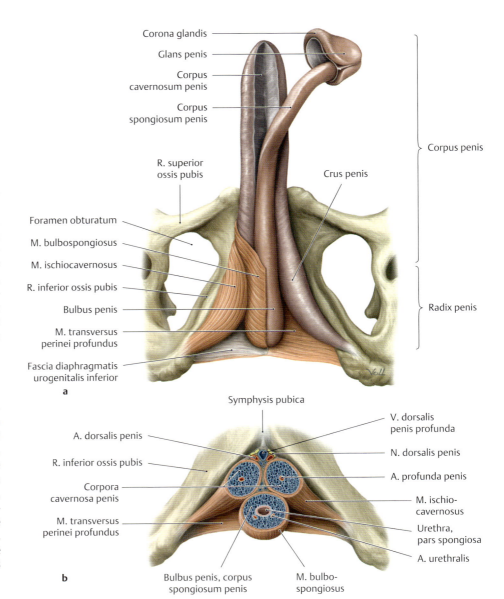

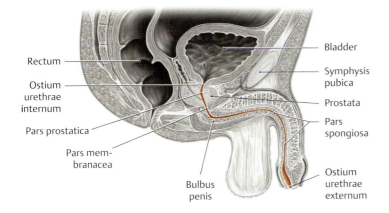

D Course of the male urethra
Midsagittal section through a male pelvis. The male urethra (urethra masculina) consists of a pars prostatica, a pars membranacea, and a pars spongiosa named for different regions of the pelvis and external genitalia (see p. 224). The pars spongiosa begins where it enters the bulbus penis and terminates at the ostium urethrae externum.

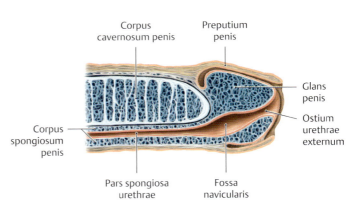

E Midsagittal section through the distal penis
The pars spongiosa urethrae undergoes an approximately 2-cm-long fusiform dilation within the glans penis. In this area of the fossa navicularis, the stratified columnar epithelium of the urethra gives way to stratified, nonkeratinized squamous epithelium. The upper cell layers of this epithelium are rich in glycogen, which—as in the vaginal milieu in females—provides a culture medium for the lactic acid bacteria that thrive there (acidic pH protects against pathogenic organisms).

13.17 Male External Genitalia: Nerves and Vessels of the Penis

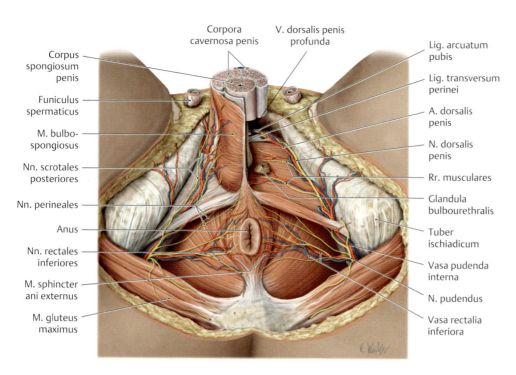
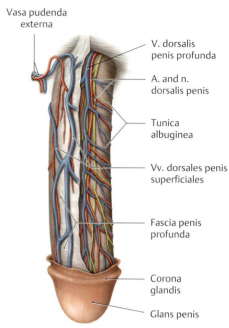

A Neurovascular structures of the male perineal region
Lithotomy position with the scrotum removed, inferior view. The spatium perinei superficialis on the right side has been opened by removing the fascia perinea superficialis. The erectile muscles and radix penis have been removed on the left side, and the spatium perinei profundum is partially exposed. The penis has been transected across the shaft, and the spermatic cords (funiculi spermatici) have been divided.

B Vessels and nerves of the Dorsum penis
The preputium penis, skin, and fascia penis superfi- cialishave been completely removed from the penile shaft. The fascia penis profunda has also been removed from the left dorsum.

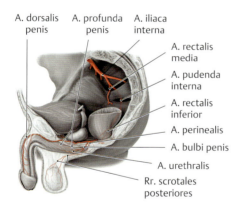
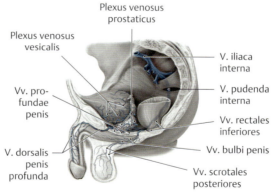
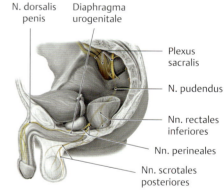

C Arterial supply to the penis and scrotum
Left lateral view. The penis and scrotum derive their arterial supply from the a. pudenda interna. This vessel enters the fossa ischio-analis and, after giving off the a. rectalis inferior to the anus, courses to the posterior border of the diaphragma urogenitale. Then, after giving off the a. perinealis, it passes through the spatium profundum perinei into the spatium superficiale perinei (see p.183), where it divides into its terminal branches: the a.dorsalis penis, a. profunda penis, a. bulbi penis, and a. urethralis.

D Venous drainage of the penis and scrotum
Left lateral view. The veins of the penis (especially the v. dorsalis penis profunda and its tributaries, the vv. profundae penis and vv. bulbi penis) open initially into the v. pudenda interna and then into the plexus venosus prostaticus. Exceptions are the vv. dorsales penis superficiales (not seen here), which drain via the vv. pudendae externae into the v. saphena magna. On its way to the plexus venosus prostaticus, the v. dorsalis penis profunda passes through a narrow space just below the symphysis between the lig. arcuatum pubis and the lig. transversum perinei (see ligaments in **A**).

E Nerve supply to the penis and scrotum
Left lateral view. The n. pudendus enters the fossa ischianalis and, after giving off the nn. rectales inferiores, courses to the m. sphincter ani externus and to the skin of the anus at the posterior border of the diaphragma urogenitale. There it divides into its terminal branches, the nn. perineales. The superficial branches pass through the spatium superficiale perinei to the skin of the perineum and posterior scrotum (rr. scrotales posteriores). The deep branches course in the spatium profundum perinei. They innervate the erectile muscles (via rr. musculares), the skin of the penis, and the erectile bodies (via the n. dorsalis penis). The course of the autonomic fibers is shown in **F**.

Trunk Wall — 13. Neurovascular Systems: Topographical Anatomy

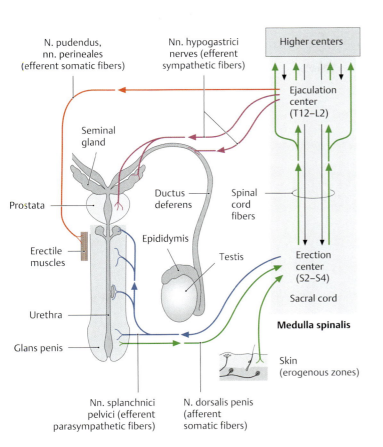

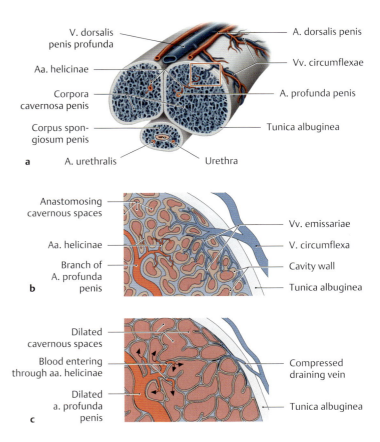

F Overview of the male sexual reflexes

The sexual reflexes in males are evoked by a variety of stimuli (e.g., tactile, visual, olfactory, acoustic, and psychogenic). Somatic and autonomic nerve pathways transmit the stimulus to the erection and ejaculation centers in the thoracolumbar and sacral spinal cord, from which it is relayed to higher centers (e.g., the hypothalamus and limbic system). For example, *tactile cutaneous stimuli* to the genitalia are transmitted to the sacral cord by *afferent somatic* fibers (n. dorsalis penis from the n. pudendus, shown in green) and are relayed in the erection center (S2–S4) to *efferent parasympathetic* fibers (pelvic splanchnic nerves, shown in blue). These impulses, which stimulate vasodilation of the arteries supplying the erectile tissues (see G), are critically influenced by descending pathways from higher centers. Conversely, the excitatory impulses evoked by increasing mechanical stimulation of the glans penis ascend from the sacral cord to the ejaculation center located at the T12–L2 levels. There they are relayed to *efferent sympathetic* fibers (nn. hypogastrici, shown in purple) and stimulate smooth muscle contractions in the epididymis, ductus deferens, prostata, and seminal glands. Simultaneous stimulation of the erectile muscles by efferent *somatic* nerve fibers (nn. perineales from the n. pudendus, shown in red) produces rhythmic contractions that expel the ejaculate from the urethra (emission). Failure to achieve an erection despite an active libido (psychological interest in sexual activity) is defined as erectile dysfunction. The recent development of successful medical treatment of *erectile dysfunction* with sildenafil (Viagra [Pfizer]) is based on its modulation of the second messenger cyclic guanosine monophosphate (cGMP). When the primary messenger nitrous oxide (NO) is released by neural stimulation, it activates the enzyme guanylate cyclase in penile erectile tissues. This enzyme generates cGMP as a second messenger, which in turn induces vasodilation and produces an erection. Sildenafil selectively inhibits cGMP breakdown by a specific phosphodiesterase (PDE5) that is prominent in erectile tissue. cGMP accumulates, vessels remain dilated, and penile erection is sustained. Sildenafil treatment thus effectively amplifies the initial neural stimuli, with the potential of prolonging normal erections and overcoming other, inhibitory physiological problems (after Klinke and Silbernagl).

G Mechanism of penile erection (after Lehnert)
a Penis in cross section, showing the blood vessels involved in erection (enlarged views in **b** and **c**).
b Corpus cavernosum penis in the flaccid state.
c Corpus cavernosum penis in the erect state.

Penile erection is based essentially on *maximum engorgement* and pressure elevation in the cavities (cavernous spaces) of the corpora cavernosa penis combined with a *constriction of venous outflow*. This mechanism raises the intracavernous blood pressure to approximately 10 times the normal systolic blood pressure (approximately 1200 mmHg in youngmen). Microscopically, the erectile tissue of the penis consists of an arborized trabecular meshwork of connective-tissue and smooth-muscle cells that is connected to the tunica albuginea. Among the trabeculae are interanastomosing cavities that are lined with endothelium. Branches of the a. profunda penis, called the aa. helicinae, open into these cavities. In the flaccid state, the aa. helicinae are more or less occluded by "intimal pads." When an erection occurs, the afferent arteries dilate, and the aa. helicinae open under the influence of the autonomic nervous system. The result is that with each pulse wave, blood is forced into the cavernous spaces, increasing the volume of the erectile tissue and raising the intracavitary pressure. The tunica albuginea, which has a limited capacity for distension, be-comes taut and compresses the veins that pass through it. This mechanism, aided by the occlusion of emissary veins (vv. emissaria), causes a constriction of venous outflow, enabling the penis to remain stiff and hard. Meanwhile, the dense venous plexuses in the corpus spongiosum penis and glans prevent excessive compression of the urethra. The flaccid phase begins with vasoconstriction of the afferent arteries. An undesired, prolonged, and painful erection is called *priapism* (from Priapus, the Greco-Roman god of procreation) and may occur, for example, in certain blood diseases or metabolic disorders. The initial treatment of this condition is medical; one option is the use of vasoconstrictors (etilefrine or norepinephrine). Surgical treatment involves making "punch anastomoses" to promote the outflow of blood.

13.18 Female External Genitalia: Overview and Episiotomy

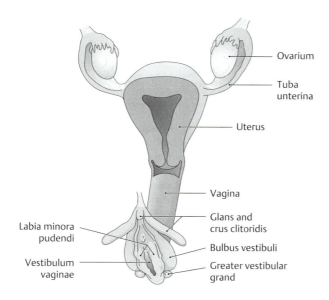

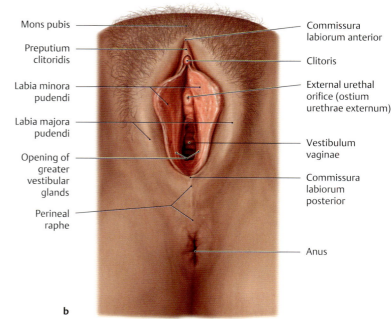

A Overview of the female genital organs
a Internal and some external genital organs.
b External genitalia, lithotomy position with the labia minora separated.

As in males, a distinction is drawn between the development and the topography of the female internal and external reproductive organs. The homology in the development of the male and female genital organs is reflected chiefly in the comparable histological features of the corresponding parts (see textbooks of histology). The female external genitalia (pudendum femininum) is also known in clinical parlance as the *vulva*. It is separated from the internal genital organs by the *hymen* (not shown here). The outer boundaries of the vulva are formed by the mons pubis, a fatty-fleshy prominence over the symphysis pubica, and the labia majora, two pigmented ridges of skin that contain smooth muscle cells as well as sebaceous glands, sweat glands, and scent glands. The labia majora are interconnected anteriorly and posteriorly by a bridge of tissue called the commissura labiorum anterior and posterior. The area be- tween the commissura labiorum posterior and anus is the raphe perinei. Specific structures are listed in the table at right.

Female internal genital organs	Female external genital organs (vulva)

• Ovarium
• Tuba uterina
• Uterus
• Vagina
	• Mons pubis
	• Labia majora pudendi
	• Labia minora pudendi
	• Vestibulum vaginae
	• Bulbus vestibuli
	• Clitoris
	• Glandulae vestibulares
	  – Glandulae vestibulares majores (Bartholin's glands)
	  – Glandulae vestibulares minores

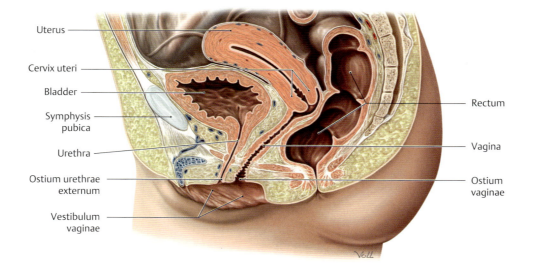

B Midsagittal section through a female pelvis
Left lateral view.
Note the close proximity of the ostium urethrae externum to the ostium vaginae, which opens into the vestibulum vaginae.

Trunk Wall — 13. Neurovascular Systems: Topographical Anatomy

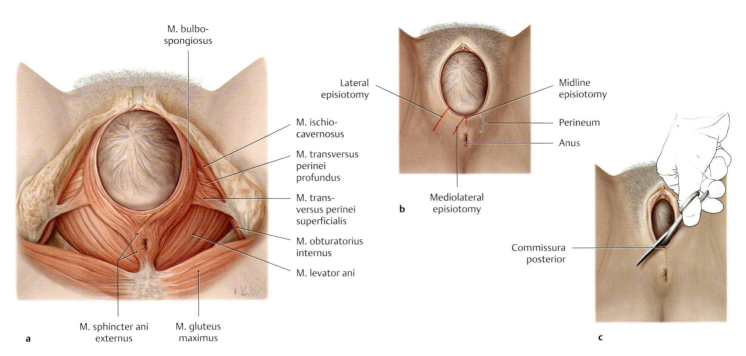

C Episiotomy: indications, types, and technique
a Indications: pelvic floor with crowning of the fetal head.
b Types of episiotomy: midline, mediolateral, lateral.
c Technique for mediolateral episiotomy performed at the height of a contraction.

Episiotomy is a common obstetric procedure utilized to enlarge the birth canal during the expulsive stage of labor (see p. 542). When the fetal head crowns through the pelvic floor, the m. levator ani in particular is passively stretched, forced downward, and rotated approximately 90°. The "levator plate" thus helps to form the wall of the distal birth canal along with the diaphragma urogenitale and m. bulbospongiosus. As such, it comes under considerable tension at the perineal body during the pushing stage of labor. To protect the perineal muscles (mm. perinei) from tearing, the obstetrician counteracts this tension by support- ing the perineum with two fingers *(perineal protection)*. An *episiotomy* is often performed to prevent uncontrolled laceration of the perineum (maternal indication). There is an imminent danger of perineal laceration during the delivery when the perineal skin is stretched to the point that it turns white, indicating diminished blood flow. The primary purpose of an episiotomy, however, is to expedite the delivery of a baby that is at risk for hypoxia during the expulsive stage. An *early episiotomy* is one that is made before the head crowns (the head is visible with contractions and pushing but recedes between contractions). A *timely episiotomy* is made after the head has crowned, when there is maximum tension on the perineal skin. Three types of episiotomy are available (see **D** for advantages and disadvantages):

- **Midline episiotomy:** straight down from the vagina toward the anus
- **Mediolateral episiotomy:** oblique incision from the commissura posterior
- **Lateral episiotomy:** lateral incision from the lower third of the vulva

After the placenta is delivered, the episiotomy is usually closed in at least three layers (vaginal suture, deep perineal suture, and cutaneous suture). Local anesthesia is generally required, especially when an early episiotomy has been done. If the episiotomy was made at the height of a contraction after crowning of the head, anesthesia is unnecessary. Local infiltration anesthesia and the pudendal block (PDB) are described on p. 542.

D Advantages and disadvantages of the different types of episiotomy (after Goerke)

Episiotomy	Divided muscles	Advantages	Disadvantages
• Midline	• None	• Easy to repair • Heals well	• May lengthen to a grade III perineal laceration (m. sphincter ani externus)
• Mediolateral	• M. bulbospongiosus • M. transversus perinei superficialis	• Gains more room • Low risk of laceration	• Heavier bleeding • More difficult to repair • More difficult healing
• Lateral*	• M. bulbospongiosus • M. transversus perinei superficialis • M. levator ani (m. puborectalis)	• Gains the most room	• Heaviest bleeding • Potential complications (e.g., anal incontinence) • Greatest postpartum complaints

*Very rarely used.

13.19 Female External Genitalia: Neurovascular Structures, Erectile Tissues, Erectile Muscles, and Vestibule

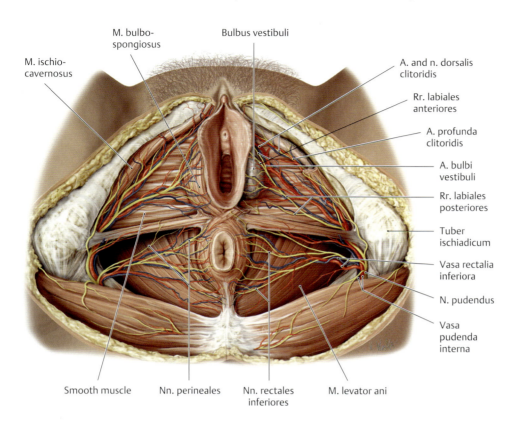

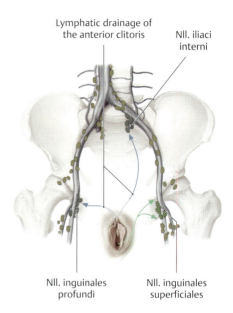

A Nerves and vessels of the female regio perinealis
Lithotomy position. The labia majora, skin, fascia perinea superficialis, and fatty tissue in the fossa ischioanalis have been removed to demonstrate the neurovascular structures. The mm. bulbospongiosus and ischiocavernosus and the fascia diaphragmatis urogenitalis inferior have also been dissected away on the left side.

B Lymphatic drainage of the female external genitalia
Female pelvis, anterior view. Lymph from the female external genitalia drains to the nll. inguinales superficiales. The only exceptions are the anterior portions of the clitoris (corpus and glans clitoridis), which drain to the nll. inguinales profundi and nll. iliaci interni.

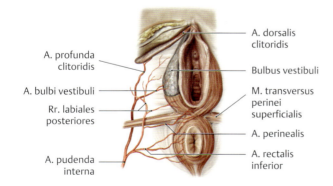

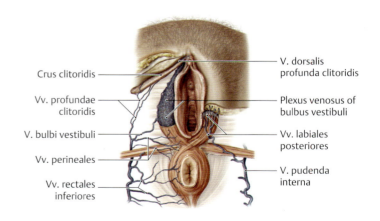

C Arterial supply to the female external genitalia
Regio perinealis, inferior view. The female external genitalia, like the penis and scrotum, are supplied by the a. pudenda interna, which enters the fossa ischioanalis (not seen here). After giving off the a. rectalis inferior to the anus, the a. pudendus interna passes anteriorly into the spatium superficiale perinei. Another branch, the a. perinealis, supplies the perineal region, the erectile muscles, and the *posterior part* of the labia majora (rr. labiales posteriores). In the spatium superficiale perinei (not visible here, see p. 183), the a. pudenda interna divides into its terminal branches, the a. bulbi vestibuli and the aa. profunda and dorsalis clitoridis (which supply the corpus cavernosum clitoridis). The *anterior part* of the labia majora is supplied by the aa. pudendae externae (rr. labiales anteriores), which arise from the a. femoralis (not seen here).

D Venous drainage of the female external genitalia
Perineal region, inferior view. Venous drainage is handled by the following vessels:

- The vv. profundae clitoridis, vv. labiales posteriores, and v. bulbi vestibuli, which drain into the *v. pudenda interna*
- The vv. dorsales superficiales clitoridis and vv. labiales anteriores, which drain into the *vv. pudendae externae* (not seen here)
- The v. dorsalis clitoridis, which drains into the *plexus venosus vesicalis*.

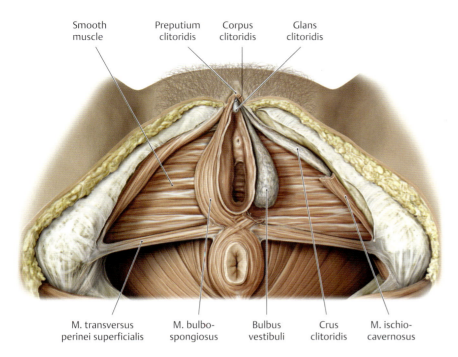

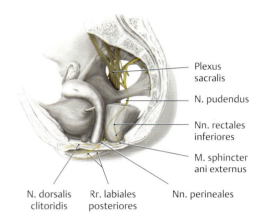

E Erectile tissues and erectile muscles in the female

Regio perinealis, lithotomy position. The labia majora and minora, skin, and fascia perinei superficialis have been removed, as well as the erectile muscles on the left side.

Erectile tissues: The erectile tissues of the *clitoris* are distributed around both of its crura and its short shaft (corpus). They correspond to the erectile tissues in the male and are named accordingly: the *corpora cavernosa clitoridis* (dextrum and sinistrum), homologous to the corpora cavernosa penis. The swelling at the end of the clitoral shaft is called the glans clitoridis, homologous to the glans penis. Its sensory innervation is like that of the penile glans, and it is mostly covered by a preputium clitoridis. The erectile tissue of the *labia minora* is located in the hairless, fat-free skin folds of the labia minora and is termed the *bulbus vestibuli,* which is the homologue of the corpus spongiosum penis in the male.

Erectile muscles: The two crura of the clitoris, one on each side, by which the clitoris arises from the ramus inferior ossis pubis are covered by the *m. ischiocavernosus*, and the dense erectile venous plexus at the base of the labia is covered by the *m. bulbospongiosus*.

F Nerve supply to the female external genitalia

Lesser pelvis, left lateral view. The n. pudendus enters the fossa ischioanalis. After giving off the nn. rectales inferiores to the m. sphincter ani externus and anal skin, it courses to the posterior border of the diaphragma urogenitale, where it divides into its terminal branches (nn. perineales). The *superficial branches* pass through the spatium perinei superficiale (not seen here) to the skin of the perineum and the posterior portions of the labia majora (rr. labiales posteriores). The *deep branches* course in the spatium profundum perinei, distributing rr. musculares to the erectile muscles and the n. dorsalis clitoridis to the clitoris. The anterior portions of the labia majora are supplied by rr. labiales anteriores from the n. ilioinguinalis (not seen here).

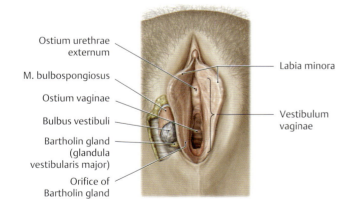

G The vestibule and vestibular glands

Lithotomy position with the labia separated. The vestibulum vaginae, bounded by the labia minora, contains the external openings of the urethra and vagina (ostium urethrae externum and ostium vaginae) and the vestibular glands. The *glandulae vestibulares minores* (not seen here) have numerous openings near the ostium urethrae externum, while the paired *glandulae vestibulares majores* each open by a 1-cm-long duct at the posterior border of the bulbus vestibuli on the inner surface of the labia minora. The glandulae vestibulares minores are homologous to the glandulae urethrales in the male, and the glandulae vestibulares majores to the glandulae bulbourethrales. The vestibular glands produce a mucous secretion that moistens the vestibulum vaginae and reduces friction during coitus, preventing epithelial injury. Next to the ostium urethrae externum are two short, blind, rudimentary excretory ducts called paraurethral ducts (not shown here). They correspond developmentally to the male prostata but have no known function in the female. Like the vestibular glands, however, they are susceptible to bacterial colonization. Bacterial colonization of the glandulae vestibulares majores (Bartholin's glands) can lead to *bartholinitis,* a painful inflammation with swelling and redness. An inflammation that occludes the excretory ducts of the glands can produce a painful *retention cyst,* which should be opened or removed.

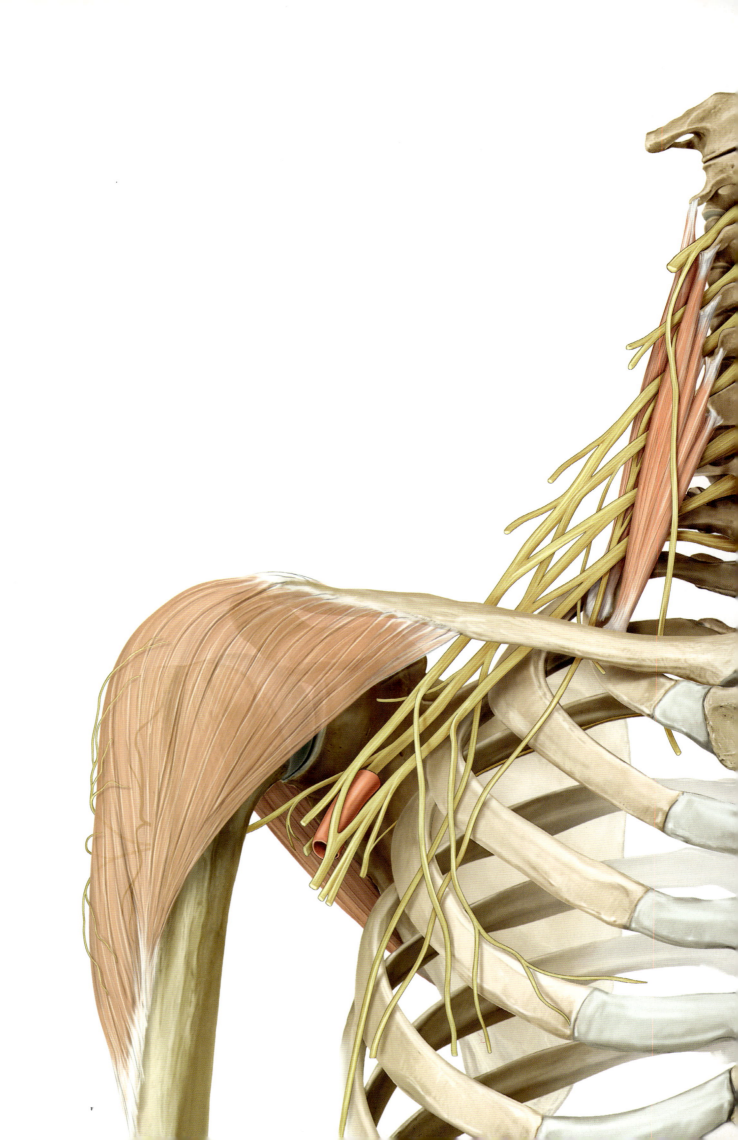

Upper Limb

14 Bones, Ligaments, and Joints 238

15 Musculature: Functional Groups 296

16 Musculature: Topographical Anatomy 330

17 Neurovascular Systems: Forms and Relations 354

18 Neurovascular Systems: Topographical Anatomy . . . 374

14.1 The Upper Limb as a Whole

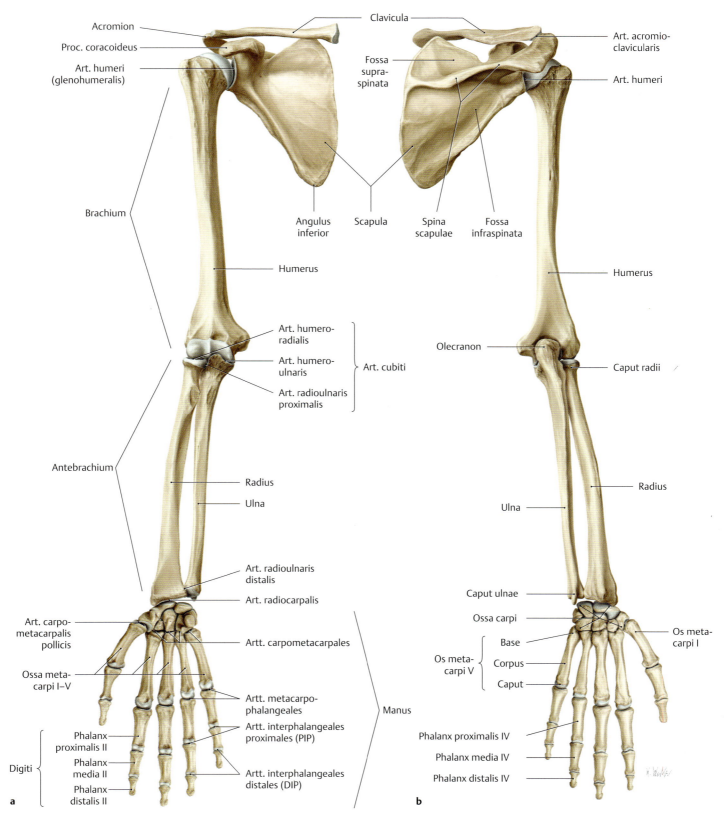

A Skeleton of the right upper limb
a Anterior view, b posterior view.
The skeleton of the upper limb consists of the shoulder girdle, arm, forearm, wrist, and hand. The shoulder girdle (clavicula and scapula) is joined to the upper limb at the shoulder joint (art. humeri [glenohumeralis]), and it joins the upper limb to the thorax at the art. sternoclavicularis (see p. 259). The limb itself consists of the

- arm (brachium),
- forearm (antebrachium), and
- hand (manus).

Upper Limb — 14. Bones, Ligaments, and Joints

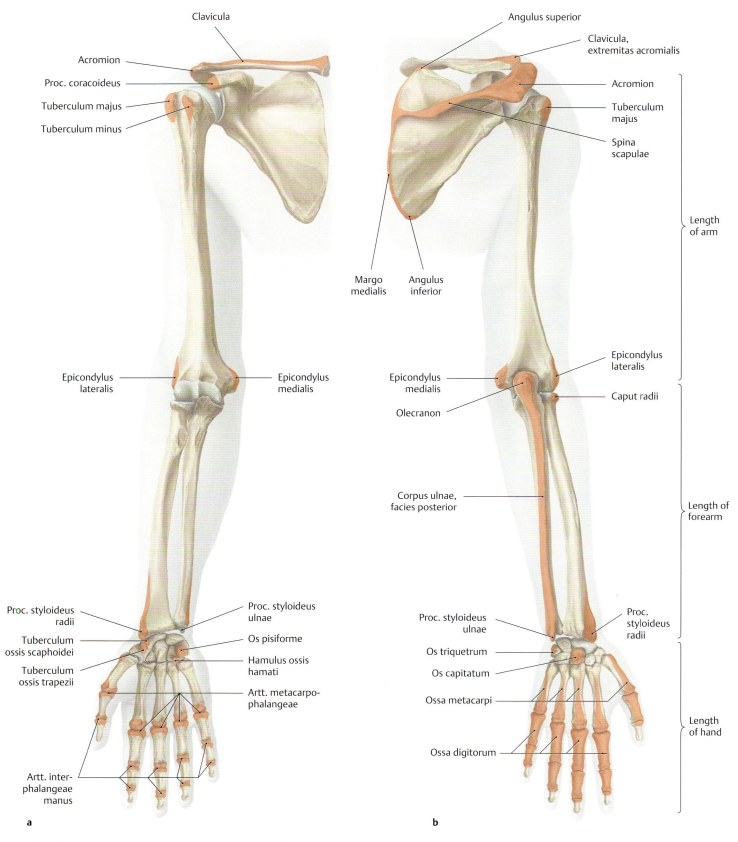

B Palpable bony prominences on the right upper limb
a Anterior view, b posterior view.
Except for the os lunatum and os trapezoideum, all of the bones in the upper limb are palpable to some degree through the skin and soft tissues. By consensus, standard reference points have been defined for use in measuring the lengths of the segments of the dependent limb (with the palm turned forward):

- Length of the arm = distance from the acromion to the epicondylus lateralis
- Length of the forearm = distance from the epicondylus lateralis to the proc. styloideus radii
- Length of the hand = distance from the proc. styloideus radii to the tip of the third finger

The segment lengths of the limb may be measured, for example, to aid in the precise evaluation of isolated pediatric growth disturbances that are confined to a particular bone.

239

14.2 Integration of the Shoulder Girdle into the Skeleton of the Trunk

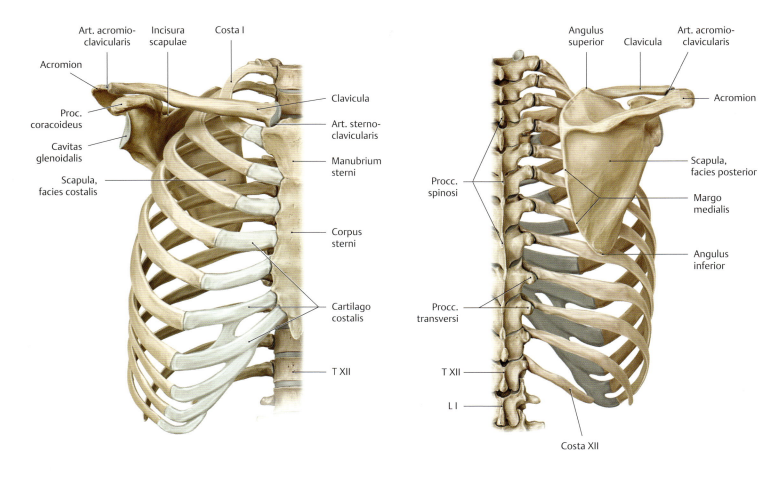

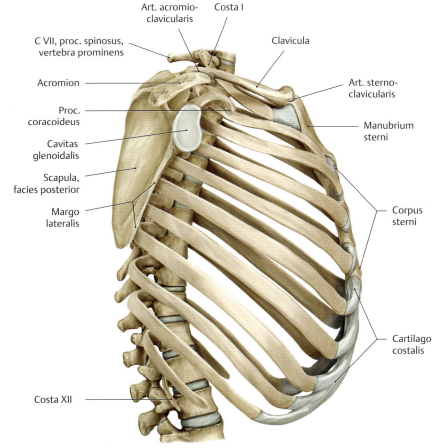

A Bones of the right shoulder girdle in relation to the trunk
a Anterior view, b posterior view, c lateral view.

The two bones of the shoulder girdle (the clavicula and scapula) are connected at the acromioclavicular joint art. acromioclavicularis (see p. 259). In its normal anatomic position, the scapula extends from the second to the seventh costa. The angulus inferior of the scapula is level with the proc. spinosus of the seventh thoracic vertebra (TVII), and the spina scapulae is level with the proc. spinosus of the third thoracic vertebra (TIII). When the scapula occupies a normal position, its long axis is angled slightly laterally, and its margo medialis forms a 3 to 5° angle with the midsagittal plane.

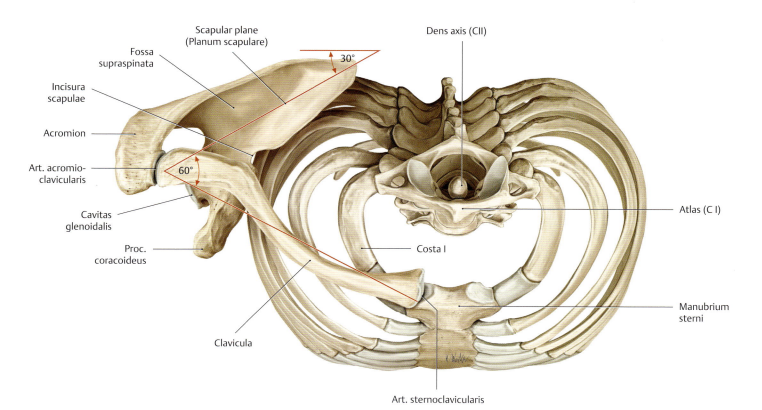

B Right shoulder girdle
Superior view. With the transition to a bipedal mode of locomotion, the human scapula moved from the more lateral placement in quadruped mammals to a more *posterior* position, and also a more frontal orientation, on the back of the thorax. Viewed from above, the scapula forms a 30° angle with the coronal plane. The scapula and clavicula subtend an angle of approximately 60°. Because of this arrangement, the two shoulder joints are angled slightly forward, shifting the range of arm movements forward into the field of vision and action. This reorientation in humans creates the opportunity for visual control of manual manipulations (hand–eye coordination).

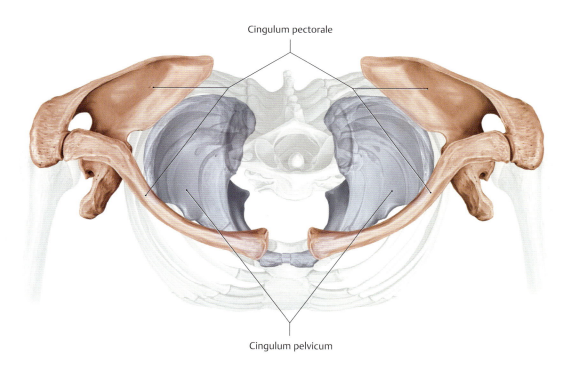

C Comparison of the shoulder girdle and pelvic girdle in their relation to the skeleton of the trunk
Superior view. Unlike the very mobile shoulder girdle, (cingulum pectorale) the pelvic girdle (cingulum pelvicum), consisting of the paired hip (coxal) bones (os coxae), is firmly integrated into the axial skeleton (skeleton axiale). As the trunk assumes an upright position, the pelvis moves over the weight-bearing surface of the feet, making it necessary for the pelvis to support the total weight of the trunk. This basically limits the lower limbs to functions of locomotion and support while freeing the upper limbs from these tasks and making them a versatile organ of movement and expression that is particularly useful for touching and grasping.

14.3 The Bones of the Shoulder Girdle

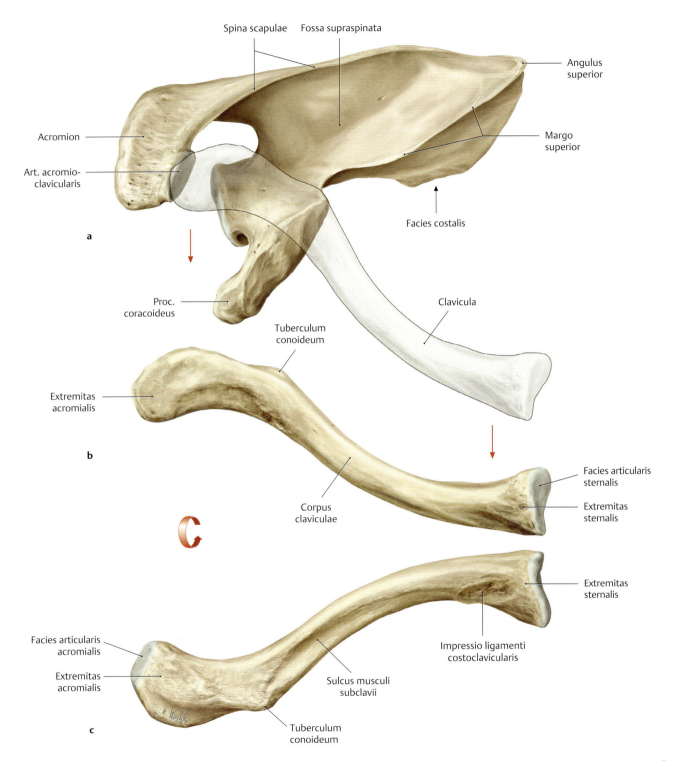

A Position and shape of the right clavicula
a Clavicula in its normal relation to the scapula, superior view.
b Isolated clavicula, superior view.
c Isolated clavicula, inferior view.

The clavicula is an S-shaped bone, approximately 12 to 15 cm long in adults, that is visible and palpable beneath the skin along its entire length. The medial or *sternal end* (extremitas sternalis) of the clavicula bears a saddle-shaped articular surface, while the lateral or *acromial end* ("extremitas acromialis") has a flatter, more vertical articular surface. The clavicula is the *only* bone in the limbs that is not preformed in cartilage during embryonic development; instead, it ossifies directly from *connective tissue* (membranous ossification). A congenital failure or abnormality in the development of this connective tissue results in an anomaly called dysostosis cleidocranialis. There may be associated ossification defects in the cranial vault, which are also formed by membranous ossification (dysostosis craniofacialis). Besides fractures due to obstetric trauma (1–2% of all newborns), fractures of the middle third of the clavicula are one of the most common fractures that are sustained by children and adults (in children, some 50% of all clavicular fractures occur before 6 years of age).

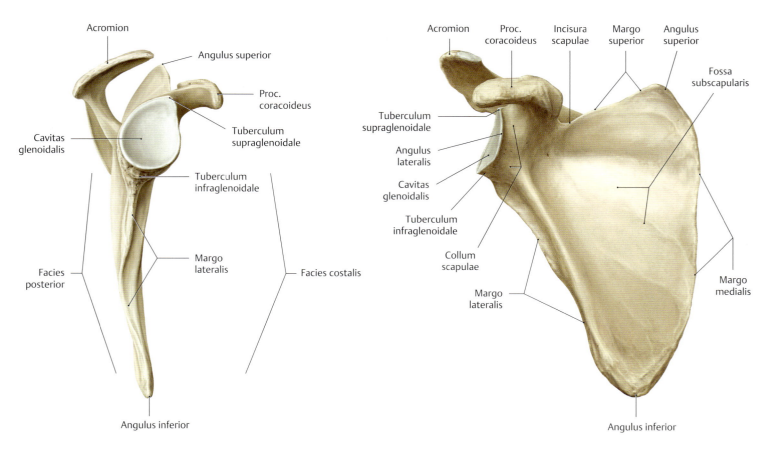

B The right scapula. Lateral view

C The right scapula. Anterior view

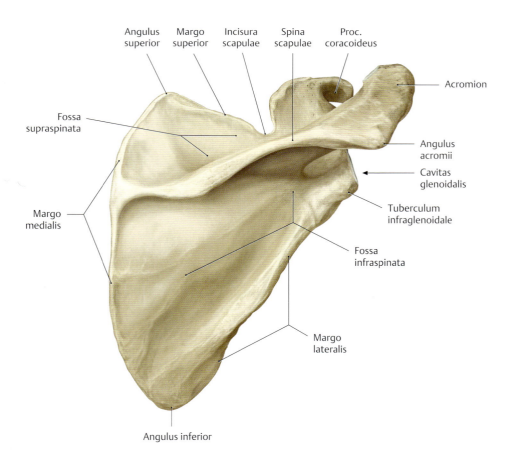

D The right scapula. Posterior view

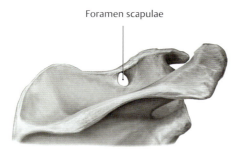

E Foramen scapulae

The lig. transversum scapulae superius (see p. 267) can become ossified, transforming the incisura scapulae into an anomalous bony canal referred to as a *foramen scapulae*. This can lead to compression of the n. suprascapularis as it passes through this canal (see p. 388). Active rotational movements of the shoulder aggravate the nerve, leading to significant symptoms (*incisura scapulae syndrome*). A common result is weakness and atrophy of the muscles—the mm. supraspinatus and infraspinatus—that the n. suprascapularis innervates (see p. 303).

14.4 The Bones of the Upper Limb: The Humerus

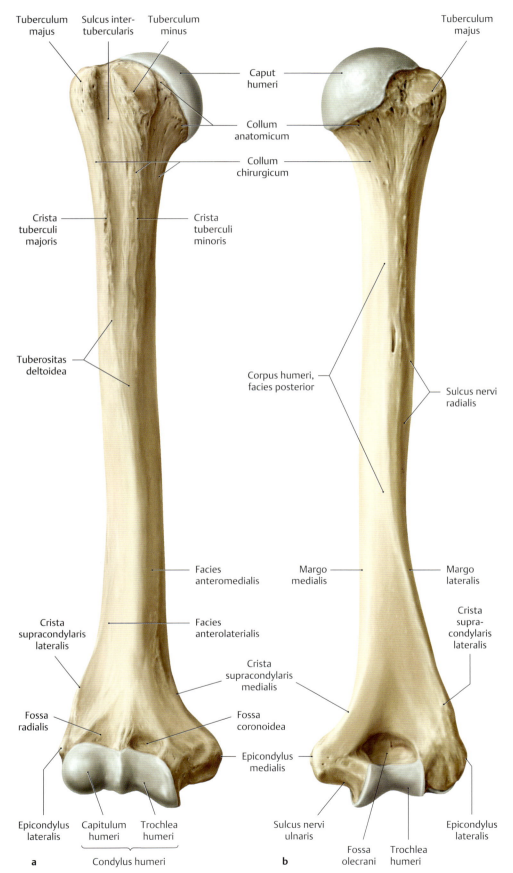

A The right humerus
a Anterior view, b posterior view.

B Processus supracondylaris
An anomaly sometimes found on the distal humerus above the epicondylus medialis is referred to as a proc. supracondylaris. This bony outgrowth is a relatively rare atavistic feature in humans that corresponds to a normal structure in other vertebrates, in which it forms part of a canalis supracondylaris (see p. 393).

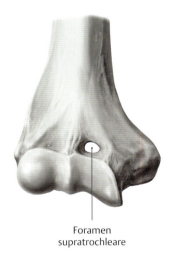

C Foramen supratrochleare
The presence of a foramen supratrochleare is another rare variant in which the two opposing fossae olecrani and coronoidea communicate through an opening.

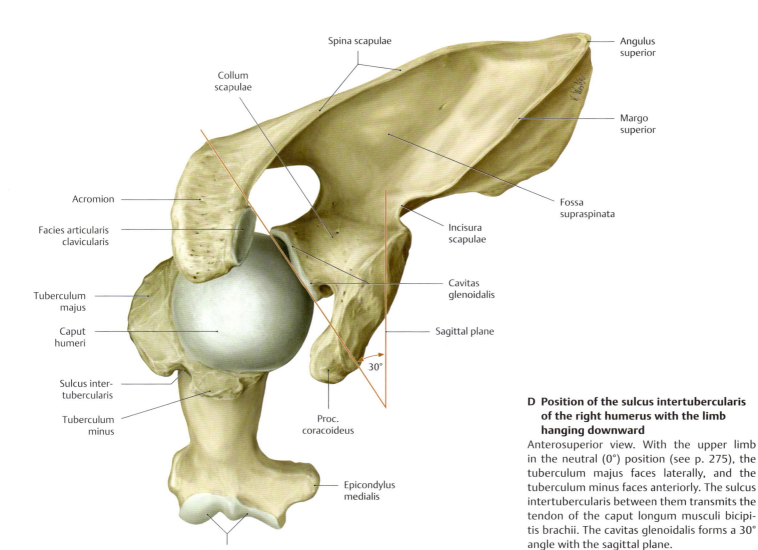

D Position of the sulcus intertubercularis of the right humerus with the limb hanging downward

Anterosuperior view. With the upper limb in the neutral (0°) position (see p. 275), the tuberculum majus faces laterally, and the tuberculum minus faces anteriorly. The sulcus intertubercularis between them transmits the tendon of the caput longum musculi bicipitis brachii. The cavitas glenoidalis forms a 30° angle with the sagittal plane.

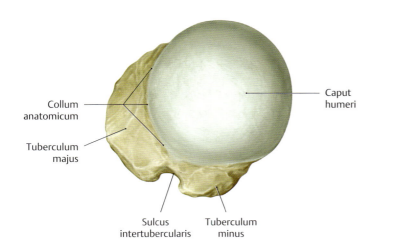

E The proximal right humerus. Superior view

F The distal right humerus. Inferior view

14.5 The Bones of the Upper Limb: Torsion of the Humerus

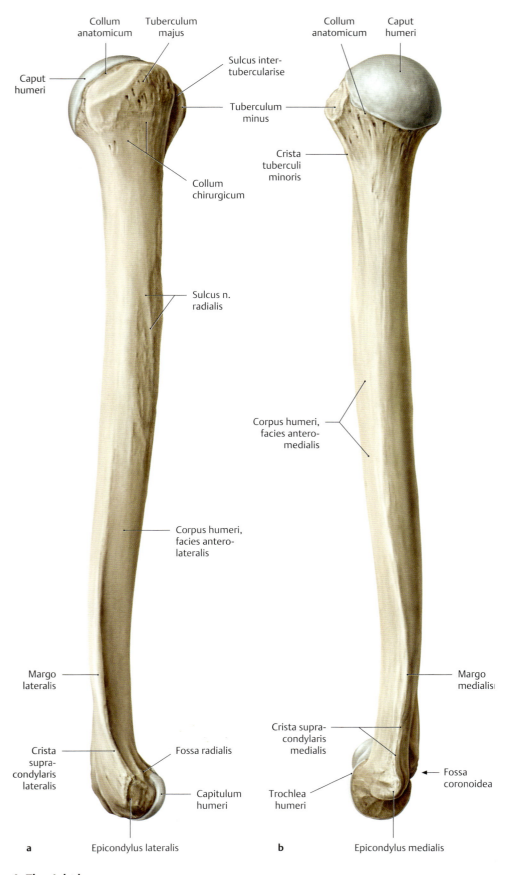

A The right humerus
a Lateral view, b medial view.

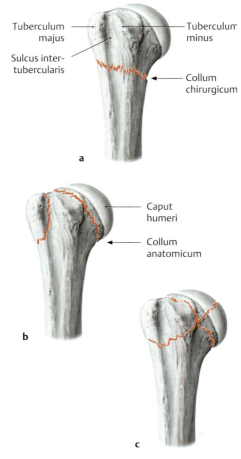

B Fractures of the proximal humerus
Anterior view. Fractures of the proximal humerus comprise approximately 4 to 5% of all fractures. They occur predominantly in older patients who sustain a fall onto the outstretched arm or directly onto the shoulder. Three main types are distinguished:

- Extra-articular fractures (**a**)
- Intra-articular fractures (**b**)
- Comminuted fractures (**c**)

Not infrequently, extra-articular fractures at the level of the *collum chirurgicum* (site of predilection for extra-articular fractures of the proximal humerus) and intra-articular fractures at the level of the *collum anatomicum* are accompanied by injuries of the blood vessels that supply the caput humeri (anterior and posterior a. circumflexa humeri, see p. 355), with an associated risk of post-traumatic avascular necrosis. Besides proximal humeral fractures, other important injuries are fractures of the humeral shaft (corpus humeri) and fractures of the distal humerus (e.g., supracondylar fractures). Fractures of the humeral shaft are frequently associated with damage to the n. radialis in its sulcus (see p. 369 for neurologic deficits following a nervus radialis lesion).

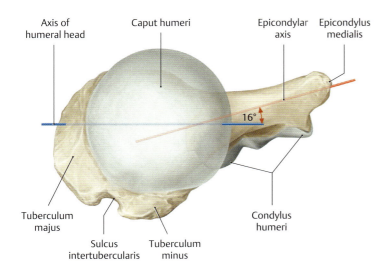

C Torsion of the humerus
Right humerus, superior view. The humeral shaft of the adult corpus humeri normally exhibits some degree of torsion, i.e., the proximal end of the humerus is rotated relative to its distal end. The degree of this torsion can be assessed by superimposing the axis of the caput humeri (from the center of the tuberculum majus to the center of the caput humeri) over the epicondylar axis of the elbow joint (art. cubiti). This *torsion angle* equals approximately 16° in an adult, compared to about 60° in a newborn. The decrease in the torsion angle with body growth correlates with the change in the position of the scapulae. Thus, while the cavitas glenoidalis in the newborn still faces anteriorly, it is directed much more laterally in the adult (see p. 241). As the position of the scapula changes, there is a compensatory decrease in the torsion angle to ensure that hand movements will remain within the visual field of the adult.

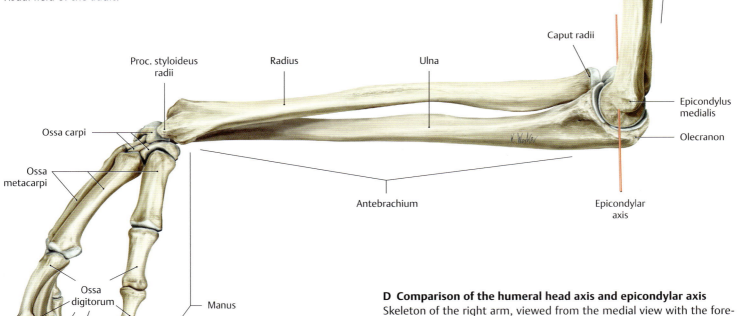

D Comparison of the humeral head axis and epicondylar axis
Skeleton of the right arm, viewed from the medial view with the forearm pronated.

14.6 The Bones of the Upper Limb: The Radius and Ulna

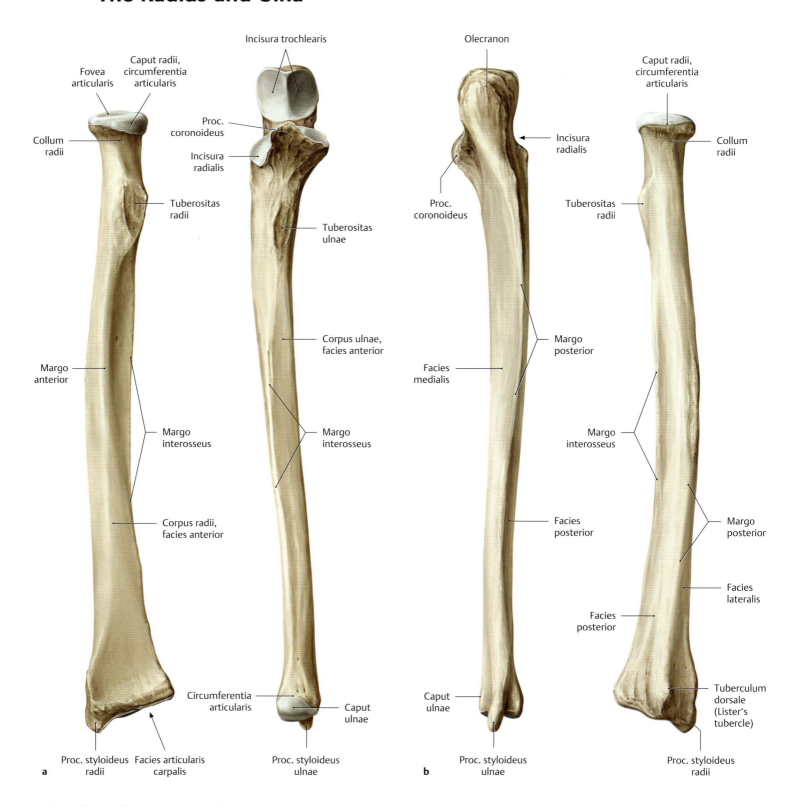

A The radius and ulna of the right forearm
a Anterior view, b posterior view.
The radius and ulna are not shown in their normal relationship; they have been separated to demonstrate the articular surfaces of the proximal and distal radioulnar joints.

Upper Limb — 14. Bones, Ligaments, and Joints

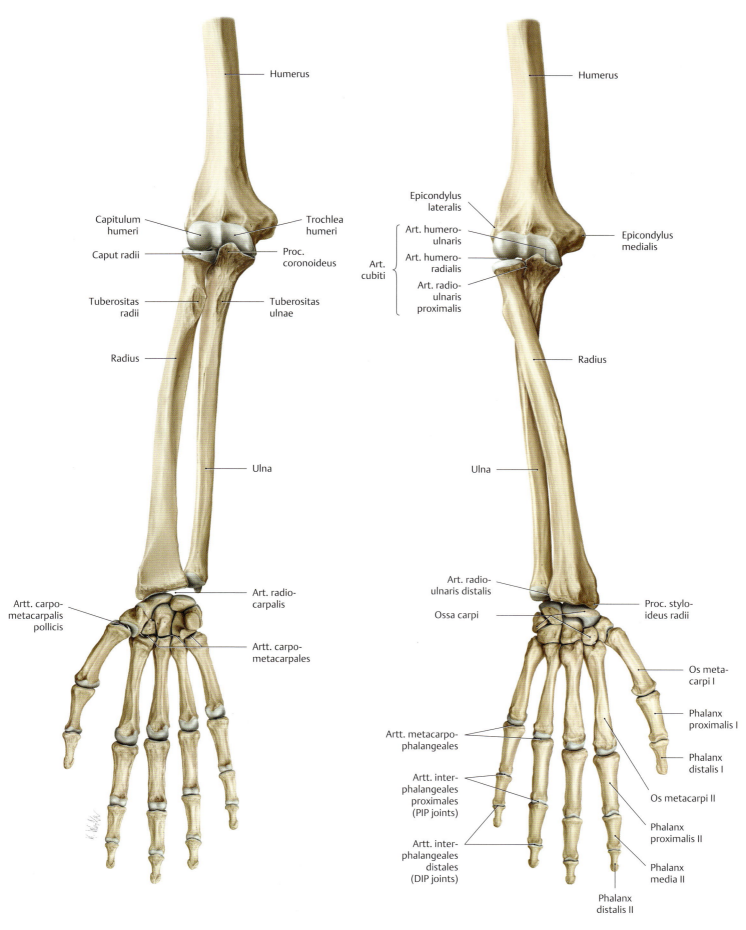

B The radius and ulna of the right arm in (a) supination and (b) pronation

The radius and ulna are parallel to each other in supination, whereas in pronation the radius crosses over the ulna. The movement of turning the palm (palma manus) upward or downward (supination/pronation) takes place at the proximal and distal radioulnar joints (see p. 280).

249

14.7 The Bones of the Upper Limb: The Articular Surfaces of the Radius and Ulna

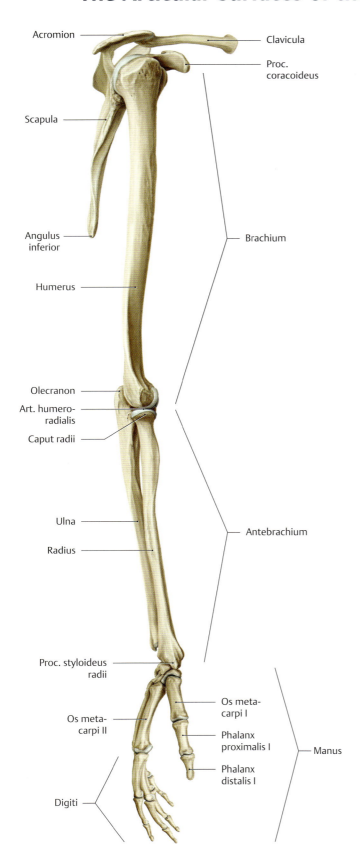

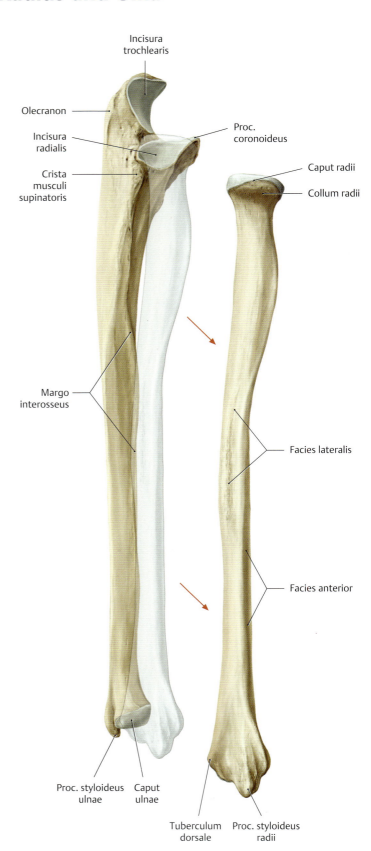

A The right upper limb
Lateral view. The forearm is supinated (the radius and ulna are parallel).

B Right forearm
Lateral view. The radius and ulna are shown disarticulated to demonstrate the articular surfaces of the ulna for the proximal and distal radioulnar joints (see **C**).

Upper Limb — 14. Bones, Ligaments, and Joints

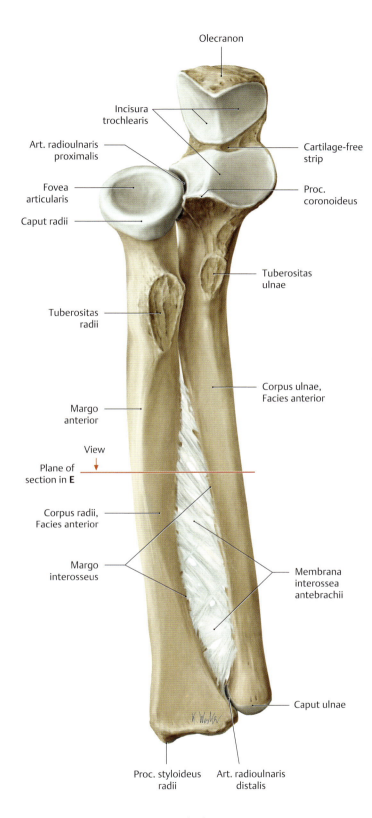

C The radius and ulna of the right forearm
Anterosuperior view. The proximal and distal radioulnar joints (art. radioulnaris proximalis and art. radioulnaris distalis) are functionally interlinked by the membrana interossea between the radius and ulna. As a result, motion in one joint is invariably combined with motion in the other (see p. 282).

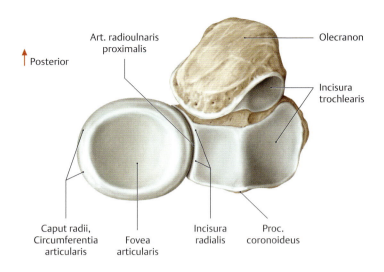

D The proximal articular surfaces of the radius and ulna of the right forearm.
Proximal view.

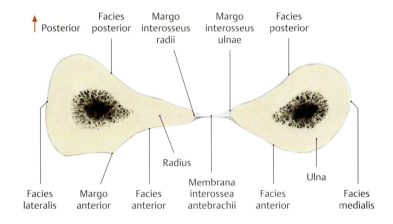

E Cross section through the right radius and ulna.
Proximal view.

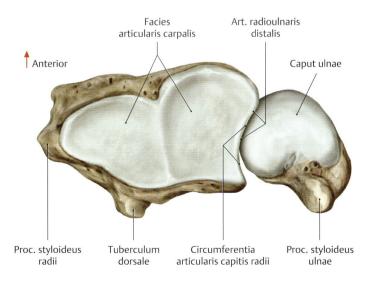

F The distal articular surfaces of the radius and ulna of the right forearm.
Distal view.

251

14.8 The Bones of the Upper Limb: The Hand

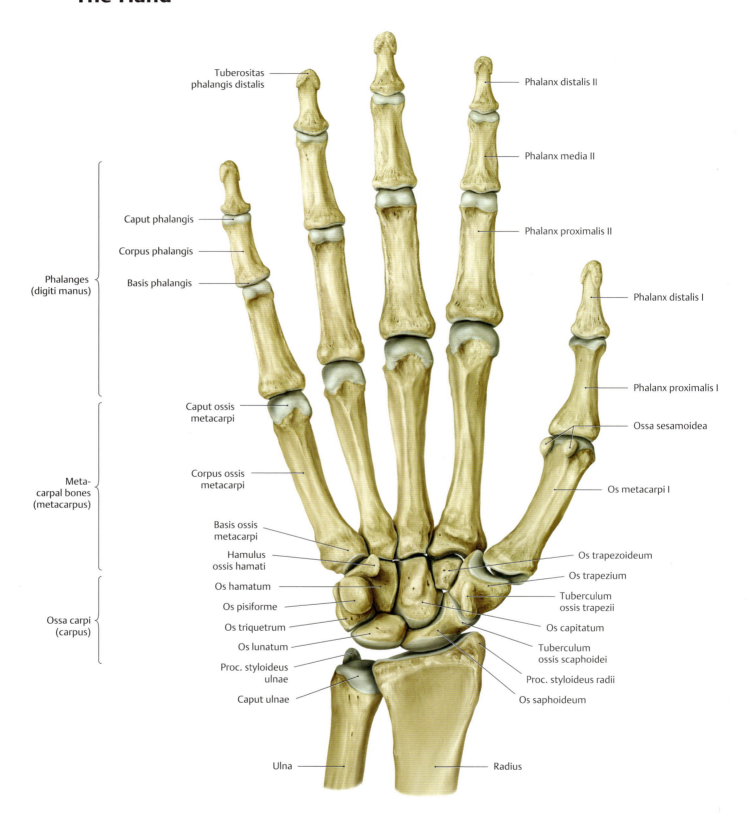

A The bones of the right hand.
Palmar view. The skeleton of the hand consists of

- the ossa carpi (carpus),
- the metacarpal bones (metacarpus), and
- the phalanges (digiti manus).

The palma manus refers to the anterior (flexor) surface of the hand, the dorsum manus to the posterior (extensor) surface. The terms of anatomic orientation in the hand are palmar or volar (toward the anterior surface), dorsal (toward the posterior surface), ulnar (toward the ulna or small finger [digitus minimus]), and radial (toward the radius or thumb [pollex]).

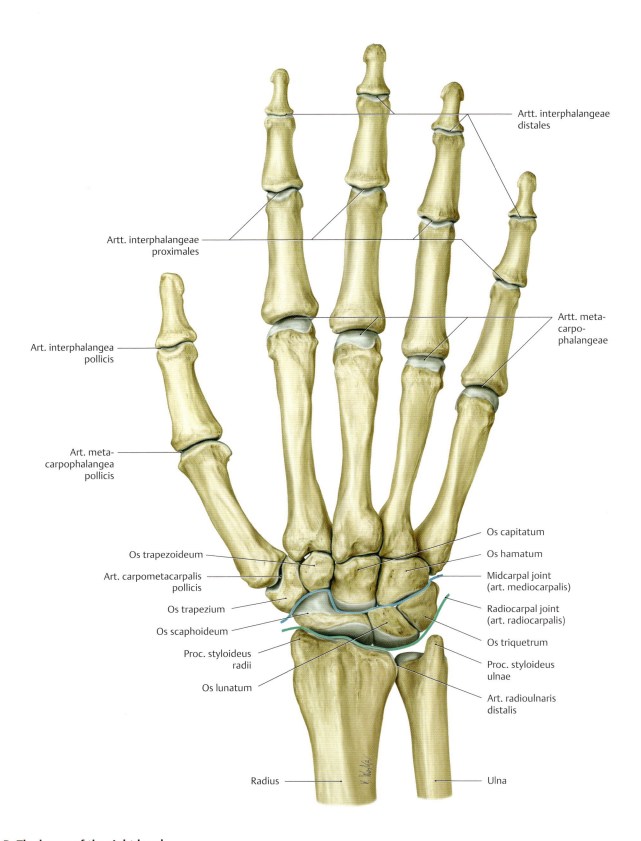

B The bones of the right hand.
Dorsal view. The radiocarpal and midcarpal joints (art. radiocarpalis and art. mediocarpalis) are indicated by blue lines.

14.9 The Bones of the Upper Limb: The Carpal Bones

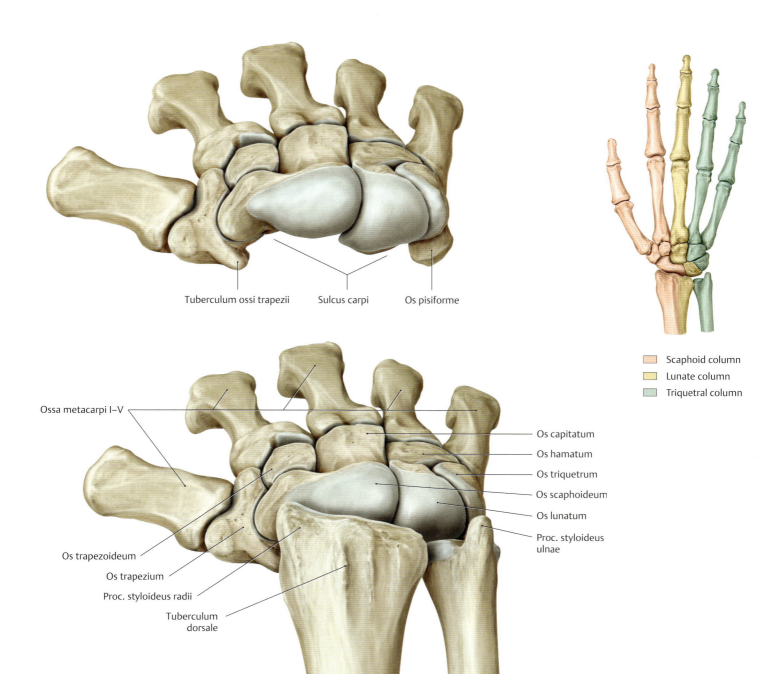

A Carpal bones (ossa carpi) of the right hand
a After removal of radius and ulna, proximal view; **b** column model of the hand, dorsal view; **c** with the wrist in flexion, proximal view.
The ossa carpi are arranged in two rows of four bones each—one proximal row and one distal row (see also **B**). From a biomechanical and clinical standpoint, the carpal bones do not form two transverse rows but are arranged in three longitudinal columns: a radial scaphoid column (consisting of the ossa scaphoideum, trapezium and trape-zoideum), a central lunate column (consisting of the ossa lunatum and capitatum), and an ulnar triquetral column (consisting of the ossa triquetrum and hamatum). In this functional classification, the os pisiforme is regarded as a sesamoid bone (os sesamoideum) embedded in the tendon of the flexor carpi ulnaris muscle (see p. 404). The bones in each row are interconnected by tight joints, their surfaces exhibiting a palmar concavity and a dorsal convexity. This creates the carpal tunnel (sulcus carpi) on the palmar surface (see p. 288), which is bounded by a bony eminence on the radial and ulnar sides.

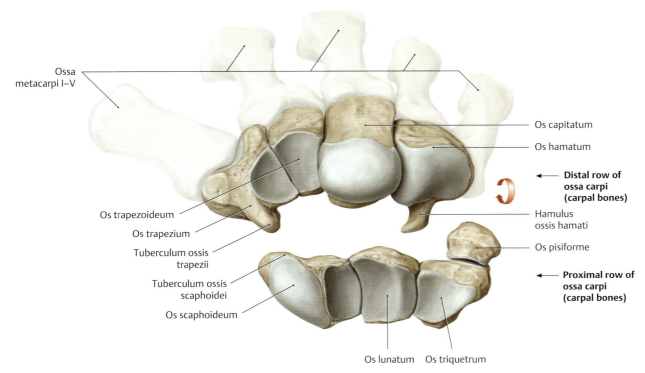

B Articular surfaces of the midcarpal joint of the right hand (articulatio mediocarpalis)
The *distal* row of ossa carpi is shown from the *proximal* view. The *proximal* row is shown from the *distal* view.

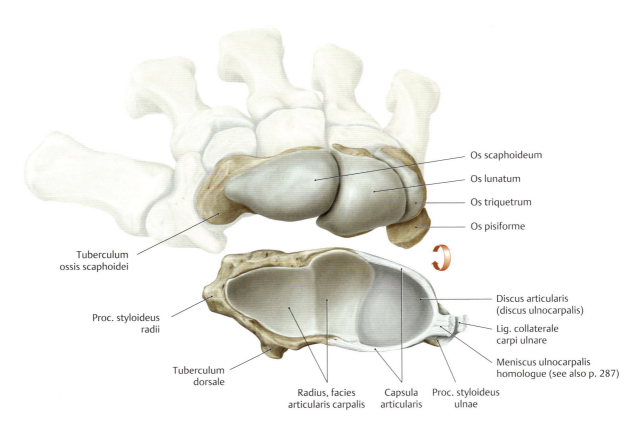

C Articular surfaces of the radiocarpal joint of the right hand (articulatio radiocarpalis)
The proximal row of ossa carpi is shown from the proximal view. The articular surfaces of the radius and ulna and the discus articularis (discus ulnocarpalis) are shown from the distal view.
Clinically, the art. radiocarpalis is subdivided into a radial compartment and an ulnar compartment. This takes into account the presence of the interposed discus ulnocarpalis, which creates a second, ulnar half of the radiocarpal joint in addition to the radial half. Accordingly, the radius articulates with the proximal row of ossa carpi in the radial compartment, while the caput ulnae and discus ulnocarpalis articulate with the proximal row of ossa carpi in the ulnar compartment.

14.10 Architecture of the Radiocarpal Junction and the Metacarpus; Distal Radius and Scaphoid Fractures

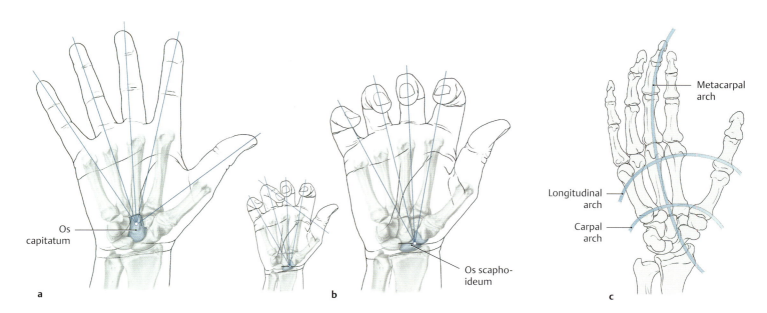

A Architecture of the metacarpus
The metacarpus is the key region for the architecture of the hand. It is where the five digital rays form and then develop into the thumb and fingers. While in neutral position, the longitudinal axis of each finger is parallel, and the longitudinal axis of the abducted thumb and the spread fingers converge to an intersection point in the os capitatum (**a**). However, when bending the finger joints, the axes converge to an intersection point in the os scaphoideum (**b**). Only the knowledge of these defined basic anatomic positions allows for the diagnosis of mal-alignments caused by injury (most notably rotational mal-alignments of the fingers, meaning "twisting" of phalanges as a result of a fracture (see inset). The five digital rays are connected by three functionally significant arches (**c**): a longitudinal arch along the third ray, a metacarpal arch, and a transverse carpal arch.

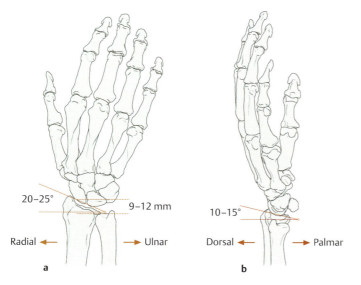 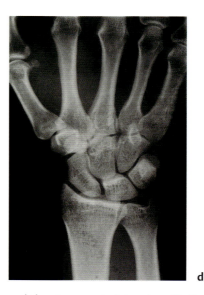 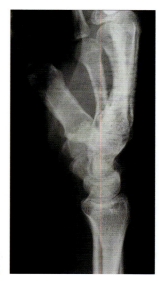

B Inclination angle of the articulating surfaces along the distal radius
a Radioulnar inclination angle, right hand, dorsal view.
b Dorsopalmar inclination angle, right hand, ulnar view.
c X-ray image of carpal region, dorsopalmar beam path and
d radioulnar beam path (**c** and **d** from Schmidt HM, Lanz U. Chirurgische Anatomie der Hand. 2nd ed. Stuttgart: Thieme; 2003).

The distal radius helps form the artt. radiocarpalis and radioulnaris distalis. In addition, it supports the discus ulnocarpalis and the strong dorsal and palmar extrinsic ligaments (see p. 284). It is primarily responsible for carpal load transmission within the longitudinally arranged columns (scaphoid column, lunate column, and triquetrum, see p. 254) and thus is prone to injury (see **C**). For a harmonious relationship between the parts forming the art. radiocarpalis in terms of optimal mobility of the hand, the position of the part of the socket formed by the radius is of importance. The facies articularis carpalis is not vertical to the longitudinal axis of the forearm, but in a radioulnar inclination angle of 20 to 25° (ulnar inclination) and a dorsopalmar inclination angle of 10 to 15° (palmar inclination). The distal radial length relative to the ulna (tip of the proc. styloideus radii to the carpal articulating surface of the ulna) is approximately 9 to 12 mm (important for optimal mobility of the hand).

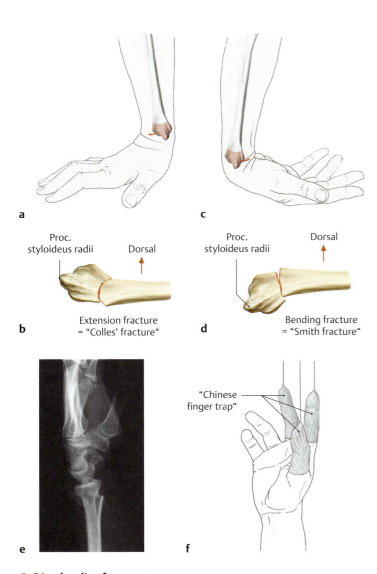
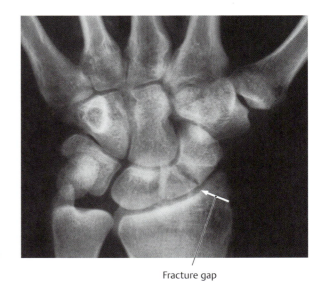
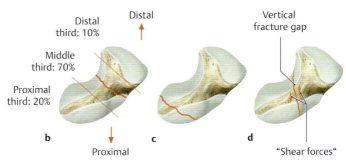

C Distal radius fractures

Accounting for 20 to 25% of all fractures, the distal (or near the wrist) radius fracture as a result of falling on the hand is the most common fracture in humans. Affected are almost 80% of women over the age of 50 (main cause: postmenopausal osteoporosis). Depending on the position of the carpus relative to the distal radius at the moment of impact, 90% of falls result in extension fractures (Colles' fracture; see **a** and **b**), and 10%, in flexion fractures (see **e**, from Henne-Bruns D, Dürig M, Kremer B. Chirurgie. 2nd ed. Stuttgart: Thieme; 2003).

In distal radius fractures, one generally differentiates between extra- and intra-articular fractures, with the extra-articular fractures typically located 3 to 4 cm proximal to the art. radiocarpalis. The standard criteria for such a diagnosis are determined by taking conventional X-ray images of the wrists in two planes (see **e** lateral view). The therapeutic procedure (conservatively using casts or surgically through osteosynthesis) depends on the angle and direction of dislocation (fracture stability), the course of the fracture line (intra-/extra-articular), and the severity of accompanying injuries (e.g., involvement of the ulna, most notably the proc. styloideus ulnae). Less complicated (nondislocated, easy-to-reposition) and primarily stable fractures are conservatively treated using "Chinese finger traps" (**f**) with the help of an image intensifier. Axial alignment, most notably of the original radial length and angle (ulnar and palmar inclination angles, see **Ba** and **Bb**), are being restored through vertical extension and joint immobilization with the help of a dorsopalmar splint. With intra-articular fractures involving large articular fragments, osteosynthetic stabilization is generally recommended.

D Scaphoid fractures

a Scaphoid fracture as seen on X-ray, dorsopalmar view (from Matzen P. Praktische Orthopädie. 3rd ed. Stuttgart: J. A. Barth Verlag Thieme; 2002).
b Frequency and distribution of scaphoid fractures.

Carpal fractures, most notably scaphoid fractures (accounting for two thirds of all cases), are another injury caused by falling on the outstretched hand extended in the dorsal direction. Unlike distal radial fractures (see **C**), young people are predominantly affected by scaphoid fractures (typical sports injury). During physical examination, the symptoms can be relatively discrete. Usually the symptoms include tenderness on palpation of the area around the anatomic snuffbox (fovea radialis) with simultaneous radial or ulnar abduction and compressive pain around the thumb and index finger. In the case of a suspected scaphoid fracture, conventional radiographs of the wrist in four different planes (so-called scaphoid quartet series) should be taken (see **a**, white arrow), in order to determine the direction of the fracture gap. If the X-ray image fails to confirm the suspected diagnosis 10 to 14 days after initially stabilizing the fracture, a follow-up X-ray must be taken (if needed using CT scans). At this point, the process of resorption in the fracture hematoma is usually completed so that the fracture gap is wider and thus more clearly visible. Depending on the location, scaphoid fractures are classified into fractures of the proximal third, central third, or distal third (see **b**). The healing process is especially lengthy in fractures of the proximal third (up to 3 months in an upper arm cast for stabilization with inclusion of the thumb basal joint, see **c**) because this part of the bone has few blood vessels (the os scaphoideum receives the majority of its blood supply via distal vessels). The healing process is just as lengthy in slanted or vertical fractures due to the resulting shearing forces pushing in opposite directions (see **d**).

Note: The os scaphoideum is involved in all movements of the hand, which makes long-term stabilization rather difficult. Thus, pseudo-osteoarthritis is a typical complication of scaphoid fracture (= false joint after failed fracture healing, see p. 88).

14.11 The Joints of the Shoulder: Overview and Clavicular Joints

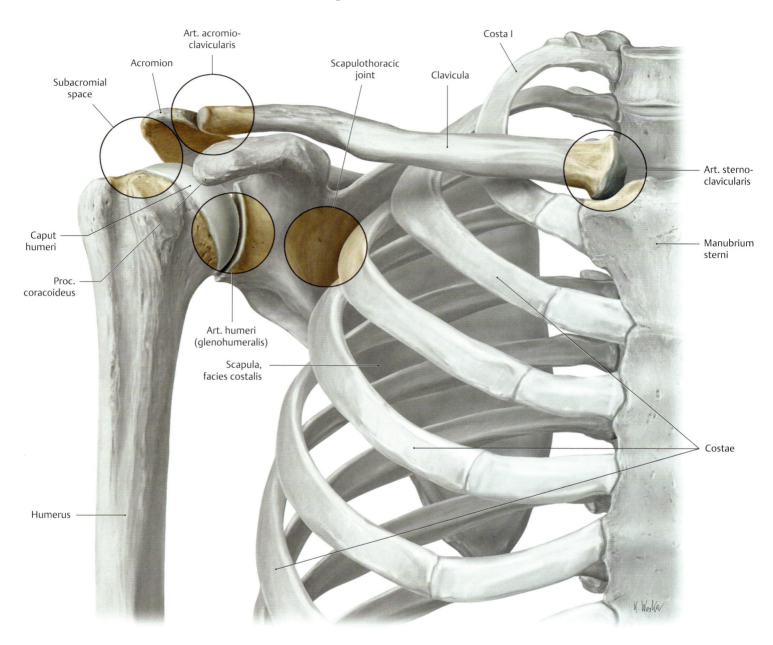

A The five joints of the shoulder
Right shoulder, anterior view. A total of five joints contribute to the wide range of arm motions at the shoulder. There are three true shoulder joints and two functional articulations:

- **True joints:**
  1. Art. sternoclavicularis (sternoclavicular joint)
  2. Art. acromioclavicularis (acromioclavicular joint)
  3. Art. humeri (glenohumeral joint)
- **Functional articulations:**
  4. Subacromial space: a space lined with bursae (bursa subacromialis and bursa subdeltoidea) that allows gliding between the acromion and the rotator cuff (= muscular cuff of the art. humeri, consisting of the mm. supraspinatus, infraspinatus, subscapularis, and teres minor, which press the caput humeri into the cavitas glenoidalis; see p. 303).
  5. Scapulothoracic joint: loose connective tissue between the mm. subscapularis and serratus anterior that allows gliding of the scapula on the chest wall.

Besides the true joints and functional articulations, the two ligamentous attachments between the clavicula and first costa (lig. costoclaviculare) and between the clavicula and proc. coracoideus (lig. costoclaviculare) contribute to the mobility of the upper limb. All of these structures together comprise a functional unit, and free mobility in all the joints is necessary to achieve a full range of motion. This expansive mobility is gained at the cost of stability, however. Since the shoulder has a loose capsule and weak reinforcing ligaments, it must rely on the stabilizing effect of the rotator cuff tendons. As the upper limb changed in mammalian evolution from an organ of support to one of manipulation, the soft tissues and their pathology assumed increasing importance. As a result, a large percentage of shoulder disorders involve the soft tissues.

Upper Limb —— 14. Bones, Ligaments, and Joints

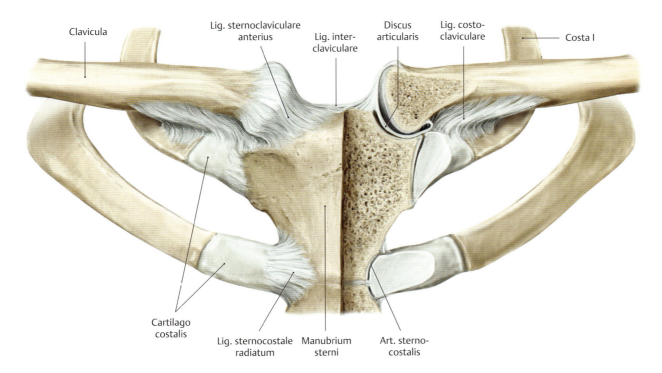

B The articulatio sternoclavicularis and its ligaments
Anterior view. The art. sternoclavicularis (also called the *medial* clavicular joint) and the art. acromioclavicularis (the *lateral* clavicular joint, see below) together make up the true joints within the shoulder girdle itself. In the figure, a coronal section has been made through the sternum and adjacent clavicula to demonstrate the inside of the left art. sternoclavicularis. A fibrocartilaginous discus articularis compensates for the mismatch of surfaces between the two saddle-shaped articular faces of the clavicula and manubrium sterni.

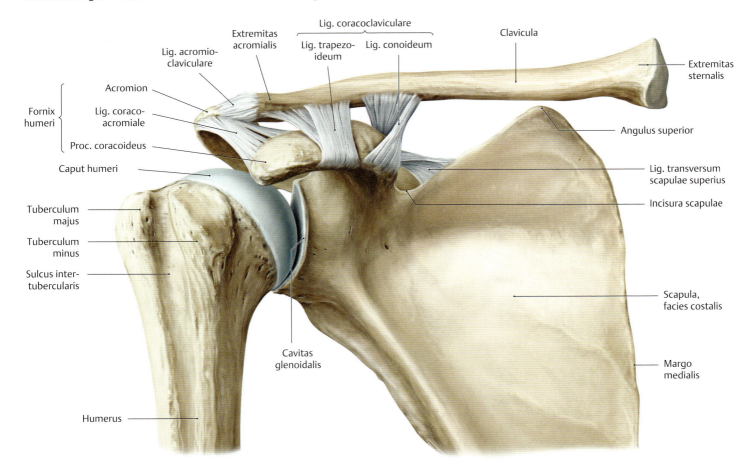

C The articulatio acromioclavicularis and its ligaments
Right shoulder, anterior view. The art. acromioclavicularis (the *lateral* clavicular joint) has the form of a *plane joint*. Because the articulating surfaces are flat, they must be held in place by strong ligaments (Ligg. acromioclaviculare, coracoacromiale and coracoclaviculare).). This greatly limits the mobility of the art. acromioclavicularis. In some individuals the art. acromioclavicularis has a variably shaped discus articularis that gives the joint greater mobility.

259

14.12 The Joints of the Shoulder: Ligaments of the Clavicular and Scapulothoracic Joints

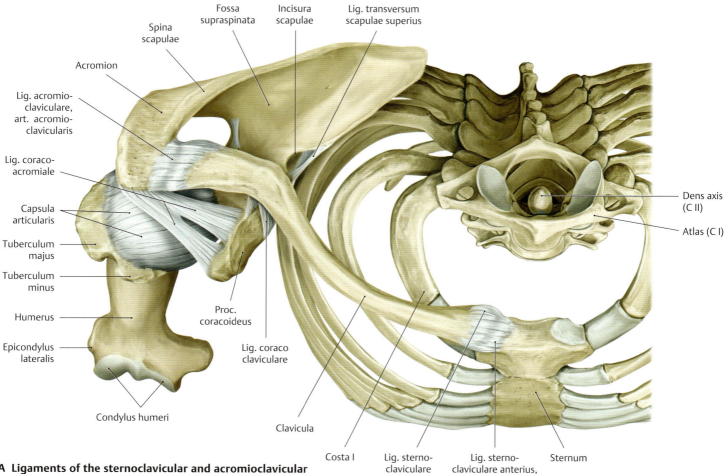

A Ligaments of the sternoclavicular and acromioclavicular joints (art. sternoclavicularis and art. acromioclavicularis)
Right side, superior view.

B Injuries to the acromioclavicular ligaments (ligg. acromioclaviculares)
These injuries often result from falling on a shoulder or an outstretched arm. According to Tossy, they can be classified into three types:

- **Tossy I** The ligg. acromioclaviculare and coracoclaviculare are stretched but still intact.
- **Tossy II** The lig. acromioclaviculare is ruptured, with subluxation of the joint.
- **Tossy III** Ligaments are all disrupted, with complete dislocation of the art. acromioclavicularis.

Rockwood added three more types that occur less frequently:

- **Rockwood IV:** Dislocation of the clavicle shifts dorsally, due to the pars clavicularis of the m. deltoideus being pulled off the clavicula.
- **Rockwood V:** Dislocation of lateral end of the clavicula is increased in the cranial direction, due to the mm. deltoideus and trapezius being pulled off the clavicula.
- **Rockwood VI:** The lateral end of the clavicula underneath the acromion or the proc. coracoideus is dislocated (very rare).

Depending on the extent of the injury, the so-called piano-key phenomenon can be triggered through palpation (caution: painful!). The lateral end of the clavicula, which is elevated due to the injury, can be reduced by applying pressure from the cranial direction, but it pops back up when pressure is released. Radiographs in different planes will show widening of the space in the art. acromioclavicularis. Comparative-stress radiographs with the patient holding approximately 10-kg weights in each hand will reveal the extent of upward displacement of the lateral end of the clavicula on the affected side (is not performed in cases of visible partial rupture of ligaments in order to avoid further damage).

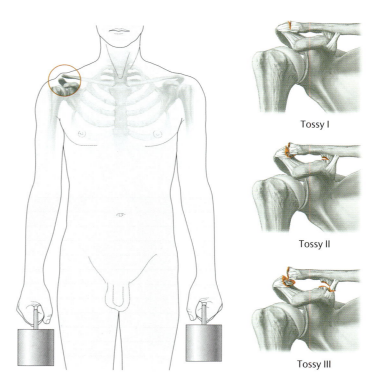

Upper Limb —— 14. Bones, Ligaments, and Joints

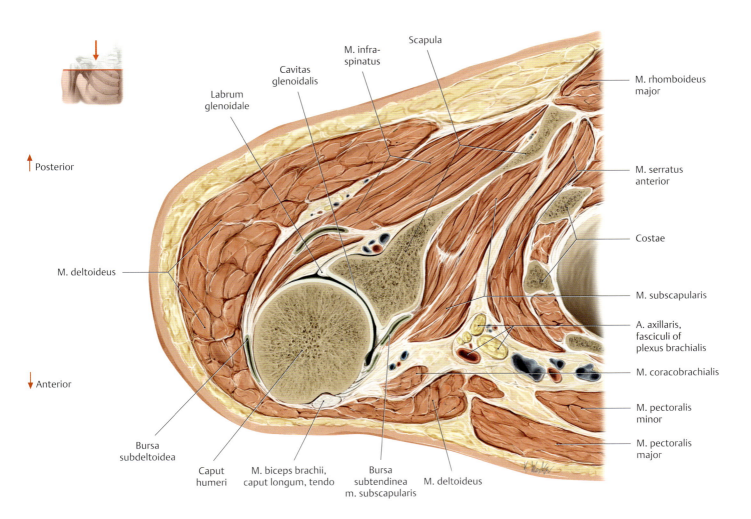

C Transverse section through the right shoulder joint
Superior view. In all movements of the shoulder girdle, the scapula glides on a curved surface of loose connective tissue between the m. serratus anterior and m. subscapularis (see **D**). This surface can be considered a "scapulothoracic" joint that allows the scapula not only to change the position of the shoulder (translational motion), but also to pivot, with the art. glenohumeralis maintained in a relatively stable position on the thorax (rotational motion) (see p. 274) (drawing based on a specimen from the Anatomical Collection at the University of Kiel).

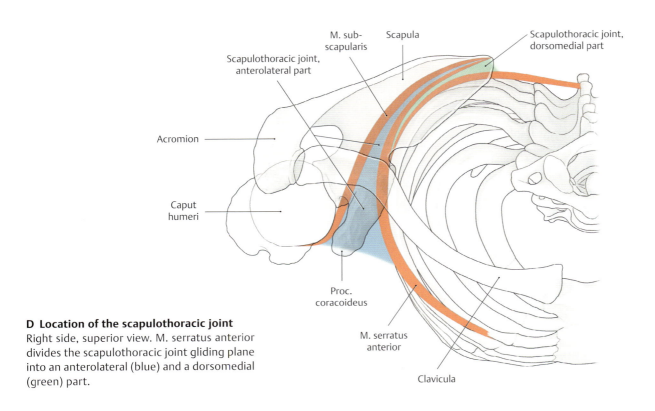

D Location of the scapulothoracic joint
Right side, superior view. M. serratus anterior divides the scapulothoracic joint gliding plane into an anterolateral (blue) and a dorsomedial (green) part.

261

14.13 The Joints of the Shoulder: Articular Surfaces, Articular Capsule, and Articular Cavity of the Articulatio humeri

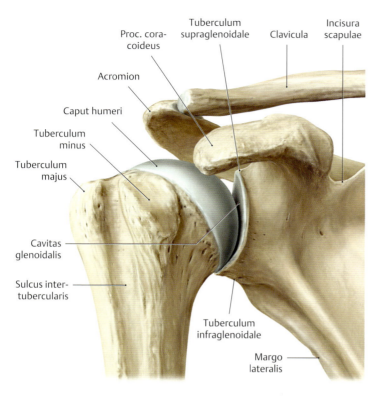
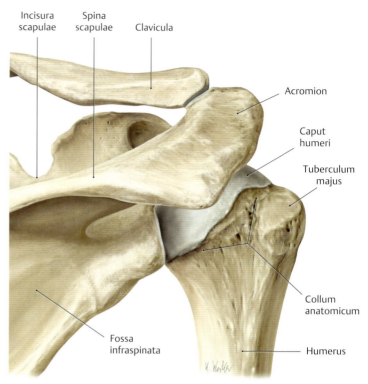

A Articulating bony elements of the right shoulder joint (art. humeri) and shoulder dislocation
a Anterior view.
b Posterior view.
c Lateral view.
d Articulating surfaces

In the shoulder joint (art. humeri)—the most mobile joint in the human body but also the most susceptible to injury—the caput humeri articulates with the cavitas glenoidalis of the scapula to form a spheroidal type of joint. The articular surface of the scapula, which is three to four times smaller than the caput humeri, is enlarged by a rim of fibrocartilage, the labrum glenoidale, which is approximately 5 mm wide at its base (see **d**). While this size discrepancy of the articulating surfaces serves to increase the range of shoulder motion, it compromises the stability of the joint. Since the joint capsule and ligaments are weak, the rotator cuff tendons are the primary stabilizers of the art. humeri (see p. 304).

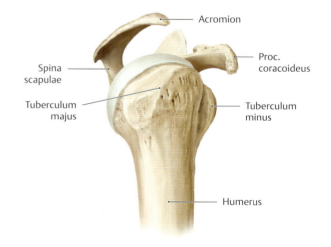

Dislocations of the shoulder joint are notoriously common. Approximately 45% of all dislocations involve the shoulder joint. In typical cases the caput humeri dislocates anteriorly or anteroinferiorly in response to forcible external rotation of the raised arm. Whereas considerable trauma is generally needed to cause the initial dislocation, certain movements of the shoulder (e.g., excessive arm rotation during sleep) may be sufficient to redislocate the caput humeri from the cavitas glenoidalis (recurrent shoulder dislocation).

In addition to physical examination (based on postion of the arm, occurring pain, and palpation of shoulder contour), diagnostic procedures for a dislocated shoulder include x-rays in two planes. In injuries caused by dislocation, especially anterior luxation, which account for most cases, mainly the rim of the socket (torn-off labrum, so-called Bankert lesion) and caput humeri (impression fracture caused by rim of socket, so-called Hill-Sachs lesion) are affected. Additional important complications include simultaneous damage to the n. axillaris (test sensitivity in the autonomous zone), the a. axillaris (test capillary refill time of the a. radialis), and, particularly in older patients, the rotator cuff (rupture).

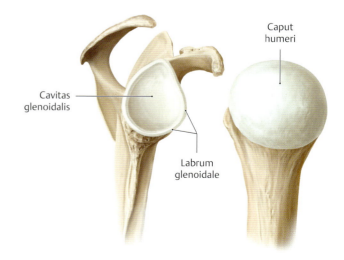

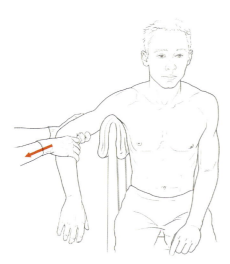
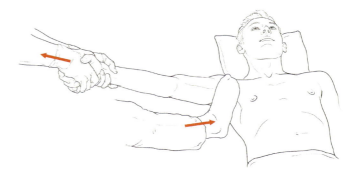

B Repositioning of shoulder dislocation
a Repositioning according to Arlt; **b** repositioning according to Hippocrates
There are several techniques to reduce a dislocated shoulder. In all cases, the repositioning should be performed with the patient under sedation or analgesia, possibly even anesthesia. According to the Arlt method, the patient sits in a chair with the affected arm hanging downward over the upholstered armrest. Repositioning is achieved through longitudinal traction, with the armrest serving as a support. The method as described by Hippocrates requires the patient to lie on the floor. Here, too, reduction is performed through longitudinal traction of the arm, with the doctor's foot placed in the axilla, thereby serving as a counter bearing.

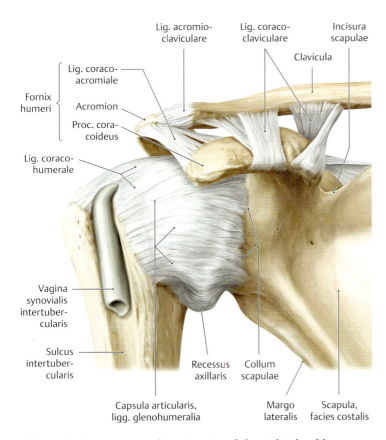
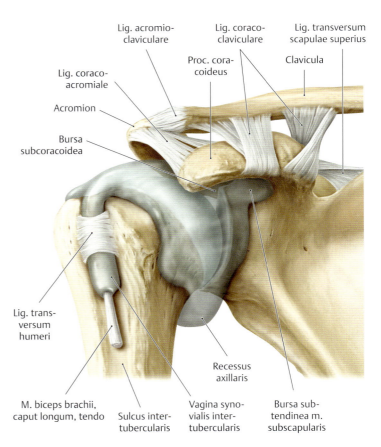

C Capsule, ligaments, and joint cavity of the right shoulder
a Anterior view.
b Joint cavity with capsule removed, anterior view.

The capsule of the shoulder joint is broad and very thin posteriorly, where it is not reinforced by ligaments. But it is strengthened anteriorly by three ligaments (the ligg. glenohumeralia superius, mediale, and inferius; see p. 264) and superiorly by the lig. coracohumerale. The lig. coracoacromiale, acromion, and proc. coracoideus together form the fornix humeri or *coracoacromial arch* (fornix humeri), which helps stabilize the caput humeri in the cavitas glenoidalis but also limits upward movement of the humerus. When the arm is hanging at the side, the lower part of the joint capsule, which is not reinforced by muscle, sags to form the recessus axillaris. This redundant fold provides a reserve capacity that is particularly useful during abduction movements of the arm. With prolonged disuse of the arm, the recessus axillaris may become atrophic or obliterated by adhesions, causing significant limitation of arm motion.

The cavity of the shoulder joint is connected to the adjacent bursae. The bursa subtendinea m. subscapularis and the bursa subcoracoidea consistently communicate with the joint cavity. The tendon sheath (vagina tendinis) of the caput longum musculi of the m. biceps brachii also communicates with the joint cavity in its passage through the sulcus intertubercularis.

Note: The ligg. glenohumeralia are usually clearly visible only at the inner side of the capsule.

14.14 The Joints of the Shoulder: Ligaments, Reinforcing Capsule, and Rotator Interval of the Articulatio humeri

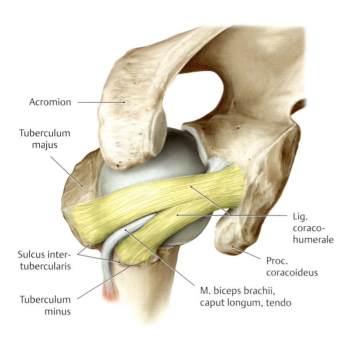

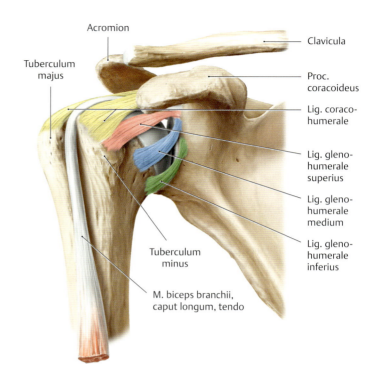

A Lig. coracohumerale
Right shoulder joint, superior view
The lig. coracohumerale arises from the base of the proc. coracoideus as a strong, broad band. Its two parts pass from there to the tubercula majus and minus of the humerus. The lig. coracohumerale stabilizes the tendon of the caput longum of the m. biceps brachii caput longum as it passes posterior to the lig. coracohumerale and before it passes through the sulcus intertubercularis.

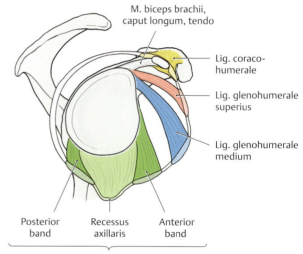

B Ligaments reinforcing capsules
a Right shoulder joint, anterior view; **b** schematic representation of the ligaments reinforcing the capsule after removal of the humeral head and cutting through both the capsule and the tendons of insertion of the rotator cuff muscles, lateral view; **c** origin and insertion of ligaments reinforcing capsule, lateral view.

The capsule of the shoulder joint is relatively loose and thin posteriorly. Anteriorly, it is reinforced by the ligg. glenohumeralia, which differ significantly in shape and are usually clearly visible only when viewed arthroscopically (see p. 270).

- **Lig. glenohumerale superius:** It passes from the upper margin of the cavitas glenoidalis to the sulcus intertubercularis and to the lesser tubercle, forming, together with the lig. coracohumerale, the rotator interval (see **D**).
- **Lig. glenohumerale medium:** It passes at almost a right angle to the tendon of insertion of the m. subscapularis, from the upper margin of the cavitas glenoidalis to the anatomic neck of the humerus.
- **Lig. glenohumerale inferius:** It consists of three parts, an anterior and posterior band and the recessus axillaris located between the bands. All three parts extend from the inferior margin of the cavitas glenoidalis to the middle of the anatomic neck of the humerus, with the middle part (recessus axillaris) extending down to the surgical neck (collum chirurgicum). The lig. glenohumerale inferius forms a hammock, especially during abduction, and is of particular significance to the anterior-inferior stability of the shoulder.

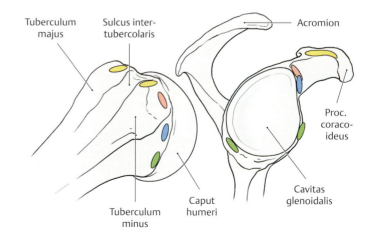

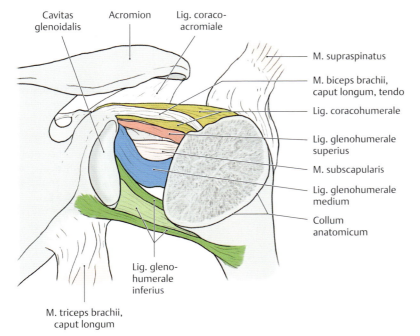

C Ligaments reinforcing the capsule
Right shoulder joint, posterior view.
View from behind the anterior articular capsule after removal of the humeral head.

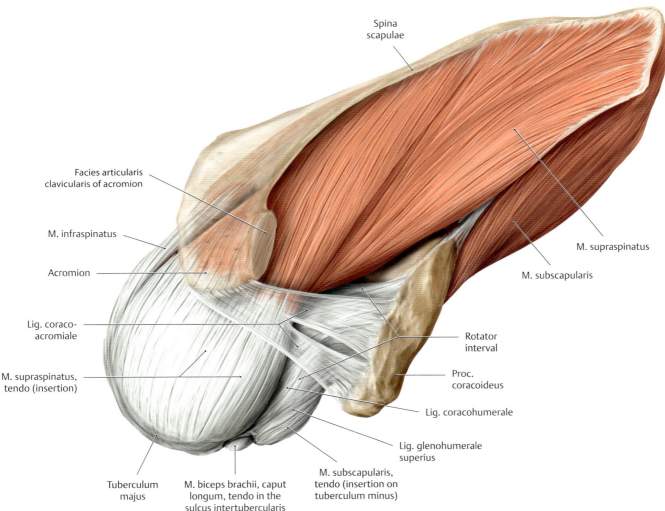

D Rotator interval
Right shoulder, superior view; clavicula and m. deltoideus have been removed. The rotator interval is the capsular area or gap between the superior edge of the m. subscapularis and the anterior edge of the m. supraspinatus. The lig. glenohumerale superius and lig. coracohumerale are primarily responsible for reinforcing the capsula articularis. The ligaments join together in the rotator interval to form the so-called biceps pulley, which surrounds the tendon of the caput longum of the m. biceps brachii and prevents it from being dislocated in an anterior-medial direction. The lig. glenohumerale superius forms the bottom and the lig. coracohumerale forms the top of this "sling" or "pulley" (see **Bb**). In addition, fibers from the tendons of insertion of the m. subscapularis and m. supraspinatus are interwoven in the rotator interval.

14.15 The Joints of the Shoulder: The Subacromial Space

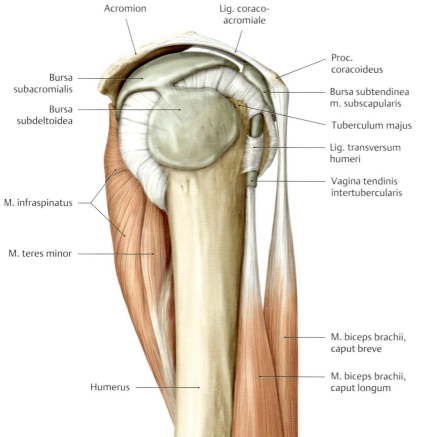

A Subacromial space, right shoulder
Lateral view. The m. deltoideus has been removed to demonstrate the following structures:

- The attachments of the rotator cuff muscles (Mm. supraspinatus, infraspinatus, teres minor, and subscapularis) to the proximal humerus (see also **B**)
- The tendons of origin of the m. biceps brachii muscle
- The subacromial space with the bursa subacromialis, which consistently communicates with the bursa subdeltoidea

The two bursae allow frictionless gliding between the humeral head and the rotator cuff tendons (especially the m. supraspinatus and the upper part of the m. infraspinatus) beneath the fornix humeri during abduction and elevation of the arm (see p. 275).

B Bursa subacromialis and cavitas glenoidalis of the right shoulder joint
Lateral view. The humeral head has been removed and the rotator cuff tendons of insertion have been divided to demonstrate the cavitas glenoidalis of the shoulder joint. The *labrum glenoidale* enlarges and deepens the glenoid only very slightly. Just before inserting into the humeral head, the muscles of the rotator cuff send tendinous expansions to the joint capsule that help to press the caput humeri into the cavitas glenoidalis. The bursa subacromialis is located between the fornix humeri and the tendons that insert on the caput humeri (see **D**).

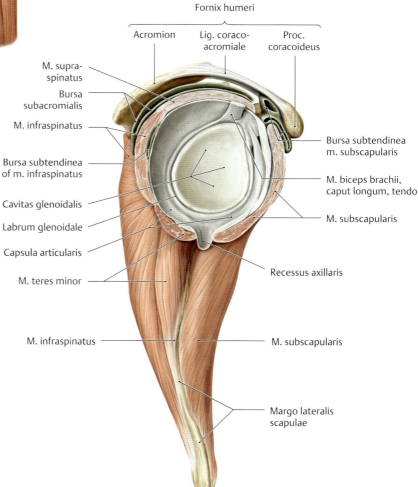

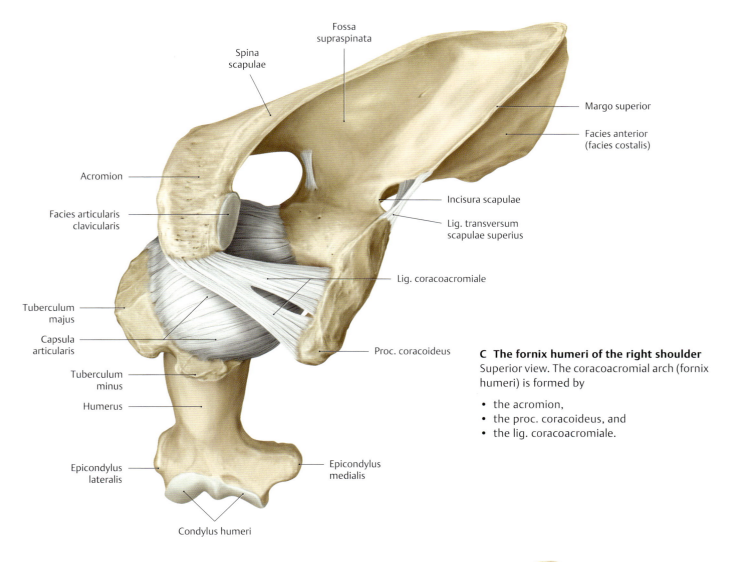

C The fornix humeri of the right shoulder
Superior view. The coracoacromial arch (fornix humeri) is formed by

- the acromion,
- the proc. coracoideus, and
- the lig. coracoacromiale.

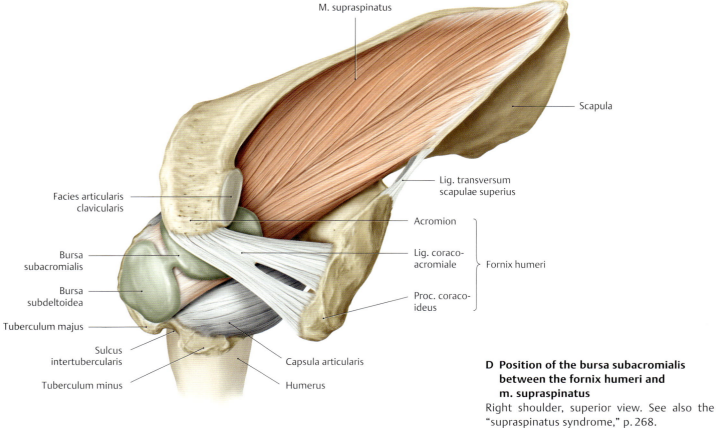

D Position of the bursa subacromialis between the fornix humeri and m. supraspinatus
Right shoulder, superior view. See also the "supraspinatus syndrome," p. 268.

14.16 The Bursa subacromialis and Bursa subdeltoidea

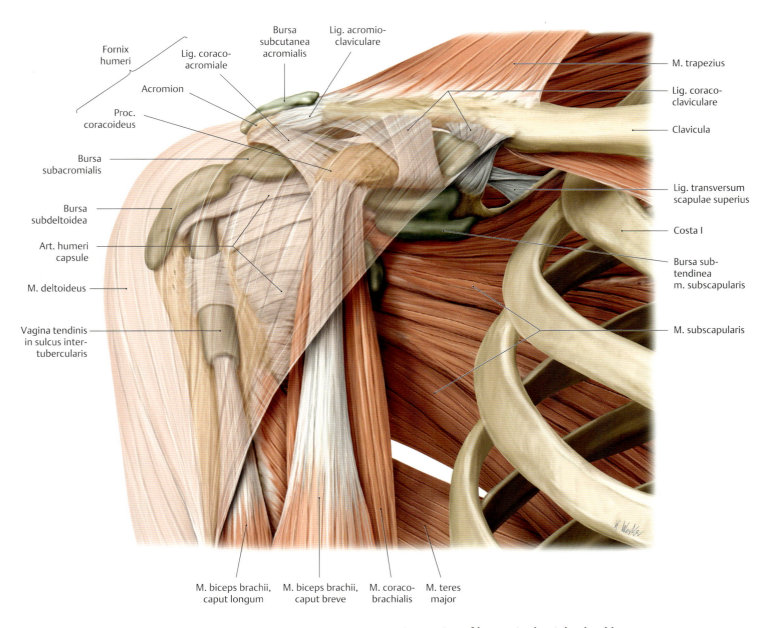

A Location of bursae in the right shoulder
Anterior view. The mm. pectorales major and minor and m. serratus anterior have been removed. The location of the bursae can be seen through the m. deltoideus, which is lightly shaded.
Note in particular the fornix humeri and the underlying bursa subacromialis.

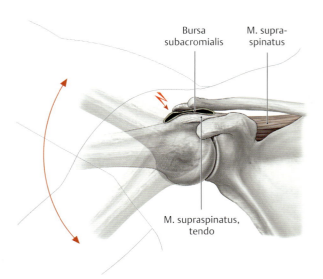

B Supraspinatus (impingement) syndrome
A supraspinatus tendon that has been thickened by calcification or other degenerative changes can be caught underneath the acromion, impinging upon the bursa subacromialis, when the arm is abducted (see **D**). Pain from this *supraspinatus*, or *impingement, syndrome* occurs in an arc of motion between 60 and 120° of abduction.

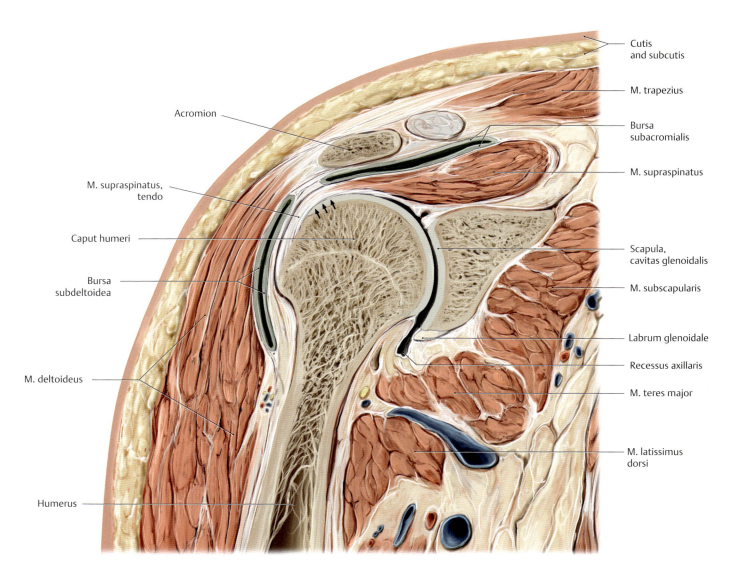

C Coronal section through the right shoulder joint
Anterior view. The tendon of insertion of the m. supraspinatus differs structurally from ordinary traction tendons. Its distal course gives it the function of a gliding tendon, which passes over the fulcrum of the humeral head (arrows). In that area, located approximately 1 to 2 cm proximal to its insertion on the tuberculum majus, the tendon tissue in contact with the humeral head is composed of fibrocartilage. This fibrocartilage zone is avascular, representing an adaptation of the tendon to the pressure loads imposed by the bony fulcrum (drawing based on a specimen from the Anatomical Collection at the University of Kiel).

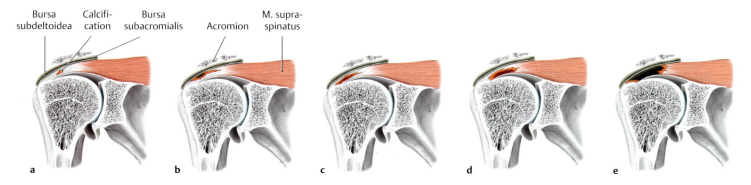

D Damage to the supraspinatus tendon

a Calcification of the tendon of insertion of the m. supraspinatus, resulting from degenerative changes in the area surrounding the fibrocartilaginous zone (see also **C**); **b–d** partial defects in the m. supraspinatus tendon (**b** bursal-sided, **c** intratendinous, and **d** articular-sided); **e** complete defect (tear) of the m. supraspinatus tendon (also called rotator cuff rupture).

Degenerative changes of the rotator cuff, primarily the m. supraspinatus tendon, together with the clinical effects on the tendon of the caput longum of the m. biceps brachii, as well as the bursae subacromialis and subdeltoidea, generally lead to a painful functional disorder in the shoulder area. The cause of this so-called subacromial syndrome (or subacromial pain syndrome) is the pathologic contact of the calcified and thickened m. supraspinatus tendon with the acromion, or subacromial impingement syndrome. After a complete rotator cuff rupture (in 95% of all cases, the m. supraspinatus tendon is involved), the bursae subacromialis and subdeltoidea communicate with the cavitas articularis. The m. supraspinatus normally contributes significantly to the first 10° of abduction (the "starter function"; see p. 302).

14.17 Shoulder Arthroscopy

A Overview of arthroscopy of the shoulder joint
Arthroscopy of the shoulder joint has taken on an increasingly important role in the field of arthroscopic surgery. Initially, it was predominantly a diagnostic procedure. Over time, as a result of technical advances, it has become a surgical technique for resection and reconstruction of the shoulder joint, increasingly replacing open shoulder surgeries. Arthroscopy is now considered primarily a therapeutic surgical procedure, which also allows a dynamic functional evaluation of the capsule ligament structures, e.g., in case of instability. Currently, all surgeries are possible as part of arthroscopy of the shoulder joint, with the exception of endoprothesis. Indications for arthroscopy of the shoulder joint can be classified into three groups:

- Shoulder instability, e.g., labral lesion (Bankert lesion), osteochondral lesion (Hill-Sachs lesion);
- Degenerative changes, e.g., partial or complete rotator cuff rupture, subacromial impingement syndrome (tendonitis calcarea), m. biceps brachii tendon; and
- Inflammatory changes, e.g., synovialitis, adhesive capsulitis.

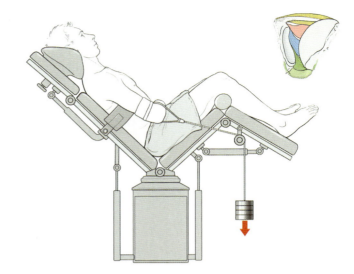

B Patient positioning for should arthroscopy
Two standard positions are available:

- Lateral position
- Half-sitting position (so-called beach-chair position)

In the half-sitting position, the patient sits upright with the back flexed at 60°. The advantage of this position is that it allows full rotation and abduction of the arm during surgery. In addition, in case the operation needs to be converted to an open procedure, there is no need to repostion the patient, nor does it require additional washing or surgical drapes. Furthermore, the arm can be extended using traction weights in order to get a better view inside the joint.

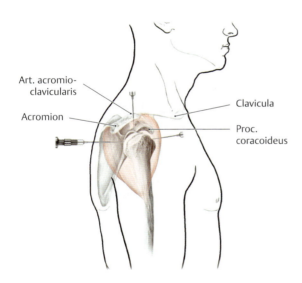

C Shoulder arthroscopy portals
a Right shoulder joint, lateral view; for better orientation, the most important anatomic landmarks have been marked: proc. coracoideus, lateral end of the clavicula, acromion, and art. acromioclavicularis; **b** arthroscopic portals, right shoulder joint (humeral head removed), lateral view. With the arthroscopic portals, one differentiates between

- Instrument portals (through which instruments are inserted) and
- Optical portals.

The standard portal for the 30°-angled arthroscope is created by making a stab incision 1 cm inferior and 1.5 cm medial to the posterolateral acromial margin (dorsal portal). Afterward, a blunt trocar is used to perforate first the subcutaneous tissue followed by the m. deltoideus and the dorsal capsule in the direction of the proc. coracoideus. Then, the trocar is replaced by a camera, and the cavitas articularis is washed. The anterior and anterior-superior portals are commonly used to insert the instruments. Depending on the type of surgery, additional portals can be created, e.g., the lateral portal for access to the subacromial space. Whereas the anterior portal is created by incising the skin lateral to the tip of the proc. coracoideus and accessing the joint above the subscapularis tendon, the anterior-superior portal is made just in front of the anterolateral angle of the acromion with the joint being accessed just behind the tendon of the caput longum of the m. biceps brachii. Moving the arthroscope to different portals allows for a complete overview of the articular space.

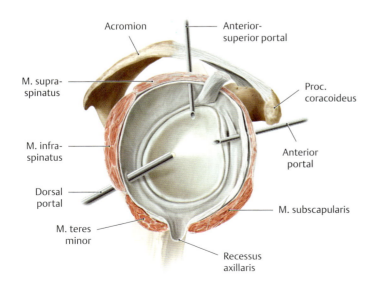

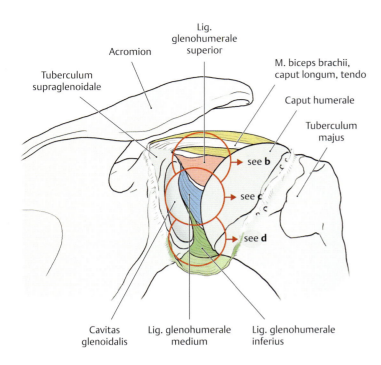
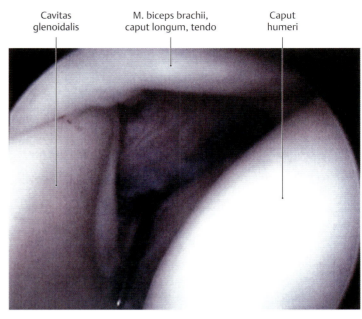
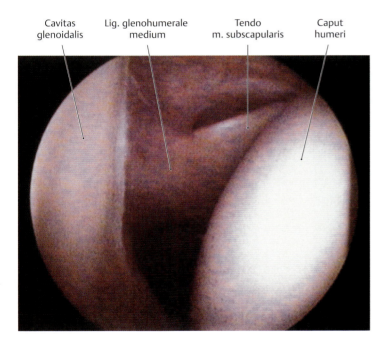
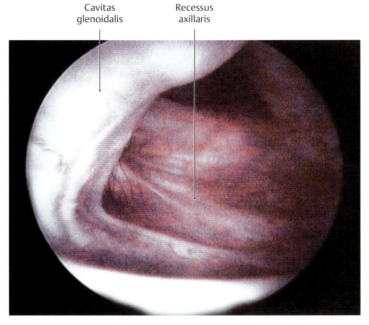

D Arthroscopic anatomy of the shoulder joint
a Schematic representation of the arthroscopically visible fields of sight; right shoulder, dorsal view; **b–d** corresponding arthroscopy images (from Merk H, Jerosch J, Hrsg. Arthroskopie des Schultergelenks. Stuttgart: Thieme; 2000).

For the arthroscopic evaluation of the intra-articular structures, a standardized test using examining hooks is recommended. The goal of this examination is the inspection of the entire articular space. In the beach-chair position (see **B**) with dorsal view, the cavitas glenoidalis is oriented vertically. Over the opposite caput humeri, the tendon of the caput longum of the m. biceps brachii passes horizontally through the superior articular space to the tuberculum supraglenoidale and is clearly visible as an anatomic landmark. The tendon of the caput longum of the m. biceps brachii together with the cavitas glenoidalis and caput humeri form the anterior triangle with the tendon of the m. subscapularis and the lig. transversum humeri (**b, c**). By tilting the lens upward, the insertion of the tendon of the caput longum of the m. biceps brachii into the sulcus intertubercularis in the rotator interval becomes visible at the anterior edge of the tendon of the m. supraspinatus. The lig. glenohumerale superius and the anterior edge of the tendon of the m. supraspinatus form a U-shaped sling, which crosses underneath the tendon of the caput longum of the m. biceps brachii. Furthermore, the undersurface of the tendons lof the m. supraspinatus and m. infraspinatus can be made visible. The lig. glenohumerale inferius can be best evaluated from the anterior portal looking toward the recessus axillaris. It extends—with its anterior and posterior fibers as well as the recessus axillaris located between them—to the inferior edge of the cavitas glenoidalis.

14.18 X-Ray and Cross-sectional Anatomy of the Shoulder

A Conventional diagnostic radiographs of the right shoulder joint
In primary diagnosis, clinical assessments and sonographic examinations (see **B**) are followed by X-ray radiography. Additionally, complicated injuries can be examined with the help of computed tomography and magnetic resonance imaging (see **C**). Similar to the X-ray radiography in evaluating bones and joints, it is recommended to take two images of two perpendicular planes:

- Anteroposterior image (**a** and **b**) and
- Axial (transverse) image (**c** and **d**).

In order to avoid overlap in X-ray images, the cavitas glenoidalis needs to be placed at a 30° angle when taking an anteroposterior image. At slight external rotation, the caput humeri and the cavitas glenoidalis are shown without overlap; the tuberculum majus forms the lateral contour. In the axial image, the patient lies on his or her back with the arm slightly rotated outward and abducted. The X-ray cassette is positioned superior to the shoulder with the beam hitting the axilla inferiorly. As a result, the caput humeri and cavitas glenoidalis are shown at a right angle to the anteroposterior image (enhancing the detection of fractures).

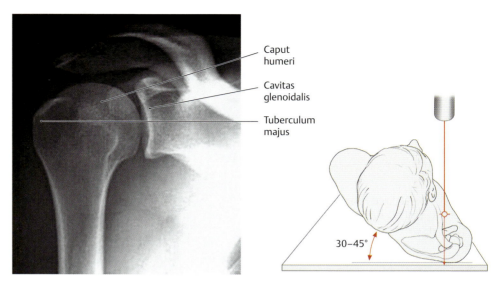

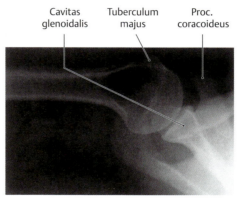

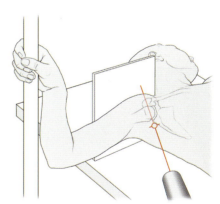

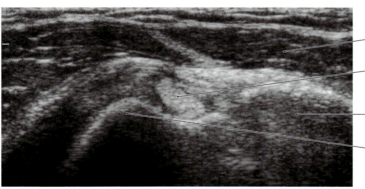

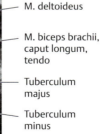

B Diagnostic ultrasound of the anterior region of the left shoulder
a Sonogram; **b** position of the transducer at the level of the sulcus intertubercularis of the left shoulder joint; **c** schematic representation of the sonogram (caudal view of the transverse section) (**a** and **b** from Konermann W, Gruber G. Ultraschalldiagnostik der Bewegungsorgane. 2nd ed. Stuttgart: Thieme; 2006).

Like diagnostic x-rays, in standardized sonographic examination of joints, two almost perpendicular planes (transverse and longitudinal) are used. Rotation of the arm, combined with different positions of the transducer, allows for an all-encompassing examination of the shoulder joint.

Upper Limb — 14. Bones, Ligaments, and Joints

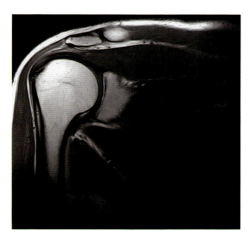

a Coronal, T1-weighted image (cross section parallel to the m. supraspinatus and perpendicular to the cavitas glenoidalis)

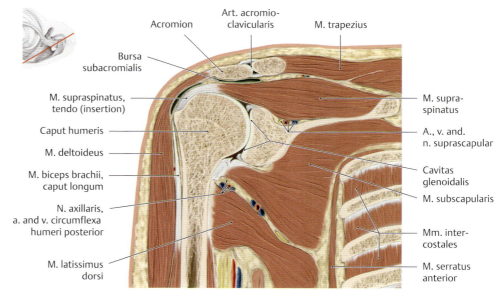

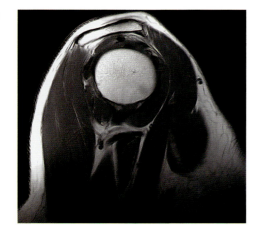

b Sagittal, T1-weighted image (cross section parallel to the cavitas glenoidalis)

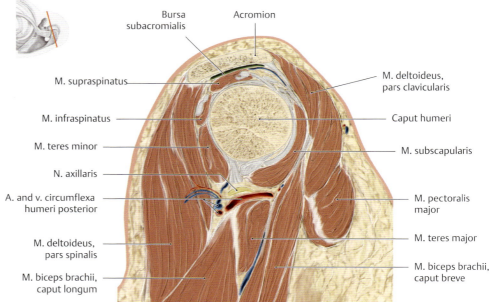

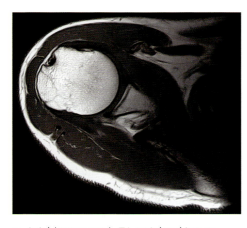

c Axial (transverse), T1-weighted image

C **MRI examination of the right shoulder joint in three planes**
(From Möller TB, Reif E. Taschenatlas der Schnittbildanatomie, Band III. Stuttgart: Thieme; 2007.) *Note:* Axial images are always viewed inferiorly.

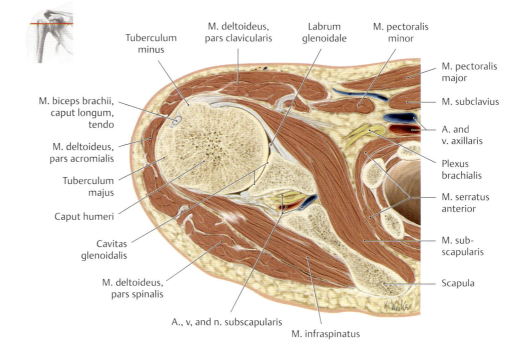

273

14.19 Movements of the Shoulder Girdle and Shoulder Joint

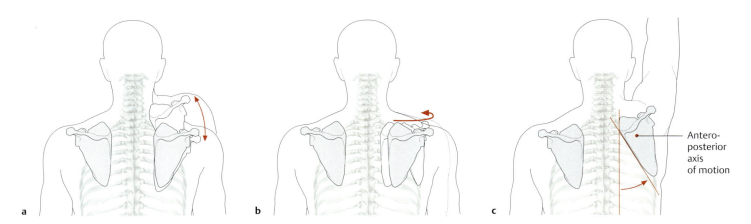

A Movements of the scapula
The artt. sternoclavicularis and acromioclavicularis are mechanically linked in such a way that all movements of the clavicula are accompanied by movements of the scapula. The scapula moves by gliding on the chest wall in the scapulothoracic joint. Both its movement and its fixation are effected by muscular slings. The following types of scapular movement are distinguished:

a Elevation and depression (during elevation and depression of the shoulder girdle): translation of the scapula in the craniocaudal direction.

b Abduction and adduction (during protraction and retraction of the shoulder girdle): horizontal translation of the scapula in the posteromedial-to-anterolateral direction.

c Lateral rotation of the angulus inferior (during abduction or elevation of the arm): rotation of the scapula about an anteroposterior axis through the center of the scapula. With an approximately 60° range of rotation, the angulus inferior of the scapula moves about 10 cm laterally, while the angulus superior moves about 2 to 3 cm inferomedially.

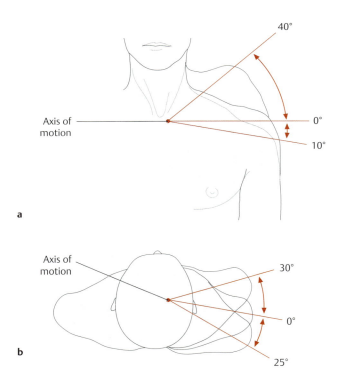

B Movements (and range of motion) in the art. sternoclavicularis
a Elevation and depression of the shoulder about a parasagittal axis.
b Protraction and retraction of the shoulder about a longitudinal (vertical) axis.

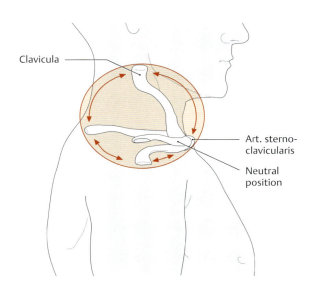

C Range of motion of the clavicula
Right clavicula, lateral view. Viewing the range of motion of the clavicula in the art. sternoclavicularis from the lateral view, we find that the clavicula moves roughly within a conical shell whose apex is directed toward the sternum, and with a slightly oval base approximately 10 to 13 cm in diameter. The clavicula also rotates about its own axis, particularly during elevation of the shoulder girdle, where its S shape significantly increases the range of shoulder elevation. The range of this rotation is approximately 45° and creates a third degree of freedom that gives the art. sternoclavicularis the function of a spheroidal joint.

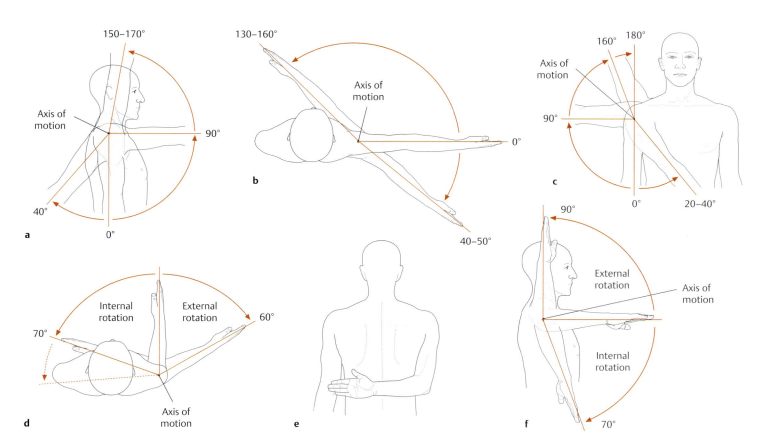

D Movements in the shoulder joint

As a typical spheroidal joint, the shoulder joint has three mutually perpendicular cardinal axes with three degrees of freedom and a total of six main directions of movement. In a very basic sense, all movements in the shoulder girdle can be classified as vertical, horizontal, or rotational. In *vertical movements*, the arm is elevated in various directions from a neutral adducted position. In *horizontal movements*, the arm is moved forward or backward while in 90° of abduction. *Rotational movements* can be performed with the arm in any position. The maximum *range* of these various movements can be achieved only by concomitant movement of the shoulder girdle, however.

a Flexion and extension occur about a horizontal axis.
b Flexion and extension of the arm raised to 90° abduction are also described as horizontal movements.
c Abduction and adduction occur about a sagittal axis, with movements past 90° often referred to as elevation. In clinical parlance, however, the term "elevation" is generally applied to all vertical movements. Past 80 to 90° of abduction, an automatic external rotation occurs that keeps the tuberculum majus from impinging on the coracoacromial arch (fornix humeri). When the arm is abducted while internally rotated, the range of abduction is decreased to approximately 60°.
d–f Internal and external rotation of the arm occurs about the longitudinal (shaft) axis of the humerus. When the arm is flexed at the elbow during these movements, the forearm can be used as a pointer. When the arm is hanging at the side, the maximum range of internal rotation is limited by the trunk. Placing the arm behind the back is equivalent to 95° of internal rotation (**e**). When the arm is abducted 90°, the range of external rotation is increased, while the maximum range of internal rotation is slightly reduced (**f**).

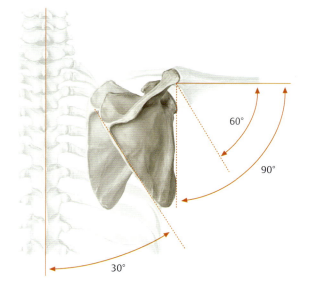

E Humeroscapular rhythm

The arm and scapula move in a 2:1 ratio during abduction. This means that when the arm is abducted 90°, for example, 60° of that movement occurs in the art. humeri (art. glenohumeralis), while 30° is accomplished by concomitant movement of the shoulder girdle. This "humeroscapular rhythm" is dependent upon freedom of movement of the scapula during abduction. Diseases of the shoulder joint can alter this rhythm, often causing the scapula to begin its rotation considerably earlier. This is best illustrated by cases involving ankylosis or arthrodesis of the art. humeri (pathological or intentional operative stiffening or fixation), in which movements in the shoulder girdle alone still permit the arm to be abducted 40 to 60° and still allow for one third the normal range of flexion/extension.

Upper Limb — 14. Bones, Ligaments, and Joints

14.20 The Elbow Joint (Articulatio cubiti) as a Whole

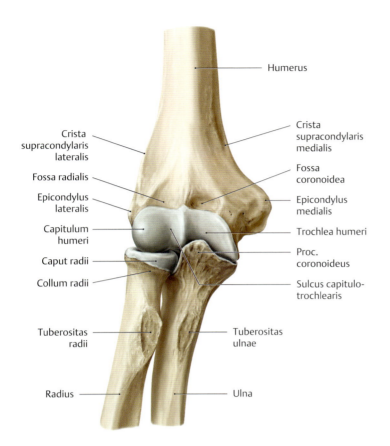

a Anterior view

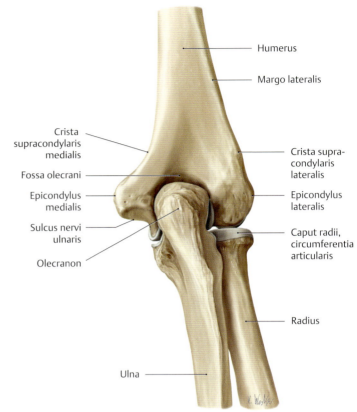

b Posterior view

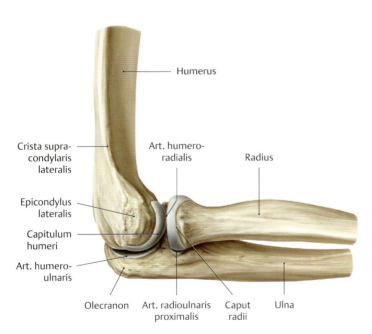

c Lateral view

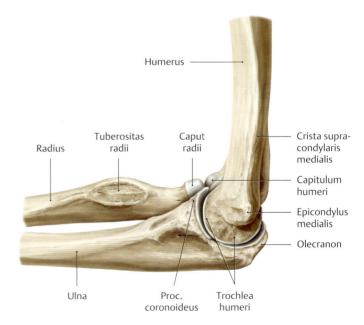

d Medial view

A The articulating skeletal elements of the right elbow joint
The *humerus, radius,* and *ulna* articulate with one another at the elbow joint (art. cubiti). The Elbow consists of three articulations:

- The humeroulnar joint (art. humeroulnaris) between the humerus and ulna
- The humeroradial joint (art. humeroradialis) between the humerus and radius
- The proximal radioulnar joint (art. radioulnaris proximalis) between the proximal ends of the ulna and radius

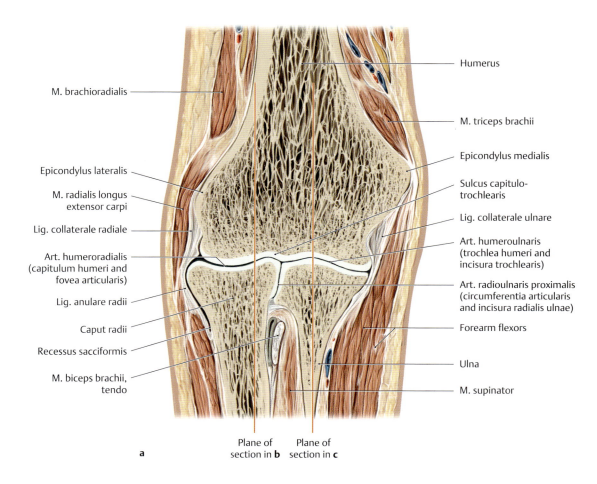
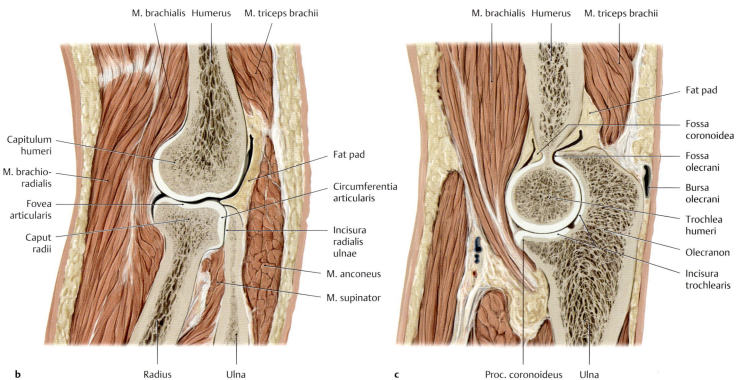

B Skeletal and soft-tissue elements of the right elbow joint
a Coronal section viewed from the front (*note* the planes of section shown in **b** and **c**).
b Sagittal section through the art. humeroradialis and art. radioulnaris proximalis, medial view.
c Sagittal section through the art. humeroulnaris, medial view.

(Drawings based on specimens from the Anatomical Collection at the University of Kiel.)

14.21 The Elbow Joint (Articulatio cubiti): Capsule and Ligaments

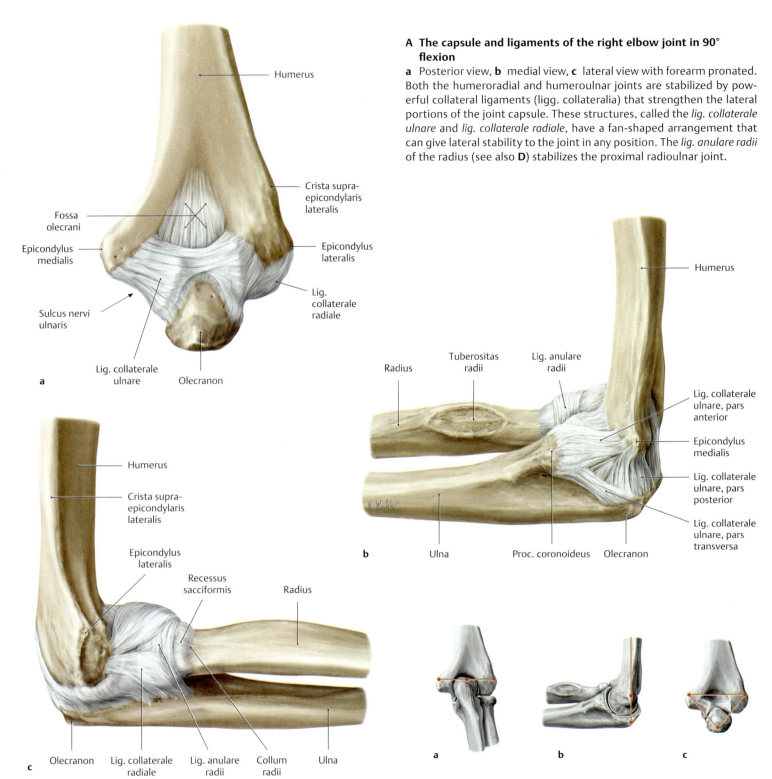

A The capsule and ligaments of the right elbow joint in 90° flexion
a Posterior view, b medial view, c lateral view with forearm pronated. Both the humeroradial and humeroulnar joints are stabilized by powerful collateral ligaments (ligg. collateralia) that strengthen the lateral portions of the joint capsule. These structures, called the *lig. collaterale ulnare* and *lig. collaterale radiale*, have a fan-shaped arrangement that can give lateral stability to the joint in any position. The *lig. anulare radii* of the radius (see also **D**) stabilizes the proximal radioulnar joint.

B The Hueter line and triangle
a Extension, posterior view; b flexion, medial view; c flexion, posterior view.
The epicondyles and the olecranon lie on a straight line when viewed from the posterior view in the extended elbow. They also lie on a straight line when viewed from the lateral view in the flexed elbow. But when the flexed elbow is viewed from behind, the two epicondyles and the tip of the olecranon form an equilateral triangle. Fractures and dislocations alter the shape of the triangle.

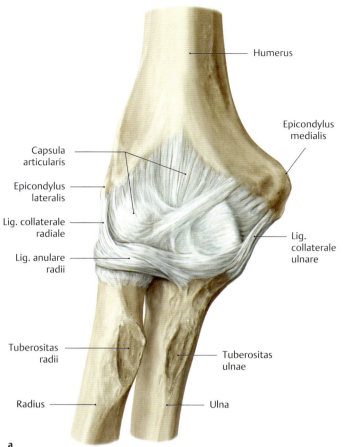
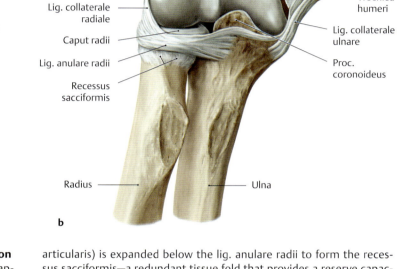

C The capsule and ligaments of the right elbow joint in extension
a Anterior view. b Anterior view with the anterior portions of the capsule removed.
The joint capsule of the elbow encloses all three articulations in the elbow joint complex. While the capsule is thin anteriorly and posteriorly, it is reinforced on each side by the collateral ligaments (ligg. collateralia) (see **A**). Over the end of the radius, the joint capsule (capsula articularis) is expanded below the lig. anulare radii to form the recessus sacciformis—a redundant tissue fold that provides a reserve capacity during pronation and supination of the forearm. During flexion and extension, the mm. brachialis and anconeus tighten the joint capsule to prevent entrapment of the capsule between the articular surfaces (see p. 312).

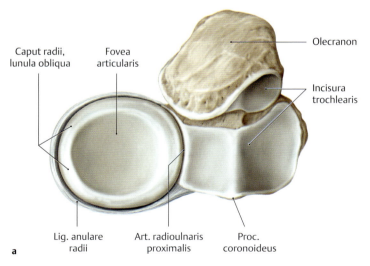
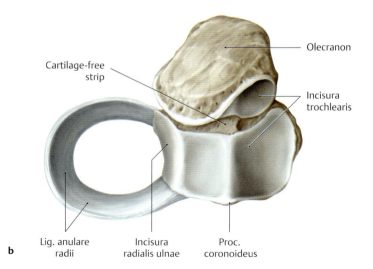

D Course of the lig. anulare radii in the right proximal radioulnar joint
a View of the proximal articular surfaces of the radius and ulna after removal of the humerus. b Same view as in **a** with the radius also removed.
The lig. anulare radii is of key importance in stabilizing the art. radioulnaris proximalis. It runs from the anterior to the posterior border of the incisura radialis ulnae (= cartilage-covered articular surface on the ulna), wrapping around the caput radii and pressing it into the ulnar articular surface. Histologically, the inner surface of the lig. anularia has the fibrocartilaginous structure of a gliding tendon, enabling it to withstand the compressive loads that are transmitted to the ligament.

14.22 The Forearm: Proximal and Distal Radioulnar Joints

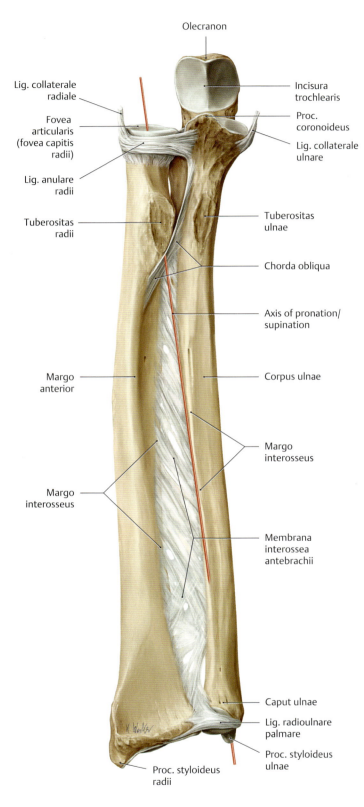
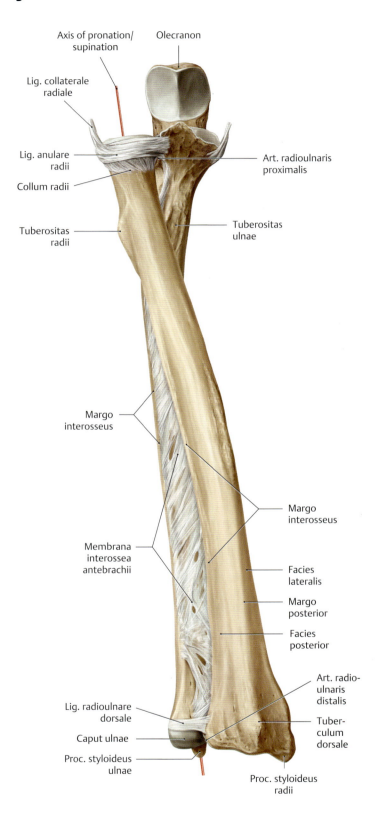

A Ligaments and axes for pronation and supination in the proximal and distal radioulnar joints
Right forearm, anterior view.

a Supination (the radius and ulna are parallel to each other).
b Pronation (the radius crosses over the ulna).

The art. radioulnaris proximalis functions together with the art. radioulnaris distalis to enable pronation and supination movements of the hand. The movements of both joints are functionally interlinked by the membrana interossea, so that the movement of one is necessarily associated with movement of the other. The axis for pronation and supination runs obliquely from the center of the capitulum humeri (not shown) through the center of the fovea capitis radii down to the proc. styloideus of the ulna (proc. styloideus ulnae).

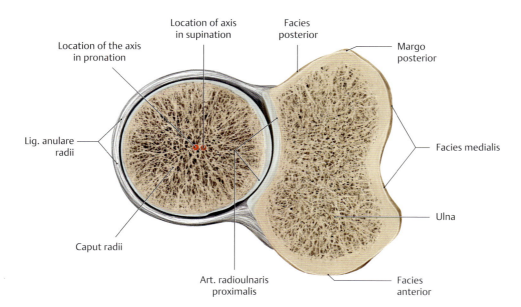

B Cross section through the right proximal radioulnar joint in pronation

Distal view. Owing to the slightly oval shape of the caput radii, the pronation/supination axis that runs through the caput radii moves approximately 2 mm radially during pronation (the long diameter of the caput radii is transverse when pronation is reached). This ensures that when the hand is pronated, there will be sufficient space for the tuberositas radii within the interosseous space (= the space between the tuberositas radii and chorda obliqua; see **Aa**, for example).

Note the thicker articular cartilage of the circumferentia articularis radii on the pronation side. This thickening occurs as an adaptation to the greater articular pressure in the art. radioulnaris proximalis in the pronated position.

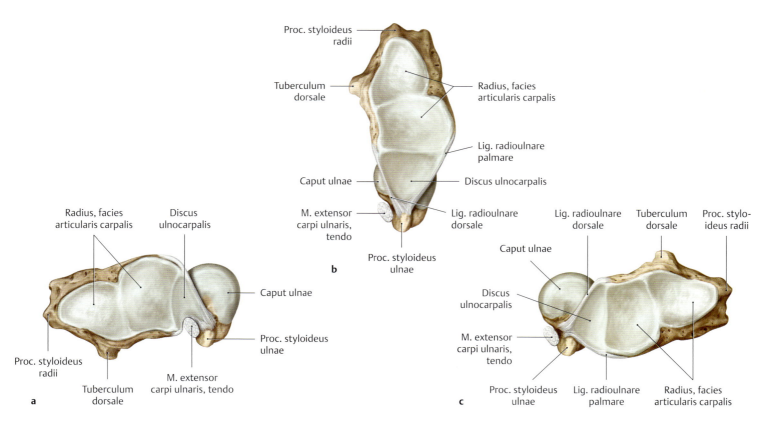

C Rotation of the radius and ulna during pronation and supination

View of the distal articular surfaces of the radius and ulna of the right forearm. For clarity, the discus ulnocarpalis is not shown.

a Supination
b Semipronation
c Pronation

The ligg. radioulnaria dorsale and palmare are part of the "ulnocarpal complex," which serves to stabilize the art. radioulnaris distalis. The mode of contact between the two distal articular surfaces varies with the position of the radius and ulna. They are in close apposition only in an intermediate (semipronated or neutral) position (after Schmidt and Lanz).

14.23 Movements of the Elbow and Radioulnar Joints (Articulationes radioulnares)

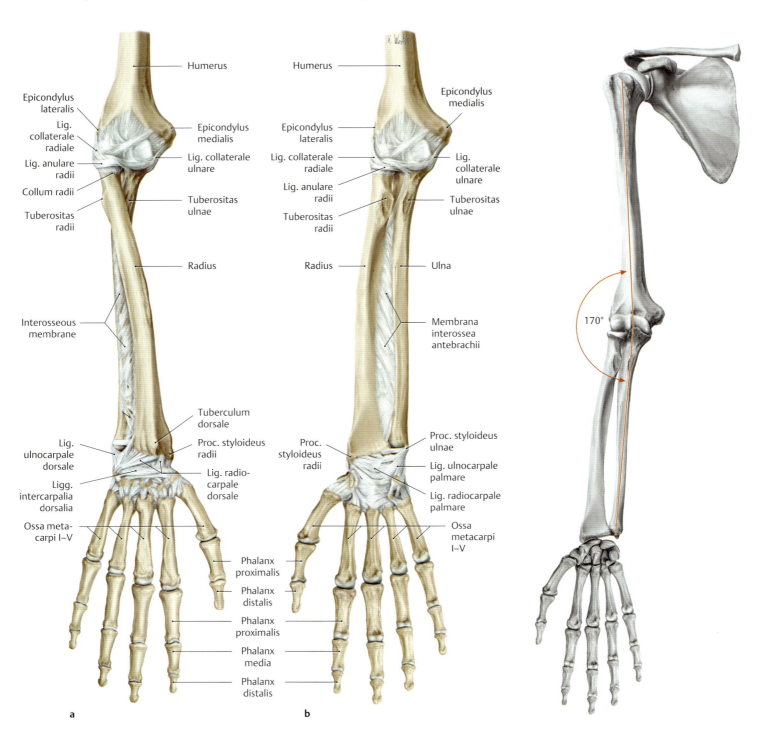

A Pronation and supination of the right hand
Anterior view.
a Pronation, **b** supination.
Pronation and supination of the hand make it possible to raise an object to the mouth for eating and to touch any area of the body for protection or cleaning. Pronation and supination are also essential for the working hand in actions such as turning a screwdriver, screwing in a light bulb, emptying a bucket, and unlocking a door. The range of hand movements can be further increased by adding movements of the shoulder girdle and trunk. This may be done, for example, to enable a full 360° twisting movement of the hand.

B Normal valgus of the elbow joint
Skeleton of the right upper limb with the forearm supinated. Anterior view.
The shape of the humeral trochlea (see p. 276) results in a normal valgus angulation between the humeral shaft and ulna (*cubitus valgus*) [valgus = bend or twist outward, away from the body axis]). This applies particularly during extension and supination. This "cubital angle" equals approximately 170°.

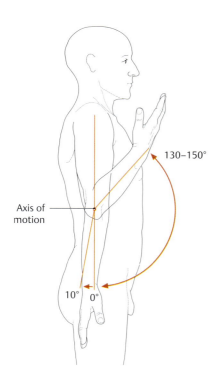

C Range of motion in the art. humeroradialis and art. humero-ulnaris of the elbow
The flexion/extension axis of the forearm runs below the epicondyles through the capitulum humeri and trochlea humeri of the humerus. Starting from the neutral (0°) position, both joints have a maximum range of 150° in flexion and approximately 10° in extension. Both movements are constrained either by soft tissues (muscles, fat, etc. = soft-tissue restraint) or by bone (olecranon = bony restraint).

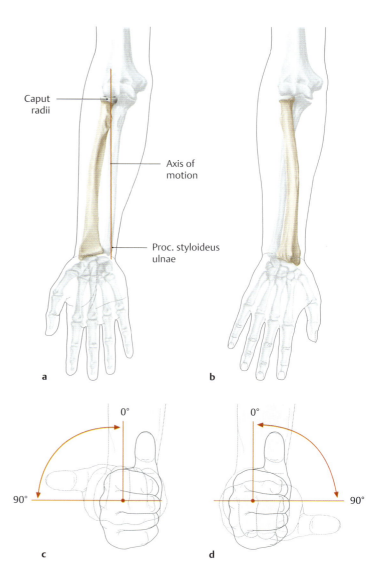

D Range and axis of pronation/supination of the right hand
The neutral (0°) position of the hand and forearm is also called semi-pronation. The axis of pronation/supination extends through the caput radii and the proc. styloideus ulnae.

a Supination (the radius and ulna are parallel to each other).
b Pronation (the radius crosses over the ulna).
c Supination of the hand with the elbow flexed, viewed from the front (the palmar surface of the hand is up).
d Pronation of the hand with the elbow flexed, viewed from the front (the palmar surface of the hand is down).

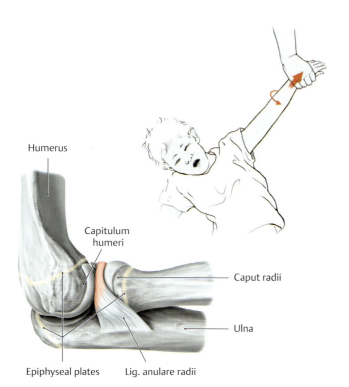

E Pronatio dolorosa, or nursemaid's elbow
Pronatio dolorosa is a very common injury in children (5- to 7-year-olds; with increasing age, ligaments become stronger, reducing the risk of injuries). The head of the radius (caput radii) slips under the lig. anulare radii (radial head subluxation), which results from a sudden pull on a child's arm. As a result, the lig. anulare radii gets stuck between the radius and the capitulum humeri, the elbow joint (art. cubiti) is locked in a slightly bent position, the arm remains turned inward (pronated position). Due to subluxation and accompanying pain, the child keeps the arm immobile and allows it to hang downward, which makes the arm appear to be paralyzed (pseudoparesis; so-called Chassaignac's paralysis). The clinical findings and the required X-ray examination in two planes to rule out any bone injuries (epiphyseal fracture of the caput radii) confirm the diagnosis. After reduction, during which the bent elbow joint is extended while firmly supinating the wrist, the child is symptom-free within minutes.

14.24 Overview of the Ligaments of the Hand

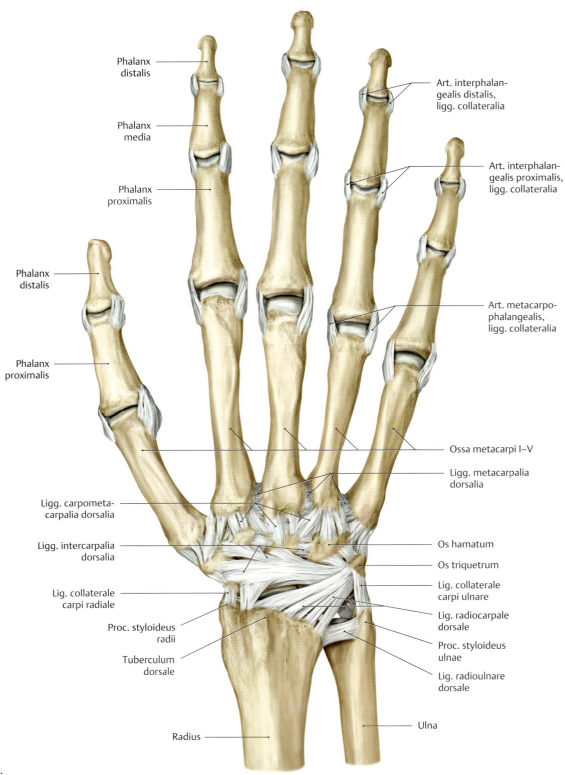

a Dorsal view.

A The ligaments of the right hand.
Carpal ligaments connect adjacent bones, limiting excess movement and providing stability to the wrist. Their courses vary. They are closely interwoven, and so they are difficult to dissect. The so-called extrinsic carpal ligaments, which run closer to the surface, are interwoven with the capsula articularis and primarily act to stabilize it. Deeper, the so-called intrinsic carpal ligaments (see p. 286) divide the joint space into individual compartments.

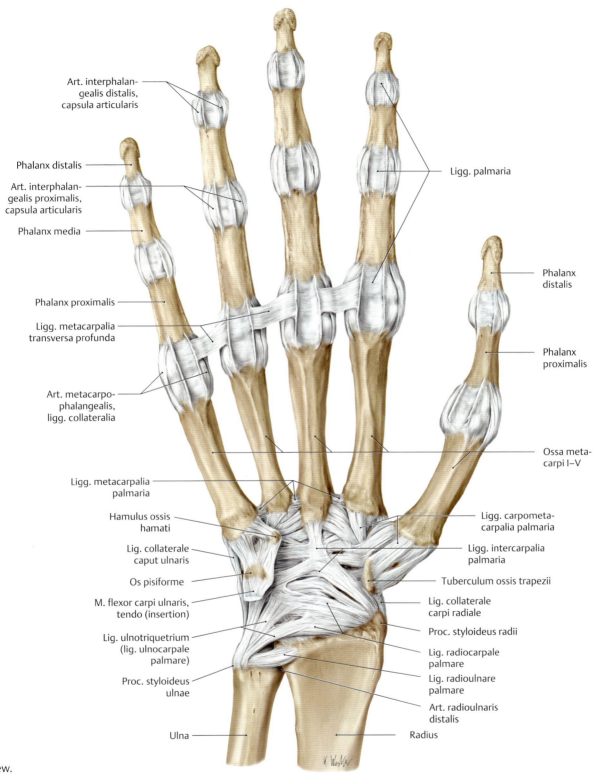

b Palmar view.

Aside from the common classification as extrinsic and intrinsic, the carpal ligaments are also classified according to their location and arrangement:

1. The ligaments between the forearm and ossa carpi (ligg. radiocarpalia and ulnocarpalia, ligg. collateralia)
2. The ligaments between individual ossa carpi (ligg. intercarpalia)
3. The ligaments between the carpal and metacarpal bones (ligg. carpometacarpalia)
4. The ligaments between the bases of the metacarpals (ligg. metacarpalia)

14.25 Intrinsic Ligaments of the Hand, Compartments of the Joint, and Ulnocarpal Complex

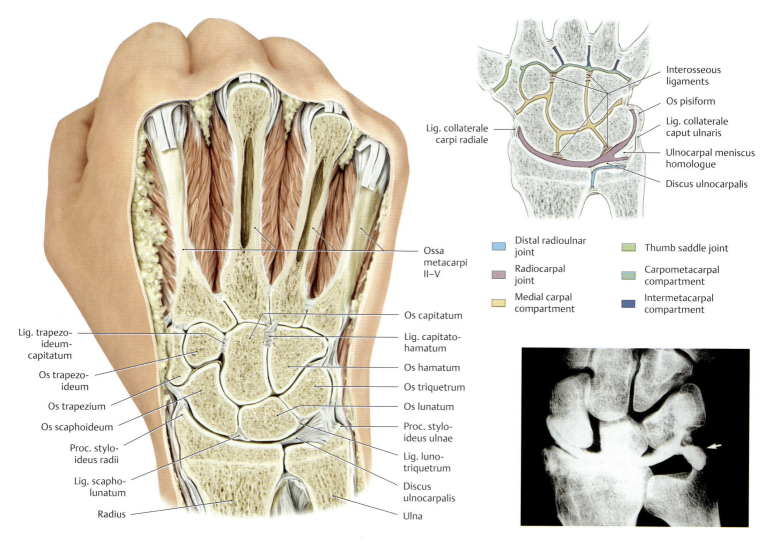
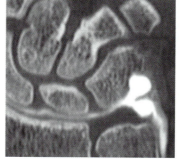
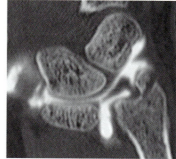

A Interosseous ligaments and compartments within the carpus
a Coronal section of the right carpus, dorsal view (drawing based on a specimen from the Anatomical Collection of the University of Kiel); **b** schematic representation of compartments (dorsal view of right hand); **c** arthrography of the art. radiocarpalis (courtesy of Dr. J. Koebke, Anatomical Institute, University of Cologne); **d** and **e** CT arthrography of the radiocarpal joint (from Bohndorf K, Imhof H, Fischer W. Radiologische Diagnostik der Knochen und Gelenke. 2nd ed. Stuttgart: Thieme; 2006): **d** intact compartment; **e** lesion of the discus ulnocarpalis with contrast agent crossing over into the radioulnar joint.

In addition to the extrinsic ligaments of the carpus (see p. 284), which reinforce the capsula articularis, there are intrinsic ligaments. These interosseous ligaments, along with the discus ulnocarpalis (discus triangularis), divide the intra-articular space into compartments, some of which are completely enclosed. These compartments include (see **b**)

- distal radioulnar joint,
- radiocarpal joint,
- medial carpal compartment,
- carpometacarpal compartment,
- intermetacarpal compartment, and
- thumb saddle joint.

Knowledge of these compartments is clinically significant for conducting and interpreting arthrographies (**c** to **e**).

The main intrinsic interosseous ligaments are those of the distal carpal row (ligg. capitatohamatum and trapeziodeumcapitatum), the proximal carpal row (ligg. lunotriquetrum and scapholunatum) (see **a**), and the discus ulnocarpalis, the most important structure of the ulnocarpal complex (see **b**). These ligament structures often succumb to degeneration or are affected in cases of carpal injuries. Degenerative changes in the discus ulnocarpalis can already be detected in people in their 20s, but degeneration of the interosseous ligaments of the proximal carpal row are more commonly seen in older patients (in 30% of all cases).

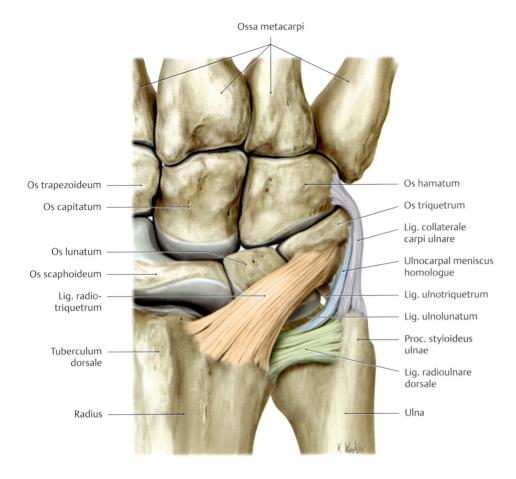

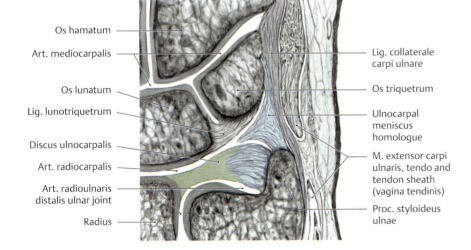

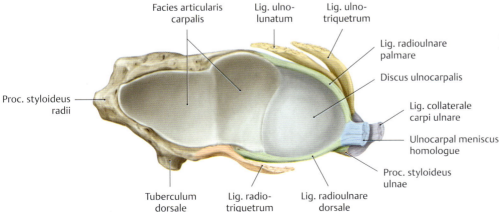

B The ulnocarpal complex

a Ulnocarpal complex of the right hand, dorsal view; **b** schematic representation of a histologic specimen of an ulnocarpal complex (from Schmidt HM, Lanz U. Chirurgische Anatomie der Hand. 2nd ed. Stuttgart: Thieme; 2003); **c** ulnocarpal complex of the right hand, distal view.

The ulnocarpal complex, which appears triangular in shape (synonym: triangular fibrocartilage complex, TFCC), is a combination of ligaments and disks and serves as a connection between the distal ulna, the distal radioulnar joint, and the proximal carpal row. In the case of injuries to the ulnocarpal complex, patients usually report ulnar-sided wrist pain. Functionally, the ulnocarpal complex is divided into

- Discus ulnocarpalis (discus triangularis)
- Ligg. radioulnare dorsale and palmare
- Ligg. ulnolunatum and ulnotriquetrum
- Meniscus ulnocarpalis
- Lig. collaterale carpi ulnare
- Lig. radiotriquetrum (a component of the lig. radiocarpale dorsale)

The discus ulnocarpalis, which consists of fibrocartilage, extends transversely and is located between the distal ulna and the os triquetrum or the os lunatum. It arises from the distal edge of the radius of the incisura ulnaris in the hyaline cartilage, and two of its fibrous bands (often) extend both to the proc. styloideus ulnae and the base of the distal ulna. The outer margins of the disk are attached to the ligg. radioulnare dorsale and palmare. The central and radial parts of the fibrocartilaginous disk are slightly vascular and thus take considerably longer to heal following injuries. Degenerative changes, too, are a common occurrence in this area. The discus ulnocarpalis is not to be confused with the meniscus ulnocarpalis, the collagen fibers of which extend from the dorsal and ulnar edges of the discus ulnocarpalis to the palmar side of the os triquetrum. The meniscus ulnocarpalis bridges the ulnar intra-articular space of the proximal wrist, which is wide in humans and especially in ulnar abduction helps to increase the force-transmitting surface.

14.26 The Carpal Tunnel (Canalis carpi)

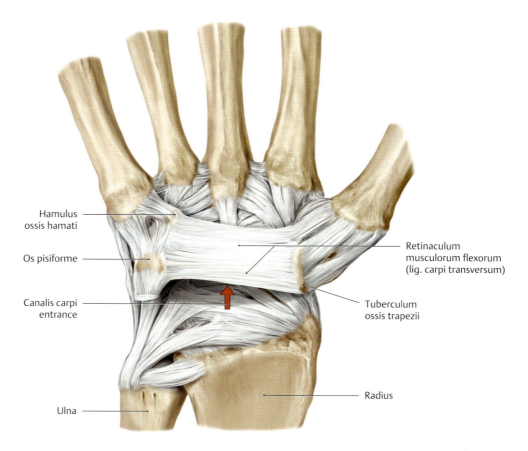

A Retinaculum musculorum flexorum (lig. carpi transversum) and carpal tunnel, right hand
Anterior view. The bony elements of the wrist form a concave groove on the palmar side (see also **C**), which is closed by the retinaculum mm. flexorum (referred to clinically as the lig. carpi transversum) to form a fibro-osseous tunnel called the carpal tunnel or carpal canal (canalis carpi). The narrowest part of this canal is located approximately 1 cm beyond the midline of the distal row of ossa carpi (see **D**). The cross-sectional area of the tunnel at that site measures only about 1.6 cm². The canalis carpi is traversed by a total of *nine flexor tendons* (enclosed in tendon sheaths and embedded in connective tissue) and the *n. medianus* (see p. 372). The tight fit of sensitive neurovascular structures with closely apposed, frequently moving tendons in this narrow space often causes problems when any of the structures swell or degenerate, leading to *carpal tunnel syndrome*. Narrowing of the tunnel can entrap or compress the n. medianus, altering its function both by direct mechanical action and by restricting the blood flow within the nerve sheath. With chronic compression, the n. medianus itself begins to degenerate beyond the site of entrapment, causing progressive pain and paresthesia and, ultimately, denervation and wasting of the muscles it serves, particularly the m. abductor pollicis brevis (see p. 372).

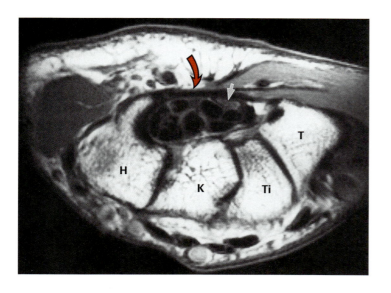

B Axial magnetic resonance image (T1-weighted) of the right hand at the level of the carpal tunnel
Proximal view. The retinaculum mm. flexorum (lig. carpi transversum) can be recognized as a band of low signal intensity (red arrow). Just below it toward the radial side is the n. medianus (small arrow), whose water and lipid contents cause it to display a higher signal intensity than the superficial and deep flexor tendons. The primary diagnosis of carpal tunnel syndrome is based on clinical signs and electrophysiological measurements such as nerve conduction velocity. While conventional radiographs and CT scans can detect bony causes of the syndrome, magnetic resonance imaging can also demonstrate soft-tissue causes (e.g., edema or swelling of the n. medianus, fibrosis, neuroma).

H = Os hamatum with hamulus ossis hamati;
K = Os capitatum;
T = Os trapezium;
Ti = Os trapezoideum.

(From Vahlensieck M, Reiser M. MRT des Bewegungsapparates. 3rd ed. Stuttgart: Thieme; 2006.)

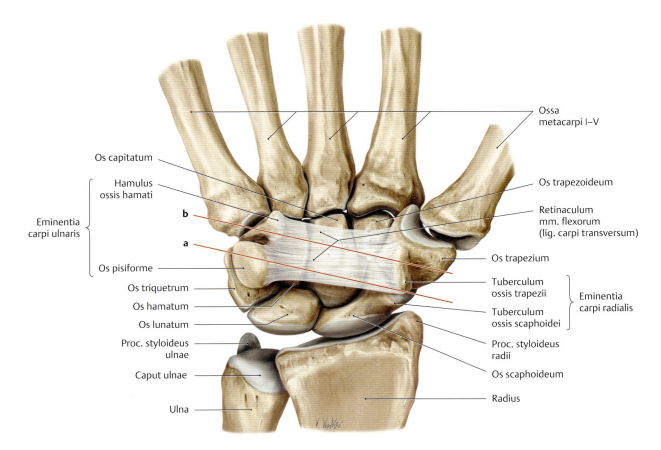

C Bony boundaries of the carpal tunnel of the right hand
Anterior view. The ossa carpi form a convex arch on the dorsal side of the wrist and a concave arch on the palmar side. This creates a *carpal groove* (sulcus carpi) on the palmar side, which is bounded by bony elevations on the radial and ulnar sides (eminentia carpi radialis and eminentia carpi ulnaris). The tubercula ossis trapezii and ossis scaphoidei form the palpable eminence on the radial side, while the hamulus ossis hamati and the os pisiforme form the eminence on the ulnar side. Stretched between them is the retinaculum mm. flexorum (lig. carpi transversum), which closes the canalis carpi on the palmar side (the planes of section marked **a** and **b** correspond to the cross sections in Fig. **D**).

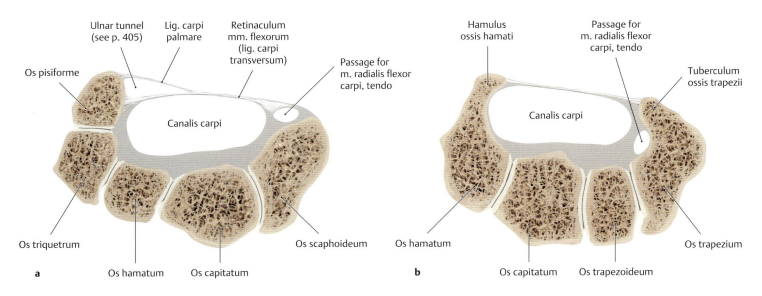

D Cross sections through the carpal tunnel
a Cross section through the proximal part of the carpal groove (sulcus carpi) (plane **a** in **C**).
b Cross section through the distal part of the carpal groove (sulcus carpi) (plane **b** in **C**).

Note: The canalis carpi is narrowest over the center of the distal row (approximately 10 mm) of ossa carpi (**b**) (after Schmidt and Lanz). In all, the canalis carpi has a median cross-sectional area of approximately 1.6 to 1.7 cm².

14.27 The Ligaments of the Fingers

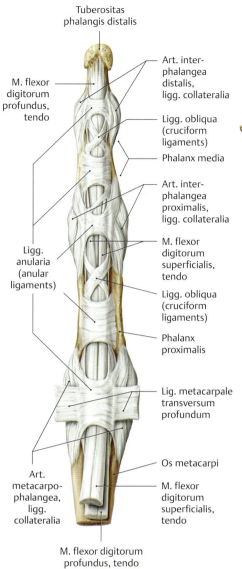

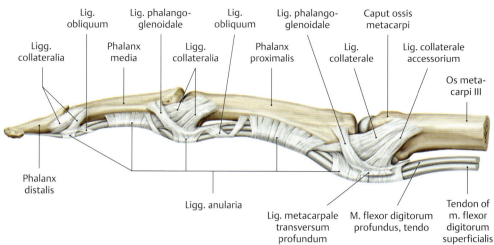

A The joint capsules, ligaments, and digital tendon sheath of the right middle finger
a Lateral view, b Palmar view.

The long flexor tendons (mm. flexores digitorum superficialis and profundus) run in a strong, common synovial tendon sheath (vagina synovialis digitorum manus; not shown here) on the palmar side of the fingers. The *tendon sheaths* (vagina tendinis) are guide mechanisms that allow for frictionless gliding of the long flexor tendons. The outer fibrous layer of the tendon sheaths, the stratum fibrosum, is strengthened by *ligg. anularia* and *ligg. obliqua* (see **B**), which also bind the sheaths to the palmar surface of the phalanx and prevent palmar deviation of the sheaths during flexion. The gaps between the ligg. anularia and ligg. obliqua (or cruciata) are necessary to allow flexion of the fingers (see also p. 344, Musculature: Topographical Anatomy).

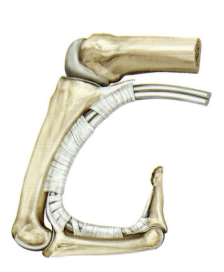

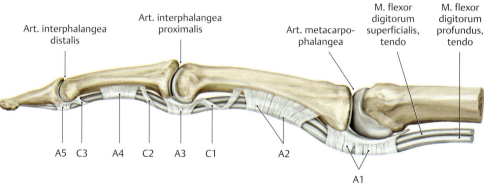

B Ligaments reinforcing the digital tendon sheath
a Lateral view in extension, b lateral view in flexion.
A1–5 = ligg. anularia, C1–3 = ligg. obliqua or cruciata.

- First lig. anularia (A1): at the level of the art. metacarpophalangea
- Second lig. anularia (A2): on the shaft of the phalanx proximalis
- Third lig. anularia (A3): at the level of the art. interphalangea proximalis
- Fourth lig. anularia (A4): on the shaft of the phalanx media
- Fifth lig. anularia (A5): at the level of the art. interphalangea distalis

The ligg. obliqua (or cruciata) are highly variable in their course.

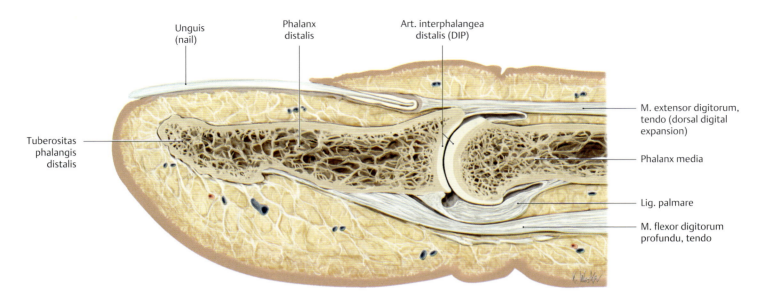

C Longitudinal section through the distal part of a finger
In the metacarpophalangeal joint, as well as the proximal and distal interphalangeal joints, the palmar articular surfaces of the phalanges are enlarged proximally by a fibrocartilaginous plate called the lig. palmare (volar plate). The ligg. palmaria also form the floor of the digital tendon sheaths at these locations (after Schmidt and Lanz).

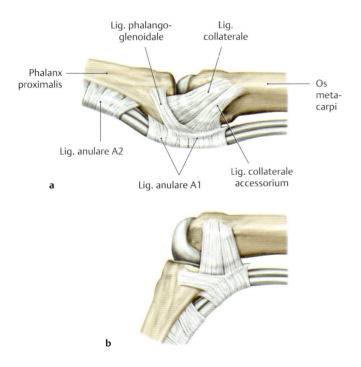

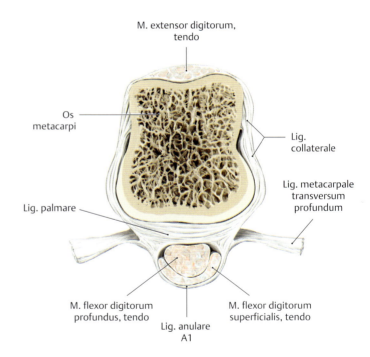

D The capsule and ligaments of the metacarpophalangeal joint
a Extension, b flexion. Lateral view.
Note: The lig. collaterale is lax in extension and taut in flexion. For this reason, the finger joints should always be placed in a "functional position" (e.g., with the artt. metacarpophalangeae flexed approximately 50 to 60°, see p. 295) if the hand is to be immobilized (e.g., in a cast) for a long period of time. If this is not done and the finger joints remain extended for a prolonged period, the lig. collaterale will shorten and create an extension deformity after the cast is removed. The ligg. collaterale accessorium and phalangoglenoidale are taut in both flexion and extension and act mainly as restraints to limit extension.

E Cross section through the head of the third metacarpal of the right hand
Proximal view. At the level of the second through fifth metacarpal heads, the volar fibrocartilage plates (ligg. palmaria) are interconnected by transverse bands, the the ligg. metacarpalia transversa profunda. By binding the ligg. palmaria to the A 1 ligg. anularia (see **B**) of the flexor tendon sheaths, they also strengthen the distal metacarpus and stabilize the transverse metacarpal arch (after Schmidt and Lanz).

14.28 The Carpometacarpal Joint of the Thumb (Articulatio carpometacarpalis pollicis)

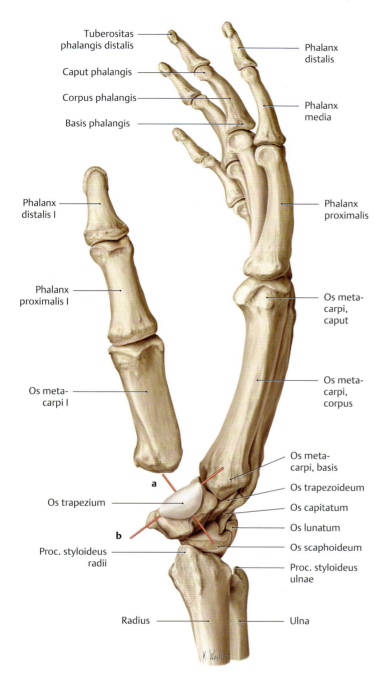

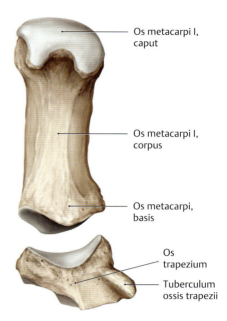

B Articulating surfaces of the carpometacarpal joint of the thumb
Palmar-ulnar view. The articular surface of the os trapezium is convex in the dorsopalmar direction and concave in the radioulnar direction. This is opposite to the curvatures found in the corresponding articular surface of the first metacarpal bone.

A Axes of motion of the carpometacarpal joint of the thumb (articulatio carpometacarpalis pollicis)
Skeleton of the right hand, radial view. The os metacarpi I has been moved slightly distally to facilitate orientation. The saddle-shaped articular surfaces of the os trapezium and os metacarpi I allow movements about two cardinal axes:

- An abduction/adduction axis (**a**)
- A flexion/extension axis (**b**)

While the axis for abduction/adduction runs approximately on a dorsopalmar line, the axis for flexion/extension runs transversely through the sellar limb of the os trapezium. When the thumb is moved toward the small finger (opposition), a rotary movement takes place about a longitudinal axis through the os metacarpi I (third degree of freedom). This oppositional movement of the thumb—essential for precision grasping movements of the hand—is made possible by the natural mismatch of the articular surfaces (see **F**).

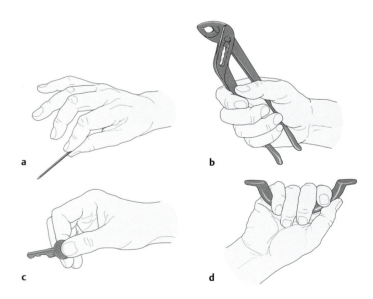

C Types of grip
Normal hand actions can be reduced to four primary types of grip:
a Pinch or precision grip, **b** power grip, **c** key grip, **d** hook grip.
The clinical examination should include function testing of the hand, giving particular attention to disturbances of fine motor skills and gross strength. It is important, for example, to evaluate the pinch grip and key grip, the pinch grip between the thumb and index finger having fundamental importance for the function of the hand. This is why, in issues pertaining to workers' compensation, loss of the thumb or index finger is considered to have a more serious impact on occupational capacity than the loss of the other long fingers.

Upper Limb — 14. Bones, Ligaments, and Joints

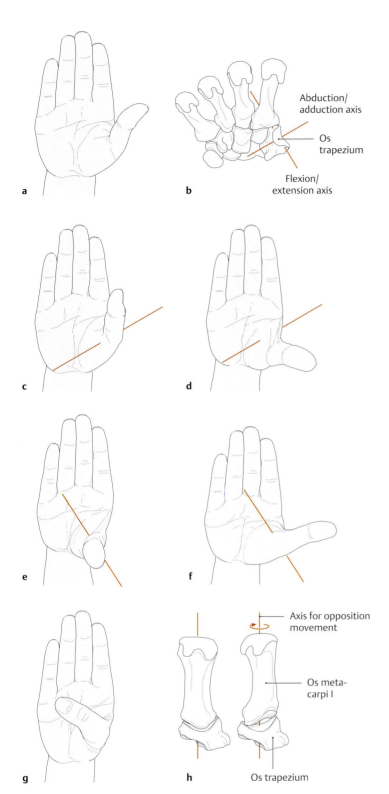

D Movements in the carpometacarpal joint of the thumb
Right hand. Palmar view.

a The neutral (0°) position.
b Axes of motion in the carpometacarpal joint of the thumb.
c Adduction.
d Abduction.
e Flexion.
f Extension.
g Opposition.
h Axis for opposition of the thumb. As the os metacarpi I rotates, its area of contact with the articular surface of the os trapezium is greatly diminished (see **F**).

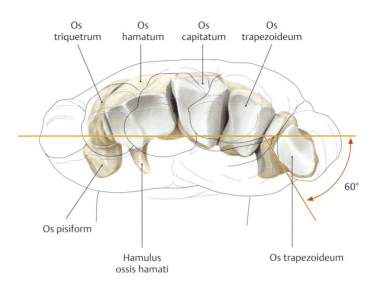

E Relationship of the thumb to the fingers in the neutral (0°) position
Right hand, distal view. Owing to the concave arch of the ossa carpi, the Os scaphoideum and Os trapezium have a markedly radiopalmar orientation. As a result, the Os metacarpi of the thumb is not placed in line with the other fingers but is rotated approximately 60° toward the palm.

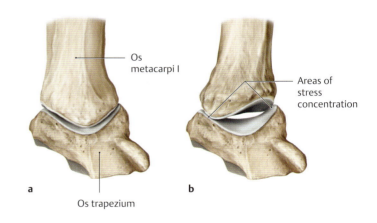

F Rotation-induced incongruity of the carpometacarpal joint during opposition of the thumb
a Neutral (0°) position, b the thumb in opposition.
As a sellar (saddle-shaped) joint, the carpometacarpal joint of the thumb is subjected to functional stresses that may promote osteoarthritis (after Koebke). The potentially harmful stresses are created by rotation of the os metacarpi I during opposition of the thumb. When the thumb is maximally opposed, this rotation greatly reduces the surface area available for stress transfer across the joint (contrast this with the large area available in **a**). This concentration of stresses in a localized area predisposes to degenerative changes in the ascending sellar limb of the os metacarpi I and the articular surface of the os trapezium (first carpometacarpal osteoarthritis).

293

14.29 Movements of the Hand and Finger Joints

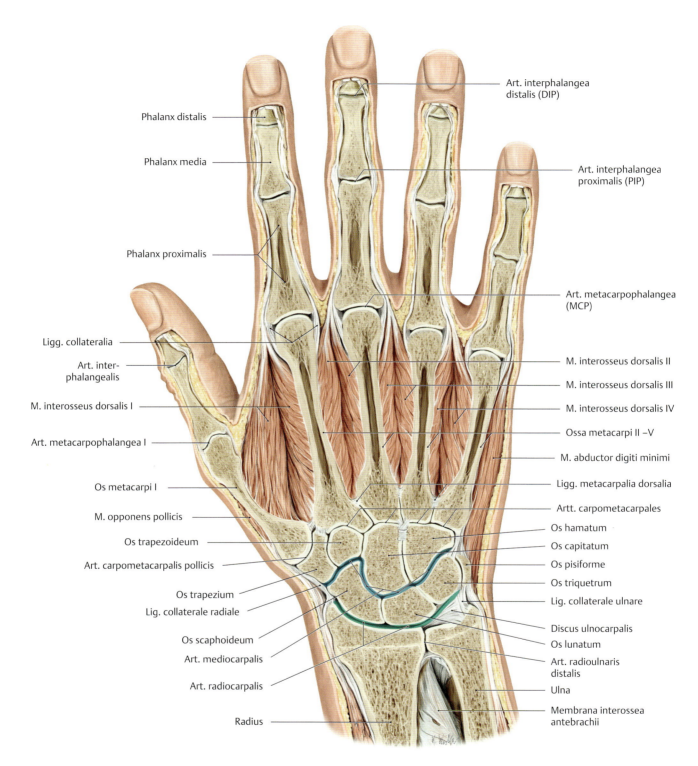

A Transverse section through the right hand
Posterior view. The hand and forearm are connected at the art. radiocarpalis and art. mediocarpalis (both indicated by blue lines in the drawing). Morphologically, the *art. radiocarpalis* is an ovoid or ellipsoidal joint, while the *art. mediocarpalis* is an interdigitating hinge joint (with an approximately S-shaped joint space between the proximal and distal rows of ossa carpi). Except for the art. carpometacarpea pollicis, the joints between the distal row of ossa carpi and the bases of the metacarpals (the artt. carpometacarpales) are *amphiarthroses* (joined by fibrocartilage) that permit very little motion.

The **finger joints** (artt. digitorum) are classified as follows:

- *Artt. metacarpophalangeae* between the ossa metacarpi and the phalanges proximales (MCP joints, spheroidal type)
- *Artt. interphalangeae proximales* between the phalanges proximales and mediales (PIP joints, hinge type)
- *Artt. interphalangeae distales* between the phalanges mediales and distales (DIP joints, hinge type)

Since the *thumb* lacks a phalanx media, it has only two joints: art. metacarpophalangea pollicis and art. interphalangea pollicis (drawing based on a specimen from the Anatomical Collection at the University of Kiel).

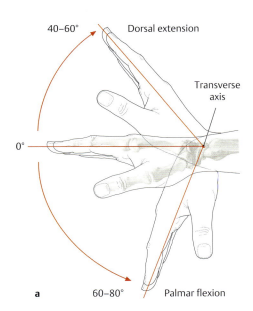
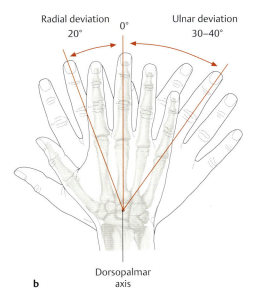
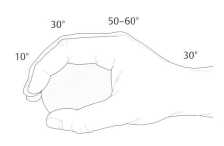

B Movements of the radiocarpal and midcarpal joints

Starting from the neutral (0°) position, palmar flexion and dorsal extension are performed about a transverse axis (**a**), while radial and ulnar deviation occurs about a dorsopalmar axis (**b**). The *transverse* axis runs through the os lunatum for the art. radiocarpalis and through the os capitatum bone for the art. mediocarpalis. The *dorsopalmar* axis runs through the os capitatum. Thus, while palmar flexion and dorsal extension can occur in both the art. radiocarpalis and art. mediocarpalis, radial and ulnar deviation can occur only in the art. radiocarpalis.

C Functional position of the hand

For postoperative immobilization of the hand, the desired position of the wrist and fingers should be considered when the cast, splint, or other device is applied. Otherwise the ligaments may shorten, and the hand can no longer assume a normal resting position.

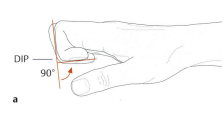
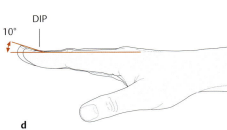
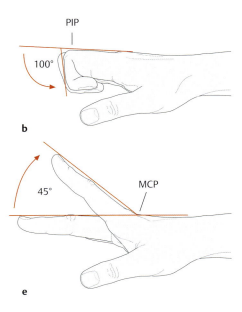
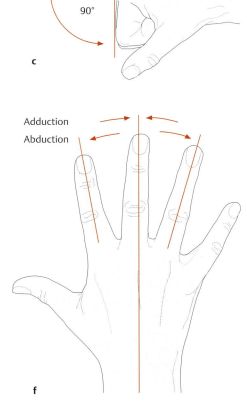

D Range of motion of the finger joints

The proximal interphalangeal (PIP) and distal interphalangeal (DIP) joints are pure hinge joints with only one degree of freedom (flexion extension). The metacarpophalangeal (MCP) joints of the second through fifth fingers are shaped like spheroidal joints with three theoretical degrees of freedom, but rotation is so limited by the collateral ligaments (ligg. collateralia) that only two degrees of freedom exist: flexion/extension and abduction/adduction. The following specific movements of the finger joints are distinguished:

a Flexion in the DIP joint.
b Flexion in the PIP joint.
c Flexion in the MCP joint.
d Extension in the DIP joint.
e Extension in the MCP joint.
f Abduction and adduction in the MCP joints (spreading the fingers apart and bringing them together about a dorsopalmar axis through the heads of the metacarpals).

Abduction/adduction movements are described in relation to the middle finger: all movements away from the middle finger are classified as abduction, all movements toward the middle finger as adduction.

15.1 Functional Muscle Groups

A Principles used in classification of the muscles of the upper limb

The muscles of the upper limb can be classified according to various criteria. An optimum system for classification should be logical and clear. The following criteria are suitable for classifying muscles:

- Origin
- Topography
- Function
- Innervation

While function and topography in the upper limb are often interrelated (muscles with the same action on a joint are often located close together), muscles that have similar actions in the shoulder region (e.g., muscles of the shoulder joint and shoulder girdle) vary considerably in their location. The following classification (**B**), then, is a compromise between topographical and functional considerations. In section **C**, a different muscle classification system, based on innervation, is presented.

The grouping of muscles by the pattern of their innervation reveals features of their embryological and phylogenetic origin and provides clinical insights into the clusters of consequences from damage to particular nerves.

B Functional-topographical classification of the muscles of the upper limb

Muscles of the shoulder girdle

Shoulder girdle muscles that have migrated from the head
- M. trapezius
- M. sternocleidomastoideus
- M. omohyoideus

Posterior muscles of the trunk and shoulder girdle
- M. rhomboideus major
- M. rhomboideus minor
- M. levator scapulae

Anterior muscles of the trunk and shoulder girdle
- M. subclavius
- M. pectoralis minor
- M. serratus anterior

Muscles of the shoulder joint

Posterior shoulder muscles
- M. supraspinatus
- M. infraspinatus
- M. teres minor
- M. subscapularis
- M. deltoideus
- M. latissimus dorsi
- M. teres major

Anterior shoulder muscles
- M. pectoralis major
- M. coracobrachialis

Muscles of the arm

Posterior arm muscles
- M. triceps brachii
- M. anconeus

Anterior arm muscles
- M. brachialis
- M. biceps brachii

Muscles of the forearm

Posterior forearm muscles
- Superficial extensors
  - M. extensor digitorum
  - M. extensor digiti minimi
  - M. extensor carpi ulnaris
- Deep extensors
  - M. supinator
  - M. abductor pollicis longus
  - M. extensor pollicis brevis
  - M. extensor pollicis longus
  - M. extensor indicis

Anterior forearm muscles
- Superficial flexors
  - M. pronator teres
  - M. flexor digitorum superficialis
  - M. flexor carpi radialis
  - M. flexor carpi ulnaris
  - M. palmaris longus
- Deep flexors
  - M. flexor digitorum profundus
  - M. flexor pollicis longus
  - M. pronator quadratus

Radial forearm muscles
- Radialis group
  - M. brachioradialis
  - M. extensor carpi radialis longus
  - M. extensor carpi radialis brevis

Muscles of the hand

Metacarpal muscles
- Mm. lumbricales I – IV
- Mm. interossei dorsales I – IV
- Mm. interossei palmares I – III

Thenar muscles
- M. abductor pollicis brevis
- M. adductor pollicis
- M. flexor pollicis brevis
- M. opponens pollicis

Hypothenar muscles
- M. abductor digiti minimi
- M. flexor digiti minimi
- M. opponens digiti minimi
- M. palmaris brevis

C Classification of the muscles of the upper limb by their innervation

Almost all the muscles of the upper limb are innervated by the plexus brachialis arising from spinal cord segments C 5–T 1. Exceptions are the mm. trapezius, sternocleidomastoideus, and omohyoideus; originating in vertebrate evolution as muscles of the head, they are supplied by n. cranialis XI (n. accessorius) and the plexus cervicalis (ansa cervicalis).

Nerve	Innervated muscles
N. accessorius	M. trapezius M. sternocleidomastoideus
Ansa cervicalis	M. omohyoideus
N. dorsalis scapulae	M. levator scapulae M. rhomboideus major M. rhomboideus minor
N. suprascapularis	M. supraspinatus M. infraspinatus
N. thoracicus longus	M. serratus anterior
N. subclavius	M. subclavius
N. subscapularis	M. subscapularis M. teres major
N. thoracodorsalis	M. latissimus dorsi
Nn. pectorales mediales and laterales	M. pectoralis major M. pectoralis minor
N. musculocutaneus	M. coracobrachialis M. biceps brachii M. brachialis
N. axillaris	M. deltoideus M. teres minor
N. radialis	M. triceps brachii M. anconeus M. supinator M. brachioradialis M. extensor carpi radialis longus M. extensor carpi radialis brevis M. extensor digitorum M. extensor digiti minimi M. extensor carpi ulnaris M. extensor pollicis longus M. extensor pollicis brevis M. extensor indicis M. abductor pollicis longus
N. medianus	M. pronator teres M. pronator quadratus M. palmaris longus M. flexor carpi radialis M. flexor pollicis longus M. flexor digitorum profundus (½) M. flexor digitorum superficialis M. abductor pollicis brevis M. opponens pollicis M. flexor pollicis brevis (caput superficiale) Mm. lumbricales I and II
N. ulnaris	M. flexor carpi ulnaris M. flexor digitorum profundus (½) M. palmaris brevis M. flexor digiti minimi M. abductor digiti minimi M. opponens digiti minimi M. adductor pollicis M. flexor pollicis brevis (caput profundum) Mm. interossei palmares and dorsales Mm. lumbricales III and IV

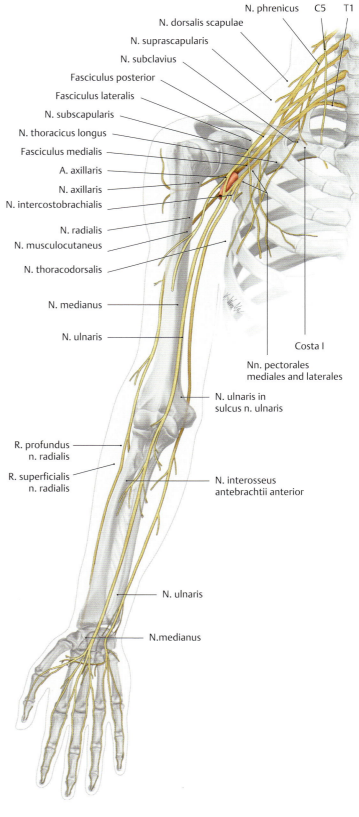

D Overview of the motor branches of the Plexus brachialis that supply the muscles of the upper limb

With the outgrowth of the limb buds from the trunk during embryonic development, the branches of the plexus brachialis follow the genetically determined *posterior* extensor muscles and *anterior* flexor muscles. The nerves for the *extensors* (nn. radialis and n. axillaris) arise from the three *posterior* divisions (divisiones posteriores) of the plexus brachialis, while the nerves for the *flexors* (n. musculocutaneous, n. ulnaris, and n. medianus) arise from the three *anterior* divisions (divisiones anteriores) (see p. 394, Neurovascular Systems: Topographical Anatomy).

15.2 The Muscles of the Shoulder Girdle: Mm. trapezius, sternocleidomastoideus, and omohyoideus

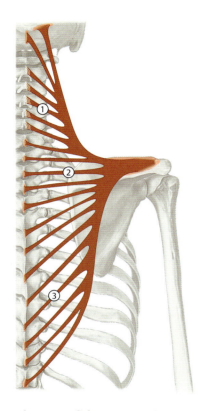

A Schematic of the M. trapezius

M. trapezius

Origin:
① Pars descendens[1]:
- os occipitale (linea nuchalis superior and protuberantia occipitalis externa)
- procc. spinosi of all cervical vertebrae via the lig. nuchae

② Pars transversa:
broad aponeurosis at the level of the procc. spinosi T I–T IV

③ Pars ascendens:
procc. spinosi of TV–TXII

Insertion:
- Lateral third of the clavicula (pars descendens)
- Acromion (pars transversa)
- Spina scapulae (pars ascendens)

Actions:
- Pars descendens:
  – Draws the scapula obliquely upward and rotates the cavitas glenoidalis inferiorly (acting with the pars inferior of the m. serratus anterior)
  – Tilts the head to the same side and rotates it to the opposite side (punctum fixum at the shoulder girdle)
- Pars transversa: draws the scapula medially
- Pars ascendens: draws the scapula medially downward (supports the rotating action of the pars descendens)
- Entire muscle: steadies the scapula on the thorax

Innervation: N. accessorius (CN XI) and plexus cervicalis (C 2–C 4)

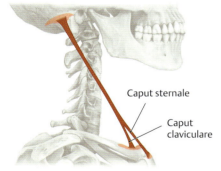

B Schematic of the M. sternocleido-mastoideus

M. sternocleidomastoideus

Origin:
- Caput sternale: manubrium sterni
- Caput claviculare: medial third of the clavicula

Insertion: Proc. mastoideus and linea nuchalis superior

Actions:
- Unilateral: – Tilts the head to the same side
              – Rotates the head to the opposite side
- Bilateral: – Extends the head
              – Assists in respiration with the punctum fixum at the head

Innervation: N. accessorius (CN XI) and the plexus cervicalis (C 1, C 2)

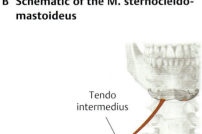

C Schematic of the M. omohyoideus

M. omohyoideus

Origin: Margo superior of the scapula
Insertion: Corpus ossis hyoidei
Actions:
- Depresses (fixes) the os hyoideum
- Moves the larynx and os hyoideum downward (for phonation and the final phase of swallowing)
- Tenses the fascia cervicalis with its tendo intermedius and maintains patency of the v. jugularis interna

Innervation: Ansa cervicalis (C1-C3) from the plexus cervicalis

[1] The tables and associated diagrams above are intended to give a systematic overview of the named muscles and their actions, while the drawings on p. 299 are intended to display the muscles as they would appear in a dissection. Not all structures listed in the tables are shown in the drawings on p. 299 because they are not all visible in those views.

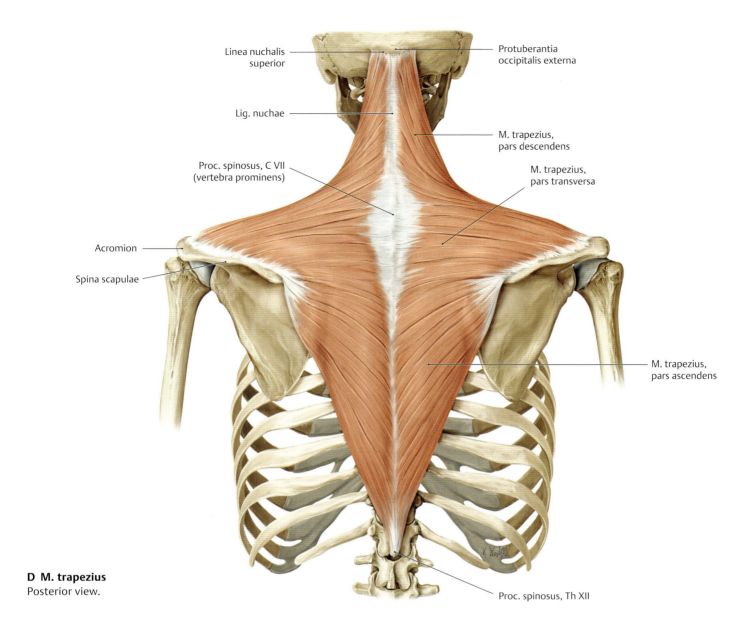

D M. trapezius
Posterior view.

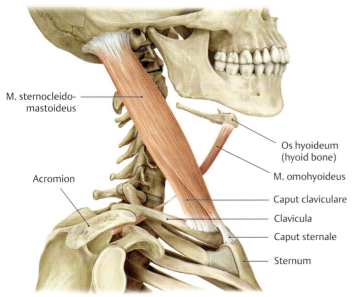

E M. sternocleidomastoideus and m. omohyoideus
Right side, lateral view.

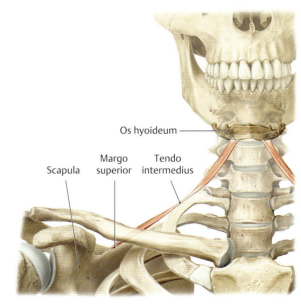

F M. omohyoideus
Right side, anterior view.

15.3 The Muscles of the Shoulder Girdle: Mm. serratus anterior, subclavius, pectoralis minor, levator scapulae, and rhomboidei major and minor

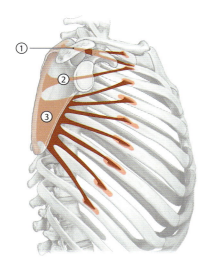

M. serratus anterior

Origin:	First through ninth ribs
Insertion:	Scapula: ① Pars superior (angulus superior)
	② Pars intermedia (margo medialis)
	③ Pars inferior (angulus inferior and margo medialis)
Actions:	• Entire muscle: draws the scapula laterally forward, elevates the ribs when the shoulder girdle is fixed (assists in respiration)
	• Pars inferior: rotates the scapula and draws its angulus inferior laterally forward (rotates cavitas glenoidalis superiorly), allowing for arm to be elevated above 90°
	• Pars superior: lowers the raised arm (antagonist to the pars inferior)
Innervation:	N. thoracicus longus (C5–C7)

A Schematic of the m. serratus anterior

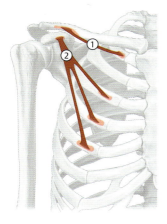

① M. subclavius

Origin:	First costa (chondro-osseous junction)
Insertion:	Inferior surface of the clavicula (lateral third)
Action:	Steadies the clavicula in the art. sternoclavicularis
Innervation:	N. subclavius (C5, C6)

② M. pectoralis minor

Origin:	Third through fifth ribs
Insertion:	Proc. coracoideus of the scapula
Actions:	• Draws the scapula downward, causing its angulus inferior to move posteromedially (lowers the raised arm), rotates glenoid inferiorly
	• Assists in respiration
Innervation:	Nn. pectorales medialis und lateralis (C8, T1)

B Schematic of the m. subclavius and m. pectoralis minor

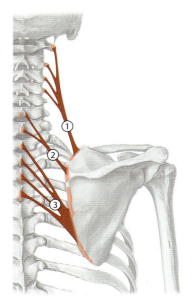

① M. levator scapulae

Origin:	Procc. transversi of the C1–C4 vertebrae
Insertion:	Angulus superior of the scapula
Actions:	• Draws the scapula medially upward while moving the angulus inferior medially (returns the raised arm to the neutral [0°] position)
	• Inclines the neck toward the same side (punctum fixum at the scapula)
Innervation:	N. dorsalis scapulae and nn. spirales cervicalis (C3, C4)

② M. rhomboideus minor

Origin:	Procc. spinosi of the CVI and CVII vertebrae
Insertion:	Margo medialis of the scapula (above the spina scapulae))
Actions:	• Steadies the scapula
	• Draws the scapula medially upward (returns the raised arm to the neutral [0°] position)
Innervation:	N. dorsalis scapulae (C4, C5)

③ M. rhomboideus major

Origin:	Procc. spinosi of the TI–TIV vertebrae
Insertion:	Margo medialis of the scapula (below the spina scapulae)
Actions:	• Steadies the scapula
	• Draws the scapula medially upward (returns the raised arm to the neutral [0°] position)
Innervation:	N. dorsalis scapulae (C4, C5)

C Schematic of the mm. levator scapulae, and rhomboidei minor and major

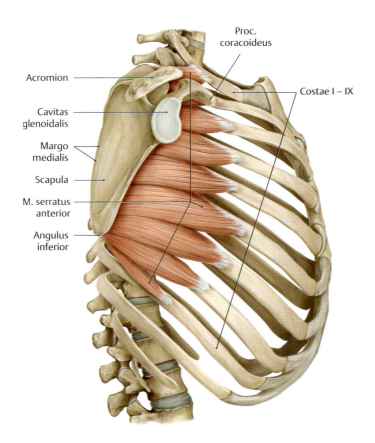

D M. serratus anterior
Right side, lateral view.

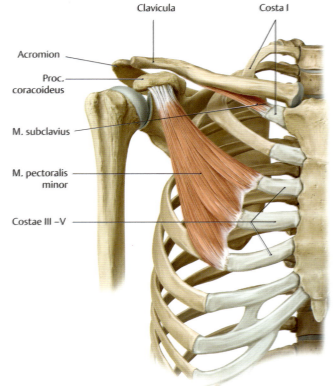

E M. pectoralis minor and m. subclavius
Right side, anterior view.

F M. levator scapulae, m. rhomboideus major, and m. rhomboideus minor
Right side, posterior view.

15.4 The Posterior Muscles of the Shoulder Joint: The Rotator Cuff

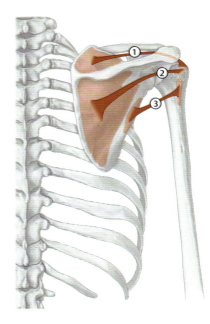

① **M. Supraspinatus**

Origin:	Fossa supraspinata of the scapula
Insertion:	Tuberculum majus of the humerus
Action:	Abduction
Innervation:	N. suprascapularis (C4–C6)

② **M. Infraspinatus**

Origin:	Fossa infraspinata of the scapula
Insertion:	Tuberculum majus of the humerus
Action:	External rotation
Innervation:	N. suprascapularis (C4–C6)

③ **M. Teres minor**

Origin:	Margo lateralis of the scapula
Insertion:	Tuberculum majus of the humerus
Action:	External rotation, weak adduction
Innervation:	N. axillaris (C5, C6)

A Schematic of the m. supraspinatus, m. infraspinatus, and m. teres minor

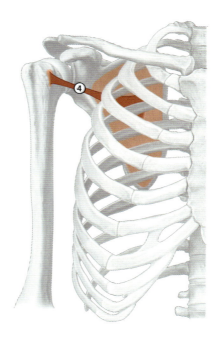

④ **M. Subscapularis**

Origin:	Fossa subscapularis of the scapula
Insertion:	Tuberculum minus of the humerus
Action:	Internal rotation
Innervation:	N. subscapularis (C5, C6)

B Schematic of the m. subscapularis

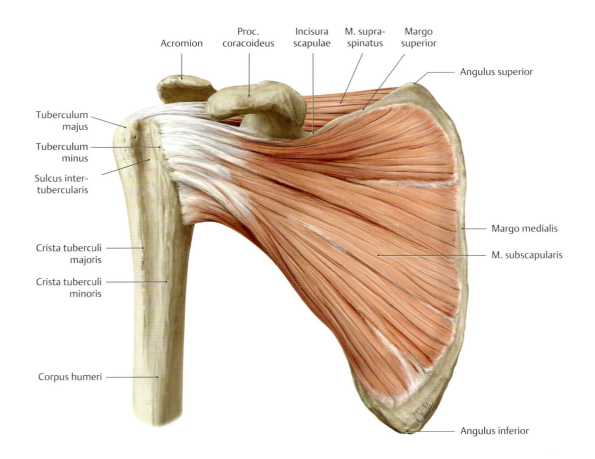

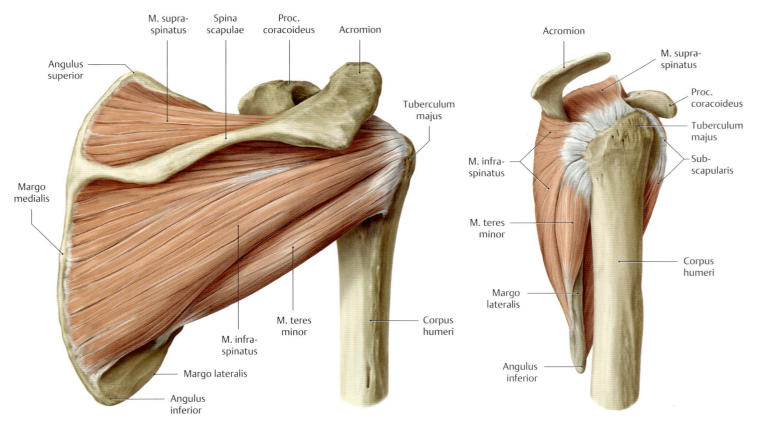

C Muscles of the rotator cuff: m. supraspinatus, m. infraspinatus, m. teres minor, and m. subscapularis
Right shoulder joint.

a Anterior view.
b Posterior view.
c Lateral view.

15.5 The Posterior Muscles of the Shoulder Joint: M. deltoideus

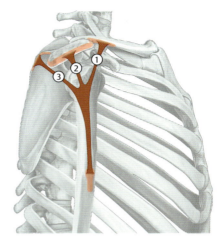

M. deltoideus

Origin: ① Pars clavicularis: lateral third of the clavicula
② Pars acromialis: acromion
③ Pars spinalis: spina scapulae
Insertion: Tuberositas deltoidea on the humerus
Actions:
- Pars clavicularis: anteversion (moves the arm and shoulder forward), internal rotation, adduction
- Pars acromialis: abduction
- Pars spinalis: retroversion (moves the arm and shoulder backward), external rotation, adduction

Between 60° and 90° of abduction, the clavicular and spinal parts of the m. deltoideus assist the pars acromialis of the muscle with abduction.
Innervation: N. axillaris (C5, C6)

A Schematic of the m. deltoideus

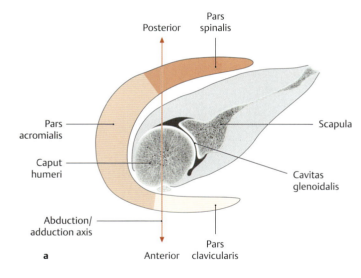
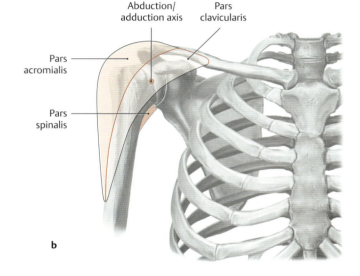
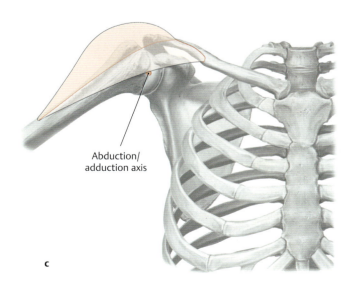

B The variable actions of the m. deltoideus
a Cross section through the right shoulder joint.
b Right shoulder joint in the neutral (0°) position, anterior view.
c Right shoulder joint in 60° of abduction, anterior view.

The actions of the three parts of the m. deltoideus (pars clavicularis, pars acromialis, and pars spinalis) depend on their relationship to the position of the humerus and its axis of motion. As a result, the parts of the m. deltoideus can act antagonistically as well as synergistically. At *less than 60°* abduction, the pars clavicularis and pars spinalis parts of the m. deltoideus act as antagonists to the acromial part, but at *more than 60°* abduction, the pars acromialis assists in the abduction. Starting from the neutral (0°) position, the pars acromialis of the m. deltoideus abducts the arm and steadies it in any position it assumes. When the arm is abducted past approximately 60°, the pars clavicularis and pars spinalis also become active as they move past the sagittal motion axis (abduction/adduction axis, **c**). This alters the action of these parts: they act as adductors below 60°, but when the arm passes 60°, they become abductors.

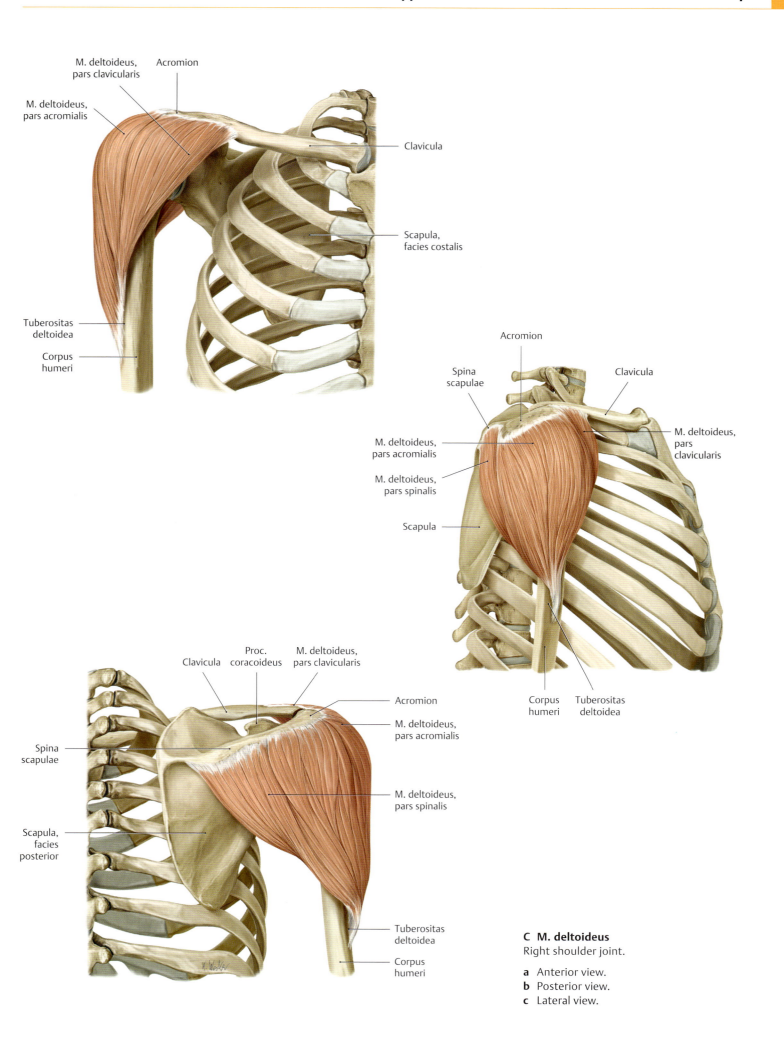

C M. deltoideus
Right shoulder joint.

a Anterior view.
b Posterior view.
c Lateral view.

15.6 The Posterior Muscles of the Shoulder Joint: Mm. latissimus dorsi and teres major

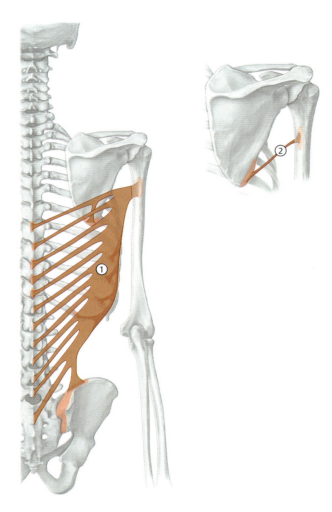

① **M. latissimus dorsi**

Origin:
- Pars vertebralis:
  - Procc. spinosi of the T VII–T XII vertebrae
  - Fascia thoracolumbalis of the procc. spinosi of all lumbar vertebrae and the os sacrum
- Pars iliaca: posterior third of the crista iliaca
- Pars costalis: 9th through 12th ribs
- Pars scapularis: angulus inferior of the scapula

Insertion: Sulcus intertubercularis humeri
Actions: Internal rotation, adduction, extension (moves the arm backward), respiration (expiration, "cough muscle")
Innervation: N. subscapularis interior (C5, C6)

② **M. teres major**

Origin: Angulus inferior of the scapula
Insertion: Crista tuberculi minoris of the humerus
Action: Internal rotation, adduction, retroversion
Innervation: N. thoracodorsalis (C6–C8)

A Schematic of the mm. latissimus dorsi and teres major

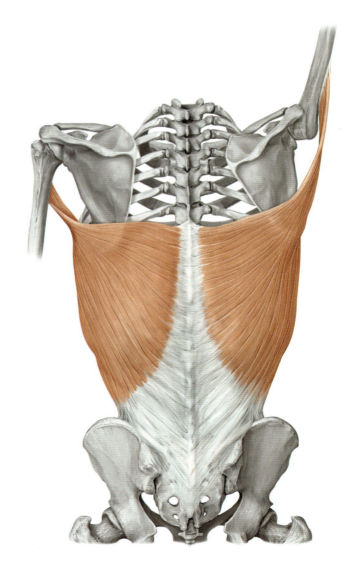

B Course of the tendon of insertion of m. latissimus dorsi in the neutral position and in elevation

Posterior view. The m. latissimus dorsi muscle is most active in the abducted or elevated arm. Raising the arm untwists the muscle fibers in the area of insertion, increasing the muscle's stretch and maximizing the force it can exert. When the position of the arm is fixed, the m. latissimus dorsi can pull the body upward, as in the act of climbing, or can depress the arm against a resistance. This makes the m. latissimus dorsi an important muscle for paraplegics, for example, who can use that muscle to raise themselves from wheelchairs.

Upper Limb — 15. Musculature: Functional Groups

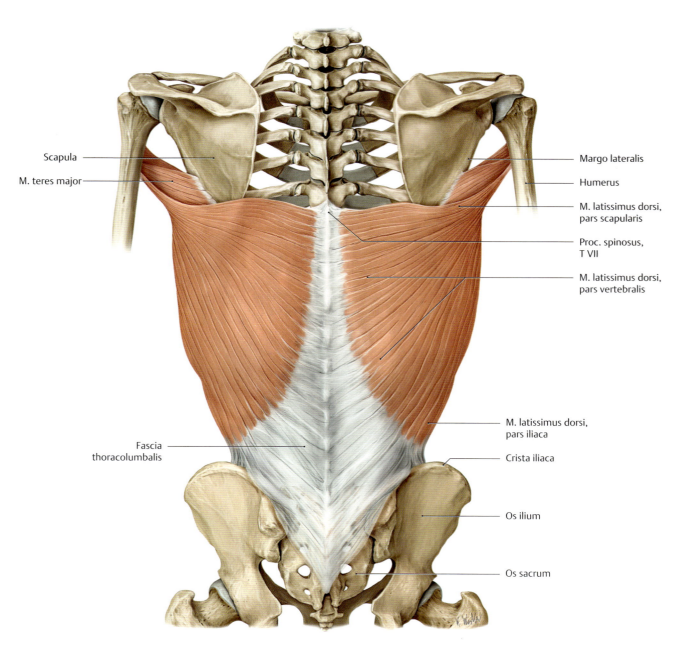

C Mm. latissimus dorsi and teres major
Posterior view.

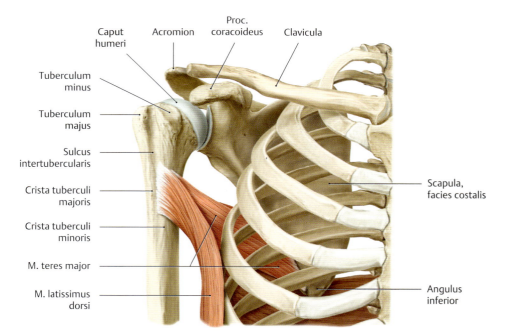

D Insertion of the m. latissimus dorsi on the sulcus intertubercularis and the m. teres major on the crista tuberculi minoris
Right side, anterior view.

307

15.7 The Anterior Muscles of the Shoulder Joint: Mm. pectoralis major and coracobrachialis

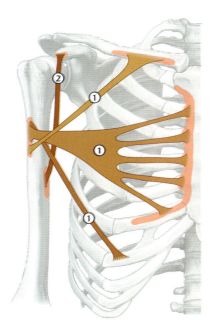

① **M. pectoralis major**

Origin:	• Pars clavicularis: medial half of the clavicula • Pars sternocostalis: sternum and the second through sixth costal cartilages • Pars abdominalis: lamina anterior of the rectus sheath
Insertion:	Crista tuberculi majoris of the humerus
Actions:	• Adduction and internal rotation (entire muscle) • Flexion (pars clavicularis and pars sternocostalis) • Assists respiration when the shoulder girdle is fixed
Innervation:	Nn. pectorales medialis and lateralis (C5–T1)

② **M. coracobrachialis**

Origin:	Proc. coracoideus of the scapula
Insertion:	Humerus (in line with the crista tuberculi minoris))
Actions:	Flexion, adduction, internal rotation
Innervation:	N. musculocutaneus (C5–C7)

A Schematic of the mm. pectoralis major and coracobrachialis

B Twisting of the tendon of insertion of the m. pectoralis major
Anterior view. The three parts of the m. pectoralis major (pars clavicularis, pars sternocostalis, and pars abdominalis) converge laterally and insert on the crista tuberculi majoris by a broad tendon that has a horseshoe-shaped cross section. The tendon fiber bundles are twisted on themselves in such a way that the pars clavicularis inserts lower on the humerus than the pars sternocostalis, which inserts lower than the pars abdominalis. As with the m. latissimus dorsi, the muscle fibers of the m. pectoralis major become untwisted and stretched with increasing elevation of the arm, increasing the force that the muscle can exert.

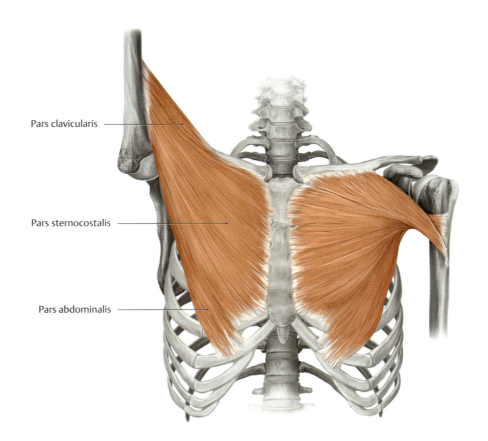

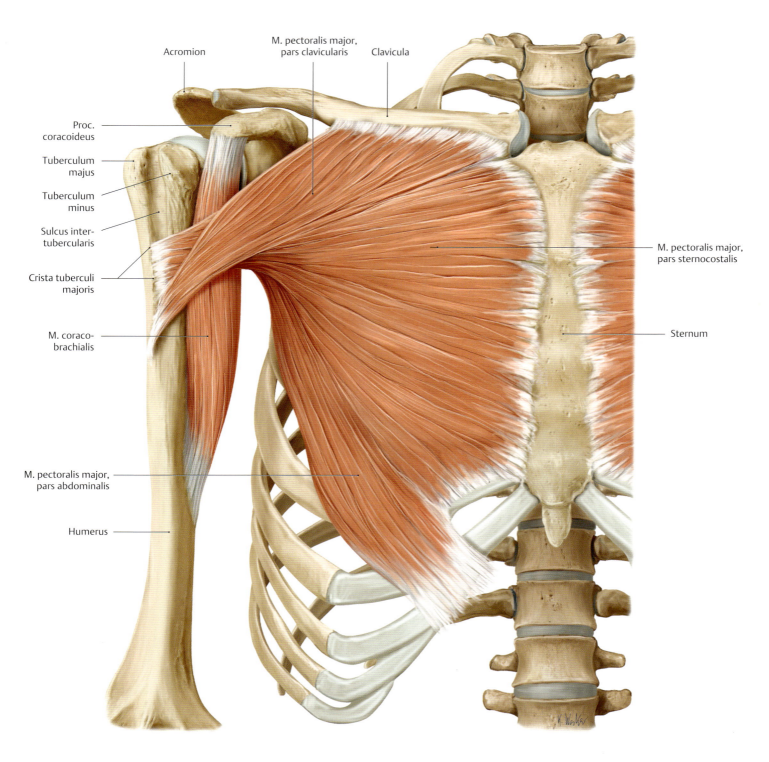

C Mm. pectoralis major and coracobrachialis
Right side, anterior view.

15.8 The Anterior Muscles of the Arm: Mm. biceps brachii and brachialis

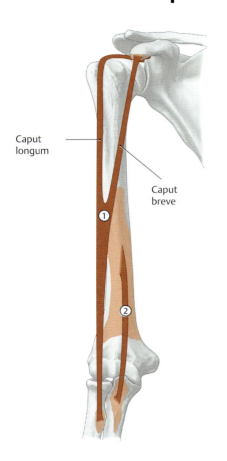

① **M. biceps brachii**

Origin:
- Caput longum: tuberculum supraglenoidale of the scapula
- Caput breve: proc. coracoideus of the scapula

Insertion: Tuberositas radii

Actions:
- Elbow joint (art. cubiti):
  – Flexion, supination (with the elbow flexed)
- Shoulder joint (art. humeri):
  – Flexion (forward motion of humerus)
  – Stabilization of caput humeri during deltoid contraction
  – Abduction and internal (medial) rotation of the humerus

Innervation: N. musculocutaneus (C5–C7)

② **M. brachialis**

Origin: Distal half of the anterior surface of the humerus, also the septa intermuscularia mediale and lateralea

Insertion: Tuberositas ulnae

Action: Flexion at the elbow joint (art. cubiti)

Innervation: N. musculocutaneus (C5, C6) and n. radialis

A Schematic of the mm. biceps brachii and brachialis

B Supinating action of the m. biceps brachii with the elbow flexed
a The forearm is pronated with the elbow flexed (right arm, medial view).
b Cross section at the level of the tuberositas radii with the forearm pronated (proximal view).
c The forearm is supinated with the elbow flexed (right arm, medial view).
d Cross section at the level of the tuberositas radii with the forearm supinated (proximal view).

When the elbow is flexed, the m. biceps brachii acts as a powerful supinator in addition to its role as a flexor, because the lever arm in that position is almost perpendicular to the axis of pronation/supination (see p. 280). This is why *supination movements* are particularly effective when the elbow is flexed. When the forearm is *pronated* (**a**), the tendon of insertion of the m. biceps brachii is wrapped around the radius. When the muscle then contracts to flex the elbow, the tendon unwraps like a rope coiled around a crank (**b**).

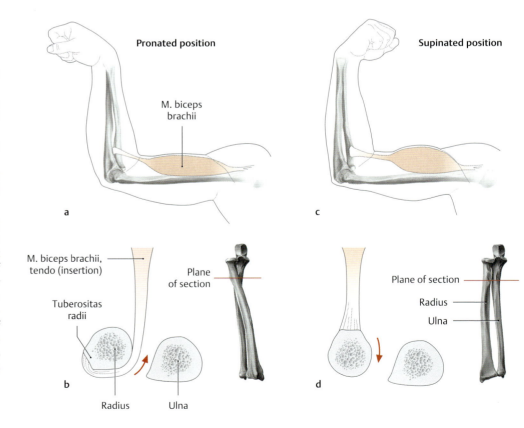

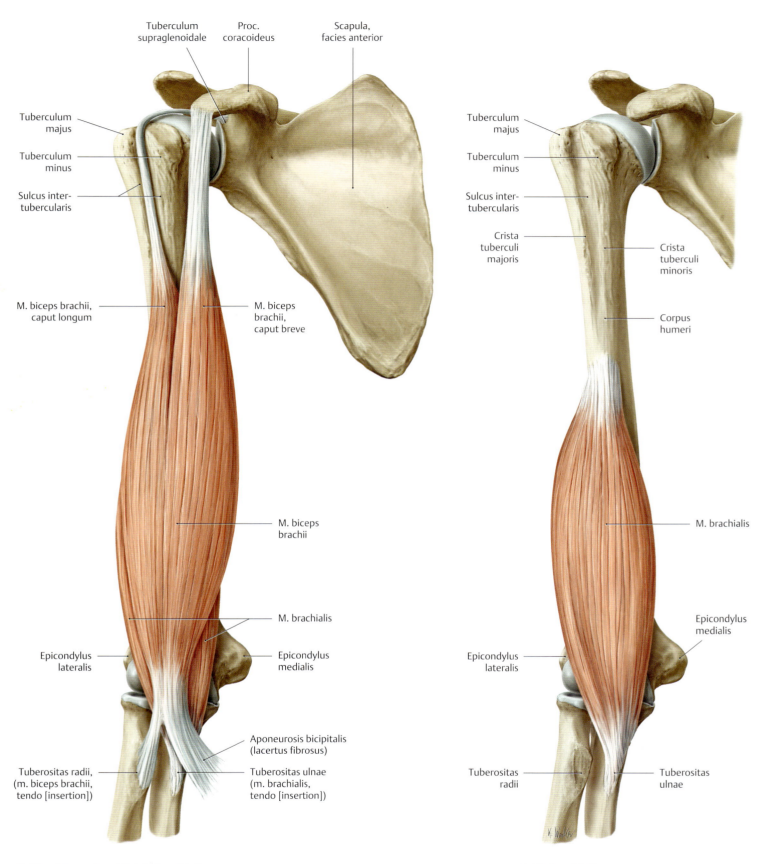

C Mm. biceps brachii and brachialis
Right arm, anterior (ventral) view.

D M. brachialis
Right arm, anterior (ventral) view.

15.9 The Posterior Muscles of the Arm: Mm. triceps brachii and anconeus

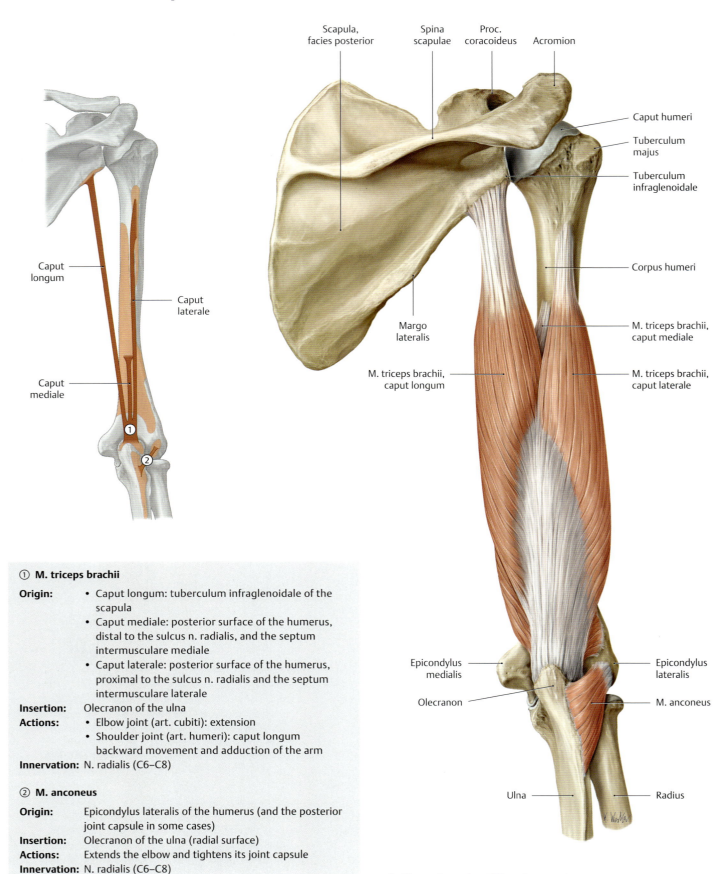

① **M. triceps brachii**

Origin:
- Caput longum: tuberculum infraglenoidale of the scapula
- Caput mediale: posterior surface of the humerus, distal to the sulcus n. radialis, and the septum intermusculare mediale
- Caput laterale: posterior surface of the humerus, proximal to the sulcus n. radialis and the septum intermusculare laterale

Insertion: Olecranon of the ulna

Actions:
- Elbow joint (art. cubiti): extension
- Shoulder joint (art. humeri): caput longum backward movement and adduction of the arm

Innervation: N. radialis (C6–C8)

② **M. anconeus**

Origin: Epicondylus lateralis of the humerus (and the posterior joint capsule in some cases)

Insertion: Olecranon of the ulna (radial surface)

Actions: Extends the elbow and tightens its joint capsule

Innervation: N. radialis (C6–C8)

A Schematic of the mm. triceps brachii and anconeus

B Mm. triceps brachii and anconeus
Right arm, posterior (dorsal) view.

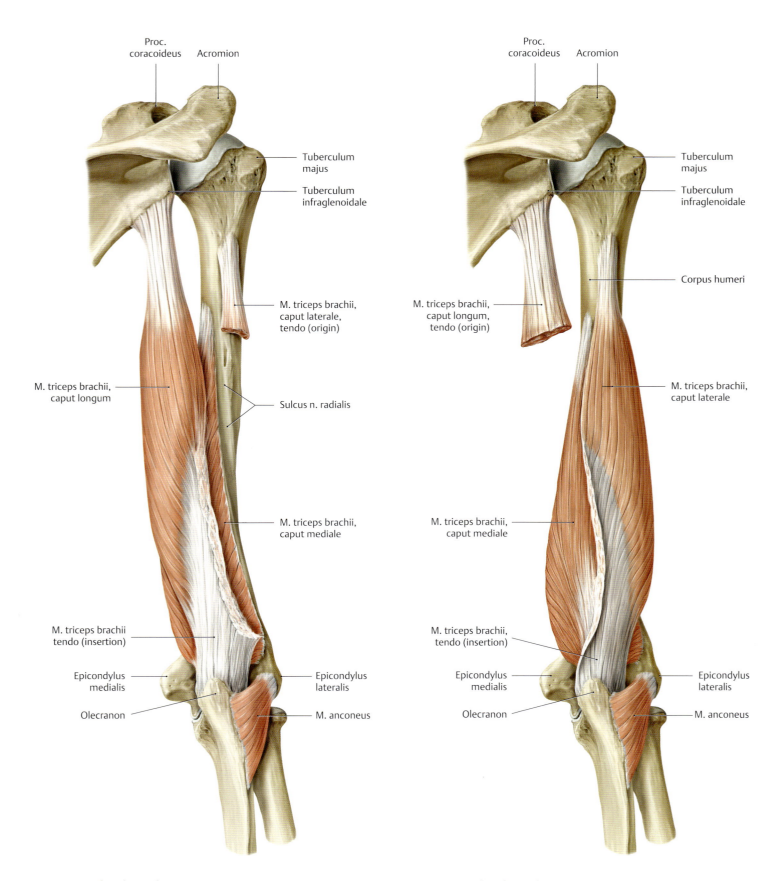

C Mm. triceps brachii and anconeus
Right arm, posterior (dorsal) view. The caput laterale of m. triceps brachii has been partially removed.

D Mm. triceps brachii and anconeus
Right arm, posterior (dorsal) view. The caput longum of m. triceps brachii has been partially removed.

15.10 The Muscles of the Anterior Forearm: The Superficial and Deep Flexors

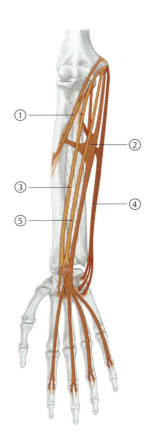

A Schematic of the superficial flexors

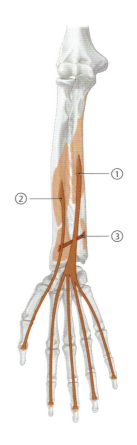

B Schematic of the deep flexors

① **M. pronator teres**

Origin:	• Caput humerale: epicondylus medialis of the humerus
	• Caput ulnare: proc. coronoideus of the ulna
Insertion:	Facies lateralis radii (distal to the m. supinator insertion)
Actions:	• Elbow joint (art. cubiti): weak flexion
	• Forearm joints (artt. radioulnaris proximalis et distalis): pronation
Innervation:	N. medianus (C6, C7)

② **M. flexor digitorum superficialis**

Origin:	• Caput humeroulnare: epicondylus medialis of the humerus and proc. coronoideus of the ulna
	• Caput radiale: distal to the tuberositas radii
Insertion:	The sides of the middle phalanges of the second through fifth digits
Actions:	Wrist joints and the MCP and PIP joints of the second through fifth digits: flexion
Innervation:	N. medianus (C8, T1)

③ **M. flexor carpi radialis**

Origin:	Epicondylus medialis of the humerus
Insertion:	Base of the os metacarpi II (and sometimes of the os metacarpi III)
Actions:	Wrist joints: flexion and abduction (radial deviation) of the hand
Innervation:	N. medianus (C6, C7)

④ **M. flexor carpi ulnaris**

Origin:	• Caput humerale: epicondylus medialis of the humerus
	• Caput ulnare: olecranon of the ulna
Insertion:	Hamulus ossis hamati, base of the os metacarpi V
Actions:	Wrist joints: flexion and adduction (ulnar deviation) of the hand
Innervation:	N. ulnaris (C8, T1)

⑤ **M. palmaris longus**

Origin:	Epicondylus medialis humeri
Insertion:	Aponeurosis palmaris
Actions:	Wrist joints: flexion tightens the aponeurosis palmaris for gripping
Innervation:	N. medianus (C7, C8)

① **M. flexor digitorum profundus**

Origin:	Proximal two-thirds of the flexor surface of the ulna and the adjacent membrana interossea
Insertion:	Palmar surface of the distal phalanges of the second through fifth digits
Actions:	Wrist joints and the MCP, PIP, and DIP joints of the second through fifth digits: flexion
Innervation:	• N. medianus (radial part, second and third digits), C8, T1
	• N. ulnaris (ulnar part, fourth and fifth digits), C8–T1

② **M. flexor pollicis longus**

Origin:	Mid-anterior surface of the radius and the adjacent membrana interossea
Insertion:	Palmar surface of the phalanx distalis of the thumb
Actions:	• Wrist joints: flexion and radial abduction of the hand
	• Carpometacarpal joint of the thumb: opposition
	• MCP and IP joints of the thumb: flexion
Innervation:	N. medianus (C8, T1)

③ **M. pronator quadratus**

Origin:	Distal one fourth of the anterior surface of the ulna
Insertion:	Distal one fourth of the anterior surface of the radius
Actions:	Pronates the hand, stabilizes the distal radioulnar joint
Innervation:	N. medianus (C8, T1)

DIP, distal interphalangeal; IP, interphalangeal; MCP, metacarpophalangeal; PIP, proximal interphalangeal.

Upper Limb — 15. Musculature: Functional Groups

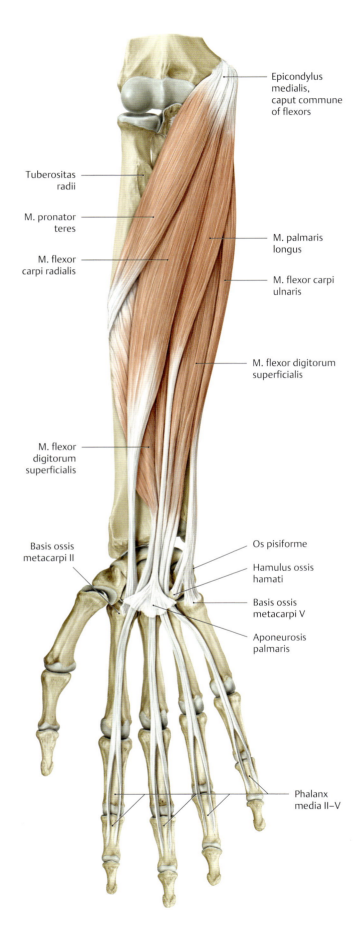

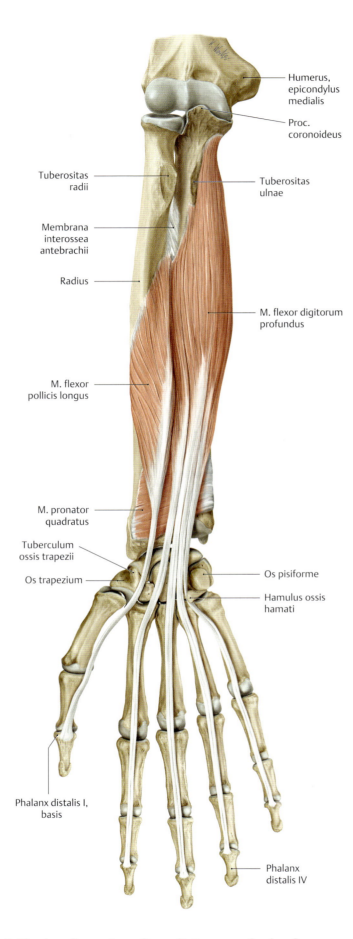

C The superficial flexors (mm. pronator teres, flexor digitorum superficialis, flexor carpi radialis, flexor carpi ulnaris, and palmaris longus))
Right forearm, anterior view.

D The deep flexors (mm. flexor digitorum profundus, flexor pollicis longus, and pronator quadratus)
Right forearm, anterior view.

315

15.11 The Muscles of the Forearm: The Radialis Muscles

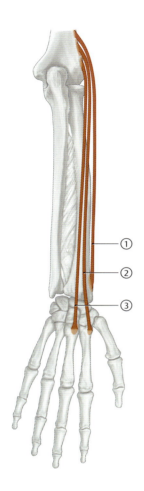

A Schematic of the radialis muscles

① **M. brachioradialis**

Origin:	Lateral surface of the distal humerus, septum intermusculare laterale
Insertion:	Proc. styloideus radii
Actions:	• Elbow joint (art. cubiti): flexion
	• Forearm joints (artt. radioulnaris proximalis et distalis): semipronation
Innervation:	N. radialis (C5, C6)

② **M. extensor carpi radialis longus**

Origin:	Lateral surface of the distal humerus (crista supracondylaris lateralis), septum intermusculare laterale
Insertion:	Dorsal base of the os metacarpi II
Actions:	Wrist joints: dorsal extension (assists in fist closure), abduction (radial deviation) of the hand
Innervation:	N. radialis (C6, C7)

③ **M. extensor carpi radialis brevis**

Origin:	Epicondylus lateralis of the humerus
Insertion:	Dorsal base of the os metacarpi III
Actions:	Wrist joints: dorsal extension (assists in fist closure), abduction (radial deviation) of the hand
Innervation:	N. radialis (C7, C8)

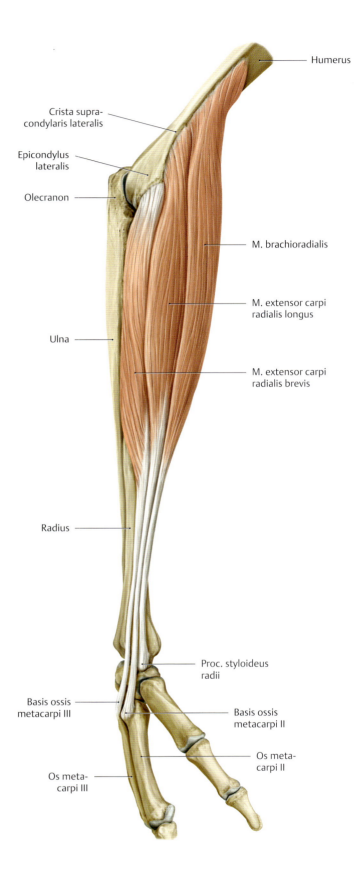
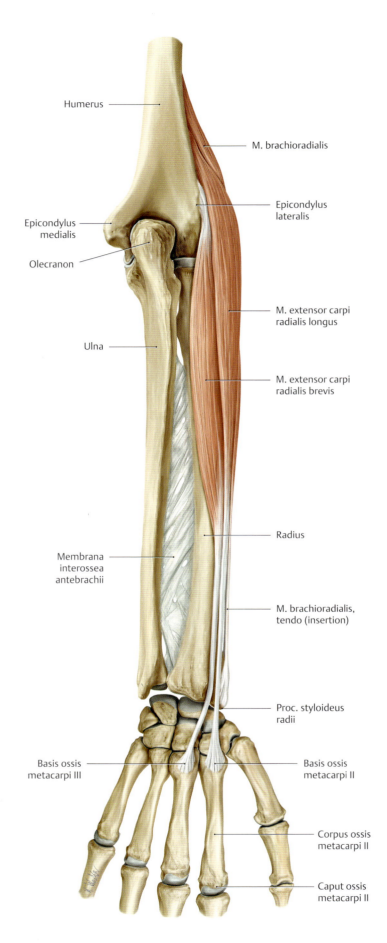

B The radialis muscles (mm. brachioradialis, extensor carpi radialis longus, and extensor carpi radialis brevis)
Right forearm.

a Lateral (radial) view.
b Posterior (dorsal) view.

15.12 The Muscles of the Posterior Forearm: The Superficial and Deep Extensors

① **M. extensor digitorum**
Origin: Caput commune (epicondylus lateralis of the humeri)
Insertion: Dorsal digital expansion of the second through fifth digits
Actions:
 • Wrist joints: extension
 • MCP, PIP, and DIP joints of the second through fifth digits: extension and abduction of the fingers
Innervation: N. radialis (C7, C8)

② **M. extensor digiti minimi**
Origin: Caput commune (epicondylus lateralis of the humerus)
Insertion: Dorsal digital expansion of the fifth digit
Actions:
 • Wrist joints: extension, adduction (ulnar deviation)
 • MCP, PIP, and DIP joints of the fifth digit: extension and abduction of the fifth digit
Innervation: N. radialis (C7, C8)

③ **M. extensor carpi ulnaris**
Origin: Caput commune (epicondylus lateralis of the humerus), caput ulnare (dorsal surface of the ulna)
Insertion: Base of the os metacarpi V
Actions: Wrist joints: extension, adduction (ulnar deviation) of the hand
Innervation: N. radialis (C7, C8)

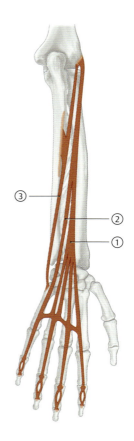

A Schematic of the superficial extensors

① **M. supinator**
Origin: Olecranon of the ulna, epicondylus lateralis of the humerus, lig. collaterale radiale, lig. anulare radii
Insertion: Radius (between the tuberositas radii and the insertion of the m. pronator teres)
Action: Radioulnar joint: supination
Innervation: N. radialis (C7, C8)

② **M. abductor pollicis longus**
Origin: Dorsal surfaces of the radius and ulna, also the membrana interossea
Insertion: Base of the os metacarpi I
Actions:
 • Radiocarpal joint: abduction (radial deviation) of the hand
 • Carpometacarpal joint of the thumb: abduction
Innervation: N. radialis (C7, C8)

③ **M. extensor pollicis brevis**
Origin: Posterior surface of the radius and the membrana interossea (distal to m. abductor pollicis longus)
Insertion: Base of the phalanx proximalis of the thumb
Actions:
 • Radiocarpal joint: abduction (radial deviation) of the hand
 • Carpometacarpal and MCP joints of the thumb: extension
Innervation: N. radialis (C7, C8)

④ **M. extensor pollicis longus**
Origin: Posterior surface of the ulna and the membrana interossea
Insertion: Base of the phalanx distalis of the thumb
Actions:
 • Wrist joints: extension and abduction (radial deviation) of the hand
 • Carpometacarpal joint of the thumb: adduction
 • MCP and interphalangeal joints of the thumb: extension
Innervation: N. radialis (C7, C8)

⑤ **M. extensor indicis**
Origin: Posterior surface of the ulna and the membrana interossea
Insertion: Posterior digital expansion of the second digit
Actions:
 • Wrist joint (art. radiocarpalis): extension
 • MCP, PIP, and DIP joints of the second digit: extension
Innervation: N. radialis (C7, C8)

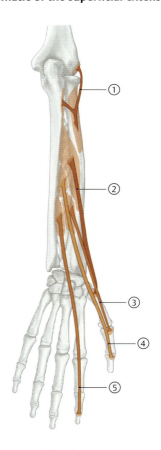

B Schematic of the deep extensors

DIP, distal interphalangeal; IP, interphalangeal; MCP, metacarpophalangeal; PIP, proximal interphalangeal.

Upper Limb —— 15. Musculature: Functional Groups

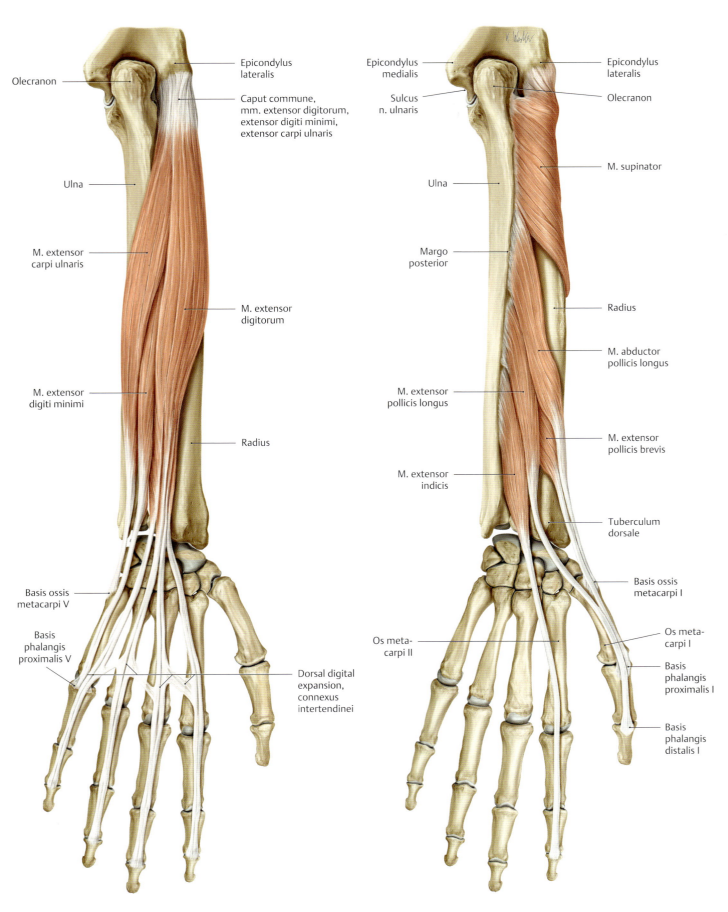

C The superficial extensors (mm. extensor digitorum, extensor digiti minimi, and extensor carpi ulnaris)
Right forearm, posterior (dorsal) view.

D The deep extensors (mm. supinator, abductor pollicis longus, extensor pollicis brevis, extensor pollicis longus, and extensor indicis)
Right forearm, posterior (dorsal) view.

319

15.13 The Intrinsic Muscles of the Hand: The Thenar and Hypothenar Muscles

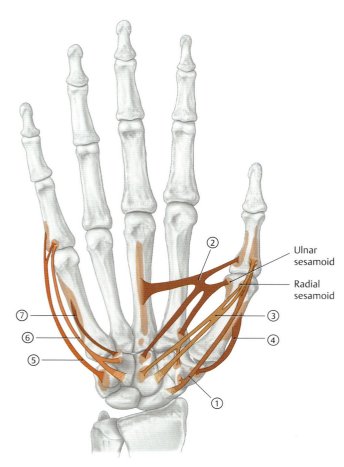

A Schematic of the thenar (①–④) and hypothenar (⑤–⑦) muscles

① **M. abductor pollicis brevis**
- Origin: Ossa scaphoideum and trapezium, retinaculum musculorum flexorum
- Insertion: Base of the phalanx proximalis of the thumb (via the radial sesamoid)
- Actions: Abduction of the thumb
- Innervation: N. medianus (C8, T1)

② **M. adductor pollicis**
- Origin:
  - Caput transversum: palmar surface of the third metacarpal
  - Caput obliquum: os capitatum, basis ossis metacarpi II
- Insertion: Base of the phalanx proximalis of the thumb (via the ulnar sesamoid)
- Actions:
  - CMC joint of the thumb: opposition
  - MCP joint of the thumb: flexion
- Innervation: N. ulnaris (C8, T1)

③ **M. flexor pollicis brevis**
- Origin:
  - Caput superficiale: retinaculum mm. flexorum
  - Caput profundum: os capitatum, os trapezium
- Insertion: Base of the phalanx proximalis of the thumb (via the radial sesamoid)
- Actions:
  - CMC joint of the thumb: flexion, opposition
  - MCP joint of the thumb: flexion
- Innervation:
  - N. medianus, C8, T1 (caput superficiale)
  - N. ulnaris, C8, T1 (caput profundum)

④ **M. opponens pollicis**
- Origin: Os trapezium
- Insertion: Radial border of the first metacarpal
- Action: CMC joint of the thumb: opposition
- Innervation: N. medianus (C8, T1)

⑤ **M. abductor digiti minimi**
- Origin: Os pisiforme
- Insertion: Ulnar base of the phalanx proximalis and the dorsal digital expansion of the fifth digit
- Actions:
  - MCP joint of the little finger: flexion and abduction of the little finger
  - PIP and DIP joints of the little finger: extension
- Innervation: N. ulnaris (C8, T1)

⑥ **M. flexor digiti minimi brevis**
- Origin: Hamulus ossis hamati, retinaculum mm. flexorum
- Insertion: Base of the phalanx proximalis of the fifth digit
- Action: MCP joint of the little finger: flexion
- Innervation: N. ulnaris (C8, T1)

⑦ **M. opponens digiti minimi**
- Origin: Hamulus ossis hamati
- Insertion: Ulnar border of the fifth metacarpal
- Action: Draws the os metacarpi in the palmar direction (opposition)
- Innervation: N. ulnaris (C8, T1)

M. palmaris brevis (not shown, see also pp. 343 and 348)
- Origin: Ulnar border of the aponeurosis palmaris
- Insertion: Skin of the hypothenar
- Action: Tightens the aponeurosis palmaris (protective function)
- Innervation: N. ulnaris (C8, T1)

CMC, carpometacarpal; DIP, distal interphalangeal; MCP, metacarpointerphalangeal; PIP, proximal interphalangeal.

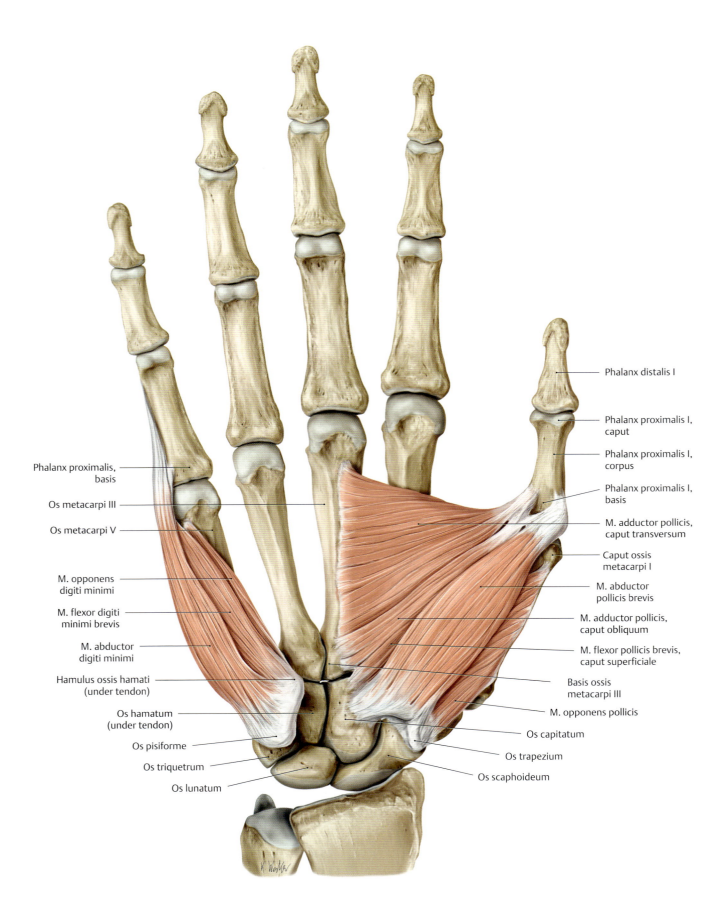

B The thenar muscles (mm. abductor pollicis brevis, adductor pollicis, flexor pollicis brevis, and opponens pollicis) and the hypothenar muscles (mm. abductor digiti minimi, flexor digiti minimi brevis, and opponens digiti minimi)
Right hand, anterior (palmar) view

15.14 The Intrinsic Muscles of the Hand: Mm. lumbricales and mm. interossei (Metacarpal Muscles)

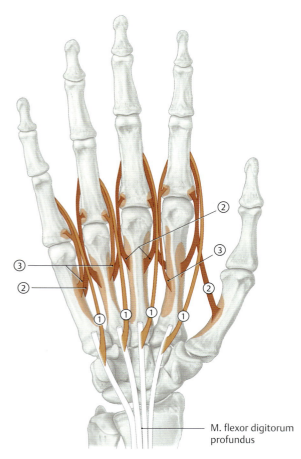

A Schematic of the metacarpal muscles

① **Mm. lumbricales I–IV**

Origin: Radial sides of the tendons of m. flexor digitorum profundus (variable)

Insertion:
- First lumbrical: dorsal digital expansion of the second digit (index finger)
- Second lumbrical: dorsal digital expansion of the third digit (middle finger)
- Third lumbrical: dorsal digital expansion of the fourth digit (ring finger)
- Fourth lumbrical: dorsal digital expansion of the fifth digit (little finger)

Actions:
- MCP joints of the second through fifth digits: flexion
- PIP and DIP joints of the second through fifth digits: extension

Innervation:
- N. medianus, C8, T1 (mm. lumbricales I + II)
- N. ulnaris, C8, T1 (mm. lumbricales III + IV)

② **Mm. interossei dorsales I–IV**

Origin: By two heads from adjacent sides of the ossa metacarpi I–V

Insertion:
- Dorsal digital expansion of the second through fourth digits, base of the phalanx proximalis
- First interosseus: radial side of the second phalanx proximalis (index finger)
- Second interosseus: radial side of the third phalanx proximalis (middle finger)
- Third interosseus: ulnar side of the third phalanx proximalis (middle finger)
- Fourth interosseus: ulnar side of the fourth phalanx proximalis (ring finger)

Actions:
- MCP joints of the second through fourth digits: flexion
- PIP and DIP joints of the second through fourth digits: extension and abduction of the fingers (abduction of the index and ring fingers from the middle finger)

Innervation: N. ulnaris (C8, T1)

③ **Mm. interossei palmares I–III**

Origin:
- First interosseus: ulnar side of the second metacarpal (index finger)
- Second interosseus: radial side of the fourth metacarpal (ring finger)
- Third interosseus: radial side of the fifth metacarpal (little finger)

Insertion: Dorsal digital expansion and base of the phalanx proximalis of the associated finger

Actions:
- MCP joints of the second, fourth, and fifth digits: flexion
- PIP and DIP joints of the second, fourth, and fifth digits: extension and adduction of the fingers (adduction of the second, fourth, and fifth digits toward the middle finger)

Innervation: N. ulnaris (C8, T1)

DIP, distal interphalangeal; MCP, metacarpointerphalangeal; PIP, proximal interphalangeal.

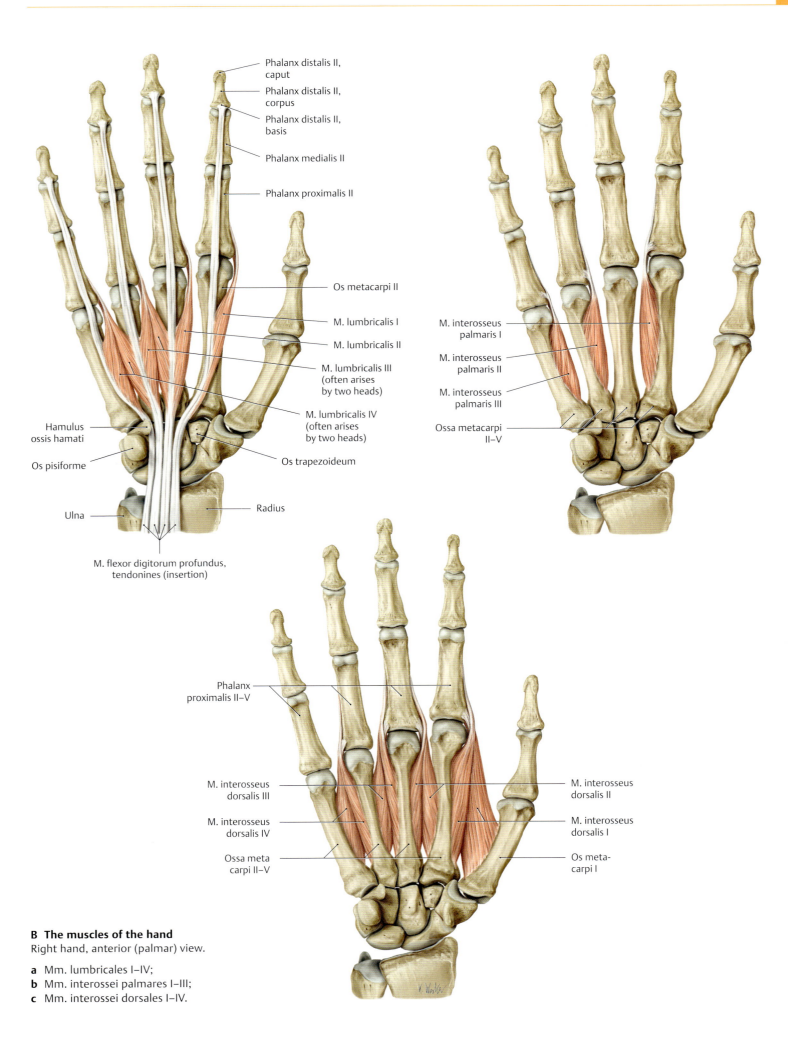

B The muscles of the hand
Right hand, anterior (palmar) view.

a Mm. lumbricales I–IV;
b Mm. interossei palmares I–III;
c Mm. interossei dorsales I–IV.

15.15 Overview of Muscle Functions: Shoulder Joint (Art. humeri)

A Movements of the shoulder joint (art. humeri [glenohumeralis])

Type of movement	Extent of movement	Muscle	Innervation	Responsible nerve segment
Flexion	90° (>90° = elevation)	• M. deltoideus pars clavicularis	• N. axillaris	• C5, C6
		• M. biceps brachii	• N. musculocutaneus	• C5–C7
		• M. pectoralis major (pars clavicularis and pars sternocostalis)	• Nn. pectorales medialis and lateralis	• C5–T1
		• M. coracobrachialis	• N. musculocutaneus	• C5–C7
Extension	40°	• M. latissimus dorsi	• N. thoracodorsalis	• C6–C8
		• M. teres major	• N. subscapularis	• C5, C6
		• M. triceps brachii, (caput longum)	• N. radialis	• C6–C8
		• M. deltoideus (pars spinalis)	• N. axillaris	• C5, C6
Abduction	90° (>90° = elevation)	• M. deltoideus (pars acromialis), above 60°, the entire muscle	• N. axillaris	• C5, C6
		• M. supraspinatus	• N. suprascapularis	• C5, C6
		• M. biceps brachii (caput longum)	• N. musculocutaneus	• C5–C7
Adduction	20–40°	• M. pectoralis major	• Nn. pectorales medialis and lateralis	• C5–T1
		• M. latissimus dorsi	• N. thoracodorsalis	• C6–C8
		• M. triceps brachii (caput longum)	• N. radialis	• C6–C8
		• M. teres major	• N. subscapularis inferior	• C5, C6
		• M. deltoideus (pars clavicularis and pars spinalis)	• N. axillaris	• C5, C6
		• M. coracobrachialis	• N. musculocutaneus	• C5–C7
Internal rotation	50–95°	• M. subscapularis	• Nn. subscapularis inferior and superior	• C5, C6
		• M. pectoralis major	• Nn. pectorales medialis and lateralis	• C5–T1
		• M. biceps brachii, (caput longum)	• N. musculocutaneus	• C5–C7
		• M. deltoideus (pars clavicularis)	• N. axillaris	• C5, C6
		• M. teres major	• N. subscapularis inferior	• C5, C6
		• M. latissimus dorsi	• N. thoracodorsalis	• C6–C8
External rotation	60–90° (depending on extent of flexion)	• M. infraspinatus	• N. suprascapularis	• C4–C6
		• M. teres minor	• N. axillaris	• C5, C6
		• M. deltoideus (pars spinalis)	• N. axillaris	• C5, C6

* The starting point of the functional approach is movement within the joint. The muscles, which are involved in all movements, are listed in the order of their strength. Remember, however, that although individual muscles may be involved in particular movements, their functional significance may be negligible. In order to describe the many types of movement in a joint, each joint (respective of its degree of freedom) is assigned clearly defined basic movements along defined axes. Only the combination of these basic movements allows for a range of movements depending on the individual joint.

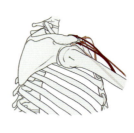
a Flexion

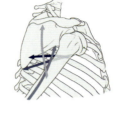
b Extension

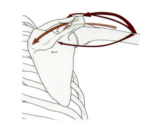
c Abduction

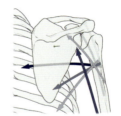
d Adduction

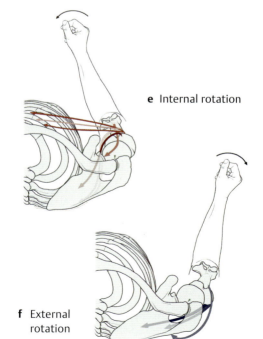
e Internal rotation

f External rotation

B Movements of the shoulder joint

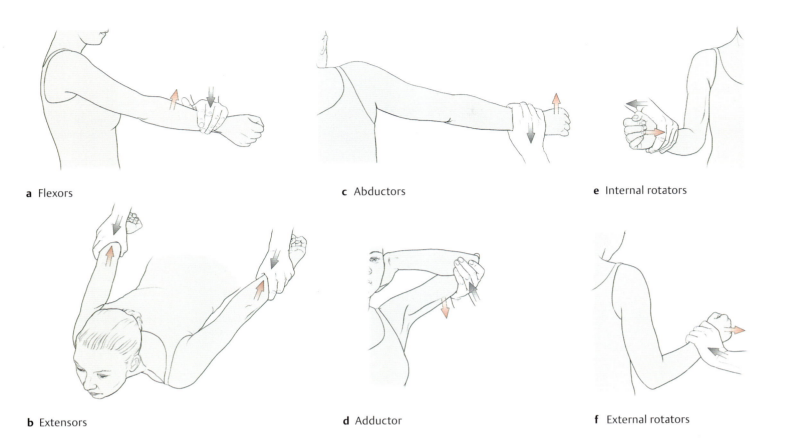

a Flexors
b Extensors
c Abductors
d Adductor
e Internal rotators
f External rotators

C Functional muscle testing in the shoulder joint
In order to test muscle strength, the patient is asked to actively move the muscle group against some resistance from the examiner. The force generated by the muscle action is measured using a scale from 0 to 5.

D Clinical symptoms of shoulder joint muscle shortening or weakness

Muscles	Symptoms of muscle shortening	Symptoms of muscle weakness
Flexors	Shoulder is commonly protracted.	In order to compensate for muscle weakness, the acromion is lifted, and the upper body, due to increased lordosis lumbalis, moves posteriorly. In case of isolated weakness of the deltoid, the shoulder contours are flattened with the acromion protuding.
Extensors	Significantly impaired movements in elevation and external rotation.	Weakness in the extensors becomes apparent when extreme loads are placed on the shoulder, e.g., propping oneself up using both arms.
Abductors	Impaired movement as a result of shortened muscles (primarily m. supraspinatus and m. biceps brachii), especially in extension, adduction, and external rotation of the shoulder joint.	The arm cannot maintain the arm in abduction against gravity. Flattening of the contours of the shoulder with typical indentation below the acromion, often subluxation of the shoulder joint with the arm hanging down.
Adductors	Protracted shoulder and increased kyphosis thoracica, along with compensatory cervical lordosis (lordosis cervicis). Significantly impaired arm movements in flexion and abduction above the head.	Dysfunction in sequence of abdominal muscles of the opposite side, meaning mainly hitting and chopping motions with the affected limb are impeded. Difficulty holding large or heavy objects at waist level.
Internal rotators	Significantly impaired movement when lifting the arm above the head with simultaneous external rotation.	An isolated weakness of the internal rotators is rare and is compensated for by the flexors (most of the daily chores are performed in flexion and internal rotation).
External rotators	Shortening of the external rotators is rare (mostly following long periods of immobilization) and leads to impaired internal rotation.	Insufficient inferior glide of the head of the humerus during elevation, resulting in compensatory activation of abductors (e.g., m. supraspinatus), pulling up of the shoulder, and increased lateral trunk flexion.

15.16 Overview of Muscle Functions: Elbow Joint

A Movements of the elbow joint (art. cubiti)*

Type of movement	Extent of movement	Muscle	Innervation	Responsible nerve segment
Flexion	130–150°	• M. biceps brachii • M. brachialis • M. brachioradialis • M. pronator teres	• N. musculocutaneus • N. musculocutaneus • N. radialis • N. medianus	• C5–7 • C5–7 • C5–7 • C6
Extension	10°	• M. triceps brachii • M. anconeus	• N. radialis • N. Radialis	• C6–C8 • C6–C8
Supination*	90°	• M. biceps brachii • M. supinator • M. brachioradialis (pronated)	• N. musculocutaneus • N. radialis • N. radialis	• C5–7 • C5–6 • C5–7
Pronation*	90°	• M. pronator quadratus • M. pronator teres • M. brachioradialis (supinated)	• N. medianus • N. medianus • N. radialis	• C8–Th1 • C6 • C5–7

*In addition to the elbow joint (art. cubiti), the distal radioulnar joint (art. radioulnaris) is involved in turning the forearm.

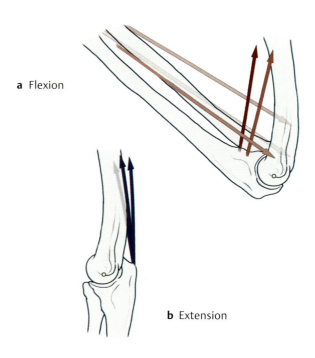

a Flexion

b Extension

c Supination

d Pronation

B Movements of the elbow joint (art. cubiti)

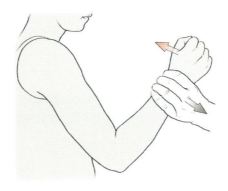
a Flexors

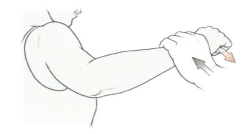
b Extensors

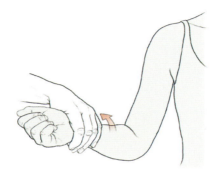
c Supinators
(elbow joint [art. cubiti] is flexed)

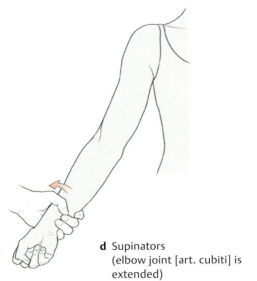
d Supinators
(elbow joint [art. cubiti] is extended)

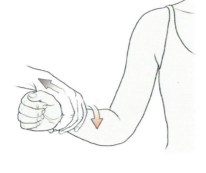
e Pronators

C Functional testing of muscles of the elbow joint (art. cubiti)

D Clinical symptoms of elbow joint muscle shortening and weakness

Muscles	Symptoms of muscle shortening	Symptoms of muscle weakness
Flexors	Flexion contracture of the elbow with involvement of the m. biceps brachii, resulting in significant functional impairment, requiring a shift from supination to pronation.	Muscle weakness, e.g., of the m. biceps brachii, leads to preponderance of pronation; thus, eating, for instance (moving a spoon to the mouth), is significantly impaired.
Extensors	Extension contracture of the elbow resulting in significant functional impairment in routine daily movements.	It is not possible to throw an object. The affected patient is not able to use a walking device since he or she cannot apply any weight on his or her hands due to inability to extend the elbow.
Supinators	Impaired movement when forearm is pronated. Compesated for by increased internal rotation and abduction of the shoulder joint.	Weakness in supinators becomes apparent in numerous routine daily movements: e.g., unlocking a door, turning off a faucet, or tightening a screw.
Pronators	Impaired movement when forearm is supinated. Compensated for by increased external rotation and adduction of the shoulder joint.	Routine daily movements are hampered (see above), such as unlocking a door or turning off a faucet.

15.17 Overview of Muscle Functions: Wrist

A Movement of the proximal and distal wrist (artt. radiocarpalis and mediocarpalis)

Type of movement	Extent of movement	Muscle	Innervation	Responsible nerve segment
Palmar flexion	60–80°	• M. flexor digitorum superficialis • M. flexor digitorum profundus • M. flexor carpi ulnaris • M. flexor pollicis longus • M. flexor carpi radialis • M. palmaris longus	• N. medianus • N. medianus • N. ulnaris • N. ulnaris • N. medianus • N. medianus • N. medianus	• C8, T1 • C8, T1 • C8, T1 • C8, T1 • C8, T1 • C6, C7 • C7, C8
Dorsal extension	40–60°	• M. extensor digitorum • Mm. extensor carpi radialis brevis et longus • M. extensor carpi ulnaris • M. extensor indicis • M. extensor pollicis longus • M. extensor digiti minimi	• N. radialis • N. radialis • N. radialis • N. radialis • N. radialis • N. radialis	• C7, C8 • C6–C8 • C7, C8 • C7, C8 • C7, C8 • C7, C8
Abduction (radial deviation)	20°	• Mm. extensor carpi radialis brevis et longus • M. abductor pollicis longus • M. extensor pollicis brevis • M. extensor pollicis longus • M. flexor pollicis longus • M. flexor carpi radialis	• N. radialis • N. radialis • N. radialis • N. radialis • N. medianus • N. medianus	• C6–C8 • C7, C8 • C7, C8 • C7, C8 • C8, T1 • C6, C7
Abduction (ulnar deviation)	30–40°	• M. extensor carpi ulnaris • M. flexor carpi ulnaris • M. extensor digiti minimi	• N. radialis • N. ulnaris • N. radialis	• C7, C8 • C8, T1 • C7, C8

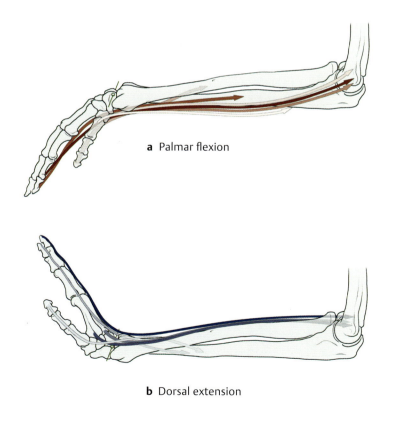

a Palmar flexion

b Dorsal extension

B Movements of the proximal and distal wrist

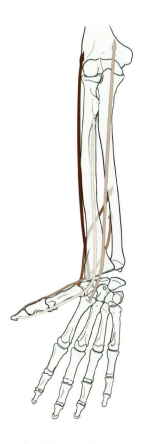

c Radial abduction

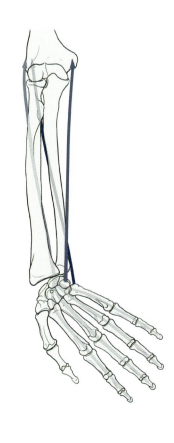

d Ulnar abduction

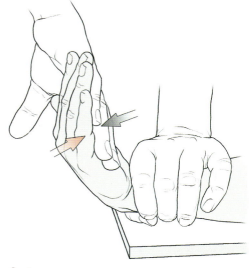

a Palmar flexion

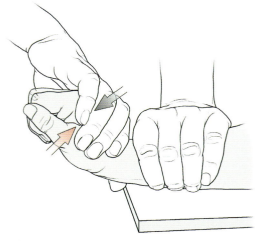

b Dorsal extension

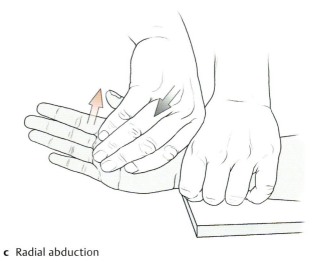

c Radial abduction

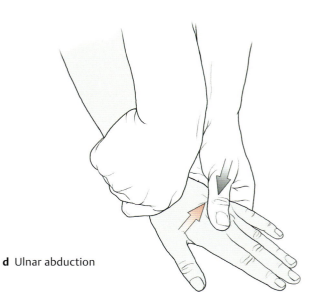

d Ulnar abduction

C Functional muscle testing of the proximal and distal wrist (artt. radiocarpalis and mediocarpalis)

D Clinical symptoms of muscle shortening and weakness of the proximal and distal wrist (artt. radiocarpalis and mediocarpalis)

Muscles	Symptoms of muscle shortening	Symptoms of muscle weakness
Palmar flexors	Impaired movement in dorsal extension of the wrist and finger joints, combined with constant overloading to palmar flexors, which leads to epicondylitis medialis (golfer's elbow).	When lifting heavy objects, the affected patient is not able to sufficiently stabilize the wrist with supinated forearm; thus, the wrist tilts in a dorsal direction. In the long run, this leads to injuries resulting from overload of the tendons of origin of the finger and hand flexors.
Dorsal extensors	Impaired movement in palmar flexion of wrist and finger joints, combined/accompanied by constant overloading to dorsal extensors, which leads to epicondylitis lateralis (tennis elbow).	When lifting heavy objects, the affected patient is not able to sufficiently stabilize the wrist with pronated forearm; thus, the wrist tilts in a palmar direction. In the long run, this leads to injuries resulting from overload of the tendons of origin of the finger and hand extensors.
Radial abductors	Weakening of ulnar abduction.	Always combined with weakened dorsal extension and palmar flexion.
Ulnar abductors	Weakening of radial abduction.	Loss of strength functionally less visible.

16.1 The Posterior Muscles of the Shoulder Girdle and Shoulder Joint

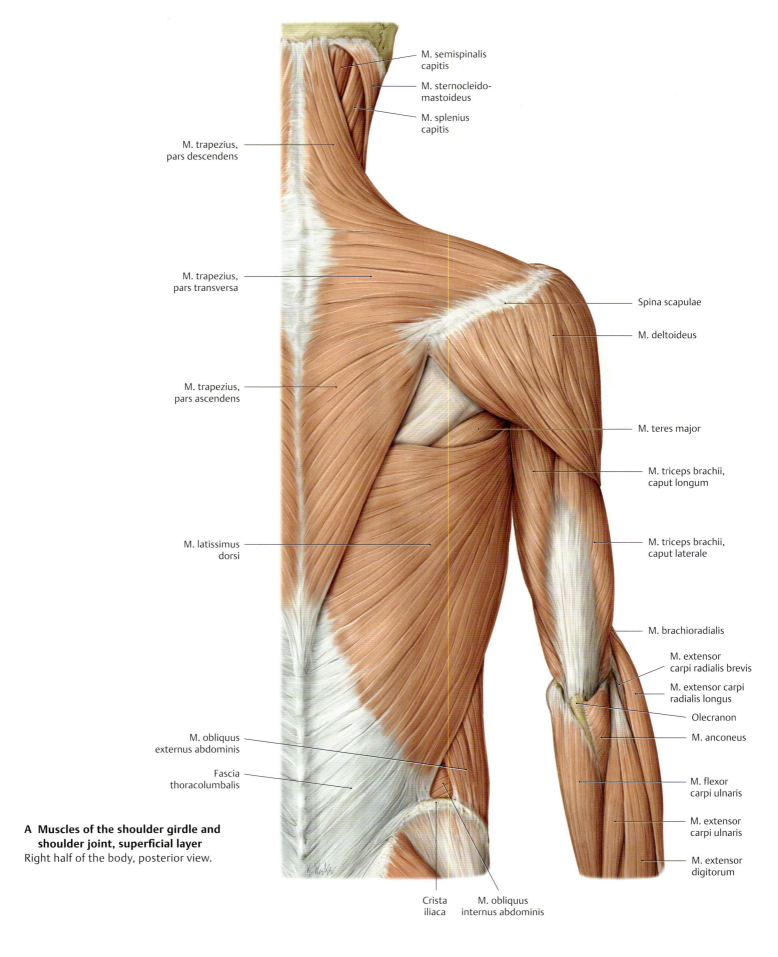

A Muscles of the shoulder girdle and shoulder joint, superficial layer
Right half of the body, posterior view.

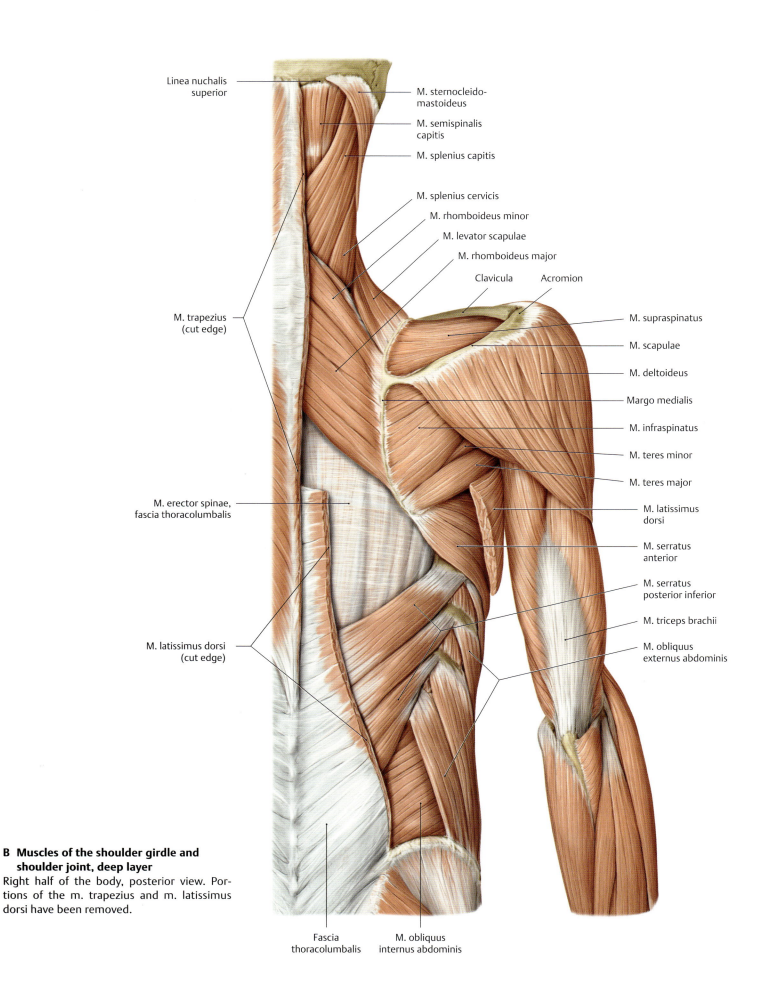

B Muscles of the shoulder girdle and shoulder joint, deep layer
Right half of the body, posterior view. Portions of the m. trapezius and m. latissimus dorsi have been removed.

16.2 The Posterior Muscles of the Shoulder Joint and Arm

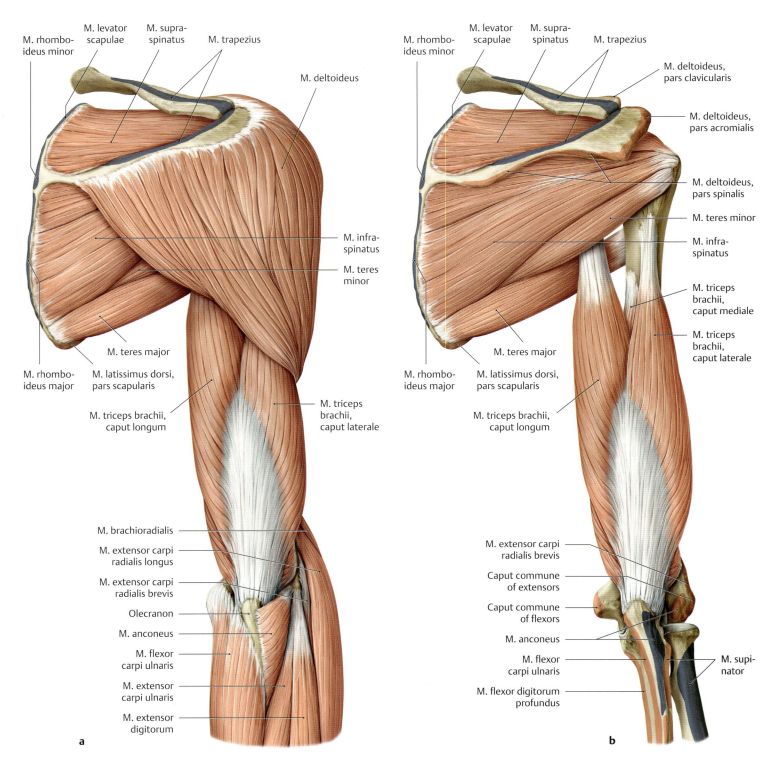

A Muscles of the shoulder and arm
Right side, posterior view. The origins and insertions of the muscles are indicated by color shading (red = origin, blue = insertion).

a After removal of the m. trapezius.
b After removal of the m. deltoideus and forearm muscles.

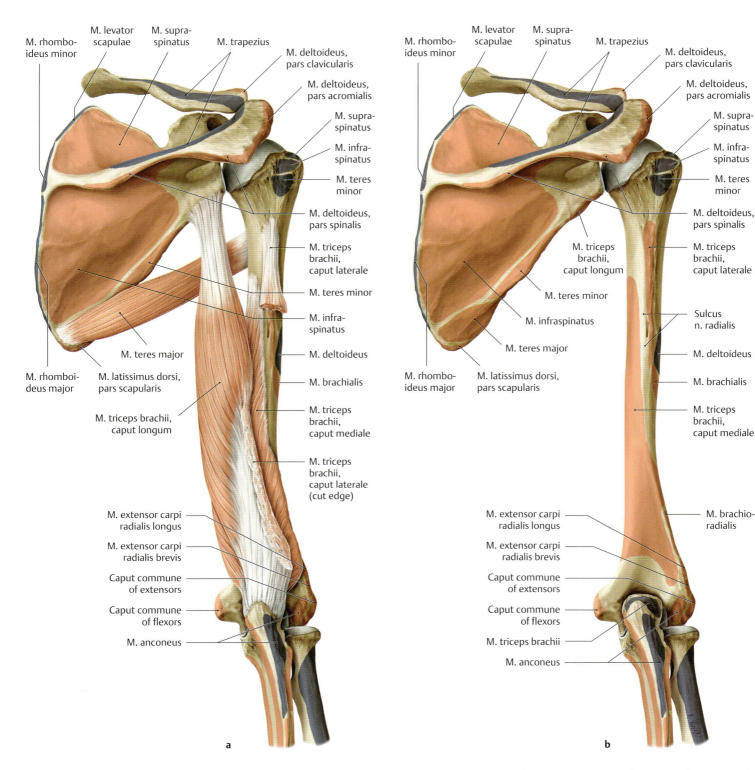

B Muscles of the shoulder and arm
Right side, posterior view. The origins and insertions of the muscles are indicated by color shading (red = origin, blue = insertion).

a The mm. supraspinatus, infraspinatus, and teres minor have been removed. The caput laterale of m. triceps brachii has been partially removed.

b All the muscles have been removed.

16.3 The Anterior Muscles of the Shoulder Girdle and Shoulder Joint

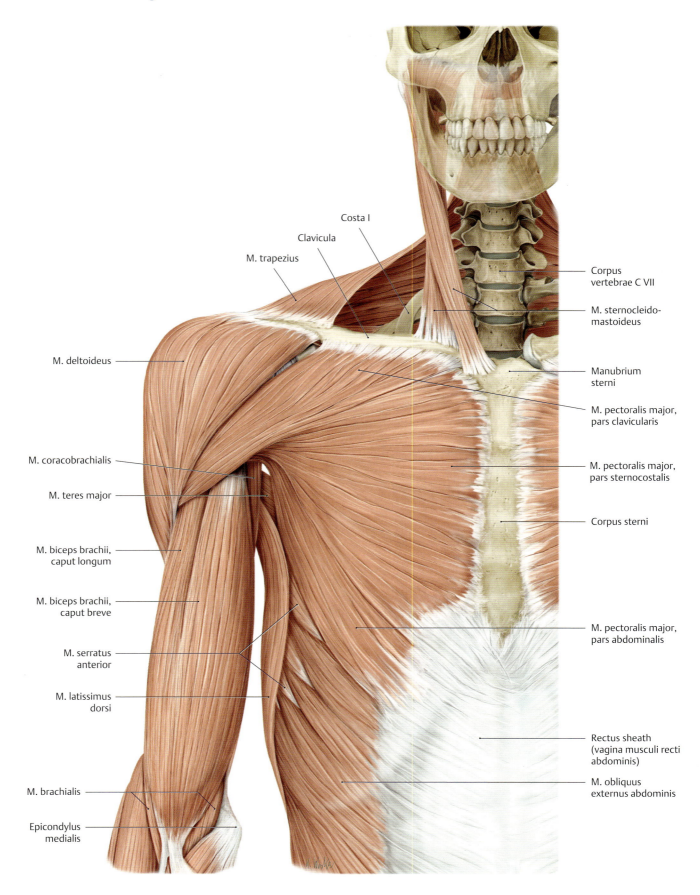

A Muscles of the shoulder and arm
Right side, anterior view.

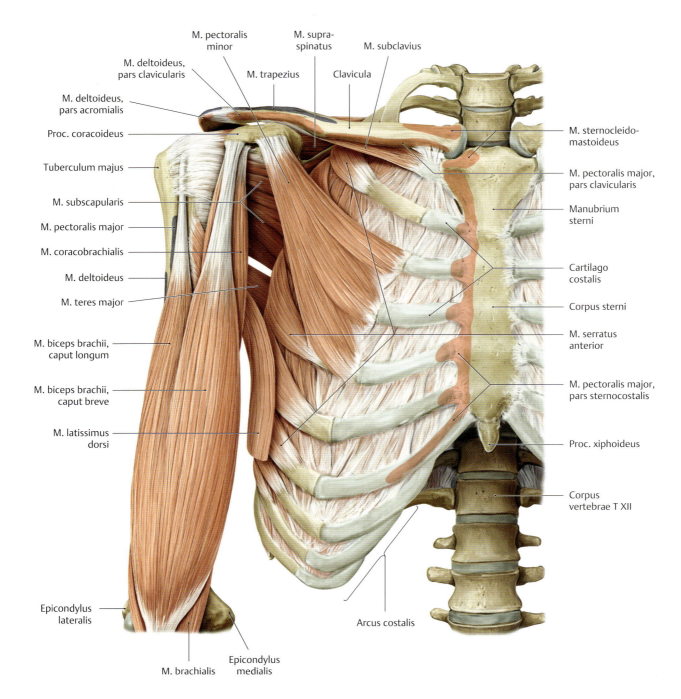

B Muscles of the shoulder and arm
Right side, anterior view. The origins and insertions of the muscles are indicated by color shading (red = origin, blue = insertion). Mm. sternocleidomastoideus, trapezius, pectoralis major, deltoideus, and obliquus externus abdominis have been completely removed. M. latissimus dorsi has been partially removed.

16.4 The Muscles of the Shoulder Joint and Anterior Arm

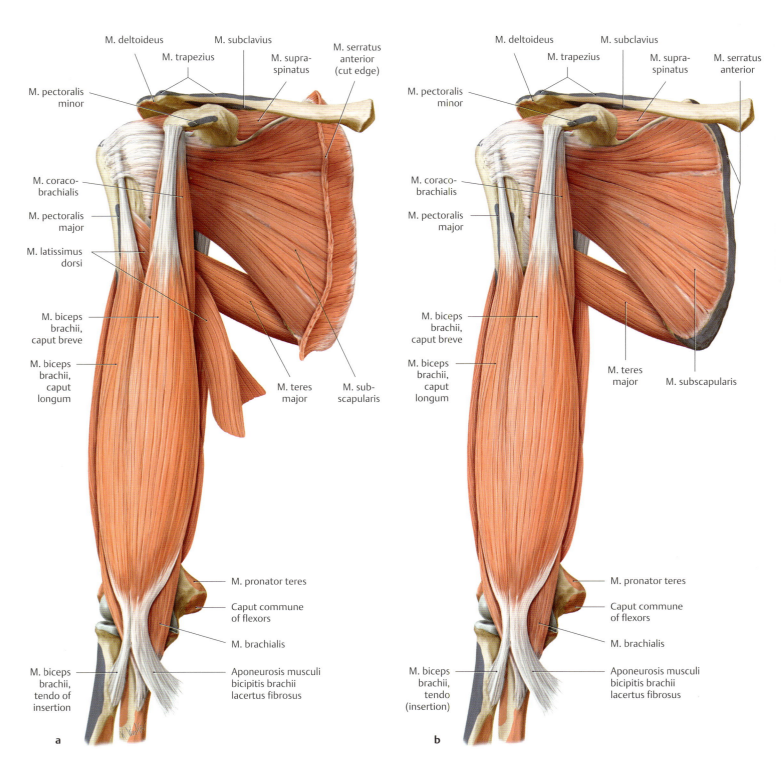

A Muscles of the shoulder and arm
Right side, anterior view. The origins and insertions of the muscles are indicated by color shading (red = origin, blue = insertion).

a M. latissimus dorsi and m. serratus anterior have been removed up to their insertions.
b Mm. latissimus dorsi and serratus anterior have been completely removed.

Upper Limb — 16. Musculature: Topographical Anatomy

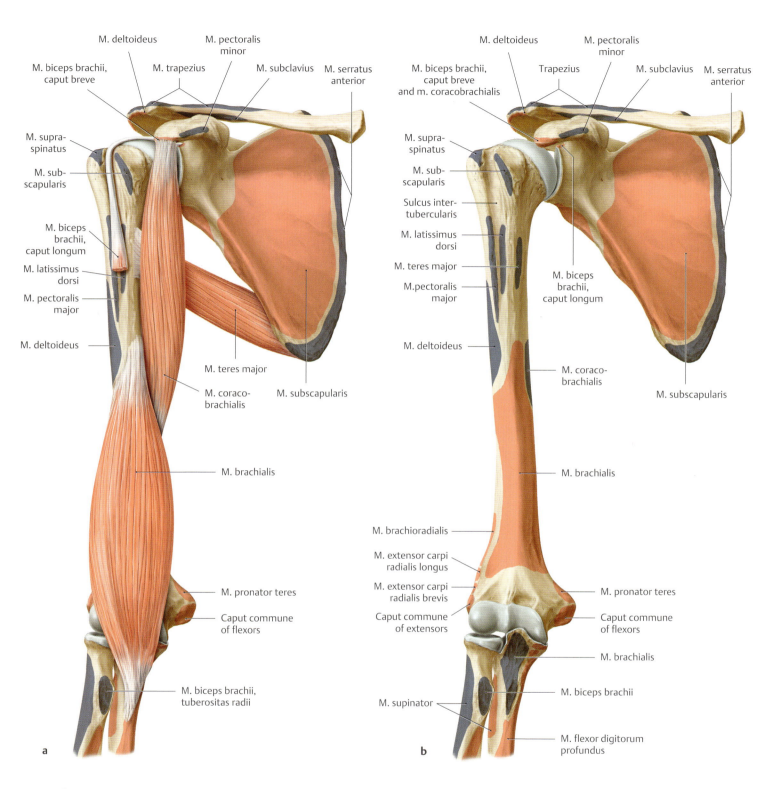

B Muscles of the shoulder and arm
Right side, anterior view. The origins and insertions of the muscles are indicated by color shading (red = origin, blue = insertion).

a The mm. subscapularis and supraspinatus have been removed. The m. biceps brachii has been removed to the tendon of origin of its caput longum (*note* its course through the sulcus intertubercularis).
b All the muscles have been removed.

337

16.5 The Muscles of the Anterior Forearm

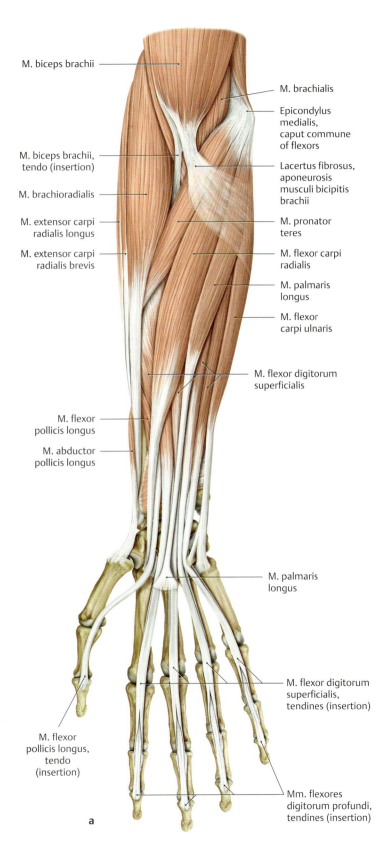
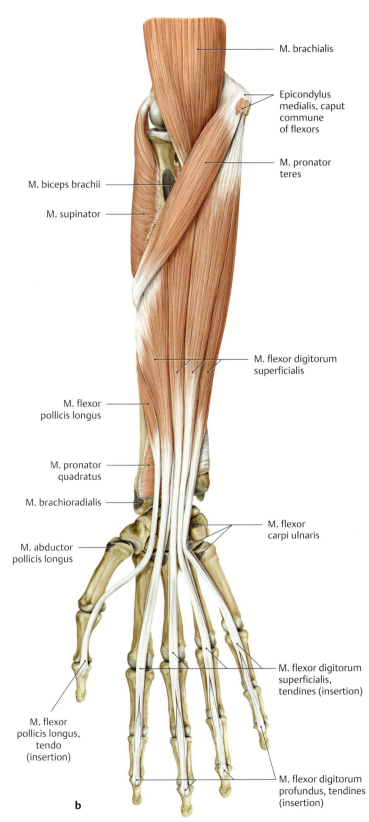

A Muscles of the forearm
Right forearm, anterior view. The origins and insertions of the muscles are indicated by color shading (red = origin, blue = insertion).

a The superficial flexors and the radialis group are shown.
b The radialis group (mm. brachioradialis, extensor carpi radialis longus, extensor carpi radialis brevis) has been completely removed, along with the mm. flexor carpi radialis, flexor carpi ulnaris, abductor pollicis longus, palmaris longus, and biceps brachii.

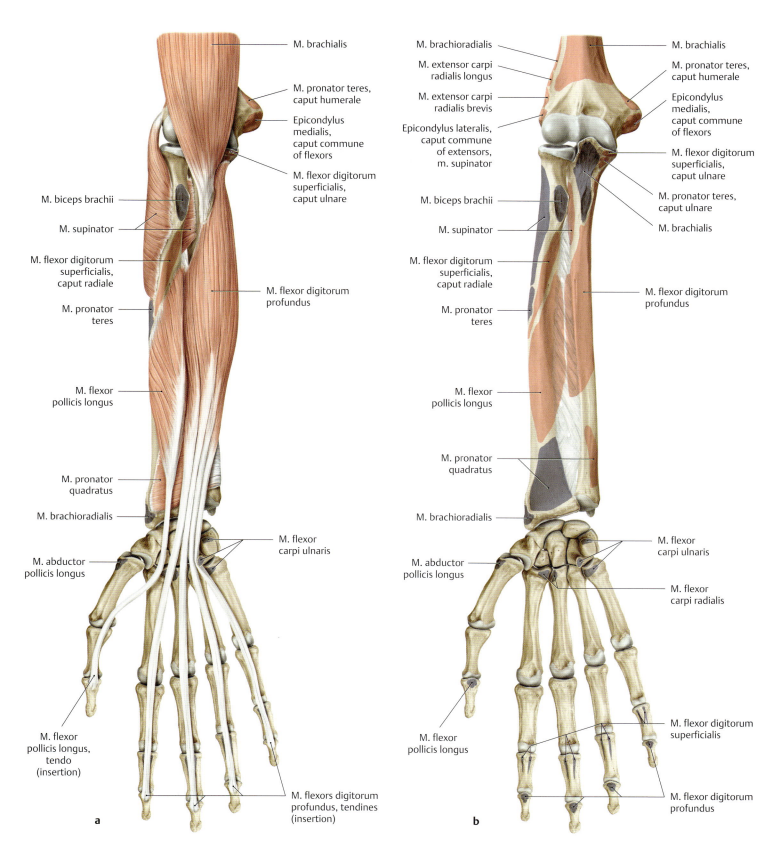

B Muscles of the forearm
Right forearm, anterior view. The origins and insertions of the muscles are indicated by color shading (red = origin, blue = insertion).

a Mm. pronator teres and flexor digitorum superficialis have been removed.
b All the muscles have been removed.

16.6 The Muscles of the Posterior Forearm

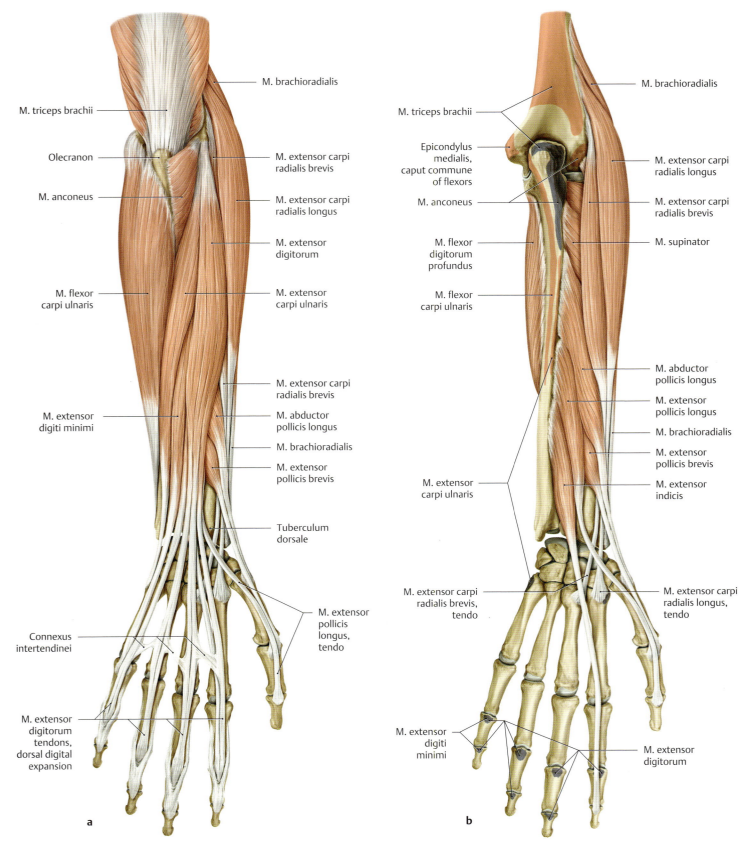

A Muscles of the forearm
Right forearm, posterior view. The origins and insertions of the muscles are indicated by color shading (red = origin, blue = insertion).

a The superficial extensors and the radialis group are shown.
b Mm. triceps brachii, anconeus, flexor carpi ulnaris, extensor carpi ulnaris, and extensor digitorum have been removed.

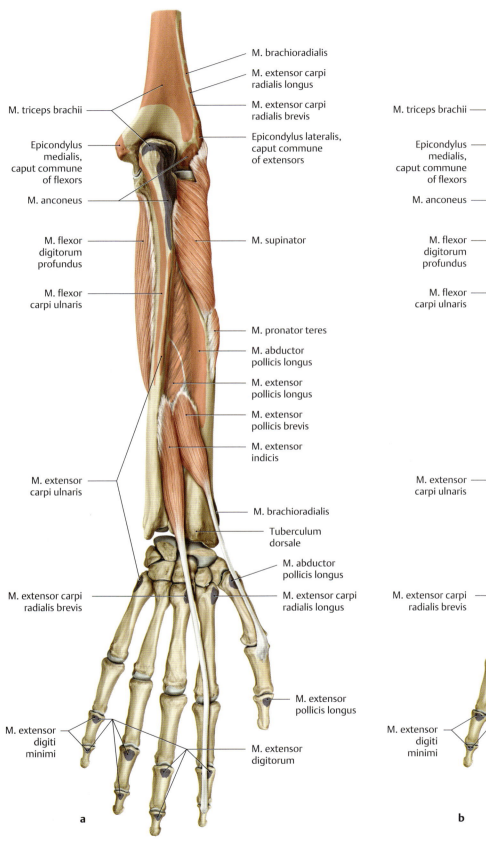
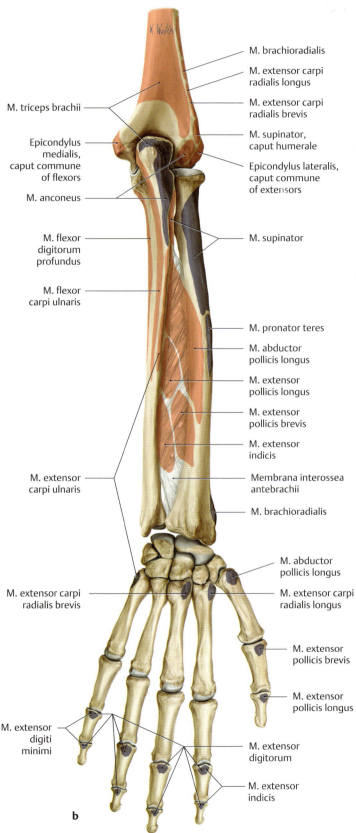

B Muscles of the forearm
Right forearm, posterior view. The origins and insertions of the muscles are indicated by color shading (red = origin, blue = insertion).
Note the membrana interossea antebrachii, which contributes to the origin of several muscles in the forearm.

a Mm. abductor pollicis longus, extensor pollicis longus, and the radialis group have been removed.
b All the muscles have been removed.

16.7 Cross-sectional Anatomy of the Arm and Forearm

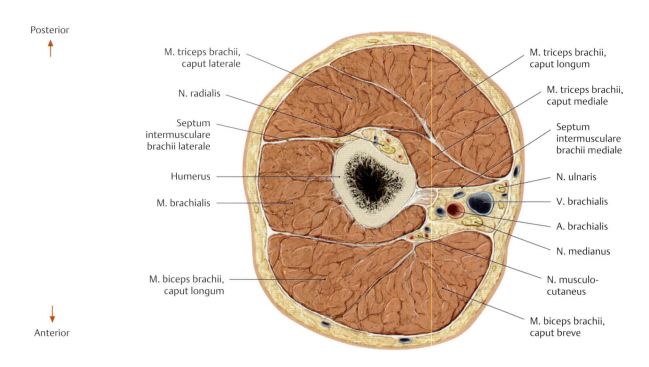

A Cross-section through the right arm
Proximal view. The location of the sectional plane is shown in **C**.

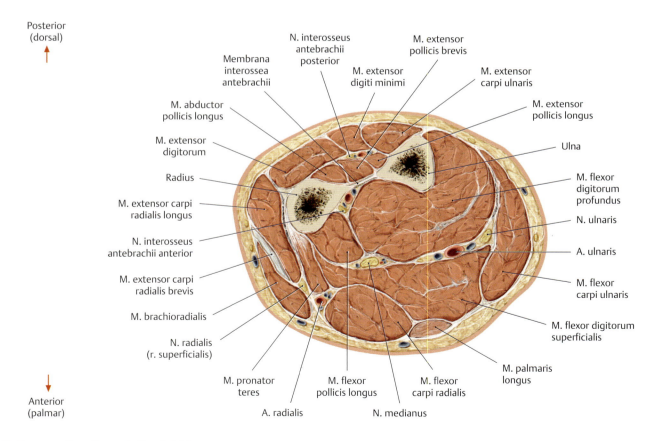

B Cross-section through the right forearm
Proximal view. The location of the plane of section is shown in **D**.

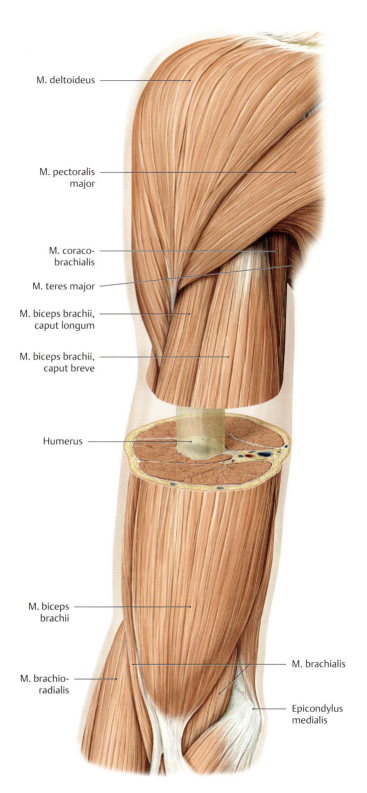

C "Windowed" dissection of the right arm
Anterior view.

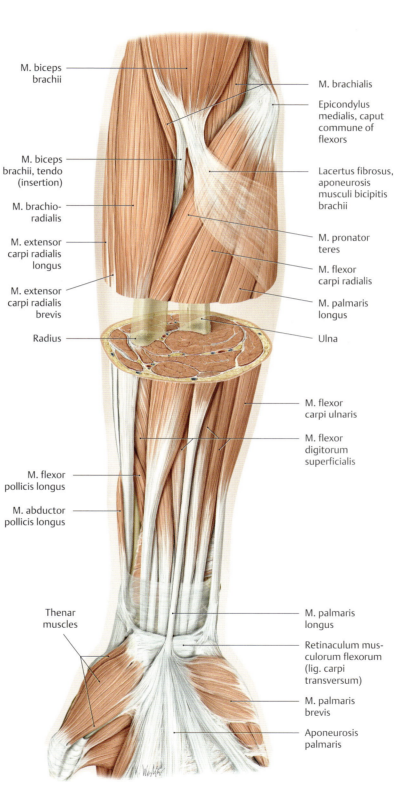

D "Windowed" dissection of the right forearm
Anterior view.

16.8 The Tendon Sheaths of the Hand

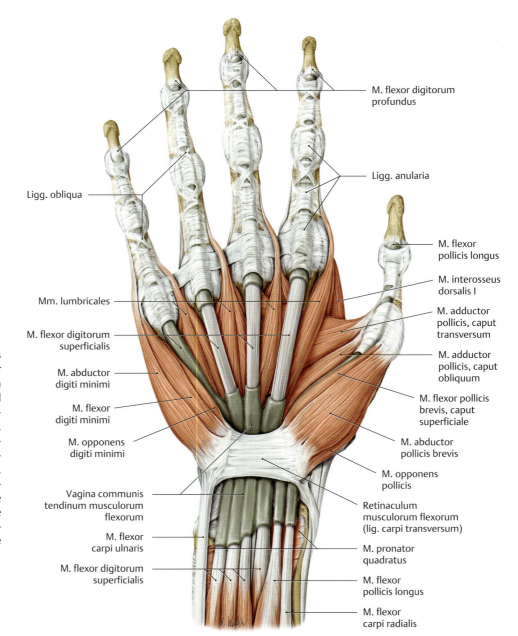

A Carpal and digital tendon sheaths on the palmar surface of the right hand
The aponeurosis palmaris (see p. 348) has been removed. The tendons of the m. flexor pollicis longus, and mm. flexores digitorum superficialis and profundus run from the distal forearm through a fibro-osseous canal (canalis carpi) to the palm, accompanied by the n. medianus and protected by the palmar carpal tendon sheaths (=vaginae tendinum carpales palmares) (see also p. 288 and p. 402). The carpal tendon sheath of the m. flexor pollicis longus is consistently continuous with the digital tendon sheath of the thumb, while the digital tendon sheaths of the remaining fingers show variable communication with the carpal tendon sheaths (see **B**).

B Communication between the carpal and digital tendon sheaths
Right hand, anterior view (after Schmidt and Lanz).

a In 71.4% of cases (Scheldrup 1951) the carpal tendon sheath of the little finger communicates directly with the digital tendon sheath, while the other tendon sheaths of the second through fourth digits extend only from the metacarpophalangeal joint to the art. interphalangea distalis.
b In 17.4% of cases the carpal tendon sheath does not communicate with the digital tendon sheath of the little finger.
c Besides being continuous with the tendon sheath of the little finger, the carpal tendon sheath may occasionally be continuous with the digital tendon sheath of the index finger (3.5%) or of the ring finger (3%; not shown).

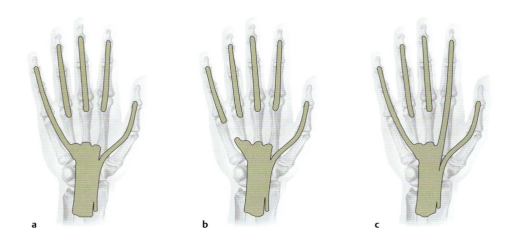

C Dorsal tendon compartments for the extensor tendons

First tendon compartment:	M. abductor pollicis longus M. extensor pollicis brevis
Second tendon compartment:	Mm. extensores carpi radialis longus and brevis
Third tendon compartment:	M. extensor pollicis longus
Fourth tendon compartment:	M. extensor digitorum M. extensor indicis
Fifth tendon compartment:	M. extensor digiti minimi
Sixth tendon compartment:	M. extensor carpi ulnaris

The location of the tendon compartments is shown in **D**.

D Retinaculum musculorum extensorum and dorsal carpal tendon sheaths (vaginae tendinum carpales dorsales)

Right hand, posterior view. The retinaculum mm. extensorum is part of the antebrachial (forearm) fascia. Its transverse fibers strengthen the stratum fibrosum of the tendon sheaths and fix them to the dorsum of the hand. Deep to the retinaculum mm. extensorum are *tendon sheath compartments*, which transmit the long extensor tendons singly or in groups. There are a total of six of these compartments, numbered 1 to 6 from the radial side to the ulnar side of the wrist (their contents are shown in **C**).

E Schematic cross-section through the forearm at the level of the distal radioulnar joint

Right forearm, proximal view. (Location of the sectional plane is shown in **D**.) Vertical connective tissue septa extend anteriorly from the deep surface of the retinaculum mm. extensorum to the bone or joint capsule (capsula articularis) and form six fibroosseous canals, the tendon sheath compartments of the extensor tendons (extensor tendon compartments).

Note also the tuberculum dorsale, which redirects the tendon of insertion of the m. extensor pollicis longus to the thumb (see also **D**).

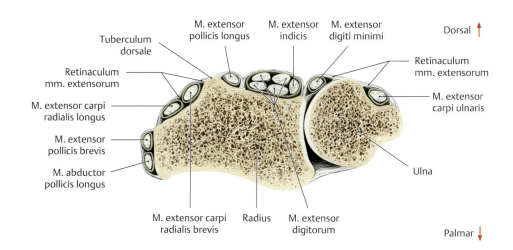

345

16.9 The Dorsal Digital Expansion

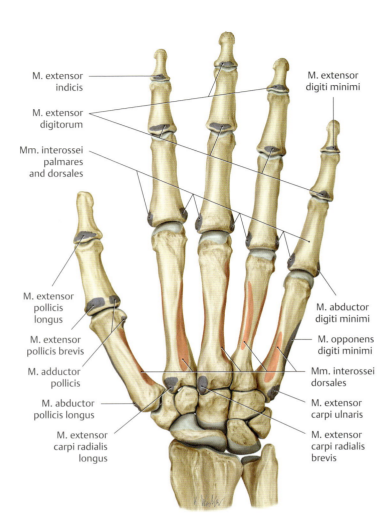

A Origins and insertions of the dorsal muscles of the right hand
The origins and insertions of the muscles are indicated by color shading (red = origin, blue = insertion).

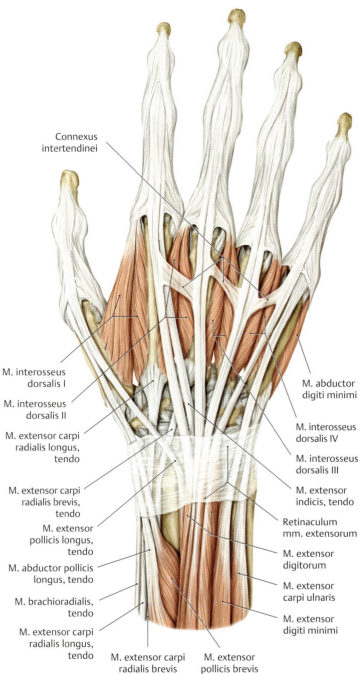

B Extensor tendons and connexus intertendinei on the dorsum of the right hand
The tendons of insertion of the m. extensor digitorum are interlinked by variable oblique bands called *connexus intertendinei*. The most proximal of the connexus intertendinei are those extending between the index and middle fingers. No such connection is present on the tendon of the m. extensor indicis. The m. extensor digitorum inserts by a variable number of tendons. Generally, all of the fingers have at least two extensor tendon elements. In addition, the *index finger* and *little finger* have their own extensor muscles (m. extensor indicis and m. extensor digiti minimi) whose tendons always run on the ulnar side of the tendons of the common m. extensor digitorum. Because the index finger and little finger have their own extensors, they can more easily be moved independently of the other fingers.

Upper Limb — 16. Musculature: Topographical Anatomy

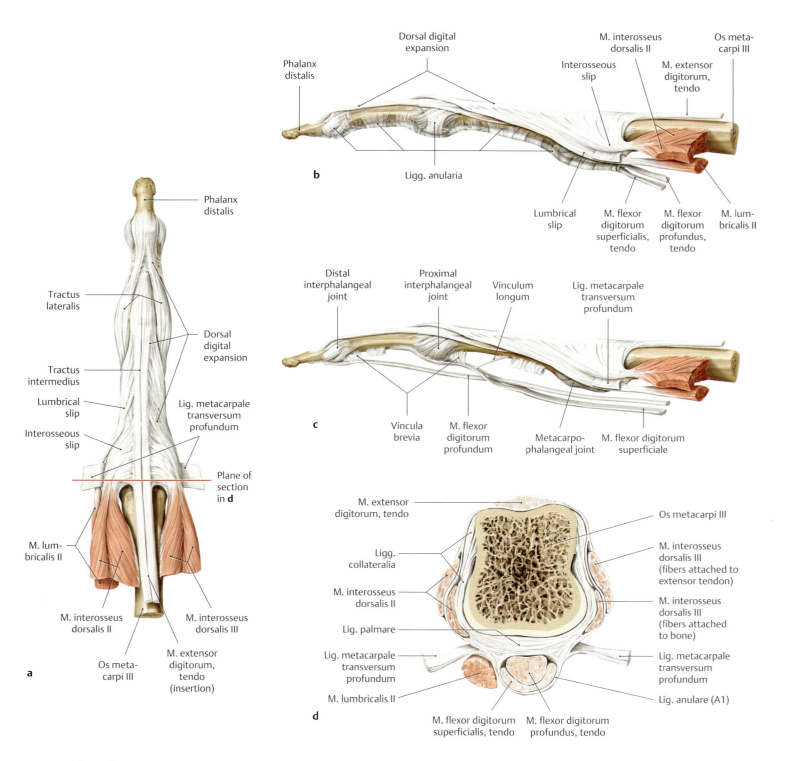

C Dorsal digital expansion
Dorsal digital expansion of the middle finger of the right hand (after Schmidt and Lanz).

a Posterior view.
b Radial view.
c Radial view, after opening the common tendon sheath of the mm. flexores digitorum superficialis and profundus.
d Cross-section at the level of the metacarpal head.

The dorsal digital expansion is more than an aponeurosis that incorporates slips from the mm. extensores digitorum, lumbricales, and interossei tendons. It is a complex system of interwoven fiber bands, joined by loose connective tissue to the periosteum of the phalanges. The dorsal digital expansion consists of a *tractus intermedius* and *tractus lateralis*, each of which has a *pars lateralis* and a *pars medialis*. The lateral part of the expansion receives slips from the tendons of the mm. lumbricales and mm. interossei (see **a**). This complex arrangement makes it possible for the long digital flexors and the short hand muscles to act on all three of the finger joints.

16.10 The Intrinsic Muscles of the Hand: Superficial Layer

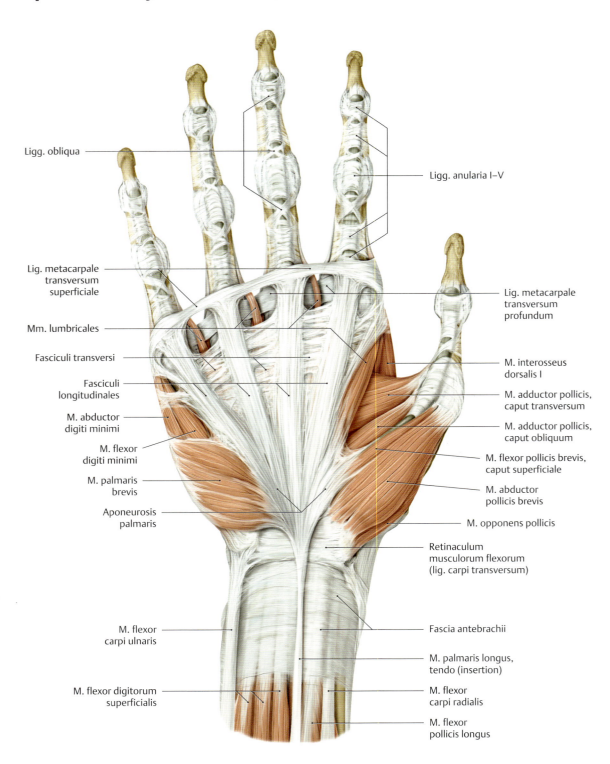

A The aponeurosis palmaris and Dupuytren's contracture
Right hand, anterior view. The muscular fascia of the palm is thickened by firm connective tissue to form the *aponeurosis palmaris*, which separates the palm from the subcutaneous fat to protect the soft tissues. It is composed mainly of longitudinal fiber bundles (*fasciculi longitudinales*), which give it a fan-shaped arrangement. The longitudinal fascicles are held together by transverse fiber bundles (*fasciculi transversi*) at the level of the metacarpal bones and by the lig. metacarpeum transversum superficiale at the level of the artt. metacarpophalangeae. Two muscles, the m. palmaris brevis and m. palmaris longus, keep the aponeurosis palmaris tense and prevent it from contracting, especially when the hand is clenched into a fist.

Gradual atrophy or contracture of the aponeurosis palmaris leads to progressive shortening of the palmar fascia that chiefly affects the little finger and ring finger (*Dupuytren's contracture*). Over a period of years, the contracture may become so severe that the fingers assume a fixed flexed position with the fingertips touching the palm; this seriously compromises the grasping ability of the hand. The causes of Dupuytren's contracture are poorly understood, but it is a relatively common condition that is most prevalent in men over 40 years of age and is associated with chronic liver disease (e.g., cirrhosis). Treatment generally consists of complete surgical removal of the aponeurosis palmaris.

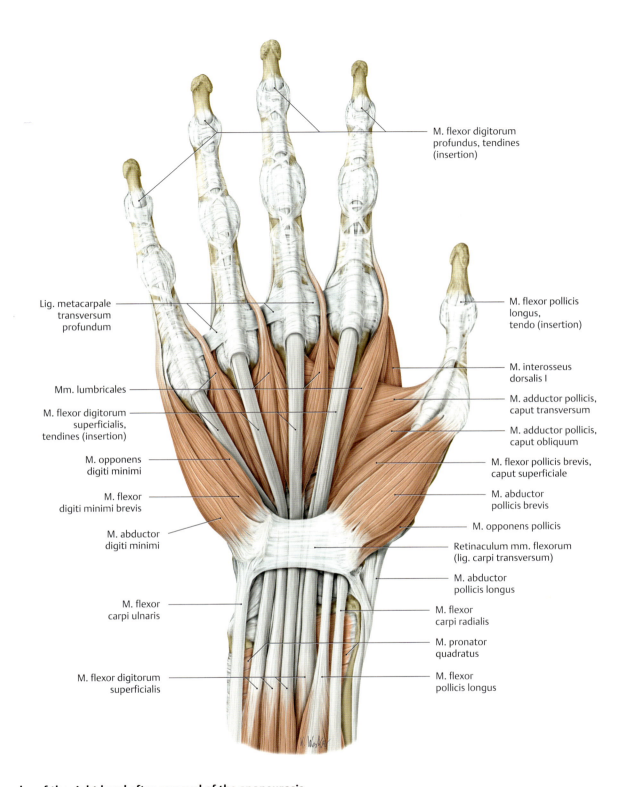

B The muscles of the right hand after removal of the aponeurosis palmaris, superficial layer

Anterior view. The aponeurosis palmaris, fascia antebrachii, and mm. palmaris brevis and longus have been removed, along with the palmar and carpal tendon sheaths.

16.11 The Intrinsic Muscles of the Hand: Middle Layer

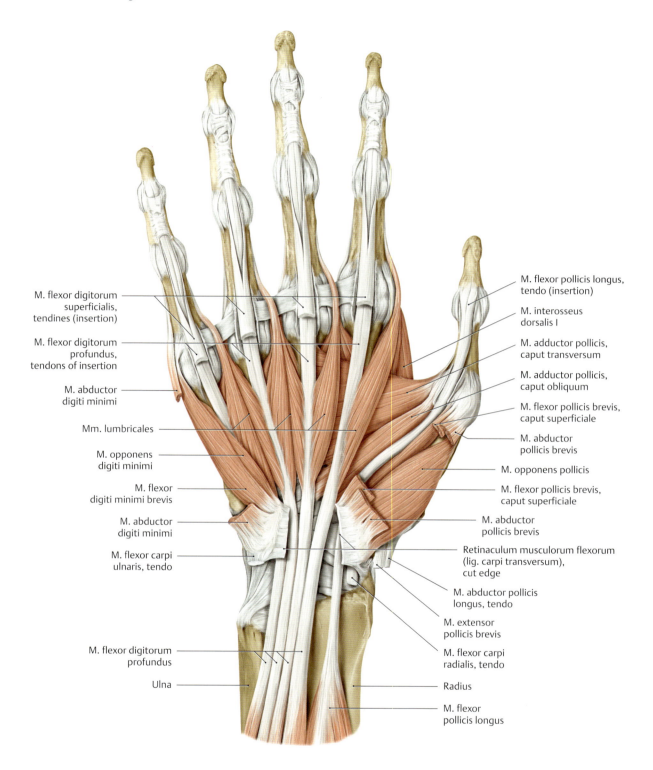

A The muscles of the right hand, middle layer
Anterior view. The m. flexor digitorum superficialis has been removed, and its four tendons of insertion have been divided at the level of the metacarpophalangeal joints (artt. metacarpophalangeae). The ligg. anularia I to III have been cut open to reveal the flexor tendons on the fingers. The retinaculum mm. flexorum (lig. carpi transversum) has been partially removed to open the carpal tunnel (canalis carpi). Of the thenar muscles, portions of the mm. abductor pollicis brevis and flexor pollicis brevis (caput superficiale) have been removed. Part of the m. abductor digiti minimi has been resected on the hypothenar side.

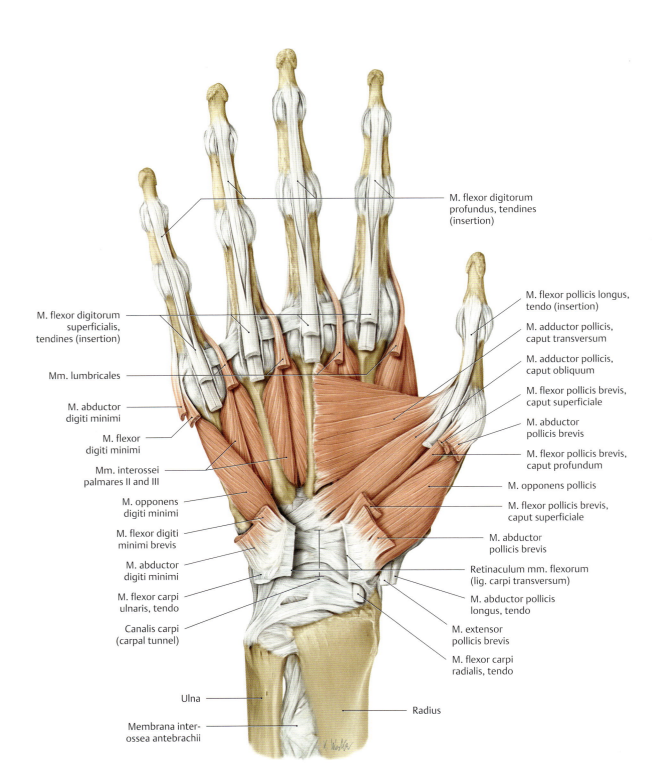

B The muscles of the right hand, middle layer
Anterior view. The m. flexor digitorum profundus has been removed, and its four tendons of insertion and the mm. lumbricales arising from them have been divided. The m. flexor pollicis longus and the m. flexor digiti minimi have also been removed.

16.12 The Intrinsic Muscles of the Hand: Deep Layer

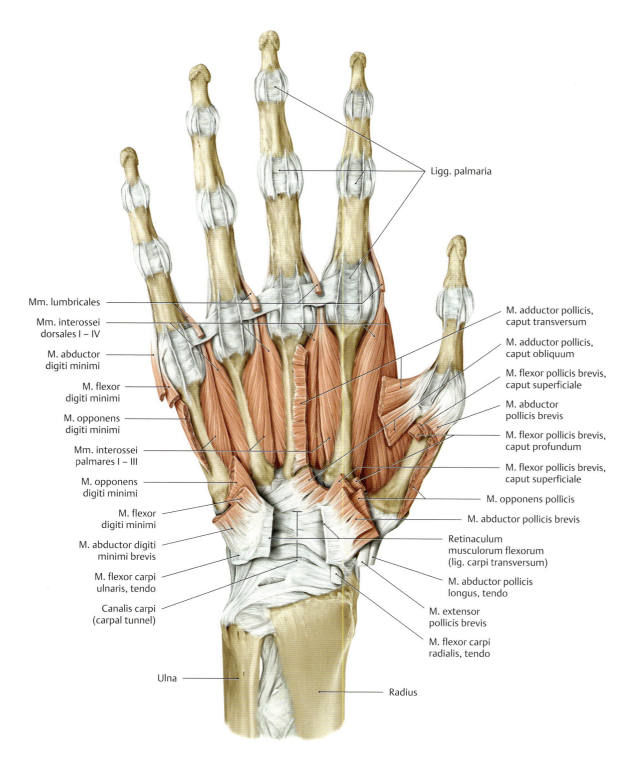

A The muscles of the right hand, deep layer
Anterior view. The tendons of insertion, tendon sheaths, and ligg. anularia of the long digital flexors have been completely removed.
Note the exposed ligg. palmaria, which combine with the tendon sheaths to form a trough that directs the long flexor tendons (see p. 291).

The m. interosseus dorsalis I and m. interosseus palmaris I have been almost completely exposed by removal of the m. adductor pollicis. Both the m. opponens pollicis and the m. opponens digiti minimi have been partially removed.

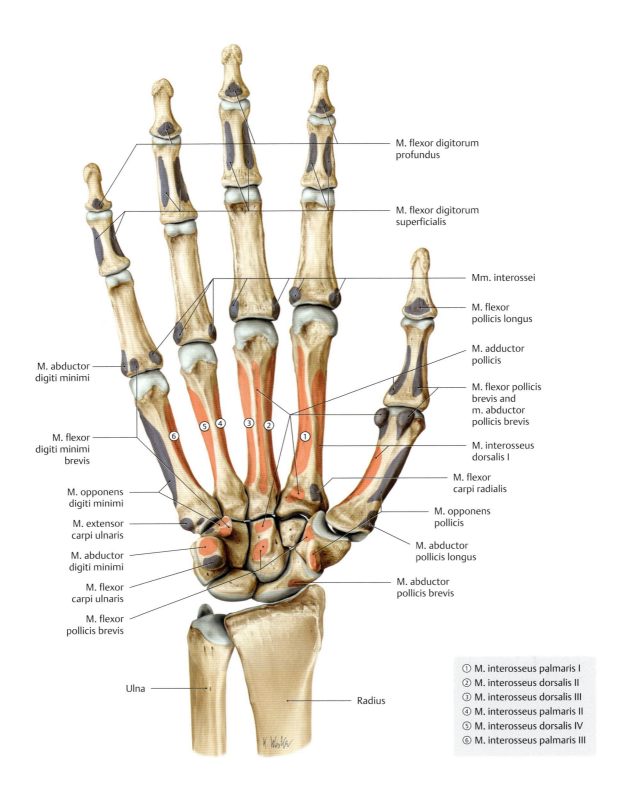

B Origins and insertions of the palmar muscles of the right hand
The origins and insertions of the muscles are indicated by color shading (red = origin, blue = insertion).

17.1 The Arteries

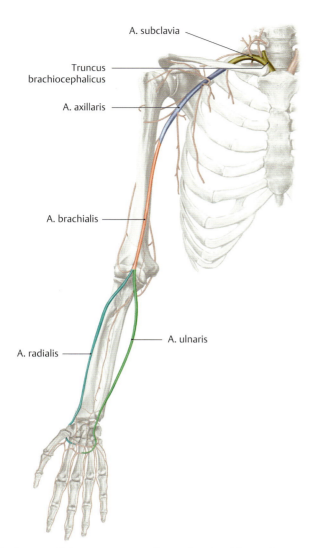

A Course of the arteries supplying the shoulder and arm
Right side, anterior view.
A. subclavia: The right a. subclavia arises from the truncus brachiocephalicus (as shown here), and the left arises directly from the aortic arch. The vessel runs over the first costa between the mm. scaleni anterior and medius (interscalene space, scalene interval) and continues as the a. axillaris (see below) on reaching the lateral border of the rib. Unlike the other arteries pictured here, the a. subclavia supplies blood not only to the upper limb (i.e., the shoulder girdle and arm) but also to

- a portion of the neck,
- the cerebral circulation, and
- the anterior chest wall.

A. axillaris: The continuation of the a. subclavia, the a. axillaris runs from the lateral border of the first rib to the inferior border of the m. teres major.
A. brachialis: The a. brachialis is the continuation of the a. axillaris. It ends at the elbow joint (art. cubiti) by dividing into the a. radialis and a. ulnaris.
A. radialis: The a. radialis runs distally on the radial side of the forearm from the division of the a. brachialis, passing between the mm. brachioradialis and flexor carpi radialis on its way to the wrist. It terminates in the arcus palmaris profundus.
A. ulnaris: This second division of the a. brachialis runs below the m. pronator teres on the ulnar side of the forearm, under cover of the m. flexor carpi ulnaris, to the arcus palmaris superficialis.

B Overview of the arteries of the shoulder and arm
The arteries of the shoulder and arm vary considerably in their origins and branching patterns (the principal variants are reviewed in Chapter 5, Neurovascular Systems: Topographical Anatomy). The branches are listed below in the order in which they arise from the parent vessels.

Branches of the a. subclavia
- A. vertebralis
- A. thoracica interna (a. mammaria interna)
- Truncus thyrocervicalis
  - A. thyroidea inferior
  - A. cervicalis ascendens
  - A. suprascapularis
  - A. transversa cervicis
- Truncus costocervicalis
  - A. cervicalis profunda
  - A. intercostalis suprema

Branches of the a. axillaris
- A. thoracica superior
- A. thoracoacromialis
  - R. acromialis
  - R. clavicularis
  - R. deltoideus
  - R. pectoralis
- A. thoracica lateralis
- A. subscapularis
  - A. thoracodorsalis
  - A. circumflexa scapulae
- A. circumflexa humeri anterior
- A. circumflexa humeri posterior

Branches of the a. brachialis
- A. profunda brachii
  - A. collateralis media
  - A. collateralis radialis
- A. collateralis ulnaris superior (rete articulare cubiti)
- A. collateralis ulnaris inferior (rete articulare cubiti)

Branches of the a. radialis
- A. recurrens radialis (rete articulare cubiti)
- R. carpalis palmaris (rete carpale palmare)
- R. palmaris superficialis (arcus palmaris superficialis)
- R. carpalis dorsalis (rete carpale dorsale)
  - Aa. metacarpales dorsales
  - Aa. digitales dorsales
- A. princeps pollicis
- A. radialis indicis
- Arcus palmaris profundus
  - Aa. metacarpales palmares
  - Rr. perforantes

Branches of the a. ulnaris
- A. recurrens ulnaris (rete articulare cubiti)
- A. interossea communis
  - A. interossea posterior
  - A. interossea recurrens
  - A. interossea anterior
- R. carpalis palmaris (rete carpale palmare)
- R. carpalis dorsalis (rete carpale dorsale)
- R. palmaris profundus (arcus palmaris profundus)
- Arcus palmaris superficialis
  - Aa. digitales palmares communes
  - Aa. digitales palmares propriae

Upper Limb — 17. Neurovascular Systems: Forms and Relations

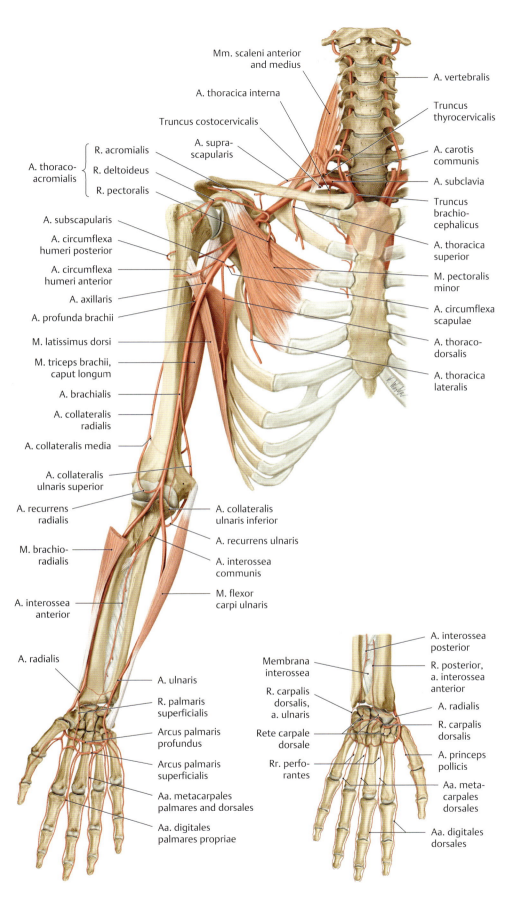

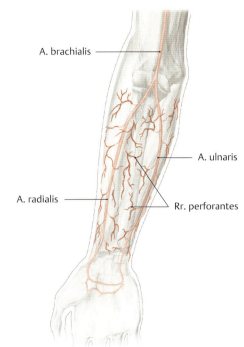

a

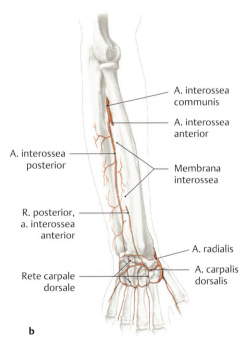

b

E Main arterial branches in the right forearm (rami perforantes)
a From the a. radialis and a. ulnaris (forearm supinated, anterior view).
b From the a. interossea posterior (forearm pronated, posterior view).

Fasciocutaneous flaps with an excellent vascular pedicle can be harvested from the thin skin of the forearm. Composed of skin, subcutaneous tissue, and fascia, these flaps are supplied by branches of the major arteries and their accompanying veins. The skin flaps carry this vascular supply with them when they are transferred to the recipient site.

C Arteries of the right upper limb
Anterior view with the forearm supinated. For clarity, some of the arteries listed in **B** are not illustrated.

D Arteries of the right hand
Posterior view.

355

17.2 The Veins

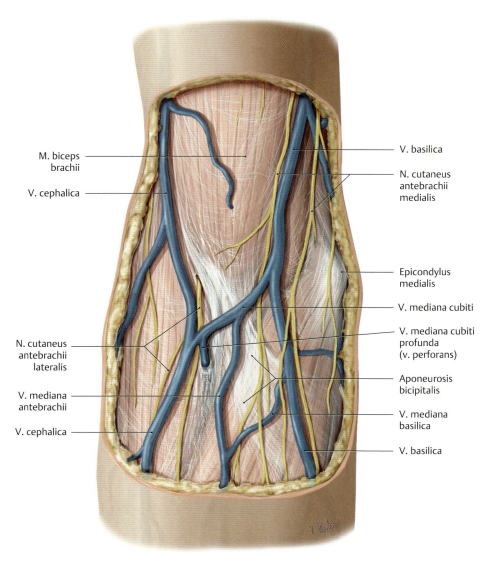

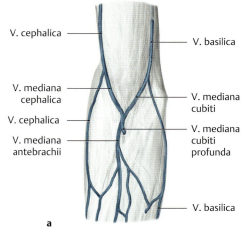

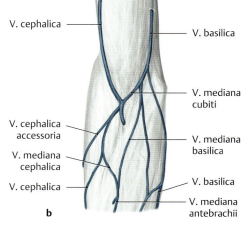

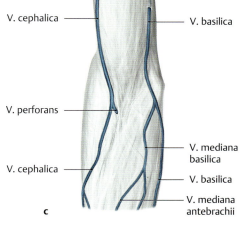

A Cutaneous veins and nerves of the right elbow

Anterior view. The subcutaneous veins of the elbow are excellent sites for administering intravenous injections and drawing blood owing to their size and accessibility and the relatively thin skin in that region. But given their close relationship to the cutaneous nerves, as illustrated by the proximity of the v. basilica to the n. cutaneus antebrachii medialis, injections into these veins may cause severe transient pain, as in cases where an accidental "paravascular" injection irritates the surrounding connective tissue. "Rolling veins" refers to a condition in which the subcutaneous veins are exceptionally mobile within the subcutaneous fat. In approximately 3% of cases the a. ulnaris may pass over the surface of the flexor muscles (a. ulnaris superficialis, see also p. 395). An unintended intra-arterial injection can have devastating consequences with certain medications. This complication can be avoided by palpating the vessel and confirming arterial-type pulsations before giving the injection and always drawing a small amount of blood back into the syringe (dark red = venous blood, bright red = arterial blood) before depressing the plunger.

B Fossa cubitalis of the right arm: variable course of the subcutaneous veins

a M-shaped venous pattern above the v. mediana antebrachii.
b Presence of an v. cephalica accessoria from the venous plexuses on the extensor side of the forearm.
c Absence of the v. mediana cubiti.

All of the illustrated variants are common.

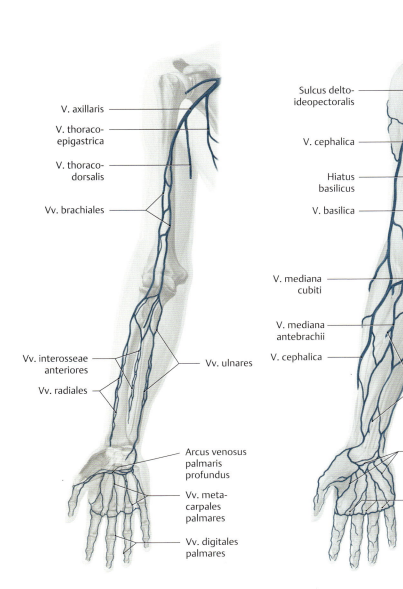

C Deep veins of the right upper limb
Anterior view.

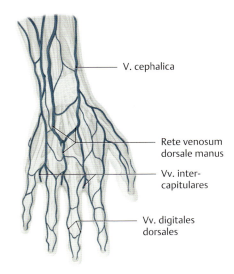

E Superficial veins of the dorsum of the right hand

D Superficial veins of the right upper limb
Anterior view. The main longitudinal trunks of the subcutaneous venous network of the arm are the v. mediana antebrachii, the v. basilica, and the v. cephalica.

V. mediana antebrachii: This vein, unlike the vv. cephalica and basilica, receives blood mainly from the cutaneous veins on the dorsum of the hand, draining the *flexor side of the forearm*. The *variable* v. mediana antebrachii opens into the corresponding longitudinal veins at the elbow, usually by way of the v. mediana cephalica and v. mediana basilica (see p. 379).

V. basilica: This vein begins at the *elbow*, first ascending in the *epifascial* plane in the sulcus bicipitalis medialis to the hiatus basilicus, where it pierces the fascia in the middle of the arm. It terminates in a *subfascial* plane at the ulnar v. brachialis.

V. cephalica: In the *arm* the v. cephalica first ascends on the lateral side of the m. biceps brachii, then enters a groove between the m. deltoideus and m. pectoralis major (the sulcus deltoideopectoralis). It finally opens into the v. axillaris in the trigonum clavipectorale (see p. 380).

F Overview of the main superficial and deep veins of the upper limb
Numerous connections exist between the deep and superficial veins of the arm—the vv. perforantes. Valves are incorporated into the veins at regular intervals, increasing the efficiency of venous return (see p. 65).

Deep veins of the upper limb (vv. profundae membri superioris)
- V. subclavia
- V. axillaris
- Vv. brachiales
- Vv. ulnares
- Vv. radiales
- Vv. interosseae anteriores
- Vv. interosseae posteriores
- Arcus venosus palmaris profundus
- Vv. metacarpales palmares

Superficial veins of the upper limb (vv. superficiales membri superioris)
- V. cephalica
- V. cephalica accessoria
- V. basilica
- V. mediana cubiti
- V. mediana antebrachii
- V. mediana cephalica
- V. mediana basilica
- Rete venosum dorsale manus
- Arcus venosus palmaris superficialis

17.3 The Lymphatic Vessels and Lymph Nodes

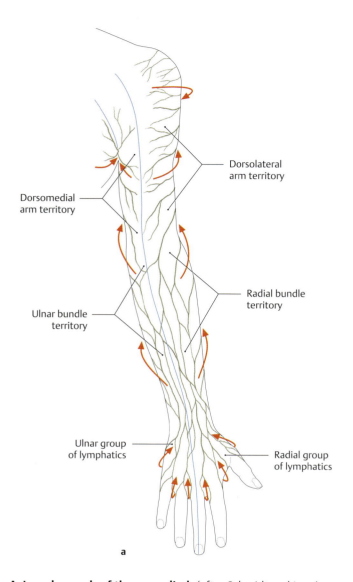
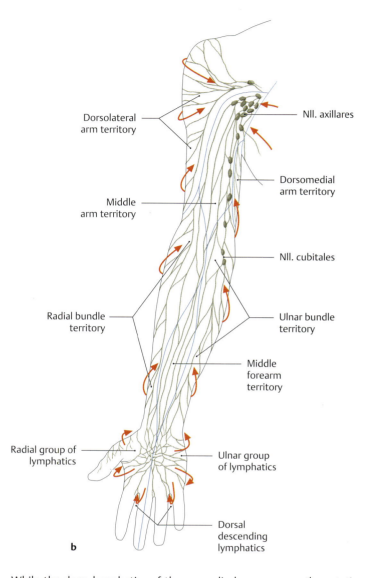

A Lymph vessels of the upper limb (after Schmidt and Lanz)
a Posterior view, **b** anterior view. The lymph vessels (lymphatics) in the upper limb are of two types:

- Superficial (epifascial) lymphatic vessels (vasa lymphatica superficialia)
- Deep lymphatic vessels (vasa lymphatica profunda)

While the deep lymphatics of the upper limb accompany the arteries and deep veins, the superficial lymphatics lie in the subcutaneous tissue (tela subcutanea). In the forearm, they are most closely related to the vv. cepha-lica and basilica. Numerous anastomoses exist between the deep and superficial systems. The arrows in the diagrams indicate the main directions of lymphatic drainage. Inflammations and infections of the hand generally incite a painful swelling of the nll. axillares. When the lymph vessels are also involved, they are visible as red streaks beneath the skin (lymphangitis).

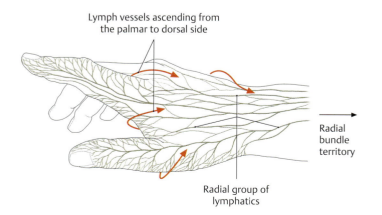

B Lymphatic drainage of the thumb, index finger, and dorsum of the hand (after Schmidt and Lanz)
The thumb, index finger, and part of the middle finger are drained by a radial group of lymph vessels that pass directly to the nll. axillares nodes. The other fingers are drained by an ulnar group of lymphatics (not shown here) that end at the nll. cubitales.

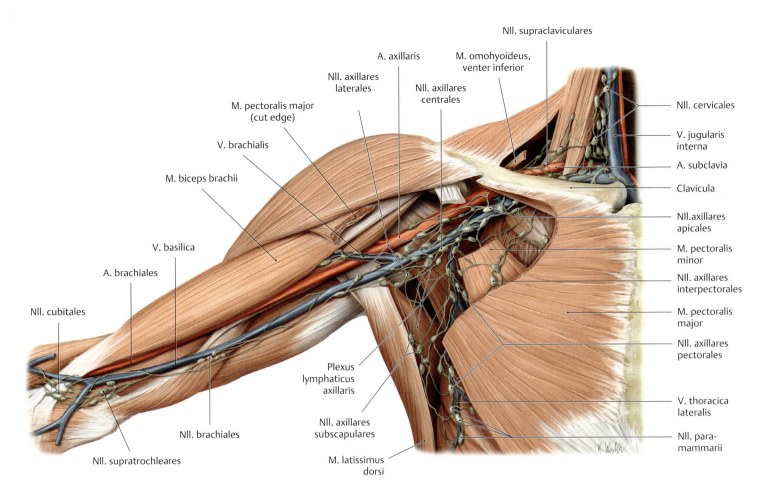

C Regional lymph nodes of the right upper limb
Anterior view. The lymph nodes of the axilla (*nodi lymphoidei axillares*) are important collecting stations for the arm, shoulder girdle, and anterior chest wall. The 30 to 60 lymph nodes of the axilla are divided into several groups or levels, numbered I to III (see **E**), which are interconnected by lymph vessels. Taken together, the lymphatics in this region form an *plexus lymphaticus axillaris* lying within the fatty tissue. Lymphatic drainage from the axilla is collected in the truncus subclavius (not shown here). On the right side, the lymph is conveyed by the truncus jugularis dexter and truncus bronchomediastinalis dexter to the ductus lymphaticus dexter, which opens into the junction of the right v. subclavia and v. jugularis interna (see p. 194).

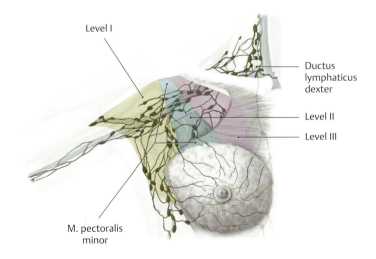

D Classification of axillary lymph nodes by levels
(after Henne-Bruns, Dürig, and Kremer)

Level I: lower axillary group
(lateral to m. pectoralis minor)
- Nll. axillares pectorales
- Nll. axillares subscapulares
- Nll. axillares laterales
- Nll. paramammarii

Level II: middle axillary group
(along m. pectoralis minor)
- Nll. axillares interpectorales
- Nll. axillares centrales

Level III: upper, infraclavicular group
(medial to m. pectoralis minor)
- Nll. axillares apicales

E Classification of axillary lymph nodes by level
The axillary lymph nodes have major clinical importance in breast cancer. A malignant breast tumor will metastasize (seed tumor cells) to the axillary nodes as it grows. As a guide for surgical removal, the axillary lymph nodes can be segregated into groups arranged in three levels, based on their relationship to the m. pectoralis minor.

- Level I: all the lymph nodes lateral to the m. pectoralis minor.
- Level II: all the lymph nodes along the m. pectoralis minor.
- Level III: all the lymph nodes medial to the m. pectoralis minor (see p. 300).

17.4 The Plexus brachialis: Structure

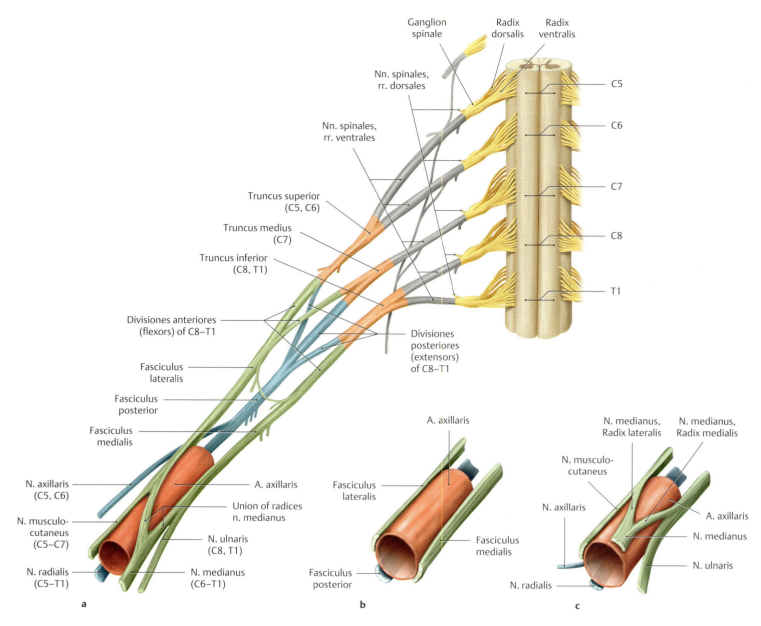

A Schematic representation of the structure of the plexus brachialis
a Names and sequence of the various components of the plexus brachialis.
b Relationship of the fasciculi of the plexus brachialis to the a. axillaris.
c Subdivision of the fasciculi of the plexus brachialis into their main branches.

B Number and location of the main components of the plexus brachialis

Components	Number	Location
1. Plexus roots (rr. ventrales of the nn. spinales from cord segments C5–T1)	5	Between m. scalenus anterior and m. scalenus medius (interscalene space)
2. The primary trunks: truncus superior, medius, and inferior	3	Lateral to the interscalene space and above the clavicula
3. The three anterior and three posterior divisions (divisiones anteriores and posteriores)	6	Posterior to the clavicula
4. The fasciculus lateralis, medialis, and posterior	3	In the axilla, posterior to the m. pectoralis minor

Upper Limb —— 17. Neurovascular Systems: Forms and Relations

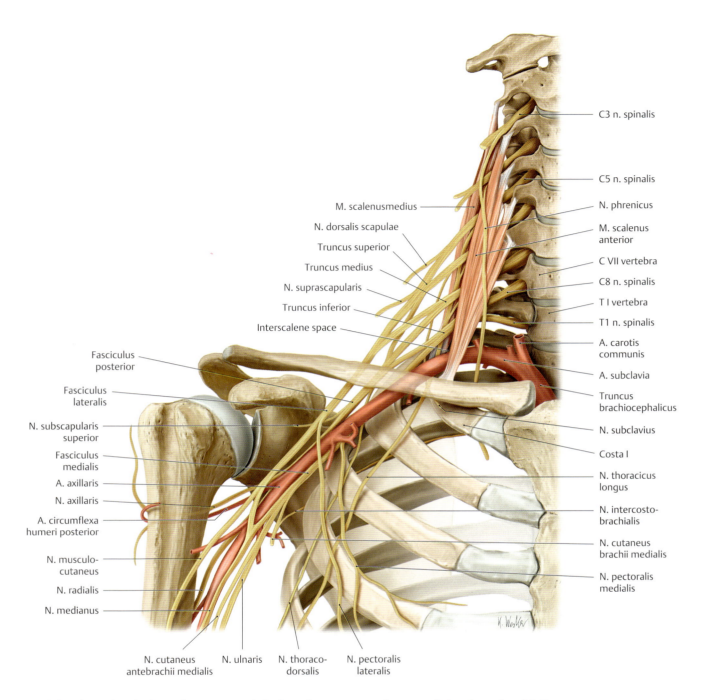

C Course of the plexus brachialis and its relation to the thorax after passing through the interscalene space
Right side, anterior view.

D Spinal cord segments and nerves of the plexus brachialis

Plexus brachialis trunci and associated spinal cord segments
- Truncus superior C5 + C6
- Truncusmedius C7
- Truncus inferior C8 + T1

Plexus brachialis fasciculi and associated spinal cord segments
- Fasciculus lateralis C5–C7
- Fasciculus medialis C8–T1
- Fasciculus posterior C5–T1

Nerves of the pars supraclavicularis of the plexus brachialis (direct branches from the rr. ventrales or trunci)
- N. dorsalis scapulae
- N. thoracicus longus
- N. suprascapularis
- N. subclavius

Nerves of the pars infraclavicularis of the plexus brachialis (short and long branches from the fasciculi)
- Fasciculus lateralis
  – N. musculocutaneus
  – N. pectoralis lateralis
  – N. medianus (radix lateralis)
- Fasciculus medialis
  – N. medianus (radix medialis)
  – N. ulnaris
  – N. pectoralis medialis
  – N. cutaneus brachii medialis
  – N. cutaneus antebrachii medialis
- Fasciculus posterior
  – N. radialis
  – N. axillaris
  – N. subscapularis superior
  – N. subscapularis inferior
  – N. thoracodorsalis

17.5 Supraclavicular Part of the Plexus brachialis (Pars supraclavicularis)

A Pars supraclavicularis of the plexus brachialis
The pars supraclavicularis of the plexus brachialis includes all nerves that arise directly from the plexus roots (rr. ventrales of nn. spinales) or from the trunci in the lateral cervical triangle between the mm. scaleni anterior and medius. The different nerves of the pars supraclavicularis are predisposed to paralysis and/or compression in varying degrees based on their location and course (see **B–D**).

Nerve	Segment	Innervated muscle
N. dorsalis scapulae	C4, C5	• M. levator scapulae • M. rhomboideus major • M. rhomboideus minor
N. suprascapularis	C4–C6	• M. supraspinatus • M. infraspinatus
N. thoracicus longus	C5–C7	• M. serratus anterior
N. subclavius	C5, C6	• M. subclavius

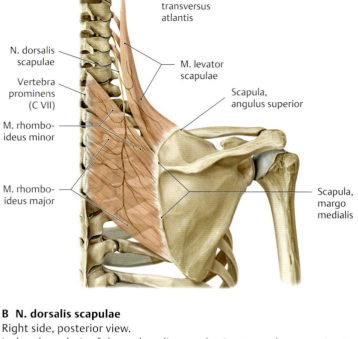

B N. dorsalis scapulae
Right side, posterior view.
Isolated paralysis of the n. dorsalis scapulae is extremely rare owing to the protected location of the nerve between the deep nuchal muscles, the m. levator scapulae, and the mm. rhomboidei.

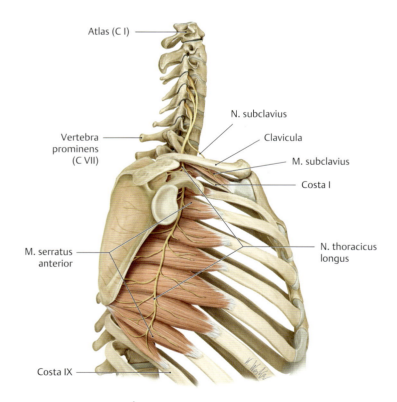

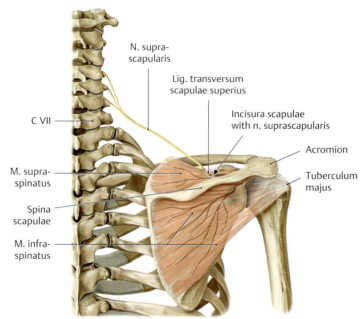

C N. thoracicus longus and n. subclavius
Right side, lateral view.
Its long, superficial course on the m. serratus anterior along the lateral chest wall makes the n. thoracicus longus susceptible to mechanical injury. The prolonged wearing of a heavy backpack is a common mechanism for this type of lesion. In iatrogenic cases, the nerve may be damaged by an axillary lymphadenectomy performed for a metastatic breast tumor. Clinically, loss of the m. serratus anterior causes the margo medialis scapulae to become elevated from the chest wall. This *"Scapula alata"* is most conspicuous when the arm is raised forward, and generally the arm cannot be elevated past 90°.

D N. suprascapularis
Right side, posterior view.
Injuries and chronic compression of the n. suprascapularis are uncommon conditions that lead to atrophy of the mm. supraspinatus and infraspinatus with weakness of arm abduction (especially during the initial phase owing to the "starter" function of the m. supraspinatus) and external rotation of the arm. Besides an isolated injury, the nerve may become compressed in the fibro-osseous canal between the scapulae and the lig. transversum scapulae superius (which is occasionally ossified to form a bony canal). The resulting symptoms are known collectively as "incisura-scapulae syndrome" (see p. 243).

E Plexus brachialis compression syndromes due to narrow anatomic passages in the shoulder region

In its course from the foramina intervertebralia to the nerves of the upper limb, the plexus brachialis must negotiate several narrow passages in which it may become compressed by surrounding structures. There are also extrinsic factors, such as carrying heavy loads, that may exert direct pressure on the plexus brachialis. Several types of compression syndrome are distinguished:

1. **Scalene syndrome or cervical rib syndrome:**
   neurovascular compression in the interscalene space caused by a cervical rib (costa cervicalis) or ligamentous structure (see F)
2. **Costoclavicular syndrome:**
   narrowing of the space between the first rib and clavicula (see G)
3. **Hyperabduction syndrome:**
   compression of the plexus brachialis by the m. pectoralis minor and coracoideus when the upper arm is raised above the head (see H)
4. Chronic heavy load on the shoulder girdle (e.g., "backpack paralysis")

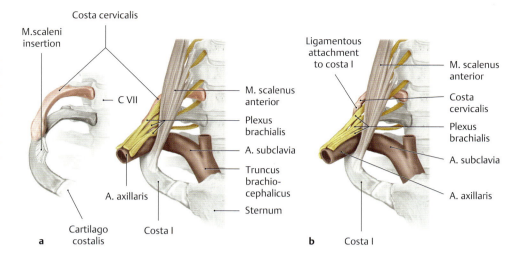

F Scalene syndrome due to narrowing of the interscalene space by a cervical rib

Right shoulder, anterior view.
In approximately 1% of the population, cervical ribs (costae cervicales) may narrow the interscalene space bounded by the mm. scaleni anterior and medius and the first rib. In this condition the trunci of the plexus brachialis that pass through the interscalene space along with the a. subclavia are compressed from behind and below, placing varying degrees of tension on the neurovascular bundle (a). If there is no bony contact between a short costa cervicalis and the first rib (b), that site is often occupied by a ligamentous structure that can also cause neurovascular compression. The main clinical manifestations are pain radiating down the arm, chiefly to the ulnar side of the hand, and circulatory impairment caused by mechanical irritation of the periarterial sympathetic plexus of the a. subclavia.

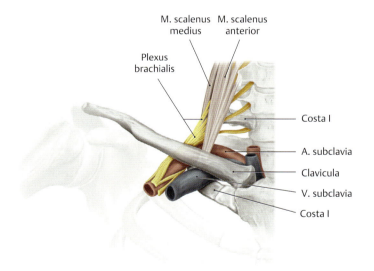

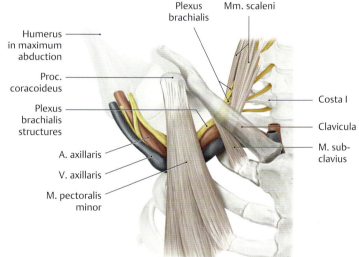

G Costoclavicular syndrome due to compression of the neurovascular bundle between the first rib and clavicula

Right shoulder, anterior view.
Narrowing of the costoclavicular space is a rare condition that is most common in persons with drooping shoulders, a flat back, retracted shoulders (from carrying heavy loads), a deformed first rib (costa I), or a previous clavicular fracture. Any narrowing of the costoclavicular space can be aggravated by lowering and retracting the shoulder girdle. The complaints are similar to those in scalene syndrome and may be accompanied by signs of venous stasis caused by impaired return through the v. subclavia.

H Hyperabduction syndrome due to compression of the neurovascular bundle below the m. pectoralis minor and proc. coracoideus

Right shoulder, anterior view.
This rare syndrome is caused by neurovascular compression beneath the tendon of the m. pectoralis minor under the proc. coracoideus. It is precipitated by maximum abduction or elevation of the arm on the affected side. A simple clinical test consists of pulling the arm upward and backward and holding it there. In normal cases, a definite radial artery pulse should still be palpable after 1 to 2 minutes, and the patient should not complain of radiating pain.

17.6 Pars infraclavicularis of the Plexus brachialis: Overview and Short Branches

A Pars infraclavicularis of the plexus brachialis
Pars infraclavicularis of the plexus brachialis includes all the nerves that leave the plexus at the level of the plexus fasciculi—the *short branches*— and those that continue down the arm as terminal branches of the individual plexus fasciculi—the *long branches*. These nerves are reviewed below, beginning with the short branches.

Nerve	Segment	Innervated muscle	Cutaneous branches
Part I: Short branches			
• N. subscapularis superior and n. subscapularis inferior	C5, C6	• M. subscapularis • M. teres major (upper only)	—
• N. thoracodorsalis	C6–C8	• M. latissimus dorsi	—
• Nn. pectorales medialis and lateralis	C5–T1	• M. pectoralis major • M. pectoralis minor	—
• N. cutaneus brachii medialis	T1	—	• N. cutaneus brachii medialis
• N. cutaneus antebrachii medialis	C8, T1	—	• N. cutaneus antebrachii medialis
• Nn. intercostobrachiales*	T2, T3	—	• Rr. cutanei laterales
Part II: Long branches			
• N. musculocutaneus (see p. 366)	C5–C7	• M. coracobrachialis • M. biceps brachii • M. brachialis	• N. cutaneus antebrachii lateralis
• N. axillaris (see p. 366)	C5, C6	• M. deltoideus • M. teres minor	• N. cutaneus brachii lateralis superior
• N. radialis (see p. 368)	C5–T1	• M. brachialis (contribution) • M. triceps brachii • M. anconeus • M. supinator • M. brachioradialis • M. extensor carpi radialis longus • M. extensor carpi radialis brevis • M. extensor digitorum • M. extensor digiti minimi • M. extensor carpi ulnaris • M. extensor pollicis longus • M. extensor pollicis brevis • M. extensor indicis • M. abductor pollicis longus	• N. cutaneus brachii lateralis inferior • N. cutaneus brachii posterior • N. cutaneus antebrachii posterior • R. superficialis n. radialis
• N. medianus (see p. 372)	C6–T1	• M. pronator teres • M. pronator quadratus • M. palmaris longus • M. flexor carpi radialis • M. flexor pollicis longus • M. flexor digitorum profundus (half) • M. flexor digitorum superficialis • M. abductor pollicis brevis • M. opponens pollicis • M. flexor pollicis brevis (caput superficiale) • Mm. lumbricales I and II	• R. palmaris n. mediani • Nn. digitales palmares communes and proprii
• N. ulnaris (see p. 370)	C8, T1	• M. flexor carpi ulnaris • M. flexor digitorum profundus (half) • M. palmaris brevis • M. flexor digiti minimi • M. abductor digiti minimi • M. opponens digiti minimi • M. adductor pollicis • M. flexor pollicis brevis (caput profundum) • Mm. interossei palmares and dorsales • Mm. lumbricales III and IV	• R. palmaris n. ulnaris • R. dorsalis n. ulnaris • Nn. digitales dorsales • Nn. digitales palmares communes and proprii

* These are the cutaneous branches of the nn. intercostales 2 and 3, which accompany the n. cutaneus brachii medialis.

B Sensory distribution of the n. cutaneus brachii medialis and n. cutaneus antebrachii medialis of the right arm
a Anterior view, b posterior view.

Upper Limb — 17. Neurovascular Systems: Forms and Relations

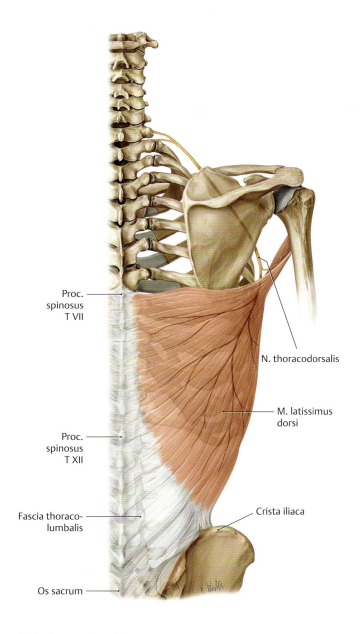

C N. thoracodorsalis
Right side, posterior view.

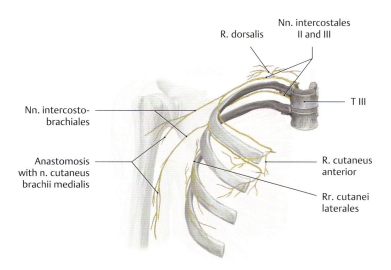

E Origin and cutaneous distribution of the Nn. intercostobrachiales of the right arm
Anterior view.

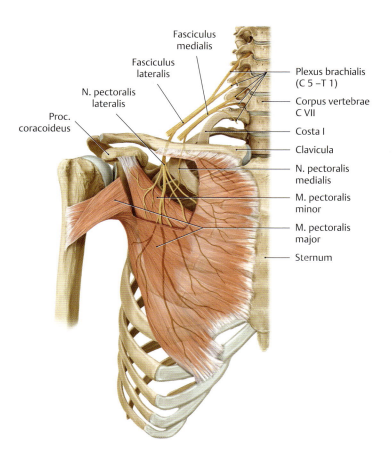

D Nn. pectorales medialis and lateralis
Right side, anterior view.

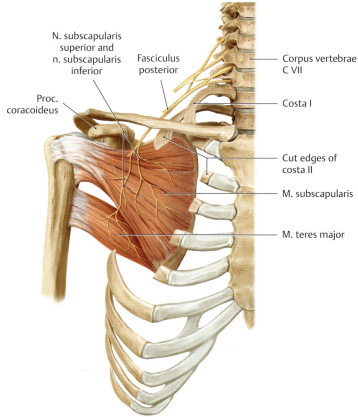

F N. subscapularis superior and n. subscapularis inferior
Right side, anterior view. The ribs have been partially removed.

365

17.7 Pars infraclavicularis of the Plexus brachialis: N. musculocutaneus and N. axillaris

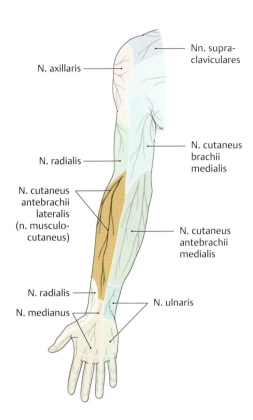

A Sensory distribution of the n.cutaneus antebrachii lateralis
Right arm, anterior view.

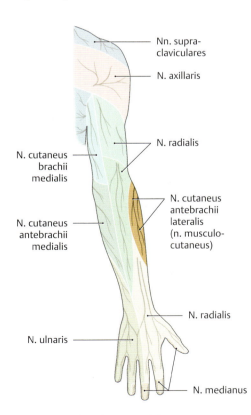

B Sensory distribution of the n.cutaneus antebrachii lateralis
Right arm, posterior view.

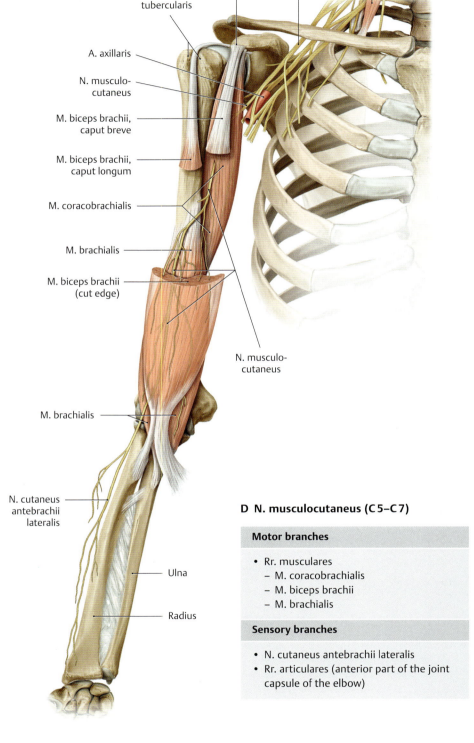

C Course of the N. musculocutaneus after leaving the fasciculus lateralis of the plexus brachialis
Right upper limb, anterior view. The n. musculocutaneus leaves the fasciculus lateralis of the plexus brachialis as a mixed nerve (n. mixtus, one with motor and sensory branches) at the level of the lateral border of the m. pectoralis minor (not shown here) and runs a short course before piercing the m. coracobrachialis. It then runs between the m. biceps brachii and m. brachialis to the elbow, where its terminal sensory branch supplies the skin on the radial side of the forearm.

D N. musculocutaneus (C5–C7)

Motor branches

• Rr. musculares
  – M. coracobrachialis
  – M. biceps brachii
  – M. brachialis

Sensory branches
• N. cutaneus antebrachii lateralis • Rr. articulares (anterior part of the joint capsule of the elbow)

Upper Limb — 17. Neurovascular Systems: Forms and Relations

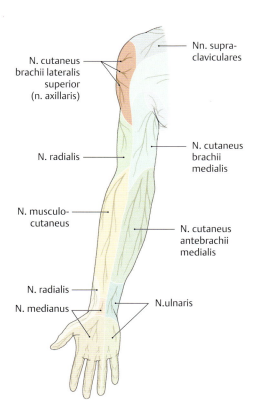

E Sensory distribution of the n. cutaneus brachii lateralis superior
Right arm, anterior view.

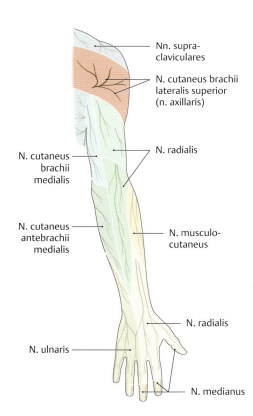

F Sensory distribution of the n. cutaneus brachii lateralis superior
Right arm, posterior view.

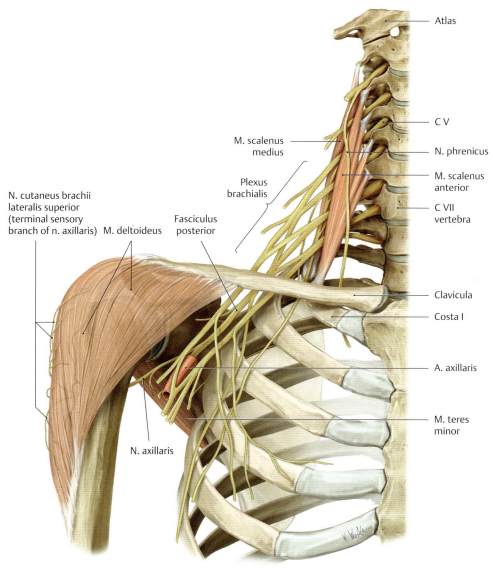

G Course of the n. axillaris after leaving the fasciculus posterior of the plexus brachialis

Right upper limb, anterior view. The n. axillaris leaves the fasciculus posterior of the plexus brachialis as a mixed nerve (n. mixtus) and runs backward through the deep part of the axilla, passing directly below the shoulder joint. It courses through the quadrangular space of the axilla (with the a. circumflexa humeri posterior) and along the collum chirurgicum to the posterior side of the proximal humerus. Its terminal sensory branch supplies the skin over the m. deltoideus. Isolated *axillary nerve palsy* may occur following an anteroinferior shoulder dislocation (or a traumatic reduction attempt), a humeral fracture at the level of the collum chirurgicum, or prolonged pressure from an improperly adjusted crutch in the axilla.

H N. axillaris (C 5 and C 6)

Motor branches

• Rr. musculares
  – M. deltoideus
  – M. teres minor

Sensory branch
• N. cutaneus brachii lateralis superior

367

17.8 Pars infraclavicularis of the Plexus brachialis: N. radialis

A N. radialis (C 5–T 1)

Motor branches

- Rr. musculares (from the n. radialis)
  - M. brachialis (contribution)
  - M. triceps brachii
  - M. anconeus
  - M. brachioradialis
  - M. extensor carpi radialis longus
  - M. extensor carpi radialis brevis
- R. profundus (terminal branch: n. interosseus posterior)
  - M. supinator
  - M. extensor digitorum
  - M. extensor digiti minimi
  - M. extensor carpi ulnaris
  - M. extensor pollicis longus
  - M. extensor pollicis brevis
  - M. extensor indicis
  - M. abductor pollicis longus

Sensory branches

- Rr. articulares (from the n. radialis)
  - Capsule of the shoulder joint
- Rr. articulares (from the n. interosseus posterior)
  - Joint capsule of the wrist and the four radial metacarpophalangeal joints
- N. cutaneus brachii posterior
- N. cutaneus brachii lateralis inferior
- N. cutaneus antebrachii posterior
- R. superficialis
  - Nn. digitales dorsales
  - R. communicans ulnaris

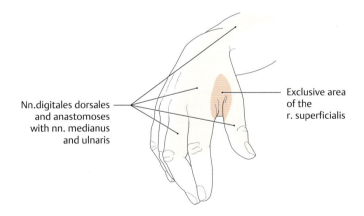

C Wrist drop due to proximal and midlevel n. radialis lesions
When the n. radialis is damaged, the patient can no longer actively extend the hand at the wrist, and *wrist drop* (drop hand) is said to be present. Besides the dropped position of the wrist, clinical examination reveals areas of sensory loss on the radial surface of the dorsum and on the extensor surface of the thumb, index finger, and the radial half of the middle finger extending to the proximal interphalangeal joint. The sensory deficits are often confined to the area of the hand that receives sensory innervation exclusively from the n. radialis (the spatium interosseum between the thumb and index finger).

B Traumatic lesions and compression syndromes involving the n. radialis
The n. radialis may be damaged anywhere in its course as a result of injury or chronic compression. The clinical features depend critically on the *site of the lesion*. As a general rule, the more proximal the site of the lesion, the greater the number of extensor muscles that are affected. The characteristic feature of a proximal ("high") n. radialis lesion is *wrist drop* (see **C**), in which the patient is unable to extend the wrist or the artt. metacarpophalangeae. Lesions at some sites may additionally cause sensory disturbances (pain, paresthesia, numbness), particularly in the exclusive sensory territory of the r. superficialis on the radial side of the dorsum (first spatium interosseum between the thumb and index finger).

Proximal n. radialis lesion

- Chronic pressure in the axilla (e.g., due to prolonged crutch use).
  Clinical features: typical dropped wrist with loss of the m. triceps brachii (and sensory disturbances).
- Traumatic lesion due to a humeral shaft fracture at the level of the sulcus n. radialis (canalis spiralis).
  Clinical features: usually a typical dropped wrist *without involvement of the m. triceps brachii,* since the rr. musculares that supply the m. triceps brachii leave the n. radialis just before it enters the sulcus n. radialis (sensory disturbances are present, however).
- Chronic compression of the n. radialis against the bony floor of the sulcus n. radialis (e.g., during sleep or due to improper positioning of the patient during general anesthesia, exuberant callus formation after a fracture, or a tendon expansion from the caput laterale of the m. triceps brachii). "Park bench palsy" is a common form caused by draping the arm over the back of a park bench.
  Clinical features: dropped wrist *without involvement of the m. triceps brachii.* Sensory disturbances are present. The prognosis is usually favorable, and the palsy should resolve in a few days.

Midlevel n. radialis lesion

- Chronic compression of the n. radialis in its passage through the septum intermusculare laterale and in the radial tunnel (e.g., by bridging vessels and connective-tissue septa).
  Clinical features: dropped wrist with sensory disturbances.

Distal n. radialis lesion

- Compression of the r. profundus of the n. radialis at its entry into the supinator canal by a sharp-edged tendon of the pars superficialis of the m. supinator: supinator syndrome or distal n. radialis compression syndrome.
  Clinical features: no typical wrist drop and no sensory disturbances involving the hand (before entering the supinator canal, the r. profundus gives off the purely sensory r. superficialis and rr. musculares for the mm. supinator, brachioradialis, and extensores carpi radialis longus and brevis). There are palsies involving the mm. extensores pollicis brevis and longus, abductor pollicis longus, extensor digitorum, extensor indicis, and extensor carpi ulnaris.
- Trauma to the r. profundus caused by a fracture or dislocation of the radius.
  Clinical features: no wrist drop and no sensory disturbances.

Upper Limb — 17. Neurovascular Systems: Forms and Relations

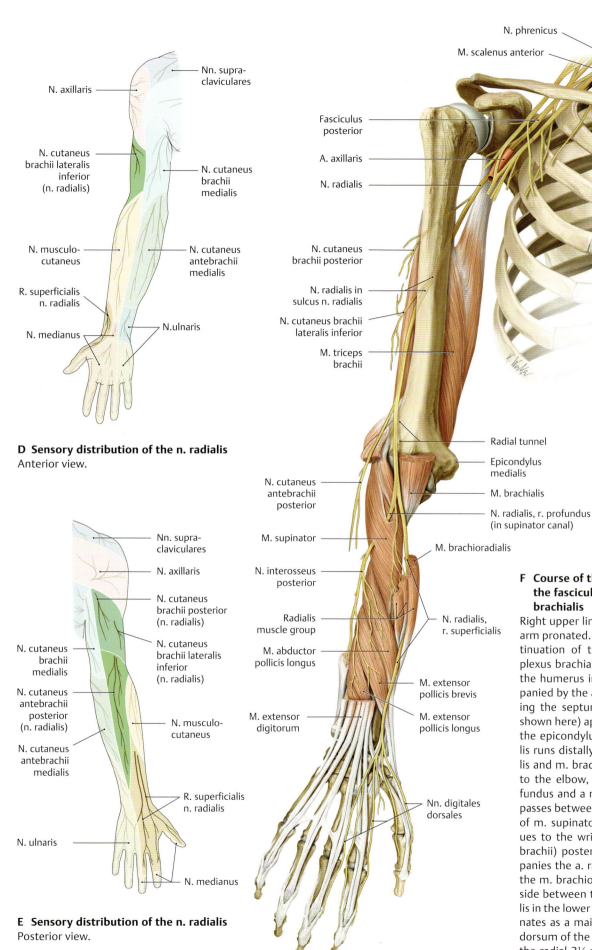

D Sensory distribution of the n. radialis
Anterior view.

E Sensory distribution of the n. radialis
Posterior view.

F Course of the n. radialis after leaving the fasciculus posterior of the plexus brachialis

Right upper limb, anterior view with the forearm pronated. The n. radialis is the direct continuation of the fasciculus posterior of the plexus brachialis. It winds around the back of the humerus in the sulcus n. radialis, accompanied by the a. profunda brachii. After piercing the septum intermusculare laterale (not shown here) approximately 10 cm proximal to the epicondylus humeri radialis, the n. radialis runs distally between the m. brachioradialis and m. brachialis (*radial tunnel*, see p. 392) to the elbow, where it divides into a r. profundus and a r. superficialis. The r. profundus passes between the superficial and deep parts of m. supinator (*supinator canal*) and continues to the wrist as the n. interosseus (antebrachii) posterior. The r. superficialis accompanies the a. radialis down the forearm along the m. brachioradialis, passes to the extensor side between the radius and m. brachioradialis in the lower third of the forearm, and terminates as a main sensory branch on the radial dorsum of the hand and the dorsal margins of the radial 2½ digits (the thumb, index finger, and radial half of the middle finger).

369

17.9 Pars infraclavicularis of the Plexus brachialis: N. ulnaris

A N. ulnaris (C8, T1)

Motor branches

- Rr. musculares (directly from the ulnar half)
  - M. flexor carpi ulnaris
  - M. flexor digitorum profundus (ulnar half)
- R. muscularis (from the r. superficialis)
  - M. palmaris brevis
- Rr. musculares (from the r. profundus)
  - M. abductor digiti minimi
  - M. flexor digiti minimi
  - M. opponens digiti minimi
  - Mm. lumbricales III und IV
  - Mm. interossei palmares and dorsales
  - M. adductor pollicis
  - M. flexor pollicis brevis (caput profundum)

Sensory branches

- Rr. articulares
  - Capsule of the art. cubiti and of the artt. intercarpeae and artt. metacarpophalangeae
- R. dorsalis n. ulnaris (terminal branches: nn. digitales dorsales)
- R. palmaris n. ulnaris
- N. digitalis palmaris proprius (from the r. superficialis)
- N. digitalis palmaris communis IV (from the r. superficialis; terminal branches: nn. digitales palmares proprii)

B Traumatic lesions and compression syndromes involving the n. ulnaris

N. ulnaris palsy is the *most common peripheral nerve paralysis*. The characteristic feature of an n. ulnaris lesion is a "*claw hand*" deformity (see C), in which *loss of the mm. interossei* causes the fingers to be hyperextended at the artt. metacarpophalangeae and slightly flexed at the proximal and distal interphalangeal joints. The deformity is least pronounced in the index and middle fingers because the mm. lumbricales I and II, which are innervated by the n. medianus, can partially compensate for the clawing of those fingers. The thumb is markedly hyperextended due to the loss of the m. adductor pollicis and the dominance of the mm. extensor pollicis longus and abductor pollicis. The mm. interossei atrophy in 2 to 3 months; this is most conspicuous in the first spatium interosseum and is accompanied by hypothenar atrophy. Sensory disturbances affect the ulnar portion of the hand, the ulnar half of the ring finger, and the entire little finger.

Proximal n. ulnaris lesion

- Traumatic lesions, usually occurring at the elbow joint due to the exposed position of the nerve in the sulcus n. ulnaris (e.g., pressure from resting on the arm), displacement of the nerve from its sulcus, or articular injuries due to fractures.
- Chronic pressure on the nerve in the sulcus n. ulnaris due to degenerative or inflammatory changes in the elbow joint, or chronic traction on the nerve caused by repetitive flexion and extension at the elbow joint (sulcus n. ulnaris syndrome).
- Possible compression between the tendons of origin of the m. flexor carpi ulnaris (cubital tunnel syndrome).
  Clinical features: claw hand and sensory disturbances.

Midlevel n. ulnaris lesion

- Traumatic lesions at the wrist (e.g., lacerations).
- Chronic compression of the nerve in the ulnar tunnel, a fibro-osseous canal between the lig. carpi palmare, os pisiforme, and retinaculum musculorum flexorum (ulnar tunnel syndrome, see p. 405).
  Clinical features: claw hand and sensory disturbances that spare the hypothenar region (r. palmaris is intact).

Distal n. ulnaris lesion

- Compression of the r. profundus of the n. ulnaris in the palm due to chronic pressure (e.g., from an air hammer or other tools).
  Clinical features: claw hand with no sensory disturbances (r. superficialis is intact).

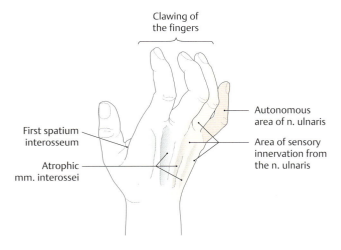

C Claw hand due to a n. ulnaris lesion

Besides the typical clawlike appearance of the hand, atrophy of the mm. interossei leads to hollowing of the interosseous spaces in the metacarpus. Sensory abnormalities are frequently confined to the little finger (exclusive sensory territory of the n. ulnaris).

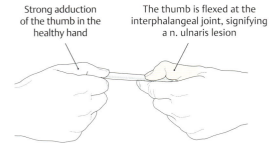

D Positive "Froment sign" in the left hand

A positive Froment sign indicates palsy of the m. adductor pollicis. When the patient is told to hold a piece of paper firmly between the thumb and index finger, he or she must use the m. flexor pollicis longus, which is innervated by the *nervus medianus*, rather than the paralyzed m. adductor pollicis, which is innervated by the *n. ulnaris*. Flexing the thumb at the interphalangeal joint signifies a positive test.

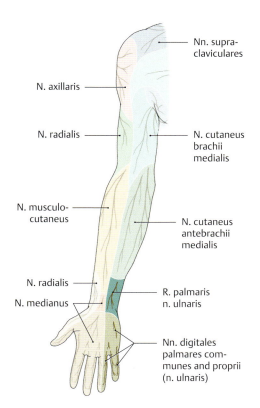

E Sensory distribution of the n. ulnaris
Right arm, anterior view.

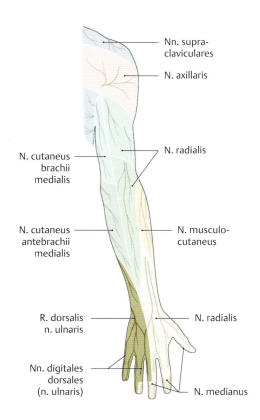

F Sensory distribution of the n. ulnaris
Right arm, posterior view.

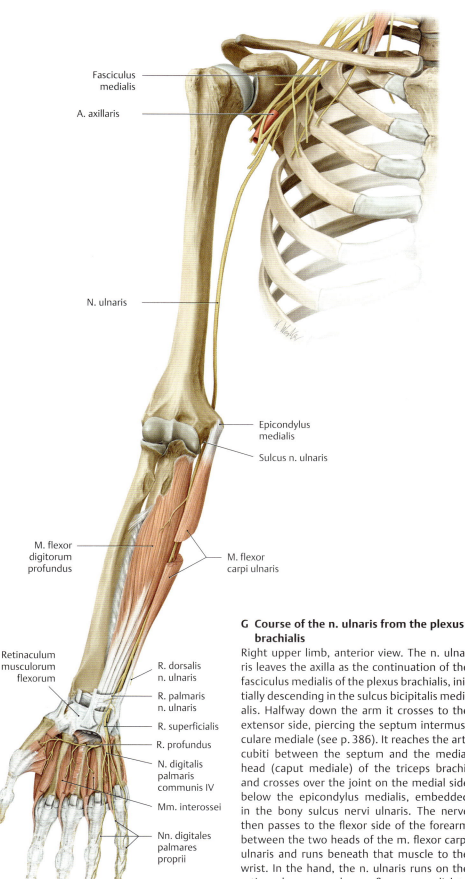

G Course of the n. ulnaris from the plexus brachialis

Right upper limb, anterior view. The n. ulnaris leaves the axilla as the continuation of the fasciculus medialis of the plexus brachialis, initially descending in the sulcus bicipitalis medialis. Halfway down the arm it crosses to the extensor side, piercing the septum intermusculare mediale (see p. 386). It reaches the art. cubiti between the septum and the medial head (caput mediale) of the triceps brachii and crosses over the joint on the medial side below the epicondylus medialis, embedded in the bony sulcus nervi ulnaris. The nerve then passes to the flexor side of the forearm between the two heads of the m. flexor carpi ulnaris and runs beneath that muscle to the wrist. In the hand, the n. ulnaris runs on the retinaculum musculorum flexorum radial to the os pisiforme, passing through the ulnar tunnel (see p. 405) to the palmar surface, where it divides into a r. superficialis and a motor r. profundus.

17.10 Pars infraclavicularis of the Plexus brachialis: N. medianus

A N. medianus (C 6–T 1)

Motor branches

- Rr. musculares (directly from the n. medianus)
  - M. pronator teres
  - M. flexor carpi radialis
  - M. palmaris longus
  - M. flexor digitorum superficialis
- Rr. musculares (from the n. interosseus antebrachii anterior)
  - M. pronator quadratus
  - M. flexor pollicis longus
  - M. flexor digitorum profundus (radial half)
- R. muscularis thenaris ("thenar branch")
  - M. abductor pollicis brevis
  - M. flexor pollicis brevis (caput superficiale)
  - M. opponens pollicis
- Rr. musculares (from the nn. digitales palmares communes)
  - Mm. lumbricales I and II

Sensory branches

- Rr. articulares
  - Capsules of the elbow joint and wrist joints
- R. palmaris n. mediani (thenar)
- R. communicans cum nervo ulnare
- Nn. digitales palmares communes
- Nn. digitales palmares proprii (fingers)

B Traumatic lesions and compression syndromes involving the n. medianus

N. medianus lesions in the arm caused by an acute injury or chronic pressure are among the most common peripheral nerve lesions. The clinical manifestations depend on the *site of the lesion*. The two main categories are proximal and distal nerve lesions, as illustrated by the *pronator teres syndrome* and the *carpal tunnel syndrome*. The hallmark of a proximal n. medianus lesion is the *"hand of benediction,"* which occurs when the patient tries to clench the hand into a fist (loss of the long digital flexors except for the part of the m. flexor digitorum profundus supplied by the n. ulnaris). This contrasts with distal n. medianus lesions, which present selectively with thenar atrophy and sensory disturbances, as in carpal tunnel syndrome.

Proximal n. medianus lesion

- Traumatic injury caused by a fracture or dislocation of the elbow joint.
- Chronic pressure injury from an anomalous proc. supracondylaris connected to the epicondylus medialis by a ligament ("Struthers' ligament," see p. 393), pressure from a tight aponeurosis bicipitalis, or a *pronator teres syndrome* in which the nerve is squeezed between the two heads of the m. pronator teres.

Clinical features: Typical "hand of benediction" (see C) when fist closure is attempted, with incomplete pronation, loss of thumb opposition, impaired grasping ability, atrophy of the thenar muscles, and sensory disturbances affecting the radial part of the palm and radial 3½ digits (also autonomic trophic disturbances, such as decreased sweat secretion and increased cutaneous blood flow). The patient also has a positive "bottle sign" in which the fingers and thumb cannot fully close around a cylindrical object due to weakness of the m. abductor pollicis brevis.

Distal n. medianus lesion

- The superficial location of the nerve in the distal forearm makes it vulnerable to cuts and lacerations (e.g., in attempted suicide).
- Chronic compression of the n. medianus in the canalis carpi (most common compression syndrome affecting the n. medianus: is carpal tunnel syndrome). Compression or entrapment of the nerve within the canalis carpi can have various causes, such as fractures and dislocations of the ossa carpi, inflammatory changes in the tendon sheaths, muscle variants (e.g., mm. lumbricales passing through the canalis carpi), and connective = tissue proliferation due to endocrine hormonal changes (diabetes mellitus, pregnancy, menopause).

Clinical features: The "hand of benediction" is not present. Initial signs consist of sensory disturbances (paresthesias and dysesthesias), chiefly affecting the tips of the index and middle fingers and thumb due to increased carpal tunnel pressure resulting from prolonged flexion or extension of the wrist during sleep ("brachialgia paraesthetica nocturna"). Chronic or severe damage leads to motor deficits involving the thenar muscles (thenar atrophy) with preservation of thenar sensation (intact r. palmaris n. mediani) and a positive bottle sign (see D).

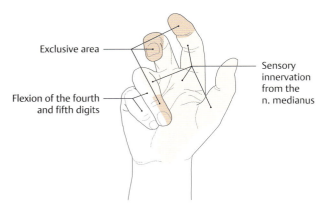

C "Hand of benediction" following a proximal n. medianus lesion
When patients try to make a fist, they can flex only the ulnar fingers. This is the "hand of benediction" deformity. There may be associated sensory disturbances, particularly in the autonomous area of the nerve (tips of the radial 3½ digits).

In a healthy hand, the thumb can be abducted to fully grasp a cylindrical object

With a proximal n. medianus lesion, the thumb cannot be fully abducted

D Positive "bottle sign" in the right hand
When proximal and distal n. medianus lesions are present, the thumb and fingers cannot completely encircle a cylindrical vessel with the affected hand due to weakness or loss of the m. abductor pollicis brevis.

Upper Limb — 17. Neurovascular Systems: Forms and Relations

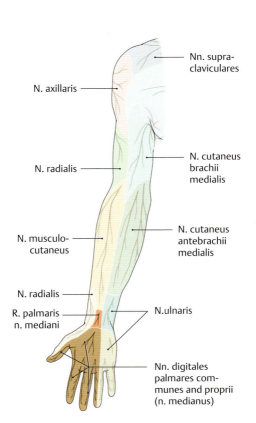

E Sensory distribution of the n. medianus.
Right arm, anterior view.

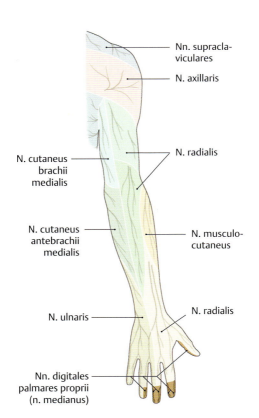

F Sensory distribution of the n. medianus
Right arm, posterior view.

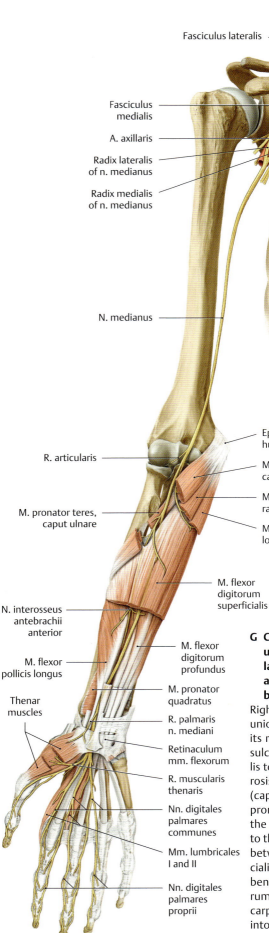

G Course of the n. medianus after the union of its radix medialis and radix lateralis from the fasciculus medialis and fasciculus lateralis of the plexus brachialis

Right upper limb, anterior view. Distal to the union of plexus brachialis's radix medialis and its radix lateralis, the n. medianus runs in the sulcus bicipitalis medialis above the a. brachialis to the elbow and passes under the aponeurosis bicipitalis and between the two heads (caput humerale and caput ulnare) of the m. pronator teres to the forearm. After giving off the n. interosseus antebrachii anterior distal to the m. pronator teres, the n. medianus runs between the mm. flexores digitorum superficialis and profundus to the wrist and passes beneath the retinaculum musculorum flexorum (lig. carpi transversum) in the canalis-carpi to the palm of the hand, where it divides into its terminal branches (a motor branch for the thenar muscles and sensory branches for the skin on the palmar side of the radial 3½ digits).

373

18.1 Surface Anatomy and Superficial Nerves and Vessels: Anterior View

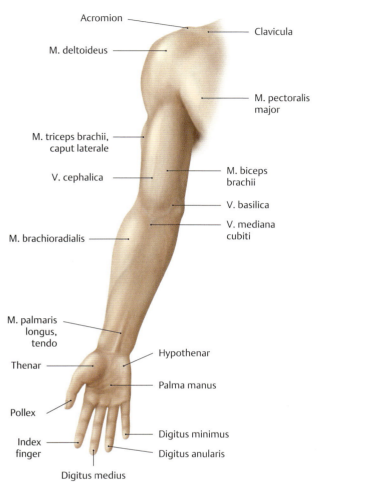

A Surface anatomy of the right upper limb
Anterior view. The palpable bony landmarks of the upper limb are reviewed on p. 239.

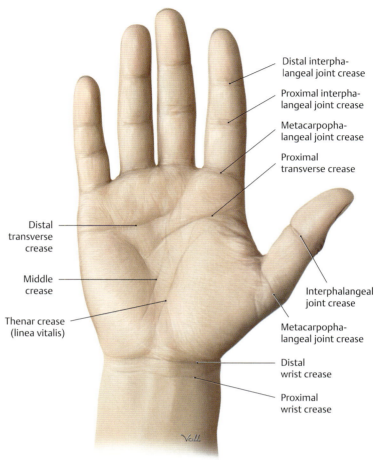

B Hand lines and flexion creases on the right palm with the wrist in slight flexion (after Schmidt and Lanz)
The *proximal wrist crease*, located approximately one finger width from the palm, coincides with the distal epiphyseal lines of the radius and ulna. The *distal wrist crease* usually overlies the art. mediocarpalis.

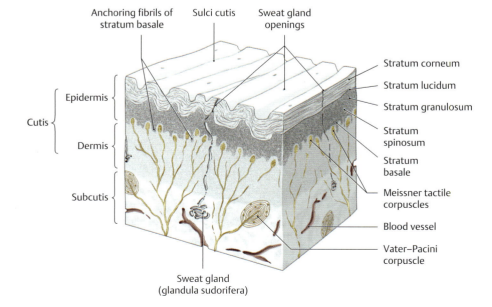

C Schematic of the structure of the ridged skin on the palm of the hand
The smooth, thin skin of the forearm gives way to the thicker, ridged skin on the palm of the hand. The papillary ridges are particularly high on the palmar skin of the fingers and, at 0.1 to 0.4 mm, are distinctly visible. The ridge pattern (dermatoglyphs) found on the bulbs of the fingers is unique for each individual. The tactile sensitivity of the fingertips is closely linked to the spatial distribution of tactile corpuscles and free nerve endings (e.g., 75–80 Vater–Pacini corpuscles per finger and approximately 100 free nerve endings per square millimeter).

Upper Limb — 18. Neurovascular Systems: Topographical Anatomy

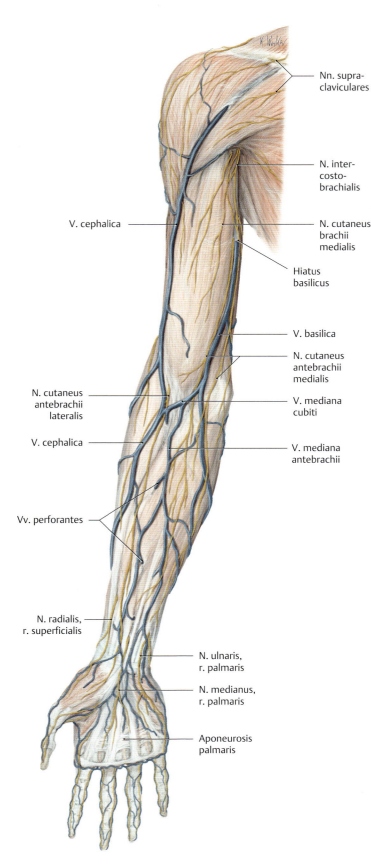

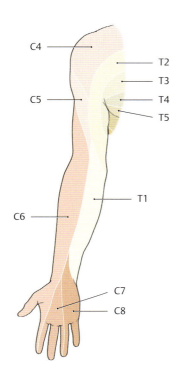

E Radicular (segmental) cutaneous innervation pattern (dermatomes) in the right upper limb
Anterior view. With the outgrowth of the upper limb during development, the sensory cutaneous segments become elongated in varying degrees to form narrow bands. In the process, segments C5–C7 become separated from the body wall.

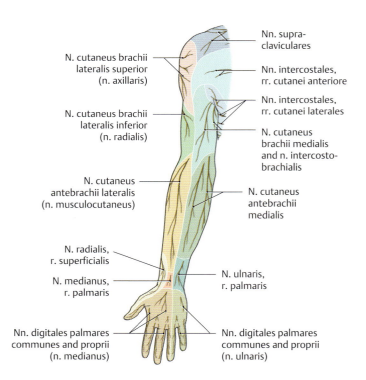

D Superficial cutaneous veins and nerves of the right upper limb
Anterior view. The arrangement of cutaneous veins about the elbow can vary considerably (see p. 356). This dissection does not show the cutaneous arteries that perforate the fascia antebrachii (particularly those arising from the a. radialis, see also p. 355).

F Peripheral sensory cutaneous innervation pattern in the right upper limb
Anterior view. The territories supplied by the peripheral cutaneous nerves (rr. cutanei) correspond to the areas of cutaneous nerve branching in the subcutaneous connective tissue that are demonstrable by dissection. The area served exclusively by a single nerve and thus rendered completely anesthetic by a lesion is much smaller because the individual sensory territories overlap extensively. *Note:* The sensory loss following damage to a *peripheral nerve* shows a completely different pattern than that of a damaged *nerve root*.

375

18.2 Surface Anatomy and Superficial Nerves and Vessels: Posterior View

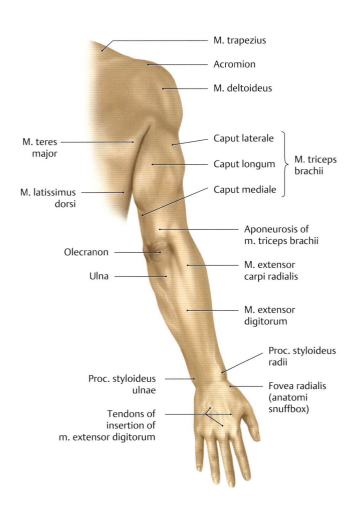

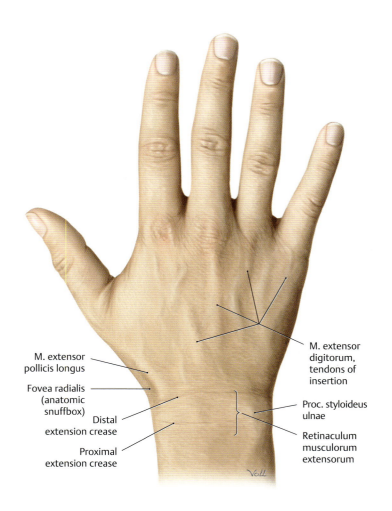

A Surface anatomy of the right upper limb
Posterior view. The palpable bony landmarks of the upper limb are reviewed on p. 239.

B Extension creases on the dorsum of the right hand (after Schmidt and Lanz)
In contrast to the palma manus, the dorsal surfaces of the hand and fingers bear indistinct extension creases that deepen with maximum dorsiflexion of the hand. The most proximal crease overlies proc. styloideus ulnae, while the most distal crease approximately overlies the distal margin of the retinaculum musculorum extensorum. Unlike the hairless ridged skin of the palm, the dorsum of the hand is covered by smooth, thin, hair-bearing skin.

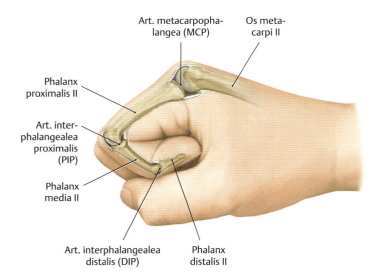

C Location of the metacarpophalangeal, proximal interphalangeal, and distal interphalangeal joint spaces
Right hand closed into a fist, radial view.

376

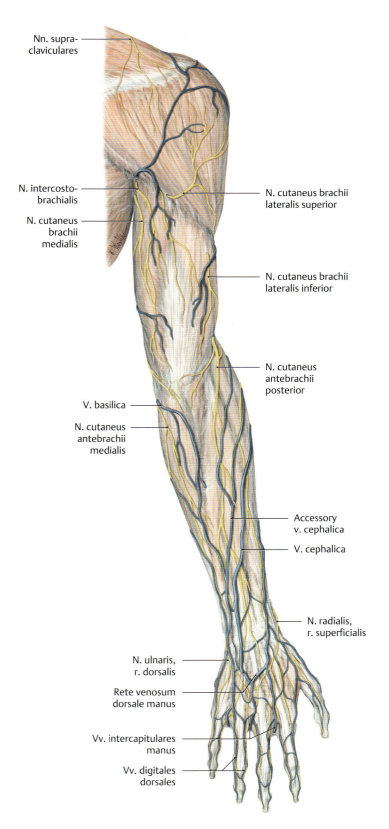

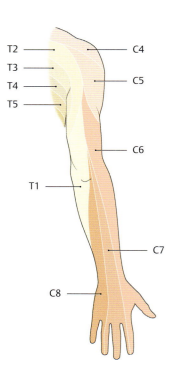

E Radicular (segmental) cutaneous innervation pattern (dermatomes) in the right upper limb

Posterior view. With the outgrowth of the limb during development, the sensory cutaneous segments become elongated in varying degrees to form narrow bands. As this occurs, segments C5–C7 become separated from the body wall.

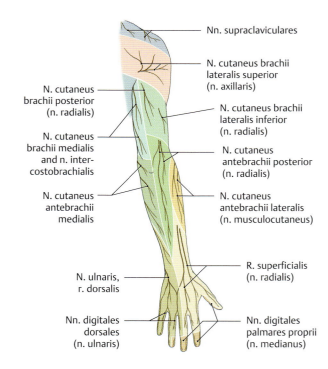

D Superficial cutaneous veins (subcutaneous veins) and cutaneous nerves of the right upper limb

Posterior view. The epifascial veins of the dorsum of the hand (rete venosum dorsale manus) display a highly variable branching pattern. Generally, the epifascial veins are clearly visible beneath the skin, receiving tributaries that include vv. perforantes from the palmar side of the hand. The v. cephalica on the radial side of the hand provides for most of the dorsal venous drainage, while the v. basilica provides for a lesser degree on the ulnar side. This dissection does not show the main branches of the a. interossea posterior that perforate the fascia antebrachii on the back of the forearm (see also p. 355).

F Peripheral sensory cutaneous innervation pattern in the right upper limb

Posterior view. The color-coded areas that are supplied by the peripheral cutaneous nerves (rr. cutanei) correspond to the areas of cutaneous nerve branching in the subcutaneous connective tissue that are demonstrable by dissection. The area of exclusive non-overlapping innervation by a specific nerve is much smaller.

Note that the sensory loss following damage to a *peripheral nerve* shows a completely different pattern from that caused by damage to a *nerve root* (see **E**).

18.3 The Shoulder Region: Anterior View

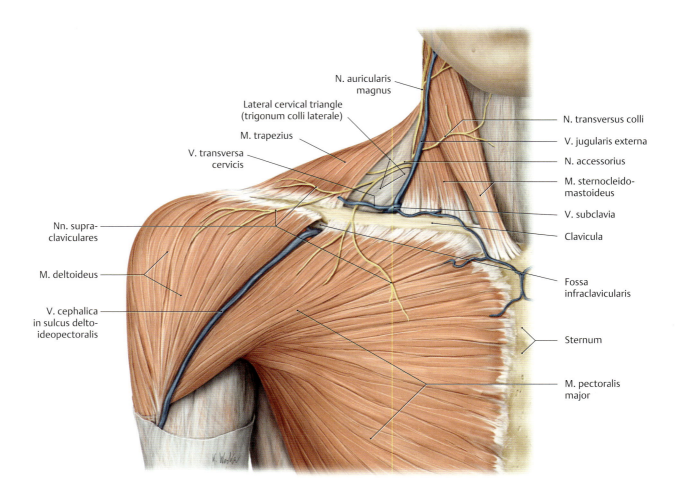

A Superficial veins and nerves of the right shoulder and neck region

Anterior view. The skin, platysma, muscle fasciae, and lamina superficialis of the fascia cervicalis have been removed in this dissection to demonstrate the branches of the plexus cervicalis (e.g., the n. auricularis magnus) and the superficial veins of the lateral and anterior neck. The v. jugularis externa and v. jugularis anterior (not shown) are visible through the skin when the patient is lying supine and the veins are well filled. When *right-sided heart failure* is present, these veins may be engorged due to the damming back of venous blood and may be visible even when the patient is sitting upright. The *v. cephalica* crosses the shoulder in the groove between the mm. pectoralis major and deltoideus (*sulcus deltoideopectoralis*) and empties into the v. axillaris. This site of entry into the v. axillaris, and thus into the deep veins, is visible and palpable on the skin as the fossa infraclavicularis. At the level of the lateral border of the first rib, the v. axillaris becomes the v. subclavia.

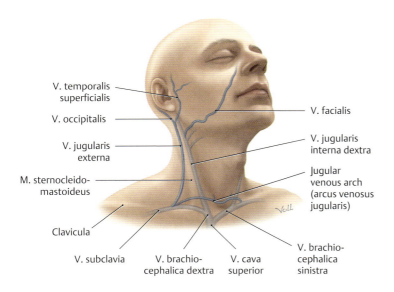

B Relationship of major superficial and deep veins in the neck to the m. sternocleidomastoideus

Anterior view. The v. jugularis interna runs almost straight downward from the foramen jugulare and unites with the v. subclavia just lateral to the art. sternoclavicularis to form the v. brachiocephalica. When its course is projected onto the side of the neck, it follows a line drawn from the earlobe to the medial end of the clavicula. The *v. jugularis interna* is crossed obliquely in its lower third by the m. sternocleidomastoideus, while the *v. jugularis externa* runs obliquely downward on the muscle and opens into the v. subclavia.

Upper Limb — 18. Neurovascular Systems: Topographical Anatomy

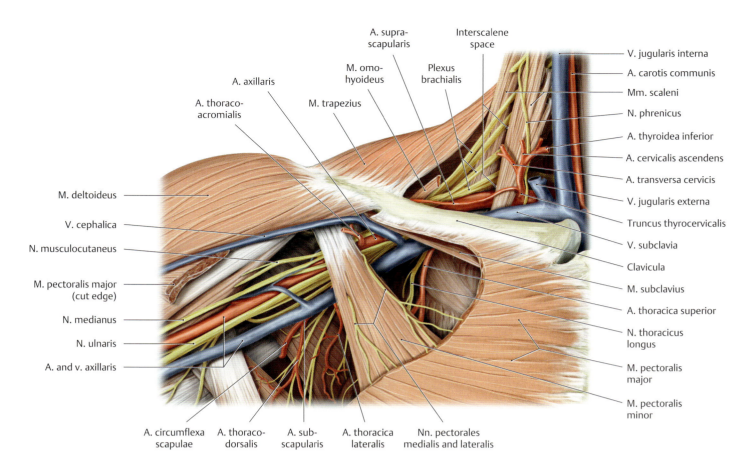

C Course of the a. subclavia dextra in the lateral neck region
Anterior view. The mm. sternocleidomastoideus and omohyoideus and all layers of the fascia cervicalis have been removed to demonstrate the deep trigonum colli laterale and the passage of the a. subclavia and plexus brachialis through the space between the mm. scaleni anterior and medius (interscalene space). At the level of the first rib the a. subclavia becomes the a. axillaris, which enters the axilla posterior to the tendon of insertion of the m. pectoralis minor.

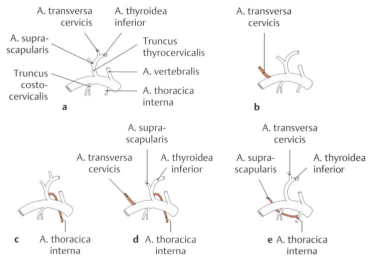

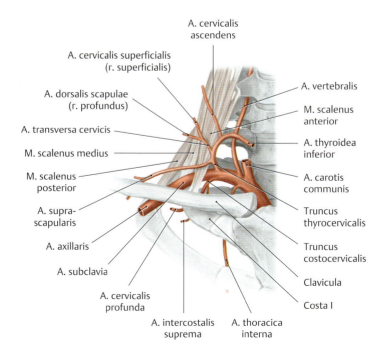

D Origin and branches of the right arteria subclavia
Anterior view.

E Branches of the a. subclavia: normal anatomy and variants
(after Lippert and Pabst)
a Normally (30% of cases), the a. subclavia gives off the following branches:
  • Truncus thyrocervicalis with the a. thyroidea inferior, a. suprascapularis, and a. transversa cervicis
  • A. vertebralis
  • A. thoracica interna
  • Truncus costocervicalis

b–e Variants:
b The a. transversa cervicis arises separately from the a. subclavia (30%).
c The a. thoracica interna arises from the truncus thyrocervicalis (10%).
d The truncus thyrocervicalis is made up of the a. thyroidea inferior, a. suprascapularis, and a. thoracica interna (8%).
e The a. subclavia gives off two main branches (4%):
  • One with the a. thyroidea inferior and a. transversa cervicis.
  • One with the a. thoracica interna and a. suprascapularis.

18.4 The Axilla (Regio axillaris): Anterior Wall

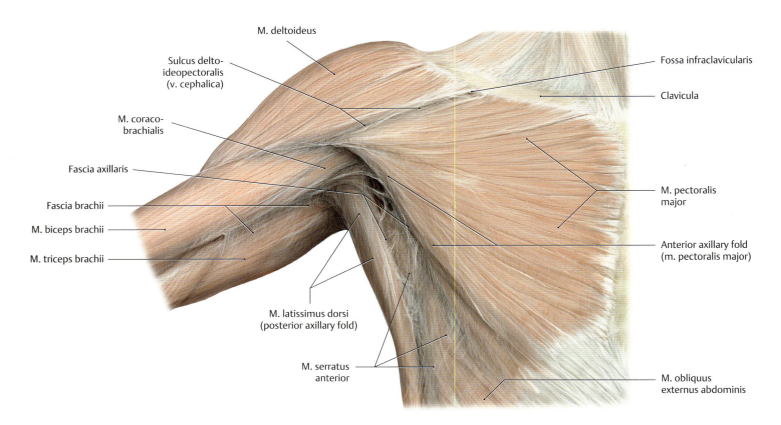

A The walls and fasciae of the right axilla
Anterior view. With the arm abducted, the axilla (Fossa axillaris) resembles a four-sided pyramid whose apex is approximately at the center of the clavicula and whose base is represented by the *fascia axillaris*. The walls of the axilla are formed by various muscles and their fasciae:

Anterior wall: The anterior wall of the axilla consists of the mm. pectorales major and minor and the fascia clavipectoralis (the m. pectoralis minor is not shown here; see **C** and **D**).
Posterior wall: This consists of the mm. subscapularis, teres major (not shown here, see p. 305), and latissimus dorsi.

Lateral wall: This is narrow and formed by the intertubercular groove (sulcus intertubercularis) of the humerus.
Medial wall: This is formed by the lateral thoracic wall (ribs 1–4 and associated intercostal muscles) and the m. serratus anterior.

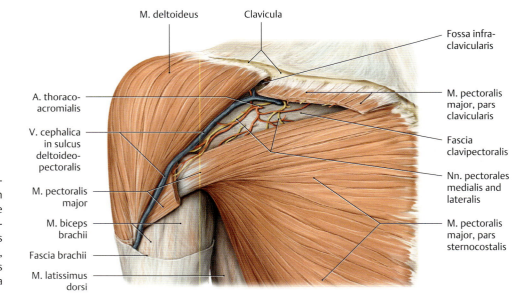

B The trigonum clavipectorale and sascia clavipectoralis
Right shoulder, anterior view. The pars clavicularis of the m. pectoralis major has been removed. In the trigonum clavipectorale bounded by the m. deltoideus, m. pectoralis major, and clavicula, the v. cephalica runs upward in the sulcus deltoideopectoralis, pierces the fascia clavipectoralis, and drains into the v. axillaris at the level of the fossa infraclavicularis.

Upper Limb — 18. Neurovascular Systems: Topographical Anatomy

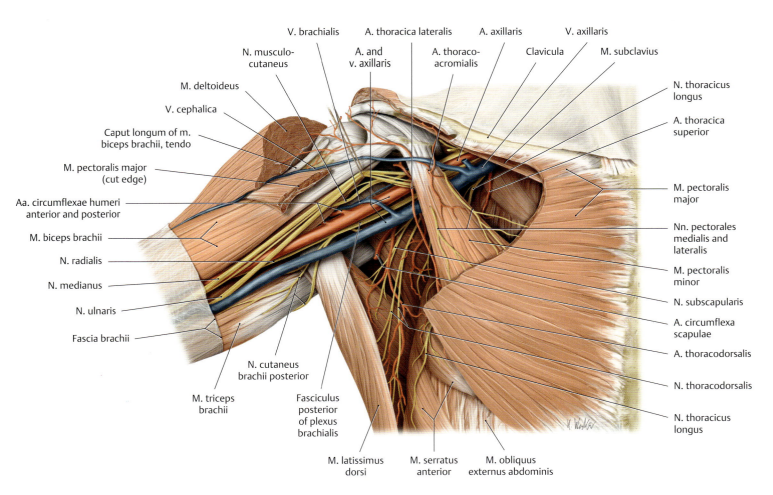

C The axilla after removal of the m. pectoralis major and fascia clavipectoralis
Right shoulder, anterior view. The a. axillaris runs approximately 2 cm below the proc. coracoideus and posterior to the m. pectoralis minor. It relates laterally to the fasciculus lateralis of the plexus brachialis and medially to the fasciculus medialis (both are retracted slightly upward in the drawing). The fasciculus posterior of the plexus brachialis, which runs behind the a. axillaris, is just visible.

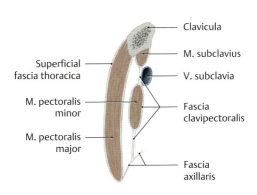

D Location of the superficial and deep thoracic fasciae
Sagittal section through the anterior wall of the right axilla. The fascia clavipectoralis, known also as the "deep" thoracic fascia, encloses the m. pectoralis minor and m. subclavius and covers the v. subclavia while being fused to its wall. The fascia is made tense by the m. pectoralis minor. The fascia clavipectoralis exerts traction on the venous wall that can keep its lumen patent, thus facilitating venous return to the vena cava superior.

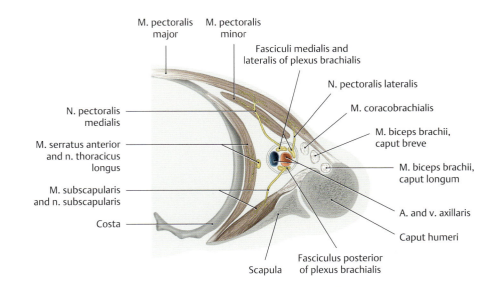

E Schematic transverse section through the right axilla
Superior view. The three muscular walls and the bony lateral wall of the axilla are clearly delineated in this view. Neurovascular structures (a. and v. axillaris plus the fasciculi medialis, lateralis and posterior of the plexus brachialis) traverse the axilla, invested by a fibrous sheath and embedded in the axillary fat.

381

18.5 The Axilla (Regio axillaris): Posterior Wall

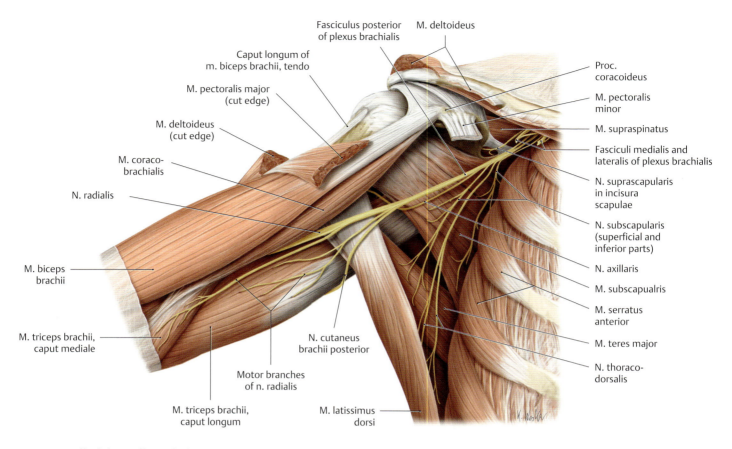

A Posterior wall of the axilla with the fasciculus posterior and its branches
Right shoulder, anterior view. The fasciculi medialis and lateralis of the plexus brachialis and the axillary vessels have been removed to demonstrate the course of the fasiculus *posterior* and its branches in the posterior axilla.

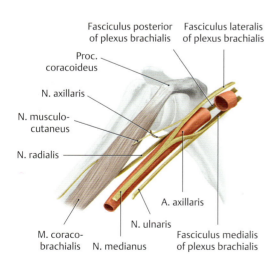

B Relationship of the fasciculi medialis, lateralis and posterior of the plexus brachialis to the a. axillaris
Note that the n. musculocutaneus passes through the m. coracobrachialis, which aids in locating the nerve. Very rarely, this nerve may be compressed as it pierces the muscle.

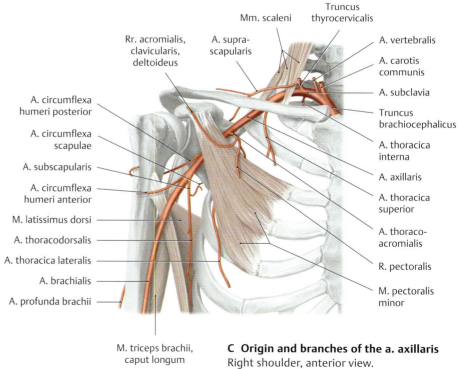

C Origin and branches of the a. axillaris
Right shoulder, anterior view.

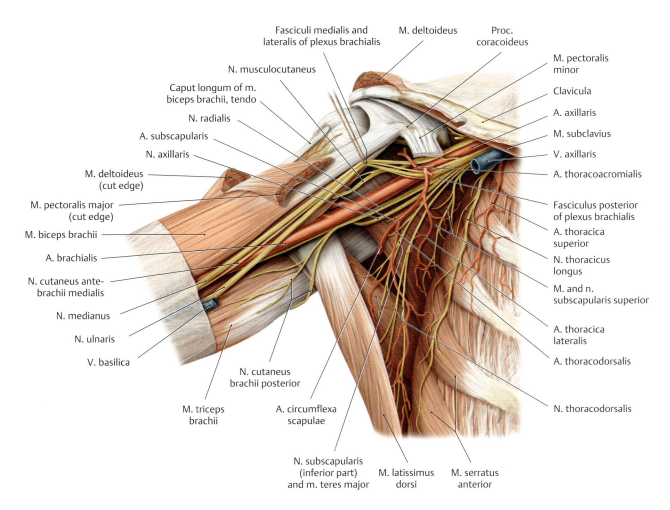

D Axilla with the entire anterior wall removed
Right shoulder, anterior view. The v. axillaris has been removed, and the fasciculus medialis and lateralis of the plexus brachialis have been retracted upward to show more clearly the location and course of the fasciculus posterior and its terminal branches, the n. radialis and n. axillaris.

Note the superficial course of the n. thoracicus longus on the m. serratus anterior.

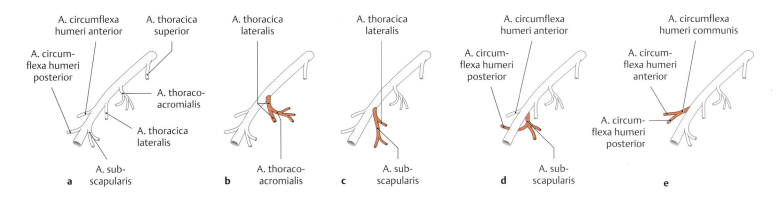

E Branches of the a. axillaris: normal anatomy and variants (after Lippert and Pabst)

a Normally (40% of cases), the a. axillaris gives off the following branches: a. thoracia superior, a. thoracoacromialis, a. thoracica lateralis, a. subscapularis, a. circumflexa humeri anterior, and a. circumflexa humeri posterior.

b–e Variants:

b The a. thoracoacromialis arises from the a. thoracica lateralis (10% of cases).

c The a. thoracica lateralis and a. subscapularis have a common origin (10% of cases).

d The a. circumflexa humeri posterior arises from the a. subscapularis (20% of cases).

e The aa. circumflexae humeri anterior and posterior have a common origin (20% of cases), the c. circumflexa humeri communis.

18.6 Conduction Anesthesia of the Plexus brachialis: Principle, Pathways, and Administering the Block

A Principle of peripheral conduction anesthesia
Peripheral conduction anesthesia is a regional anesthetic technique that blocks conduction of action potentials. The area under anesthesia is thus distal to the puncture site. It is possible to anesthetize both individual peripheral nerves and the entire plexus.

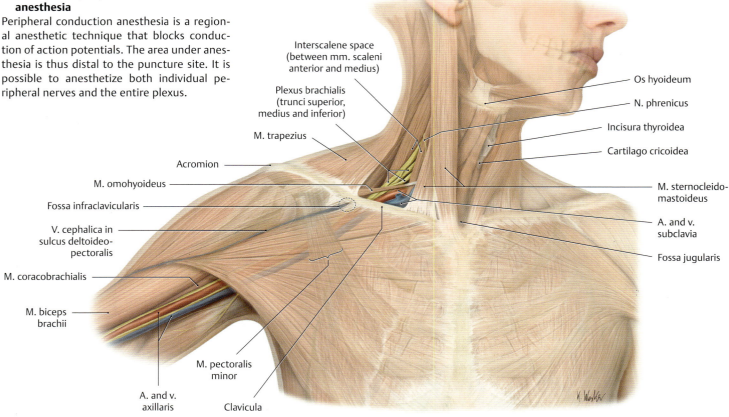

B Topography of the plexus brachialis and anatomic landmarks
The plexus brachialis is responsible for the motor and sensory nerve supply of the upper limbs. It is formed from the rr. ventrales of the nn. spinales C5–T1 (see p. 360). In the course of the plexus, first the trunci are formed, followed by the divisions, and then the fasciculi. The trunci are located at the level of the interscalene space. The divisions are above and behind the clavicula. The fasciculi run infraclavicular and start cranial or lateral to the a. axillaris and at the level of the axilla, posterior (fasciculus posterior), lateral (fasciculus lateralis), and medial (fasciculus medialis) to the a. axillaris.

Note the important anatomic landmarks for the individual pathways: m. sternocleidomastoideus, cartilago cricoidea, thyroidea, mm. scaleni anterior and medius (interscalene space), clavicula, acromion, Fossa jugularis, Fossa infraclavicularis (Mohrenheim's fossa), m. coracobrachialis, and a. axillaris. In addition, it is recommended to memorize the topographic anatomy in relation to the following structures, which could potentially get damaged: n. phrenicus, n. laryngeus recurrens, cervical or thoracic and cervical sympathetic ganglia (e.g., ganglion stellatum), a. vertebralis, cervical epidural and subarachnoid space, and pleural dome.

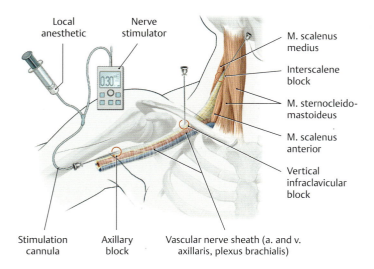

C Brachial plexus sheath and electrical nerve stimulation
During passage through the interscalene space to the axillary region, the plexus brachialis, and the accompanying a. and v. axillaris, are enclosed in a connective-tissue sheath. Within this sheath, local anesthetic can dissipate more or less evenly, anesthetizing all nerves in this area. In order to target and effectively block one particular nerve, electrical nerve stimulation is performed. A stimulation cannula, only the tip of which is not isolated, emits a defined current pulse. This current pulse can trigger action potentials at the respective motor axons (for more details about the possible responses, see **E**). With correct positioning of the cannula, ensuing injection of 1 to 2 mL of a suitable local anesthetic should lead to immediate suppression of muscle function (extinction phenomenon).

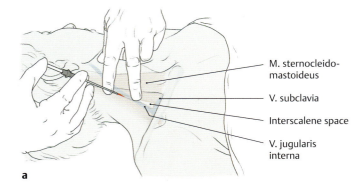

a

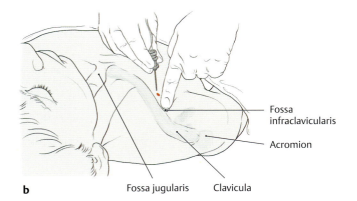

b

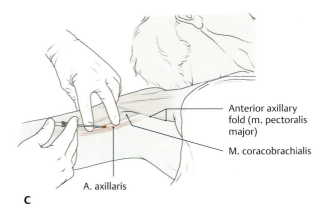

c

D Clinically important pathways for administering a brachial plexus block

In order to avoid mechanical lesions, usually the use of an electrical nerve stimulation device (see **C**) and atraumatic needles is recommended. Among common contraindications for all procedures mentioned here are infections of the puncture site and manifest coagulation disorders. Among more specific contraindications are contralateral n. phrenicus paresis and contralateral paresis of n. laryngeus recurrens.

a **Interscalene approach according to Meier:** Allows for the most proximal brachial plexus block and thus procedures involving the neck and shoulder. Anatomic landmark is the posterior edge of the m. sternocleidomastoid at the level of the incisura thyroidea 2 cm above the cartilageo cricoidea. Puncture is made from the cranial direction down toward the interscalene space with the needle at 30° to the skin.

b **Verticalinfraclavicular block according to Kilka, Geiger, and Mehrkens:** Advantage over the axillary block (see below): in addition to blocking the fasciculi, it safely blocks the n. musculocutaneus. The anatomic landmarks are the anterior edge of the acromion and the center of the fossa jugularis. The midpoint of the connecting line between those two points marks the puncture site and the lower edge of the clavicula. For better orientation, the index finger is placed in the fossa infraclavicularis (Mohrenheim's fossa). Puncture is made from medial of the index finger and strictly vertical to the exam table cover. A medial direction of puncture and puncturing too deep is to be avoided in any case due to the risk of pneumothorax.

c **Axillary brachial plexus anesthesia:** Most established, technically easiest, and least risky pathway to the plexus brachialis. Among the indications are all procedures involving the hand, forearm, and distal upper arm. Anatomic landmark is the a. axillaris medial to the m. coracobrachialis, which is usually easily palpable. Use two fingers to palpate the gap between the a. axillaris and m. coracobrachialis, which is located directly distal to the anterior axillary fold (lateral edge of the m. pectoralis major). The puncture is made in this gap with the needle being held at a 30 to 45° angle parallel to the artery. After penetrating the vessel nerve sheath, which is felt as a sturdy, elastic resistance, the needle is lowered and advanced tangentially in the proximal direction all the way. By using a nerve stimulation device, the position of the tip of the needle can be optimized. "Problematic nerves" (= those difficult to anesthetize) include the n. radialis, which runs behind the a. axillaris, and the n. musculocutaneus, which has left the vessel nerve sheath significantly more proximal.

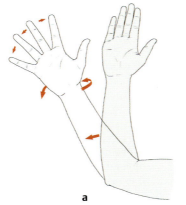
a

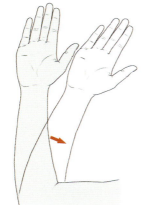
b

c

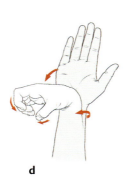
d

E Motor response of individual nerves of the upper limb following electrical nerve stimulation

a **N. radialis:** elbow joint extension (m. triceps brachii), extension and radial abduction of the wrist, supination of forearm, and finger extension.

b **N. musculocutaneus:** elbow joint flexion (m. biceps brachii).

c **N. ulnaris:** ulnar flexion of the wrist, flexion of artt. metacarpophalangeae II–V.

d **N. medianus:** flexion and pronation of the wrist, flexion of the phalanges, including the thumb.

18.7 The Anterior Brachial Region (Regio brachialis anterior)

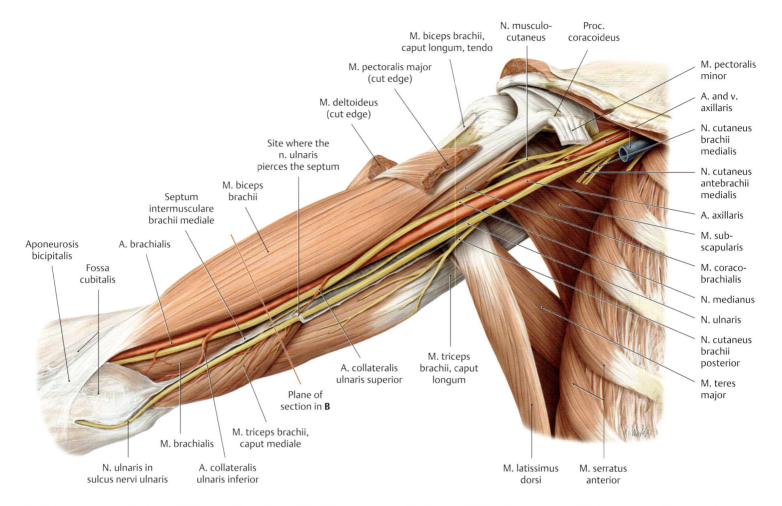

A Main neurovascular tract of the arm: the sulcus bicipitalis medialis

The right arm has been abducted and slightly rotated externally and is viewed from the anterior view. The mm. deltoideus, pectoralis major, and pectoralis minor have been removed. The sulcus bicipitalis medialis is a subcutaneous longitudinal groove on the medial side of the arm that is bounded deeply by the mm. biceps brachii and brachialis and the septum intermusculare brachii mediale. It marks the location of the main neurovascular tract of the arm extending from the axilla to the fossa cubitalis. The *most superficial* structure of the tract is the *n. cutaneus antebrachii medialis*, which leaves the sulcus bicipitalis medialis at the hiatus basilicus in company with the v. basilica (see p. 375). The *most medial* structure is the *n. ulnaris*, which initially courses on the septum intermusculare brachii mediale. In the lower third of the arm, the n. ulnaris pierces the septum intermusculare and passes to the back of the septum, entering the sulcus nervi ulnaris on the epicondylus medialis humeri. The *deep part of the sulcus bicipitalis medialis* transmits the principal artery of the arm, the *a. brachialis*, which extends from the axilla to the elbow accompanied by the *n. medianus*.

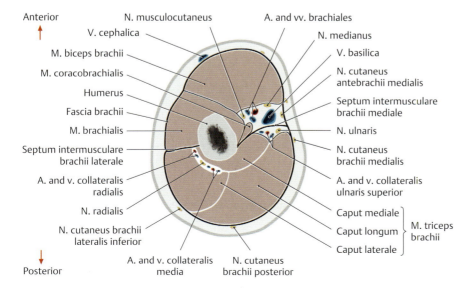

B Cross section through the middle third of the right arm

Distal view.

Note for orientation that the hiatus basilicus (where the v. basilica perforates the deep fascia medial to the m. biceps brachii) is distal to (below) the level of this section. Thus, the v. basilica and n. cutaneus antebrachii medialis are subfascial. The n. ulnaris and a. collateralis ulnaris superior have left the sulcus bicipitalis medialis and have pierced the septum intermusculare brachii mediale, and thus lie posterior to it, in this section. Proximal to (above) this level, the a. profunda brachii has split into its two terminal branches, the a. collateralis radialis and the a. collateralis media, which are seen here posterior to the humerus.

Upper Limb — 18. Neurovascular Systems: Topographical Anatomy

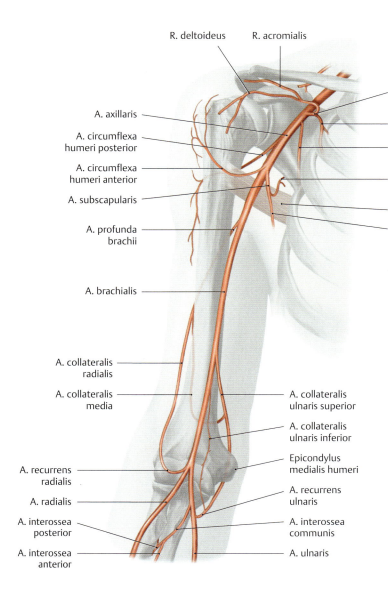

C Course of the a. brachialis in the arm
Right arm, anterior view. The a. brachialis arises from the a. axillaris at the level of the m. teres major and descends in the sulcus bicipitalis medialis to the elbow. There it divides into the a. radialis and a. ulnaris. On its way down the arm, it gives off branches to the arm muscles as well as the a. profunda brachii, which runs on the extensor side of the arm and divides distal to the sulcus nervi radialis into the a. collateralis media (to the caput mediale of the m. triceps brachii) and the a. collateralis radialis (to the rete articulare cubiti). The a. brachialis supplies the rete articulare cubiti through the aa. collaterales ulnares superior and inferior. It is significant clinically that the a. brachialis can be ligated distal to the origin of the a. profunda brachii with little risk (e.g., to control heavy posttraumatic bleeding) because the rete articulare cubiti (see **C**, p. 393) can establish an adequate collateral circulation. The m. biceps brachii is a useful landmark for locating the a. brachialis, whose pulse is palpable all along the ulnar border of that muscle.

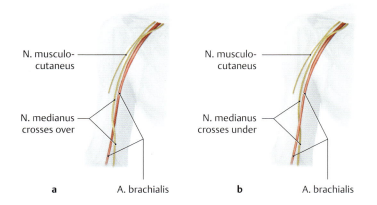

D Course of the a. brachialis in the arm:
normal anatomy and variants (after von Lanz and Wachsmuth)
Right shoulder, anterior view.

a Usually (74% of cases) the n. medianus *crosses over* the a. brachialis in the lower third of the arm.

b–d Variants:

b The n. medianus *crosses under* the a. brachialis (very rare, 1% of cases).

c, d The a. brachialis divides into a a. brachialis *superficialis* and a a. brachialis while still in the arm ("high division" pattern, 25% of cases). In cases where both of these arteries are well developed and flank the union of the n. medianus roots and the n. medianus itself, the a. radialis arises from the a. brachialis superficialis ("high origin of the a. radialis"), and the a. ulnaris is the continuation of the a. brachialis (see p. 395).

18.8 The Shoulder Region: Posterior and Superior Views

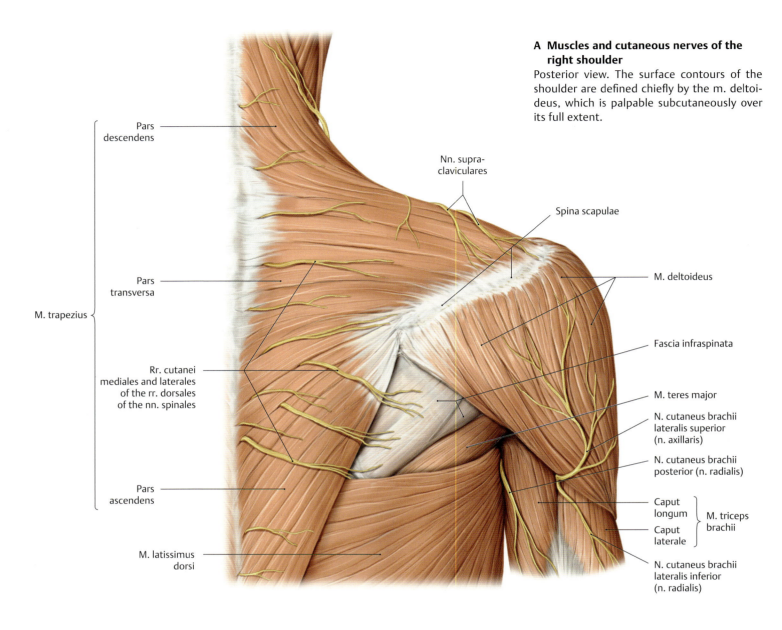

A Muscles and cutaneous nerves of the right shoulder
Posterior view. The surface contours of the shoulder are defined chiefly by the m. deltoideus, which is palpable subcutaneously over its full extent.

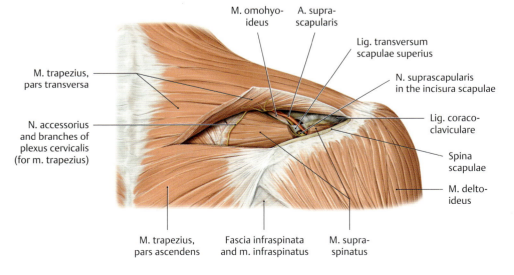

B Suprascapular region of the right shoulder
Posterior view. A flap from the pars transversa of the m. trapezius has been raised to reveal the suprascapular region, and the central portion of the m. supraspinatus has been removed.
Note the course of the *n. suprascapularis* in its fibro-osseous canal *below* the lig. transversum scapulae superius in the incisura scapulae. Compression of this nerve in its canal, especially on extreme external rotation of the shoulder, can lead to paralysis of the mm. supra- and infraspinatus (*incisura scapulae syndrome*). Ossification of the lig. transversum scapulae superius creates a foramen scapulae that can also cause n. suprascapularis compression (see p. 243).

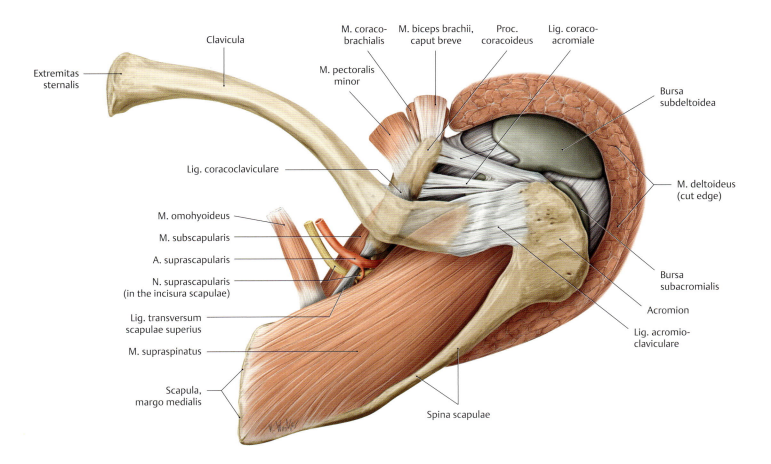

C Suprascapular region of the right shoulder
Superior view. The mm. trapezius and deltoideus have been removed to demonstrate the *m. supraspinatus*, which originates in the fossa supraspinata and passes laterally beneath the bursa subacromialis in the subacromial space to insert on the tuberculum majus. The A. and n. suprascapularis run along the anterior border of the m. supraspinatus at the level of the lig. transversum scapulae superius, just lateral to the m. omohyoideus insertion—the artery above the ligament and the nerve below it (see **B** and **D**).

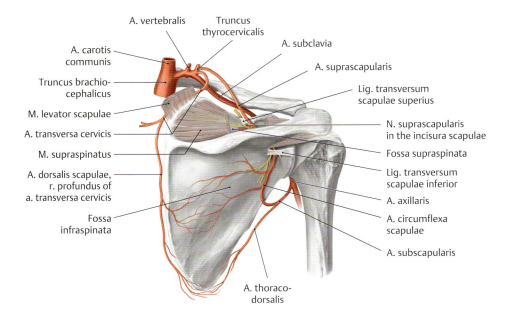

D Scapular arcade
Right scapula, posterior view. The a. suprascapularis arises from the truncus thyrocervicalis and passes *over* the lig. transversum scapulae *superius* to enter the fossa supraspinata. From there it runs past the collum scapulae, passing *under* the lig. transversum scapulae *inferius* (often absent), and enters the fossa infraspinata, where it communicates with the a. circumflexa scapulae (from the a. subscapularis) and the r. profundus (a. dorsalis scapulae) of the a. transversa cervicis.
Note the anastomosis between the a. suprascapularis and the a. circumflexa scapulae (*scapular arcade*). It is important clinically because it can provide a collateral circulation in response to ligation or occlusion of the a. axillaris (see also p. 390).

18.9 The Posterior Brachial Region (Regio brachialis posterior)

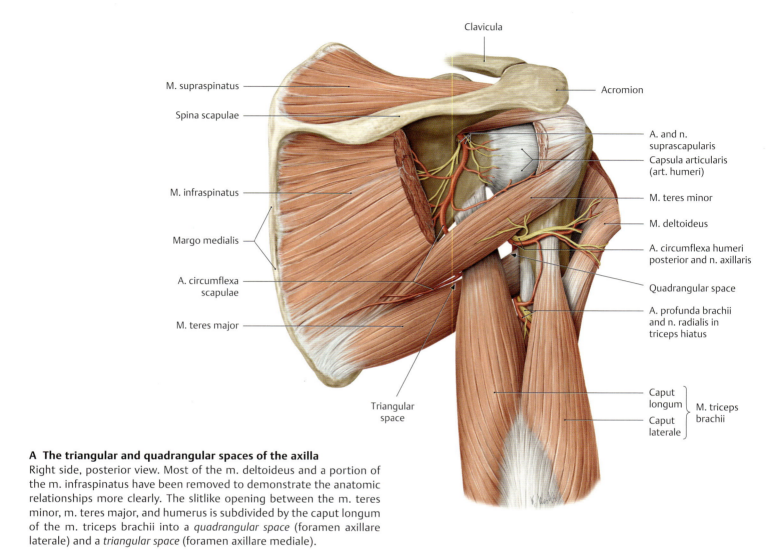

A The triangular and quadrangular spaces of the axilla
Right side, posterior view. Most of the m. deltoideus and a portion of the m. infraspinatus have been removed to demonstrate the anatomic relationships more clearly. The slitlike opening between the m. teres minor, m. teres major, and humerus is subdivided by the caput longum of the m. triceps brachii into a *quadrangular space* (foramen axillare laterale) and a *triangular space* (foramen axillare mediale).

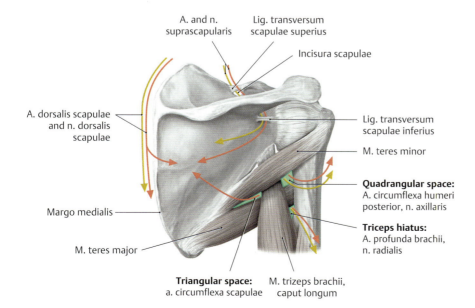

B Neurovascular tracts associated with the scapula
Right side, posterior view. The triangular and quadrangular spaces of the axilla and the triceps hiatus provide important passageways that transmit neurovascular structures from the anterior to the posterior scapular region.

Passageways	Structures transmitted
• Triangular space	A. circumflexa scapulae
• Quadrangular space	A. circumflexa humeri posterior and n. axillaris
• Triceps hiatus	A. profunda brachii and n. radialis

Upper Limb — 18. Neurovascular Systems: Topographical Anatomy

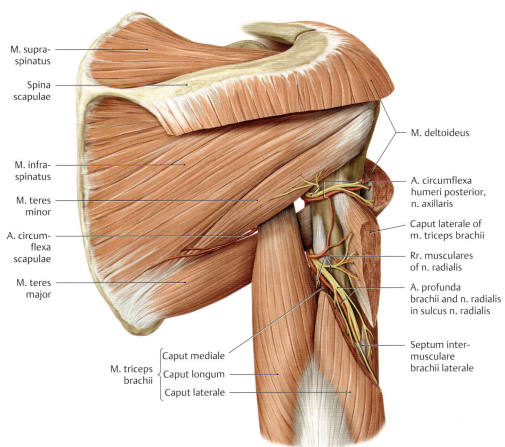

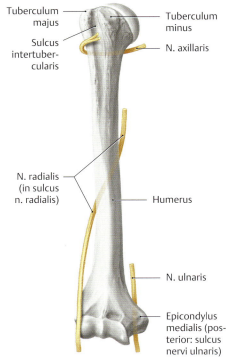

E Nerves that are closely related to the humerus
Right humerus, anterior view.

C Course of the n. radialis in the sulcus nervi radialis
Right shoulder and upper arm, posterior view. The caput laterale of the m. triceps brachii has been divided to show how the n. radialis spirals around the humerus. The dissection shows the bony sulcus nervi radialis between the origins of the caput mediale and caput laterale of the m. triceps brachii. At the distal end of the sulcus, the n. radialis passes through the septum intermusculare brachii laterale to the front of the humerus and continues in the radial tunnel to the fossa cubiti (not shown here; see also p. 392).
Note that the rr. musculares of the n. radialis for the m. triceps brachii arise proximal to the sulcus n. radialis. Thus, the m. triceps brachii may still be functional after a humeral shaft fracture at the level of the sulcus n. radialis, even though the n. radialis has been damaged, because the rr. musculares to the m. triceps brachii arise proximal to the site of the lesion.

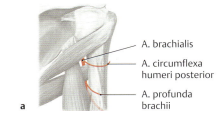

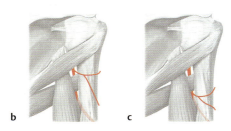

F Branches of the a. brachialis: normal anatomy and variants (after von Lanz and Wachsmuth)
a Typically (77% of cases), the a. profunda brachii and the a. circumflexa humeri posterior arise from the a. brachialis.

b, c Variants:
b The a. profunda brachii arises from the a. circumflexa humeri posterior (7%).
c As in b, but the a. circumflexa humeri posterior runs through the triceps hiatus rather than the quadrangular space (16%).

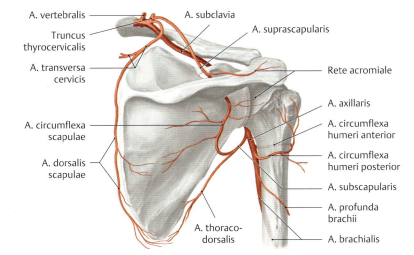

D Arterial supply to the scapular region
Right shoulder, posterior view.

391

18.10 The Elbow (Regio cubitalis)

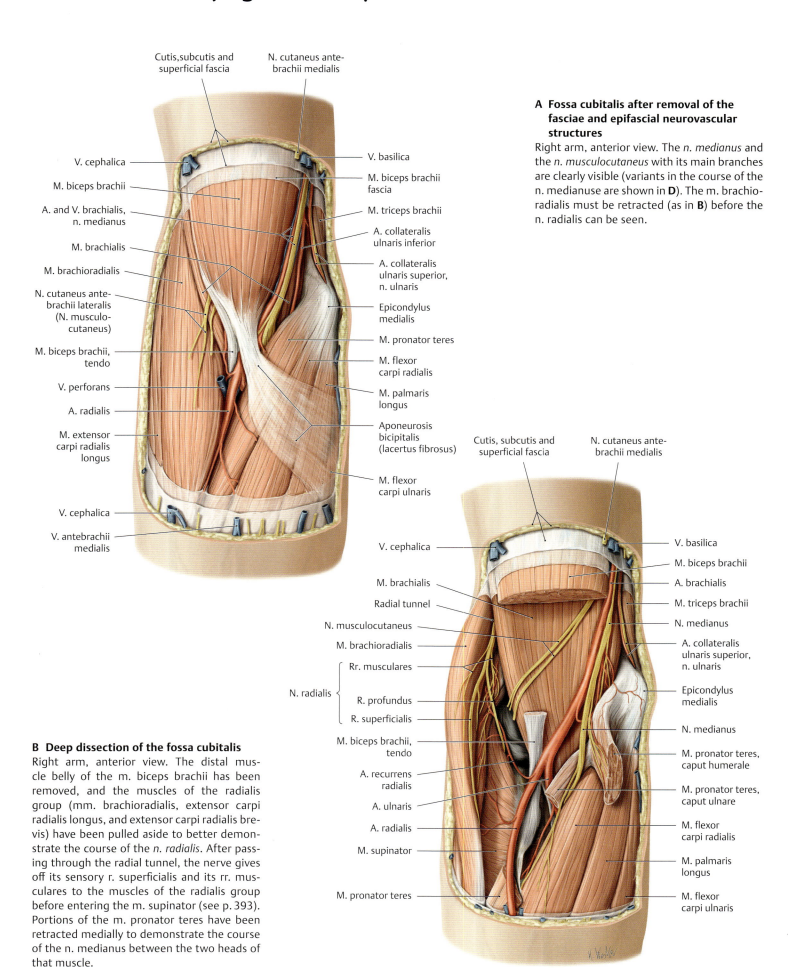

A Fossa cubitalis after removal of the fasciae and epifascial neurovascular structures
Right arm, anterior view. The *n. medianus* and the *n. musculocutaneus* with its main branches are clearly visible (variants in the course of the n. medianuse are shown in **D**). The m. brachioradialis must be retracted (as in **B**) before the n. radialis can be seen.

B Deep dissection of the fossa cubitalis
Right arm, anterior view. The distal muscle belly of the m. biceps brachii has been removed, and the muscles of the radialis group (mm. brachioradialis, extensor carpi radialis longus, and extensor carpi radialis brevis) have been pulled aside to better demonstrate the course of the *n. radialis*. After passing through the radial tunnel, the nerve gives off its sensory r. superficialis and its rr. musculares to the muscles of the radialis group before entering the m. supinator (see p. 393). Portions of the m. pronator teres have been retracted medially to demonstrate the course of the n. medianus between the two heads of that muscle.

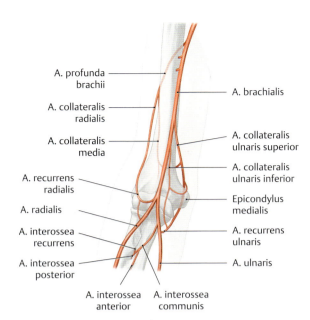

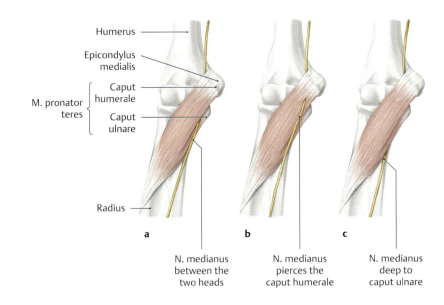

a N. medianus between the two heads

b N. medianus pierces the caput humerale

c N. medianus deep to caput ulnare

C Arterial anastomoses in the elbow region: the rete articulare cubiti

Right arm, anterior view. The arterial anastomoses in the elbow region form a vascular network (rete articulare cubiti) that is fed by several arteries:

- The a. collateralis media and a. collateralis radialis from the a. profunda brachii (communicate with the a. radialis via the a. recurrens radialis and a. interossea recurrens)
- The a. collateralis ulnaris superior and a. collateralis ulnaris inferior from the A. brachialis (communicate with the a. ulnaris via the a. recurrens ulnaris)

Because of this arterial network, the a. brachialis can be ligated distal to the origin of the a. profunda brachii without compromising the blood supply to the elbow region.

D Relationship of the n. medianus to the m. pronator teres: normal anatomy and variants (after von Lanz and Wachsmuth)
Right arm, anterior view.

a In the great majority of cases (95%) the n. medianus runs between the two heads of the m. pronator teres.

b, c Variants:
b The n. medianus pierces the caput humerale of the m. pronator teres (2% of cases).
c The n. medianus runs on the bone beneath the caput ulnare of the m. pronator teres (3% of cases).

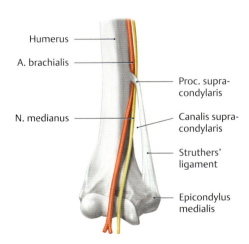

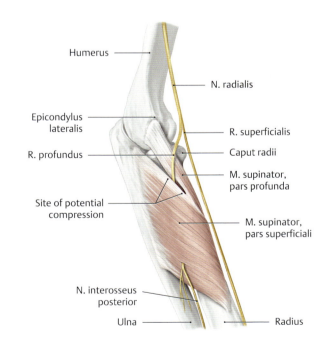

E Processus supracondylaris of the humerus
Distal humerus, right arm, anterior view. The *proc. supracondylaris* is an unusual anomaly (0.7% of the population), a bony outgrowth above the epicondylus medialis (see p. 214). When present, it can serve as an attachment for a connective tissue band referred to as Struthers' ligament, which ends on the epicondylus medialis. The resulting fibroosseous *canalis supracondylaris* can entrap and compress the a. brachialis and n. medianus.

F Relationship of the n. radialis to the m. supinator
Right elbow region, radial view. Just proximal to the m. supinator, the n. radialis divides into its motor r. profundus and sensory r. superficialis. This arrangement can lead to entrapment and compression of the motor r. profundus, with resulting selective palsy of the extensor muscles (and m. abductor pollicis longus) served by this nerve.

18.11 The Anterior Forearm Region (Regio antebrachialis anterior)

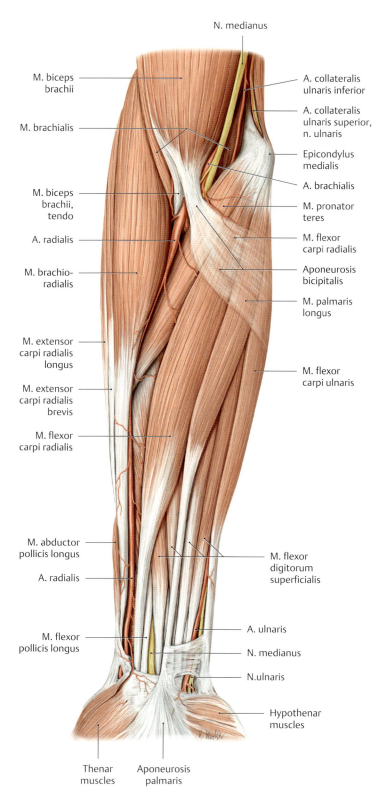

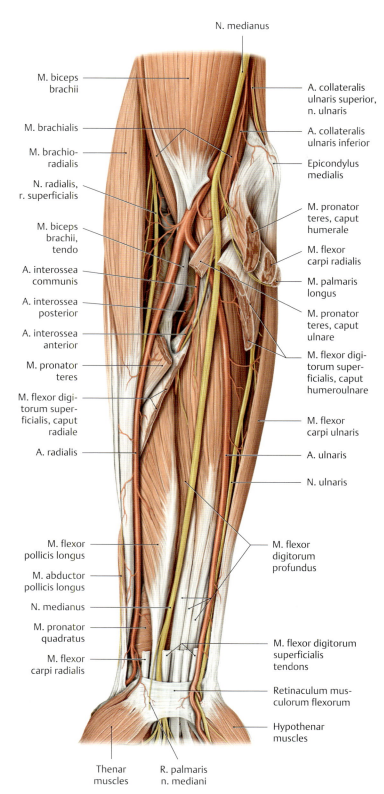

A Right forearm, superficial layer
Anterior view. The fasciae and superficial neurovascular structures have been removed. Most of the forearm's neurovascular structures are obscured in this view. (The superficial veins are shown in **D**, p. 375.)

B Right forearm, deep layer
Anterior view. The mm. pronator teres, flexor digitorum superficialis, palmaris longus, and flexor carpi radialis have been partially removed to demonstrate the n. medianus, the r. superficialis of the n. radialis, and the a. radialis and a. ulnaris (variants in the course of the arteries are shown in **D**).

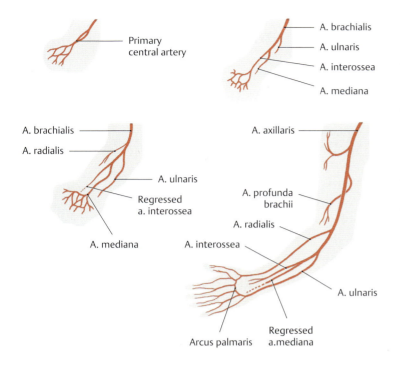

C Development of the arteries of the upper limb (after Stark)
The vascular system in the embryonic arm does not develop directly into the mature anatomy but undergoes several basic changes. The early limb bud is supplied by a central vascular trunk that develops distally into the *a. interossea communis*. As development proceeds, a second longitudinal trunk, called the *a. mediana*, is formed parallel to the n. medianus. That vessel provides most of the blood supply to the forearm and hand, while the a. interossea regresses. Finally, the initially small muscular branches enlarge on the ulnar and radial sides to form the *a. ulnaris* and the *a. radialis*, which replace the a. mediana in primates and assume its functions. The a. interossea remains the principal vessel of the arm in nonmammals, while the a. mediana is the dominant vessel in lower mammals. The a. interossea and a. mediana may persist in humans as well-developed atavistic anomalies (see **D**), providing most of the blood supply to the palm of the hand.

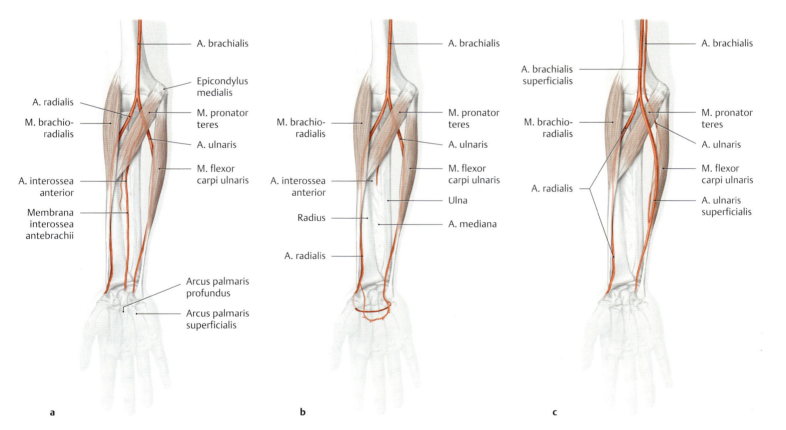

D Arteries of the forearm: normal anatomy and variants
(after Lippert and Pabst)
Right forearm, anterior view.

a Typical arterial anatomy in the forearm (84% of cases).

b, c Variants:
b A persistent a. mediana is present, often arising from the a. ulnaris distal to the origin of the a. interossea communis (8% of cases).
c Accessory superficial arteries are present in the forearm (aa. antebrachiales superficiales, 8% of cases), such as an a. ulnaris superficialis arising from an a. brachialis superficialis, which runs over the surface of the flexor muscles and may unite distally with the a. ulnaris. The presence of this vessel is a potential hazard during intravenous injections in the cubital area (see p. 356). Accessory superficial arteries most commonly develop in cases where the a. brachialis divides in the arm into an a. brachialis superficialis (which becomes the a. radialis in the forearm) and an a. brachialis (which becomes the a. ulnaris in the forearm; see "high division" pattern, p. 387).

18.12 The Posterior Forearm Region (Regio antebrachialis posterior) and the Dorsum of the Hand (Dorsum manus)

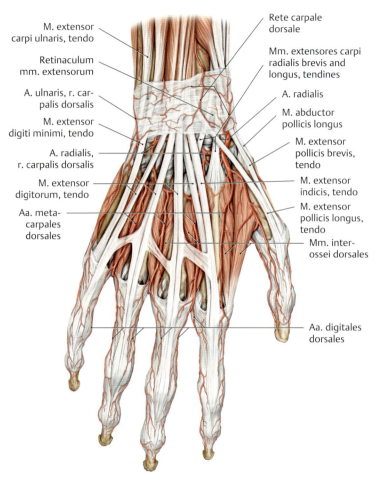

A Arteries of the dorsum of the hand (dorsum manus) and the extensor sides of the fingers in the right hand
The skin, subcutaneous tissue, and dorsal fascia of the hand have been removed to demonstrate the dorsal arteries (for clarity, the veins and nerves have also been removed). The dorsum of the hand receives most of its blood supply from the a. radialis, while the a. ulnaris contributes only one small vessel (the r. carpalis dorsalis). The rr. perforantes, however, create numerous connections between the palmar and dorsal arteries of the hand. In the fingers, these connections are provided by lateral anastomoses between the aa. digitales dorsales and the aa. digitales palmares propriae (not seen here).

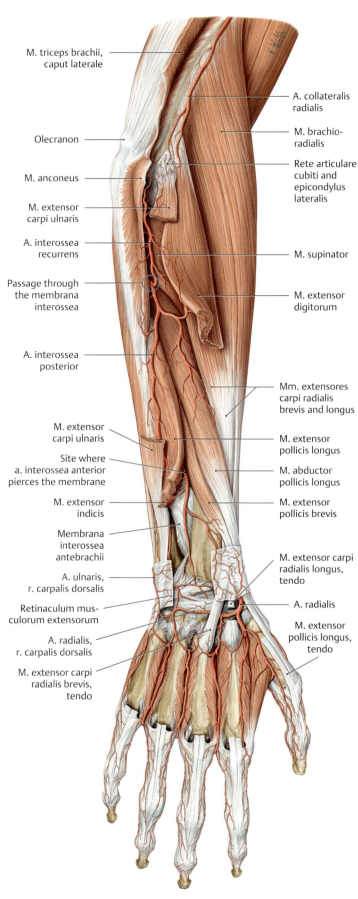

B Deep dissection of the arteries on the extensor side of the forearm and the dorsum of the hand in the right arm
In the elbow region the m. anconeus has been released from its origin and reflected to the side. The m. triceps brachii has also been released from its origin at a more proximal level. On the extensor side of the forearm, the m. extensor carpi ulnaris and m. extensor digitorum have been partially resected.
Note how the a. interossea posterior pierces the membrana interossea antebrachii just below the lower border of the m. supinator and enters the extensor compartment of the forearm. In the distal forearm, parts of the m. extensor pollicis longus and m. extensor indicis have been removed to demonstrate the site where the a. interossea anterior pierces the membrana interossea to reach the back of the forearm. Both arteries are important sources of blood to the extensor compartment.

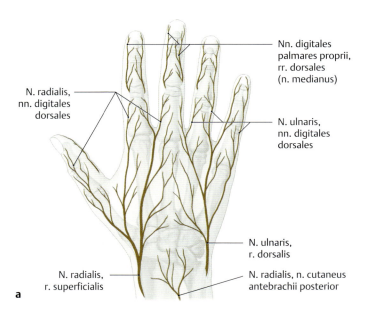
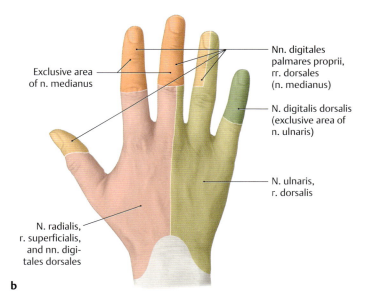

C Nerve supply to the dorsum of the hand
Right hand, posterior view.

a The cutaneous nerves on the dorsum of the hand. *Note* that the index and middle fingers and the radial part of the ring finger are supplied by different nerves in their proximal and distal portions.

- *Distal:* by the rr. dorsales of the nn. digitales palmares from the *n. medianus*.
- *Proximal:* by the nn. digitales dorsales from the *n. radialis* (to about the artt. interphalangeae proximales of the index and middle fingers) and from the *n. ulnaris* (also to the level of the artt. interphalangeae proximales of the middle and ring fingers).

b Exclusive (dark shaded) and overlapping (light shaded) areas of sensory innervation on the dorsum of the hand, for n. ulnaris, n. medianus, and n. radialis. The light shaded areas are those that receive sensory innervation mostly but not exclusively from the nerve indicated. Because the sensory territory of each nerve overlaps extensively with that of the adjacent sensory territories of other nerves, an isolated nerve lesion does not render that nerves territory without sensation. Instead extensive or complete loss of sensation is restricted to the dark shaded areas, where this is little or no overlap.

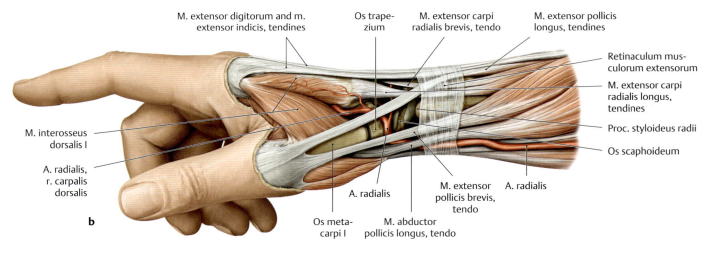

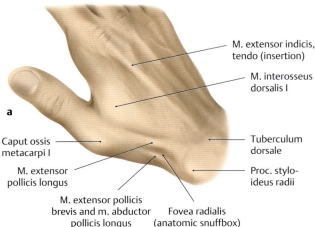

D Boundaries of the anatomic snuffbox (Foveavradialis)
a Surface anatomy of the dorsum of the right hand, posterolateral view.
b Muscles and tendons of the dorsum of the right hand, radial view. The three-sided fovea radialis ("anatomic snuffbox") is bounded on the palmar side by the tendons of insertion of the mm. abductor pollicis longus and extensor pollicis brevis and dorsally by the tendon of insertion of the m. extensor pollicis longus. The floor is formed mostly by the os scaphoideum and os trapezium. Fractures of the os scaphoideum are thus often associated with deep tenderness in the snuffbox. The snuffbox is bounded proximally by the retinaculum mm. extensorum.

Note that the a. radialis runs deep in the snuffbox between the os trapezium and os scaphoideum, providing a landmark for dissection.

397

18.13 The Palm of the Hand (Palma manus): Epifascial Nerves and Vessels

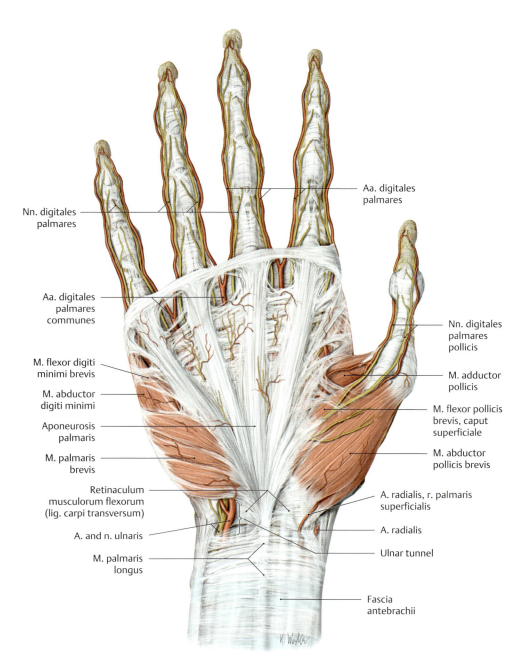

A Superficial arteries and nerves of the palm

Right hand, anterior view. All of the fasciae except the aponeurosis palmaris have been removed to demonstrate the superficial neurovascular structures of the palm. The lig. carpi palmare has also been removed to demonstrate the neurovascular structures that pass through the ulnar tunnel (the a. and n. ulnaris; see p. 405).

Note the r. palmaris superficialis of the a. radialis, which is highly variable in its course. In the case shown, it runs between the origins of the mm. abductor and flexor pollicis brevis to the palm of the hand. In approximately 30% of cases it combines with the a. ulnaris to form the arcus palmaris superficialis (not shown here; see also p. 400).

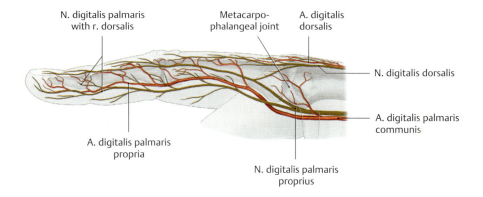

B Arteries and nerves of the right middle finger

Lateral view. The arteries of the palm are anterior to the nerves, but they are dorsal to the nerves in the fingers (generally crossing at the level of the metacarpophalangeal joint). The lateral and distal dorsal surfaces of the fingers are supplied by branches of the n. digitalis palmaris proprius (from the n. medianus).

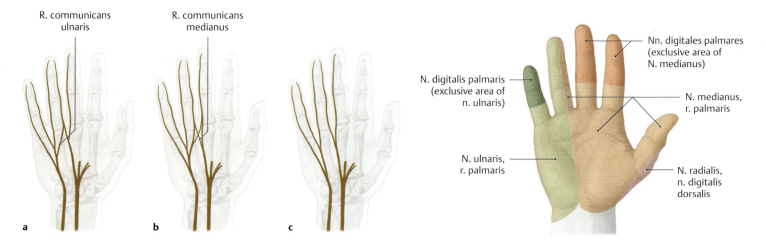

C Nerve supply to the palm of the hand
Right hand, anterior view.

a–c Innervation patterns in the palm of the hand (after Schmidt and Lanz). The sensory innervation pattern is marked by the presence of connecting branches between the n. medianus and n. ulnaris. The following innervation patterns are most frequently encountered:

a Most commonly (46% of cases) the n. medianus and n. ulnaris are interconnected by an r. communicans ulnaris.

b Variant 1 (20% of cases): The n. medianus and n. ulnaris are cross-connected by an r. communicans ulnaris and a r. communicans medianus.

c Variant 2 (20% of cases): There are no communicating branches between the n. medianus and n. ulnaris.

d Exclusive (dark shaded) and overlapping (light shaded) areas of sensory innervation of the hand, palmar view. Non-overlapping areas of exclusive innervation are indicated by darker shading. Compare with the dorsal aspect (**C**, p. 397).

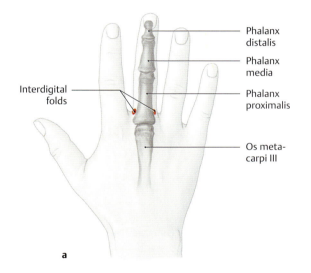
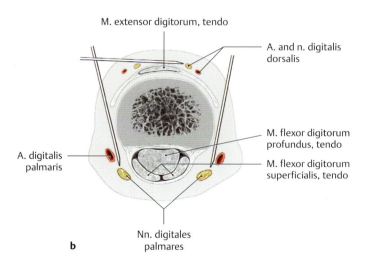

D The Oberst nerve block
Right hand, posterior view. This type of local anesthesia is clinically useful for injuries of the fingers, specifically for wounds that require suturing.

a The injection sites are located in the interdigital folds.
b After the dorsal nerve branches have been numbed, the needle is advanced on both the radial and ulnar sides toward the palmar nerves, and a subcutaneous bolus of 1 to 2 mL of local anesthetic is injected at each site.

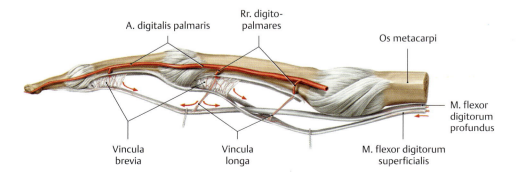

E Blood supply to the flexor tendons of the finger within the tendon sheath
(after Lundborg)

Right middle finger, lateral view. The flexor tendons are supplied within their sheath by branches of the aa. digitales palmares that are transmitted to the tendons through the mesotendineum (vinculum longum and vinculum breve).

18.14 The Palm of the Hand (Palma manus): Vascular Supply

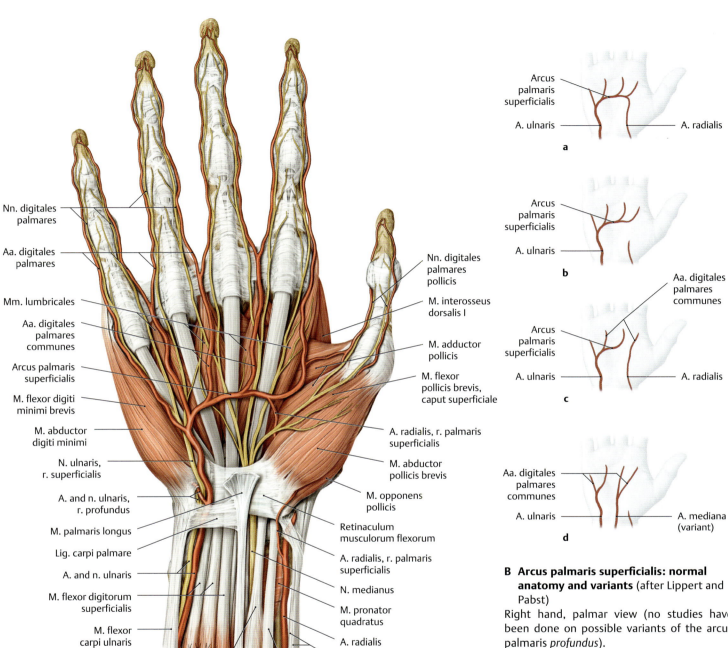

A The superficial palmar arch (arcus palmaris superficialis) and its branches
Right hand, anterior (palmar) view. The aponeurosis palmaris and other fasciae have been removed to demonstrate the *arcus palmaris superficialis* (variants are shown in **B**).

B Arcus palmaris superficialis: normal anatomy and variants (after Lippert and Pabst)
Right hand, palmar view (no studies have been done on possible variants of the arcus palmaris *profundus*).

a Normally (37% of cases), the a. radialis and a. ulnaris contribute equally to the arcus palmaris superficialis.

b–d Variants:
b The arcus palmaris arises entirely from the a. ulnaris (37% of cases).
c All of the aa. digitales palmares communes arise from the a. ulnaris except for the first, which arises from the a. radialis a. mediana (13%).
d The a. ulnaris and an a. mediana give off the aa. digitales palmares communes (very rare).

Upper Limb — 18. Neurovascular Systems: Topographical Anatomy

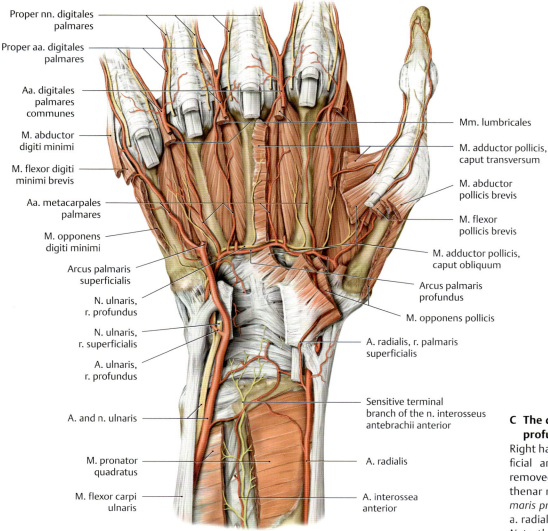

C The deep palmar arch (arcus palmaris profundus) and its branches

Right hand, anterior (palmar) view. The superficial and deep flexor tendons have been removed along with the thenar and hypothenar muscles to demonstrate the *arcus palmaris profundus* as the terminal branch of the a. radialis.

Note the end branch of the n. interosseus antebrachii anterior, which extends between the membrana interossea and m. pronator quadratus distally and supplies sensory innervation to the capsule of the wrists.

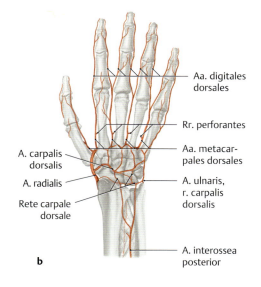

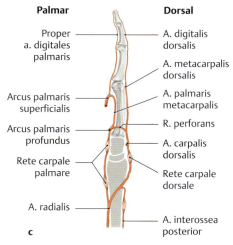

D Arterial anastomoses in the hand

The a. ulnaris and a. radialis are interconnected by the arcus palmaris superficialis and profundus, the rr. perforantes, and the rete carpale dorsale.

a Right hand, anterior (palmar) view.
b Right hand, posterior (dorsal) view.
c Right middle finger, lateral view.

401

18.15 The Carpal Tunnel (Canalis carpi)

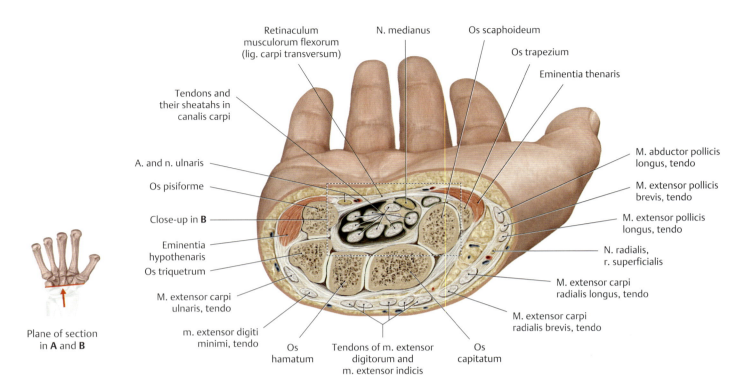

A Cross section through the right wrist (see also **B**)
Distal view. The *carpal tunnel (canalis carpi)* is a fibro-osseous canal (see p. 288) through which pass the n. medianus and the tendons of insertion of the Mm. flexor digitorum superficialis, flexor digitorum profundus, and flexor pollicis longus. Its dorsal boundary is formed by the sulcus carpi on the anterior surface of the ossa carpi, and its palmar boundary is the retinaculum mm. flexorum (also known clinically as the lig. carpi transversum). The a. ulnaris and n. ulnaris pass through the *ulnar tunnel* on the palmar side of the retinaculum mm. flexorum (see p. 405).

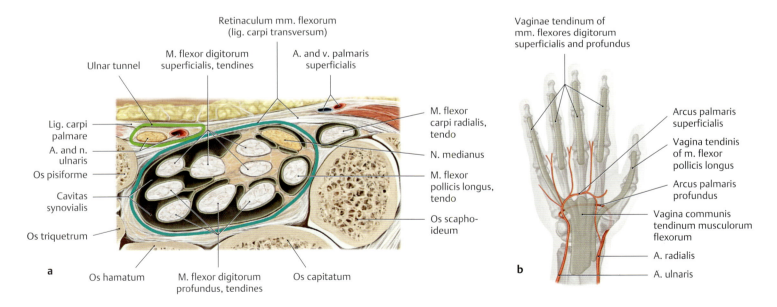

B Relationship of the palmar arches to the carpal and digital tendon sheaths (after Schmidt and Lanz)
a Tendon sheaths in the canalis carpi (detail from **A**). The long flexor tendons pass through the canalis carpi, encased in their palmar tendon sheaths. The tendons of the mm. flexores digitorum superficialis and profundus are contained in their own ulnar synovial sheath. Radial to this sheath is the tendon of the m. flexor pollicis longus. The common mesotendineum of all the digital flexor tendons is attached to the radial and palmar walls of the canalis carpi. The n. medianus generally occupies a separate space just deep to the retinaculum mm. flexorum (the variable course of the tendon sheaths is described on p. 344).
b Relationship of the carpal and digital tendon sheaths to the palmar arches.
Right hand, anterior view.

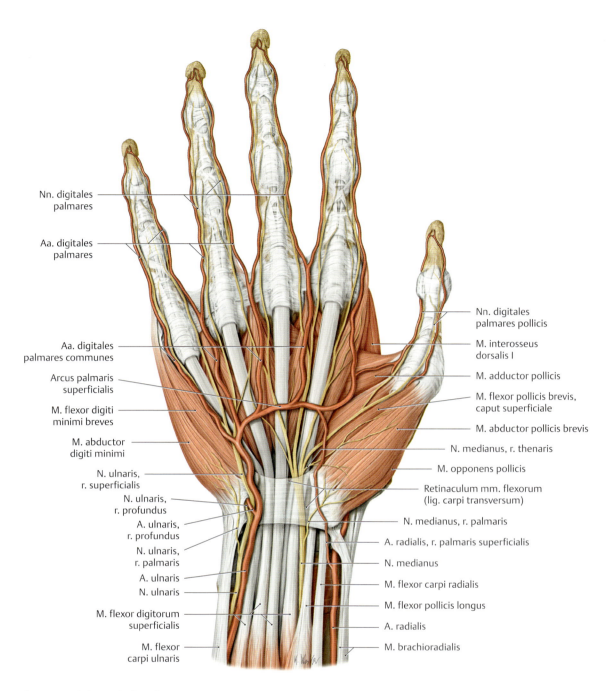

C View into the canalis carpi of the right hand
Anterior (palmar) view. The lig. carpi transversum is transparent to reveal the contents of the canalis carpi. The ulnar tunnel has been opened to expose the a. ulnaris and n. ulnaris.
Note the superficial course of the n. medianus in the canalis carpi and the origin of its motor r. thenaris just distal to the retinaculum mm. flexorum (variants are shown in **D**). During surgical division of the lig. carpi transversum for carpal tunnel syndrome, the hand surgeon must be aware of its variable course to avoid cutting the r. thenaris.
The r. palmaris superficialis of the a. radialis runs on the lig. carpi transversum in the case shown here, but frequently it passes through the thenar muscles (see p. 320).

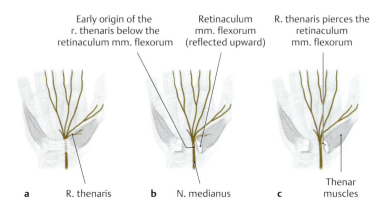

D Origin of the motor r. thenaris of the n. medianus: normal anatomy and variants (after Schmidt and Lanz)
a Normally (46% of cases) the n. medianus gives off its r. thenaris distal to the retinaculum mm. flexorum (lig. carpi transversum).

b, c Variants:
b The r. thenaris has a subligamentatous origin and course (31% of cases).
c The r. thenaris pierces the retinaculum mm. flexorum (lig. carpi transversum; approximately 23% of all cases), making it vulnerable during surgical division of the retinaculum.

403

18.16 The Ulnar Tunnel and Anterior Carpal Region (Regio carpalis anterior)

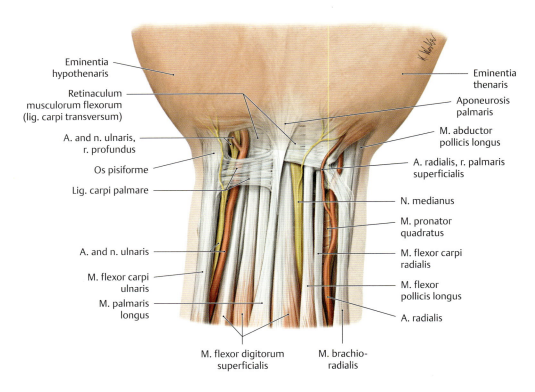

A Superficial structures of the regio carpalis anterior
Right hand, anterior view. The skin and fascia antebrachii have been removed to demonstrate the superficial structures of the regio carpalis anterior, which is bounded distally by the retinaculum mm. flexorum (lig. carpi transversum). The tendons of insertion of the mm. flexor carpi ulnaris, palmaris longus, and flexor carpi radialis in particular are clearly visible and palpable beneath the skin, especially when the fist is tightly clenched and the wrist is in slight flexion (see **B**). The tendon of the m. flexor carpi radialis is a useful landmark for locating the *a. radialis pulse*. The m. flexor carpi ulnaris tendon is palpable proximally over the os pisiforme.

Note: Due to their superficial course, the nn. medianus and ulnaris and the aa. radialis and ulnaris are particularly susceptible to injury from lacerations of the wrist.

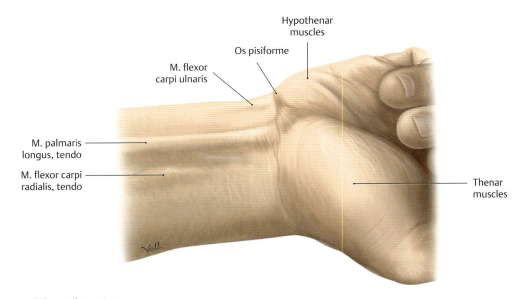

B Surface anatomy of the right wrist
Anterior view.

Upper Limb — 18. Neurovascular Systems: Topographical Anatomy

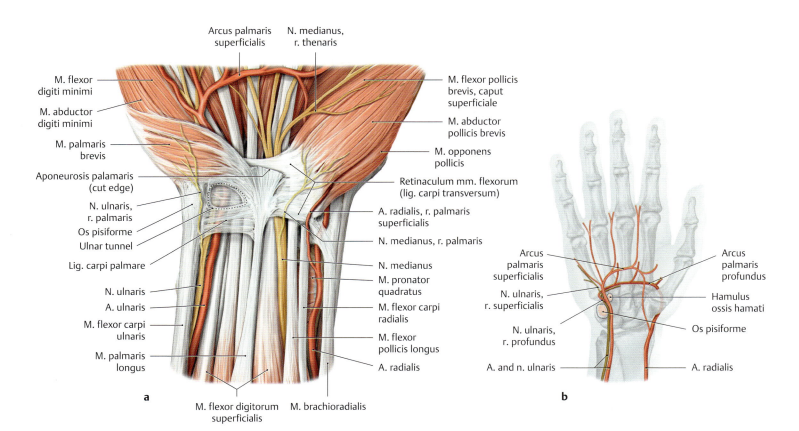

C Course of the a. and n. ulnaris in the ulnar tunnel and deep palm
Right hand, anterior view.

a The aponeurosis palmaris and fascia antebrachii have been removed to demonstrate the course of the a. and n. ulnaris through the ulnar tunnel.

b Bony landmarks within the ulnar tunnel.
Right hand, anterior view. The os pisiforme on the ulnar side of the wrist and the hamulus ossis hamati, which lies more distally and radially, provide the bony landmarks between which the a. and n. ulnaris pass through the ulnar tunnel.

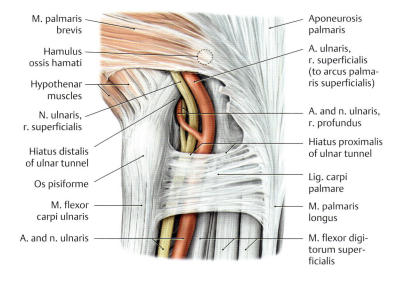 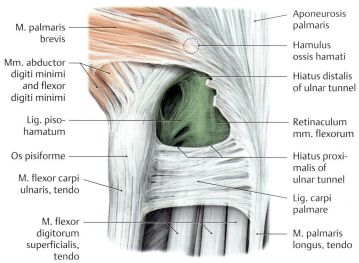

D Apertures and walls of the ulnar tunnel with (a) and without (b) the nerves and vessels (after Schmidt and Lanz)
Right hand, anterior view. The *palmar* roof of the ulnar tunnel is formed by skin and subcutaneous fat, the lig. carpi palmare (proximal), and the m. palmaris brevis (distal). The ulnar tunnel is bounded *dorsally* by the retinaculum mm. flexorum (lig. carpi transversum) and the lig. pisohamatum. The *entrance* to the tunnel begins at the level of the os pisiforme below the lig. carpi palmare (hiatus proximalis). The *outlet* is at the level of the hamulus ossis hamati, marked by a taut, transverse, crescent-shaped arcus tendineus between the os pisiforme and the hamulus ossis hamati (hiatus distalis), the latter giving attachment to the m. flexor digiti minimi. The *r. profundus* of the a. and n. ulnaris reach the central compartment of the palm on the lig. pisohamatum, passing deep to the arcus tendineus. The *r. superficialis* of the a. and n. ulnaris run distally above the arcus tendineus, passing deep to the m. palmaris brevis.

405

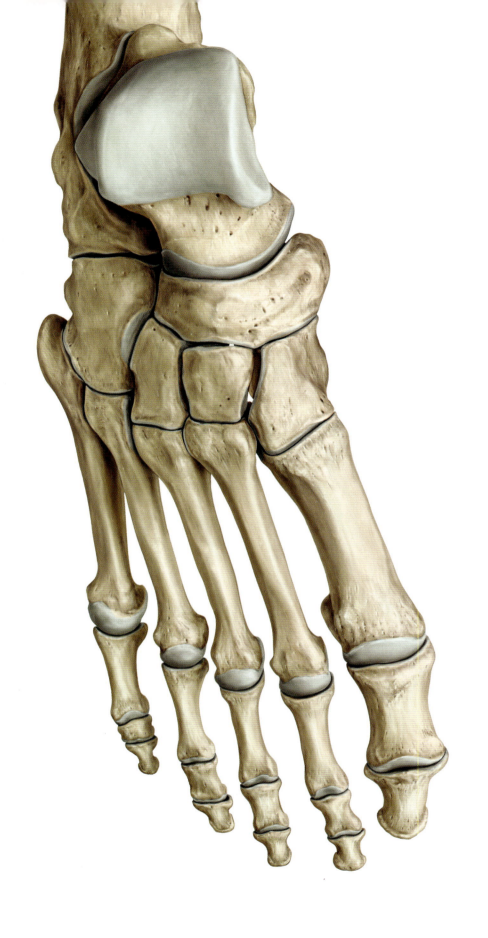

Lower Limb

19 Bones, Ligaments, and Joints 408
20 Musculature: Functional Groups 474
21 Musculature: Topographical Anatomy 502
22 Neurovascular Systems: Forms and Relations 524
23 Neurovascular Systems: Topographical Anatomy . . 544

19.1 The Lower Limb: General Aspects

A Unique features and specialized function of the human lower limb

The evolution of the lower limb into a mechanism specifically adapted for bipedal locomotion, along with the specialization of the upper limb for visually guided manipulation, is a distinctive feature of the anatomy of the human primate. The uniquely human conformation of shapes and proportions is the end result of a process that rearranged the primate center of gravity and the positions of internal organs, dramatically altering the form and biomechanics of the trunk to produce a progressively more efficient bipedal gait. Other primates have the capacity to assume an erect body posture and to walk upright, but only for short periods and at a much greater relative expenditure of energy. The habitual upright gait of humans has been achieved through a series of anatomic adaptations of the musculoskeletal system. The most critical of these adaptations occurred in the vertebral column and pelvis. The design of the human vertebral column differs markedly from that of other primates: the simple "arch and cord" construction of the chimpanzee spine has been abandoned in favor of the human double-S-shaped curve, which allows the human axial skeleton (skeleton axiale) to act as a shock-absorbing spring (see p. 103), while shifting the entire weight of the trunk over the load-bearing surface of the feet. This shift to an upright posture has imposed the full weight of the abdominal viscera upon the pelvis. Concomitantly, the iliac wings of the pelvis have spread farther apart, and the sacrum has broadened, to generate a structure in humans that is now specialized for bearing the load of the viscera. The efficiency of upright gait has been improved further by stabilization of the pelvis and secure anchoring to the spine via the sacrum. The unique proportions of the human lower limb provide a dramatic demonstration of the extent of this specialization. Because their function is more exclusively directed toward support and locomotion, the legs are exceptionally long and powerful in humans. While the leg length is 111% of trunk length in orangutans and 128% in chimpanzees, it measures 171% of trunk length in humans. The specialization of the human lower limb for bipedal gait is also reflected in the substantial changes in function of certain muscles, particularly the gluteal muscles, the knee-joint extensors, and the muscles of the calf.

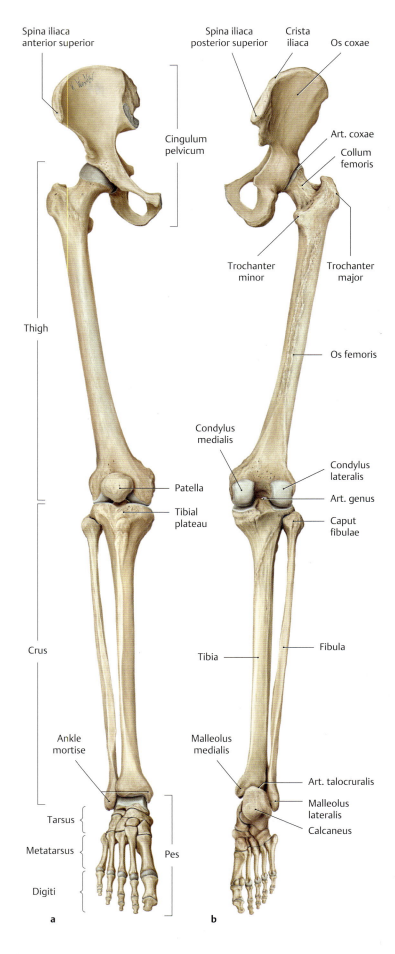

B Skeleton of the lower limb
a Right lower limb, anterior view.
b Right lower limb, posterior view (the foot is in maximum plantar flexion in both views).

As in the upper limb, the skeleton of the lower limb consists of a limb girdle and the attached free limb.

- The pelvic girdle (cingulum pelvicum/cingulum membri inferioris) in adults is formed by the paired hip bones (ossa coxae, innominate bones). They differ from the shoulder girdle in that they are firmly integrated into the axial skeleton through the sacroiliac joints (see p. 142). The two hip bones (ossa coxae) combine with the os sacrum and symphysis pubica to form the *pelvic ring* (see p. 413).
- The free limb (pars libera membri inferioris) consists of the thigh (femur), the leg (crus, with tibia and fibula), and the foot (pes). It is connected to the pelvic girdle by the joint (art. coxae).

Lower Limb — 19. Bones, Ligaments, and Joints

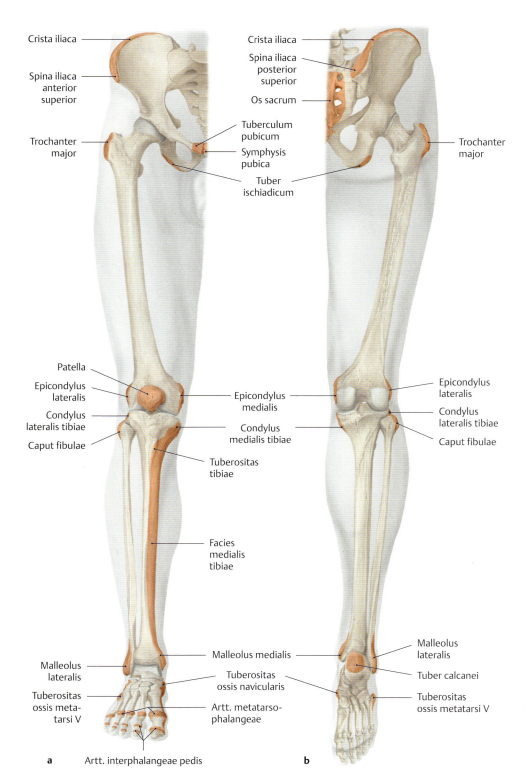

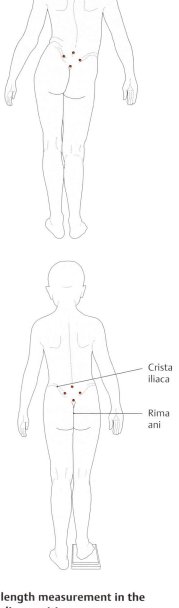

C Palpable bony prominences of the right lower limb

a Anterior view, **b** posterior view.

Almost all the skeletal elements of the lower limb have bony prominences, margins, or surfaces (e.g., the facies medialis of the Tibia) that can be palpated through the skin and soft tissues. The only exceptions are structures that are largely covered by muscle, such as the hip joint, the neck and shaft of the femur, and large portions of the shaft of the fibula. Several standard anatomic landmarks have been defined in the lower limb for use in measuring the length of the leg and certain skeletal elements. They are the spina iliaca anterior superior, the trochanter major, the medial joint space of the knee (superior margin of the condylus medialis tibiae), and the malleolus medialis. The clinical evaluation of leg length discrepancy is important because "true" shortening of the leg (a disparity of anatomic leg lengths), as well as functional leg shortening (e.g., due to muscle contractures), can lead to *pelvic tilt* and scoliotic deformity of the spine (see p. 133).

D Leg length measurement in the standing position

Leg length discrepancy can be measured with reasonable accuracy in the standing patient by placing wooden blocks of known thickness (0.5 cm, 1 cm, 2 cm) beneath the foot of the shorter leg until the pelvis is horizontal. Horizontal position is confirmed when noting that both cristae iliacae are at the same level when palpated from behind and the rima ani is vertical. If the pelvis cannot be leveled by placing blocks under the apparently shorter limb, then a "functional" leg length discrepancy is present rather than a "true" discrepancy. Most cases of this kind are caused by a fixed pelvic tilt secondary to a hip joint contracture or scoliosis. The measured leg lengths in these cases may actually be equal, and the pelvic tilt only mimics a length discrepancy.

19.2 The Anatomic and Mechanical Axes of the Lower Limb

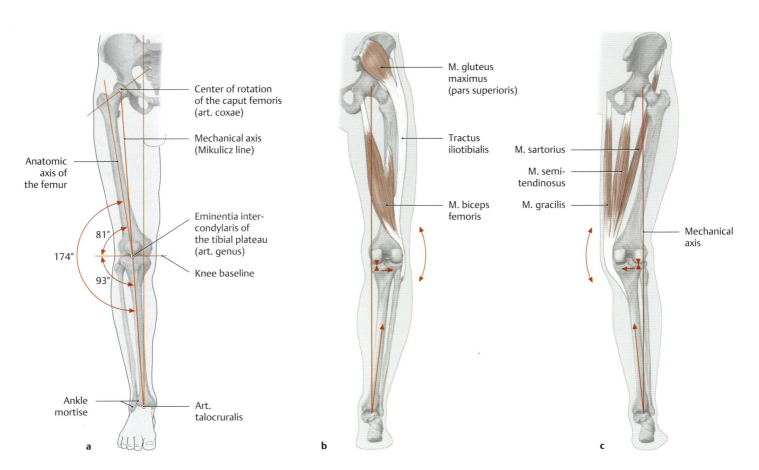

A The mechanical axis of the leg (the Mikulicz line)
a Mechanical axis in normal alignment, anterior view.
b Mechanical axis in genu varum, posterior view.
c Mechanical axis in genu valgum, posterior view.

In an individual with normal axial alignment, the large joints of the lower limb (the hip, knee, and ankle) lie on a straight line that represents the mechanical longitudinal axis of the leg (the *Mikulicz line*). This mechanical axis extends from the center of rotation of the caput femoris through the eminentia intercondylaris of the tibial plateau and down through the center of the ankle *mortise* (the pocket created by the fibula and tibia for the talus in the art. talocruralis, from Arabic, *murtazz*, fastened). While the mechanical axis and anatomic axis coincide in the *tibial shaft*, the anatomic and mechanical axes of the *femoral shaft* diverge at a 6° angle. Thus, the longitudinal anatomic axes of the femur and tibia do not lie on a straight line but form a laterally open angle of 174° at the level of the knee joint in the coronal plane (the *femorotibial angle*). In genu varum (**b**) the center of the knee joint is lateral to the mechanical axis, and in genu valgum (**c**) it is medial to the mechanical axis. Both conditions impose abnormal, unbalanced loads on the joints (see **B**) that gradually cause degenerative changes to develop in the bone and cartilage (osteoarthritis of the knee) accompanied by stretching of the associated joint capsule, ligaments, and muscles. In genu varum (**b**), for example, the medial joint complex of the knee is subjected to abnormal pressure, while the lateral joint structures (e.g., the lig. collaterale fibulare), tractus iliotibialis, and m. biceps femoris are subjected to abnormal tension. Genu varum also places greater stress on the lateral border of the foot, resulting in a fallen pedal arch.

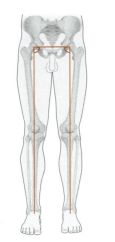

B Position of the mechanical axes with the feet slightly apart and together
Anterior view.

a In upright stance with the feet placed slightly apart, the mechanical axis runs almost vertically through the center of the three large joints.
b The legs are generally considered "straight" if, when the feet are together, the opposing medial malleoli and knees are touching. Accordingly, the intercondylar distance and the intermalleolar distance between the legs provide an index for the measurement of genu varum and genu valgum. When this stance is attempted, an intercondylar distance greater than 3 cm or an intermalleolar distance greater than 5 cm is considered abnormal (see **C**).

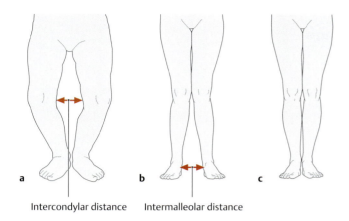

a Intercondylar distance   b Intermalleolar distance   c

C The normal leg axes at different ages
a Infant, b small child, c school-age child.
Up to about 20° of genu varum is considered normal during the first year of life. Up to about 10° of genu valgum is also considered normal through 2 years of age. By the time the child enters school, the legs are essentially straight as a result of musculoskeletal growth.

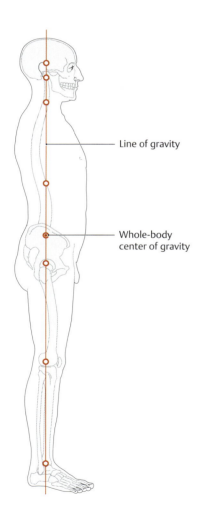

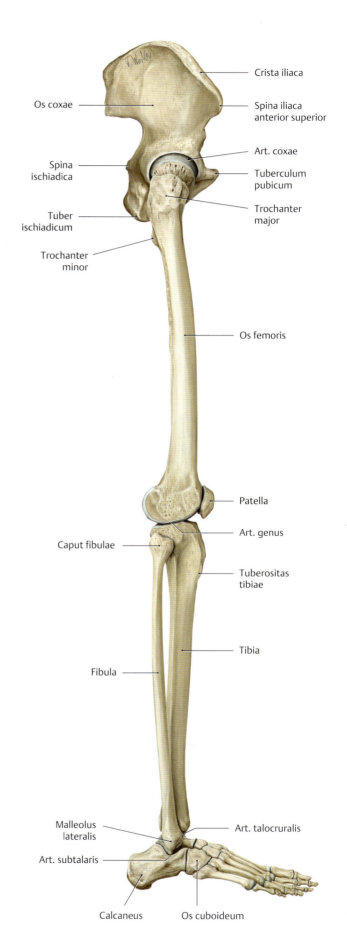

D Normal anatomic position in relation to the line of gravity
Right lateral view. The *line of gravity* runs vertically from the whole-body center of gravity to the ground. In normal upright humans, it intersects the external auditory canal, the dens of the axis (C II), the inflection points between the normal curves in the columna vertebralis (between the cervical and thoracic curves, and thoracic and lumbar curves), the whole-body center of gravity, and the hip, knee, and ankle joints (art.coxae, art. genus, and art. talocruralis). Chronic deviation of any reference point from this line imposes abnormal stresses on different clusters of musculoskeletal elements.

E Skeleton of the right lower limb
Right lateral view.

19.3 The Bones of the Pelvic Girdle

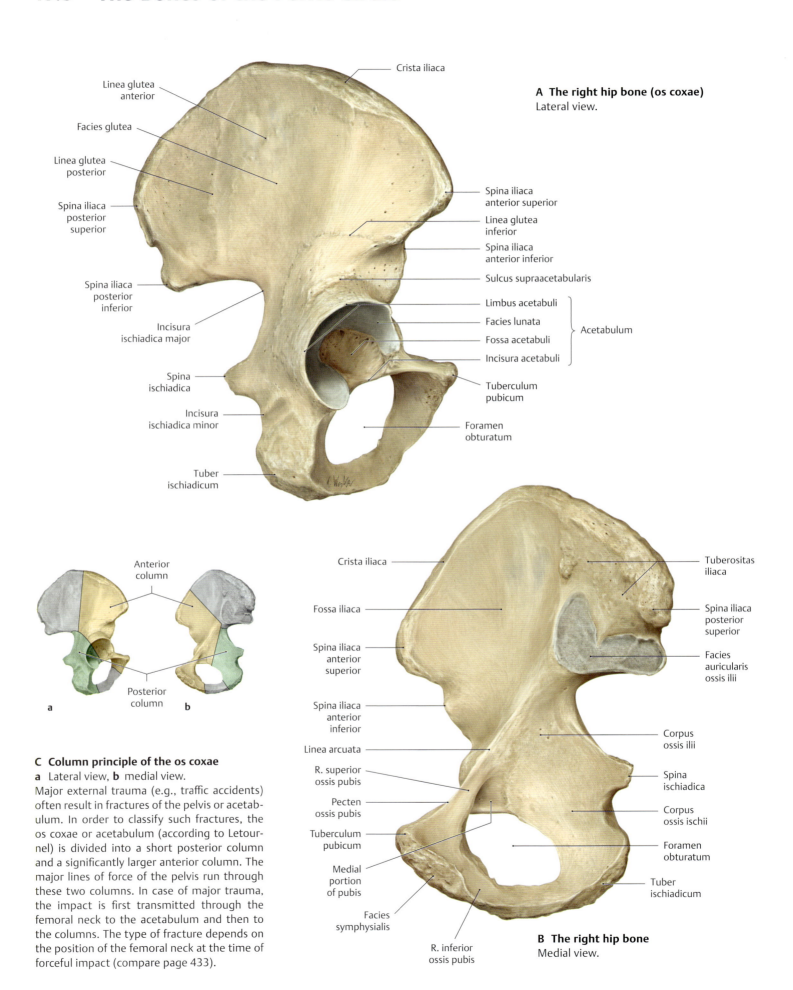

A The right hip bone (os coxae) Lateral view.

B The right hip bone Medial view.

C Column principle of the os coxae
a Lateral view, **b** medial view.
Major external trauma (e.g., traffic accidents) often result in fractures of the pelvis or acetabulum. In order to classify such fractures, the os coxae or acetabulum (according to Letournel) is divided into a short posterior column and a significantly larger anterior column. The major lines of force of the pelvis run through these two columns. In case of major trauma, the impact is first transmitted through the femoral neck to the acetabulum and then to the columns. The type of fracture depends on the position of the femoral neck at the time of forceful impact (compare page 433).

Lower Limb — 19. Bones, Ligaments, and Joints

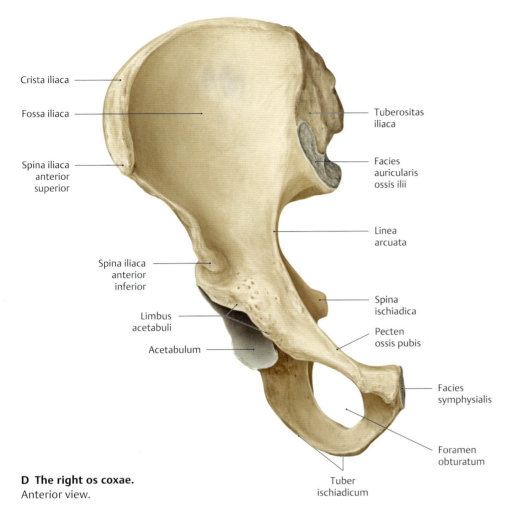

D The right os coxae.
Anterior view.

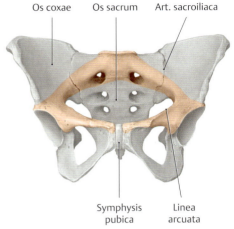

E The pelvic girdle and pelvic ring
Anterior view. The paired ossa coxae that make up the cingulum pelvicum (pelvic girdle) are connected to each other at the cartilaginous symphysis pubica and to the os sacrum at the sacroiliac joints (see p. 142). This creates a stable ring, the bony pelvic ring (shaded in red), that permits very little motion. This stability throughout the pelvic ring is an important prerequisite for the transfer of trunk loads to the lower limb necessary for normal gait.

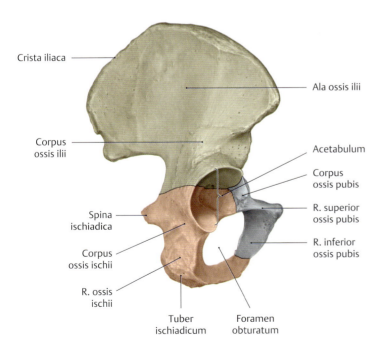

F The triradiate cartilage of a right hipbone(Oscoxae): the junction of the os ilium, os ischii, and os pubis.
Lateral view.

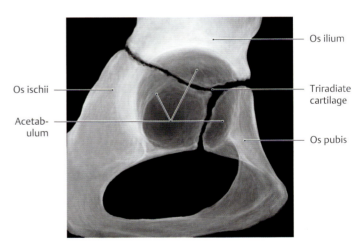

G Radiograph of the right acetabulum of a child
Lateral view (lateral projection). The bony elements of the os coxae come together in the acetabulum, with the os ilium and os ischium each comprising two fifths of the acetabulum and the os pubis one fifth. Definitive fusion of the Y-shaped growth plate (triradiate cartilage) occurs between the 14th and 16th years of life.

19.4 The Femur (Os femoris): Importance of the Femoral Neck Angle

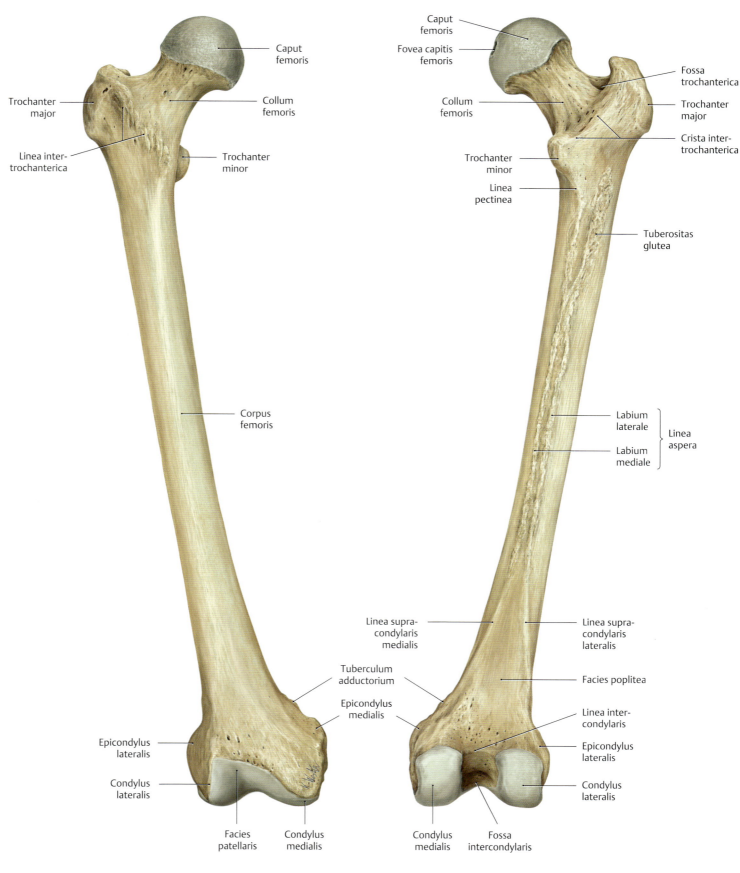

A The right os femoris
Anterior view.

B The right os femoris
Posterior view.

Lower Limb — 19. Bones, Ligaments, and Joints

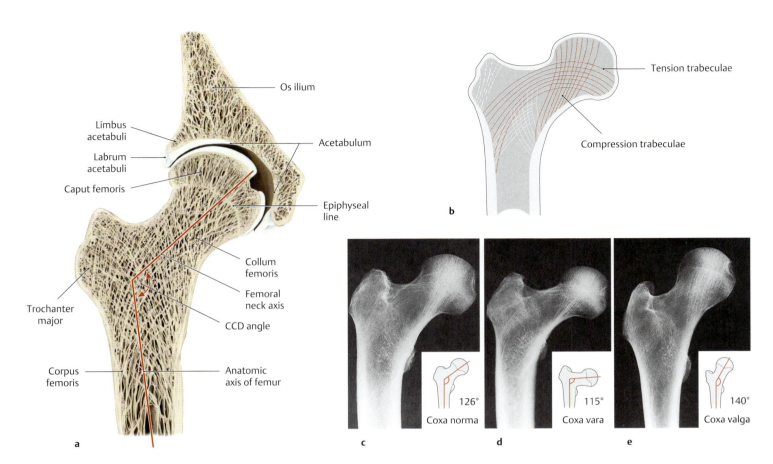

C The arrangement and prominence of tension trabeculae and compression trabeculae as a function of the femoral neck angle
Right femur, anterior view.

a Coronal section through the right hip joint at the level of the fovea capitis femoris. The angle between the longitudinal axis of the femoral neck (collum femoris) and the axis of the femoral shaft (corpus femoris) is called the femoral neck angle or CCD angle (centrum-collum-diaphysis angle). This angle (coxa norma) normally measures approximately 126° in adults and 150° in newborns. It decreases continually during growth due to the constant bone remodeling that occurs in response to the changing stress patterns across the hip.

b The trabecular pattern associated with a normal femoral neck angle.
c–e Radiographs in the sagittal projection.
c Normal femoral neck angle with a normal bending load.
d A *decreased* femoral neck angle (coxa vara) leads to a greater bending load with higher tensile stresses, thereby stimulating the formation of more tension trabeculae.
e An *increased* femoral neck angle (coxa valga) leads to a greater pressure load with higher compressive stresses, stimulating the formation of more compression trabeculae.

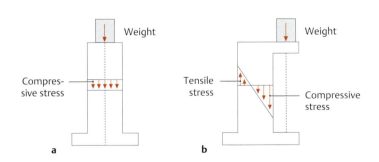

D Compressive and tensile stresses in a bone model
a An *axial* (centered) weight placed atop a Plexiglas model of a pillar creates a uniform pressure load that is evenly distributed over the cross section of the pillar and whose sum is equal to the applied weight.
b A *nonaxial* (eccentric) weight placed on an overhang creates a bending load that generates both tensile and compressive stresses in the pillar.

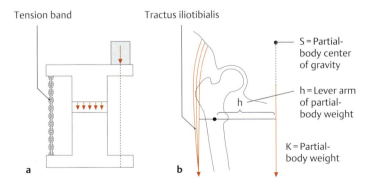

E The principle of the tension band (after Pauwels)
a The bending load acting on an I-beam model can be reduced by placing a high-tensile-strength member (chain) on the side opposite the bending force. This added member transforms the bending load into a pure compressive load.
b In the leg, the fascia lata on the lateral side of the thigh is thickened to form the tractus iliotibialis (see p. 479). By functioning as a tension band, the tractus iliotibialis reduces the bending loads on the proximal femur.

415

19.5 The Femoral Head and Deformities of the Femoral Neck

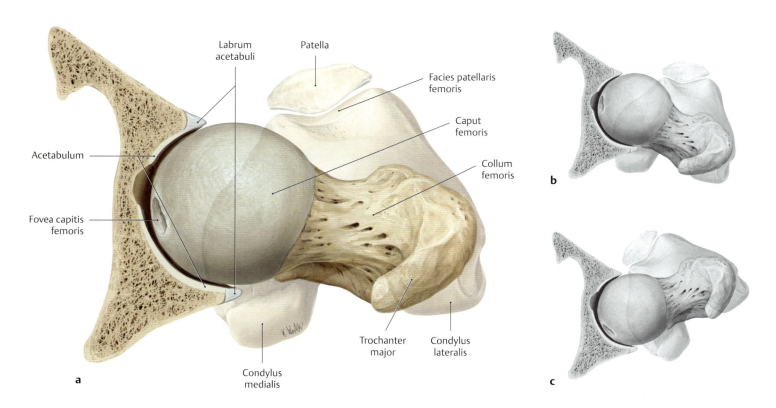

A The right femur
Proximal view. For clarity, the acetabulum has been sectioned in the horizontal plane. The distal end of the femur (with patella) has been added in light shading.
a Hip joint with centered femoral head (caput femoris), **b** hip joint externally rotated, **c** hip joint internally rotated.
Note the orientation of the acetabulum, which is angled forward by approximately 17°. This anterior angle affects the stability and "seating" of the femoral head in the hip joint (see p. 427). With the caput femoris centered in the acetabulum and the femur slightly medially rotated (**a**), the distal femur and thus the knee joint point slightly inward (physiologic internal rotation of the knee). The position of the foot is additionally influenced by the external rotation of the tibia (see p. 421) (see **D**).

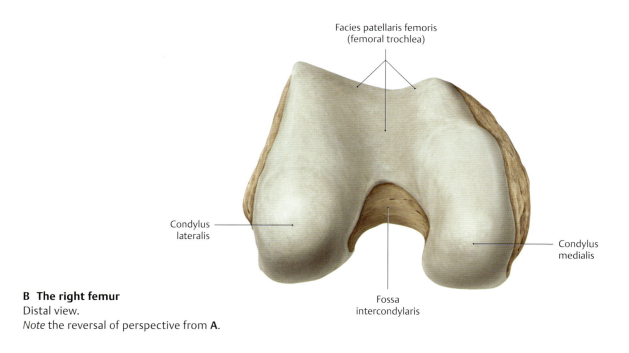

B The right femur
Distal view.
Note the reversal of perspective from **A**.

Lower Limb —— *19. Bones, Ligaments, and Joints*

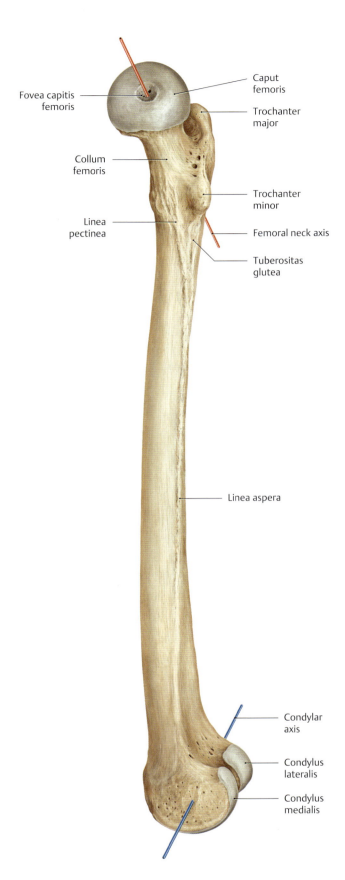

C The right femur.
Medial view.
Note the transverse condylar axis and the femoral neck axis. When the axes are superimposed, the two lines intersect each other at a 12° angle in adults (anteversion angle, see also **D** and **A**). This angle is considerably larger at birth, measuring 30° to 40°, but decreases to the normal adult value by the end of the 2nd decade.

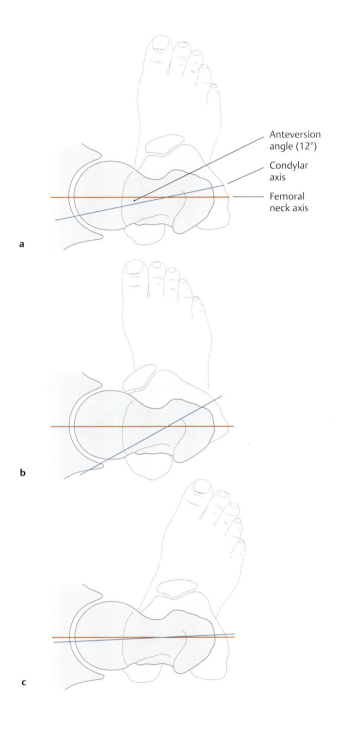

D Rotational deformities of the femoral neck (collum femoris)
Right hip joint, superior view. Increased or decreased torsion of the femoral shaft (corpus femoris) results in torsion angles of varying size. When the hip is centered, this leads to increased internal or external rotation of the leg with a corresponding change in gait (a "toeing-in" or "toeing-out" gait). When the condylar axis is taken as the reference point, femoral torsion may be described as normal (**a**), increased (**b**), or decreased (**c**).

a A normal anteversion angle of approximately 12° with the foot directed forward (taking into account tibial torsion of 23°, see p. 421).
b An increased anteversion angle (*coxa anteverta*) typically leads to a toeing-in gait accompanied by a pronounced limitation of external rotation.
c The collum femoralis is retroverted (points backward in relation to the condylar axis). The result is *coxa retroverta* with a toeing-out gait.

19.6 The Patella

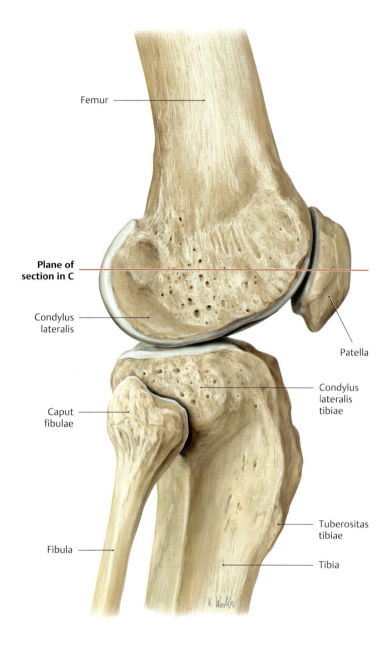

A Location of the patella
Right knee joint, lateral view. The red line indicates the plane of section in **C**.

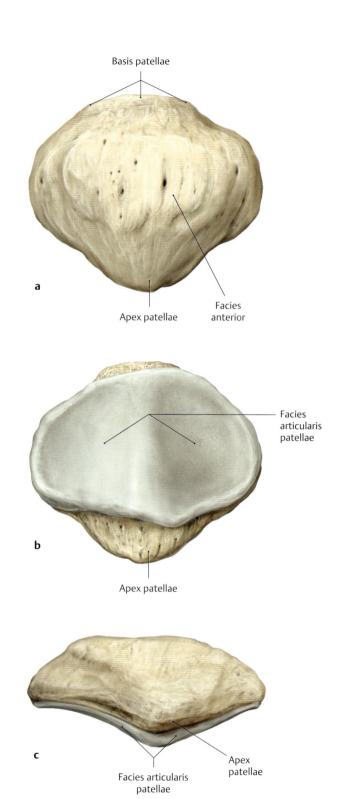

B Right patella
a Anterior view, **b** posterior view, **c** distal view.
Note that the *apex* of the patella points downward.

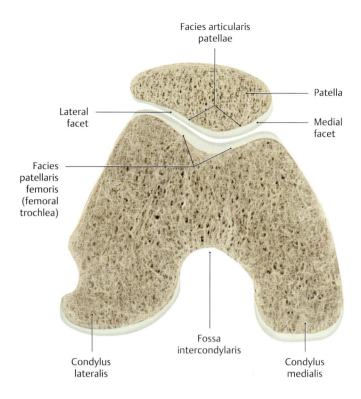
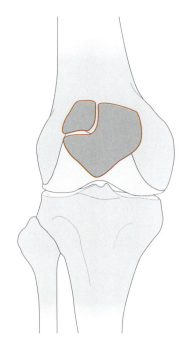

C Cross section through the femoropatellar joint
Right knee, distal view. The level of the cross section is shown in **A**. The femoropatellar joint is the site where the facies patellaris femoris, often called the trochlea femoris (by analogy with the distal humerus), articulates with the posterior articular surface of the patella (facies arti- cularis patellae). The patella is a sesamoid bone (the largest of the ossa sesamoidea) embedded in the m. quadriceps femoris tendon. The patella is well centered when the ridge on the undersurface of the patella is seated within the groove of the femoral trochlea (fossa intercondylaris femoris). The main functional role of the patella is to lengthen the effective lever arm of the m. quadriceps femoris (the only extensor muscle of the knee), thereby reducing the force required to extend the knee joint (see also p. 482).

D Patella bipartita (patella bipartita)
Because the patella develops from multiple ossification centers, the failure of an ossification center to fuse results in a two-part (bipartite) patella. The upper lateral quadrant of the patella is most commonly affected. A fracture should always be considered in the radiographic differential diagnosis of a bipartite patella.

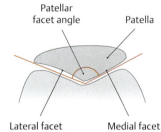

a   Lateral facet   Medial facet   b   c   d

E The evaluation of patellar shape
Diagrams of tangential radiographs of the patella ("sunrise" view: supine position, knee flexed 60°, caudocranial beam directed parallel to the posterior patellar surface). Each diagram shows the relation of the patella to the femoral trochlea in a horizontal plane through the right knee joint. The posterior articular surface of the patella bears a vertical ridge dividing it into a lateral facet and a medial facet. Generally, the lateral facet is slightly concave, while the medial facet is slightly convex. The angle between the lateral and medial facets, called the patellar facet angle, is normally 130° ± 10°. Wiberg, Baumgart, and Ficat devised the following scheme for the classification of patellar shape based on the facet angle:

a   Patella with medial and lateral facets of approximately equal size and a facet angle within the normal range.
b   Most common patellar shape with a slightly smaller medial facet.
c   A distinctly smaller medial facet ("medial hypoplasia").
d   Patellar dysplasia with a very steep medial facet ("hunter's hat") configuration.

Besides the various patellar shapes, the facies patellaris femoris (the trochlea femoris) has a variable morphology (described in the Hepp classification system). Developmental dysplasias of the patella and femoral trochlea lead to patellar instability marked by recurrent lateral or medial subluxation or dislocation of the patella.

19.7 The Tibia and Fibula

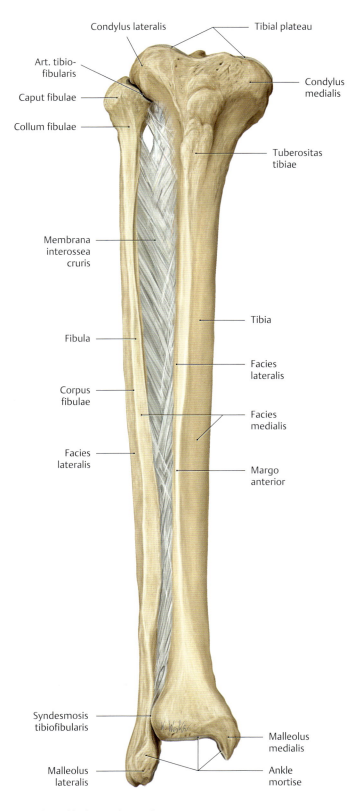

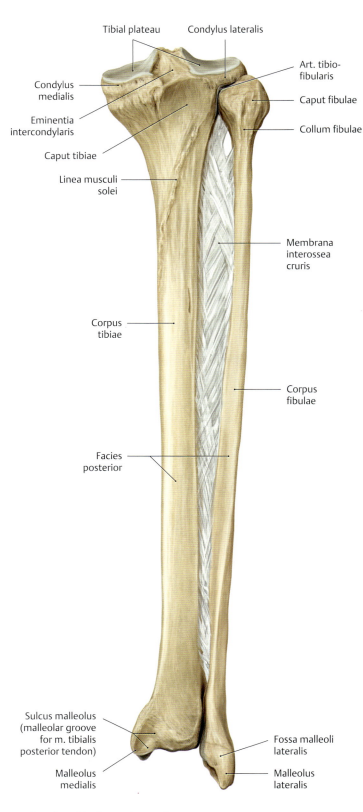

A The tibia, fibula, and membrana interossea cruris
Right leg, anterior view. The tibia bond fibula articulate at two joints that allow only very limited motion (rotation). Proximally, near the knee, is the synovial art. tibiofibularis; distally, at the ankle, is the syndesmosis tibiofibularis (fibrous joint with bony elements united by ligaments). The membrana interossea cruris (see also **F**) is a sheet of tough connective tissue that serves as an origin for several muscles in the leg. Additionally, it acts with the syndesmosis tibiofibularis to stabilize the ankle mortise.

B The tibia, fibula, and membrana interossea cruris
Right leg, posterior view.

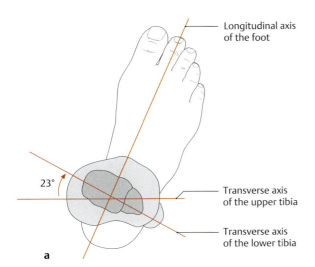
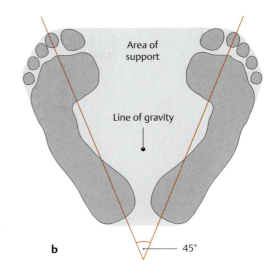

C Normal orientation of the tibia and its role in stability
When the transverse axes of the upper tibia (tibial plateau) and lower tibia (ankle mortise) are superimposed, they form an angle of approximately 23°; i.e., the transverse axis of the ankle joint is rotated 23° laterally relative to the transverse axis of the tibial plateau (*normal tibial orientation*, **a**). As a result of this, the longitudinal anatomic axis of the foot does not lie in the sagittal plane, and the toes point outward when the upper tibia is directed forward (**b**). This significantly improves the stability of bipedal stance by placing the line of gravity close to the center of the area of support (see p. 411).

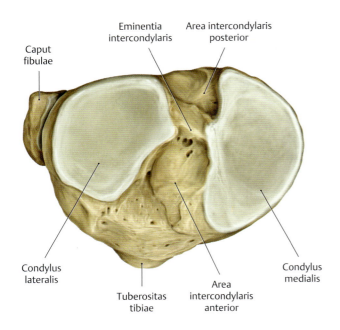

D The right tibial plateau
Proximal view.

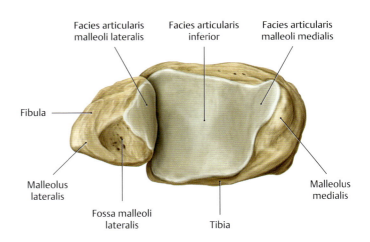

E The right ankle mortise
Distal view.

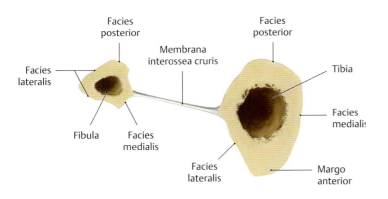

F Cross section through the middle third of the right leg
Proximal view.

19.8 The Bones of the Foot from the Dorsal and Plantar Views

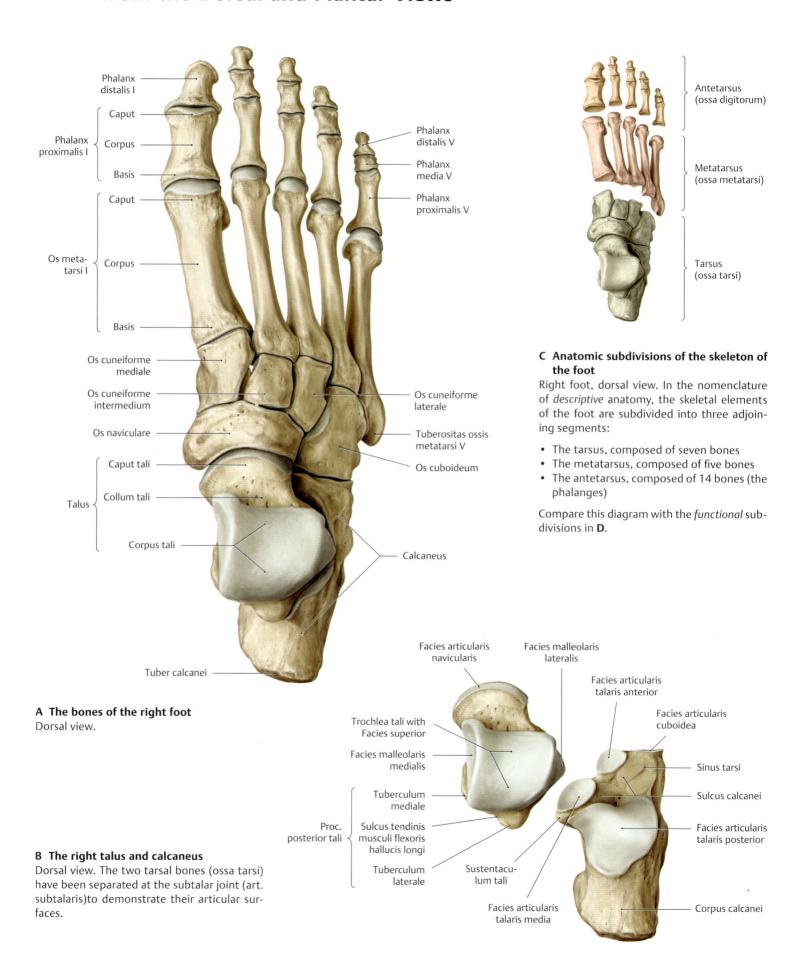

A The bones of the right foot
Dorsal view.

B The right talus and calcaneus
Dorsal view. The two tarsal bones (ossa tarsi) have been separated at the subtalar joint (art. subtalaris) to demonstrate their articular surfaces.

C Anatomic subdivisions of the skeleton of the foot
Right foot, dorsal view. In the nomenclature of *descriptive* anatomy, the skeletal elements of the foot are subdivided into three adjoining segments:

- The tarsus, composed of seven bones
- The metatarsus, composed of five bones
- The antetarsus, composed of 14 bones (the phalanges)

Compare this diagram with the *functional* subdivisions in **D**.

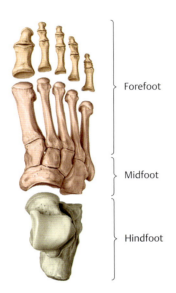

D Functional subdivisions of the pedal skeleton

Right foot, dorsal view. The skeleton of the foot is often subdivided as follows based on functional and clinical criteria:

- The hindfoot (calcaneus and talus)
- The midfoot (os cuboideum, os naviculare, and ossa cuneiformia)
- The forefoot (the phalanges proximales, medius, and distales)

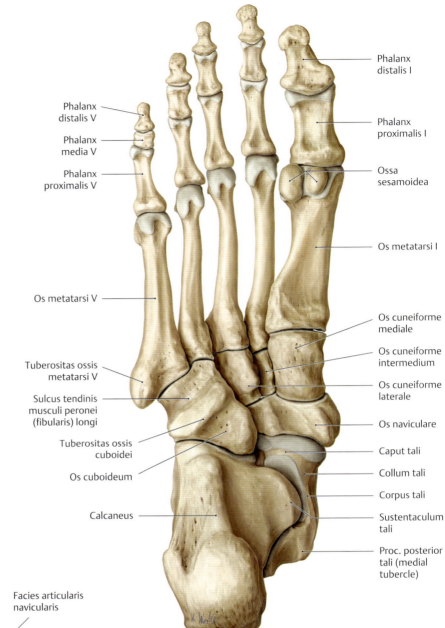

E The bones of the right foot
Plantar view.

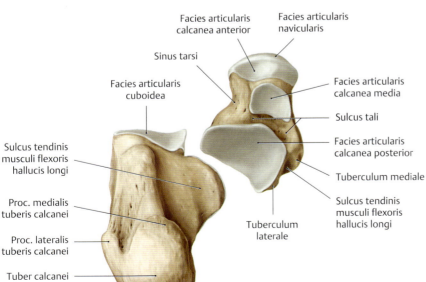

F The right talus and calcaneus
Plantar view. The two tarsal bones (ossa tarsi) have been separated at the art. subtalaris to demonstrate their articular surfaces.

19.9 The Bones of the Foot from the Lateral and Medial Views; Accessory Tarsal Bones

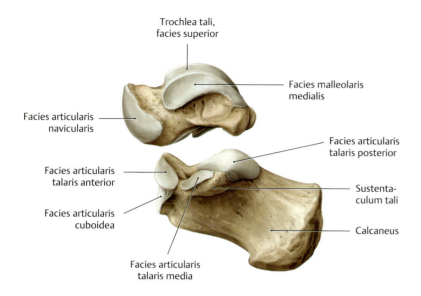

A The right talus and calcaneus
Medial view. The two tarsal bones (ossa tarsi) have been separated at the subtalar joint (art. subtalaris) to demonstrate their articular surfaces.

C Sustentaculum tali ("supports talus")

- The sustentaculum tali is a bony protuberance of the medial calcaneus.
- It is palpable approximately 1.5 cm below the tip of the malleolus medialis.
- It supports the talus on its balcony-like protrusion.
- It marks the end of the tarsal canal (canalis tarsi), which divides the two chambers of the lower ankle joint (see p. 459).
- It serves as a support for the tendon of the m. flexor hallucis longus (thereby supporting the calcaneus in upright position, see p. 466).
- It has a longitudinal groove for the m. flexor digitorum longus (see p. 466).
- Two ligaments insert at the sustentaculum tali: the spring ligament and part of the lig. deltoideum (see p. 460).
- It is a common location for fractures associated with snowboarding.

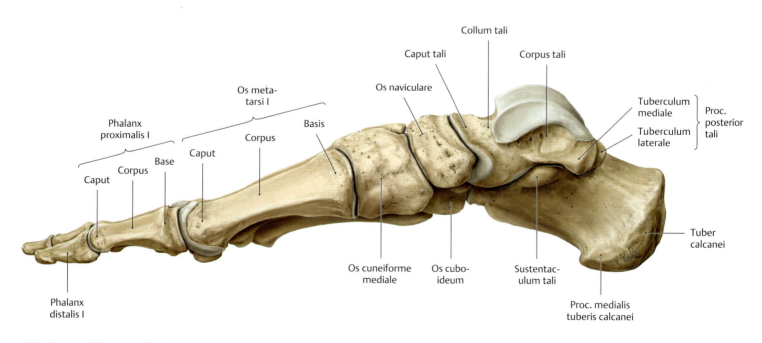

B The bones of the right foot
Medial view.

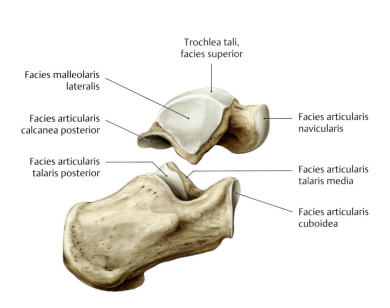

C The right talus and calcaneus
Lateral view. The two tarsal bones (ossa tarsi) have been separated at the subtalar joint to demonstrate their articular surfaces.

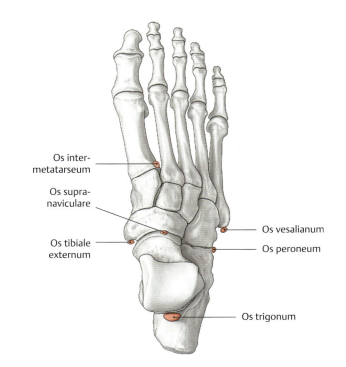

E Accessory tarsal bones
Right foot, dorsal view. A number of accessory (inconstant) ossicles are sometimes found in the foot. While they rarely cause complaints, they do require differentiation from fractures. A clinically important accessory bone is the os tibiale externum, which can be a source of discomfort when tight shoes are worn.

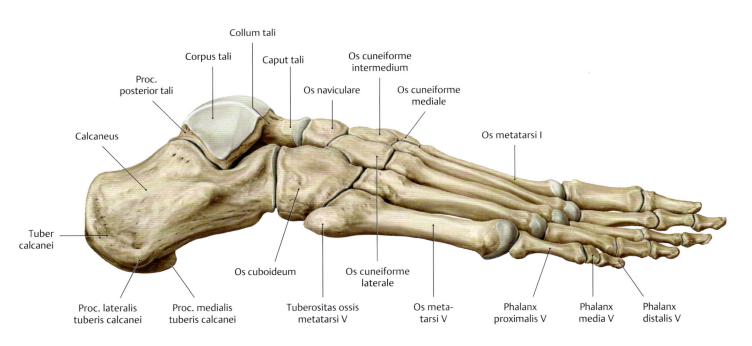

D The bones of the right foot
Lateral view.

19.10 The Hip Joint (Articulatio coxae): Articulating Bones

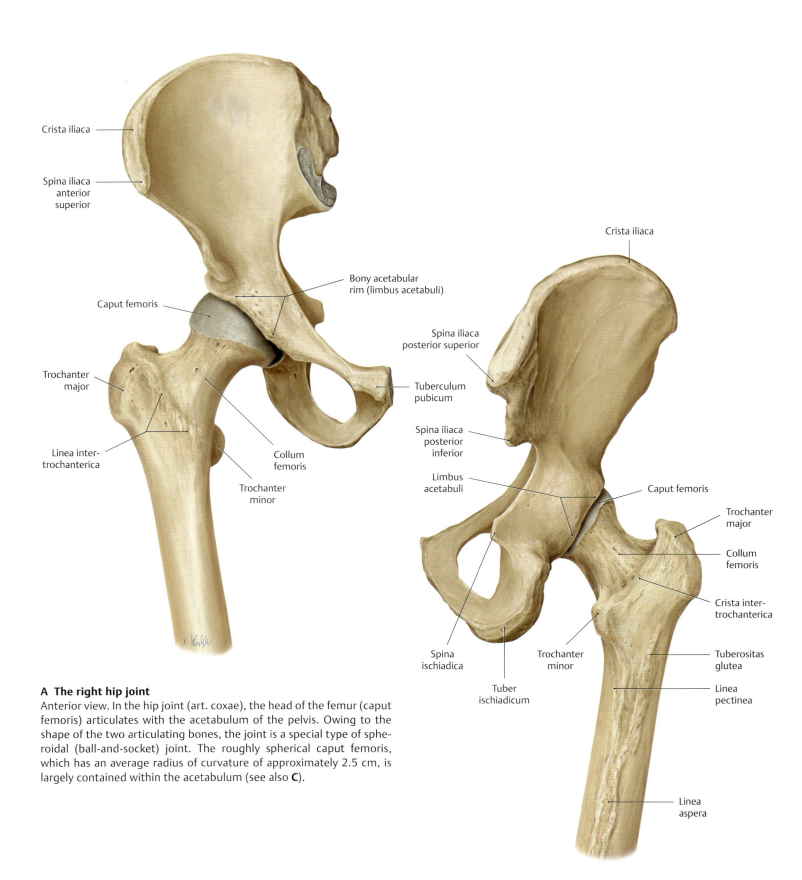

A The right hip joint
Anterior view. In the hip joint (art. coxae), the head of the femur (caput femoris) articulates with the acetabulum of the pelvis. Owing to the shape of the two articulating bones, the joint is a special type of spheroidal (ball-and-socket) joint. The roughly spherical caput femoris, which has an average radius of curvature of approximately 2.5 cm, is largely contained within the acetabulum (see also **C**).

B The right hip joint
Posterior view.

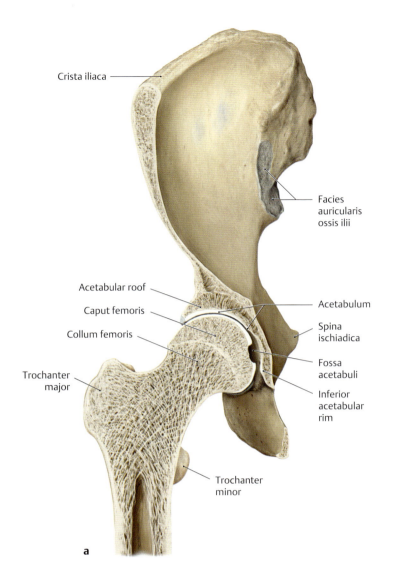

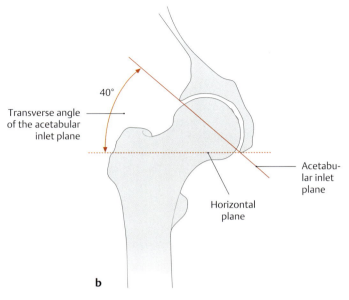

b

C Transverse angle of the acetabular inlet plane in the adult
Right hip joint, anterior view. Coronal section at the level of the fossa acetabuli. The acetabular inlet plane, or bony acetabular rim, faces inferolaterally (*transverse angle*) and also anteroinferiorly (*sagittal angle*; see **D**). The inferolateral tilt of the acetabulum can be determined by drawing a line from the superior acetabular rim to the inferior acetabular rim (lowest point of the incisura acetabuli) and measuring the angle between that line and the true horizontal. This transverse angle normally measures approximately 51° at birth, 45° at 10 years of age, and 40° in adults (after Ullmann and Sharp). The value of the transverse angle affects several parameters, including the degree of lateral coverage of the caput femoris by the acetabulum (the *center-edge angle* of Wiberg, see p. 439).

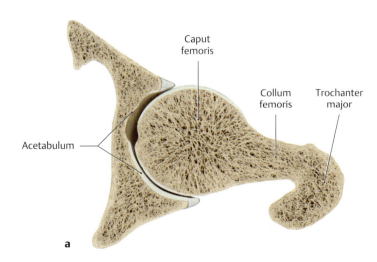

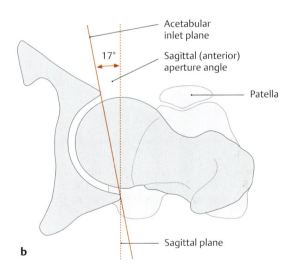

b

D Sagittal angle of the acetabular inlet plane in the adult
Right hip joint, superior view. Horizontal section through the center of the caput femoris.

The bony acetabular rim is angled anteroinferiorly relative to the sagittal plane (compare this with the horizontal plane in **C**). This aperture angle measures approximately 7° at birth and increases to 17° by adulthood (after Chassard and Lapine).

19.11 The Ligaments of the Hip Joint: Stabilization of the Femoral Head (Caput femoris)

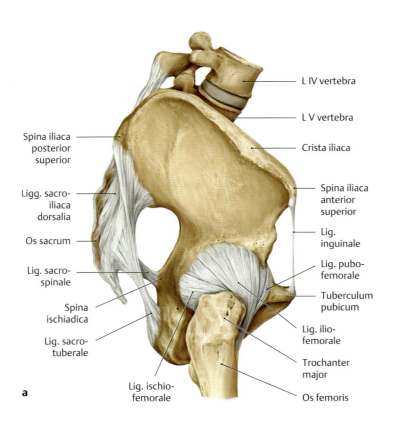

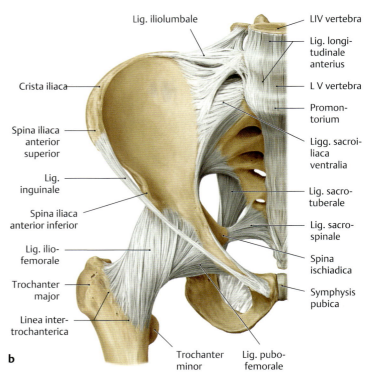

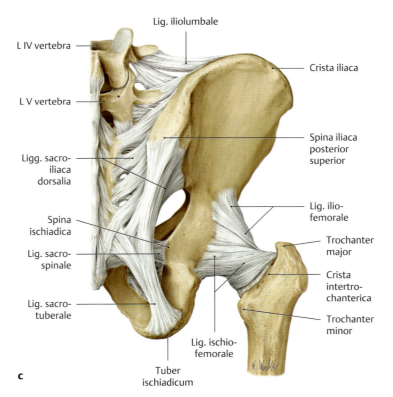

A The ligaments of the right art. coxae
a Lateral view, b anterior view, c posterior view.
The strongest of the three ligaments, the lig. iliofemorale, arises from the spina iliaca anterior inferior and fans out at the front of the hip, attaching along the linea intertrochanterica (see **b**). With a tensile strength greater than 350 N, it is the most powerful ligament in the human body and provides an important constraint for the hip joint: it keeps the pelvis from tilting posteriorly in the upright stance, without the need for muscular effort. It also limits adduction of the extended limb (particularly the lateral elements of the ligament) and stabilizes the pelvis on the stance side during gait; i.e., it acts with the small gluteal muscles to keep the pelvis from tilting toward the swing side.

B The ligaments of the hip joint

- Lig. iliofemorale
- Lig. pubofemorale
- Lig. ischiofemorale
- Zona orbicularis (anular ligament)*
- Lig. capitis femoris**

* Not visible externally, it encircles the femoral neck (collum femoris) like a buttonhole (see p. 431, **C**).
** Has no mechanical function, but transmits vessels that supply the femoral head (caput femoris)(see also p. 431).

Lower Limb — 19. Bones, Ligaments, and Joints

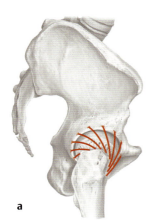
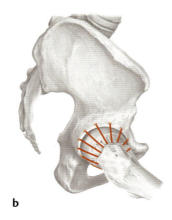

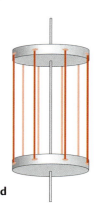

C Actions of the ligaments as a function of joint position

a Right hip joint in extension, lateral view. The capsular ligaments (ligg. capsularia) of the hip joint (see facing page) form a ringlike collar that encircles the femoral neck. When the hip is extended, these ligaments become twisted upon themselves (as shown here), pushing the caput femoris more firmly into the acetabulum (joint-stabilizing function of the ligaments).

b Right hip joint in flexion, lateral view. During flexion (anteversion), the ligament fibers are lax and press the caput femoris less firmly into the acetabulum, allowing a greater degree of femoral mobility.

c, d The twisting mechanism of the ligg. capsularia can be represented by a model consisting of two disks interconnected by parallel bands. The situation in **c** represents the position of the ligaments when the hip joint is extended. When one of the two disks rotates (blue arrow), the bands become twisted and draw the two disks closer together (red arrows). Panel **d** models the situation in the flexed hip. The ligaments are no longer twisted, so the distance between the two disks is increased (after Kapandji).

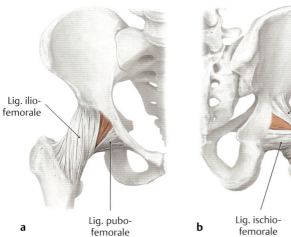
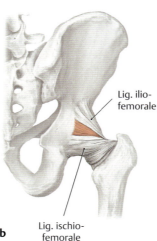
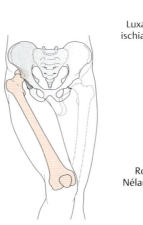
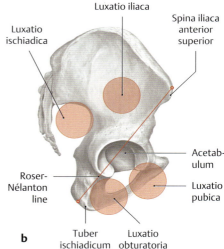

D Weak spots in the capsule of the right hip joint

a Anterior view, **b** posterior view.

There are weak spots in the joint capsule (color-shaded areas) located between the ligaments that strengthen the membrana fibrosa (see **A**). External trauma may cause the caput femoris to dislocate from the acetabulum at these sites (see **E**).

The combination of great ligament strength and the close congruity of the caput femoris in the acetabulum makes the hip joint very stable and dislocations relatively rare. The situation is different, however, following a hip replacement arthroplasty. The hip joint ligaments must be at least partially divided to implant the prosthesis, and the risk of dislocation is markedly increased.

E Traumatic dislocation of the hip

a It is most common for the caput femoris to dislocate upward and backward from the acetabulum (luxatio iliaca) between the lig. iliofemorale and lig. ischiofemorale. Typically, this is caused by a fall from a great height, a motor vehicle accident (front-end collision), etc. In this type of dislocation, the leg assumes a position of adduction and slight internal rotation.

b Lateral view. Position of the caput femoris in various types of dislocation (red-shaded area). The trochanter major may be above or below the Roser–Nélaton line.

19.12 The Ligaments of the Hip Joint: Nutrition of the Femoral Head (Caput femoris)

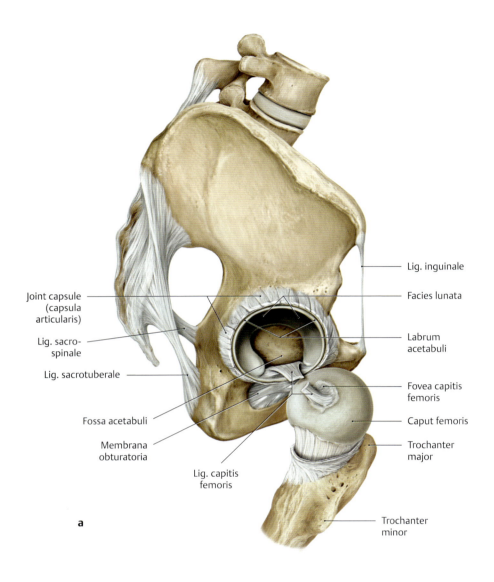

A The ligaments of the right hip joint
a Lateral view. The joint capsule (capsula articularis) has been divided at the level of the labrum acetabuli, and the caput femoris has been dislocated to expose the divided lig. capitis femoris. This ligament transmits important nutrient blood vessels for the caput femoris.
b Anterior view. The membrana fibrosa of the capsula articularis has been removed at the level of the femoral neck to show the conformation of the membrana synovialis. This membrane extends laterally from the acetabular rim, and about 1 cm proximal to the attachment of the membrana fibrosa, it is reflected onto the femoral neck within the joint cavity. It continues up the femoral neck to the chondro-osseous junction of the caput femoris (see also the coronal section in **C**).
c Posterior view.

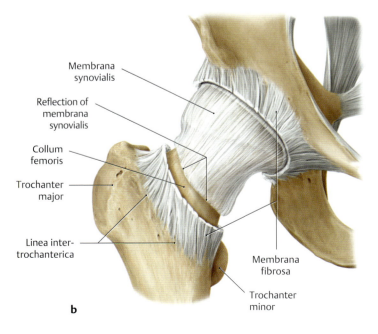

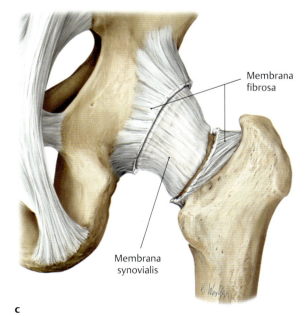

430

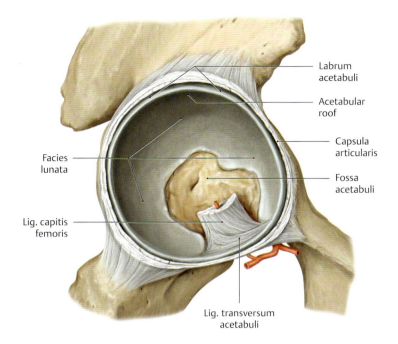

B The acetabulum of the right art. coxae with the femoral head removed

Lateral view. The cartilage-covered articular surface of the acetabulum is crescent-shaped (facies lunata) and is broadest and thickest over the acetabular roof. The facies lunata is bounded externally by the slightly protruding bony rim of the acetabulum, which is extended by a lip (the labrum acetabuli) composed of tough connective tissue and fibrocartilage. The cartilaginous articular surface lines much of the fossa acetabuli, which is occupied by loose, fibrofatty tissue and is bounded inferiorly by the lig. transversum acetabuli in the area of the incisura acetabuli (not visible here). The lig. capitis femoris, which has been sectioned in the drawing, transmits blood vessels that nourish the caput femoris (see **C**).

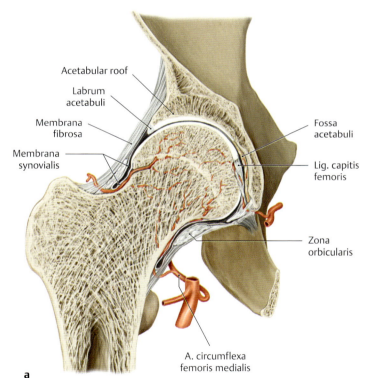
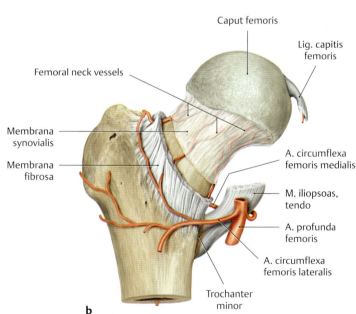

C The blood supply to the caput femoris
a Coronal section through the right hip joint, anterior view.
b Course of the femoral neck vessels in relation to the capsula articularis (right femur, anterior view).

The caput femoris derives its blood supply from the aa. circum flexae femoris lateralis and medialis and the a. lig. capitis femoris, which branches from the a. obturatoria (see p. 550). If the anastomoses between the vessels of the lig. capitis femoris and the femoral neck vessels are absent or deficient due to the avulsion of blood vessels caused by a dislocation or femoral neck fracture, the bony tissue in the head of the femur may become necrotic (avascular necrosis of the femoral head).

19.13 Cross-sectional and X-Ray Anatomy of the Hip Joint. Typical Medical Condition of the Elderly: Femoral Neck Fractures

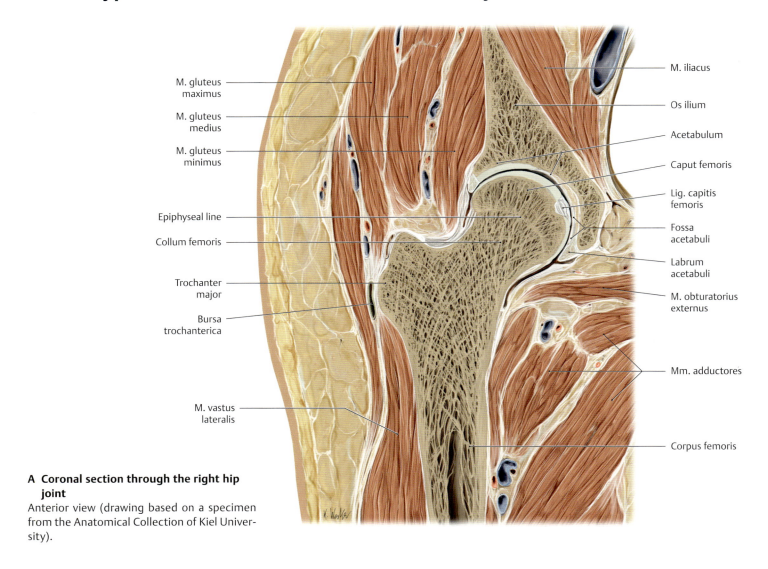

A Coronal section through the right hip joint
Anterior view (drawing based on a specimen from the Anatomical Collection of Kiel University).

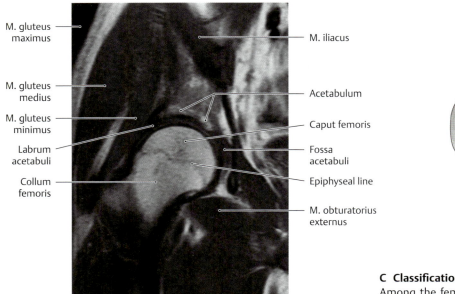

B MRI of the hip region: coronal T1-weighted spin-echo (SE) image at the level of the fossa acetabuli (from Vahlensieck M, Reiser M. MRT des Bewegungsapparates. 3rd ed. Stuttgart: Thieme; 2006).

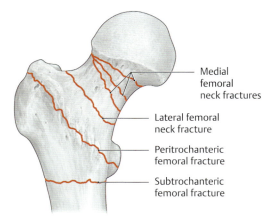

C Classification of proximal femoral fractures
Among the femoral fractures close to the hip joint, the medial femoral fractures (see **F**) are typical injuries of the osteoporotic bones of the elderly. Often, the cause is minimal trauma, such as falling on the trochanter major or the outstretched leg.

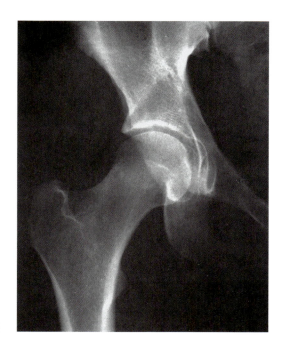
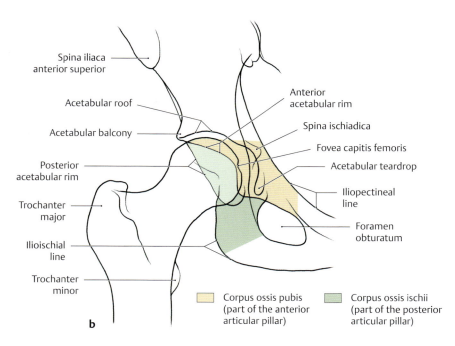

D X-ray of the hip joint and diagnostic reference levels in case of acetabular fractures

a X-ray of the hip joint in a sagittal beam path (section of a pelvis survey radiograph; from Möller TB, Reif E. Taschenatlas der Röntgenanatomie. 2nd ed. Stuttgart: Thieme; 1998); **b** reference lines important for diagnostic radiology, mainly X-rays of the hip socket.

The pelvis survey radiograph is supplemented only when needed (i.e., when it is not sufficent for making a diagnosis, e.g., in undisplaced femoral fractures) with the help of particular positioning (e.g., ala and obturator view in which the healthy or injured side, respectively, is lifted at 45°) or MRI (see **E**) or CT. The CT is required as soon as an acetabular fracture has been diagnosed.

Particular reference lines in the anteroposterior (AP) view of the pelvis are important for diagnosis or for therapeutic surgical procedures in cases of acetabular fractures: anterior and posterior acetabular rim, acetab- ular roof, acetabular teardrop (equivalent to the depth of the fossa acetabuli), and iliopectineal or ilioischial line.

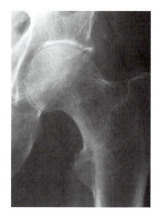
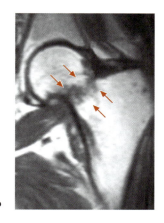
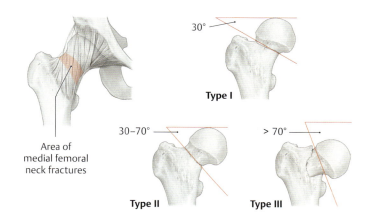

E Radiologic diagnosis of proximal femoral fractures (from Bohndorf K, Imhof H, Fischer W. Radiologische Diagnostik der Knochen und Gelenke. 2nd ed. Stuttgart: Thieme; 2006)

a Normal X-ray of a medial, undisplaced femoral fracture in a sagittal beam path; **b** MRI identifies fracture: T1-weighted coronal cross section of the same patient with fracture-related edema (red arrows).

Whereas diagnosing displaced factures usually is not a problem since they are visible in AP views of the hip (see **Da**), both undisplaced femoral fractures and stress fractures often appear in conventional X-rays only as slight abnormalities of the trabecular bone structure, so an MRI is needed to identify a fracture (because fracture-related edema can be identified through the lesser signals it emits).

F Medial femoral fractures (classification according to Pauwels)

Medial femoral fractures are much more common than lateral femoral fractures (95% of cases compared to 5% of cases). Medial femoral fractures are always intracapsular and due to common complications associated with them (e.g., ischemic femoral head necrosis, delayed fracture healing, and development of pseudarthrosis) are of particular clinical relevance. In particular, the damage to epiphyseal vessels as a result of intracapsular femoral head fractures leads to impaired blood flow to the caput femoris. Medial femoral fractures are classified according to Pauwels by the inclination angle of the fracture line relative to the horizontal line (type I: 0–30°, type II: 30–70°, and type III: > 70°). The steeper the angle, meaning the steeper the course of the fracture line, the greater the risk of the caput femoris sliding off and the greater the danger of pseudarthrosis.

19.14 Cross-sectional Anatomy of the Hip Joint (Art. coxae): Sonographic Representation of Hip Joint Effusion

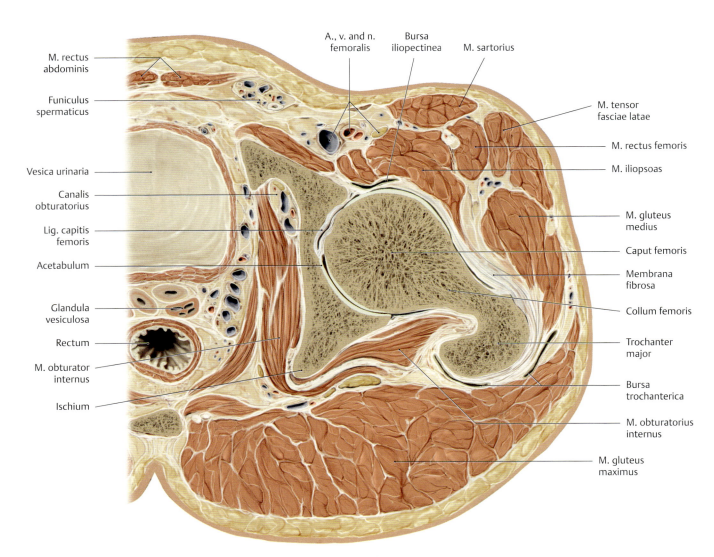

A Transverse section through the right hip joint (art. coxae)
Superior view (drawing based on a specimen from the Anatomical Collection of Kiel University).

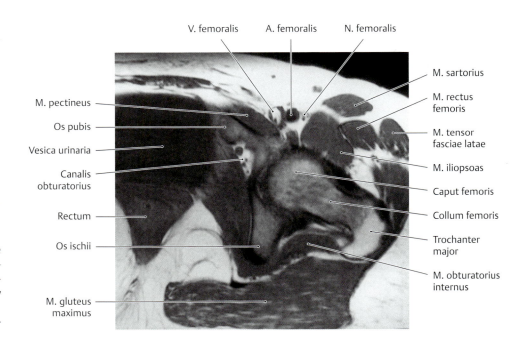

B MRI of the hip region (regio coxae): axial (transverse) T1-weighted SE image at the level of the collum femoris (from Vahlensieck and Reiser. MRT des Bewegungsapparates. 3rd ed. Stuttgart: Thieme; 2006).
The synovial bursae shown in **A** are not visible because in T1-weighted MRI, they always appear as low signal intensity and are barely distinguishable from muscles, which are also low signal intensity.
Note: With MRI, axial (transverse) cross sections are always inferior views.

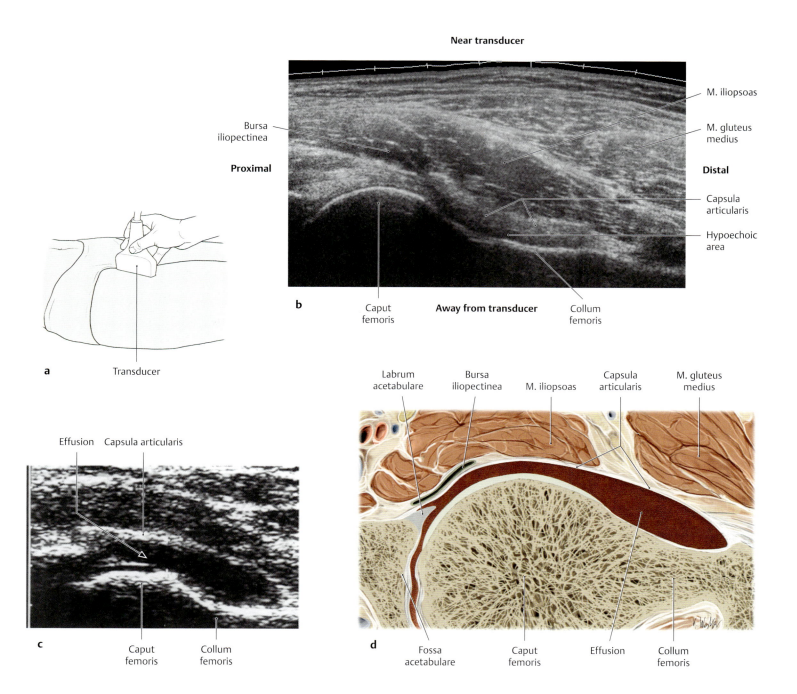

C Diagnostic ultrasonography of a longitudinal section of the hip: normal and showing hip joint effusion

a Position of the transducer from the front and in the longitudinal direction to the collum femoris; **b** sonogram of a normal finding (from Konermann W, Gruber G. Ultraschalldiagnostik der Bewegungsorgane. 2nd ed. Stuttgart: Thieme; 2006); **c** sonogram of hip joint effusion (from Niethard FU, Pfeil J. Orthopädie. 5th ed. Stuttgart: Thieme; 2005); **d** schematic view of the sonogram. Ultrasonography delivers real-time pictures. Structures at the top of the screen are close to the transducer; those at the bottom are farther away; those on the left side are proximal, those on the right side are distal.

In addition to the diagnostic ultrasound of the embryonic hip joint, the standardized diagnostic ultrasound of the hip joint is of great significance as it is cost-effective and quickly done. Just as in diagnostic x-rays, two almost perpendicular sectional planes (transverse and longitudinal relative to the collum femoris) are recorded. The sonographic examination of the hip joint is performed with the patient lying down. The hip and knee joints are in neutral position (**a**). The longitudinal section allows for a very good evaluation of the anterior parts of the articular capsule and the osseous and periarticular structures of the art. coxae. The surface contours of the anterior margin, the crescent-shaped caput femoris, and the collum femoris appear as echoic structures. The joint capsule runs parallel to the caput femoris and collum femoris and is generally separated from the collum femoris by a narrow Rypoechoic area (**b**). Hip disorders, which are accompanied by increased intra-articular volume (e.g., effusion as part of synovialitis or bacterial coxitis), can be very well shown with the help of an anterior longitudinal section because the effusion appears as a capsular distention at the anterior side of the collum femoris (see **c**). A lateral difference of more than 2 mm between the healthy and affected area (between the joint capsule and the collum femoris) is viewed as significant and implies increased intra-articular volume.

Note: Diagnostic ultrasound helps to show periarticular fluid buildup, e.g., in the case of bursitis trochanterica (= above the trochanter major).

19.15 The Movements and Biomechanics of the Hip Joint

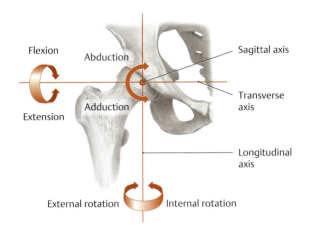

A   The axes of motion in the art. coxae
Right art. coxae, anterior view. As a spheroidal joint, the hip has three principal axes of motion, all of which pass through the center of the caput femoris (the rotational center of the hip) and are mutually perpendicular. Accordingly, the joint has three degrees of freedom, allowing movement in six principal directions:

1. Transverse axis: flexion and extension
2. Sagittal axis: abduction and adduction
3. Longitudinal axis: internal rotation and external rotation

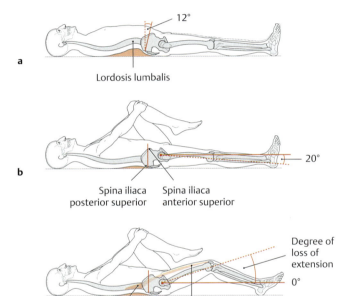

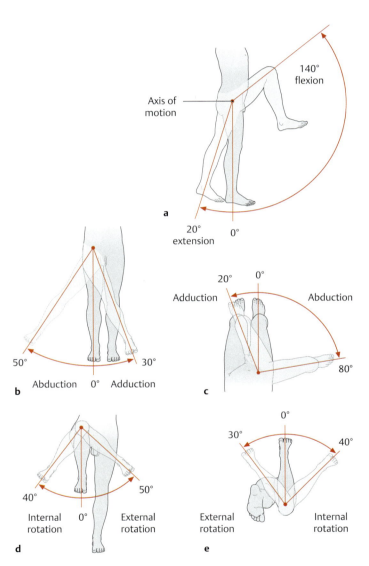

B   Range of extension of the right hip joint determined with the Thomas maneuver
The Thomas maneuver is used to measure the range of hip extension while the patient lies supine on a hard surface.

a Starting position with a slight degree of anterior pelvic tilt (approximately 12°). In this position it cannot be determined whether or not there is a flexion contraction of the hip joint. This is because the patient can compensate for any limitation of extension by increasing lordosis lumbalis (arching the lower back to hyperlordosis) and increasing the degree of anterior pelvic tilt.
b Pelvic tilt can be temporarily eliminated by drawing the opposite hip joint (in this case, the left hip) into a position of maximum flexion. If the right thigh remains flat on the table, the right hip joint will be in approximately 20° of extension (*normal extension*).
c If hip extension is limited (e.g., due to a shortened m. rectus femoris or m. iliopsoas), and the opposite (left) hip is placed in maximum flexion, the femur of the affected leg will be raised from the table by an amount equal to the loss of extension. Increased lordosis lumbalis is generally present when hip extension is limited and is easily detected clinically by palpating the lower back.

C   Range of motion of the hip joint from the neutral (0°) position (after Debrunner)
The range of motion of the art. coxae is measured using the neutral zero method (see p. 50).

a Range of flexion/extension.
b Range of abduction/adduction with the hip extended.
c Range of abduction/adduction with the hip flexed 90°.
d Range of internal rotation/external rotation with the hip flexed 90°.
e Range of internal rotation/external rotation in the prone position with the hip extended (when measuring rotation, the examiner uses the leg, flexed 90°, as a pointer to determine the range of motion).

D Hip joint load depending on the femoral neck angle

Each illustration shows a one-legged stance, standing on the right leg, anterior view. **a** Normal femoral neck angle, around 126°, see p. 415 (the femoral neck angle is also known as the CCD angle—centrum-collum-diaphysis angle, with centrum standing for the center of the caput femoris); **b** increased femoral neck angle (coxa valga); **c** decreased femoral neck angle (coxa vara).

The resultant force vector *R* determines the hip joint load when walking or standing on one leg. It is calculated using the partial body weight *K*, muscle force *M*, and length of the lever arms (see p. 51). With a normal CCD angle, the lever arm of the body weight is around 3 times more than the one of the muscle force, meaning in a one-legged stance, the hip joint load is 4 times ($R = 4$) larger than the body weight *K*. Since the lever arm of muscle force in coxa valga (coxa vara) decreases (increases), the hip joint load increases in coxa valga ($R = 7$) and decreases in coxa vara ($R = 3$). *S* is the partial-body center of gravity.

Note: Due to this relationship, surgically changing the femoral neck angle (osteotomy) influences the hip joint load.

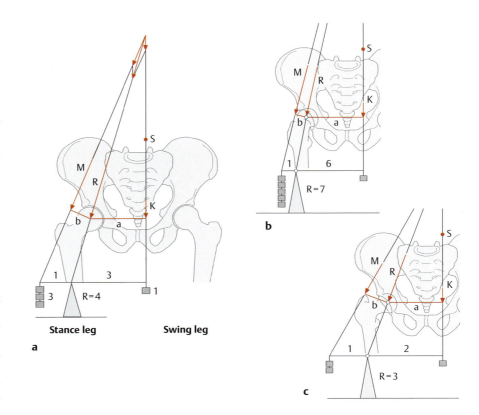

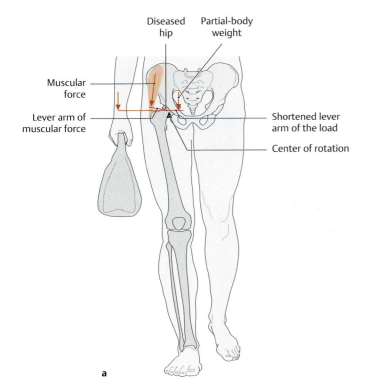

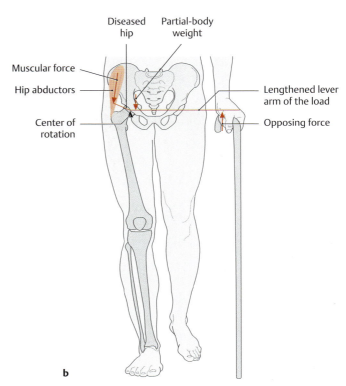

E Reducing the stresses on an osteoarthritic right hip

Anterior view. In patients with advanced osteoarthritis of the hip, various measures can be taken to alleviate the stresses, and thus the pain, on the affected side.

a Shift the partial-body center of gravity (see above) toward the affected side. One way to do this is by carrying a shopping bag *on the affected right side*, as shown here. This moves the partial-body center of gravity closer to the center of the caput femoris, thereby *shortening* the lever arm of the load (in this case, the partial-body weight) and also reducing the torque generated by the partial-body weight. The same effect is produced by adopting a *Duchenne limp*—an unconscious response in which the patient leans over the affected side with the upper body during the stance phase of gait (see also p. 536).

b Use a cane *on the unaffected (left) side*. While this *lengthens* the lever arm of the load (the partial-body weight), it also provides a force (the cane) that counteracts the body load at the end of that lever arm. This reduces the torque generated by the load (as in **a**).

19.16 The Development of the Hip Joint

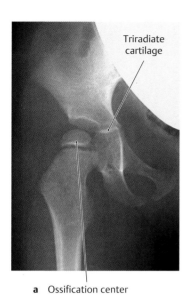
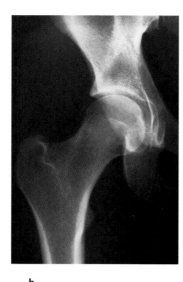

a Ossification center   b

A Radiographic appearance of the right hip joint
Anteroposterior projection.

a Boy, 2 years of age (original film from the Department of Diagnostic Radiology, Schleswig-Holstein University Hospital, Kiel Campus, Prof. S. Müller-Hülsbeck, M. D.).
Note: The ossification center for the caput femoris is already visible.
b Man 25 years of age (from Möller TB, Reif E. Taschenatlas der Röntgenanatomie. 2nd ed. Stuttgart: Thieme; 1998).

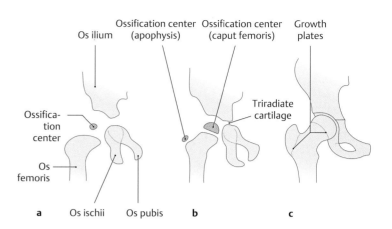

B Stages in the radiographic development of the hip joint
Schematic representations of AP radiographs taken at various stages in the development of the right hip joint. The ossification centers are indicated by dark shading.

a The ossification center for the caput femoris (caput femoris) can be identified at 6 months of age.
b Ossification centers for the caput femoris and trochanter major are visible at 4 years of age.
c At 15 years of age, the growth plates have not yet fused.

The anatomic differentiation of all structures that comprise the hip joint is largely complete by the 12th week of development (crown–rump length 80 mm). Whereas ossification of the acetabulum begins between the 3rd and 6th months of fetal development, the ossification center for the epiphysis capitis femoris (femoral head) does not appear until about 5 to 6 months after birth. The ossification center for the apophysis of the trochanter major appears during the 4th year of life. Fusion of the growth plates takes place between 16 and 18 years of age in the proximal femur and at about 15 years of age in the triradiate cartilage.

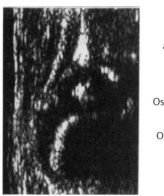
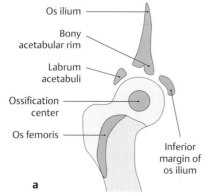
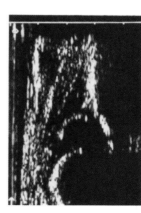
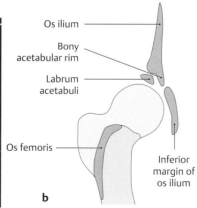

C Ultrasound evaluation of the infant hip (from Niethard FU, Pfeil J. Orthopädie. 5th ed. Stuttgart: Thieme; 2005).
a Normal hip joint in a 5-month-old child.
b Hip dislocation (type IV) in a 3-month-old child.

Ultrasonography is the most important imaging method for screening the infant hip, demonstrating potential morphological changes during the 1st year of life without exposure to ionizing radiation. The child is examined in the lateral decubitus position (on his or her side, other hip on the examination table) with the ultrasound transducer placed longitudinally over the art. coxae and perpendicular to the skin. The key landmark is the superior acetabular rim, i. e., the bony and cartilaginous roof of the acetabulum (see **D**). Infant hips are classified into four types based on their sonographic features:

- Type I   Normal hip
- Type II   Physiologically immature hip
- Type III   Subluxated hip
- Type IV   Dislocated hip

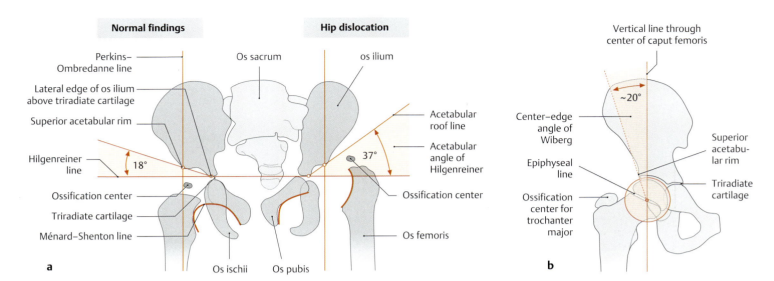

D Radiographic evaluation of the pediatric hip
Schematic representations of AP pelvic radiographs. Radiographic evaluation of the infant hip is feasible after 3 months of age, by which time there has been sufficient ossification of the joint. Both hips should always be imaged on the same radiograph.

a Normal findings (left half of figure) contrasted with findings in congenital hip dislocation (right half of figure) in a 2-year-old child. The following standard reference lines are used in the radiographic analysis of the infant hip:

- *Hilgenreiner line:* connects the inferolateral edge of the os ilium above the triradiate cartilage on both sides.
- *Perkins–Ombredanne line:* drawn from the most lateral edge of the acetabular roof, perpendicular to the Hilgenreiner line.
- *Menard–Shenton line:* curved line drawn from the superior border of the foramen obturatorium along the medial border of the femoral neck.
- *Acetabular angle of Hilgenreiner (AC angle):* angle formed by the intersection of the Hilgenreiner line and a line connecting the superior acetabular rim with the lowest part of the os ilium at the triradiate cartilage (see p. 413). This angle measures approximately 35° at birth, approximately 25° at 1 year of age, and should be less than 10° by 15 years of age.

Typically, the acetabular angle of Hilgenreiner is increased on the affected (left) side, while the center–edge angle of Wiberg (see below) is decreased. Additionally, there is a discontinuity in the Ménard–Shenton line, and the Perkins–Ombredanne line runs medial to the femoral shaft.

b Evaluation of lateral caput femoris coverage based on the center–edge angle of Wiberg (drawing of a radiograph of the right hip in a 5-year-old child). The angle is formed by a vertical line through the center of the caput femoris (within the future epiphyseal line [lineaepiphysealis]) and a line drawn from the center of the caput femoris to the superior acetabular rim. The center–edge angle should not be less than 10° between 1 and 4 years of age, and it should be in the range of 15 to 20° at 5 years of age.

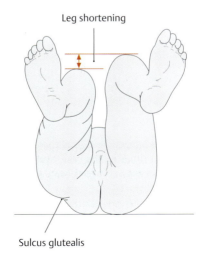

E Clinical examination of congenital dysplasia and dislocation of the hip
Hip dysplasia is characterized by an abnormal development of the acetabulum (acetabular dysplasia) in which the steep, flattened acetabular roof provides insufficient coverage for the caput femoris (see also **D**). The principal complication is dislocation of the hip, since the caput femoris is poorly contained in the dysplastic acetabulum and may be displaced upward and backward as a result of muscular traction or external loads. The *etiology* of hip dysplasia and dislocation is related to endogenous factors (familial disposition, maternal hormone status), as well as exogenous factors.

The following *clinical signs* may direct attention to a dysplastic and dislocated hip:

- Instability of the art. coxae: paucity of kicking activity or a positive Ortolani click caused by subluxation of the caput femoris. The Ortolani test requires a very experienced examiner. While still considered part of the clinical examination, it is performed less often today owing to the availability of ultrasound.
- Leg shortening with asymmetry of the posterior leg folds and gluteal folds.
- Limitation of abduction due to increased reflex tension from the hip adductors.

19.17 The Knee Joint (Articulatio genus): Articulating Bones

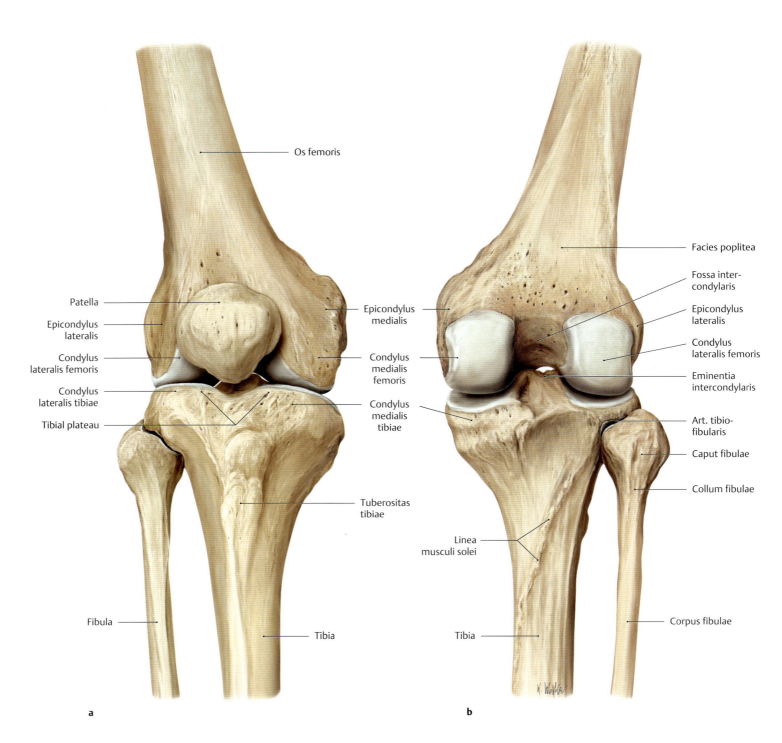

A The right knee joint (art. genus)
(a) Anterior view and (b) posterior view.
Three bones articulate at the knee joint (art.genus): the femur, tibia, and patella. The femur and tibia form the *femorotibial joint* (art .femorotibialis), while the femur and patella form the *femoropatel-* *lar joint* (art. femoropatellaris). Both joints are contained within a common capsule and have communicating articular cavities (see p. 450). Contrasting with the elbow joint (art. cubiti), where the forearm bones articulate with the humerus, the fibula is *not* included in the knee joint. It forms a separate, rigid articulation with the tibia called the *tibiofibular joint* (art. tibiofibularis).

Lower Limb —— 19. Bones, Ligaments, and Joints

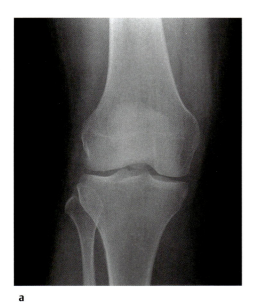
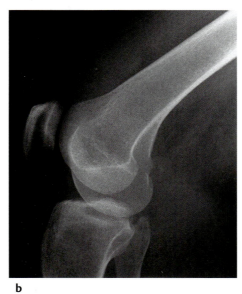

B The knee joint (art. genus)
Right art. genus, **a** anteroposterior projection and **b** lateral projection (original films from the Department of Diagnostic Radiology, Schleswig-Holstein University Hospital, Kiel Campus, Prof. S. Müller-Hülsbeck, M.D.). There are three standard radiographic views of the knee that demonstrate the joint in three planes: anteroposterior, lateral, and tangential. The *anteroposterior (AP) view* is particularly useful for evaluating the width of the joint space and the contours of the tibial plateau. *Lateral views* are good for evaluating the shape of the femoral condyles and the height of the patella. The *tangential (sunrise) view* is used mainly for examining the femoropatellar joint and evaluating the position of the patella in the femoral trochlea (see **C**).

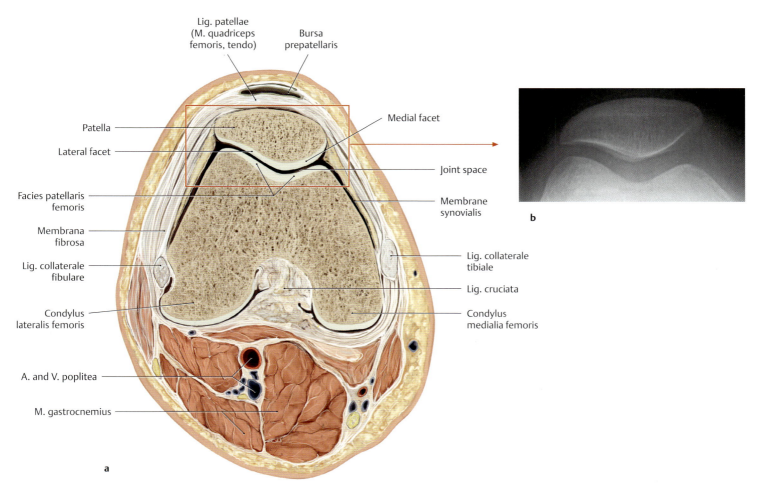

C The femoropatellar joint (art. femoropatellaris)
a Transverse section at the level of the femoropatellar joint. Right knee joint in slight flexion, distal view (drawn from a specimen in the Anatomical Collection of Kiel University).
b Tangential radiographic view of the patella and femoral trochlea ("sunrise" view of the right knee joint in 60° flexion with the beam parallel to the posterior patellar surface). This view is excellent for evaluating the articular surface of the patella and the femoral trochlea. The radiographic "joint space" appears particularly wide owing to the relatively thick articular cartilage in this region (articular cartilage is not visible on radiographs). (Original film from the Department of Diagnostic Radiology, Schleswig-Holstein University Hospital, Kiel Campus, Prof. S. Müller-Hülsbeck, M.D.).

19.18 The Ligaments of the Knee Joint: An Overview

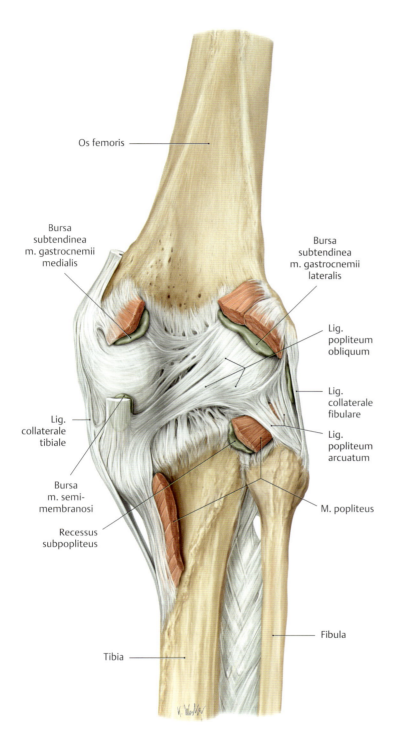

A The capsule, ligaments, and periarticular bursae of the popliteal fossa (fossa poplitea)

Right knee, posterior view. Besides the ligaments that reinforce the capsula articularis (lig. popliteum obliquum and lig. popliteum arcuatum), the capsule is strengthened posteriorly by the tendinous attachments of the muscles in the popliteal region. There are several sites where the joint cavity communicates with periarticular bursae—these include the recessus subpopliteus, bursa m.semimembranosi, and the bursa subtendinea m. gastrocnemii medialis.

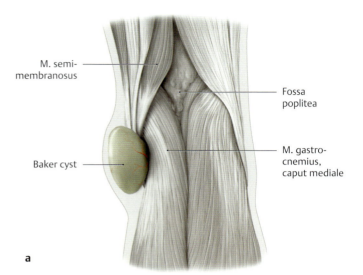

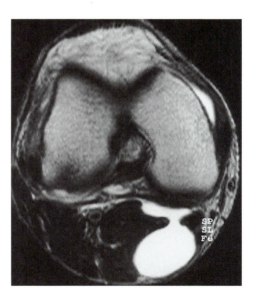

B Bursagastrocnemio-semimembranosa ("Baker cyst") in the popliteal region

a Depiction of a Baker cyst in the right fossa poplitea. A painful swelling behind the knee may be caused by a cystic outpouching of the capsula articularis ("synovial popliteal cyst"). This frequently results from a joint effusion (e.g., in rheumatoid arthritis) causing a rise of intra-articular pressure. A common Baker cyst is a cystic protrusion occurring in the medial part of the fossa poplitea between the m. semimembranosus tendon and the of the caput mediale m. gastrocnemius at the level of the posterior part of condylus medialis femoris (bursa gastrocnemio-semimembranosa = communication between the bursa semimembranosa and the bursa subtendinea m. gastrocnemii medialis).

b Axial MR image of a knee with a Baker cyst. The cystic mass in the fossa poplitea and its communication with the joint cavity appear as conspicuous areas of high signal intensity in the T 2-weighted image (from Vahlensieck M, Reiser M. MRT des Bewegungsapparates. 3rd ed. Stuttgart: Thieme; 2006).

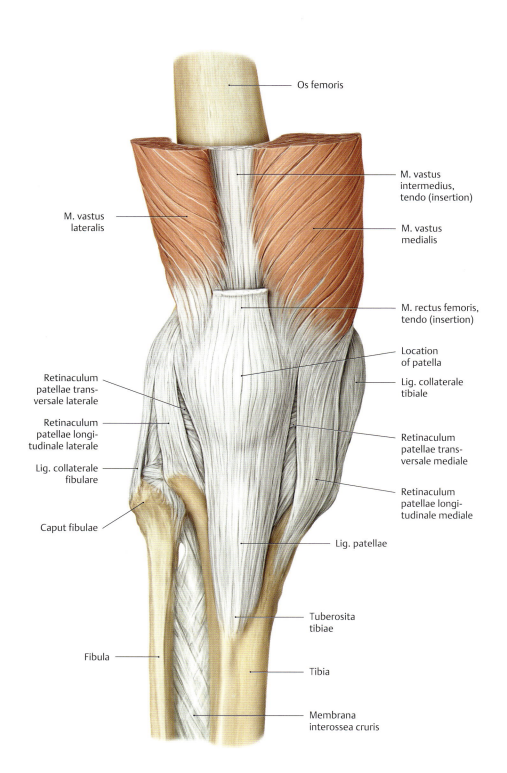

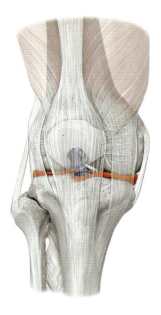

D Location of the cruciate ligaments and menisci in the knee joint
Anterior view of the right knee, in which the capsule and patella are shown in light shading. The cruciate ligaments are colored dark blue, the menisci red.

E Overview of the ligaments of the knee joint
Because its articulating bony surfaces are not closely apposed over a large area, the knee must rely upon a group of strong and extensive ligaments for stability. These ligaments of the knee joint can be segregated into two groups, extrinsic and intrinsic.

Extrinsic ligaments

• Anterior side
  – Lig. patellae
  – Retinaculum patellae longitudinale mediale
  – Retinaculum patellae longitudinale laterale
  – Retinaculum patellae transversale mediale
  – Retinaculum patellae transversale laterale
• Medial and lateral sides
  – Lig. collaterale tibiale
  – Lig. collaterale fibulare
• Posterior side
  – Lig. popliteum obliquum
  – Lig. popliteum arcuatum

Intrinsic ligaments
– Lig. cruciatum anterius – Lig. cruciatum posterius – Lig. transversum genus – Lig. meniscofemorale posterius

C The anterior and lateral capsule and ligaments of the right knee joint
Anterior view. The capsule and ligaments at the front of the knee serve mainly to stabilize the patella. The key stabilizers for this purpose are the tendons of insertion of the m. rectus femoris and mm. vasti medialis and lateralis, the retinacula patellae longitudinalia and transversalia, and, at a deeper level, the lig. patellae.

19.19 The Knee Joint: The Cruciate and Collateral Ligaments (Ligg. cruciata et collateralia)

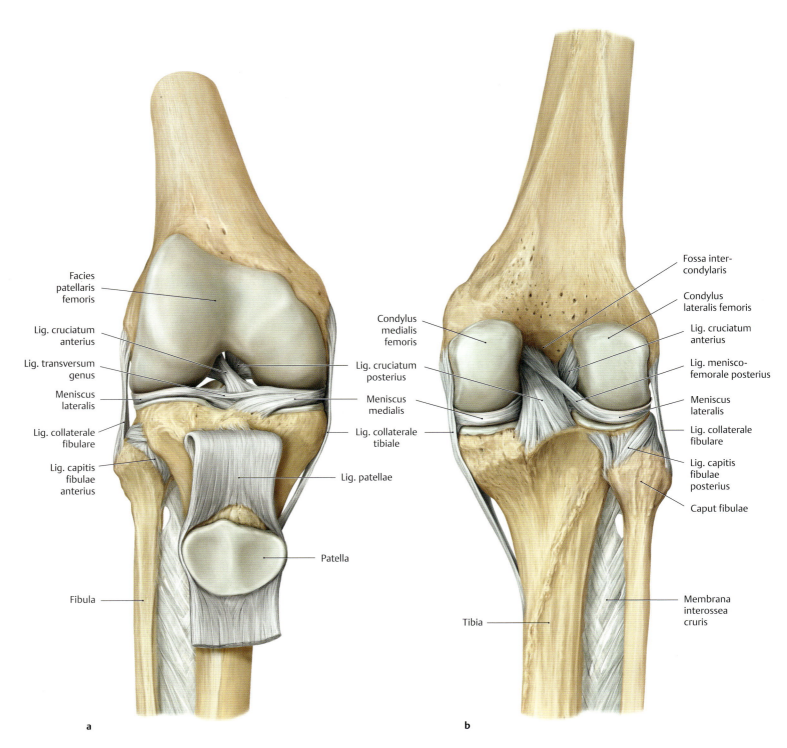

A The cruciate ligaments of the right knee joint
a Anterior view. The lig. patellae has been reflected downward with the attached patella.
b Posterior view.

The cruciate ligaments of the knee stretch between the areae intercondylares anterior and posterior of the tibia (not visible here; see p. 446) and the fossa intercondylaris of the femur.

- The *lig. cruciatum anterius* runs from the area intercondylaris anterior of the tibia to the medial surface of the condylus lateralis femoris.

- The *lig. cruciatum posterius* is thicker than the anterior ligament and runs approximately at right angles to it, passing from the area intercondylaris posterior to the lateral surface of the condylus medialis femoris.

The cruciate ligaments keep the articular surfaces of the femur and tibia in contact while stabilizing the knee joint primarily in the sagittal plane. Some portions of the cruciate ligaments are taut in every position of the joint (see p. 448).

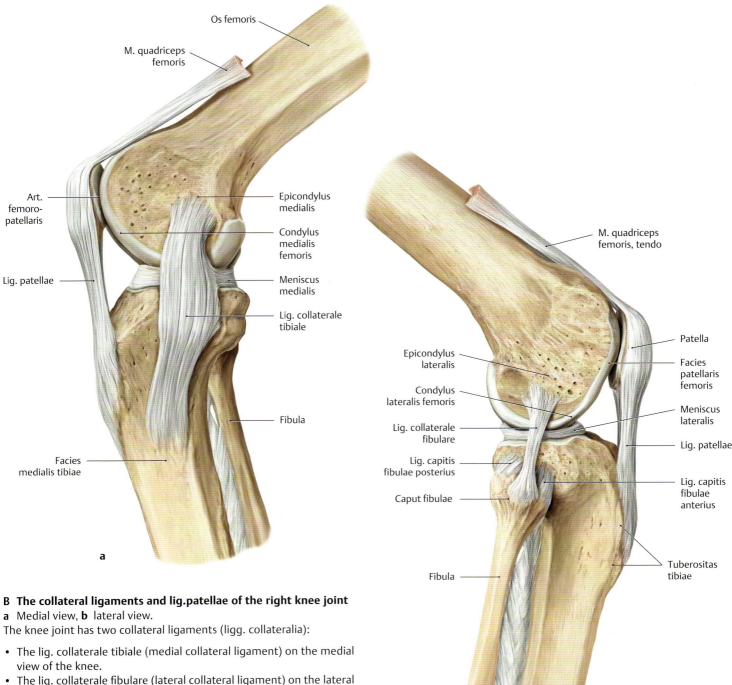

B The collateral ligaments and lig.patellae of the right knee joint
a Medial view, b lateral view.
The knee joint has two collateral ligaments (ligg. collateralia):

- The lig. collaterale tibiale (medial collateral ligament) on the medial view of the knee.
- The lig. collaterale fibulare (lateral collateral ligament) on the lateral view of the knee.

The *lig. collaterale tibiale* (medial collateral ligament) is the broader of the two ligaments. It runs obliquely downward and forward from the epicondylus medialis of the femur to the facies medialis tibiae approximately 7 to 8 cm below the tibial plateau. The *lig. collaterale fibulare* (lateral collateral ligament) is a round cord that runs obliquely downward and backward from the condylus lateralis femoris of the femur to the caput fibulae. Both of the ligg. collateralia are taut when the knee is in *extension* (see **A**). When the knee is in *flexion*, the radius of curvature is decreased, and the origins and insertions of the ligg. collateralia move closer together, causing the ligaments to become lax. Both ligg. collateralia stabilize the knee joint in the coronal plane. Thus, damage or rupture of these ligaments can be diagnosed by examining the mediolateral stability of the knee and the extent of medial and lateral opening of the joint space with manipulation.

Note the different relationship of each collateral ligament to the joint capsule and associated meniscus. The lig. collaterale *tibiale* is firmly attached both to the capsule and the meniscus medialis, whereas the lig. collaterale *fibulare* has no direct contact with the capsule or the meniscus lateralis. As a result, the meniscus medialis is less mobile than the meniscus lateralis and is thus far more susceptible to injury (see also p. 447).

19.20 The Knee Joint: The Menisci

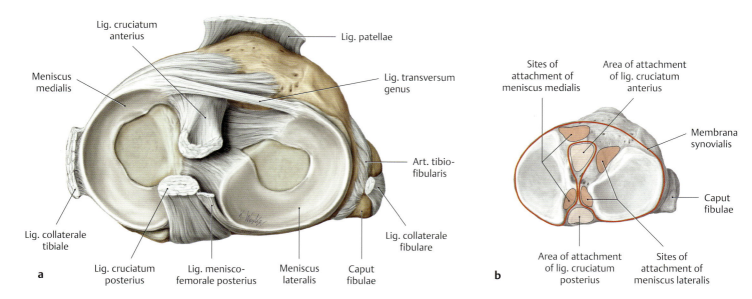

A The tibial plateau with the medial and lateral menisci and the sites of attachment of the menisci and cruciate ligaments
Right tibial plateau, proximal view with the ligg. cruciata (cruciate ligaments) and ligg. collateralia (collateral ligaments) divided and the femur removed.

a Shape and attachments of the menisci: The medial and lateral menisci are both crescent-shaped when viewed from above (L. *meniscus* = crescent). Their ends (the anterior and posterior horns) are attached by short ligaments to the bone of the areae intercondylares anterior and posterior of the tibia. The *lateral meniscus (meniscus lateralis)* forms almost a complete ring, while the *medial meniscus (meniscus medialis)* has a more semicircular shape. On the whole, the meniscus medialis is less mobile than the meniscus lateralis because its points of attachment to the bone are spaced farther apart (see **b**), and it is also firmly attached peripherally to the medial collateral ligament (lig. collaterale tibiale). The meniscus lateralis, by contrast, has no attachment to the lig. collaterale fibulare (see **E**).

b Sites of attachment of the medial and lateral menisci and the cruciate ligaments: The red line indicates the tibial attachment of the membrana synovialis, which covers the cruciate ligaments anteriorly and at the sides. The cruciate ligaments lie in the subsynovial connective tissue of the capsula articularis and are covered posteriorly by the heavy membrana fibrosa. Because the cruciate ligaments migrate forward into the knee joint during development, they are extracapsular but intra-articular in their location (see also p. 450), and they derive their blood supply from the fossa poplitea (a. media genus, see p. 561).

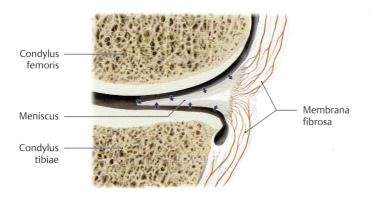

B Blood supply to the menisci
Schematic coronal section through the femorotibial joint. The fibrous portions of the menisci located adjacent to the capsula articularis have a rich blood supply (the aa.articulares inferiores mediales and lateralis from the a. poplitea, see p. 561). But the more central portions of the menisci, composed of fibrocartilage, are avascular and are nourished entirely by the synovial fluid blue (arrows).

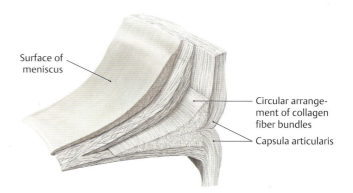

C Structure of the meniscus (after Petersen and Tillmann)
The meniscus has a wedge-shaped cross section, the base of the wedge (right) is directed toward the periphery and attached to the joint capsule. The lower surface facing the tibial plateau is flat, while the upper surface facing the femoral condyles is concave. The central, inner two thirds of the menisci are composed of fibrocartilage and the outer third of tough connective tissue. The bundles of collagen fibers in both the fibrocartilage and the connective tissue have a predominantly circular arrangement, reflecting the high tensile stresses that develop in the menisci. The ability of the meniscal tissue to move outward in response to loading is similar to that found in disci intervertebrales (converting pressure to tensile forces).

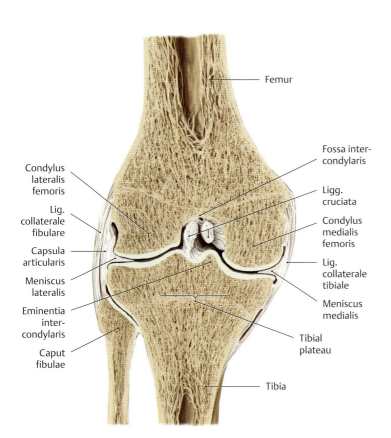

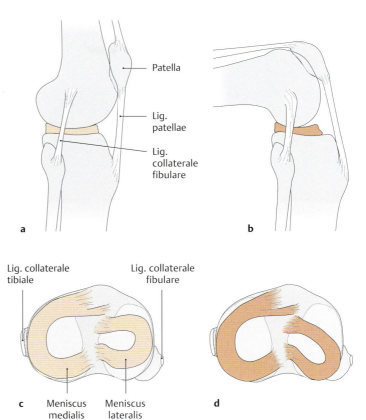

D Coronal section through the femorotibial joint
Right knee, anterior view. An essential task of the menisci is to increase the surface area available for load transfer across the knee joint. With their different curvatures, the menisci compensate for the mismatch in the articulating surfaces of the femur and tibia. They absorb approximately one third of the loads imposed on the knee, and they help to distribute the pressures more evenly within the femorotibial joint.

E Movements of the menisci during flexion of the knee
The drawings show a right art. genus from the lateral view in extension (**a**) and flexion (**b**) and the associated tibial plateau viewed from above in extension (**c**) and flexion (**d**).
Note that the meniscus medialis, which is anchored more securely than the meniscus lateralis, undergoes considerably less displacement during knee flexion.

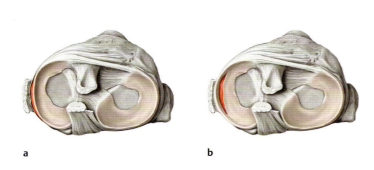

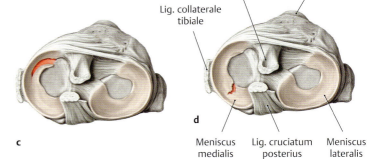

F Different patterns of meniscal tears
Right tibial plateau, proximal view.

a Peripheral tear.
b Bucket-handle tear.
c Longitudinal or flap tear of the anterior horn.
d Radial tear of the posterior horn.

The meniscus medialis, being less mobile, is injured far more often than the meniscus lateralis. Meniscal injuries most commonly result from sudden extension or rotational movements of the flexed knee (external and internal rotation) while the leg is fixed, as may occur while skiing or playing soccer. The resultant shearing forces can tear the substance of the meniscus or avulse it from its peripheral attachment. The cardinal feature of a fresh meniscal injury is a painful limitation of active and passive knee extension immediately after the trauma, while the patient favors the knee by keeping it slightly flexed. Degenerative changes in the menisci occur with aging and are exacerbated by excessive loads and angular deformities of the knee (genu varum or valgum, see p. 410).

19.21 The Movements of the Knee Joint

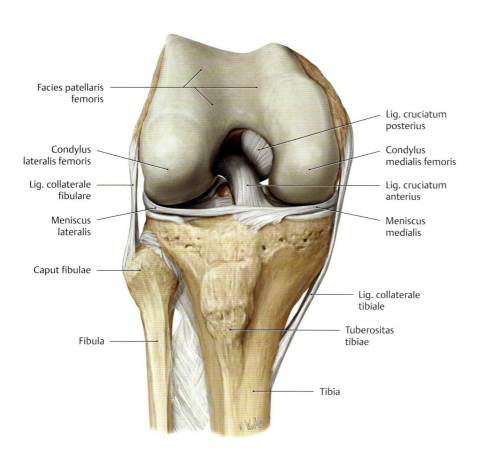

A The right knee joint in flexion
Anterior view with the capsula articularis and patella removed.

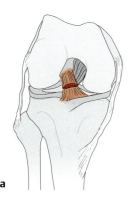
a

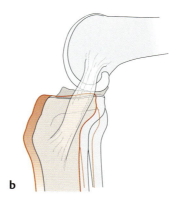
b

B Condition after rupture of the anterior cruciate ligament (lig. cruciatum anterius)
a Right knee joint in flexion, anterior view.
b Right knee joint in flexion, medial view.

The instability that results from a cruciate ligament rupture allows the tibia to be moved forward or backward like a drawer relative to the femur, depending on whether the anterior or lig. cruciatum posterius has been torn ("anterior or posterior drawer sign," elicited with the Lachman test). A rupture of the lig. cruciatum anterius, as shown in the diagram, is approximately 10 times more common than a rupture of the posterior cruciate ligament. The most common mechanism of injury is an internal rotation trauma with the leg fixed (see **D**). A lateral blow to the fully extended knee, with foot planted, tends to cause concomitant rupture of the lig. cruciatum anterius and the medial collateral ligament (lig. collaterale tibiae), and tearing of the attached medial meniscus (meniscus medialis), referred to colloquially as the "unhappy triad."

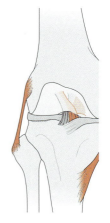
a

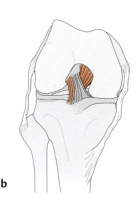
b

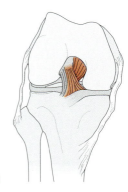
c

C Behavior of ligg. cruciata and ligg. collateralia in flexion and extension
Right knee, anterior view. Ligament fibers that are taut are colored red.

a Extension.
b Flexion.
c Flexion and internal rotation.

While the collateral knee ligaments are taut *only in extension* (**a**), the cruciate ligaments, or at least portions of them, are taut *in every joint position*: the medial portions of both cruciate ligaments in extension (**a**), the lateral part of the lig. cruciatum anterius and the entire lig. cruciatum posterius in flexion (**b**), the medial part of the lig. cruciatum anterius and the entire lig. cruciatum posterius in flexion and internal rotation (**c**). The cruciate ligaments thus help stabilize the knee in any joint position.

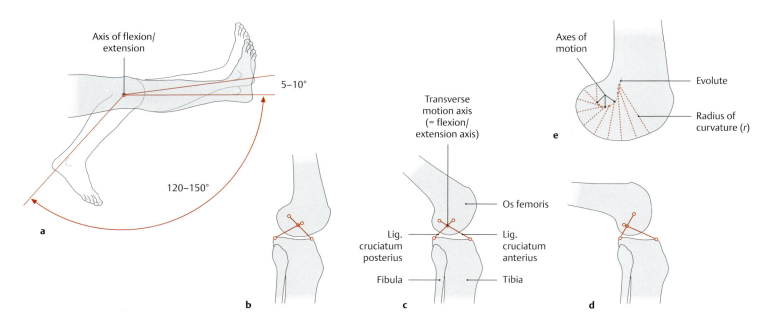

D Flexion and extension of the knee joint (art. genus)
Right knee joint, lateral view.
Flexion and extension of the art. genus take place about a transverse axis (**a**) that passes through the dynamic center of rotation in any joint position. That center is located at the point where both the ligg. collateralia and the cruciate ligaments intersect (**b**). With increasing flexion of the knee (**c, d**), the dynamic flexion axis moves upward and backward along a curved line (the evolute, **e**). The momentary distance from that curve to the articular surface of the femur is equal to the changing radius of curvature (*r*) of the condyli femoris. The total range of motion, especially in flexion, depends on various parameters (soft-tissue restraints, active insufficiency or hamstring tightness, see p. 485).

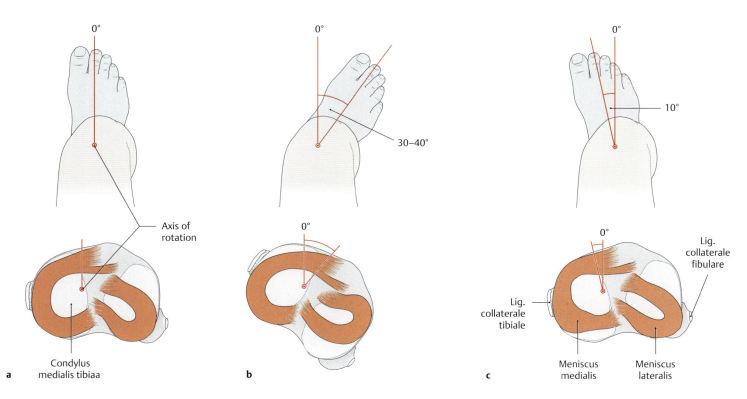

E Rotational movements of the tibia relative to the femur with the knee flexed 90°
Right knee joint, proximal view of the flexed knee and corresponding tibial plateau.

a Neutral (0°) position.
b External rotation.
c Internal rotation.

The axis of tibial rotation runs vertically through the medial part of the condylus medialis tibiae. Because the cruciate ligaments (not shown here) twist around each other during internal rotation, the range of internal rotation in the knee (approximately 10°) is considerably smaller than the range of external rotation (30–40°). As a result, the majority of cruciate ligament tears occur during internal rotation and involve the lig. cruciatum anterius.
Note the different degrees of displacement of the lateral and medial menisci.

19.22 The Knee Joint (Art. genus): Capsule and Joint Cavity

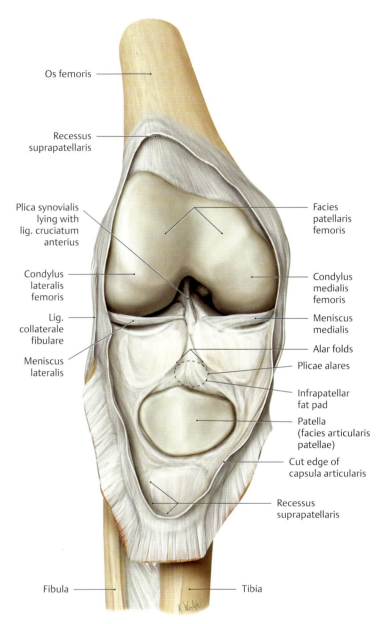

A The right knee with opend joint capsule (capsula articularis)
The patella has been reflected downward. In the anterior meniscofemoral portion of the capsula articularis, variable folds of the capsule project into the joint cavity (plicae alares on both sides of the infrapatellar fat pad [Corpus adiposum infrapatellare]), increasing its capacity.

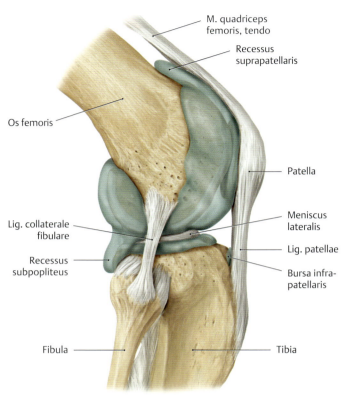

B Extent of the joint cavity (cavitas articularis)
Right knee joint, lateral view. The joint cavity was demonstrated by injecting a liquid plastic into the knee joint and later removing the capsule after the plastic had cured.

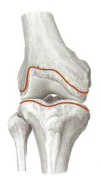

C Anterior femoral and tibial attachments of the joint capsule
Right knee joint, anterior view.

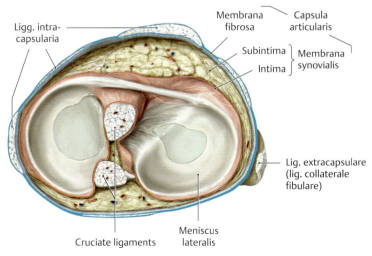

D Structure of the joint capsule
Right knee after removal of femur (joint capsule and ligaments severed), superior view.
Note: The cruciate ligaments run in the subintima of the synovial membrane and are thus intracapsular but extra-articular (compare p. 43). In contrast, the menisci are intra-articular because they are not covered by a synovial intima and have direct contact with synovial fluid.

Lower Limb — 19. Bones, Ligaments, and Joints

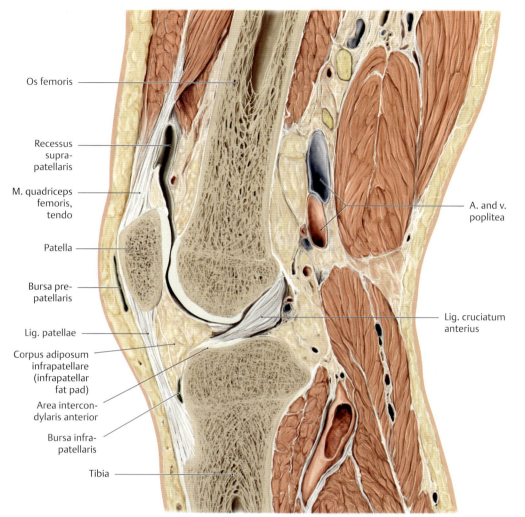

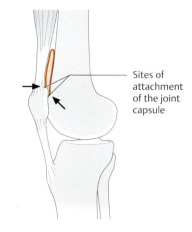

a

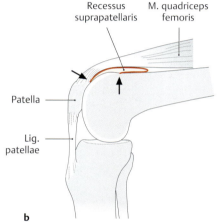

b

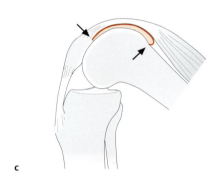

c

E Mid-sagittal section through the right knee joint (art. genus)
Note the extent of the recessus suprapatellaris (also called the *bursa* suprapatellaris) and compare it with **F**. Note also the placement of the infrapatellar fat pad (corpus adiposum infrapatellare) between the area intercondylaris anterior and the deep surface of the lig. patellae. A fall onto the knee or chronic mechanical irritation due to frequent kneeling can cause pain and inflammation of the bursae about the patella: bursitis infrapatellaris ("clergyman's knee") and bursitis prepatellaris (drawn from a specimen in the Anatomical Collection of Kiel University).

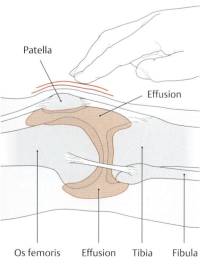

F The "ballottable patella sign" of knee effusion
When an effusion develops in the knee joint due to inflammatory changes or injury, various degrees of joint swelling may be seen. To differentiate an intra-articular effusion from swelling of the capsula articularis itself, the leg is placed in a position of maximum extension. This will force the (potentially increased) intra-articular fluid out of the recessus suprapatellaris and into the space between the patella and femur. The examiner then pushes the patella downward with the index finger. If there is excessive fluid in the joint, the patella will rebound when released, signifying a positive test.

G Unfolding of the recessus suprapatellaris during flexion
Right knee joint, medial view.

a Neutral (0°) position.
b 80° of flexion.
c 130° of flexion.

The recessus suprapatellaris extends proximally from the superior pole of the patella, turns back distally, and inserts at the chondro-osseous junction on the facies patellaris femoris. This redundant fold provides a reserve capacity when the knee is flexed, opening up completely past about 130° of flexion.

19.23 Cross-sectional Anatomy of the Knee

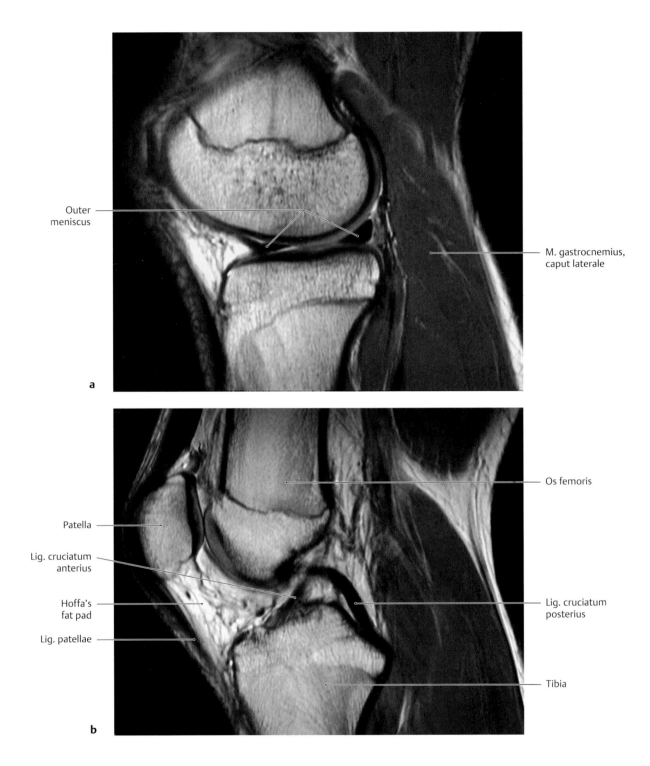

A Magnetic resonance imaging of the knee joint
(from Vahlensieck M, Reiser M. MRT des Bewegungsapparates. 3rd ed. Stuttgart: Thieme; 2006)
T1-weighted sagittal sequences at the level of the outer meniscus (**a**) and the cruciate ligaments (**b**).
The strength of MRI when evaluating joints includes the differentiated representation of the joint's internal structure, the capsular ligament apparatus, the surrounding soft tissue, and the subchondral bone. In addition to the T1-weighted images, which are primarily suited for anatomic orientation and tissue characterization, the T2-weighted proton-density fat saturation sequences (PDFS) play an important role in the diagnosis of joint damage because they are very sensitive to edematous changes (e.g., fractures) and allow for the high signal intensity of the hyaline articular cartilage (see **B**). (In MRI, fat emits a high signal, which, however, is often adverse and thus needs to be suppressed.)
Note: In sagittal cross sections, the lig. cruciatum posterius is clearly visible in its entirety, even its archlike course from posterior to superior. In contrast, the lig. cruciatum anterius is only visible in its entirety when the knee joint, as shown here, is externally rotated at 15 to 20°.

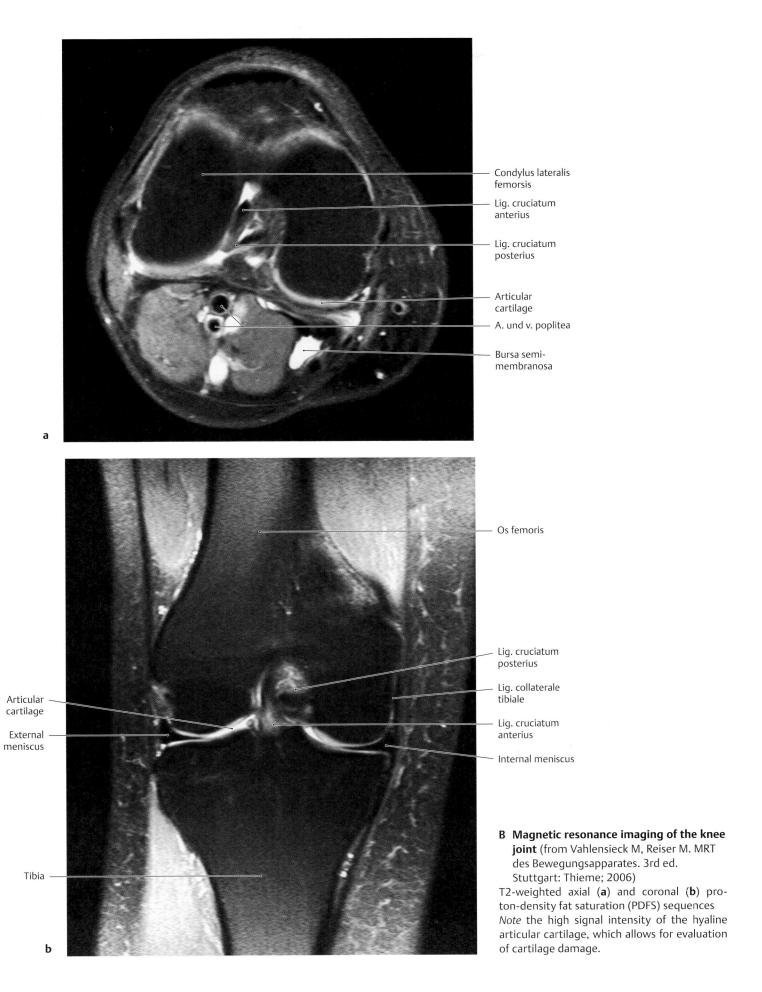

B Magnetic resonance imaging of the knee joint (from Vahlensieck M, Reiser M. MRT des Bewegungsapparates. 3rd ed. Stuttgart: Thieme; 2006)
T2-weighted axial (**a**) and coronal (**b**) proton-density fat saturation (PDFS) sequences
Note the high signal intensity of the hyaline articular cartilage, which allows for evaluation of cartilage damage.

19.24 The Joints of the Foot: Overview of the Articulating Bones and Joints

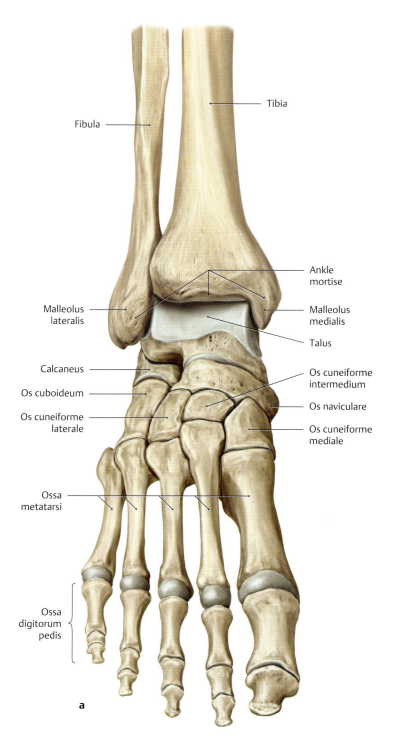
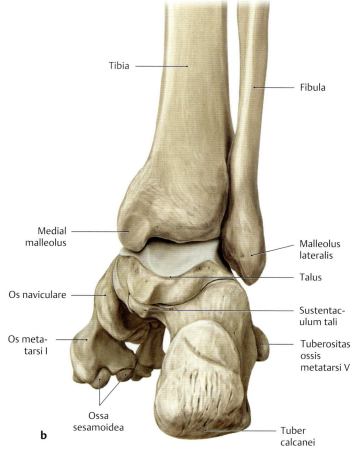

A The articulating bones in different joints of the right foot

a Anterior view with the talocrural joint (art. talocruralis) in plantar flexion.
b Posterior view with the foot in the neutral (0°) position.

B Overview of the joints in the foot (Articulationes pedis)

- Art. talocruralis (ankle joint)
- Subtalar joint (art. subtalaris and art. talocalcaneonavicularis)*
- Art. calcaneocuboidea (between the calcaneus and cuboid bone)
- Art. talonavicularis (between the talus and navicular bone)
- Art. tarsi transversa**
- Art. cuneonavicularis (between the ossa cuneiformia and os naviculare)
- Artt. intercuneiformes (between the cuneiform bones)
- Art. cuneocuboidea (between the os cuneiforme laterale and os cuboideum)
- Artt. tarsometatarsales
- Artt. intermetatarsales (between the bases of the metatarsal bones)
- Artt. metatarsophalangeae
- Artt. interphalangeae proximales
- Artt. interphalangeae distales

* In the subtalar joint, the talus articulates with the calcaneus and the navicular bone to form two separate articulations: the art. subtalaris posteriorly and the art. talocalcaneonavicularis anteriorly. Both are often referred to collectively as the "subtalar joint."

** Consists of the art. calcaneocuboidea and art. talonavicularis.

Lower Limb — 19. Bones, Ligaments, and Joints

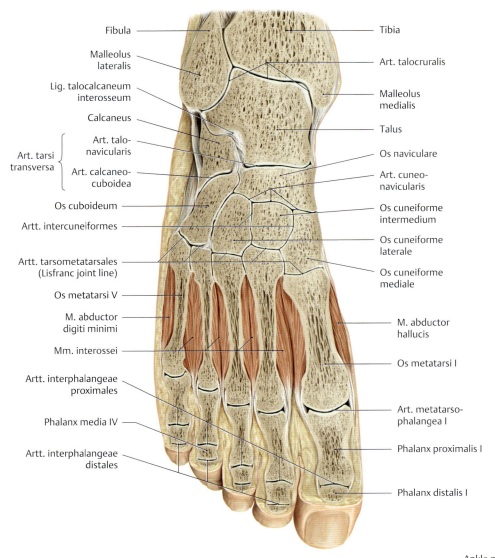

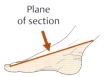

C Oblique transverse section through the foot
Right foot, superior view. The foot is plantar flexed at the art. talocruralis (drawn from a specimen in the Anatomical Collection of Kiel University).

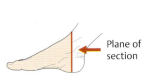

D Coronal section through the talocrural and subtalar joint
Right foot, proximal view. The art. talocruralia is plantar flexed, and the art. subtalaris is sectioned through its posterior compartment (drawn from a specimen in the Anatomical Collection of Kiel University).

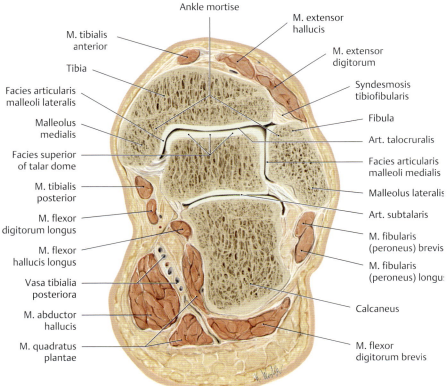

455

19.25 The Joints of the Foot: Articular Surfaces

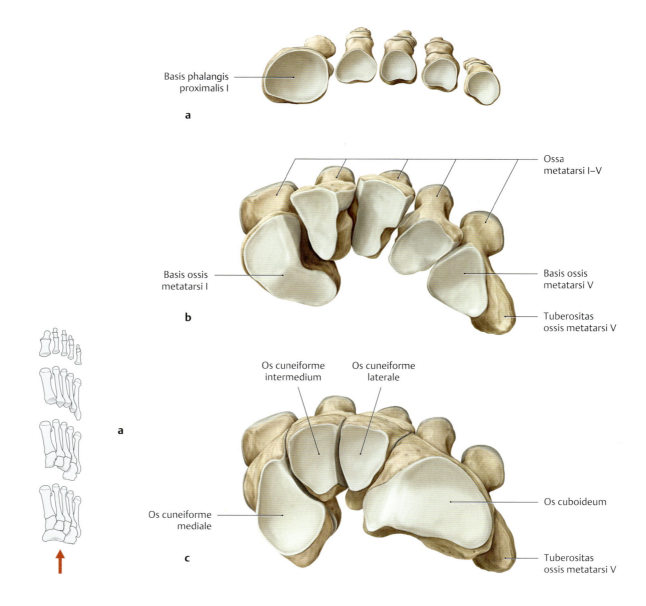

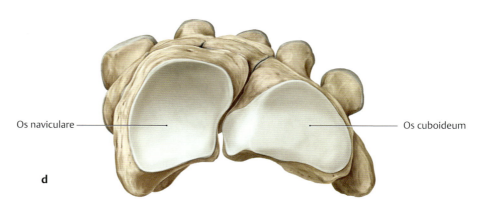

A Proximal articular surfaces
Right foot, proximal view.

a Artt. metatarsophalangeae: bases of the phalanges proximales I to V.
b Artt. tarsometatarsales: bases of the ossa metatarsi I–V.
c Art. cuneonavicularis and art. calcaneocuboidea: proximal articular surfaces of the ossa cuneiformia mediale, intermedium, and laterale and the os cuboideum.
d Art. talonavicularis and art. calcaneocuboidea: proximal articular surfaces of the os naviculare and os cuboideum.

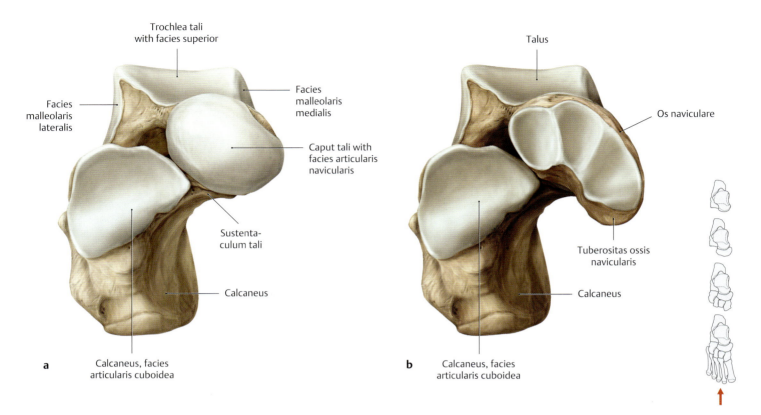
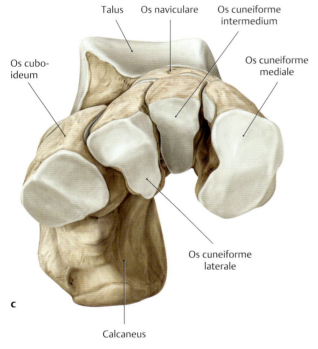
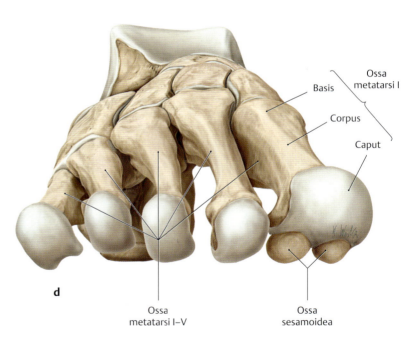

B Distal articular surfaces
Right foot, distal view.

a The art. talonavicularis and art. calcaneocuboidea: distal articular surfaces of the calcaneus and talus.
b The art. cuneonavicularis and art. calcaneocuboidea: distal articular surfaces of the os naviculare and calcaneus.
c The artt. tarsometatarsales: distal articular surfaces of the ossa cuneiformia mediale, intermedium, and laterale and the os cuboideum.
d Artt. metatarsophalangeae: heads of the ossa metatarsil–V.

19.26 The Joints of the Foot: The Talocrural and Subtalar Joints

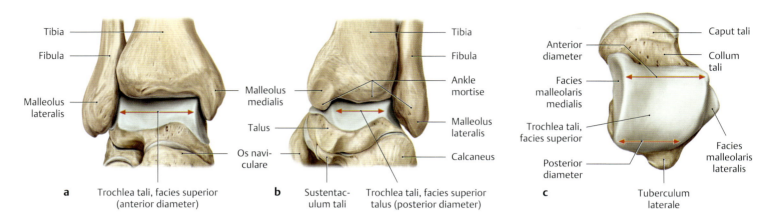

A The articulating skeletal elements of the talocrural joint (art. talocruralis)
a Right foot, anterior view.
b Right foot, posterior view.
c Right trochlea tali, superior view.

The art. talocruralis, called also the ankle joint, is formed by the distal ends of the tibia and fibula (the ankle mortise, see also **B**) articulating with the trochlea tali. This provides the art. talocruralis with good bony and ligamentous stability and helps to stabilize the body in an erect posture. However, due to the shape of the trochlea tali (the anterior part, the facies superior, is approximately 5 to 6 mm broader than the posterior part), the bony stability of the art. talocruralis differs in flexion and extension. When the broader anterior part of the trochlea articulates with the ankle mortise in *dorsiflexion* (where the foot moves closer to the leg, as in squatting), the syndesmotic ligaments (see p. 460) are tightly stretched, and there is excellent bony stability. But when the narrower posterior part of the trochlea comes in contact with the ankle mortise in *plantar flexion* (e.g., standing on the toes), the talus no longer provides a high degree of bony stability within the ankle mortise.

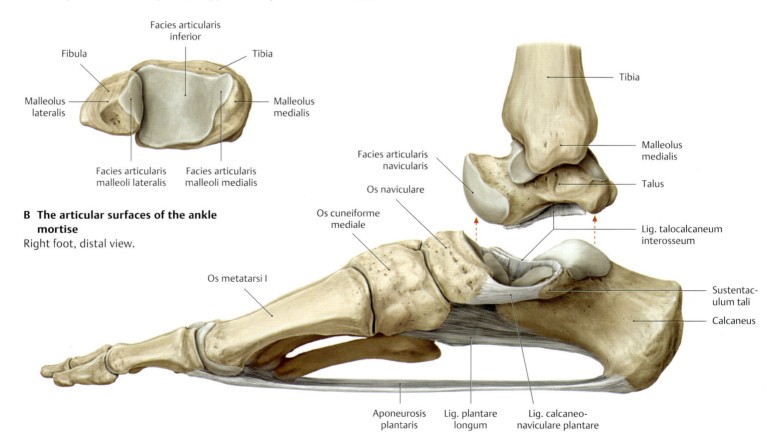

B The articular surfaces of the ankle mortise
Right foot, distal view.

C Overview of an opened art. subtalaris
Right foot, medial view. The lig. talocalcaneum interosseum has been divided, and the talus has been displaced upward to demonstrate the articular surfaces of the subtalar joint.

Note the course of the plantar calcaneonavicular ligament (lig. calcaneonaviculare plantare), which functions with the lig. plantare longum and aponeurosis plantaris to support the longitudinal arch of the foot (see also **D** and p. 467).

Lower Limb — 19. Bones, Ligaments, and Joints

D Course of the lig. calcaneonaviculare plantare
Right foot, plantar view. The plantar calcaneonavicular (spring) ligament stretches between the sustentaculum tali and the os naviculare. It completes the bony socket of the talocalcaneal joint from the plantar side.

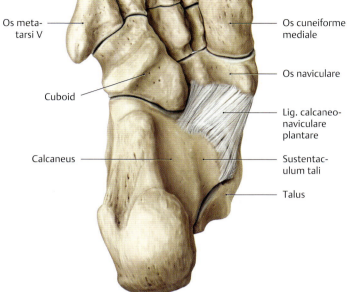

E The articular surfaces of an opened subtalar joint
Right foot, dorsal view (after separation of the talus). In the subtalar joint, the talus articulates with the calcaneus and the os naviculare. It consists of two completely separate articulations:

- a posterior compartment (the art. subtalaris) and
- an anterior compartment (the art. talocalcaneonavicularis).

The boundary between the two compartments is formed by the lig.talocalcaneum interosseum located in the canalis tarsi (bony canal formed by the sulcus tali and sulcus calcanei; its entrance is the sinus tarsi)). The lig. calcaneonaviculare plantare, which has cartilage cells in its medial surface, loops like a tendon around the plantar head of the talus, which acts as a fulcrum. It stabilizes the position of the talus on the calcaneus and helps to support the apex of the longitudinal arch (see p. 467). Overstretching of the lig. calcaneonaviculare plantare due to flattening of the plantar vault promotes the development of flat foot.

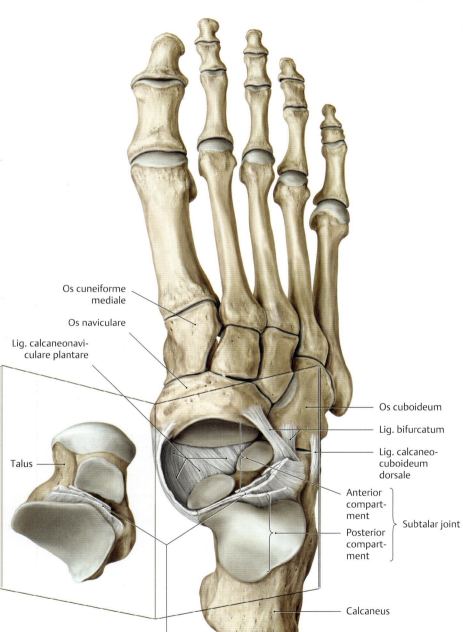

459

19.27 The Ligaments of the Foot

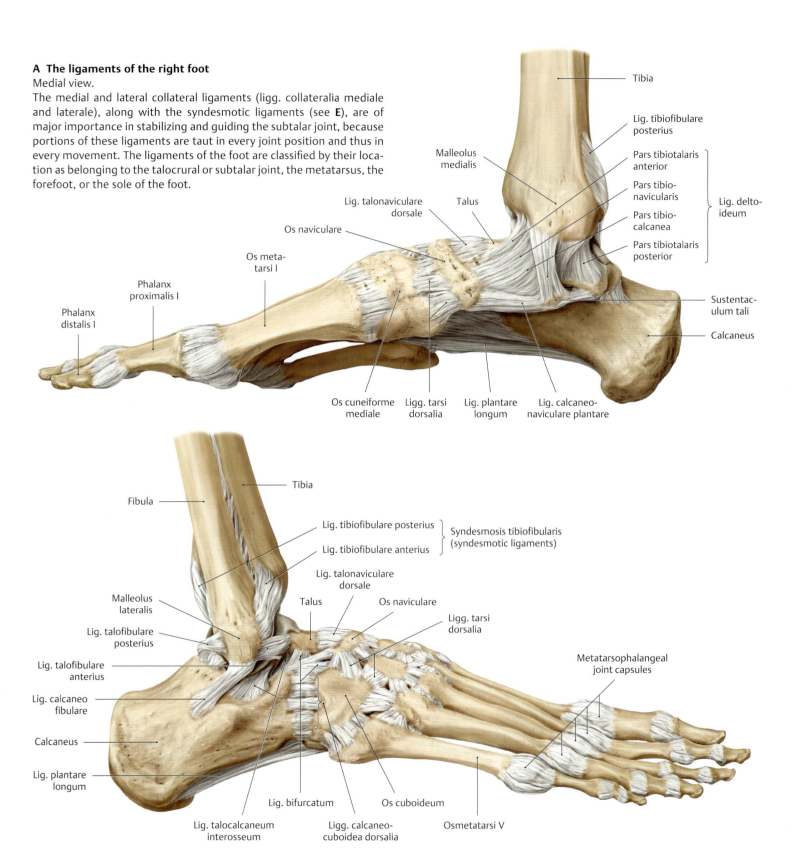

A The ligaments of the right foot
Medial view.
The medial and lateral collateral ligaments (ligg. collateralia mediale and laterale), along with the syndesmotic ligaments (see **E**), are of major importance in stabilizing and guiding the subtalar joint, because portions of these ligaments are taut in every joint position and thus in every movement. The ligaments of the foot are classified by their location as belonging to the talocrural or subtalar joint, the metatarsus, the forefoot, or the sole of the foot.

B The ligaments of the right foot
Lateral view.
Sprains of the ankle joint and especially of its lateral ligaments (usually supination trauma = buckling of the ankle in a supinated position) are extremely common injuries. They often occur during plantar flexion of the foot, a position that provides less bony stability to the art. talocruralis (see p. 458). Most of these injuries occur during sports and other leisure activities when the ankle gives way on uneven ground. Typically, the trauma will cause stretching or tearing of the lig. talofibulare anterius, the lig. calcaneofibulare, or both. If the leg is twisted violently while the foot is fixed, there may also be separation of the ankle mortise with disruption of the syndesmosis tibiofibularis (see **D**).

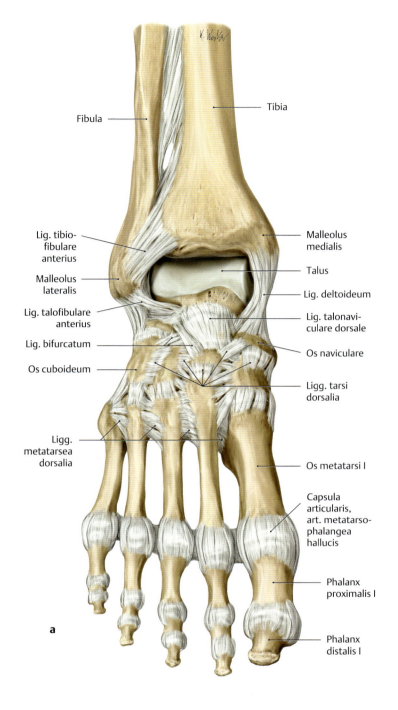

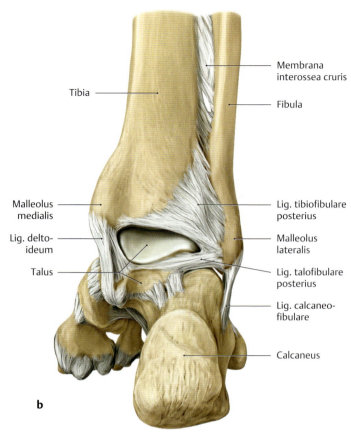

C The ligaments of the right foot
a Anterior view (art. talocruralis in plantar flexion).
b Posterior view (plantigrade foot position).

The anterior and posterior portions of the talocrural joint capsule have been removed to demonstrate more clearly the placement of the ligaments.

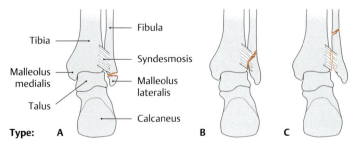

D Weber fractures
A Weber fracture is an avulsion fracture of the malleolus lateralis of the fibula. Weber fractures are classified as type A, B, or C depending on whether the fibula is fractured below, level with, or above the syndesmosis. The syndesmosis *may* be torn in a Weber type B fracture (as shown here), but it is *always* torn in a Weber type C fracture.

E The ligaments of the art. talocruralis (the ligaments of the subtalar joint are reviewed on p. 459).

Ligg. collateralia laterale
- Lig. talofibulare anterius
- Lig. talofibulare posterius
- Lig. calcaneofibulare

Ligg. collateralia mediale
- Lig. deltoideum
  – Pars tibiotalaris anterior
  – Pars tibiotalaris posterior
  – Pars tibionavicularis
  – Pars tibiocalcanea

Syndesmotic ligaments of the ankle mortise
- Lig. tibiofibulare anterius
- Lig. tibiofibulare posterius

19.28 The Movements of the Foot

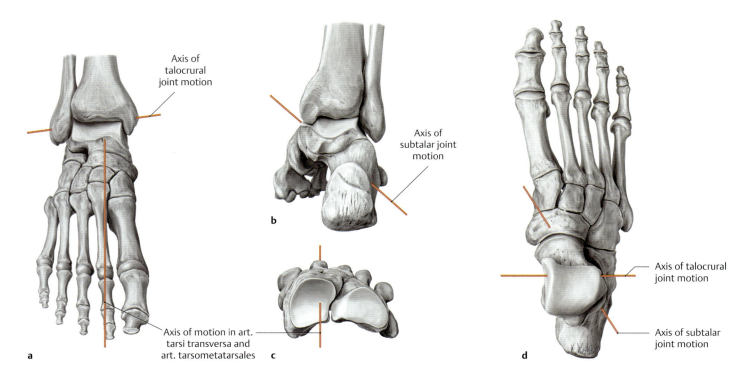

A The principal axes of motion in the right foot
a Anterior view with the talocrural joint in plantar flexion.
b Posterior view in the functional position (see **B**).
c Isolated right forefoot, proximal view.
d Superior view.

The axes of articular motion in the foot are complex, and the descriptions of movements in the pedal joints are often inconsistent and confusing. The following axes of motion are important in clinical parlance and for the testing of joint motion (compare with the facing page):

- **Axis of talocrural joint motion (plantar flexion/dorsiflexion):** This axis runs almost transversely through the lateral and medial malleoli. It forms an approximately 82° angle with the tibial shaft axis in the frontal plane, and it forms a 10° angle with the frontal plane on the medial side (**a, d**).
- **Axis of subtalar joint motion (inversion/eversion):** This axis runs obliquely upward through the foot in a posterolateral-to-anteromedial direction, i.e., from the lateral calcaneus through the medial portion of the canalis tarsi to the center of the os naviculare. It forms an approximately 30° angle with the horizontal plane and a 20° angle with the sagittal plane (**b, d**).
- **Axis of forefoot motion in the art. tarsi transversa and artt. tarsometatarsales (pronation/supination):** This axis lies approximately in the sagittal plane, running from the calcaneus through the os naviculare and along the second ray (**a, c**).

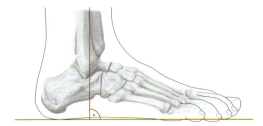

B The functional position of the foot
Right foot, lateral view. In the neutral (0°) position, the skeleton of the foot is angled approximately 90° relative to the skeleton of the leg. This *plantigrade foot position* is termed the "functional position" and is an important basis for normal standing and walking.

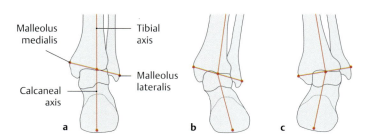

C Axis of the hindfoot
Distal right leg and hindfoot, posterior view.

a With normal axial alignment in the hindfoot, the tibial axis and calcaneal axis lie on a vertical line (pes rectus). The calcaneal axis bisects a line drawn between the two malleoli.
b Pes valgus: the foot is in an everted position.
c Pes varus: the foot is in an inverted position.

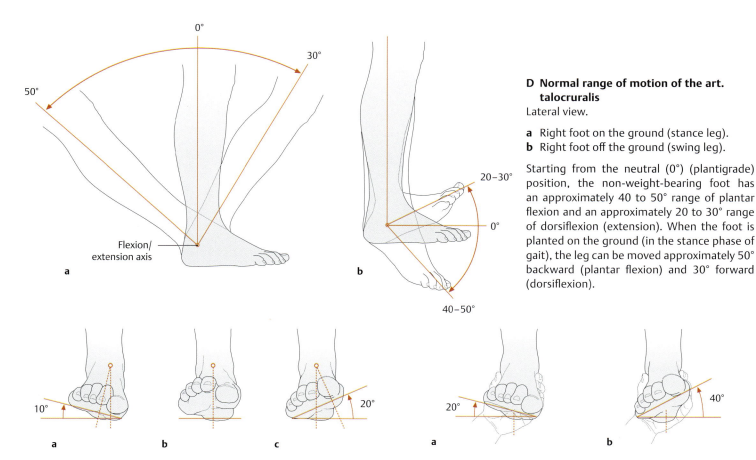

D Normal range of motion of the art. talocruralis
Lateral view.

a Right foot on the ground (stance leg).
b Right foot off the ground (swing leg).

Starting from the neutral (0°) (plantigrade) position, the non-weight-bearing foot has an approximately 40 to 50° range of plantar flexion and an approximately 20 to 30° range of dorsiflexion (extension). When the foot is planted on the ground (in the stance phase of gait), the leg can be moved approximately 50° backward (plantar flexion) and 30° forward (dorsiflexion).

E Range of motion of the subtalar joint
Right foot, anterior view.

a Everted by 10°.
b Neutral (0°) position.
c Inverted by 20°.

Rotation of the calcaneus medially (inversion) and laterally (eversion) is measured from the neutral (0°) position. This is done clinically by holding the leg stationary and moving the calcaneus back and forth. Estimation of the range of inversion/eversion is based on the calcaneal axis.

F Range of pronation/supination of the transverse tarsal and tarsometatarsal joints (art. tarsi transversa and artt. tarsometatarsales).
Right foot, anterior view.

a Range of pronation of the forefoot: 20°.
b Range of supination of the forefoot: 40°.

Range of motion is tested with the hindfoot fixed. Pronation/supination of the forefoot is tested by rotating the forefoot outward relative to the hindfoot (raising the lateral border of the foot) or inward (raising the medial border of the foot).

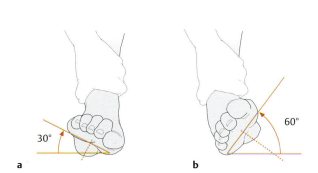

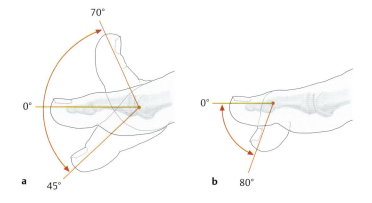

G Total range of motion of the forefoot and hindfoot
Right foot, anterior view.

a Eversion and pronation of the forefoot: 30°.
b Inversion and supination of the forefoot: 60°.

Because the movements in the joints are complex, and different joint movements are almost always mechanically coupled, the range of all joint movements can be assessed by holding the leg stationary and raising the entire foot in the medial and lateral directions.

H Range of motion of the joints of the great toe
Lateral view.

a Flexion/extension of the first metatarsophalangeal joint.
b Flexion of the first interphalangeal joint.

The toes, especially the great toe, can be passively extended to approximately 90°. This is an important prerequisite for walking, especially during the phase between heel take-off and toe strike.

19.29 Overview of the Plantar Vault and the Transverse Arch (Arcus pedis transversus)

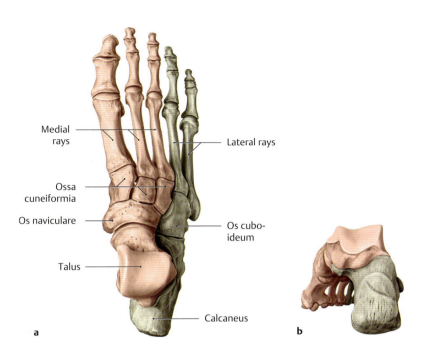

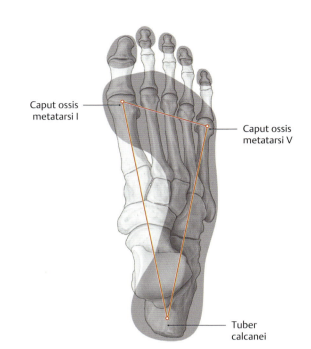

A The plantar vault
a Right foot, superior view.
b Right foot, posteromedial view.

From the perspective of structural engineering, the forces borne by the foot are distributed among two lateral (fibular) rays and three medial (tibial) rays. The lateral rays extend across the os cuboideum to the calcaneus, while the medial rays extend across the ossa cuneiformia and os naviculare to the talus. The arrangement of these rays—adjacent distally and overriding proximally—creates a longitudinal arch (arcus pedis longitudinalis) and a transverse arch (arcus pedis transversus) in the sole of the foot (planta). These plantar arches enable the foot to adapt optimally to uneven terrain, ensuring that the compressive forces can be transmitted under optimum mechanical conditions in any situation. The arches thus perform a kind of shock-absorber function, creating a springy flexibility that helps the foot absorb vertical loads. The deficient arches in a flatfoot or a splayfoot, for example, can lead to considerable pain during walking.

B The plantar architecture of the right foot
Superior view showing the bony points of support for the plantar vault (connected by red lines) and the associated footprint (dark gray shading). The area that is outlined by interconnecting the bony supports (the tuber calcanei and the heads of the first and fifth metatarsals) has the shape of a triangle. By contrast, the area of ground contact defined by the plantar soft tissues (the footprint or *podogram*) is considerably larger. The calluses typically found on the heel and the balls of the great and little toes confirm that these areas bear the brunt of the loads.

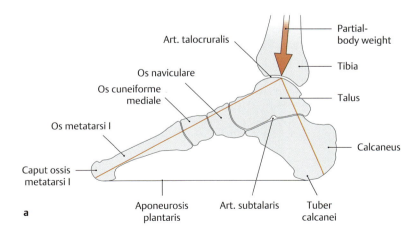

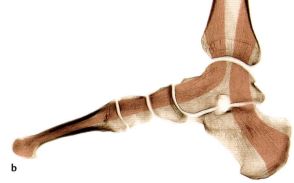

C Transfer of compressive stresses in the weight-bearing foot
Schematic sagittal section at the level of the first ray, medial view.

a During stance, the partial body weight on the talocrural joint is transferred across the talus to the forefoot and hindfoot.

b A schematic radiograph illustrates the parallel arrangement of the cancellous bony trabeculae. The pattern conforms to the compressive stresses (indicated by color shading) that result from the loads acting on the forefoot and hindfoot (**a**) (after Rauber and Kopsch).

Lower Limb — **19. Bones, Ligaments, and Joints**

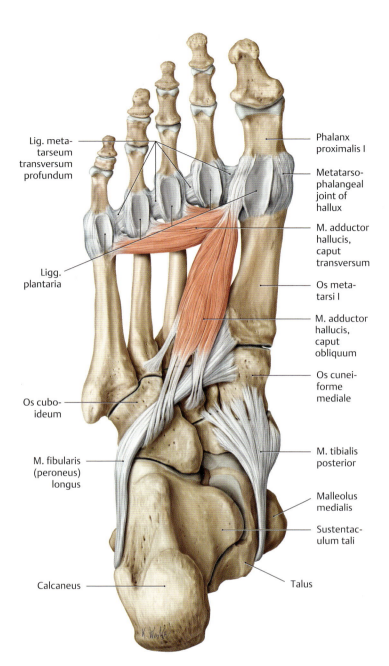

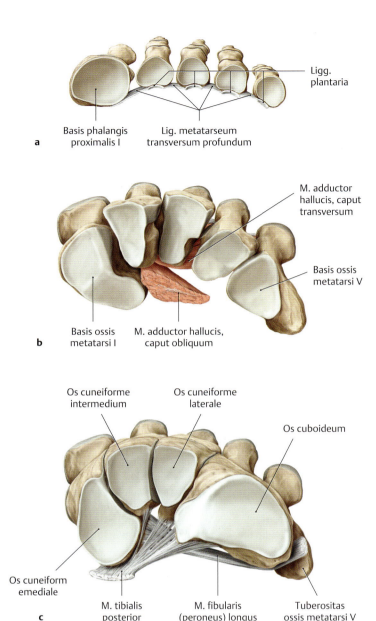

D Active and passive stabilizers of the transverse arch
Right foot, plantar view. Both active and passive stabilizing structures maintain the curvature of the transverse arch (arcus pedis transversus). The passive stabilizers are ligaments, and the active stabilizers are muscles. In the foot, ligamentous structures are usually able to maintain the pedal arches without assistance from the muscles. But when the loads on the foot are increased, as during walking or running on uneven ground, active muscular forces are recruited to give additional support.

E Active and passive stabilizers of the transverse arch
Right foot, proximal view.

a Arch stabilizers in the forefoot (anterior arch). The lig. metatarseum transversum profundum stabilizes the anterior arch at the level of the metatarsal heads. The arch of the forefoot thus relies entirely on *passive* stabilizers, while the arches of the metatarsus and tarsus (**b, c**) have only *active* stabilizers.

b Stabilizers of the metatarsal arch. The caput transversum of m. adductor hallucis is the primary muscular stabilizer of the metatarsal arch.

c The principal arch-supporting muscle in the tarsal region is the m. fibularis (peroneus) longus. After winding around the cuboid, its tendon of insertion runs from the lateral border of the foot and across the sole (planta) to the os cuneiforme mediale and the base of the first metatarsal. Another active stabilizer in this region is the m. tibialis posterior, whose tendon of insertion gives off expansions to the cuneiform bones. Like the m. fibularis (peroneus) longus, its oblique course enables it to support the longitudinal arch in addition to its transverse component.

465

19.30 The Longitudinal Arch (Arcus pedis longitudinalis) of the Foot

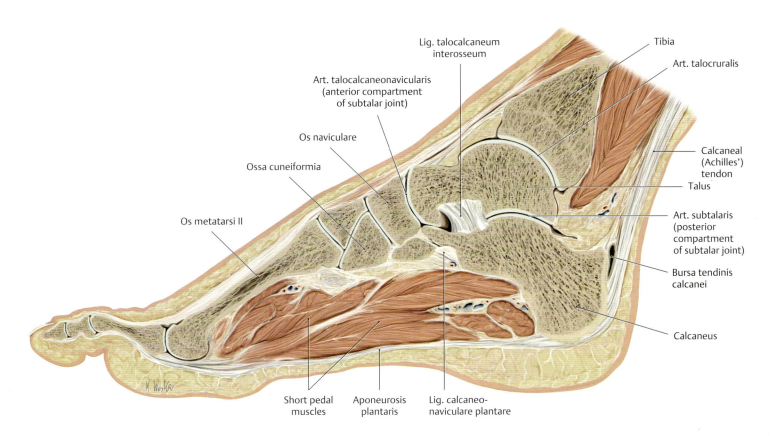

A Active stabilizers of the longitudinal arch
Sagittal section at the level of the second ray of the right foot, medial view. The second ray (consisting of the second toe, os metatarsi II, os cuneiforme intermedium, os naviculare, and calcaneus) forms the highest arch within the overall longitudinal plantar arch, its height diminishing laterally. The main active stabilizers of the longitudinal arch (arcus pedis longitudinalis) are the *short muscles of the foot*: mm. abductor hallucis, flexor hallucis brevis, flexor digitorum brevis, quadratus plantae, and abductor digiti minimi (drawn from a specimen in the Anatomical Collection of Kiel University).

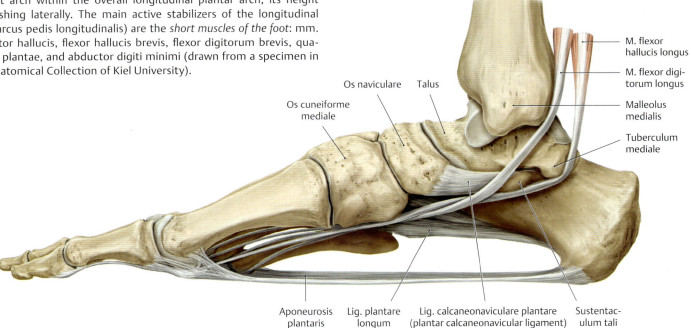

B Passive stabilizers of the longitudinal arch
Right foot, medial view. The main passive stabilizers of the longitudinal arch (arcus pedis longitudinalis) are the aponeurosis plantaris, the lig. plantare longum, and the lig. calcaneonaviculare plantare. The aponeurosis plantaris is particularly important owing to its long lever arm, while the lig. calcaneonaviculare plantare is the weakest component (shortest distance from the apex of the longitudinal arch). The *tendons of insertion of the long flexors* of the foot (m. flexor hallucis longus and m. flexor digitorum longus) also help to prevent sagging of the longitudinal arch. The m. flexor hallucis longus, which runs beneath the sustentaculum tali, is particularly effective in tightening the longitudinal arch like the chord of an arc.

Lower Limb — 19. Bones, Ligaments, and Joints

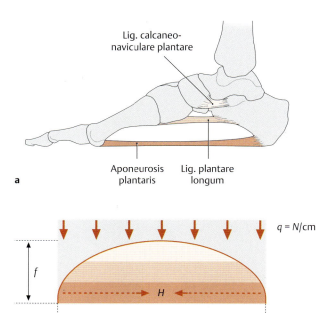

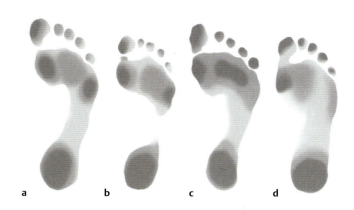

D Footprints (podograms) of right feet (after Rauber and Kopsch)
a Normal plantar arches (pes rectus).
b Increased height of the longitudinal arch (pes cavus).
c Loss of the transverse arch (splayfoot = pes transversoplanus).
d Loss of the longitudinal arch (flatfoot = pes planus).

Foot deformities—deviations from a normal, healthy foot shape—may be congenital, or acquired through paralysis or trauma. Structural abnormalities caused by chronic loads imposed on the foot by the body weight are referred to specifically as *static deformities*.

C Support of the longitudinal arch
a Ligamentous support of the longitudinal arch (right foot, medial view).
b Calculating the constraining force (*H*) needed to maintain the longitudinal arch (after Rauber and Kopsch).

Comparing the longitudinal arch of the foot (arcus pedis longitudinalis) to a theoretical parabolic arch, we see that a constraining force (*H*) must be applied to maintain the arch curvature. The magnitude of this force depends on the load (*q*), the chord length of the arch (*l*), and the height of the arch (*f*). As a result, the structures that are most effective in maintaining the pedal arch are those closest to the ground, since the long lever arm of those structures requires the least expenditure of force. The formula also dictates that the constraining force must increase as the distance *l* between the points of support increases, or as the arch becomes flatter (smaller *f*).

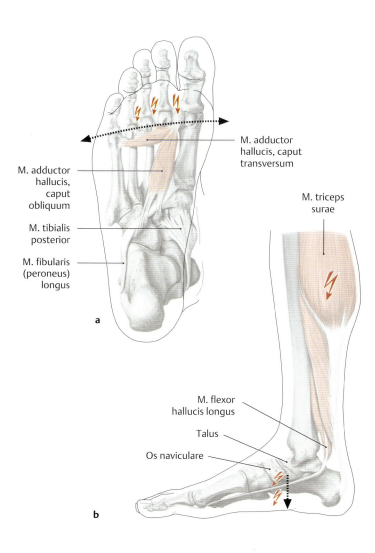

E Location of pain associated with splayfoot and flatfoot (after Loeweneck)
a Right splayfoot, plantar view. The collapse of the transverse arch (arcus pedis transversus) results in a broadened forefoot (arrows) with greater pressure acting on the heads of the second through fourth metatarsals and the associated artt. metatarsophalangeae. Typically, very painful calluses will form between the balls of the great and little toes in this situation.
b Right flatfoot, medial view. With collapse of the longitudinal arch, marked by downward displacement of the talus and os naviculare (arrow), weight bearing often incites a diffuse foot pain that is most intense in the area of the stretched lig. calcaneonaviculare plantare. Calf pain may also develop as a result of sustained, increased tension on the calf muscles (and the pedal muscles as well, which must compensate for the deficient passive stabilizers).

19.31 The Sesamoid Bones (Ossa sesamoidea) and Deformities of the Toes

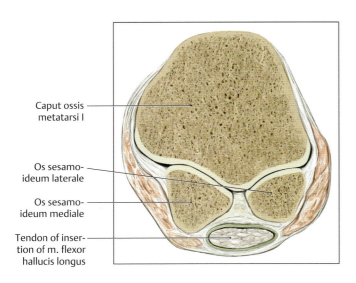

A Cross section through the head of the first metatarsal at the level of the ossa sesamoidea
Great toe of right foot, proximal view. The plane of section is indicated in **B**. The lateral and medial sesamoids are hemispheric bones, each presenting a slightly convex dorsal articular surface that articulates with the grooved plantar articular surfaces on the head of the first metatarsal (caput ossis metatarsi I). Sesamoids protect the tendons from excessive friction. They are important functionally for their ability to lengthen the effective lever arm of the muscle, so that muscular forces can be applied more efficiently. The development of sesamoids can be interpreted as a functional adaptation to the presence of pressure tendons.

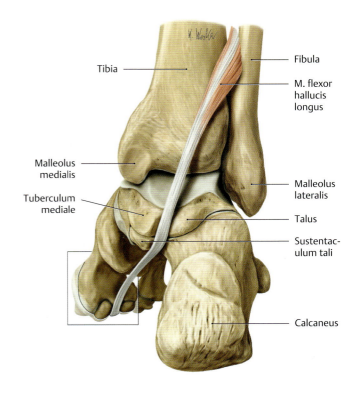

B Ossa sesamoidea of the hallux (great toe)
Right foot, posteromedial view. The tendon of the m. flexor hallucis longus runs between the two sesamoids. The box around the sesamoids indicates the plane of section shown in **A**.

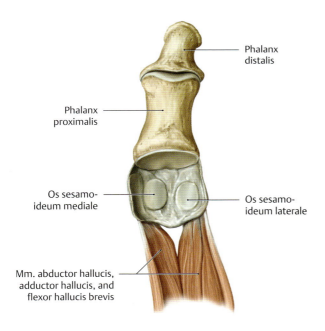

C The articular surfaces of the ossa sesamoidea
Right foot, dorsal view with the first metatarsal (os metatarsi I) removed.

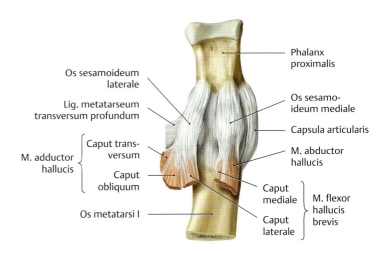

D The capsule and ligaments of the ossa sesamoidea and muscular attachments
First metatarsophalangeal joint of the right foot, plantar view. Both sesamoids are attached to the capsula articularis and to the ligg. collateralia of the art. metatarsophalangea. They are embedded in the tendons of insertion of the following muscles:

- Os sesamoideum mediale
  – M. abductor hallucis
  – Caput mediale of m. flexor hallucis brevis

- Os sesamoideum laterale
  – Caput laterale of m. flexor hallucis brevis
  – Caput transversum of m. adductor hallucis
  – Caput obliquum of m. adductor hallucis

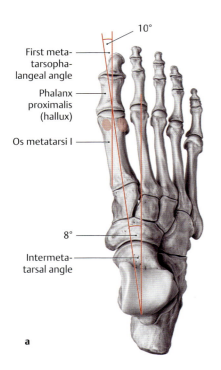
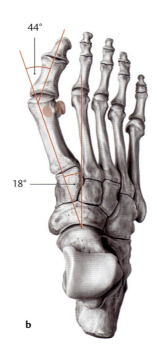
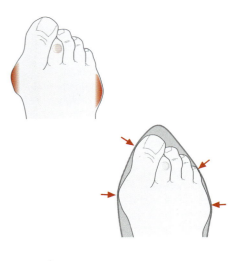

E Change in the first intermetatarsal angle and first metatarsophalangeal angle in hallux valgus
Right foot, superior view.

a Skeleton of a normal right foot.
b Lateral deviation of the first ray with subluxation of the metatarsophalangeal joint in hallux valgus.

In a normal foot, the *first intermetatarsal angle* (angle between the longitudinal axes of the first and second metatarsals) should not exceed 8°. The *first metatarsophalangeal angle* (angle between the longitudinal axes of the phalanx proximalis of the great toe and the first metatarsal) should be less than 20°. In hallux valgus and also in splayfoot, which generally precedes it, both the intermetatarsal angle and metatarsophalangeal angle are significantly increased.

G Etiology of hallux valgus
Hallux valgus usually develops secondary to splayfoot. When a broadened forefoot is forced into a narrow, pointed shoe, the outer toes are crowded against the middle toes. This results in the pressure points and pain that are typical of hallux valgus and predominantly affect the medial side of the head of the first metatarsal, with chronic irritation of the first metatarsophalangeal joint and the overlying bursa (bursitis), in addition to reactive bone changes (exostosis). The middle toes are squeezed together anteriorly and become clawed (hammer toes, claw toes) (after Debrunner).

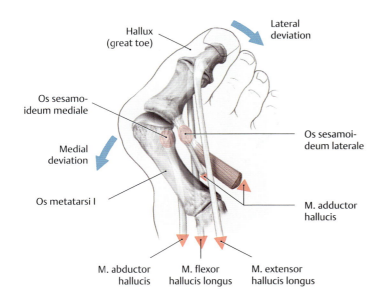
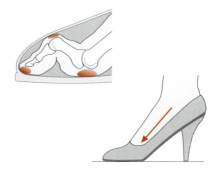

F Pathogenic mechanism of hallux valgus
Right forefoot, superior view. As the os metatarsi I deviates medially and the great toe deviates laterally, a muscular imbalance develops marked by a change in the direction of tendon pull, which perpetuates and exacerbates the deformity. Most notably, the m. abductor hallucis moves laterally with the medial sesamoid, causing it to become an *adductor*. Meanwhile, the tendons of m. flexor hallucis longus and m. extensor hallucis longus move laterally, reinforcing the lateral angulation at the first metatarsophalangeal joint.

H Claw toes and hammer toes
Toe deformities are a very common associated feature of hallux valgus and splayfoot. When the foot is placed into a tight shoe with a high heel, it tends to slide forward and downward, and the resulting pressure leads to a typical cramped deformity with degenerative changes in the toe joints and painful callus formation. The claw toe deformity is characterized by marked hyperextensibility of the metatarsophalangeal joint with flexion of the proximal and distal interphalangeal joints. With hammer toes, there is less pronounced dorsiflexion of the metatarsophalangeal joint.

19.32 X-Ray and Cross-sectional Anatomy of the Foot

A X-ray anatomy of the foot
a Upper ankle joint in anterior-posterior beam path, b hindfoot in lateral beam path, c forefoot in dorsal plantar beam path (from Möller TB, Reif E. Taschenatlas der Röntgenanatomie. 2nd ed. Stuttgart: Thieme; 1998).
The conventional X-ray is still the basis for medical imaging of the foot. It provides the highest spatial resolution and a good overview of the overall shape of the foot.

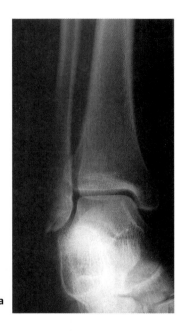
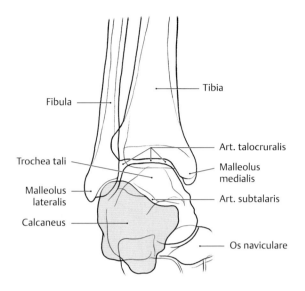
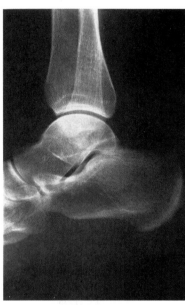
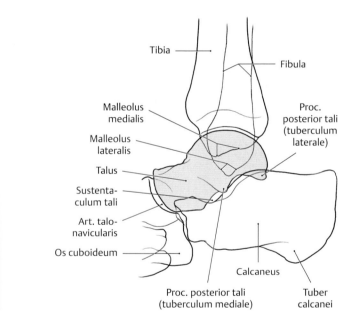
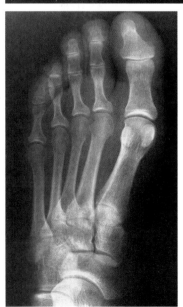
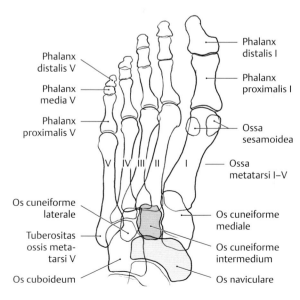

Lower Limb — 19. Bones, Ligaments, and Joints

B Magnetic resonance imaging of the right foot (from Vahlensieck M, Reiser M. MRT des Bewegungsapparates. 3rd ed. Stuttgart: Thieme; 2006)
T1-weighted **a** sagittal, **b** coronal, and **c** axial (transverse) sequences.

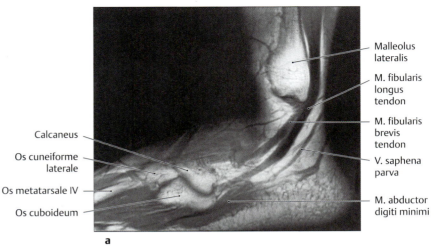

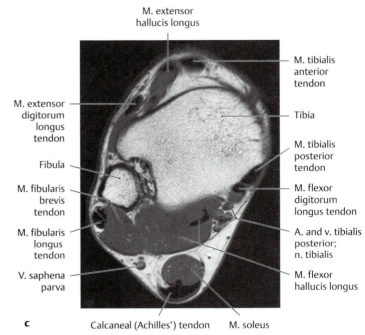

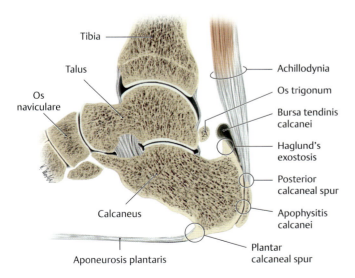

C Anatomic localization of clinically important disorders of the regio calcanea (heel region)

Heel pain is a very common symptom in patients of doctors specializing in foot surgery. In addition to the physical examination, only a few medical imaging techniques are necessary to make a diagnosis and treatment plan. Aside from typical bone diseases (osteoporosis and tumors), the following causes for heel pain need to be considered:

- Plantar heel spur: osseous exostosis at the calcaneal insertion of the aponeurosis plantaris (often as a result of chronic plantar fascitis).
- Apophysitis calcanei: defect in the bony apophysis of the tuber calcanei during growth; pain results from overload of the cartilaginous growth plate.
- Posterior heel spur: prominent exostosis at the insertion on the calcaneus of the calcaneal (Achilles') tendon (insertional tendinopathy of the chondral apophyseal enthesis).
- Haglund's exostosis: proximally located dorsal calcaneal tuberosity (benign cartilaginous growth); pain results from inflammation of the adjacent bursa (bursa tendinis calcanei).
- Os trigonum: accessory bone of the foot at the posterior talus.
- Achillodynia: painful peridentinitis of the calcaneal (Achilles') tendon; symptoms include hyperthermia, swelling, and impaired function.

19.33 Human Gait

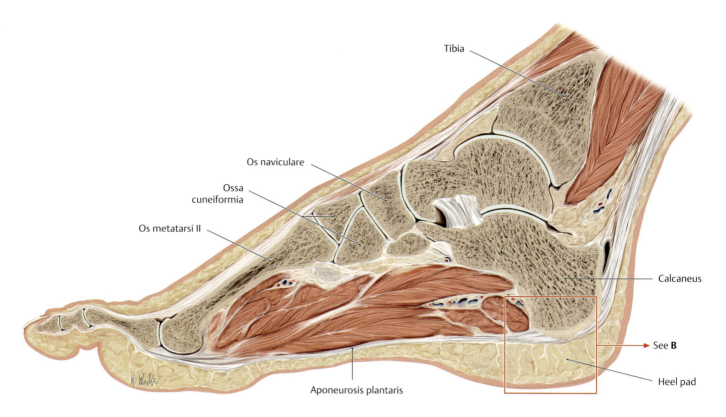

A The pressure chamber system in the sole of the foot
Sagittal section through the right foot at the level of the second ray, medial view (see detail in **B**).
During walking and particularly in the stance phase, large compressive forces are exerted on the heel pad and on the balls of the great and little toes. To distribute these concentrated forces more evenly over a larger area, the sole of the foot (planta) is covered by a layer of subcutaneous connective tissue up to 2 cm thick. As a functional adaptation to these demands, the tissue has a "pressure chamber" construction that acts as a shock absorber while also enhancing the mechanical stability of the sole. Without this pressure chamber construction, the loads on the foot would generate very high, localized stresses that would result in pressure necrosis (drawn from a specimen in the Anatomical Collection of Kiel University).

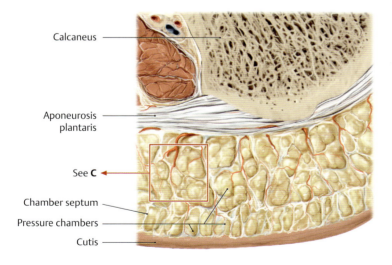

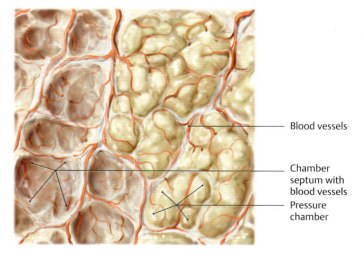

B The plantar pressure chambers
Detail from **A**.
Each of the pressure chambers contains an internal fibrofatty tissue covered externally by tough connective tissue made up of collagen fibers. These fibrous septa are firmly attached between the aponeurosis plantaris and plantar skin and are supplied by an extensive network of blood vessels that further stabilize the walls of the pressure chambers (see close-up in **C**).

C Structure of the pressure chambers
Detail from **B**.
The fatty tissue has been removed from the chambers on the left side of the drawing to demonstrate the blood vessels that permeate the septa. (The sole of the foot is one of the most highly vascularized regions of the body surface.)

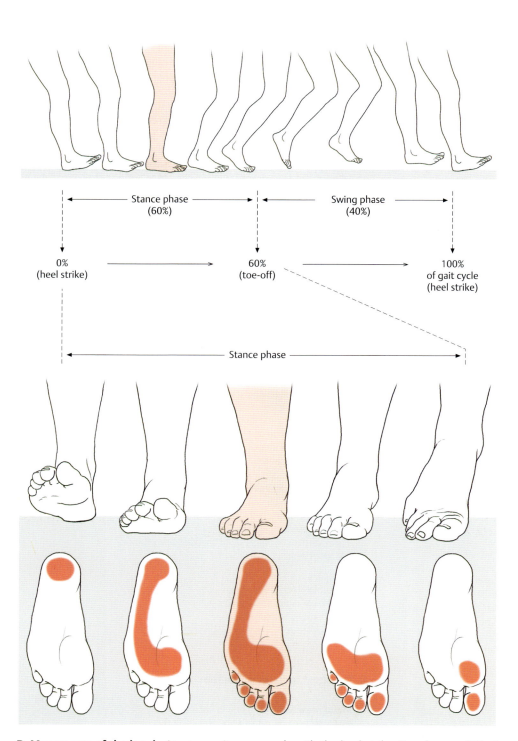
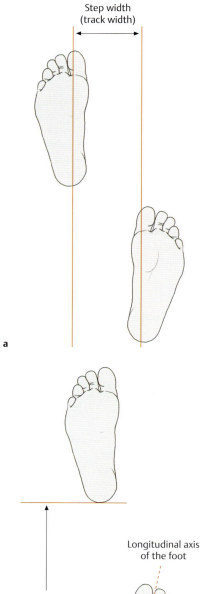

D Movements of the leg during one gait cycle

During normal walking, each leg functions alternately as the stance leg and swing leg. The stance phase begins when the heel contacts the ground (heel strike) and ends when the toes push off from the ground (toe-off). This phase makes up 60% of the gait cycle. The swing phase begins with the toe-off and ends with the heel strike. It makes up 40% of the gait cycle (100% of gait cycle = the period between two heel strikes of the same foot). In the neutral position, the longitudinal axis of the foot relative to the line of direction shows an external rotation of 7°. This slight external rotation of the foot occurs also during walking.

E Step width (a) and step length (b)

The step width (track width) is evaluated from behind. Generally, it is narrower than the distance between the two hip joints. The step length (evaluated from the side) equals approximately 2 to 3 foot lengths.

The step width and step length define the area of support and thus play a critical role in stability. This is particularly important in hemiplegic patients, for example, in whom impaired proprioception can lead to instabilities of gait and stance.

20.1 The Muscles of the Lower Limb: Classification

In most mammals, the upper and lower limbs share many functions and have analogous functional groups of muscles. In humans, however, the specializations of the upper limb for manipulation and the lower limb for ambulation have imposed radically different requirements upon their respective muscle groups. For instance, the shoulder girdle has considerable freedom of motion on the trunk and is acted upon by an array of muscles, but the pelvic ring is firmly fixed to the columna vertebralis and changes position very little relative to the trunk, and has no comparable muscles to move it. In contrast, the hip and gluteal muscles have evolved into massive and powerful movers and stabilizers of the femur, counteracting the loads imposed by support of the whole body weight on two limbs and maintaining balance and stability during bipedal locomotion; these muscles are, in aggregate, larger than their counterparts that act upon the humerus, with a significantly different arrangement and orientation.

As with the upper limb (see p. 296), the muscles of the lower limb can be classified on the basis of origin, topography, function, and innervation. Each such classification system has advantages and disadvantages, so several schemes are presented here. Segregation of muscles that act at the hip into specific functional groups is valid only for a particular joint position, because the axis of motion changes relative to the muscles as the joint is dynamically reoriented, causing abductors to become adductors, for example. Muscles surrounding the hip can be categorized topographically into an inner and outer group, relative to the pelvic girdle (see **A**). Muscles acting on the knee and foot can be grouped logically in an arrangement that uses both functional and topographical criteria, because these muscles tend to be clustered by functional groups into discrete compartments and act in a consistent way on joints with restricted ranges of motion. As with the upper limb, it is also instructive to categorize the lower limb's muscles by the pattern of their innervation (see **E**), a pattern that reveals the underlying logic of different clinical syndromes involving nerve damage.

A The hip and gluteal muscles

Inner hip muscles
- M. psoas major
- M. psoas minor
- M. iliacus

} Act in unison as the m. iliopsoas

Outer hip muscles
- M. gluteus maximus
- M. gluteus medius
- M. gluteus minimus
- M. tensor fasciae latae
- M. piriformis
- M. obturatorius internus
- Mm. gemelli
- M. quadratus femoris

Muscles of the adductor group*
- M. obturatorius externus
- M. pectineus
- M. adductor longus
- M. adductor brevis
- M. adductor magnus
- M. adductor minimus
- M. gracilis

* For functional reasons, the muscles of the adductor group, all of which are located on the medial side of the thigh, are classified as hip muscles because they act mainly on the hip joint.

B The thigh muscles

Anterior thigh muscles
- Sartorius
- Quadriceps femoris
  - Rectus femoris
  - Vastus medialis
  - Vastus lateralis
  - Vastus intermedius
  - (Articularis genus, the "fifth head" of m.quadriceps femoris, see p. 482)

Posterior thigh muscles
- Biceps femoris
- Semimembranosus
- Semitendinosus

} The ischiocrural muscles (hamstrings)

C The leg muscles

Anterior compartment
- M. tibialis anterior
- M. extensor digitorum longus
- M. extensor hallucis longus
- M. fibularis tertius

Lateral compartment
- M. fibularis longus
- M. fibularis brevis

Posterior compartment
Superficial part
- M. triceps surae
  - M. soleus
  - M. gastrocnemius (caput mediale and laterale)
- M. plantaris

Deep part
- M. tibialis posterior
- M. flexor digitorum longus
- M. flexor hallucis longus
- M. popliteus

D The short muscles of the foot

Dorsal muscles
- M. extensor digitorum brevis
- M. extensor hallucis brevis

Plantar muscles
Medial compartment
- M. abductor hallucis
- M. flexor hallucis brevis (caput mediale and laterale)

Lateral compartment
- M. abductor digiti minimi
- M. flexor digiti minimi brevis
- M. opponens digiti minimi

Central compartment
- M. flexor digitorum brevis
- M. adductor hallucis (caput transversum und obliquum)
- M. quadratusbplantae
- Mm. lumbricales I–IV
- Mm. interossei plantares I–III
- Mm. interossei dorsales I–IV

E Classification of muscles based on their motor innervation

All the muscles of the lower limb are supplied by branches of the plexus lumbalis (T12–L4) and the plexus sacralis (L5–S3). They may be supplied by short, direct branches or by long nerves emanating from the corresponding plexus (see p. 530). See also **F**.

Nerve or plexus	Muscles supplied
Plexus lumbalis	
Direct branches (Rr. musculares) (L1–L3)	Mm. psoas major and minor
Nerves of the lower limb arising from the plexus lumbalis	
N. femoralis (L2–L4)	M. iliacus, m. pectineus; M. sartorius, m. quadriceps femoris
N. obturatorius (L2–L4)	M. obturatorius externus, m. pectineus, mm. adductores longus, brevis, magnus (deep part), and minimus, m. gracilis
Plexus sacralis	
Direct branches (Rr. musculares) (L5–S2)	M. piriformis, m. obturatorius internus Mm. gemelli, m. quadratus femoris
Nerves of the lower limb arising from the plexus sacralis	
N. gluteus superior (L4–S1)	M. tensor fasciae latae, mm. glutei medius and minimus
N. gluteus inferior (L5–S2)	M. gluteus maximus
N. ischiadicus (L4–S3) (see **F**)	M. adductor magnus (superficial part, tibial part); m. biceps femoris (caput longum, tibial part); m. biceps femoris (caput breve, fibular part); m. semimembranosus, m. semitendinosus (tibial part)
• N. fibularis* communis (L4–S2)	
– N. fibularis profundus	M. tibialis anterior, mm. extensores digitorum longus and brevis, m. fibularis tertius, mm. extensores hallucis longus and brevis
– N. fibularis superficialis	Mm. fibulares longus and brevis
• N. tibialis (L4–S3)	M. popliteus, m. triceps surae, m. plantaris, m tibialis posterior, m. flexor digitorum longus, m. flexor hallucis longus
– N. plantaris medialis	M. abductor hallucis, m. flexor hallucis brevis (caput mediale), m. flexor digitorum brevis, mm. lumbricales I + II
– N. plantaris lateralis	M. flexor hallucis brevis (caput laterale), m. adductor hallucis, m. abductor digiti minimi, m. flexor digiti minimi brevis, m. opponens digiti minimi, m. quadratus plantae, mm. lumbricales III + IV, mm. interossei plantares I–III; mm. interossei dorsales II–IV

* The N. fibularis communis, for example n. peroneus communis and its divisions are also referred to as "peroneal nerves"

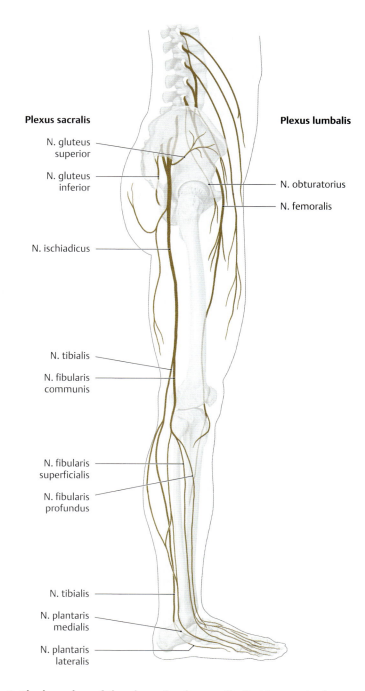

F The branches of the plexus lumbosacralis that innervate the muscles of the lower limb

Right leg, lateral view. The rami anterior of the nn. lumbales and nn. sacrales, with contributions from the n. subcostalis and n. coccygeus (not shown here), combine to form the plexus lumbosacralis. While the branches arising from the plexus lumbalis run *anterior to* the hip joint and mainly supply the muscles on the anterior and medial views of the thigh, the branches from the plexus sacralis run *behind* the hip joint to supply the posterior thigh muscles and all the muscles of the leg and foot. The grossly visible division of the n. ischiadicus into its two terminal branches (the n. tibialis and n. fibularis communis) is generally located just above the knee joint, as pictured here (*low division*). But the nerve fibers that make up the two terminal branches become organized into bundles at a much more proximal level, where they already appear as separate nerve branches within their common fibrous sheath. In the *high division* pattern, the nerve divides into its terminal branches while still in the lesser pelvis (see p. 553).

20.2 The Hip and Gluteal Muscles: The Inner Hip Muscles

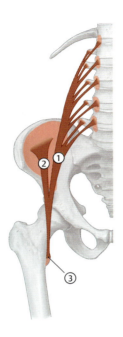

	M. iliopsoas
Origin:	• ① M. psoas major (superficial layer): lateral surfaces of the T XII corpus vertebrae, the L I–L IV vertebral bodies, and the associated disci intervertebrales • ① M. psoas minor (deep layer): procc. costales of the L I–L V vertebrae • ② M. iliacus: the fossa iliaca
Insertion:	Common insertion on the trochanter minor of the femur as the m. iliopsoas ③ (m. psoas minor inserts into the arcus iliopectineus, (not depicted; see **B**)
Action:	• Hip joint: flexion and external rotation • Lumbar spine: unilateral contraction (punctum fixum on the femur) bends the trunk laterally to the same side, bilateral contraction raises the trunk from the supine position
Innervation:	N. femoralis (m. iliacus) and direct branches from the plexus lumbalis (psoas) (L I–L III)

A Schematic of the inner hip muscles

Properties and clinical aspects of the m. iliopsoas
The m. iliopsoas is classified as a hip flexor, along with the mm. rectus femoris, sartorius, and tensor fasciae latae. It is the most powerful flexor, its long vertical travel making it an important muscle for standing, walking, and running. As a typical postural muscle with a preponderance of slow-twitch red (type I) fibers, however, the m. iliopsoas is inherently susceptible to pathological shortening (particularly in older patients with a sedentary lifestyle or chronic immobilization conditions) and requires regular stretching to maintain normal tone (see pp. 56 and 499).
Shortening (contracture) of the hip flexors leads to

- increased anterior pelvic tilt,
- increased lordosis lumbalis, and
- limitation of hip extension.

Unilateral shortening of the m. iliopsoas, in which the os ilium on the affected side is tilted forward, can be diagnosed with the Thomas maneuver (see p. 436). This condition leads to pelvic torsion, in which the pelvis becomes twisted upon itself. This mainly alters the function of the sacroiliac joints but also compromises the intervertebral joints and the lumbosacral junction (increased lordosis of the lumbar spine with degenerative changes in the vertebral bodies, see p. 130). Patients with *bilateral* iliopsoas weakness or paralysis are unable to raise the trunk from the supine position, despite intact abdominal muscles, without using their arms and are greatly limited in their ability to walk and climb stairs without assistance.

Lower Limb —— 20. Musculature: Functional Groups

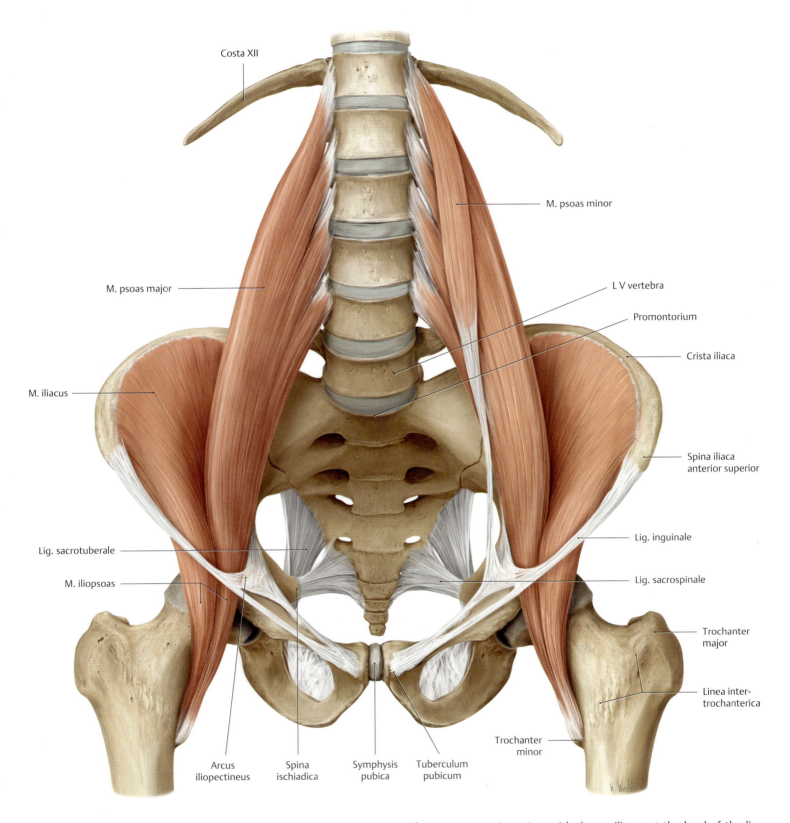

B The inner hip muscles
Anterior view.
Note: In the lower extremity, the directional terms "posterior" and "anterior" are better used than "dorsal" and "ventral"; for background, see rotation of the extremities on p. 21.

The m. psoas major unites with the m. iliacus at the level of the lig. inguinale to form a conjoined muscle, the m. iliopsoas. Approximately 50% of the population also has a m. psoas minor muscle (as shown here), which arises from the T XII and L I vertebrae and inserts into the arcus iliopectineus (fascia iliaca).

20.3 The Hip and Gluteal Muscles: The Outer Hip Muscles

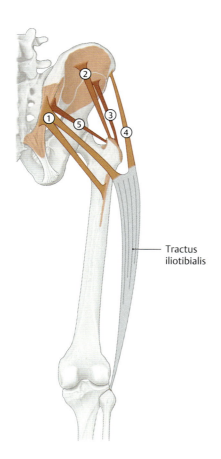

A Schematic of the vertically oriented outer hip muscles

① **M. gluteus maximus**
- Origin: Lateral part of the dorsal surface of the os sacrum, posterior part of the facies glutea of the os ilium (behind the Linea glutea posterior), also from the fascia thoracolumbalis and lig. sacrotuberale
- Insertion:
  • Upper fibers: tractus iliotibialis
  • Lower fibers: tuberositas glutea
- Action:
  • Entire muscle: extends and externally rotates the hip, stabilizes the hip in both the sagittal and coronal planes
  • Upper fibers: abduction
  • Lower fibers: adduction
- Innervation: N. gluteus inferior (L5–S2)

② **M. gluteus medius**
- Origin: Facies glutea of the os ilium (below the crista iliaca between the lineae gluteae anterior and posterior)
- Insertion: Lateral surface of the trochanter major of the os femoris
- Action:
  • Entire muscle: abducts the hip, stabilizes the pelvis in the coronal plane
  • Anterior part: flexion and internal rotation
  • Posterior part: extension and external rotation
- Innervation: N. gluteus superior (L4–S1)

③ **M. gluteus minimus**
- Origin: Facies glutea of the os ilium (below the origin of m. gluteus medius)
- Insertion: Anterolateral surface of the trochanter major of the femur
- Action:
  • Entire muscle: abducts the hip, stabilizes the pelvis in the coronal plane
  • Anterior part: flexion and internal rotation
  • Posterior part: extension and external rotation
- Innervation: N. gluteus superior (L4–S1)

④ **M. tensor fasciae latae**
- Origin: Spina iliaca anterior superior
- Insertion: Tractus iliotibialis
- Action:
  • Tenses the fascia lata
  • Art. coxae: abduction, flexion, and internal rotation
- Innervation: N. gluteus superior (L4–S1)

⑤ **M. piriformis**
- Origin: Facies pelvica of the os sacrum
- Insertion: Apex of the trochanter major of the femur
- Action:
  • External rotation, abduction, and extension of the hip joint
  • Stabilizes the hip joint
- Innervation: Direct branches from the plexus sacralis (S1, S2)

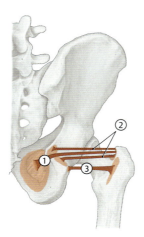

B Schematic of the horizontally oriented outer hip muscles

① **M. obturatorius internus**
- Origin: Inner surface of the membrana obturatoria and its bony boundaries
- Insertion: Facies medialis of the trochanter major
- Action: External rotation, adduction, and extension of the hip joint (also active in abduction, depending on the position of the joint)
- Innervation: Direct branches from the plexus sacralis (L5, S1)

② **Mm. gemelli**
- Origin:
  • M. gemellus superior: spina ischiadica of the os ischii
  • M. gemellus inferior: tuber ischiadicum of the os ischii
- Insertion: Jointly with m. obturatorius internus tendon (facies medialis, trochanter major)
- Action: External rotation, adduction, and extension of the hip joint (also active in abduction, depending on the position of the joint)
- Innervation: Direct branches from the plexus sacralis (L5, S1)

③ **M. quadratus femoris**
- Origin: Lateral border of the tuber ischiadicum of os ischii
- Insertion: Crista intertrochanterica of the os femoris
- Action: External rotation and adduction of the hip joint
- Innervation: Direct branches from the plexus sacralis (L5, S1)

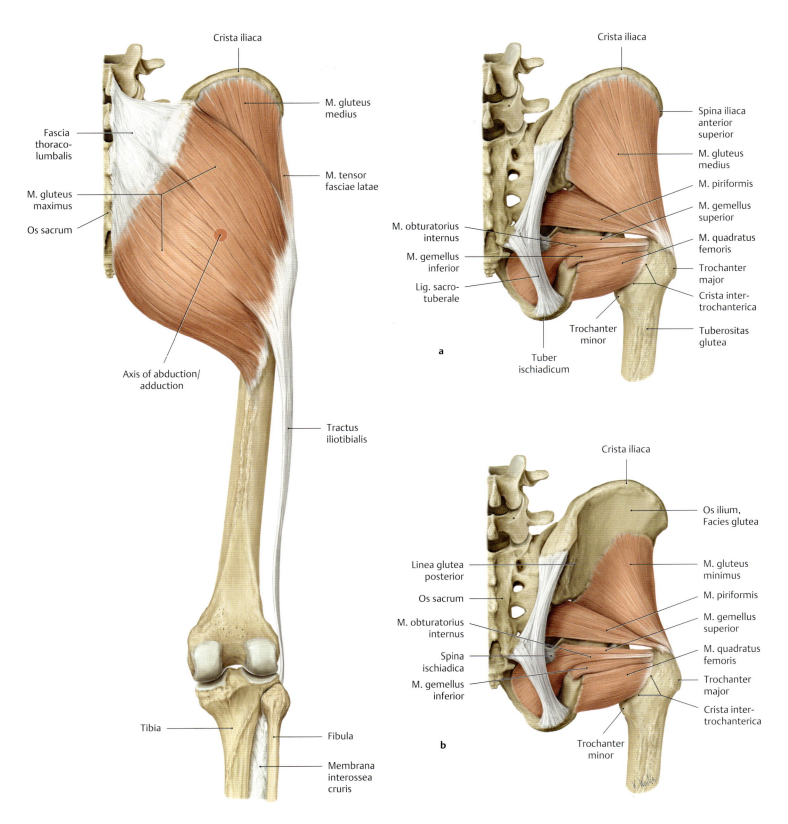

C The outer hip muscles: superficial layer
Right side, posterior view.
Note the position of the m. gluteus maximus muscle in relation to the axis of hip abduction and adduction. While the fibers of the m. gluteus maximus that run *above* the axis and insert on the tibia via the tractus iliotibialis are active in abducting the hip joint, the muscle fibers that run *below* the axis are active in adduction.

D The outer hip muscles: deep layer
Right side, posterior view.

a With the m. gluteus maximus removed.
b With the m. gluteus medius removed.

If there is weakness or paralysis of the small gluteal muscles (mm. glutei medius and minimus), the pelvis can no longer be stabilized in the coronal plane and will tilt toward the unaffected side (positive Trendelenburg sign, see also p. 536).

20.4 The Hip and Gluteal Muscles: The Adductor Group

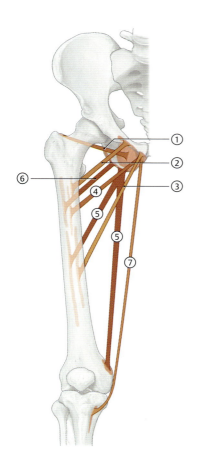

A Schematic of the adductors

① **M. obturatorius externus**

- Origin: Outer surface of the membrana obturatoria and its bony boundaries
- Insertion: Fossa trochanterica of the os femoris
- Action:
  - Adduction and external rotation of the hip joint
  - Stabilizes the pelvis in the sagittal plane
- Innervation: N. obturatorius (L3, L4)

② **M. pectineus**

- Origin: Pecten ossis pubis
- Insertion: Linea pectinea and the proximal linea aspera of the os femoris
- Action:
  - Adduction, external rotation, and slight flexion of the hip joint
  - Stabilizes the pelvis in the coronal and sagittal planes
- Innervation: N. femoralis, n. obturatorius (L2, L3)

③ **M. adductor longus**

- Origin: R. superior of os pubis and anterior side of the symphysis pubica
- Insertion: Linea aspera: labium mediale in the middle third of the os femoris
- Action:
  - Adduction and flexion (up to 70°) of the hip joint (extends the hip past 80° of flexion)
  - Stabilizes the pelvis in the coronal and sagittal planes
- Innervation: N. obturatorius (L2–L4)

④ **M. adductor brevis**

- Origin: R. inferior of os pubis
- Insertion: Linea aspera: labium mediale in the upper third of the os femoris
- Action:
  - Adduction and flexion (up to 70°) of the hip joint (extends the hip past 80° of flexion)
  - Stabilizes the pelvis in the coronal and sagittal planes
- Innervation N. obturatorius (L2, L3)

⑤ **M. adductor magnus**

- Origin: R. inferior of os pubis, r. ossis ischii, and tuber ischiadicum
- Insertion:
  - Deep part ("fleshy insertion"): labium mediale of linea aspera
  - Superficial part ("tendinous insertion"): epicondylus medialis of the os femoris
- Actions:
  - Adduction, external rotation, and extension of the hip joint (the tendinous insertion is also active in internal rotation)
  - Stabilizes the pelvis in the coronal and sagittal planes
- Innervation:
  - Deeppart: n. obturatorius (L2–L4)
  - Superficial part: n. tibialis (L4)

⑥ **M. adductor minimus (upper division of m. adductor magnus)**

- Origin: R. inferior of os pubis
- Insertion: Labium mediale of linea aspera
- Action: Adduction, external rotation, and slight flexion of the hip joint
- Innervation: N. obturatorius (L2–L4)

⑦ **M. gracilis**

- Origin: R. inferior of os pubis below the symphysis pubica
- Insertion: Medial border of the tuberositas tibiae (along with the tendons of mm. sartorius and semitendinosus)
- Action:
  - Art. coxae: adduction and flexion
  - Art. genus: flexion and internal rotation
- Innervation: N. obturatorius (L2, L3)

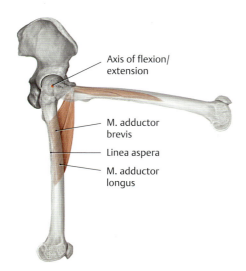

B Reversal of muscle actions, illustrated for the mm. adductor brevis and longus

Right hip joint, lateral view. The os femoris in 80° of flexion is shown in lighter shading. In addition to their primary action as adductors, both muscles may be active in flexion and extension, depending on the joint position.

- They assist in flexion from the neutral (0°) position to approximately 70°.
- Their actions reverse past approximately 80° of flexion, and they become active in extension.

The flexor components of both muscles are transformed into extensor components as soon as their insertion (the linea aspera) moves higher than their origin (the r. inferior or superior of the os pubis).

Lower Limb — 20. Musculature: Functional Groups

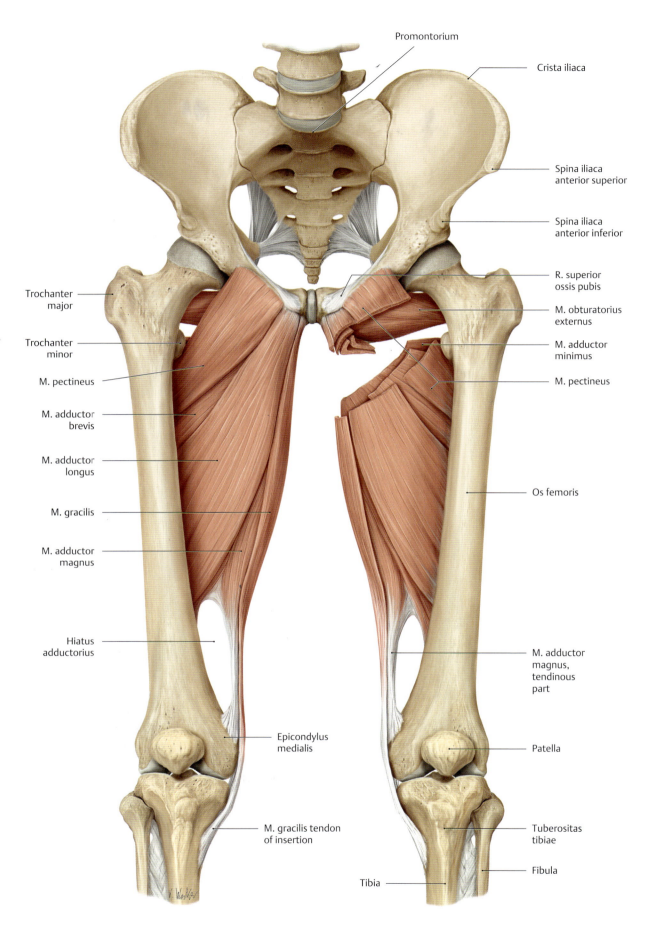

C The adductors (m. obturatorius externus; m. pectineus; mm. adductores longus, brevis, magnus, and minimus; and m. gracilis)
Anterior view. A portion of the mm. adductores, m. pectineus, and m. gracilis muscles on the left side have been removed just past their origins to demonstrate the course of the m. obturatorius externus more clearly.
Note: Unilateral shortening of the adductors leads to functional leg shortening on the affected side.

481

20.5 The Anterior Thigh Muscles: The Extensor Group

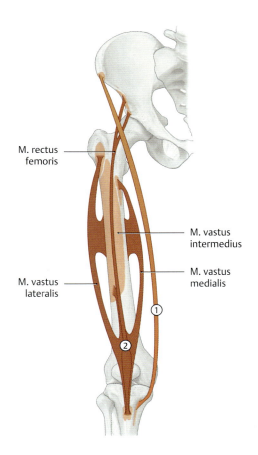

① **M. sartorius**

Origin:	Spina iliaca anterior superior
Insertion:	Medial to the tuberositas tibiae (together with m. gracilis and m. semitendinosus)
Action:	• Hip joint: flexion, abduction, and external rotation
	• Knee joint: flexion and internal rotation
Innervation:	N. femoralis (L2, L3)

② **M. quadriceps femoris**

Origin:
- M. rectus femoris: spina iliaca anterior inferior (caput rectum), acetabular roof of the hip joint (caput reflexum)
- M. vastus medialis: labium mediale of the linea aspera, distal part of the linea intertrochanterica
- M. vastus lateralis: labium laterale of the linea aspera, lateral surface of the trochanter major
- M. vastus intermedius: anterior side of the femoral shaft
- M. articularis genus (distal fibers of the m. vastus intermedius): anterior side of the femoral shaft at the level of the recessus suprapatellaris

Insertion:
- On the tuberositas tibiae via the lig. patellae (entire muscle)
- Both sides of the tuberositas tibiae on the condylus medialis and lateralis via the retinaculum patellae mediale and laterale (m. vastus medialis and lateralis)
- The recessus suprapatellaris of the knee joint capsule (m. articularis genus)

Action:
- Hip joint: flexion (m. rectus femoris)
- Knee joint: extension (all parts), prevents entrapment of the capsule (m. articularis genus)

Innervation: N. femoralis (L2–L4)

A Schematic of the extensors

B Deficient stabilization of the knee joint due to weakness or paralysis of m. quadriceps femoris
Right lower limb, lateral view.

a When the m. quadriceps femoris is intact and the knee is in slight flexion, the line of gravity falls *behind* the transverse axis of knee motion. As the only extensor muscle of the knee joint, the m. quadriceps femoris keeps the body from tipping backward and ensures stability.

b With weakness or paralysis of the m. quadriceps femoris, the knee joint can no longer be actively extended. In order to stand upright, the patient must hyperextend the knee so that the line of gravity, and thus the whole-body center of gravity, is shifted forward, in front of the knee, to utilize gravity as the extending force. The joint is stabilized in this situation by the posterior capsule and ligaments of the knee.

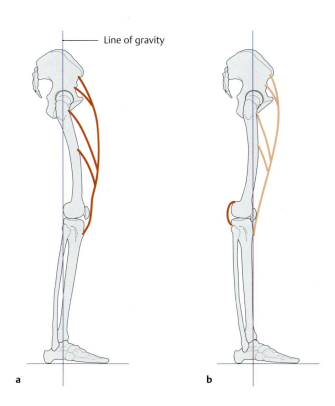

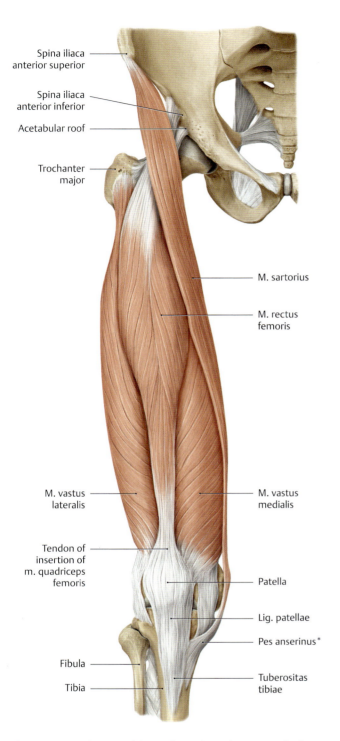
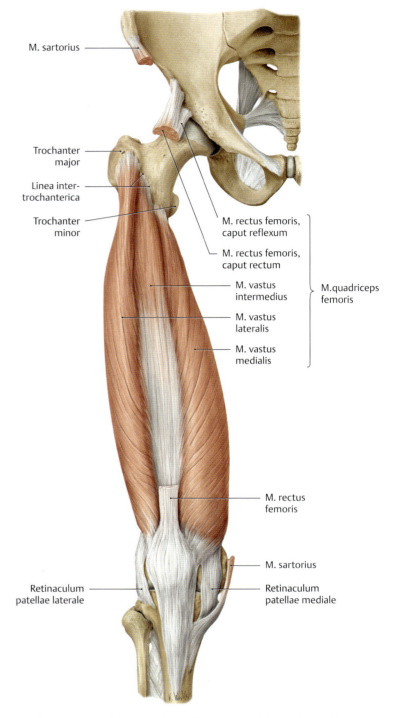

C The extensors (m. quadriceps femoris and m. sartorius)
Right side, anterior view. As its name implies, the m. quadriceps femoris is basically a four-headed muscle consisting of the m. rectus femoris and the mm. vasti medialis, lateralis, and intermedius (the m. vastus intermedius, covered here by the m. rectus femoris, is visible in **D**). It may also be considered as having a fifth head, the m. articularis genus. The latter is composed of distal fibers of the m. vastus intermedius and so does not constitute a separate muscle. But because its fibers insert in the recessus suprapatellaris (not shown), unlike the other four parts, which all attach to the lig. patellae, the m. articularis genus is often regarded as the fifth head of the m. quadriceps femoris.
Note: The only biarticular part of the m. quadriceps femoris is the m. rectus femoris, which acts on both the hip and knee joints.

* The pes anserinus is the common tendinous expansion for the m. gracilis, m. sartorius, and m. semitendinosus.

D The extensors (deep portion of the m. quadriceps femoris)
Right side, anterior view. The m. sartorius and m. rectus femoris have been removed to their origins and insertions.
The area of origin of the m. rectus femoris is intimately related to the anterior aspect of the capsule of the hip joint, a relation with functional and clinical consequences. Pathological swelling of the joint capsule can cause pain that induces reflex reactions when the m. rectus femoris is used for knee flexion; such reflex reactions are the basis of a useful test. With the patient lying prone, the examiner flexes the patient's knee. This causes passive stretching of the m. rectus femoris and adds significant pressure to a hip joint capsule already distended by effusion. The patient reflexively "escapes" the painful stimulus by raising the buttock, a positive "rectus sign."

20.6 The Posterior Thigh Muscles: The Flexor Group

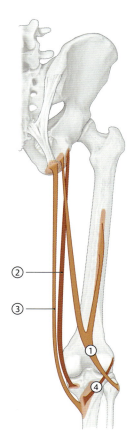

A Schematic of the flexors

① **M. biceps femoris**
Origin:
- Caput longum: tuber ischiadicum, lig. sacrotuberale (caput commune with m. semitendinosus)
- Caput breve: labium laterale of the linea aspera in the middle third of the os femoris

Insertion: Caput fibulae
Action:
- Hip joint (caput longum): extends the hip, stabilizes the pelvis in the sagittal plane
- Knee joint (entire muscle): flexion and external rotation

Innervation:
- N. tibialis, L5–S2 (caput longum)
- N. fibularis (peroneus) communis, L 5–S 2 (caput breve)

② **M. semimembranosus**
Origin: Tuber ischiadicum
Insertion: Condylus medialis tibiae, lig. popliteum obliquum, m. popliteus fascia
Action:
- Hip joint: adduction, extends the hip, stabilizes the pelvis in the sagittal plane
- Knee joint: flexion and internal rotation

Innervation: N. tibialis (L5–S2)

③ **M. semitendinosus**
Origin: Tuber ischiadicum and lig. sacrotuberale (caput commune with caput longum of m. biceps femoris)
Insertion: Medial to the tuberositas tibiae in the pes anserinus (along with the tendons of mm. gracilis and sartorius)
Action:
- Hip joint: extends the hip, stabilizes the pelvis in the sagittal plane
- Knee joint: flexion and internal rotation

Innervation: N. tibialis (L5–S2)

④ **M. popliteus**
Origin: Condylus femoralis lateralis, posterior horn of the meniscus lateralis
Insertion: Facies posterior tibiae (above the origin of m. soleus)
Action: Flexion and unlocking knee joint by internal rotation of os femoris on the fixed head of the tibia by 5°
Innervation: N. tibialis (L4–S1)

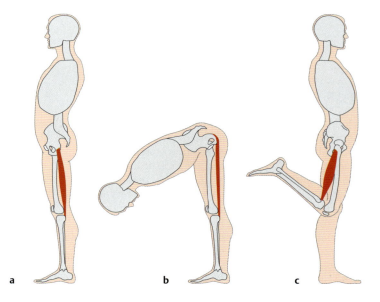

B Passive and active muscle insufficiency illustrated for ischiocrural muscles

a The ischiocrural (hamstring) muscles extend from the ischium over the hip joint and knee joint to the lower leg.

b Passive muscle insufficiency (lack of muscle length): With the knee outstretched, the ischiocrural muscles do not stretch enough to allow for maximal bending of the hip joint.

c Active muscle insufficiency (lack of muscle strength): With hip extended, the ischiocrural muscles cannot shorten enough to allow for maximal bending of the knee joint (see also muscle and soft tissue suspension, p. 50).

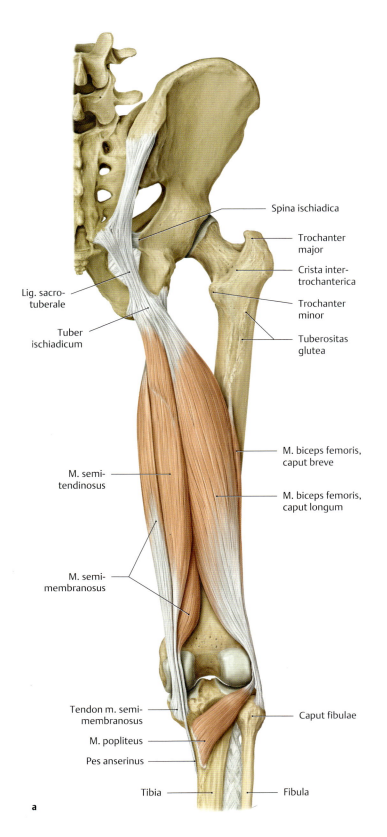
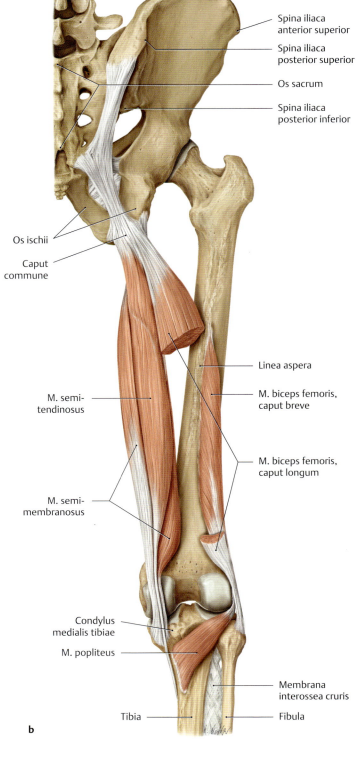

C The flexors (ischiocrural muscles and m. popliteus)
Right side, posterior view.

a The ischiocrural muscles (*hamstrings*) are the posterior thigh muscles that arise from the os ichii and insert on the leg: m. biceps femoris, m. semimembranosus, and m. semitendinosus. All but the caput breve of the m. biceps femoris are "biarticular," spanning both the art. coxae and art. genus.

b A portion of the caput longum of the m. biceps femoris has been removed to display the caput breve and its origin from the labium laterale of the linea aspera.

20.7 The Leg Muscles: The Anterior and Lateral Compartments (Extensor and Fibularis* Group)

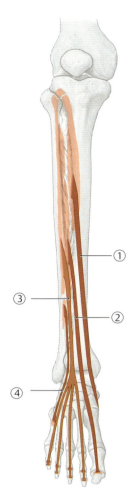

A Schematic of the anterior compartment

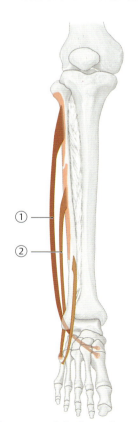

B Schematic of the lateral compartment

① **M. tibialis anterior**
Origin: Upper two thirds of the facies lateralis tibiae, the membrana interossea cruris, and the highest part of the fascia cruris superficialis
Insertion: Medial and plantar surface of the os cuneiforme mediale, the medial base of the os metatarsi I
Action:
- Talocrural joint: dorsiflexion
- Subtalar joint: inversion (supination)

Innervation: N. fibularis profundus (L4, L5)

② **M. extensor digitorum longus**
Origin: Condylus lateralis tibiae, caput fibulae, margo anterior fibulae, and membrana interossea cruris
Insertion: By four slips to the dorsal aponeuroses of the second through fifth toes and the bases of the phalanges distales of the second through fifth toes
Action:
- Talocrural joint: dorsiflexion
- Subtalar joint: eversion (pronation)
- Extends the metatarsophalangeal and interphalangeal joints of the second through fifth toes

Innervation: N. fibularis profundus (L4, L5)

③ **M. extensor hallucis longus**
Origin: Middle third of the facies medialis fibulae, membrana interossea cruris
Insertion: Dorsal aponeurosis of the big toe and the base of its phalanx distalis
Action:
- Talocrural joint: dorsiflexion
- Subtalar joint: active in both eversion and inversion (pronation/supination), depending on the initial position of the foot
- Extends the metatarsophalangeal and interphalangeal joints of the big toe

Innervation: N. fibularis profundus (L4, L5)

④ **Fibularis tertius (part of m. extensor digitorum longus) (see also pp. 510 and 514)**
Origin: Margo anterior of the distal fibula
Insertion: Basis ossis metatarsi
Action:
- Talocrural joint: dorsiflexion
- Subtalar joint: eversion (pronation)

Innervation: N. fibularis profundus (L5, S1)

① **M. fibularis longus**
Origin: Caput fibulae, proximal twothirds of the facies lateralis fibulae (arising partly from the septa intermuscularia)
Insertion: Plantar side of os cuneiforme mediale, base of the os metatarsi I
Action:
- Talocrural joint: plantar flexion
- Subtalar joint: eversion (pronation)
- Supports the transverse arch of the foot

Innervation: N. fibularis superficialis (L5, S1)

② **M. fibularis brevis**
Origin: Distal half of the facies lateralis fibulae, septa intermuscularia
Insertion: Tuberositas ossis metatarsi V (with an occasional division to the dorsal aponeurosis of the fifth toe)
Action:
- Talocrural joint: plantar flexion
- Subtalar joint: eversion (pronation)

Innervation: N. fibularis superficialis (L5, S1)

③ **M. fibularis tertius (part of m. extensor digitorum longus)**
Origin: Margo anterior of the distal fibula
Insertion: Base of the os metatarsi V
Action:
- Talocrural joint: dorsiflexion
- Subtalar joint: eversion (pronation)

Innervation: N. fibularis profundus (L5, S1)

* also called "peroneus"

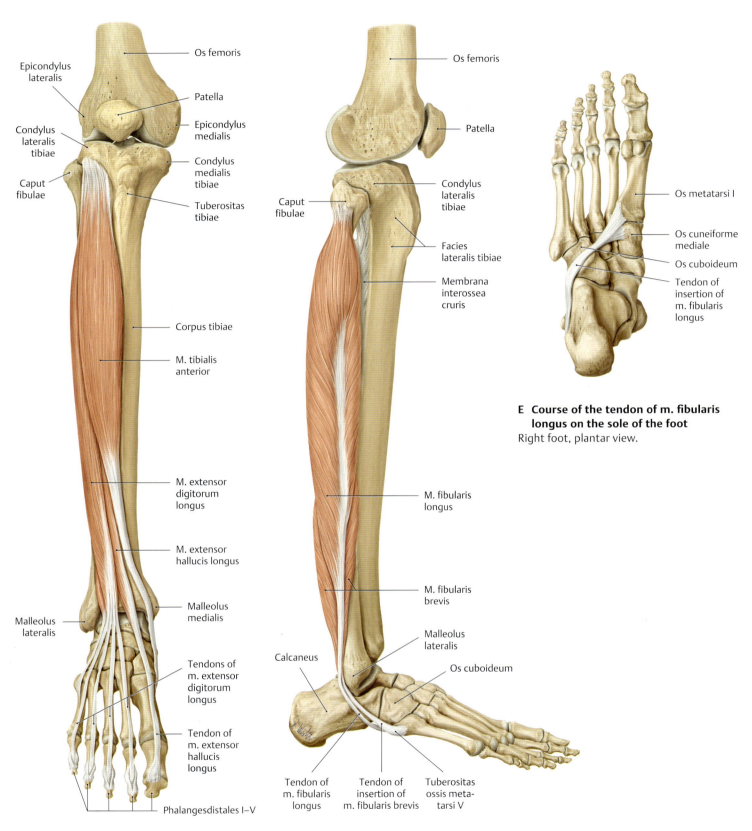

C The anterior compartment (m. tibialis anterior, m. extensor digitorum longus, and m. extensor hallucis longus)
Right leg, anterior view.

D The lateral compartment (mm. fibularis longus and brevis)
Right leg, lateral view.

E Course of the tendon of m. fibularis longus on the sole of the foot
Right foot, plantar view.

20.8 The Leg Muscles: The Posterior Compartment (Superficial Flexor Group)

① **M. triceps surae**
Origin:
- M. soleus: posterior surface of the caput and collum fibulae; attached to the linea musculi solei of the tibia via an arcus tendineus
- M. gastrocnemius, caput mediale: epicondylus medialis femoris
- M. gastrocnemius, caput laterale: epicondylus lateralis femoris

Insertion: The Tuber calcanei via the calcaneal (Achilles') tendon
Action:
- Talocrural joint: plantar flexion
- Subtalar joint: inversion (supination)
- Knee joint: flexion (m. gastrocnemius)

Innervation: N. tibialis (S1, S2)

② **M. plantaris**
Origin: Epicondylus lateralis proximal to the caput laterale of m. gastrocnemius
Insertion: The Tuber calcanei via the tendo calcaneus
Action: Negligible due to its small cross section; may act to prevent compression of the posterior leg musculature during knee flexion
Innervation: N. tibialis (S1, S2)

A Schematic of the superficial flexors

B Rupture of the calcaneal (Achilles') tendon (tendo calcaneus)
Right leg, posterior view. The calcaneal (Achilles') tendon is the common tendon of insertion of the muscles that comprise the m. triceps surae (the m. soleus and both heads of the m. gastrocnemius). The tendon has an average length of 20 to 25 cm, a mean cross-sectional area of approximately 70 to 80 mm², and a breaking strength of 60 to 100 N/mm². A healthy tendon can thus bear a load of nearly 1 ton. It is very unlikely, then, that the tendo calcaneus will rupture unless it has been subjected to chronic excessive loads (in high jumpers, for example). Repetitive microtrauma can compromise the blood supply to the tendon, causing it to degenerate and gradually lose its strength. This is particularly damaging in the area where the tendon already has the least blood flow: approximately 2 to 6 cm proximal to its insertion on the tuber calcanei. This is the most common site of a degenerative tendo calcaneus rupture, which is eventually precipitated by a trivial injury. The rupture is accompanied by a whiplike snapping sound. Afterward the patient loses active plantar flexion and has only residual flexion from the deep flexor muscles (see also pp. 510–514).

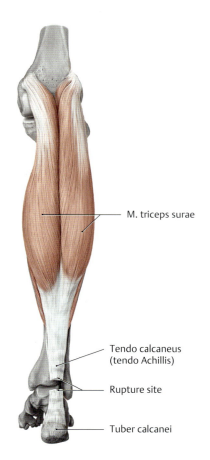

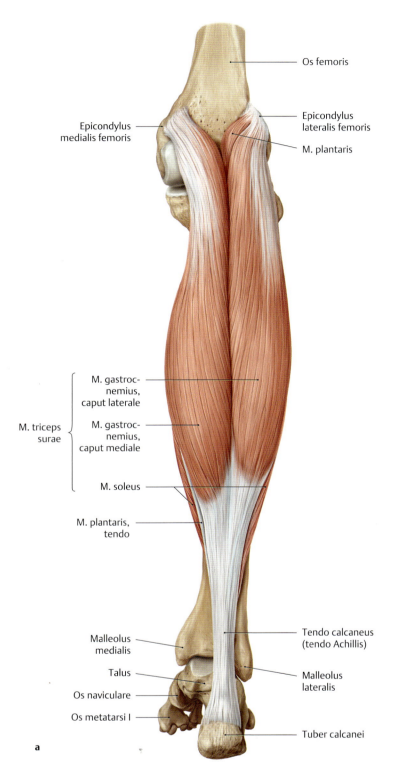
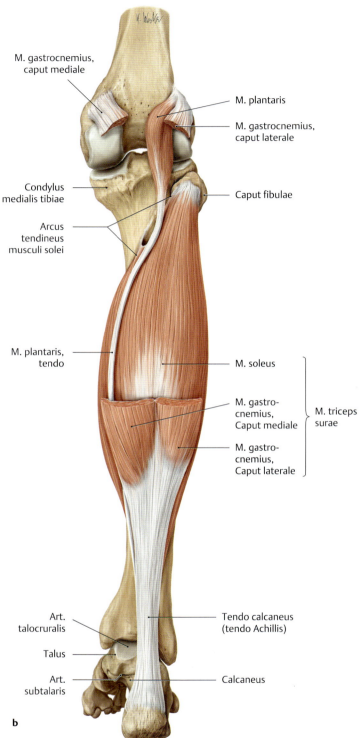

C The superficial flexors (m. triceps surae and m. plantaris)
Right leg, posterior view.

a The three heads of the m. triceps surae are clearly distinguishable: the caput laterale and mediale of the m. gastrocnemius and the m. soleus. The m. plantaris, which arises proximal to the caput laterale of the m. gastrocnemius, is often viewed as the fourth head of the m. triceps surae.

b Portions of the caput laterale and mediale of the m. gastrocnemius have been removed to expose the m. soleus and the m. plantaris with its long, narrow tendon of insertion.

20.9 The Leg Muscles: The Posterior Compartment (Deep Flexor Group)

① **M. tibialis posterior**

Origin: Membrana interossea cruris and the adjacent borders of the tibia and fibula
Insertion: Tuberositas ossis navicularis, ossa cuneiformia mediale, intermedium, and laterale, bases of the ossa metatarsi II–IV
Action:
- Talocrural joint: plantar flexion
- Subtalar joint: inversion (supination)
- Supports the longitudinal and transverse arches of the foot

Innervation: N. tibialis (L4, L5)

② **M. flexor digitorum longus**

Origin: Middle third of the facies posterior of the tibia
Insertion: Bases of the second through fifth distal phalanges
Action:
- Talocrural joint: plantar flexion
- Subtalar joint: inversion (supination)
- Metatarsophalangeal and interphalangeal joints of the second through fifth toes: plantar flexion

Innervation: N. tibialis (L5–S2)

③ **M. flexor hallucis longus**

Origin: Distal twothirds of the facies posterior fibulae, adjacent membrana interossea cruris
Insertion: Base of the phalanx distalis hallucis
Action:
- Talocrural joint: plantar flexion
- Subtalar joint: inversion (supination)
- Metatarsophalangeal and interphalangeal joints of the big toe: plantar flexion
- Supports the medial longitudinal arch of the foot

Innervation: N. tibialis (L5–S2)

④ **M. popliteus**

Origin: Condylus lateralis femoris, posterior horn of meniscus lateralis
Insertion: Facies posterior tibiae (above the origin of the soleus)
Action: Flexes and unlocks the knee by internally rotating the femur on the fixed tibia 5°
Innervation: N. tibialis (L4–S1)

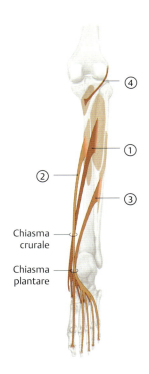

A Schematic of the deep flexors

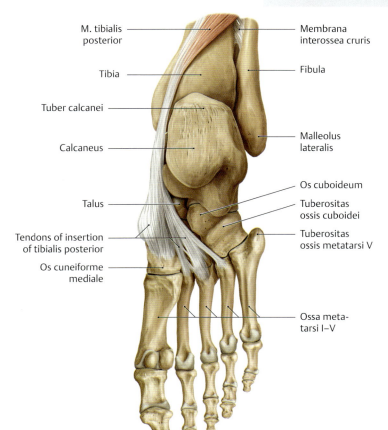

B Insertion of the m. tibialis posterior
Right foot in plantar flexion, plantar view. With its fan-shaped insertion, the m. tibialis posterior assists in stabilizing both the longitudinal and transverse arches of the foot.

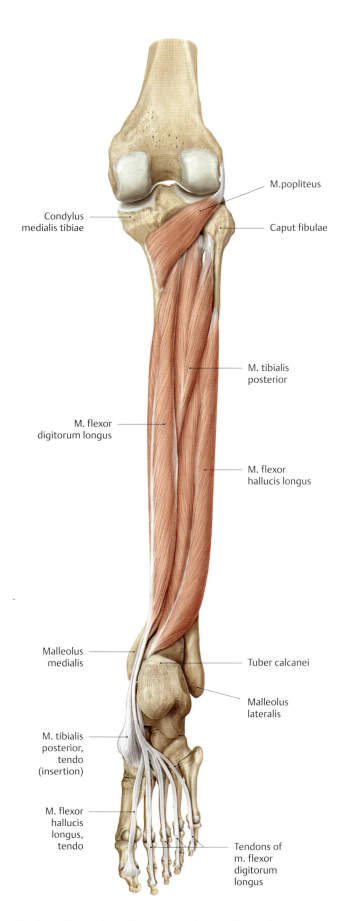
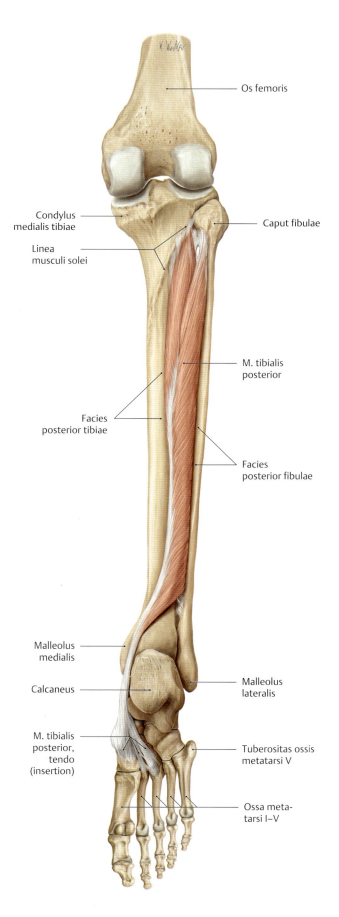

C The deep flexors (m. tibialis posterior, m. flexor digitorum longus, and m. flexor hallucis longus)
Right leg with the foot in plantar flexion, posterior view.

D The m. tibialis posterior
Right leg with the m. flexor digitorum longus and m. flexor hallucis longus removed, foot in plantar flexion, posterior view.

20.10 The Short Muscles of the Foot: Dorsum, and Medial and Lateral Compartments of the Plantar Surface

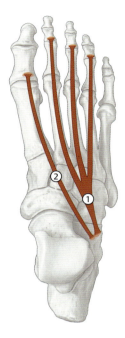

A Dorsal view

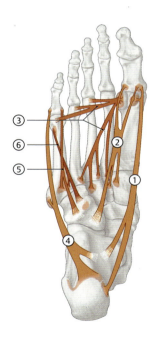

B Plantar view of the medial and lateral compartments

① **M. extensor digitorum brevis**
- Origin: Dorsal surface of the calcaneus
- Insertion: Dorsal aponeurosis of the second through fourth toes, bases of the middle phalanges of these toes
- Action: Extension of the metatarsophalangeal and proximal interphalangeal joints of the second through fourth toes
- Innervation: N. fibularis profundus (L5, S1)

② **M. extensor hallucis brevis**
- Origin: Dorsal surface of the calcaneus
- Insertion: Dorsal aponeurosis of the big toe, base of the phalanx proximalis of the big toe
- Action: Extension of the metatarsophalangeal joint of the big toe
- Innervation: N. fibularis profundus (L5, S1)

① **M. abductor hallucis**
- Origin: Proc. medialis of the tuber calcanei, aponeurosis plantaris
- Insertion: Base of the phalanx proximalis of the big toe via the medial sesamoid
- Action: First metatarsophalangeal joint: flexion and medial abduction of the first toe; supports the longitudinal arch
- Innervation: N. plantaris medialis (S1, S2)

② **M. flexor hallucis brevis**
- Origin: Os cuboideum, os cuneiforme laterale, lig. calcaneocuboideum plantare
- Insertion:
  - Caput mediale: base of the phalanx proximalis of the big toe via the medial sesamoid
  - Caput laterale: base of the phalanx proximalis of the big toe via the lateral sesamoid
- Action: Flexes the first metatarsophalangeal joint, supports the longitudinal arch
- Innervation: N. plantaris medialis (S1, S2)

③ **M. adductor hallucis** (for clarity, the m. adductor hallucis is depicted here, although it is located in the central compartment)
- Origin:
  - Caput obliquum: bases of the ossa metatarsi II–IV, os cuboideum, os cuneiforme laterale
  - Caput transversum: artt. metatarsophalangeae of the third through fifth toes, lig. metatarseum transversum profundum
- Insertion: Base of the first phalanx proximalis by a common tendon via the lateral sesamoid
- Action: Flexes the first metatarsophalangeal joint, adducts the big toe; caput transversum supports the transverse arch, caput obliquum supports the longitudinal arch
- Innervation: N. plantaris lateralis (S2, S3)

④ **M. abductor digiti minimi**
- Origin: Proc. lateralis and inferior surface of the tuber calcanei, aponeurosis plantaris
- Insertion: Base of the phalanx proximalis of the little toe, tuberositas ossis metatarsi V
- Action: Flexes the metatarsophalangeal joint of the little toe, abducts the little toe, supports the longitudinal arch
- Innervation: N. plantaris lateralis (S1–S3)

⑤ **M. flexor digiti minimi brevis**
- Origin: Basis ossis metatarsi V, lig. plantare longum
- Insertion: Base of the phalanx proximalis of the little toe
- Action: Flexes the metatarsophalangeal joint of the little toe
- Innervation: N. plantaris lateralis (S2, S3)

⑥ **M. opponens digiti minimi** (often included with m. flexor digiti minimi brevis)
- Origin: Lig. plantare longum, plantar tendon sheath of the m. fibularis longus
- Insertion: Os metatarsi V
- Action: Pulls the os metatarsi V slightly in the plantar and medial direction
- Innervation: N. plantaris lateralis (S2, S3)

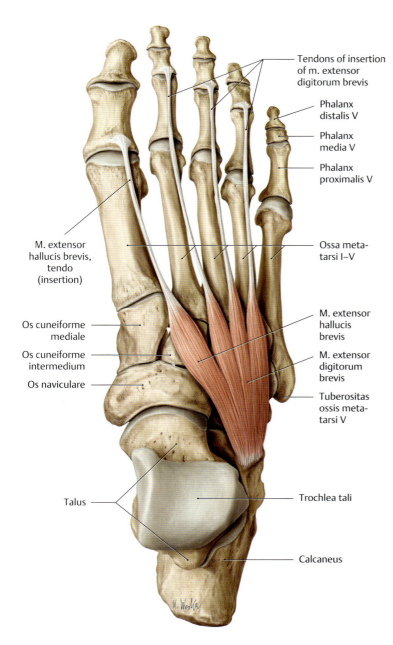

C The dorsal muscles of the foot (mm. extensor digitorum brevis and extensor hallucis brevis)
Right foot, dorsal view.

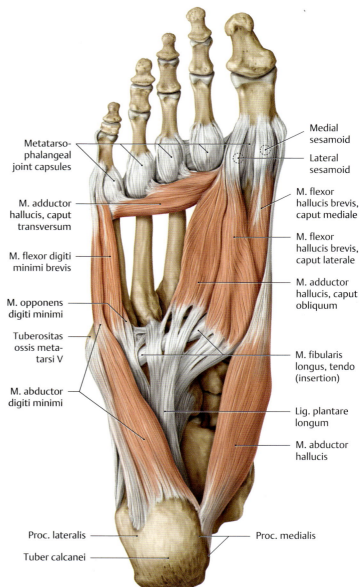

D The plantar muscles of the medial and lateral compartments (mm. abductor hallucis, adductor hallucis*, flexor hallucis brevis, abductor digiti minimi, flexor digiti minimi brevis, and opponens digiti minimi)
Right foot, plantar view.

* The m. adductor hallucis is considered part of the central compartment (see p. 494).

20.11 The Short Muscles of the Foot: Central Compartment of the Plantar Surface

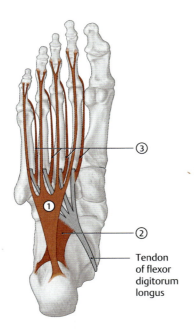

A Plantar view of m. flexor digitorum brevis, m. quadratus plantae, and the mm. lumbricales I–IV

① **M. flexor digitorum brevis**
- **Origin:** Proc. medialis tuberis calcanei, aponeurosis plantaris
- **Insertion:** The sides of the middle phalanges of the second through fifth toes
- **Action:**
  - Flexes the metatarsophalangeal and proximal interphalangeal joints of the second through fifth toes
  - Supports the longitudinal arch of the foot
- **Innervation:** N. plantaris medialis (S1, S2)

② **M. quadratus plantae**
- **Origin:** Medial and plantar borders on the plantar side of the tuber calcanei
- **Insertion:** Lateral border of the tendon of the m. flexor digitorum longus
- **Action:** Redirects and augments the pull of m. flexor digitorum longus
- **Innervation:** N. plantaris lateralis (S1–S3)

③ **Mm. lumbricales I–IV**
- **Origin:** Medial borders of the tendon of the m. flexor digitorum longus
- **Insertion:** Dorsal aponeuroses of the second through fifth toes
- **Action:**
  - Flexes the artt. metatarsophalangeae of the second through fifth toes
  - Extension of the interphalangeal joints of the second through fifth toes
  - Moves the toes closer together (adducts the second through fifth toes toward the big toe)
- **Innervation:**
  - M. lumbricalis: n. plantaris medialis (S2, S3)
  - Mm. lumbricales II–IV: n. plantaris lateralis (S2, S3)

④ **Mm. interossei plantares I–III**
- **Origin:** Medial border of the ossa metatarsi III–V
- **Insertion:** Medial base of the phalanx proximalis of the third through fifth toes
- **Action:**
  - Flexes the artt. metatarsophalangeae of the third through fifth toes
  - Extension of the interphalangeal joints of the third through fifth toes
  - Moves the toes closer together (adducts the third through fifth toes toward the second toe)
- **Innervation:** N. plantaris lateralis (S2, S3)

⑤ **Mm. interossei dorsales I–IV**
- **Origin:** By two heads from opposing sides of the ossa metatarsi I–V
- **Insertion:**
  - First interosseus: medial base of the second phalanx proximalis, aponeurosis dorsalis of the second toe
  - Second through fourth interossei: lateral base of the second through fourth proximal phalanges, dorsal aponeurosis of the second through fourth toes
- **Action:**
  - Flexes the artt. metatarsophalangeae of the second through fourth toes
  - Extension of the interphalangeal joints of the second through fourth toes
  - Spreads the toes apart (abducts the third and fourth toes from the second toe)
- **Innervation:** N. plantaris lateralis (S2, S3)

B Plantar view of the mm. interossei plantares I-III and the mm. interossei dorsales I-IV

*The m. adductor hallucis, though part of the central compartment, is not pictured here (see p. 493).

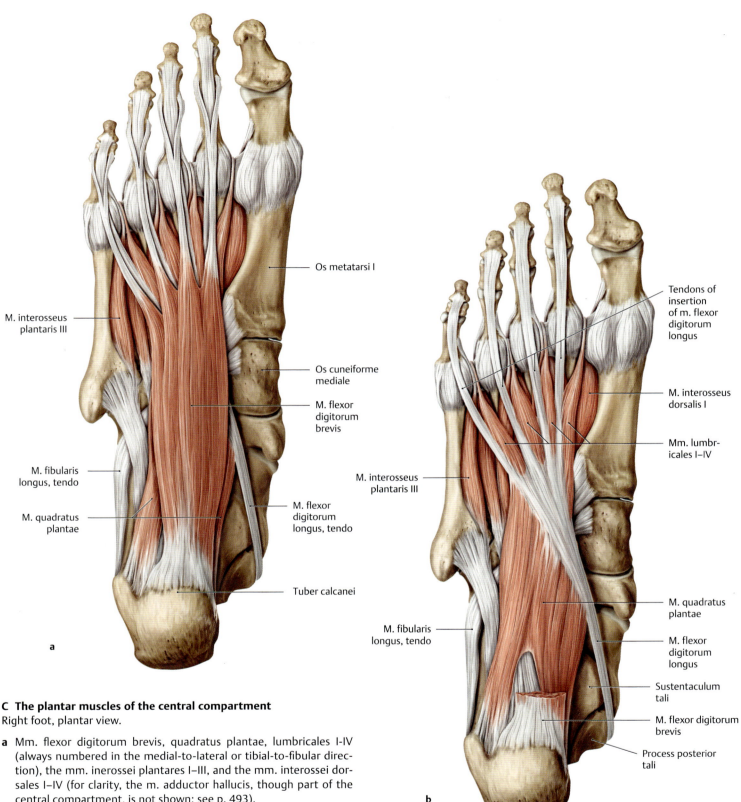

C The plantar muscles of the central compartment
Right foot, plantar view.

a Mm. flexor digitorum brevis, quadratus plantae, lumbricales I–IV (always numbered in the medial-to-lateral or tibial-to-fibular direction), the mm. inerossei plantares I–III, and the mm. interossei dorsales I–IV (for clarity, the m. adductor hallucis, though part of the central compartment, is not shown; see p. 493).

b M. lexor digitorum brevis has been removed to its origin to display more clearly the insertion of quadratus plantae on the lateral margin of the m. flexor digitorum longus tendon.

Note: The "movable origins" of the mm. lumbricales I–IV form the medial borders of the m. flexor digitorum longus tendons. When the m. flexor digitorum longus contracts and therefore shortens, the origins of the mm. lumbricales move proximally. This "prestretching" of the mm. lumbricales improves their ability to contract, enabling them to develop greater force.

20.12 Overview of Muscle Functions: Hip Joint

A Movements of the hip joint

Type of movement	Range of motion	Muscle	Innervation	Nerve segment
Flexion	120–140°	• M. iliopsoas (m. psoas major und m. iliacus) • M. rectus femoris • M. tensor fasciae latae • M. sartorius • M. pectineus • M. adductor longus • M. adductor brevis • M. gracilis • Mm. glutei medius and minimus, anterior parts	• Direct branches from plexus lumbalis and n. femoralis • N. femoralis • N. gluteus superior • N. femoralis • N. obturatorius • N. obturatorius • N. obturatorius • N. obturatorius • N. gluteus superior	• L1–L3 • L2–L4 • L4–S1 • L2, L3 • L2, L3 • L2–L4 • L2, L3 • L2, L3 • L4–S1
Extension	20°	• M. gluteus maximus • M. semitendinosus • M. semimembranosus • M. biceps femoris, caput longum • Mm. glutei medius and minimus, posterior parts • M. adductor magnus • M. piriformis • M. obturatorius internus	• N. gluteus inferior • N. tibialis • N. tibialis • N. tibialis • N. gluteus superior • N. obturatorius • N. tibialis • Direct branches from plexus sacralis • Direct branches from plexus sacralis	• L5–S2 • L5–S2 • L5–S2 • L5–S2 • L4–S1 • L2–L4 • L4 • S1, S2 • L5, S1
Abduction	50–80°	• M. gluteus medius • M. tensor fasciae latae • M. gluteus maximus (upper fibers) • M. gluteus minimus • M. piriformis • M. sartorius	• N. gluteus superior • N. gluteus superior • N. gluteus inferior • N. gluteus superior • Direct branches from plexus sacralis • N. femoralis	• L4–S1 • L4–S1 • L5–S2 • L4–S1 • S1, S2 • L2, L3
Adduction	20–30°	• M. adductor magnus • M. adductor longus • M. adductor brevis • M. gluteus maximus (lower fibers) • M. pectineus • M. gracilis • M. semitendinosus • M. semimembranosus • M. biceps femoris, caput longum • M. quadratus femoris • M. obturatorius internus • M. obturatorius externus	• N. obturatorius • N. tibialis • N. obturatorius • N. obturatorius • N. gluteus inferior • N. obturatorius • N. obturatorius • N. tibialis • N. tibialis • N. tibialis • N. gluteus inferior • Direct branches from plexus sacralis • N. obturatorius	• L2–L4 • L4 • L2–L4 • L2, L3 • L5–S2 • L2, L3 • L2, L3 • L5–S2 • L5–S2 • L5–S2 • L5, S1 • L5, S1 • L3, L4
Internal rotation	40°	• Mm. glutei medius and minimus, anterior parts • M. tensor fasciae latae • M. adductor magnus (tendinous insertion on epicondylus medialis)	• N. gluteus superior • N. gluteus superior • N. tibialis	• L4–S1 • L4–S1 • L4
External rotation	30–50°	• M. gluteus maximus • M. obturatorius internus • M. quadratus femoris • M. obturatorius externus • Mm. glutei medius and minimus, posterior parts • M. adductor magnus • M. adductor longus • M. adductor brevis • M. pectineus • M. sartorius • M. iliopsoas (m. psoas major and m. iliacus)	• N. gluteus inferior • Direct branches from plexus sacralis • N. gluteus inferior • N. obturatorius • N. gluteus superior • N. obturatorius • N. tibialis • N. obturatorius • N. obturatorius • N. obturatorius • N. femoralis • Direct branches from plexus lumbalis and n. femoralis	• L5–S2 • L5, S1 • L5, S1 • L3, L4 • L4–S1 • L2–L4 • L4 • L2–L4 • L2, L3 • L2, L3 • L2, L3 • L1–L3

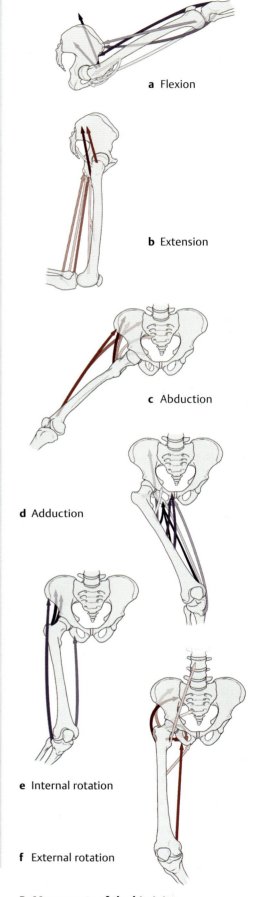

a Flexion

b Extension

c Abduction

d Adduction

e Internal rotation

f External rotation

B Movements of the hip joint

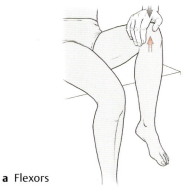

a Flexors

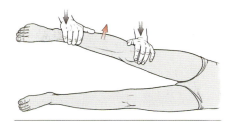

c Abductors

g Internal rotators

b Extensors

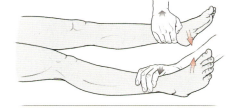

d Side-by-side comparison of adductors

e Adductors in side-lying position with strength unimpaired

f External rotators

C Functional muscle testing of the hip joint

In order to test the strength of each muscle, the patient is asked to resist (push back against) the examiner. The force generated by the muscle action is measured using a scale from 0 to 5.

D Clinical symptoms of hip muscle shortening and weakness

Muscles	Symptoms of muscle shortening	Symptoms of muscle weakness
Flexors	Shortened hip flexors lead to the pelvis tilting forward with increased lordosis lumbalis and impaired hip extension range. One-sided shortening leads to pelvic torsion with ensuing functional impairment mainly in the sacroiliac joint of the affected side.	Activities such as climbing stairs, walking uphill, shifting from supine position to sitting position, or moving the upper body forward when sitting up are considerably impaired with weak flexors. When walking, instead of hip joint flexion, the leg is moved forward by circumduction or through movements of the pelvis.
Extensors	Contractions of the m. gluteus maximus are rare. Shortened ischiocrural muscles lead to a characteristic posture with increased hip extension and flexion of the lumbar spine.	A lack of stability in stance phase is due to weak hip joint extensors. It is compensated for through shifting of the upper body (to achieve support from the lig. iliofemorale).
Abductors	Shortened abductors lead to pelvic shift in the frontal plane and functional leg elongation of the affected side. The body compensates for leg length discrepancy through increased flexion of the knee joint.	Due to the lack of muscular support, the pelvis cannot be kept in place in stance phase. It drops to the healthy side (positive Trendelenburg sign). In case of slight weakness, the upper body shifts to the affected side, thus preventing the pelvis from dropping (Duchenne sign).
Adductors	Shortened adductors also lead to pelvic shift in the frontal plane and functional leg elongation of the affected side.	Generally, weakness becomes noticeable during extreme movements, e.g., horse riding, skiing (e.g., rider cannot keep himself on horse).
Internal rotators	The hip joint can no longer rotate externally to the full extent, so sitting cross-legged is no longer possible.	Leads to a significant predominance of external rotation (= change of gait pattern: forward section of foot increasingly points outward when walking).
External rotators	The one-sided pull at the sacrum, especially with shortening of the short external rotators, leads to functional impairment of the sacroiliac joint of the affected side.	Increased internal rotation of the affected leg (forward section of foot increasingly points inward when walking).

20.13 Overview of Muscle Functions: Knee Joint

A Movements of the knee joint

Type of movement	Range of motion	Muscle	Innervation	Nerve segment
Flexion	120–150°	• M. semimembranosus • M. semitendinosus • M. biceps femoris, caput longum • M. biceps femoris, caput breve • M. gracilis • M. sartorius • M. gastrocnemius (caput mediale and caput laterale) • M. popliteus • M. popliteus	• N. tibialis • N. tibialis • N. tibialis • N. fibularis communis • N. obturatorius • N. femoralis • N. tibialis • N. tibialis • N. tibialis	• L5–S2 • L5–S2 • L5–S2 • L5–S2 • L2, L3 • L2, L3 • S1, S2 • L4–S1 • S1, S2

Extension	5–10°	• M. quadriceps femoris
		  – M. rectus femoris
		  – M. vastus lateralis
		  – M. vastus medialis
		  – M. vastus inter medius
			• N. femoralis	• L2–L4

Internal rotation	10°	• M. semimembranosus • M. semitendinosus • M. gracilis • M. sartorius • M. popliteus	• N. tibialis • N. tibialis • N. obturatorius • N. femoralis • N. tibialis	• L5–S2 • L5–S2 • L2, L3 • L2, L3 • L4–S1
External rotation	30–40°	• M. biceps femoris, caput longum • M. biceps femoris, caput breve	• N. tibialis • N. fibularis communis	• L5–S2 • L5–S2

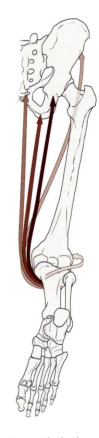

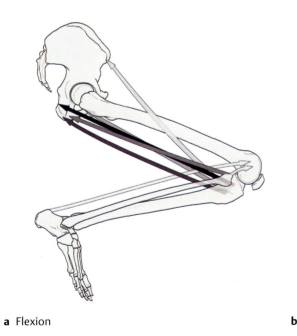

a Flexion

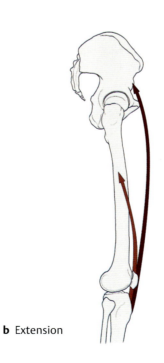

b Extension

c Internal rotation with the knee joint flexed

B Movements of the knee joint

d External rotation with the knee joint extended

C Clinical symptoms of knee muscle shortening and weakness

Muscles	Symptoms of muscle shortening	Symptoms of muscle weakness
Flexors	Impaired knee extension with bent hip or impaired hip flexion with hip outstretched. Shortened flexors, especially the ischiocrural (hamstring) muscles on both sides, lead to straightening of the pelvis and flattening of the lumbar spine. If the shortening affects only one side, the differing pull causes pelvic tilt and functional disorder of the sacroiliac joint.	Weakness of the ischiocrural muscles makes itself felt in both the knee and the hip joint. When standing, the result is increased pelvic tilt forward with simultaneous knee hyperextension. The clinical symptoms are similar to those of extensor weakness.
Extensors	In the case of muscle shortening, the m. rectus femoris is most commonly affected since it is the only part of the m. quadriceps femoris that crosses two joints. The result is impaired knee flex-ion or impaired hip extension with increased lordosis lumbalis.	In case of muscle weakness, essential functions are significantly impaired, e.g., climbing stairs, walking uphill, standing up, and sitting down. Compensation occurs through shifting the center of gravity forward via knee hyperextension (genu recurvatum): "Gravity turns into stretching force."
Internal rotators		Increased external rotation of the lower leg
External rotators		Increased internal rotation of the lower leg

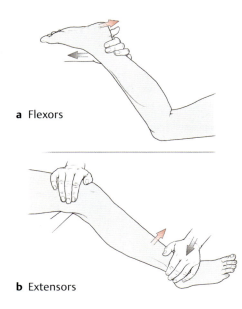

a Flexors

b Extensors

D Functional muscle testing of the knee joint

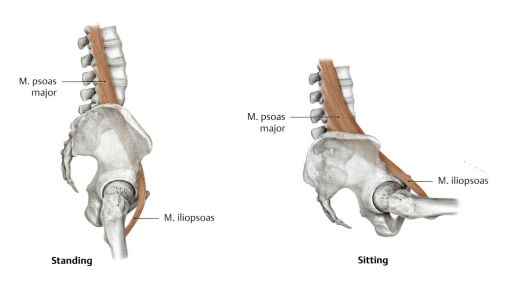

Standing Sitting

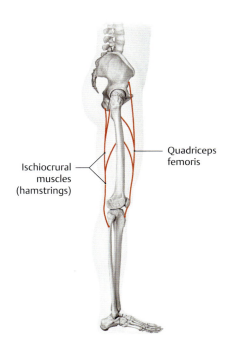

E Muscular dysbalance

Iliacus not shown. Special testing of every joint includes not only movement tests (neutral zero method, see p. 50) but also functional tests. The primary goal is not to test a muscle's force and ability to stretch but to detect muscular imbalances and coordination impairment. If there is an imbalance between a muscle and its antagonist, it is called a muscular dysbalance. What is commonly meant by this is an imbalance between tonic and phasic muscles.

Tonic muscles primarily consist of slow-twitch muscle fibers and in case of disorders tend to shorten (see p. 56). **Phasic muscles** consist of fast-twitch fibers and tend to weaken (atrophy) in case of disorders. Those disorders are the result of the insufficient adaptation of the musculoskeletal system to the civilized way of life, which includes prolonged sitting, inactivity, habitual movements, and abnormal postures in daily life and at work. Increased variety of motion and better quality of motion are easy ways to counteract the lack of stimuli to muscles.

Examples of the development of muscular dysbalances:
- Prolonged sitting leads to increased structural shortening of the mm. iliopsoae (these muscles reach their original length in standing position) and significant weakening of the hip extensors (mainly the m. gluteus maximus and ischiocrural muscles [hamstrings]) and thus indirectly leads to increased susceptibility to disorders of the hyperlordotic lumbar spine.
- High heels lead to increased load on the m. quadriceps femoris with corresponding increased tonus; the ischiocrural muscles, as antagonists, however, shorten due to the constant lack of stimuli.

20.14 Overview of Muscle Functions: Ankle Joints

A Movements of the art. talocruralis (upper ankel joint), the art. subtalaris (lower ankle joint) and the art. tarsi transversa

Type of movement	Range of motion	Muscle	Innervation	Nerve segment
Plantar flexion	40–50°	• M. triceps surae	• N. tibialis	• S1, S2
		• M. fibularis longus	• N. fibularis superficialis	• L4, L5
		• M. fibularis brevis	• N. fibularis superficialis	• L4, L5
		• M. flexor hallucis longus	• N. tibialis	• L5–S2
		• M. flexor digitorum longus	• N. tibialis	• L5–S2
		• M. tibialis posterior	• N. tibialis	• L4, L5
		• M. plantaris	• N. tibialis	• S1, S2
Dorsiflexion	20–30°	• M. tibialis anterior	• N. fibularis profundus	• L4, L5
		• M. extensor digitorum longus	• N. fibularis profundus	• L4, L5
		• M. extensor hallucis longus	• N. fibularis profundus	• L4, L5
Inversion and supination	60°	• M. tibialis posterior	• N. tibialis	• L4, L5
		• M. flexor hallucis longus	• N. tibialis	• L5–S2
		• M. flexor digitorum longus	• N. tibialis	• L5–S2
		• M. tibialis anterior	• N. fibularis profundus	• L4, L5
		• (M. extensor hallucis longus)	• N. fibularis profundus	• L4, L5
Eversion and pronation	30°	• M. fibularis longus	• N. fibularis superficialis	• L5, S1
		• M. fibularis brevis	• N. fibularis superficialis	• L5, S1
		• M. extensor digitorum longus	• N. fibularis profundus	• L4, L5
		• (M. extensor hallucis longus)	• N. fibularis profundus	• L4, L5

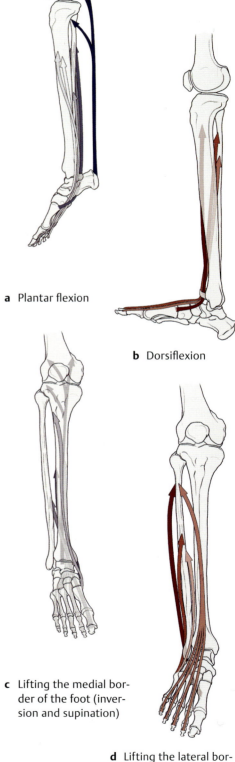

a Plantar flexion

b Dorsiflexion

c Lifting the medial border of the foot (inversion and supination)

d Lifting the lateral border of the foot (eversion and pronation)

B Movements of the art. talocruralis, the art. subtalaris and the art. tarsi transversa

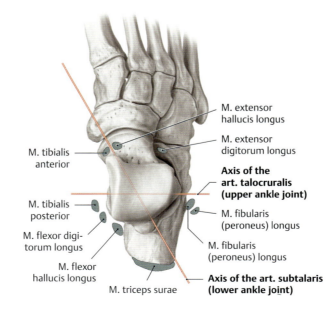

C Location of tendons of the long muscles of the foot relative to the axes of the upper and lower ankle joint

Superior view of both axes of the upper and lower ankle joint (art. talocruralis and art. subtalaris). Depending on the course of the tendons, the muscles can plantar flex and dorsal extend while at the same time pronating (everting) and supinating (inverting).

Labels: M. tibialis anterior; M. tibialis posterior; M. flexor digitorum longus; M. flexor hallucis longus; M. triceps surae; M. extensor hallucis longus; M. extensor digitorum longus; Axis of the art. talocruralis (upper ankle joint); M. fibularis (peroneus) longus; M. fibularis (peroneus) longus; Axis of the art. subtalaris (lower ankle joint)

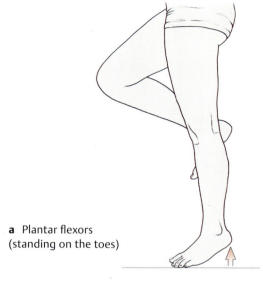
a Plantar flexors (standing on the toes)

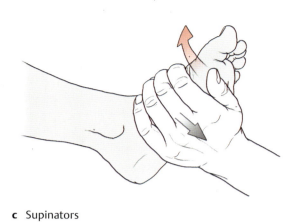
c Supinators

b Dorsiflexors

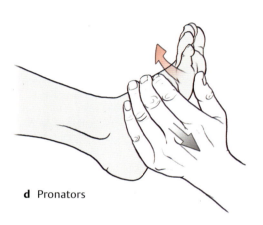
d Pronators

D Functional muscle testing of the ankle joint

E Clinical symptoms of muscle shortening and weakness of the upper and lower ankle joint

Muscles	Symptoms of muscle shortening	Symptoms of muscle weakness
Plantar flexors	Shortening of plantar flexors, mainly the m. triceps surae, leads to club foot (pes equinovarus). The resulting functional leg elongation when standing is compensated for by increased knee flexion. When walking, during swing phase, increased hip flexon is required in order to allow for the functionally longer leg to swing forward.	Tiptoe position is usually not possible, and jumping power is considerably impaired. Also common is development of talipes calcaneus (no rollover during walking). In stance phase, there is increased knee extension.
Dorsi-flexors	Shortening of dorsiflexors leads to impaired plantar flexion and impaired rollover during walking. Often development of club foot (no rollover during walking) develops.	The forward section of the foot cannot be sufficiently lifted during swing phase. To compensate for it, there is increased bending of the hip and knee. This characteristic gait pattern is also known as stork legs.
Supinators	Shortening leads to pes equinovarus. There is increased load on the lateral bottom of the foot during walking.	Weakened supinators result in valgus position (talipes valgus) with increased load on the inside edge of the foot.
Pronators	If the pronators (e.g., the fibularis muscles) are affected by shortening, the foot is plantar flexed and pronated (pes equinovalgus).	Weakened pronators often result in distortion of the upper and lower ankle joint (so-called supination trauma).

21.1 The Muscles of the Medial and Anterior Thigh, Hip, and Gluteal Region

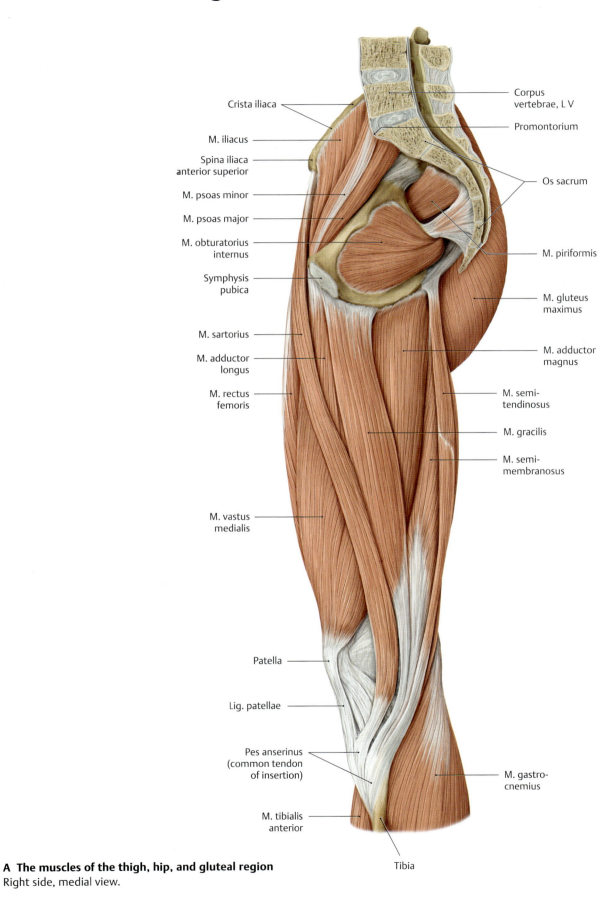

A The muscles of the thigh, hip, and gluteal region
Right side, medial view.

Lower Limb —— 21. Musculature: Topographical Anatomy

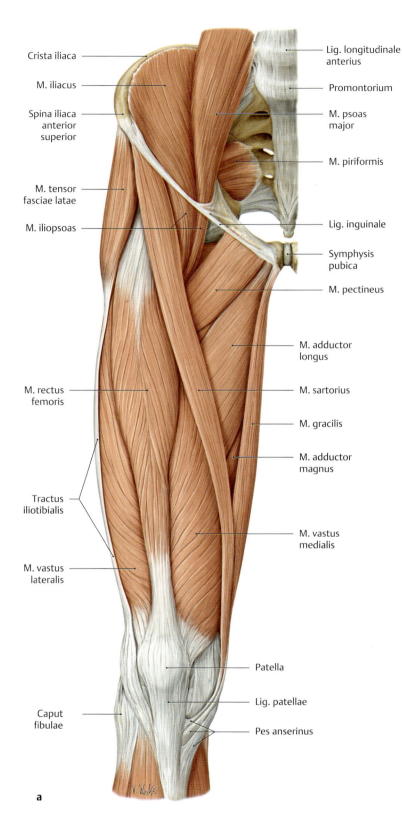
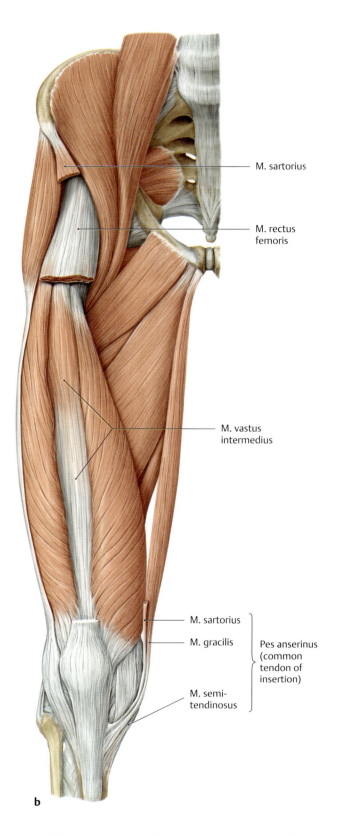

B The muscles of the thigh, hip, and gluteal region
Right side, anterior view.
a The fascia lata of the thigh (see p. 545) has been removed as far as the lateral tractus iliotibialis.
b Portions of the m. sartorius and m. rectus femoris have also been removed.

503

21.2 The Muscles of the Anterior Thigh, Hip, and Gluteal Region: Origins and Insertions

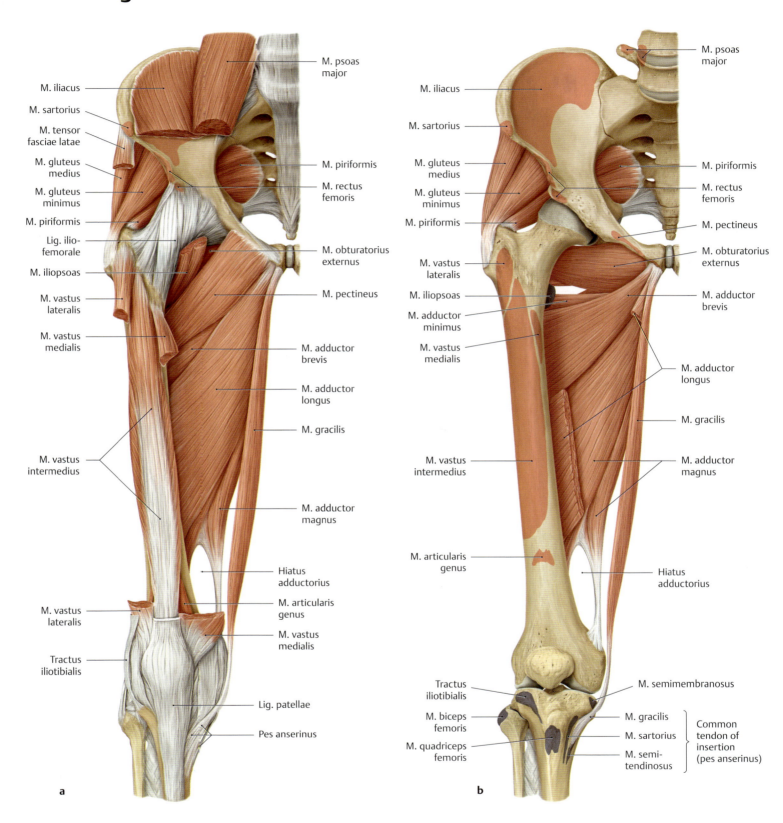

A The muscles of the thigh, hip, and gluteal region
Right side, anterior view. The origins and insertions of the muscles are indicated by color shading (red = origin, blue = insertion).

a The Mm. iliopsoas and tensor fasciae latae have been partially removed. The mm. sartorius, rectus femoris, vastus lateralis, and vastus medialis have been completely removed.

b The mm. quadriceps femoris, iliopsoas, tensor fasciae latae, and pectineus have been completely removed. The midportion of the m. adductor longus has been removed.

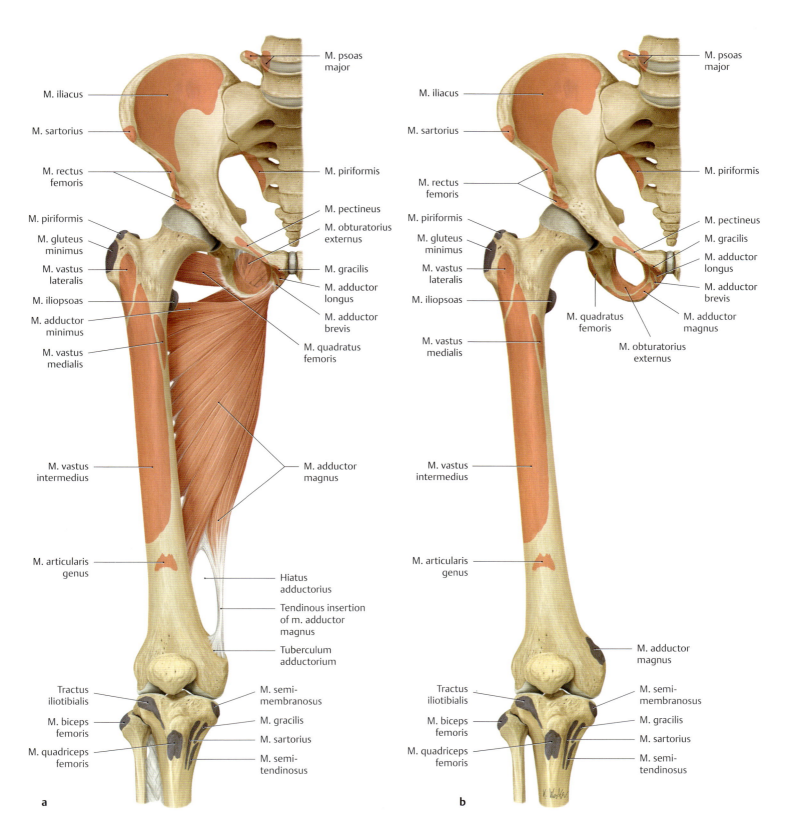

B The muscles of the thigh, hip, and gluteal region
Right side, anterior view. The origins and insertions of the muscles are indicated by color shading (red = origin, blue = insertion).

a All of the muscles have been removed except for m. adductor magnus and m. quadratus femoris.
b All of the muscles have been removed.

Note the hiatus adductorius, through which the a. and v. femoralis enter the fossa poplitea of the leg.

21.3 The Muscles of the Lateral and Posterior Thigh, Hip, and Gluteal Region

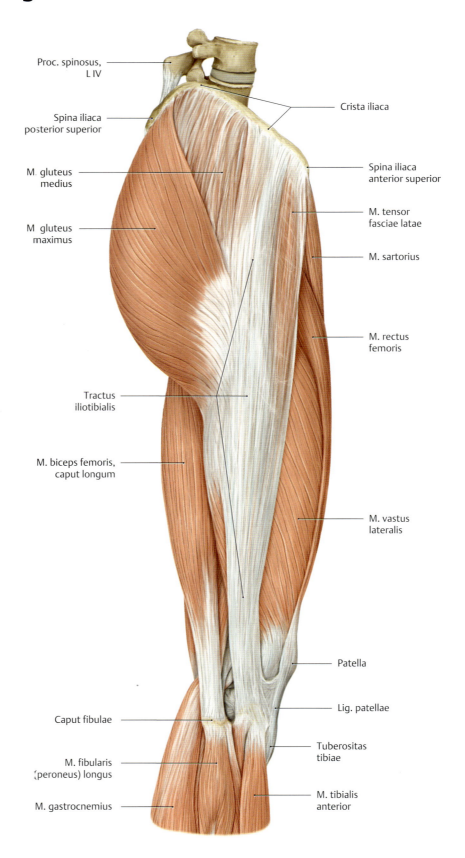

A The muscles of the thigh, hip, and gluteal region
Right side, lateral view. *Note* the mm. tensor fasciae latae and gluteus maximus, whose tendons of insertion strengthen and thicken the lateral part of the fascia lata. This thickened band, called *the Tractus ilio-* *tibialis* because it runs between the crista *iliaca* and the lateral side of the upper *tibia*, functions mechanically as a tension band to reduce the bending loads on the proximal femur (after Pauwels, see also p. 415).

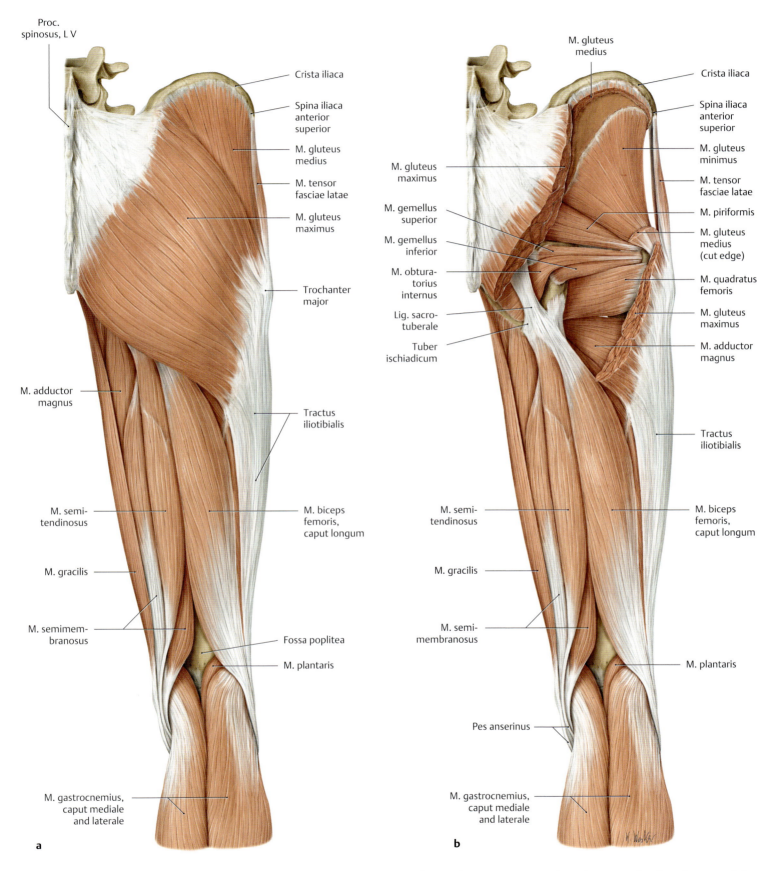

B The muscles of the thigh, hip, and gluteal region
Right side, posterior view.
a The fascia lata has been removed as far as the tractus iliotibialis (the portion over the buttock is called the fascia glutaea).
b The m. gluteus maximus and m. gluteus medius have been partially removed.

21.4 The Muscles of the Posterior Thigh, Hip, and Gluteal Region: Origins and Insertions

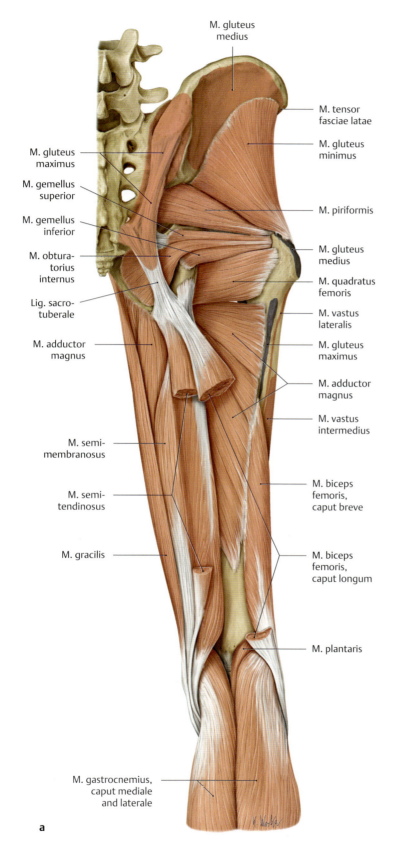
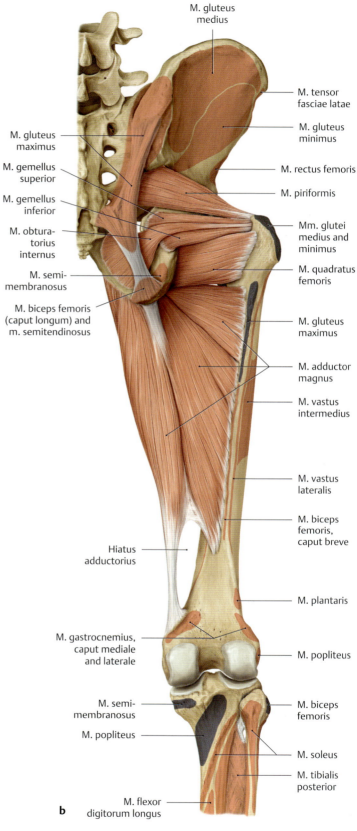

A The muscles of the thigh, hip, and gluteal region
Right side, posterior view. The origins and insertions of the muscles are indicated by color shading (red = origin, blue = insertion).

a The mm. semitendinosus and biceps femoris have been partially removed. The mm. gluteus maximus and gluteus medius have been completely removed.

b The hamstrings (also called ischiocrural muscle; mm. semitendinosus, semimembranosus, and biceps femoris) and m. gluteus minimus have been completely removed.

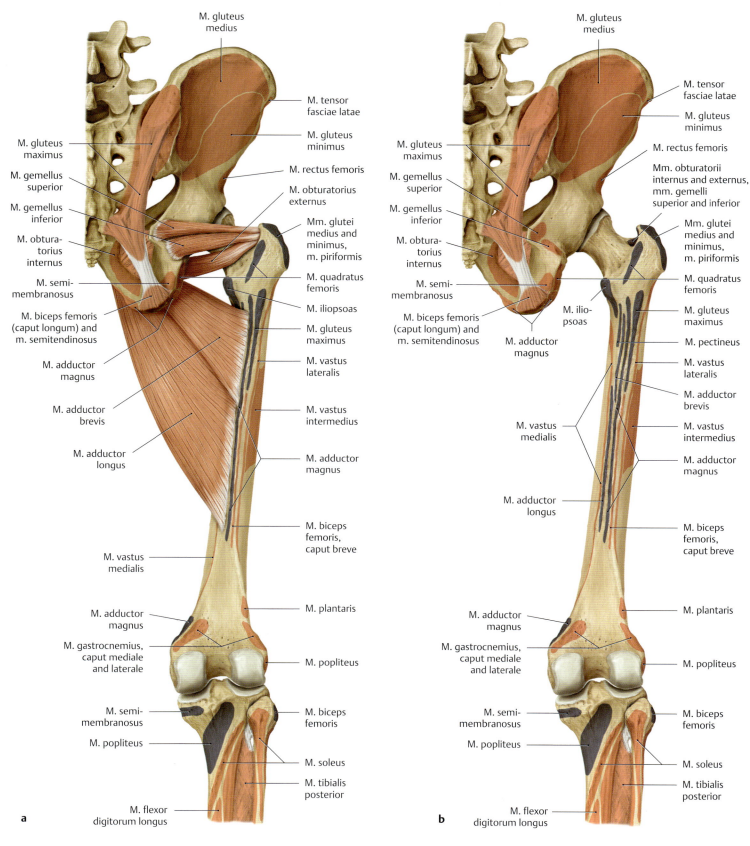

B The muscles of the thigh, hip, and gluteal region
Right side, posterior view. The origins and insertions of the muscles are indicated by color shading (red = origin, blue = insertion).

a All muscles have been removed except for the mm. adductor brevis, adductor longus, gemelli superior and inferior, and obturatorius externus.

b All of the muscles have been removed.

21.5 The Muscles of the Lateral and Anterior Leg: Origins and Insertions

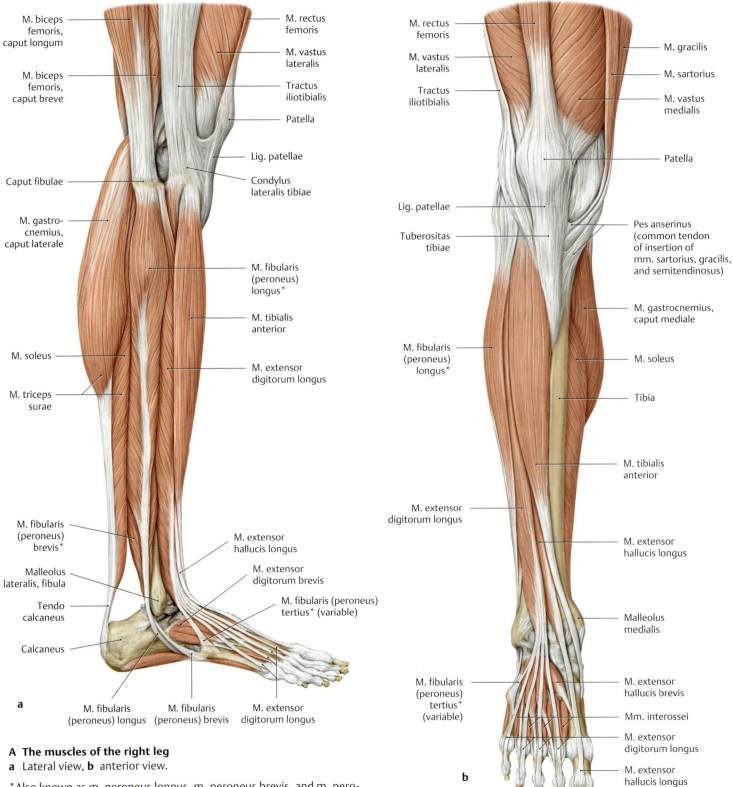

A The muscles of the right leg
a Lateral view, b anterior view.

*Also known as m. peroneus longus, m. peroneus brevis, and m. peroneus tertius.

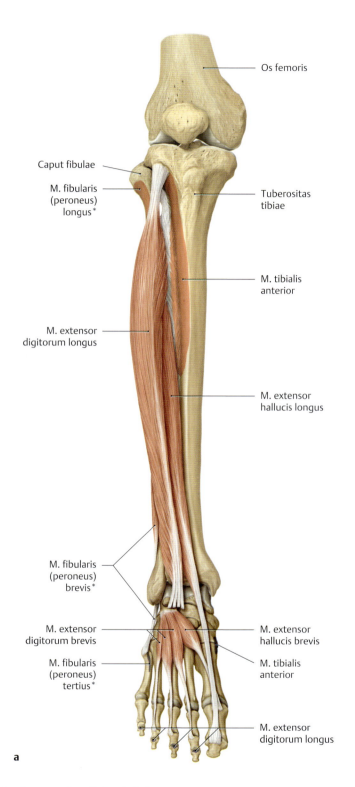
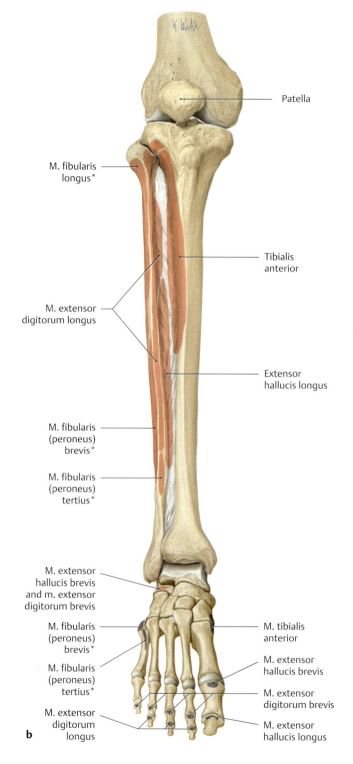

B The muscles of the right leg
Anterior view. The origins and insertions of the muscles are indicated by color shading (red = origin, blue = insertion).

a The mm. tibialis anterior and fibularis (peroneus) longus have been completely removed, as have the distal portions of the m. extensor digitorum longus tendons. The m. fibularis (peroneus) tertius is a division of the m. extensor digitorum longus.
b All of the muscles have been removed.

21.6 The Muscles of the Posterior Leg: Origins and Insertions

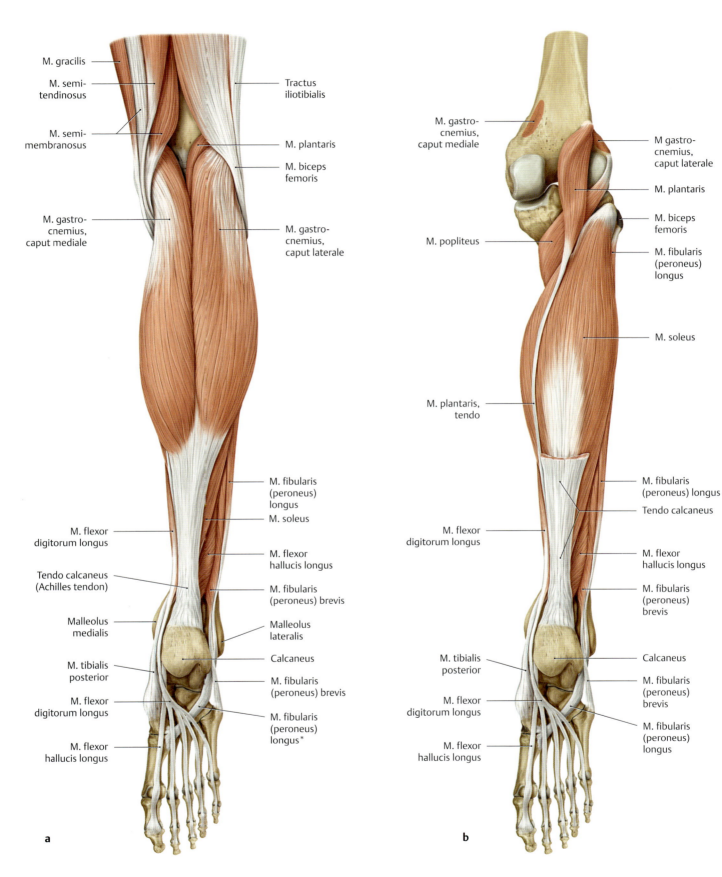

A The muscles of the right leg
Posterior view. The origins and insertions of the muscles are indicated by color shading (red = origin, blue = insertion). The foot is shown in a plantar-flexed position to better demonstrate the plantar tendons.

a The bulge of the calf (sura) is produced mainly by the m. triceps surae (= m. soleus plus the two heads of the m. gastrocnemius).
b Both heads of the m. gastrocnemius have been removed.

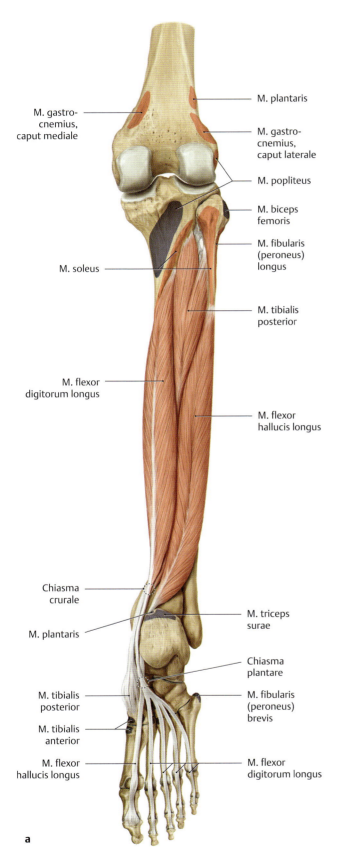
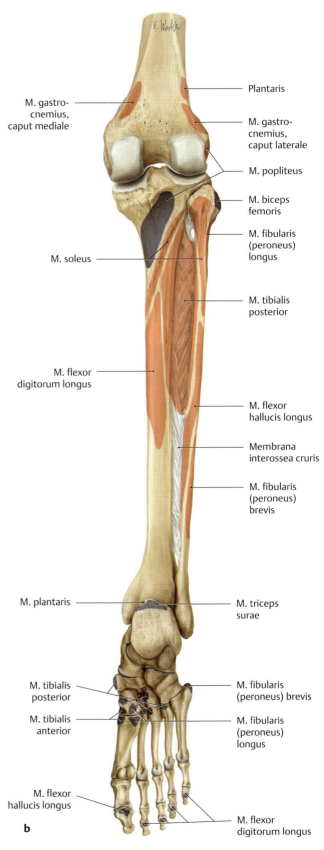

B The muscles of the right leg
Posterior view. The origins and insertions of the muscles are indicated by color shading (red = origin, blue = insertion). The foot is shown in a plantar-flexed position to better demonstrate the plantar tendons.

a The mm. triceps surae, plantaris, and popliteus have been removed.
b All of the muscles have been removed.

21.7 The Tendon Sheaths and Retinacula of the Foot

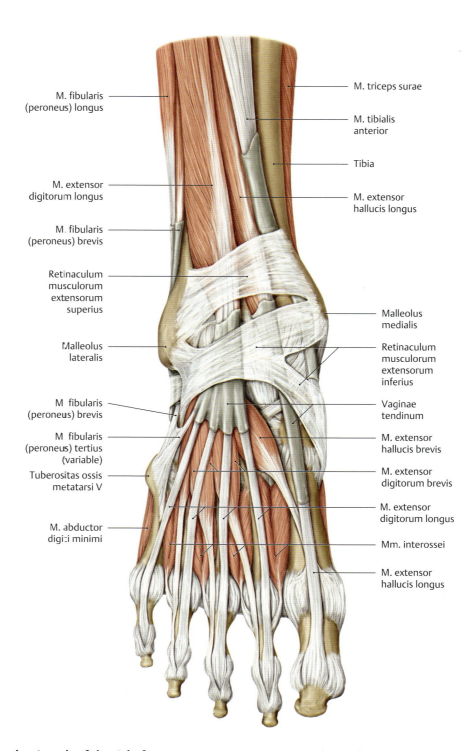

A The tendon sheaths and retinacula of the right foot
Anterior view. The foot is plantar-flexed, with fascia cruris superficialis removed, to display the deep fascial bands—*retinacula*—that hold in place the tendon sheaths (vaginae tendineum) of the long foot extensors and flexors. The superior and inferior extensor retinacula retain the long extensor tendons, allowing efficient redirection of the forces generated by their muscles (mm. tibialis anterior, extensor digitorum longus, extensor hallucis longus, and fibularis tertius) while preventing the tendons from rising away from the bones of the ankle when the foot is dorsiflexed. Similarly, the fibular retinacula, laterally, hold the fibular muscle tendons in place posterior to the malleolus lateralis (see **B**, part **a**), and the flexor retinaculum (retinaculum musculorum flexorum) retains the long flexor tendons behind the malleolus medialis (see **B**, part **b**), preventing displacement of these tendons while enabling them to operate smoothly regardless of the orientation of the art. talocruralis.

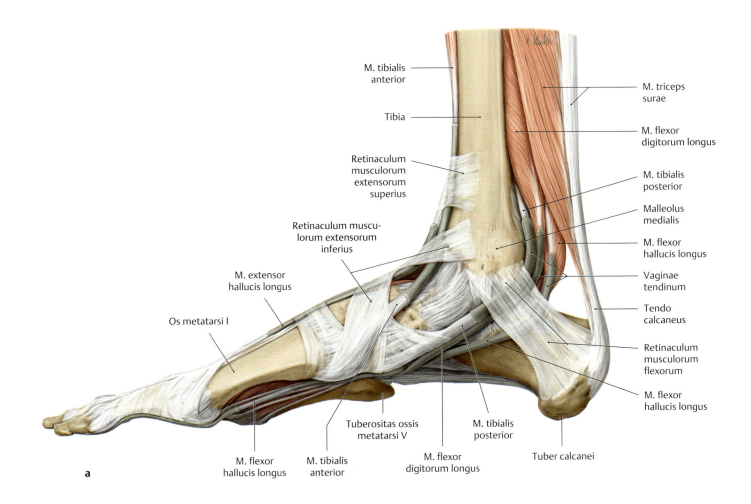
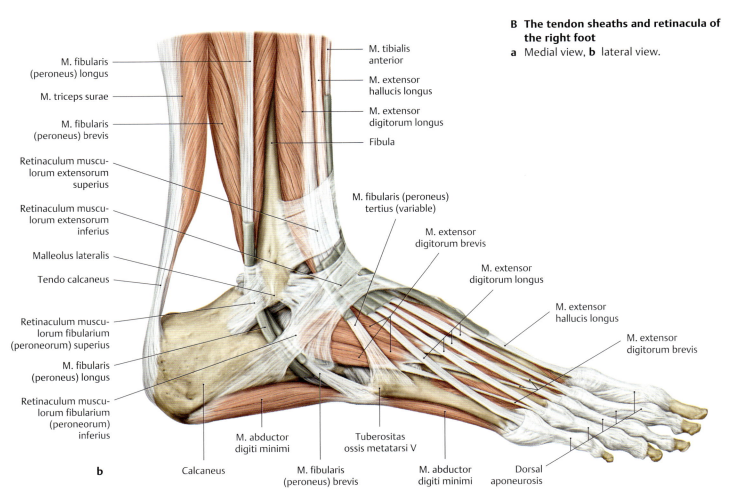

B The tendon sheaths and retinacula of the right foot
a Medial view, b lateral view.

21.8 The Intrinsic Foot Muscles from the Plantar View: The Aponeurosis plantaris and Superficial Layer

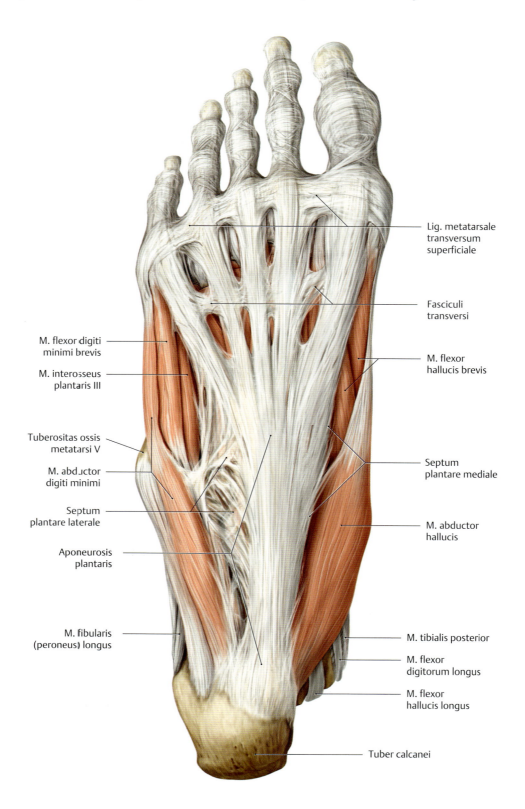

A The aponeurosis plantaris of the right foot
Plantar view. The *aponeurosis plantaris* is a tough aponeurotic sheet that is thicker centrally than medially and laterally and blends with the fascia dorsalis pedis (not shown here) at the borders of the foot. Two sagittal expansions of the thick *central* aponeurosis (the septum plantare mediale and septum plantare laterale) extend deep to the bones of the foot, defining the boundaries of three muscle compartments in the plantar region: the medial compartment, lateral compartment, and central compartment (not labeled here, see p. 494). The main function of the aponeurosis plantaris is to give *passive* support to the longitudinal arch of the foot (see also p. 467).

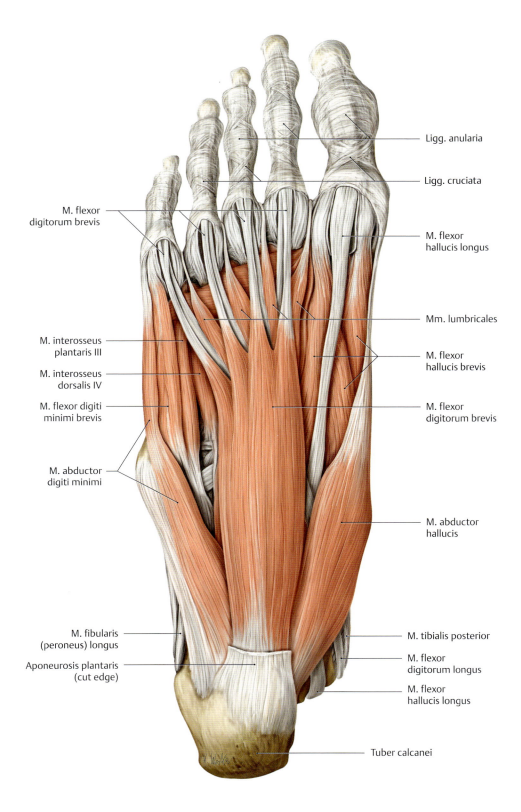

B The short muscles of the right foot, superficial layer
Plantar view. The entire aponeurosis plantaris, including the lig. metatarsale transversum superficiale, has been removed.

Note the ligg. anularia on the plantar side of the toes. Together with the oblique ligg. cruciata, they strengthen the tendon sheaths and help to hold the tendons in position.

21.9 The Intrinsic Foot Muscles from the Plantar View: Middle Layer

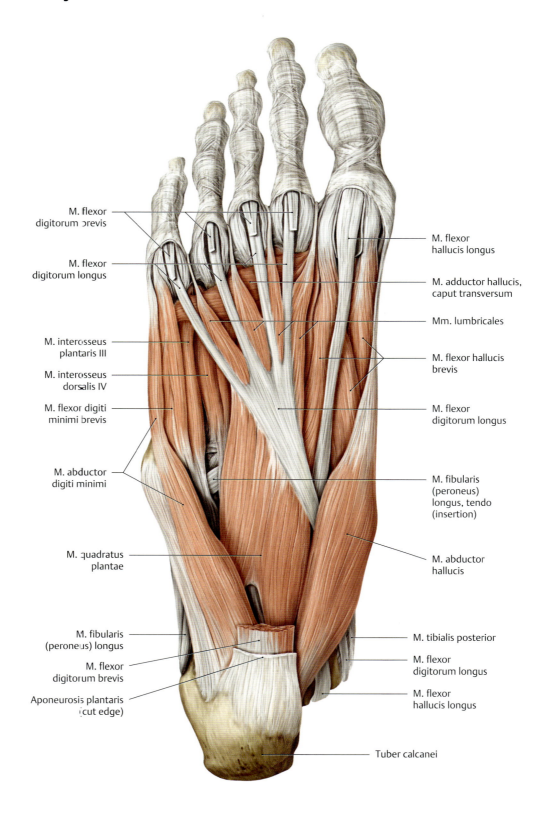

A The short muscles of the right foot, middle layer
Plantar view. The aponeurosis plantaris and m. flexor digitorum brevis have been removed.

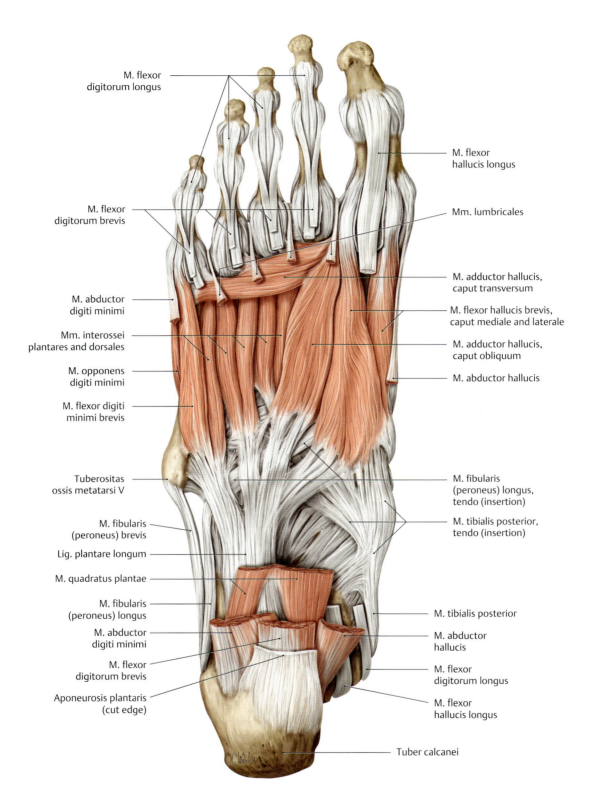

B The short muscles of the right foot, middle layer
Plantar view. The aponeurosis plantaris has been removed in addition to the following muscles: mm. flexor digitorum brevis, abductor digiti minimi, abductor hallucis, quadratus plantae, and lumbricales, and the tendons of insertion of the mm. flexor digitorum longus and flexor hallucis longus.

Note that each of the four tendons of insertion of the m. flexor digitorum brevis divides into two slips, and that the tendons of the m. flexor digitorum longus pass between these slips to insert on the distal phalanges.

21.10 The Intrinsic Foot Muscles from the Plantar View: Deep Layer and Origins and Insertions

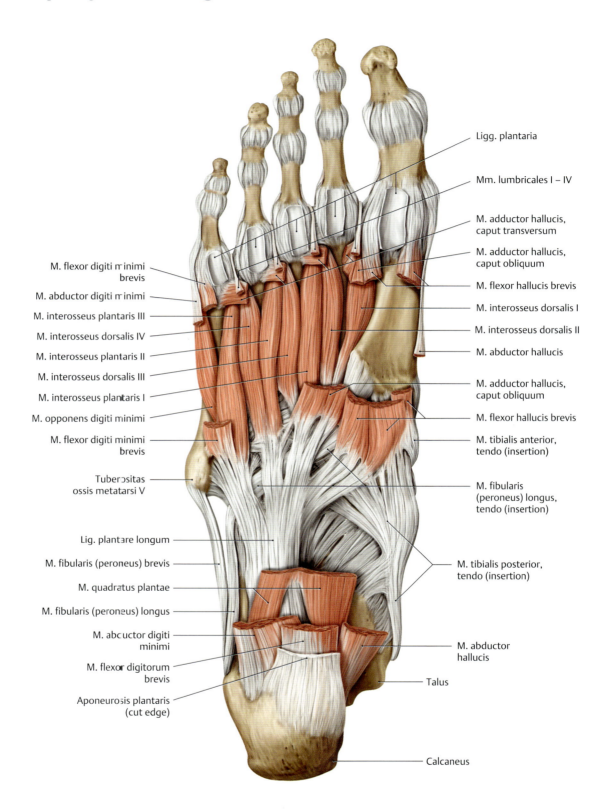

A The short muscles of the right foot, deep layer
Plantar view. All of the short foot muscles except for the mm. interossei dorsales and plantares have been removed, leaving behind their origins and insertions.

Note the course of the m. tibialis posterior and m. fibularis (peroneus) longus tendons of insertion, both of which help to support the transverse arch of the foot.

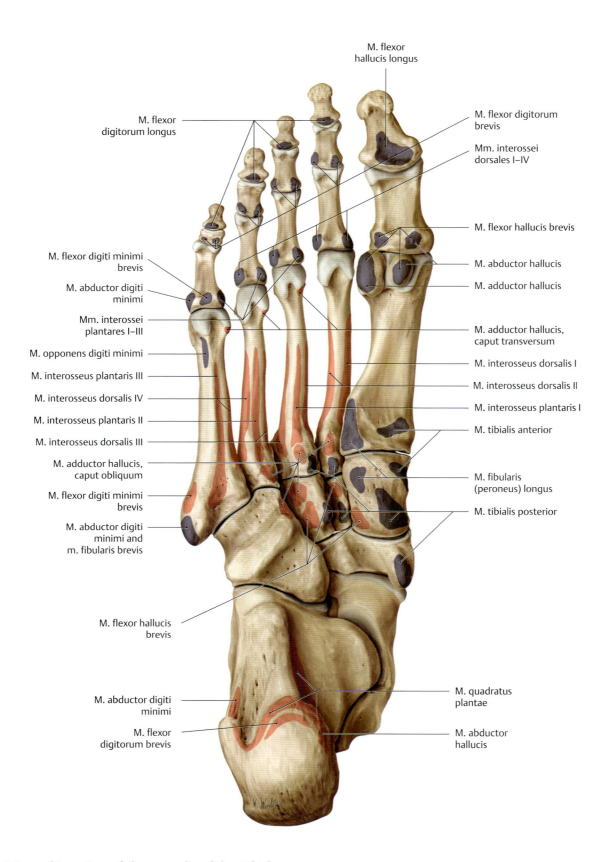

B Muscle origins and insertions of short muscles of the right foot
Plantar view. The origins and insertions of the muscles are indicated by color shading (red = origin, blue = insertion).

21.11 Cross-sectional Anatomy of the Thigh, Leg, and Foot

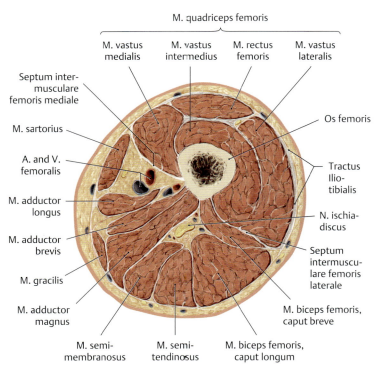

A Cross section through the right thigh
Proximal view. The level of the section is shown in **C**.

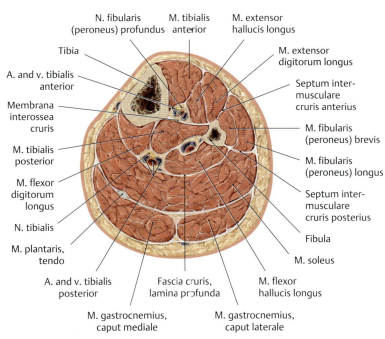

B Cross section through the right leg
Proximal view. The level of the section is shown in **C**.

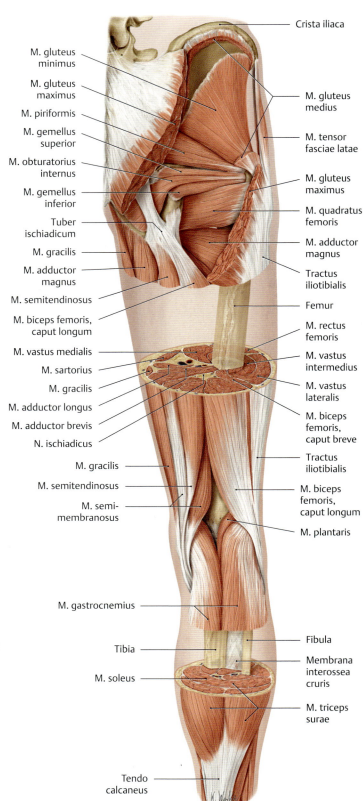

C "Windowed" dissection of the right lower limb
Posterior view. Portions of the mm. glutei maximus and medius have been removed (the removed cross sections are shown in **A** and **B**). The lower limb is one of the body regions most frequently examined by tomographic methods, and a knowledge of its cross-sectional anatomy is critically important in the identification of landmarks in both radiological and magnetic resonance (MR)–based images.

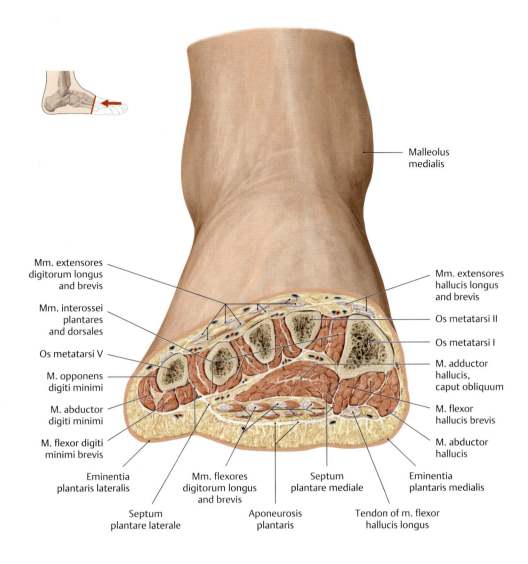

D Cross section through the right foot at the level of the metatarsals

View of the distal cut surface. The muscle compartments of the foot are formed mainly by the aponeurosis plantaris, the septa plantaria mediale and laterale, and the fascia plantaris profunda (see also **E**). Foot injuries, such as fracture-dislocations of the tarsus and metatarsus, may lead to *compartment syndromes of the foot*. They are caused by increased tissue pressure in the affected compartment due to the local extravasation of blood. The raised pressure in the compartment leads to impaired venous drainage and diminished capillary perfusion, manifested clinically by swelling and pain. This leads in turn to neuromuscular dysfunction with circulatory compromise that may culminate in muscle necrosis (drawn from a specimen in the Anatomical Collection of Kiel University).

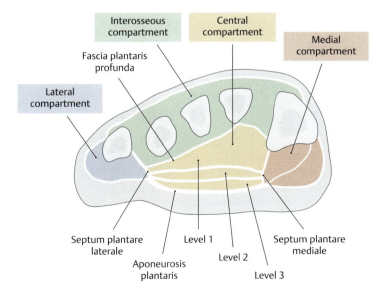

E Location of the compartments of the foot
Schematic cross section through a right foot, distal view. The different muscle compartments are indicated by color shading.

F The four compartments of the foot and their muscular contents (see also **E**)

Interosseous compartment
- Mm. interossei dorsales and plantares

Medial compartment
- M. abductor hallucis
- M. flexor hallucis brevis
- Tendon of insertion of M. flexor hallucis longus

Lateral compartment
- M. abductor digiti minimi
- M. flexor digiti minimi brevis
- M. opponens digiti minimii

Central compartment, consisting of three levels
- Level 1: M. adductor hallucis
- Level 2: M. quadratus plantae, mm. lumbricales and m. flexor digitorum longus tendons
- Level 3: M. flexor digitorum brevis

(after Mubarak and Hargens)

22.1 The Arteries

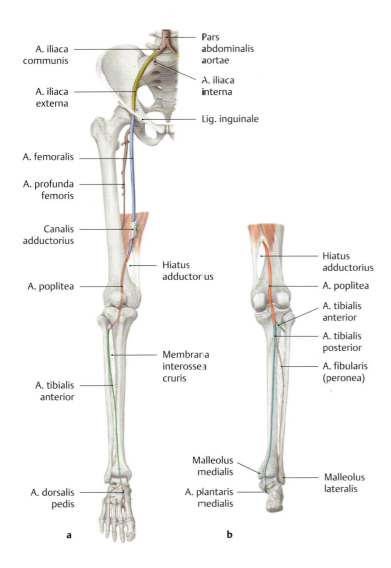

A Different segments of the arteries of the lower limb
a Right leg, anterior view; **b** right leg, posterior view. The different arterial segments are shown in different colors.

A. iliaca externa arises with the a. iliaca interna from the a. iliaca communis and descends along the medial border of the m. psoas major through the lacuna vasorum (see p. 549). It becomes the a. femoralis at the level of the lig. inguinalis.

A. femoralis, the continuation of the a. iliaca externa, runs down the medial side of the thigh to the canalis adductorius, through which it passes from the anterior to the posterior side of the limb. On leaving the hiatus adductorius, it becomes the a. poplitea.

A. poplitea runs from the hiatus adductorius through the fossa popliteal to the m. popliteus, dividing at the inferior border of that muscle into its terminal branches, the aa. tibiales anterior and posterior.

A. tibialis anterior enters the extensor compartment of the leg at the upper border of the membrana interossea and descends between the m. tibialis anterior and m. extensor hallucis longus. Distal to the Retinaculum musculorum extensorum, it continues onto the dorsum of the foot as the *a. dorsalis pedis*.

A. tibialis posterior, the direct continuation of the a. poplitea, enters the flexor compartment of the leg and passes behind the malleolus medialis. At that level it divides into its two terminal branches, the *aa. plantares medialis* and *lateralis* (the latter is shown in **D**), which continue onto the plantar side of the foot. The a. tibialis posterior also gives rise to the a. fibularis (peronea).

B Overview of the principal arteries of the lower limb
The arteries of the lower limb vary considerably in their origins and branching patterns (the main variants are reviewed in Chapter 5, Neurovascular Systems: Topographical Anatomy). The branches are listed in the order in which they arise from the parent vessels.

Branches of the a. iliaca externa
- A. epigastrica inferior
  - A. cremasterica
  - A. ligamenti teretis uteri
  - R. pubicus
- A. circumflexa ilium profunda

Branches of the a. femoralis (superficialis*)
- A. epigastrica superficialis
- A. circumflexa ilium superficialis
- A. pudenda externa superficialis
- A. pudenda externa profunda
- A. profunda femoris
  - A. circumflexa femoris medialis
  - A. circumflexa femoris lateralis
  - Aa. perforantes
- A. descendens genus

Branches of the a. poplitea
- A. superior medialis genus and a. superior lateralis genus
- Aa. surales
- A. mediagenus
- A. inferior medialis genus and a. inferior lateralis genus
  Note that the paired a. superior genus and a. inferior genus form the *arterial anastomotic network (rete articulare genus)* around the knee.

Branches of the a. tibialis anterior
- A. recurrens tibialis anterior
- A. malleolaris anterior lateralis
- A. malleolaris anterior medialis
- A. dorsalis pedis
  - A. tarsalis lateralis
  - A. tarsalis medialis
  - A. arcuata with the aa. metatarsales dorsales
    (→ aa. digitales dorsales)

Branches of the a. tibialis posterior
- A. recurrens tibialis posterior (rete articulare genus)
- A. fibularis (peronea)
  - R. perforans
  - R. communicans
  - Rr. malleolares laterales
  - Rr. calcanei
- Rr. malleolares mediales
- Rr. calcanei
- A. plantaris medialis
  - R. superficialis
  - R. profundus (→ arcus plantaris profundus)
- A. plantaris lateralis (→ arcus plantaris profundus)
- Aa. metatarsales plantares
- Aa. digitales plantares communes

* Often referred to clinically as the a. femoralis superficialis.
→ = is continuous with
Note: In addition, the lower limb is supplied by branches of the a. iliaca interna (e.g., the a. obturatoria).

Lower Limb — 22. Neurovascular Systems: Forms and Relations

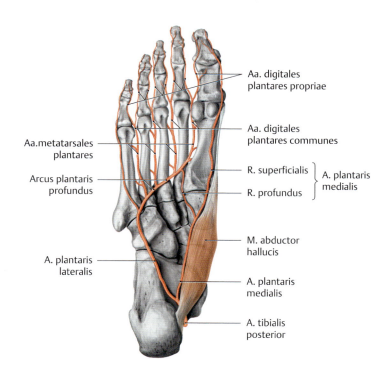

D The arteries of the sole of the foot
Right foot, plantar view.

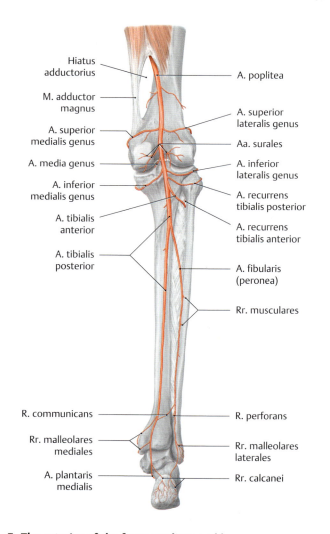

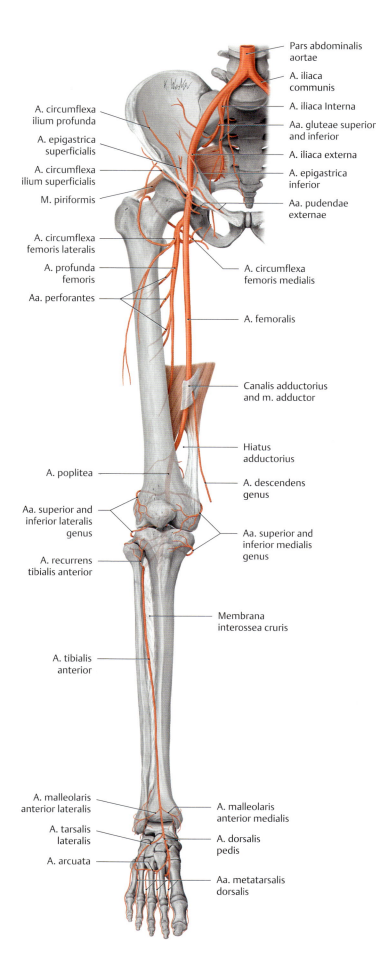

C The arteries of the lower limb
Right leg, anterior view with the foot in plantar flexion.

E The arteries of the fossa poplitea and leg
Right leg, posterior view.

525

22.2 The Veins

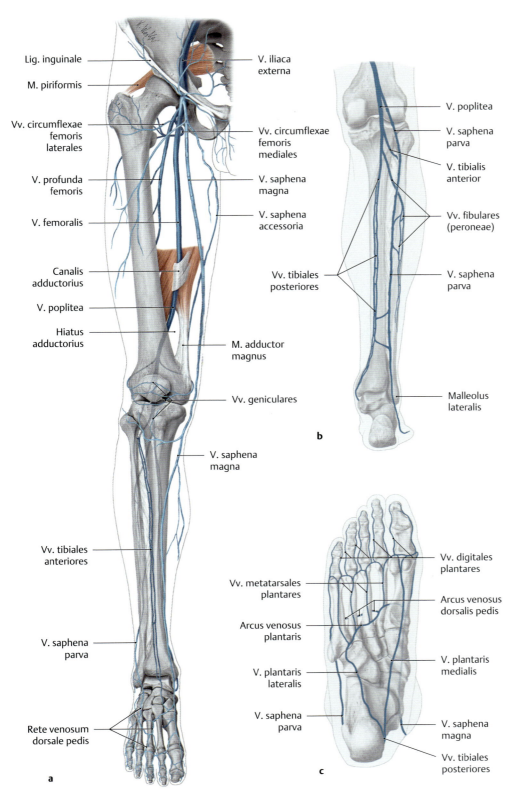

B Overview of the principal veins of the lower limb

The veins of the lower limb are subdivided into three systems: a superficial (epifascial) system, a deep (intermuscular) system, and a *perforating* system that interconnects the superficial and deep veins. The upright human body posture places an exceptional load on the veins of the lower limb, which must act against the force of gravity in returning the blood to the heart (the deep venous system handles approximately 85% of the venous return, the superficial veins approximately 15%). A series of venous valves (valvulae venosae) help to maintain the normal superficial-to-deep direction of blood flow (compare **E**). Note that, for the sake of clarity, not all of the veins in the table below have been depicted in these illustrations.

Deep veins of the lower limb (vv. profundae membri inferioris)
- V. femoralis
- V. profunda femoris
- Vv. circumflexae femoris mediales and laterales
- V. poplitea
- Vv. surales
- Vv. geniculares
- Vv. tibiales anteriores and posteriores
- Vv. fibulares (peroneae)
- Vv. metatarsales dorsales and plantares (see **Ac**)
- Vv. digitales plantares (see **Ac**)

Superficial lower limb veins (vv. superficiales membri inferioris)
- V. saphena magna
- Vv. pudendae externae
- V. circumflexa ilium superficialis
- V. epigastrica superficialis
- V. saphena accessoria
- V. arcuata cruris posterior
- V. saphena parva (see **Cb**)
- V. femoropoplitea (see **Cb**)
- Rete venosum dorsale pedis (see **Ca**)
- Arcus venosus dorsalis pedis
- Rete venosum plantare
- Arcus venosus plantaris

Perforating veins (vv. perforantes)
Of the many perforating veins in the leg, three groups have the greatest clinical importance (see **E**):
- The Dodd group (medial side of the thigh, middle third)
- The Boyd group (medial side of the leg below the knee)
- The Cockett group (medial side of the distal leg)

A The deep and superficial veins of the right lower limb
a Thigh, leg, and dorsum of the foot, anterior view.
b Leg, posterior view.
c Sole of the foot (planta), plantar view.

For clarity, only the most important veins are demonstrated here.

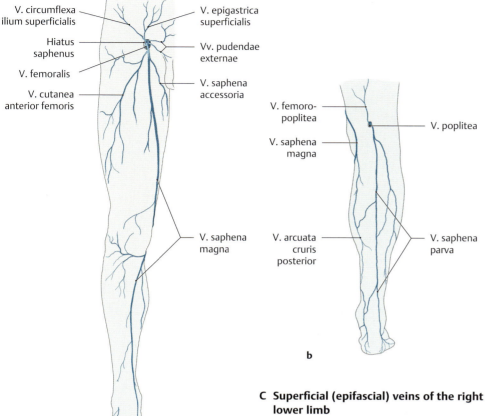
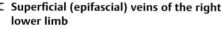
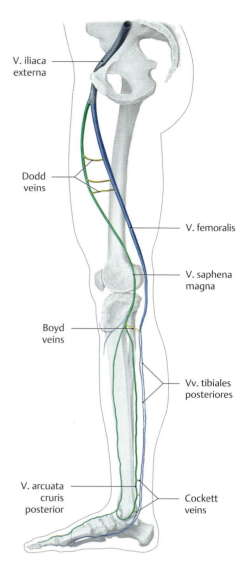

C Superficial (epifascial) veins of the right lower limb
a Thigh, leg, and dorsum of the foot, anterior view.
b Leg, posterior view.

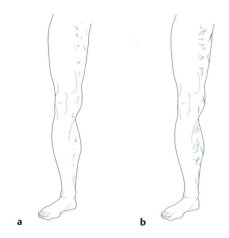
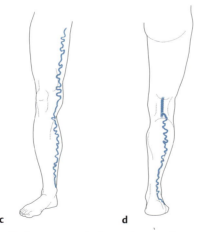

D Varices of the superficial leg veins
a Spider veins (tiny intradermal varices).
b Reticular varices (weblike dilations of small subcutaneous veins).
c Vena-saphena-magna varicosity.
d Vena-saphena-parva varicosity.

Varicose disease of the *superficial* leg veins is the most common *chronic* venous disease, affecting 15% of the adult population. Varicose veins can be classified as primary idiopathic varices (75%) or as secondary symptomatic varices. **Primary varices** generally result from degeneration of the venous wall leading to incompetence of the venous valves (valvulae venosae). **Secondary varices** result from chronic occlusion of the *deep* venous system with incompetence of the perforating veins (vv. perforantes) and a reversal in the direction of venous flow. Besides chronic conditions, there are also important *acute* diseases that may affect the superficial venous system (e.g., thrombophlebitis) and deep venous system (e.g., venous thrombosis).

E Clinically important perforating veins
Right leg, medial view. Numerous perforating veins (vv.perforantes) interconnect the deep and superficial venous systems of the leg. Their venous valves (valvulae venosae) normally prevent blood flow from the deep veins to the superficial cutaneous veins. The clinically important members of this system are located between the deep veins and the tributary region of the v. saphena magna:

- **Dodd veins:** located between the v. saphena magna and v. femoralis at the level of the canalis adductorius.
- **Boyd veins:** located between the v. saphena magna and vv. tibiales posteriores on the medial side of the proximal leg.
- **Cockett veins (I–III):** located between a curved branch of the v. saphena magna behind the malleolus medialis (the v. arcuata cruris posterior) and the vv. tibiales posteriores. The Cockett veins on the medial side of the distal leg are of special clinical importance because of this region's susceptibility to ulceration.

22.3 The Lymphatic Vessels and Lymph Nodes

A The superficial lymphatic system of the right lower limb

a Anterior view, **b** posterior view. (The arrows indicate the main directions of lymphatic drainage.)

The lymph in the lower limb is drained by a superficial (epifascial) system and a deep (subfascial) system, similar to the arrangement in the arm. The largest lymph vessels, called collectors, basically follow the course of the superficial veins (v. saphena magna, v. saphena parva) and deep veins (v. poplitea, v. femoralis) and are interconnected by anastomoses located mostly in the popliteal and inguinal regions. While the superficial lymph vessels (vasa lymphatica superficialia) primarily drain the cutis and subcutis, the deep system drains lymph from the muscles, joints, and nerves. The vasa lymphatica superficialia consist of an anteromedial bundle and a posterolateral bundle. The **anteromedial bundle** runs along the v. saphena magna to the superficial *inguinal* lymph nodes (nll. inguinales superficiales). It drains all of the cutis and subcutis of the lower limb except for the lateral border of the foot and a narrow strip on the calf. Those areas are drained by the **posterolateral bundle** (see **b**), which thus receives drainage from a considerably smaller region. The lymph in the posterolateral bundle first passes along the v. saphena parva to the superficial *popliteal* lymph nodes (nll. popliteales superficiales) and then drains through the nll. popliteales profundi to the deep inguinal lymph nodes (nll. inguinales profundi).

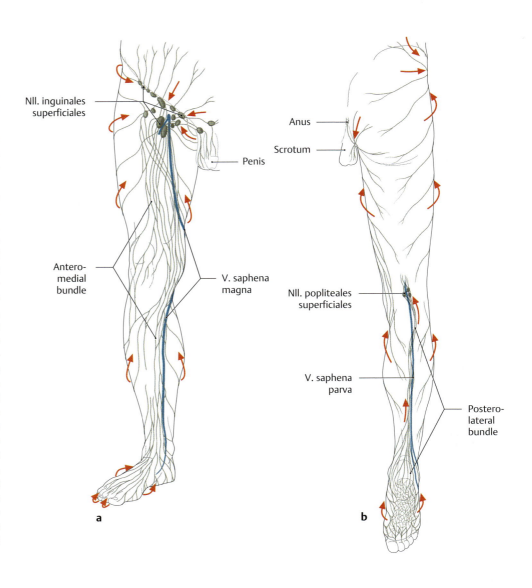

B The deep lymph nodes of the inguinal region

Right inguinal region after removal of the fascia cribrosa about the hiatus saphenus, anterior view. The veins and lymphatic system above the lig. inguinale are shown in light shading. The deep inguinal lymph nodes (nll. inguinales profundi) are located near the termination of the v. saphena magna, medial to the v. femoralis. They are important because *all the lymph from the limb* filters through them before reaching the nll. iliaci. The largest lymph node of this group (*Rosenmüller's lymph node*) is also the highest, placed at the level of the canalis femoralis. The group of nll. pelvis, which includes the nll. iliaci externi, begins just above the lig. inguinale.

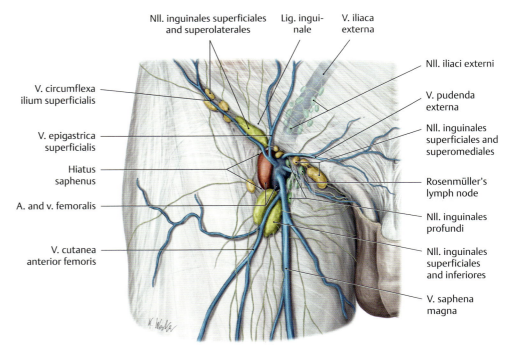

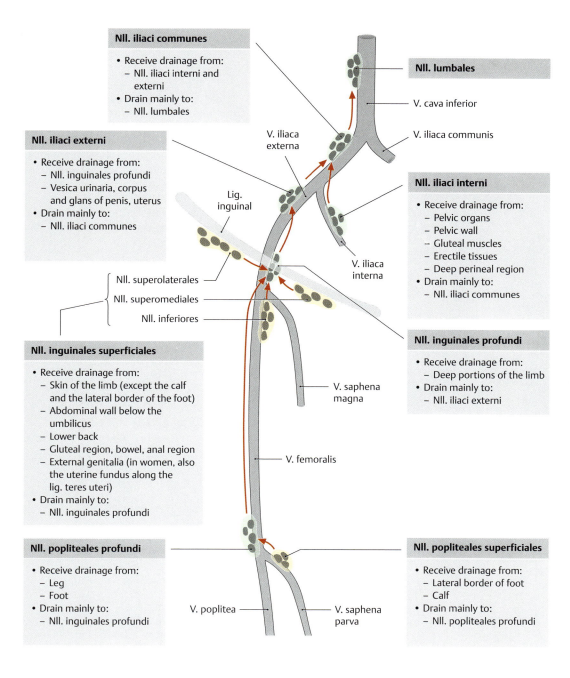

C The lymph node groups and drainage pathways in the lower limb

Right limb, anterior view. The arrows indicate the main directions of lymph flow in the superficial and deep lymphatic systems.

Note: Lymph from the cutis and subcutis of the calf and the lateral border of the foot passes through the nll. popliteales superficiales and profundi along the deep system of lymphatics *directly* to the deep inguinal lymph nodes (nll. inguinales profundi). By contrast, lymph from the rest of the skin of the lower limb first drains through the anteromedial bundle along the v. saphena magna to the superficial inguinal lymph nodes (see also **A**). The *superficial inguinal lymph nodes (nll. inguinales superficiales)*, located on the fascia lata, consist of the following:

- Lymph nodes arranged parallel to the lig. inguinale (nll. inguinales superomediales and superolaterales)
- Lymph nodes distributed vertically along the terminal segment of the v. saphena magna (nll. inguinales inferiores)

These nodes first drain into the deep inguinal lymph nodes (nll. inguinales profundi) (see **B**) and then along the v. iliaca externa to the iliac lymph nodes (nll. iliaci externi and communes), finally reaching the lumbar lymph nodes (nll. lumbales).

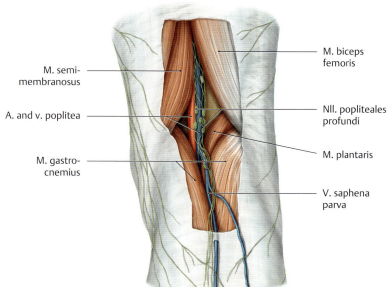

D The deep lymph nodes of the popliteal region (regio genus posterior)

The fossa poplitea of the right leg, posterior view. Lymph from the deep lymphatics of the leg drains (through the nll. popliteales profundi between the posterior knee joint capsule and vasa poplitea) along the v. femoralis and then anteriorly through the hiatus adductorius to the nll. inguinales profundi.

22.4 The Structure of the Plexus lumbosacralis

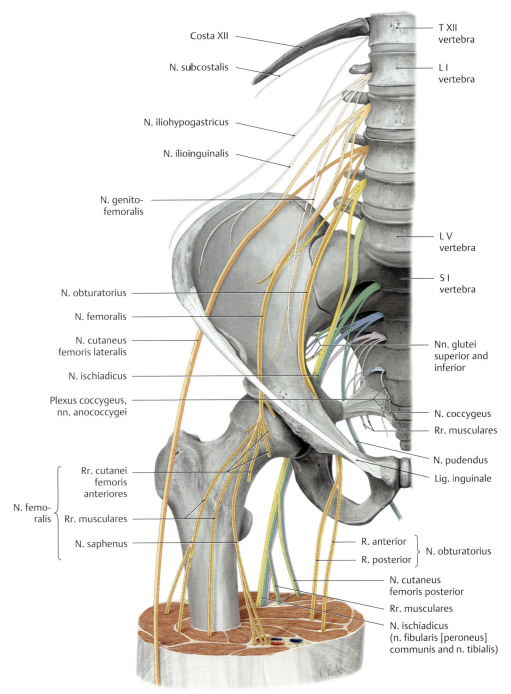

A The plexus lumbosacralis and its branches
Right side, anterior view. For clarity, the muscles of the pelvis and lumbar spine have been removed. Lateral to the foramina intervertebralia of the lumbar spine, the rr. ventrales of the first four *lumbar* spinal nerves (L1–L4) form the plexus lumbalis and pass through the m. psoas major. The smaller rr. musculares are distributed directly to the m. psoas major. The larger branches emerge from the muscle at various sites and pass sharply downward to reach the abdominal wall and thigh, except for the n. obturatorius, which runs down the lateral wall of the lesser pelvis to the thigh. The rr. ventrales of the first four *sacral* spinal nerves (S1–S4) emerge from the foramina sacralia pelvina and unite on the anterior surface of the m. piriformis with the r. ventralis of L5 to form the plexus sacralis. The nerves from the plexus sacralis are distributed to the back of the thigh, the leg, and the foot (after Mumenthaler).

B Spinal cord (medulla spinalis) segments and nerves of the plexus lumbosacralis
The plexus lumbosacralis supplies sensory and motor innervation to the lower limb. It is formed by the rr. ventrales of the lumbar and sacral spinal nerves, with contributions from the n. subcostalis (T 12) and n. coccygeus (Co1) (see **D**). The plexus lumbosacralis is subdivided into the plexus lumbalis and plexus sacralis based on its distribution and topography.

Plexus lumbalis (T12–L4)
- N. iliohypogastricus (L1)
- N. ilioinguinalis (L1)
- N. genitofemoralis (L1, L2)
- N. cutaneus femoris lateralis (L2, L3)
- N. obturatorius (L2–L4)
- N. femoralis (L2–L4)
- Short, direct muscular branches (rr. musculares) to specific hip muscles

Plexus sacralis (L5–S4)*
- N. gluteus superior (L4–S1)
- N. gluteus inferior (L5–S2)
- N. cutaneus femoris posterior (S1–S3)
- N. ischiadicus (L4–S3) with its two large branches:
  - N. tibialis (L4–S3)
  - N. fibularis (peroneus) communis (L4–S2)
- N. pudendus (S2–S4)
- Short, direct muscular branches (rr. musculares) to specific hip muscles

* Often the plexus sacraliss is further subdivided into a plexus ischiadicus and a plexus pudendus. The main branch of the plexus pudendus, the n. pudendus, supplies the skin and muscles of the pelvic floor, perineum, and external genitalia.

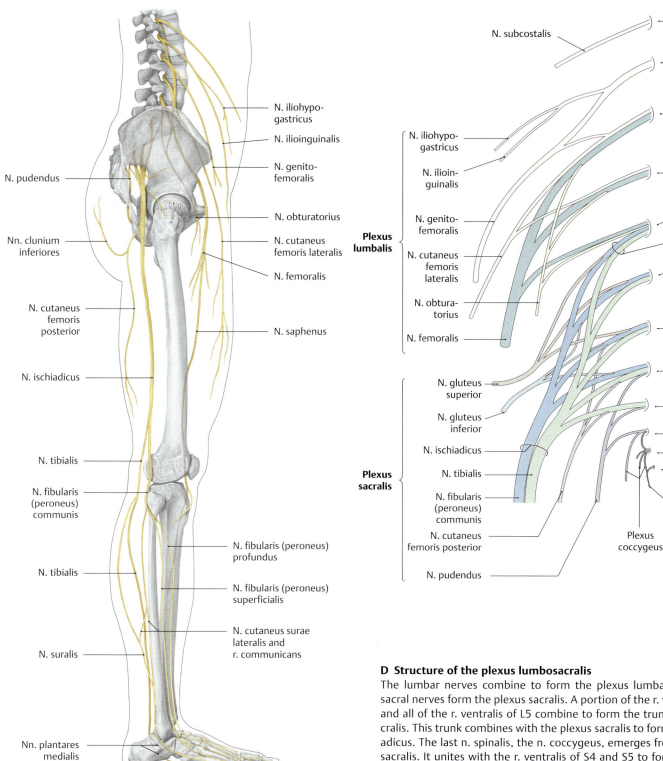

C Topography of the plexus lumbosacralis
Right lower limb, lateral view. The nerves of the plexus lumbalis reach the lower limb *in front of* the art. coxae and mainly supply the *anterior side of the thigh*, while the nerves of the plexus sacralis descend *behind* the art. coxae and innervate the *posterior side of the thigh*, most of the *leg*, and the entire *foot*.

D Structure of the plexus lumbosacralis
The lumbar nerves combine to form the plexus lumbalis, while the sacral nerves form the plexus sacralis. A portion of the r. ventralis of L4 and all of the r. ventralis of L5 combine to form the truncus lumbosacralis. This trunk combines with the plexus sacralis to form the n. ischiadicus. The last n. spinalis, the n. coccygeus, emerges from the hiatus sacralis. It unites with the r. ventralis of S4 and S5 to form the plexus coccygeus (see p. 542).

22.5 The Nerves of the Plexus lumbalis: Nn. iliohypogastricus, ilioinguinalis, genitofemoralis, and cutaneus femoris lateralis

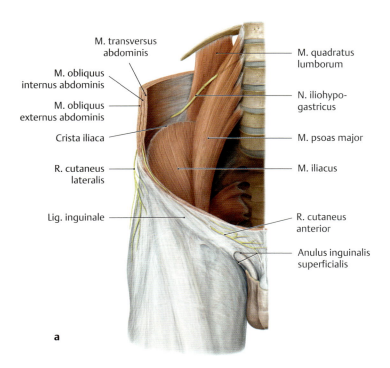
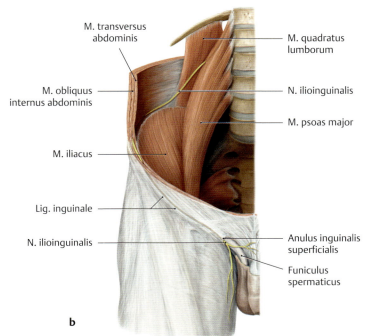

A Course of the nn. iliohypogastricus, ilioinguinalis, genitofemoralis, and cutaneus femoris lateralis after emerging from the plexus lumbalis (after Mumenthaler).
Right lateral and posterior abdominal wall region, anterior view.

a The **n. iliohypogastricus** generally emerges with the n. ilioinguinalis (see **b**) at the lateral border of the m. psoas major and runs laterally and obliquely on the anterior surface of the m. quadratus lumborum. Approximately 3 to 4 cm past the lateral border of that muscle, it pierces the m. transversus abdominis and runs anteriorly above the crista iliaca, passing between the m. transversus abdominis and m. obliquus internus abdominis. After giving off several rr. musculares to both of these muscles and a sensory r. cutaneus lateralis to the skin of the lateral hip region, the terminal branch of the n. iliohypogastricus courses medially, running parallel to the lig. inguinale. Above the external inguinal ring (anulus inguinalis superficialis) it pierces the aponeurosis of the m. obliquus externus abdominis and supplies a sensory r. cutaneus anterior to the skin above the lig. inguinale.

b The **n. ilioinguinalis** generally courses with the n. iliohypogastricus (see **a**) on the m. quadratus lumborum but soon separates from it and runs at the level of the crista iliaca to the lateral abdominal wall, which it pierces at a variable location. It runs medially at the level of the lig. inguinale between the mm. transversus abdominis and obliquus internus abdominis, supplying twigs to both muscles, and it distributes sensory fibers through the anulus inguinalis superficialis to the skin over the symphysis and to the lateral portion of the labia majora or scrotum.

B Overview of the nerves of the plexus lumbalis

Nerve	Segment	Innervated muscles	Cutaneous branches (to the region receiving sensory innervation, see **C** and pp. 534 and 535)
• N. iliohypogastricus	L1	• M. transversus abdominis, m. obliquus internus abdominis (the inferior portions of each)	• R. cutaneus anterior • R. cutaneus lateralis
• N. ilioinguinalis	L1	• M. transversus abdominis, m. obliquus internus abdominis (the inferior portions of each)	• Nn. scrotales anteriores in males, nn. labiales anteriores in females
• N. genitofemoralis	L1, L2	• M. cremaster in males (r. genitalis)	• R. genitalis, r. femoralis
• N. cutaneus femoris lateralis	L2, L3		• N. cutaneus femoris lateralis
• N. obturatorius (see p. 534) – R. anterior – R. posterior	L2–L4	• M. obturatorius externus • M. adductor longus, m. adductor brevis, m. gracilis, m. pectineus • M. adductor magnus	• R. cutaneus
• N. femoralis (see p. 535)	L2–L4	• M. iliopsoas, m. pectineus, m. sartorius, m. quadriceps femoris	• Rr. cutanei anteriores, n. saphenus
• Short, direct muscular branches (rr. musculares) (see p. 534)	T12–L4	• M. psoas major, m. quadratus lumborum, m. iliacus, mm. intertransversarii lumborum	

Lower Limb — 22. Neurovascular Systems: Forms and Relations

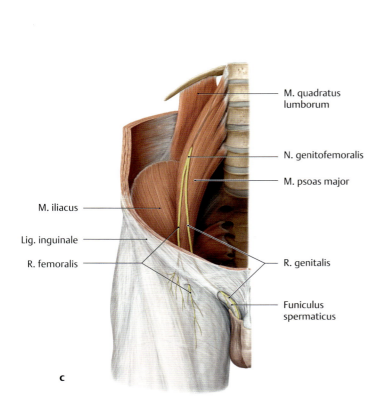

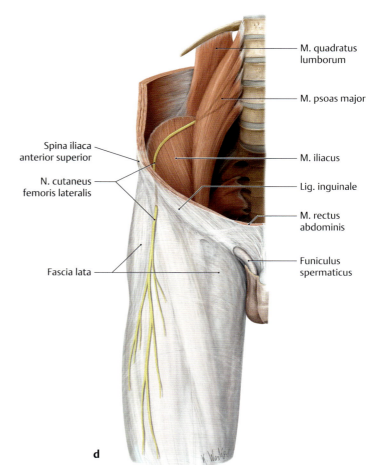

c The **n. genitofemoralis** pierces the m. psoas major and descends upon its anterior surface, dividing into its two terminal branches: the r. genitalis and r. femoralis:

- The *purely sensory r. femoralis* pierces the lacuna vasorum in the area of the hiatus saphenus (see p. 549) and becomes superficial, supplying the skin below the lig. inguinale in both sexes.
- The *mixed r. genitalis* runs in the funiculus spermaticus in males. In females it initially passes through the canalis inguinalis accompanied by the lig. teres uteri. In its further course it distributes sensory fibers to the scrotal skin in males and to the skin of the labia majora in females. It also supplies motor fibers to the m. cremaster in males (see p. 176).

d The **n. cutaneus femoris lateralis** emerges from the lateral border of the m. psoas major and runs obliquely downward and laterally beneath the m. iliacus fascia toward the spina iliaca anterior superior. Medial to the spina iliaca, the nerve leaves the pelvis through the lateral lacuna musculorum (see p. 549) and first runs beneath the fascia lata and then upon it to the skin of the anterior thigh, piercing the fascia approximately 2 to 3 cm below the spina iliaca anterior superior. The nerve is susceptible to occasional mechanical injury at its site of emergence from the pelvis below the lig. inguinale, as it makes an approximately 80° angle at that site and is susceptible to stretching, especially on extension of the hip. The nerve also has only scant coverage by fatty tissue at that location. Stretch injuries are manifested by sensory disturbances (paresthesias) or pain in the *lateral* part of the thigh.

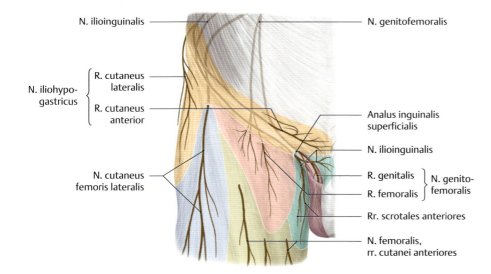

C Sensory innervation of the inguinal region (regio inguinalis) and thigh
Right regio inguinalis of a male, anterior view. The territories of the various sensory nerves are indicated by different colors.
Note: Both the n. ilioinguinalis and the r. genitalis of the n. genitofemoralis pass through the anulus inguinalis superficialis. The two nerves are frequently confused. The r. genitalis in males is located by first opening the funiculus spermaticus. In females, the r. genitalis accompanies the lig. teres uteri to supply the skin of the labia majora (see also **Ac**).

22.6 The Nerves of the Plexus lumbalis: Nn. obturatorius and femoralis

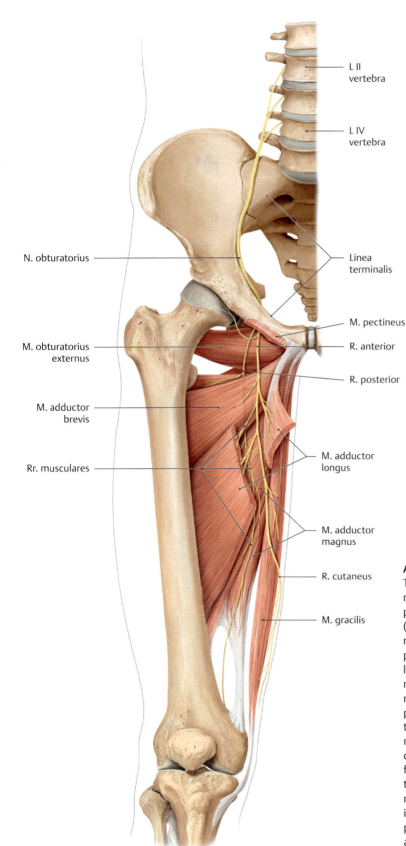

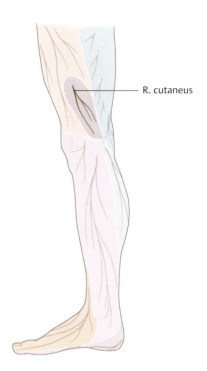

B Sensory distribution of the n. obturatorius
Right leg, medial view.

A Course of the n. obturatorius
The right regio inguinales and thigh, anterior view. The n. obturatorius receives fibers from the L 2–L 4 spinal segments. After leaving the plexus lumbalis, it descends behind and medial to the m. psoas major (not shown here) toward the lesser pelvis and enters the canalis obturatorius (not shown here, see p. 558) below the linea terminalis, accompanied by the vasa obturatori. Farther distally it distributes rr. musculares to the m. obturatorius externus and subsequently divides into an r. anterior and a r. posterior. These branches continue distally, passing respectively anterior and posterior to the m. adductor brevis, and supply motor innervation to the rest of the adductor muscles (mm. pectineus, adductor longus, adductor brevis, adductor magnus, adductor minimus, and gracilis). The r. anterior gives off a terminal, sensory r. cutaneus at the anterior border of the m. gracilis, which pierces the fascia lata to supply a palm-sized area of skin on the medial view of the distal thigh. In evaluating motor deficits that are associated with n. obturatorius injuries (e.g., intrapartum or due to pelvic fractures), it is important to know that the n. femoralis contributes to the supply of the m. pectineus, while the n. ischiadicus helps to supply the m. adductor magnus.

Lower Limb — 22. Neurovascular Systems: Forms and Relations

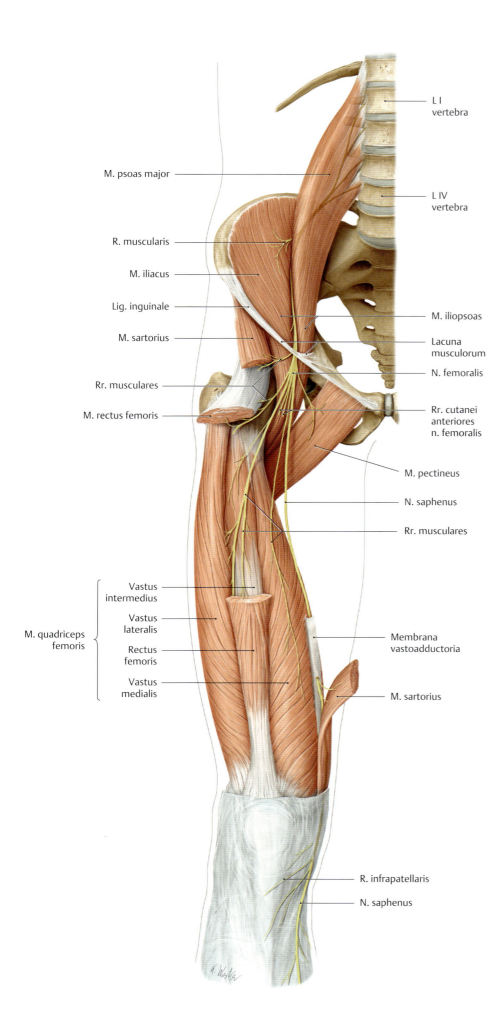

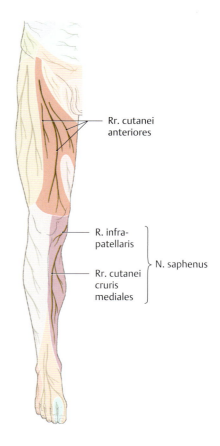

D Sensory distribution of the n. femoralis
Right leg, anterior view.

C Course of the n. femoralis
The right regio inguinalis and thigh, anterior view. As the largest and longest nerve of the plexus lumbalis, the n. femoralis receives fibers from the second through fourth lumbar segments of the spinal cord. It supplies *motor* innervation to the mm. iliopsoas, pectineus, sartorius, and quadriceps femoris and *sensory* innervation to the skin of the anterior thigh, medial leg, and hindfoot (see **D**). The nerve, under cover of the m. psoas fascia (pars psoatica fasciae iliopsoaticae), runs in a groove between the m. psoas major and m. iliacus to the medial lacuna musculorum while giving off branches to both muscles. Approximately 8 cm below the lig. inguinalis, the n. femoralis divides divides into numerous cutaneous branches (rr. cutanei anteriores) and muscular branches (rr. musculares), as well as a long, terminal sensory branch that continues to the foot, the n. saphenus. Initially, the n. saphenus enters the canalis adductorius with the vasa femoralia (below the membrana vastoadductoria) but then leaves the canal through the membrana vastoadductoria and passes with the m. sartorius toward the medial side of the knee. After giving off a sensory r. infrapatellaris to the skin of the medial knee, it follows the v. saphena magna to the skin of the medial leg and foot.

22.7 The Nerves of the Plexus sacralis: N. gluteus superior, N. gluteus inferior, and N. cutaneus femoris posterior

A Nerves of the plexus sacralis (part I)
(For nerves of the plexus sacralis, parts II and III, see pp. 538 and 540.)

Nerve	Segment	Innervated muscles	Cutaneous branches
• N. gluteus superior	L4–S1	• M. gluteus medius • M. gluteus minimus • M. tensor fasciae latae	
• N. gluteus inferior	L5–S2	• M. gluteus maximus	

• N. cutaneus femoris posterior	S1–S3		• N. cutaneus femoris posterior
			  – Nn. clunium inferiores
			  – Rr. perineales (see **F** for sensory distribution)

• *Direct branches from the plexus:*			
– N. musculi piriformis	S1, S2	• M. piriformis	
– N. musculi obturatorii interni	L5, S1	• M. obturatorius internus • Mm. gemelli	
– N. musculi quadrati femoris	L5, S1	• M. quadratus femoris	

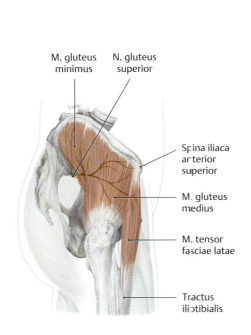

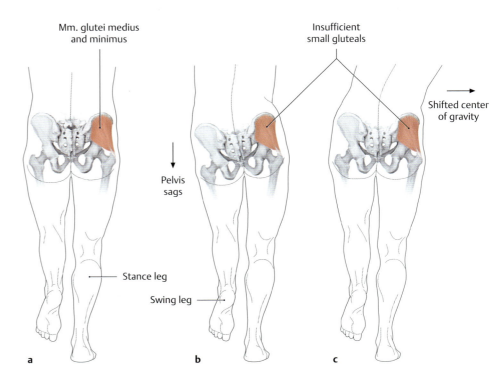

B Motor distribution of the n. gluteus superior
Right hip region, lateral view. Accompanied by blood vessels with the same name, the n. gluteus superior leaves the pelvis minor through the foramen suprapiriforme in foramen ischiadicum majus (see p. 554), runs in the spatium intergluteale, and supplies motor fibers to the small gluteal muscles (mm. glutei medius and minimus) and m. tensor fasciae latae.

C Clinical indicators of small gluteal muscle weakness: the Trendelenburg sign and the Duchenne limp
Lower half of body, posterior view.

a In normal one-legged stance, the small gluteal muscles on the stance side can stabilize the pelvis in the coronal plane.
b Weakness or paralysis of the small gluteal muscles (e.g., due to a faulty intramuscular injection causing damage to the n. gluteus superior) is manifested by weak abduction of the affected hip joint and an inability to stabilize the pelvis in the coronal plane. In a positive Trendelenburg test, the pelvis sags toward the *normal* unsupported side.
c Tilting the upper body toward the affected side shifts the center of gravity onto the stance side, thereby elevating the pelvis on the swing side (*Duchenne limp*). With bilateral loss of the small gluteals, the patient exhibits a typical waddling gait.

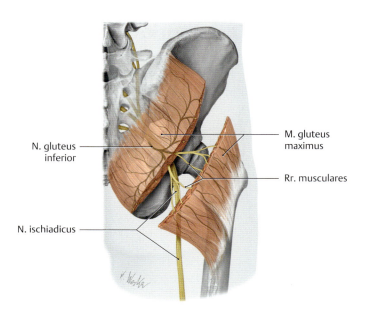
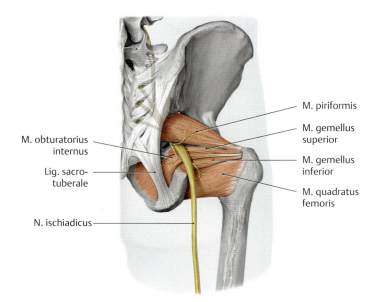

D Motor distribution of the n. gluteus inferior
Right half of the pelvis, posterior view. The n. gluteus inferior leaves the lesser pelvis with the n. ischiadicus through the greater sciatic foramen infrapiriforme (see p. 554) and supplies numerous rr. musculares to the m. gluteus maximus. Paralysis of the m. gluteus maximus causes little impairment of normal gait on even ground because the deficit is well compensated by the hamstrings (see p. 484). The affected patient is unable to run, jump, or climb stairs, however.

E Muscles that are supplied by direct branches from the plexus sacralis
Right half of the pelvis, posterior view. The direct branches of the plexus sacralis are listed in **A**.

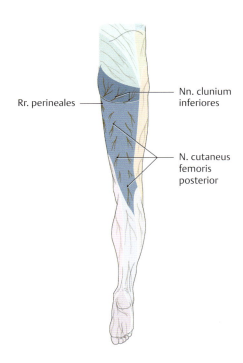
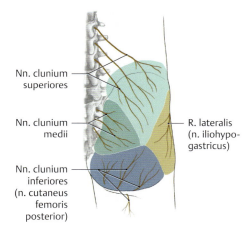
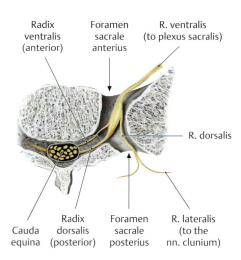

F Sensory distribution of the n. cutaneus femoris posterior
Right lower limb, posterior view. Besides the skin of the posterior thigh, the n. cutaneus femoris posterior distributes several branches to the skin of the gluteal sulcus (nn. clunium inferiores), and its rr. perineales supply the skin of the perineal region (darker shading indicates the exclusive area).

G Sensory innervation of the gluteal region
Right buttock, posterior view.
The gluteal region (regio glutealis) receives sensory innervation from portions of the Plexus sacralis and plexus lumbalis (rr. ventrales of the nn. spinales) and also from rr. dorsales:

- From the plexus sacralis: nn. clunium inferiores (from the n. cutaneus femoris posterior).
- From the plexus lumbalis: the r. lateralis of the n. iliohypogastricus.
- From rr. dorsales of the nn. spinales: the nn. clunium superiores (rr. dorsales of L1–L3) and clunium medii (rr. dorsales of S1–S3).

H An emerging sacral nerve
Horizontal section through the right half of the os sacrum at the level of the foramina sacralia. While the r. ventralis of a sacral nerve emerges from the os sacrum through an *anterior* foramen sacralis, the corresponding r. dorsalis passes through the *posterior* foramen sacralis to supply the skin of the buttock.

22.8 The Nerves of the Plexus sacralis: The N. ischiadicus (Overview and Sensory Distribution)

A Nerves of the plexus sacralis (part II)
The largest and longest of the peripheral nerves, the n. ischiadicus, leaves the lesser pelvis through the foramen infrapiriforme and passes below the m. gluteus maximus to the back of the thigh. It divides into its two main branches, the n. tibialis and n. fibulari communis, at a variable level but generally before entering the fossa poplitea. The *rr. musculares* of the n. ischiadicus, however, can already be identified as consisting of a *fibular part* (Fib) and a *tibial part* (Tib) while still proximal to the bifurcation (see also p. 540). Injuries of the n. ischiadicus may be caused by compression of the nerve at its emergence inferior to the m. piriformis (usually by extrinsic pressure, such as sitting). Other potential causes are misdirected intramuscular injections (in which the nerve is accidentally pricked), pelvic fractures, and surgical procedures (e.g., hip replacement).

Nerve	Segment	Innervated muscles	Cutaneous branches

N. ischiadicus	L4–S3	• M. semitendinosus (Tib)
		• M. semimembranosus (Tib)
		• M. biceps femoris
		  – Caput longum (Tib)
		  – Caput breve (Fib)
		• M. adductor magnus (Tib), medial part	

N. fibularis (peroneus) communis	L4–S2		• N. cutaneus surae lateralis • R. communicans fibularis
– N. fibularis (peroneus) superficialis		• M. fibularis (peroneus) longus • M. fibularis (peroneus) brevis	• N. cutaneus dorsalis medialis • N. cutaneus dorsalis intermedius
– N. fibularis (peroneus) profundus		• M. tibialis anterior • M. extensor digitorum longus • M. extensor digitorum brevis • M. extensor hallucis longus • M. extensor hallucis brevis • M. fibularis tertius	• N. cutaneus hallucis lateralis • N. cutaneus digiti secundi medialis
• N. tibialis	L4–S3	• M. triceps surae • M. plantaris • M. popliteus • M. tibialis posterior • M. flexor digitorum longus • M. flexor hallucis longus	• N. cutaneus surae medialis (→ n. suralis) • Rr. calcanei laterales • Rr. calcanei mediales • N. cutaneus dorsalis lateralis
– N. plantaris medialis		• M. abductor hallucis • M. flexor digitorum brevis • M. flexor hallucis brevis, caput mediale • M. lumbricalis I	• Nn. digitales plantares proprii
– N. plantaris lateralis		• M. flexor hallucis brevis, caput laterale • M. quadratus plantae • M. abductor digiti minimi • M. flexor digiti minimi brevis • M. opponens digiti minimi • Mm. lumbricales II + IV • Mm. interossei plantares I–III • Mm. interossei dorsales I–IV • M. adductor hallucis	• Nn. digitales plantares proprii

→ = is continuous with

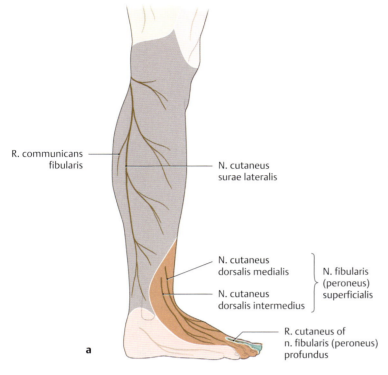
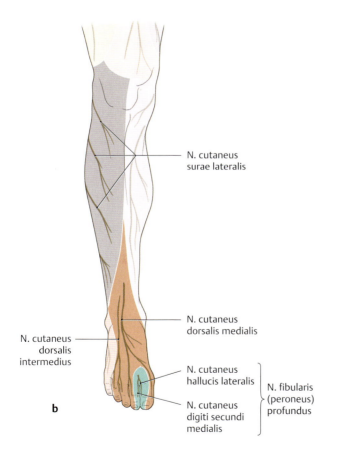
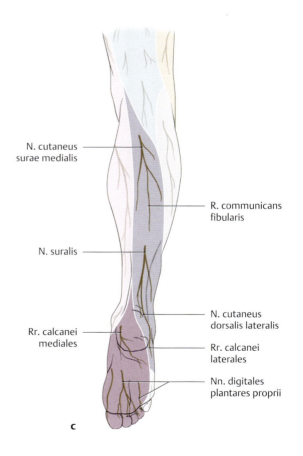

B Sensory distribution of the n. ischiadicus
Right leg. **a** Lateral view, **b** anterior view, **c** posterior view.

22.9 The Nerves of the Plexus sacralis: N. ischiadicus (Course and Motor Distribution)

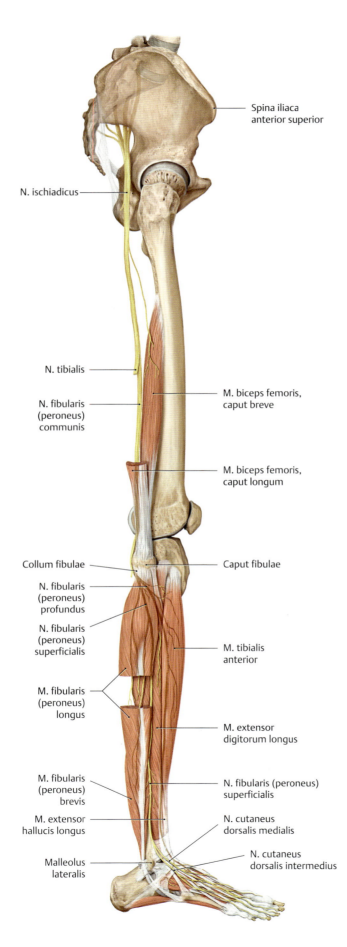

A Course and motor distribution of the n. ischiadicus: the fibular part (n. fibularis (peroneus) communis)

Right lower limb, lateral view. After giving off several rr. musculares from its fibular part (to the caput breve of m. biceps femoris), the n. ischiadicus consistently divides in the lower third of the thigh into the n. tibialise and n. fibularis (peroneus) communis. The n. fibularis (peroneus) communis then follows the medial border of the m. biceps femoris to the caput fibulae and winds around the collum fibulae to the front of the leg. Immediately after entering the m. fibularis (peroneus) longus, it divides into its two terminal branches, the n. fibularis (peroneus) profundus and n. fibularis (peroneus) superficialis. The n. fibularis (peroneus) superficialis supplies the mm. fibulares and runs between the m. fibularis (peroneus) longus and fibula to the dorsum of the foot. The n. fibularis (peroneus) profundus runs through the membrana interossea to enter the extensor compartment. After supplying the mm. tibialis anterior, extensor digitorum longus, and extensor hallucis longus, it runs in a groove between the m. tibialis anterior and m. extensor hallucis longus on the membrana interossea cruris, accompanied by the vasa tibialia anteriora, to the dorsum of the foot.

- If the nerve is damaged at the level of the collum fibulae (a very exposed location!) before dividing into its two terminal branches, the result is weakness or paralysis of the anterior and lateral compartment muscles, resulting in foot drop with some inversion.
- If the nerve is damaged after dividing into its terminal branches, the result may be an isolated weakness or paralysis of the anterior compartment or the lateral muscles, depending on whether the n. fibularis (peroneus) profundus or the n. fibularis (peroneus) superficialis is affected. Accordingly, the result may be weak dorsiflexion or weakness of eversion. An isolated lesion of the n. fibularis (peroneus) superficialis generally affects only the sensory terminal branch, with pain involving the distal leg and dorsum of the foot. Gait disturbance will occur only with an isolated lesion of the n. fibularis (peroneus) profundus (as in compartment syndrome caused by anterior compartment hemorrhage, see p. 567), resulting in foot drop and a "steppage gait." Increased flexion of the hip and knee joints is necessary to keep the toe from dragging the ground during the swing phase of gait.

Lower Limb — 22. Neurovascular Systems: Forms and Relations

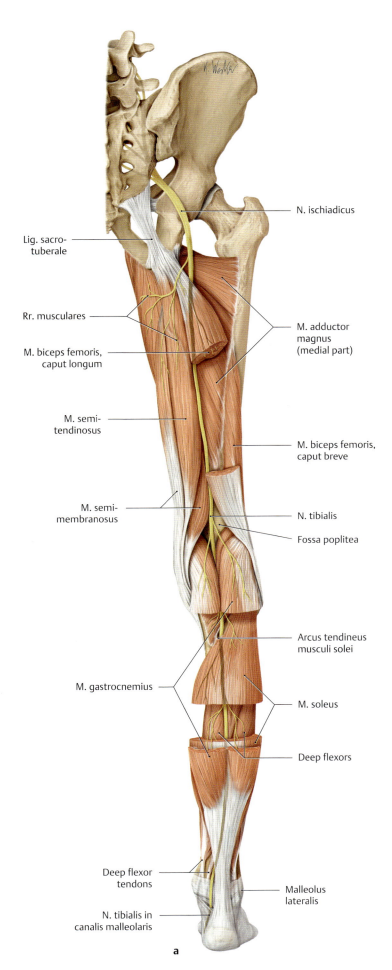
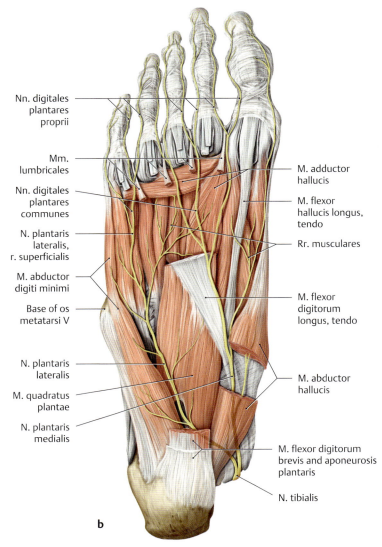

B Course and motor distribution of the n. ischiadicus: the tibial part (n. tibialis)

a Right lower limb, posterior view; **b** right foot, plantar view. While still in the thigh, the tibial part of the n. ischiadicus distributes several rr. musculares to the mm. semitendinosus, semimembranosus, biceps femoris (caput longum), and adductor magnus (medial part). After division of the n. ischiadicus, the n. tibialis continues straight down through the center of the fossa poplitea and runs below the arcus tendineus musculi solei to the superficial and deep plantar flexors, which it supplies. In the deep posterior compartment, the n. tibialis continues distally in a neurovascular bundle with the vasa tibialia posteriora (not shown here) and passes through the canalis tarsalis, accompanied by the deep flexor tendons, to the plantar side of the foot (**b**). In passing through the canalis tarsalis, the n. tibialis divides into its two terminal branches (nn. plantares lateralis and medialis), which supply all the muscles on the plantar side of the foot. Compression of the n. tibialis or its terminal branches at this site leads to an entrapment syndrome (= tarsal tunnel syndrome). This can result in pain and sensory disturbances affecting the sole of the foot or even palsies of the intrinsic foot muscles, particularly following severe nerve trauma in connection with a fracture of the tibial shaft or malleolus medialis.

22.10 The Nerves of the Plexus sacralis: N. pudendus and N. coccygeus

A Nerves of the plexus sacralis (part III)
The n. pudendus, the lowest branch of the plexus sacralis, arises from a separate small plexus formed by the rr. ventrales of S1–S4; hence it is occasionally referred to as the *plexus pudendus*.

Nerve	Segment	Innervated muscles	Cutaneous branches
• N. pudendus (plexus pudendus)	S2–S4	• Pelvic floor muscles – M. levator ani – M. transversus perinei superficialis – M. transversus perinei profundus – M. bulbospongiosus – M. sphincter ani externus – M. sphincter urethrae	• Nn. rectales inferiores • Nn. perineales – Nn. labiales posteriores in females – Nn. scrotales posteriores in males – N. dorsalis clitoridis in females
• N. coccygeus (plexus coccygeus)	S5–Co2	• M. coccygeus	• Nn. anococcygei (rr. ventrales) • Rr. dorsales

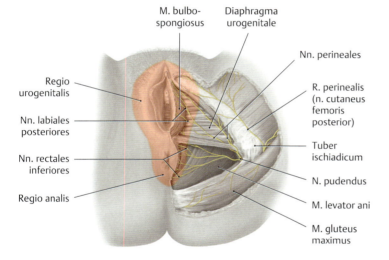

B The cutaneous branches of the n. pudendus and its sensory distribution in the female
Lithotomy position, inferior view. The skin layers have been removed on the left side to demonstrate the terminal branches of the n. pudendus in the fossa ischioanalis (see p. 556). The area of cutaneous sensory innervation is color-shaded. Large portions of the regio urogenitalis and regio analis receive their sensory supply from the n. pudendus. The skin area supplied by the n. pudendus can be anesthetized during childbirth by infiltration anesthesia or a nerve block, enabling the obstetrician to perform and repair an episiotomy without pain (see p. 232). This can be done either by infiltrating the perineum between the anus and posterior fornix (fornix vaginae, pars posterior) with a local anesthetic or by temporarily blocking the n. pudendus with local anesthetic injected near the spina ischiadica (before the nerve has branched; see diagram in **C**).

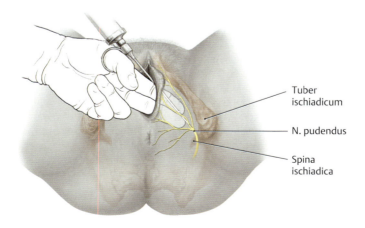

C Technique of a left-sided n. pudendus block
Lithotomy position, inferior view. The most common type of conduction anesthesia used in vaginal deliveries is the n. pudendus block, which renders the perineum, vulva, and lower third of the vagina insensitive to pain. In the *transvaginal approach*, a special guide cannula is introduced into the vagina, and 10 mL of a local anesthetic solution is injected approximately 1 cm above and 1 cm lateral to the palpable spina ischiadica on each side. An injection at this site will block the N. pudendus *before* it has entered the canalis pudendalis (Alcock's canal) and *before* it has divided into its terminal branches. Often the nerve block is administered at the end of the expulsion stage to relieve stretch pain in the perineal region (see p. 232).

Lower Limb — 22. Neurovascular Systems: Forms and Relations

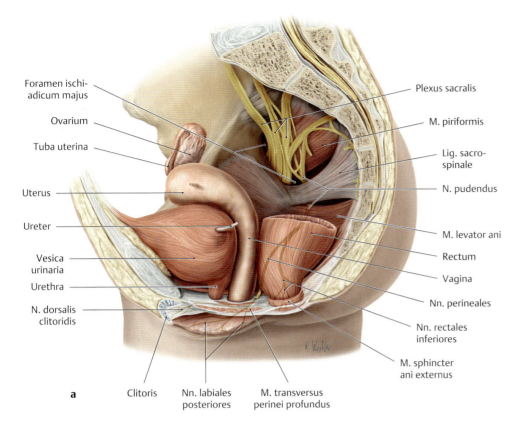

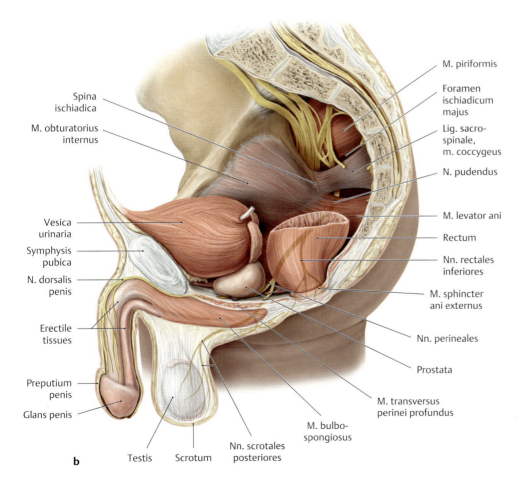

D Course of the n. pudendus and n. coccygeus in the female and male
a Sagittal section through the female pelvis, left lateral view.
b Sagittal section through the male pelvis, left lateral view.

The **n. pudendus** emerges from the lesser pelvis through the foramen ischiadicum *majus*. It then courses around the spina ischiadica and lig. sacrospinale and passes through the foramen ischiadicum *minus* into the fossa ischioanalis (ischiorectalis) (see p. 556). It runs forward in the lateral wall of the fossa, embedded in a duplication of the m. obturatorius internus fascia (canalis pudendalis = Alcock's canal) and accompanied by the vasa pudenda interna (see p. 555). Below the symphysis it passes to the dorsum of the penis or the clitoris. The n. pudendus gives off numerous *branches* from within the perineum:

- The *nn. rectales inferiores* supply motor innervation to the m. sphincter ani externusand sensory innervation to the skin around the anus.
- The *nn. perineales* distribute motor branches to the muscles of the perineum (see p. 162) and sensory branches to the skin of the posterior scrotum or labia majora and minora, to the skin of the penis or clitoris, and to the glans of the penis or clitoris, prepuce, and erectile tissues.

Damage to the n. pudendus (e.g., resulting from perineal injuries during childbirth) leads to loss of function of the muscles of the perineum, especially the sphincter muscles of the bladder and bowel, causing urinary and fecal incontinence. N. pudendus lesions can also lead to sexual dysfunction (e.g., male impotence).

The rr. ventrales of the fifth sacral nerve and the first or second n. coccygeus form the **n. coccygeus** (synonym: *plexus coccygeus*). This nerve and its terminal sensory branches, the nn. anococcygei, pass along the lig. anococcygeum to supply the skin between the os coccygis and anus.

23.1 Surface Anatomy and Superficial Nerves and Vessels: Anterior View

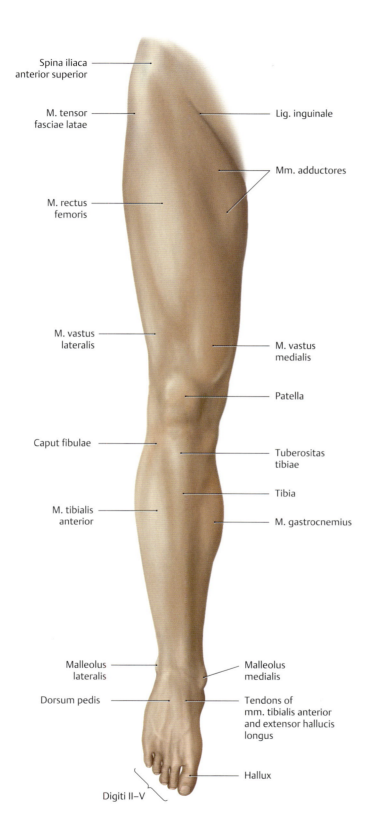

A Surface anatomy of the right lower limb
Anterior view. Palpable *bony prominences* on the lower limb are reviewed on p. 409.

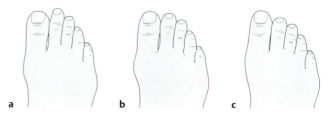

B The most common variants of the forefoot and toes (after Debrunner and Lelievre)
Three types of foot shape are distinguished based on the relative lengths of the first and second toes:

a The "Greek" type, in which the second toe is longer than the first.
b The square type, in which the first and second toes are of equal length.
c The "Egyptian" type, in which the first toe is longer than the second.
In the "Greek" type, the os metatarsi II of the second toe is generally longer than the os metatarsi of the first toe (great toe; hallux). As a result, the caput of the second os metatarsale of the second toe is often subject to painful overloading, especially when high heels are worn.

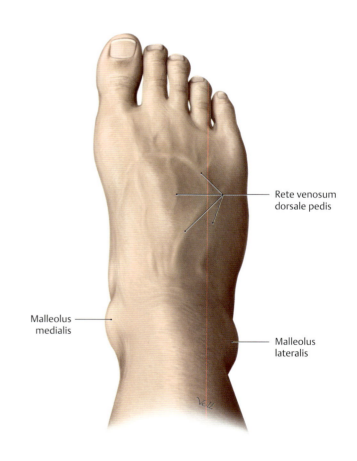

C The dorsum pedis of the right foot
The rete venosum dorsale pedis is visible on the dorsum of the foot (compare with **D**).

Lower Limb — 23. Neurovascular Systems: Topographical Anatomy

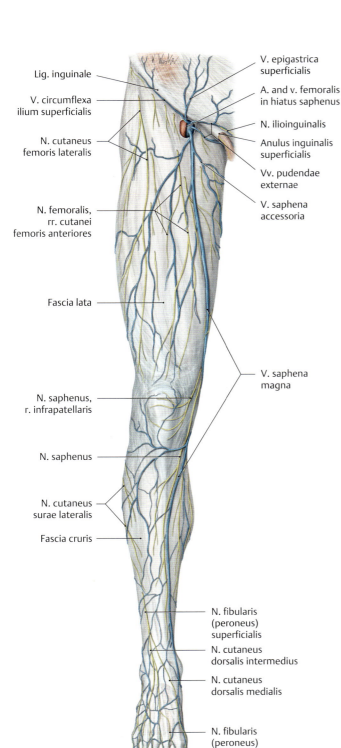

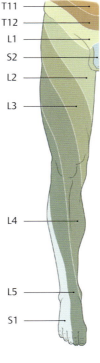

E Radicular (segmental) cutaneous innervation pattern (dermatomes) in the right lower limb

Anterior view. As in the arm, the outgrowth of the lower limb during development causes the sensory cutaneous segments to become elongated and drawn out into narrow bands. The L4, L5, and S1 segments in particular move so far peripherally that they no longer have any connection with the corresponding segments of the trunk.

Note that the dermatomes of the lumbar trunk segments lie mostly on the front of the leg, while those of the sacral segments are mostly on the back of the leg (see p. 86). This can be of diagnostic importance in patients with a herniated disk, for example, in order to determine the level of the herniation (after Mumenthaler).

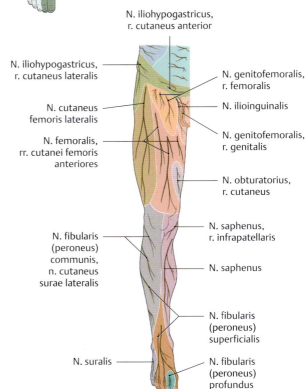

D Superficial cutaneous veins and nerves of the right lower limb

Anterior view. The dorsal venous network of the foot (rete venosum dorsale pedis) is drained by two large venous trunks (v. saphena magna and v. saphena parva), which receive a variable pattern of cutaneous veins. While the v. saphena parva (see p. 547) enters the v. poplitea at the level of the fossa poplitea, the v. saphena magna extends up the medial side of the leg to a point just below the lig. inguinale, where it passes through the hiatus saphenus of the fascia lata to enter the v. femoralis. The superficial veins of the lower limb are commonly affected by varicosity, causing them to become thickened, tortuous, and distinctly visible and palpable (see also p. 527).

F Peripheral sensory cutaneous innervation pattern in the right lower limb

Anterior view. As in the arm, the sensory distribution in the lower limb corresponds to the branching patterns of the peripheral cutaneous nerves (rr. cutanei) in the subcutaneous connective tissue. The territories of the individual peripheral nerves overlap, especially at their margins. Hence the clinically determined *exclusive area* of a particular cutaneous nerve (the area supplied by that nerve alone) tends to be considerably smaller than the *maximum area* that can be demonstrated anatomically. For this reason, the traumatic disruption of a nerve causes a complete loss of sensation (anesthesia) in the exclusive area but often will cause only diminished sensation (hypoesthesia) at the perimeter of that area.

Note that the sensory loss resulting from a peripheral nerve injury presents a completely different pattern from that caused by injury to a nerve root (see p. 88) (after Mumenthaler).

23.2 Surface Anatomy and Superficial Nerves and Vessels: Posterior View

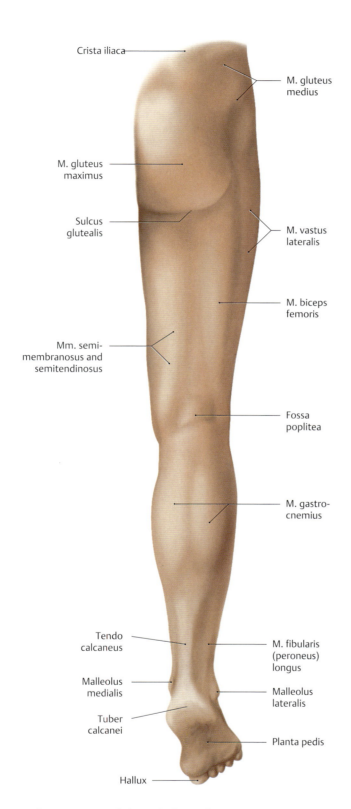

A Surface anatomy of the right lower limb
Posterior view. The foot is in plantar flexion. (Palpable *bony prominences* on the lower limb are reviewed on p. 409.)

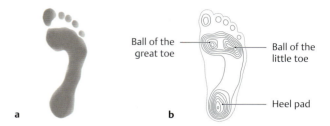

B Footprints (podograms) of the normal right foot in an adult
A podogram provides a graphic representation of the loads borne by the foot. Besides visual inspection of the sole of the foot, analysis of the podogram supplies the most useful information on the weight-bearing dynamics of the foot.

a Footprint created with an ink pad.
b Pressure podogram showing a normal weight-bearing pattern on the foot. The concentric lines indicate that the pressure is evenly distributed over all the major points of support. These three areas are clearly defined, while the intervening plantar arches bear essentially no weight (see p. 464).

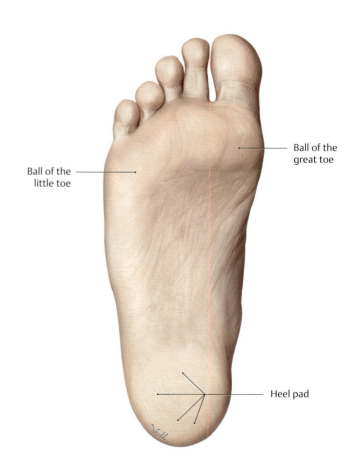

C The planta pedis of the right foot
The skin on the plantar surface of the foot serves as a sensory organ for contact with the ground, perceiving its consistency during stance and locomotion by means of receptors in the sole (planta) of the foot. Stresses acting on the heel pad and the balls of the great and little toes generate high local compressive forces at those sites, to which the subcutaneous connective tissue has adapted functionally by developing a system of *pressure chambers* (see p. 472).

Lower Limb — 23. Neurovascular Systems: Topographical Anatomy

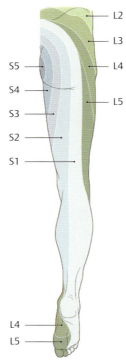

E Radicular (segmental) cutaneous innervation pattern (dermatomes) of the right lower limb

Posterior view, foot in plantar flexion. As in the arm, the outgrowth of the lower limb during development causes the sensory cutaneous segments to become elongated into narrow bands. The L4, L5, and S1 segments in particular move so far peripherally that they no longer have any connection with the corresponding segments of the trunk.

Note that the dermatomes of the lumbar trunk segments lie mostly on the front of the leg, while those of the sacral segments are mostly on the back of the leg (see p. 86). This can be of diagnostic importance in patients with a herniated disk, for example, in order to determine the level of the herniation (after Mumenthaler).

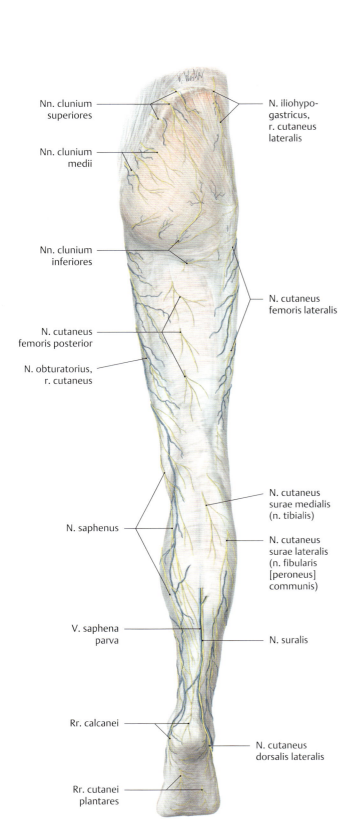

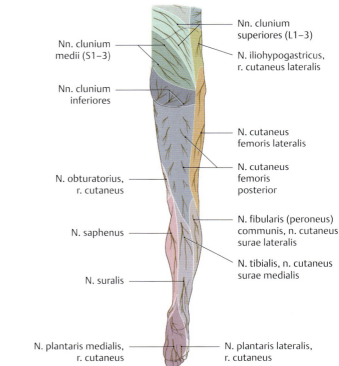

D The superficial (epifascial) cutaneous veins and nerves of the right lower limb
Posterior view.

F Peripheral sensory cutaneous innervation pattern in the right lower limb

Posterior view, foot in plantar flexion. As in the arm, the sensory distribution in the lower limb corresponds to the branching patterns of the peripheral cutaneous nerves (rr. cutanei) in the subcutaneous connective tissue. The territories of the individual peripheral nerves overlap, especially at their margins. Hence the clinically determined *exclusive area* of a particular cutaneous nerve (the area supplied by that nerve alone) tends to be considerably smaller than the *maximum area* that can be demonstrated anatomically. For this reason, the traumatic disruption of a nerve causes a complete loss of sensation (anesthesia) in the exclusive area but often will cause only diminished sensation (hypoesthesia) at the perimeter of that area.

Note that the sensory loss resulting from a peripheral nerve injury presents a completely different pattern from that caused by injury to a nerve root (see p. 88) (after Mumenthaler).

23.3 The Anterior Femoral Region (Regio femoralis anterior) Including the Trigonum femorale

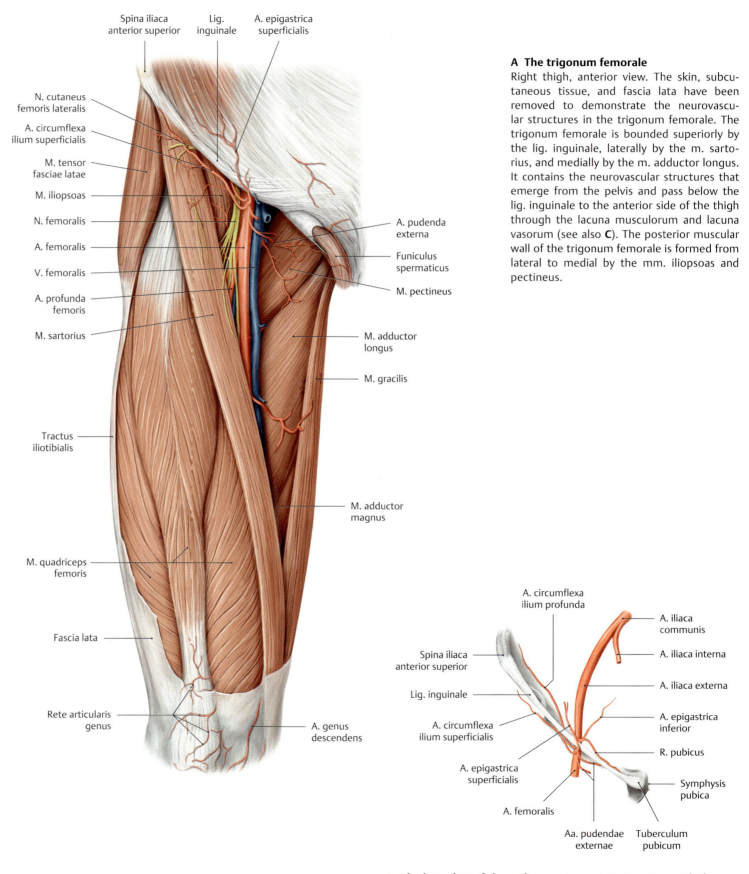

A The trigonum femorale
Right thigh, anterior view. The skin, subcutaneous tissue, and fascia lata have been removed to demonstrate the neurovascular structures in the trigonum femorale. The trigonum femorale is bounded superiorly by the lig. inguinale, laterally by the m. sartorius, and medially by the m. adductor longus. It contains the neurovascular structures that emerge from the pelvis and pass below the lig. inguinale to the anterior side of the thigh through the lacuna musculorum and lacuna vasorum (see also **C**). The posterior muscular wall of the trigonum femorale is formed from lateral to medial by the mm. iliopsoas and pectineus.

B The branches of the a. iliaca externa at its junction with the a. femoralis in the region of the lig. inguinale

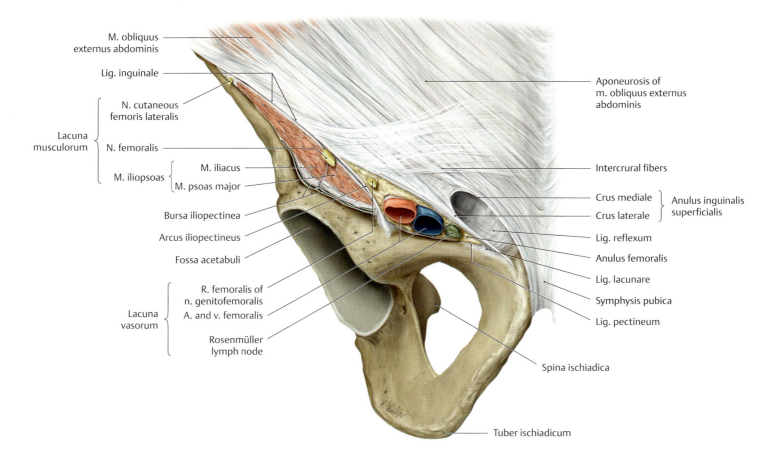

C Inguinal region and the contents of the lacuna musculorum and lacuna vasorum

Anterior view. The drawing shows a portion of the right hip bone and the adjacent anterior inferior abdominal wall with the anulus inguinalis superficialis and the contents of the lacuna musculorum and lacuna vasorum below the lig. inguinale. The site of emergence of the muscles and vessels, bounded by the lig. inguinale and the superior pelvic rim, is subdivided by the fibrous arcus iliopectineus.

The **lacuna vasorum** is located medial to the arcus iliopectineus and is traversed from lateral to medial by the r. femoralis of the n. genitofemoralis, the a. and v. femoralis, and the vasa lymphatica inguinalia profunda (only one lymph node is shown here). The part of the lacuna vasorum that lies medial to the v. femoralis is called the *anulus femoralis*. The lymph vessels from the thigh pass through that ring to enter the pelvis. The anulus femoralis is covered by a thin sheet of connective tissue called the septum femorale (not shown here), which usually contains a lymph node (the Rosenmüller node) belonging to the group of nll. inguinales profundi (see also p. 528).

The **lacuna musculorum** is lateral to the arcus iliopectineus and is traversed by the m. iliopsoas, n. femoralis, and n. cutaneus femoris lateralis.

Note the bursa iliopectinea located below the m. iliopsoas. It is the largest bursa of the hip region and communicates in 15% of cases with the joint cavity of the hip. For this reason, an inflammatory disease of the hip joint may incite inflammation of this bursa (bursitis). When inflamed, the bursa iliopectinea is frequently painful and swollen and may occasionally be mistaken for a neoplasm on MR images.

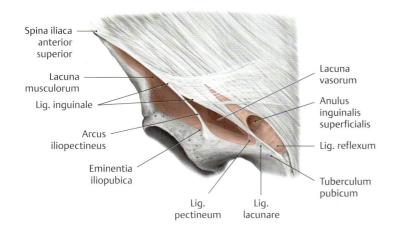

D The connective-tissue and bony boundaries of the lacuna musculorum and lacuna vasorum

Diagram of the right inguinal region, anterior view. The connective-tissue boundary between the lacuna musculorum and lacuna vasorum is formed by the arcus iliopectineus, a thickened band in the medial portion of the m. iliacus fascia. It extends between the lig. inguinale and the eminentia iliopubica. The fibrous band that curves downward from the medial attachment of the lig. inguinale is called the lig. lacunare and extends further medially as the lig. pectineum (Cooper's ligament) and runs along the ramus superior ossis pubis. This sharp-edged ligament defines the medial boundary of the lacuna *vasorum* (femoral ring) and may entrap the hernial sac in patients with a femoral hernia (see p. 214). Above the lig. inguinale is the external anulus inguinalis superficialis, which is the external opening of the canalis inguinalis (see p. 210). The lacuna *musculorum* is bounded laterally by the spina iliaca anterior superior.

23.4 Arterial Supply to the Thigh

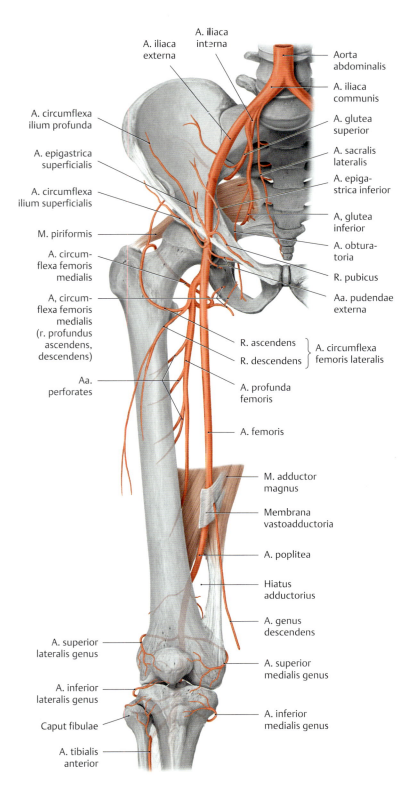

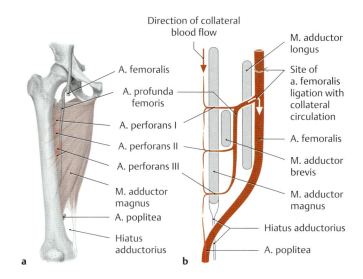

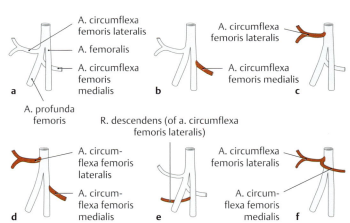

A Course and branches of the a. femoralis
The a. femoralis, the distal continuation of the a. iliaca externa, runs along the medial side of the thigh to the adductor canal (canalis adductorius), through which it passes to the back of the leg. After emerging from the hiatus adductorius, it becomes the a. poplitea. In clinical parlance the a. femoralis is often called the a. femoralis *superficialis* because of its superficial course down the front of the thigh, distinguishing it from the more deeply placed *a. profunda femoris* that arises from it (see **D**).

B Course of the a. profunda femoris and sites where the aa. perforantes pierce the adductor muscles
a Right thigh, anterior view; **b** schematic longitudinal section through the adductor muscles at the level of the aa. perforantes. The a. profunda femoris has approximately three to five terminal branches that pass from the front to the back of the thigh through the femoral insertions of the adductor muscles (= aa. perforantes I–III) to supply the hamstring muscles (mm. biceps femoris, semitendinosus, and semimembranosus). Generally, the arteries pierce the adductor muscles above and below the m. adductor brevis and just above the hiatus adductorius. Ligation of the a. femoralis proximal to the origin of the a. profunda femoris is relatively well tolerated owing to a good collateral supply from branches of the a. iliaca interna (a. glutea superior and a. obturatoria).

C Variants in the a. femoralis branching pattern (after Lippert and Pabst)
a Usually the a. profunda femoris and aa. circumflexae femoris medialis and lateralis arise from the a. femoralis by a common trunk (58% of cases, also shown in the other figures on this page).
b The a. circumflexa femoris medialis arises directly from the a. femoralis (18% of cases).
c The a. circumflexa femoris lateralis arises directly from the a. femoralis (15% of cases).
d The aa. circumflexae arise separately from the a. femoralis (4% of cases).
e The r. descendens of the a. circumflexa femoris lateralis springs directly from the a. femoralis (3% of cases).
f The aa. circumflexae arise by a common trunk (1% of cases).

Lower Limb — 23. Neurovascular Systems: Topographical Anatomy

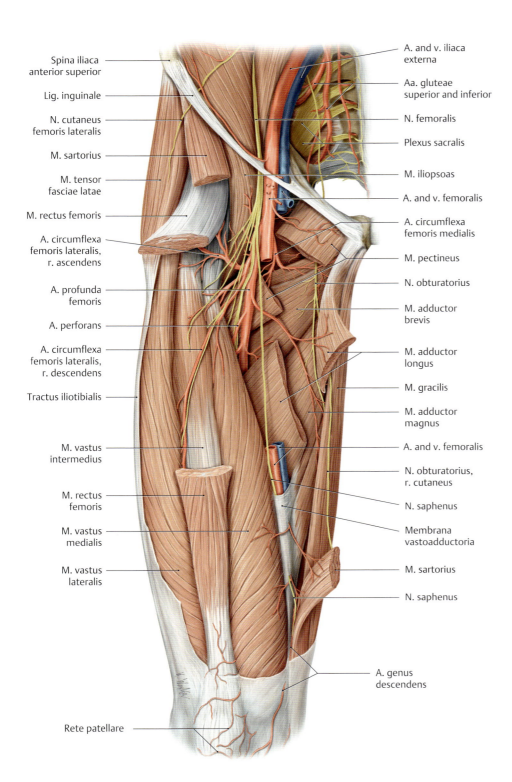

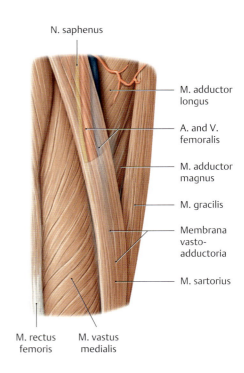

E The location of the canalis adductorius
Right thigh, anterior view. The n. saphenus passes down the canalis adductorius on the anterior side of the thigh, accompanied by the a. and v. femoralis. While both vessels continue toward the fossa poplitea through the hiatus adductorius, the n. saphenus pierces the membrana vastoadductoria along with the a. genus descendens and passes to the medial side of the art. genus (see also **F**).

F The boundaries and contents of the canalis adductorius (canalis adductorius)

Boundaries
- Mm. adductores longus and magnus (posterior)
- M. sartorius (medial)
- Membrana vastoadductoria (anterior)
- M. vastus medialis (lateral and anterior)

Contents
- A. femoralis
- V. femoralis
- N. saphenus ⎫ Pierce the
- A. genus   ⎬ membrana
  descendens ⎭ vastoadductoria

D The blood supply to the thigh from the a. profunda femoris
Right thigh, anterior view. The mm. sartorius, rectus femoris, adductor longus, and pectineus have been partially removed, along with the central portion of the a. femoralis, to demonstrate the course of the a. profunda femoris. For clarity, the veins have also been removed to the level of the v. iliaca externa. This dissection does not show the anterior abdominal wall or the pelvic and abdominal organs above the level of the lig. inguinale. While the branches of the Aa.circumflexae medialis and lateralis mainly supply blood to the art. coxae and extensors and adductors of the thigh, the terminal branches of the a. profunda femoris (the aa. perforantes I–III, see **B**) on the medial side of the femur pass to the back of the thigh through gaps in the insertions of the adductor muscles and supply the hamstrings (mm. biceps femoris, semitendinosus, and semimembranosus).
Note: The membrana vastoadductoria is pierced by the a. genus descendens and n. saphenus (see **F**).

551

Lower Limb — 23. Neurovascular Systems: Topographical Anatomy

23.5 The Gluteal Region (Regio glutealis): Overview of Its Vessels and Nerves

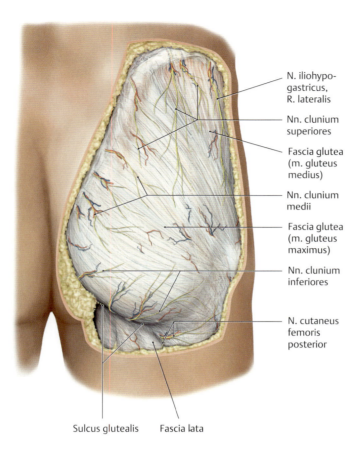

A The fasciae and cutaneous nerves of the superficial gluteal region

Right gluteal region, posterior view. The regio glutealis is covered by the fascia glutea, which is part of the fascia lata (although the term "fascia lata" strictly refers only to the part below the mm. gluteus medius and maximus). The fascia covering the m. gluteus maximus forms septa-like invaginations between the muscle bundles. At the junction of the regio glutealis with the back of the upper thigh is the gently curved sulcus glutealis, in which thickened fiber tracts of the fascia lata run transversely across the thigh at the level of the tuber ischiadicum.

Note on surface anatomy: The oblique inferior border of the m. gluteus maximus (see **B**) *crosses* the sulcus glutealis. Its course, then, is not identical to that of the sulcus glutealis.

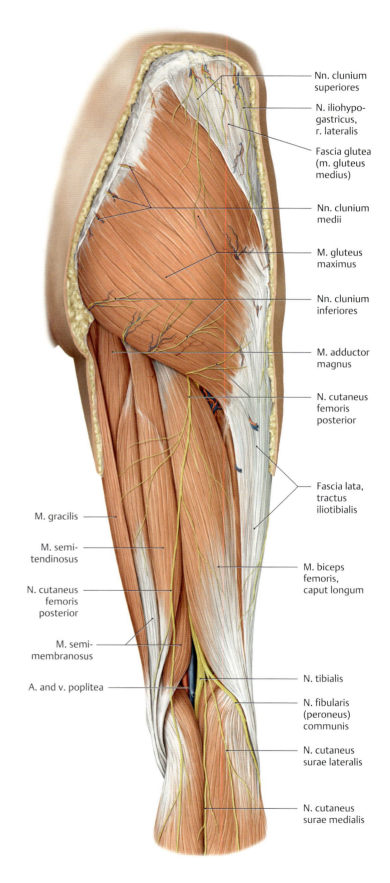

B The gluteal region and thigh with the fasciae removed

Right side, posterior view. With the fascia lata removed, the main trunk of the n. cutaneus femoris posterior, which is subfascial over much of its course, can be traced into the fossa poplitea.

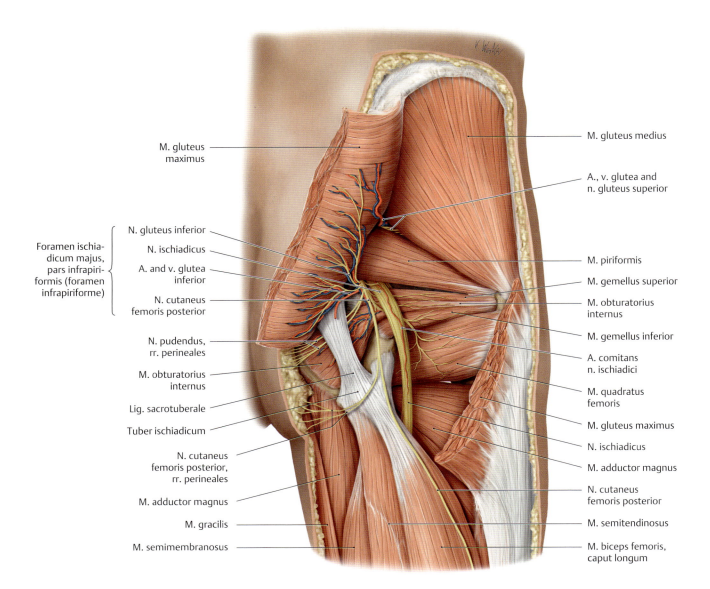

C The vessels and nerves of the deep gluteal region
Right side, posterior view, with the m. gluteus maximus partially removed.
The neurovascular structures of the deep gluteal region traverse an extensive fatty and connective-tissue space below the m. gluteus maximus. The floor of this space is formed by the mm. piriformis, obturatorius internus, gemellus superior, and quadratus femoris. It communicates through the foramina ischiadica with the connective-tissue spaces of the lesser pelvis and fossa ischioanalis (not shown here). A useful topographic landmark is the m. piriformis, which extends from the facies pelvica of the sacrum through the foramen ischiadicum majus to the tip of the trochanter major (see **A**, p. 554).

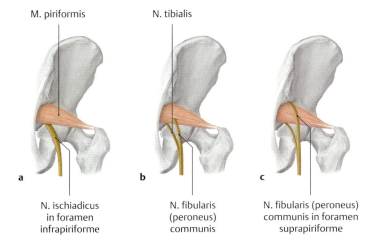

D Variable course of the n. ischiadicus in relation to the m. piriformis (after Rauber/Kopsch)
a The n. ischiadicus leaves the lesser pelvis inferior to the foramen infrapiriforme (almost 85% of cases).
b This variant illustrates a *high division* of the n. ischiadicus (approximately 15% of cases). In this pattern the fibular division (n. fibularis [peroneus] communis) and sometimes the n. cutaneus femoris posterior pass through the m. piriformis and may become compressed at that location, causing a "piriformis syndrome." Usually this term refers to the complaints that may develop following trauma to the gluteal region and are marked by severe gluteal pain. It is still uncertain whether these complaints are actually referable to the compression of n. ischiadicus segments.
c In this variant the fibular part of the n. ischiadicus leaves the lesser pelvis above the m. piriformis (rare, only about 0.5% of cases).

23.6 The Gluteal Region (Regio glutealis): The Foramina ischiadica and N. ischiadicus

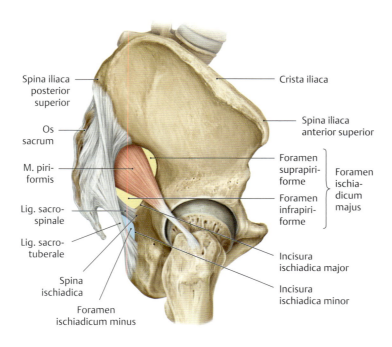

A Location of the greater and lesser sciatic foramina
Right os coxae, lateral view.

B The boundaries of the foramina ischiadica and the structures that traverse them
The subgluteal connective-tissue space communicates through the foramina ischiadica with the connective-tissue spaces of the lesser pelvis and the fossa ischioanalis.

Foramen	Boundaries	Transmitted structures
• Foramen ischiadicum majus	• Incisura ischiadica major • Lig. sacrospinale • Os sacrum	• *Foramen suprapiriforme* – a. and v. glutea superior – N. gluteus superior • *Foramen infrapiriforme* – A. and v. glutea inferior – N. gluteus inferior – A. and v. pudenda interna – N. pudendus – N. ischiadicus – N. cutaneus femoris posterior
• Foramen ischiadicum minus	• Incisura ischiadica minor • Lig. sacrospinale • Lig. sacrotuberale	– A. and v. pudenda interna – N. pudendus – M. obturatorius internus

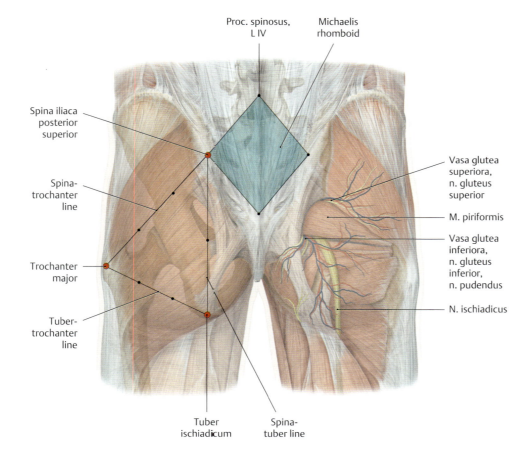

C Reference lines used for locating neurovascular structures in the regio glutealis
Right and left gluteal regions, posterior view. The reference lines are drawn between the following points: the spina iliaca posterior superior (lateral point of the Michaelis rhomboid), the tuber ischiadicum, and the trochanter major.

- **Spina-trochanter line:** The vasa glutea superiora emerge from the foramen suprapiriforme between the middle and upper thirds of this line.
- **Tuber-trochanter line:** The n. ischiadicus runs downward between the middle and medial thirds of this line.
- **Spina-tuber line:** The n. ischiadicus, n. gluteus inferior, and n. pudendus and the vasa glutea inferiora emerge from the foramen infrapiriforme at the midpoint of this line.

Lower Limb — 23. Neurovascular Systems: Topographical Anatomy

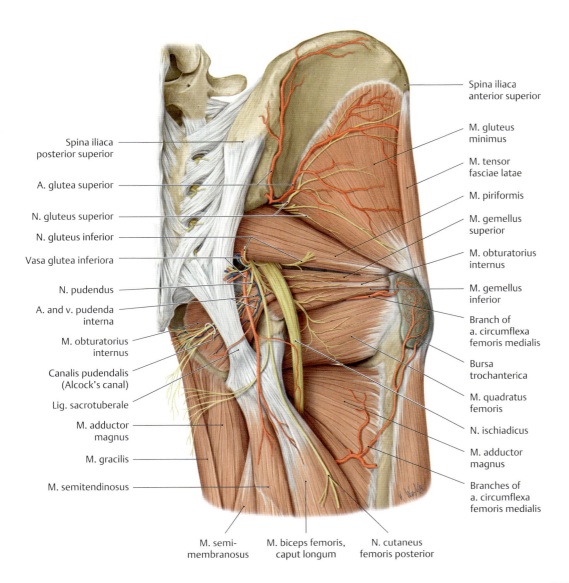

D The vessels and nerves of the regio glutealis and fossa ischioanalis

Right gluteal region, posterior view, with the mm. gluteus maximus and medius removed.
Note the course of the vasa pudenda interna and n. pudendus in the lateral wall of the fossa ischioanalis. They run in the canalis pudendalis (Alcock's canal), which is formed by the m. obturatorius internus fascia (see p. 478).

E Location of the nn. ischiadicus and gluteus superior and their protection during intragluteal injections

Right gluteal region, lateral view.

a Two very important nerves are found in the regio glutealis: the n. ischiadicus and the n. gluteus superior. To avoid jeopardizing these nerves during intramuscular injections, the needle should be inserted with the greatest possible safety margin with respect to these structures. Making the injection within the "von Hochstetter triangle" ensures that this safety margin is maintained.

b Locating the von Hochstetter triangle: The target site is located in the anterolateral gluteal region (accounting for the term "ventrogluteal injection"). To give an intramuscular injection on the right side, for example, place the palm of the left hand on the trochanter major and the tip of the index finger on the spina iliaca anterior superior. Keeping the hand in place, abduct the middle finger away from the index finger and introduce the needle perpendicular to the skin surface within the triangular zone between the two fingers and the crista iliaca.

555

23.7 The Fossa ischioanalis

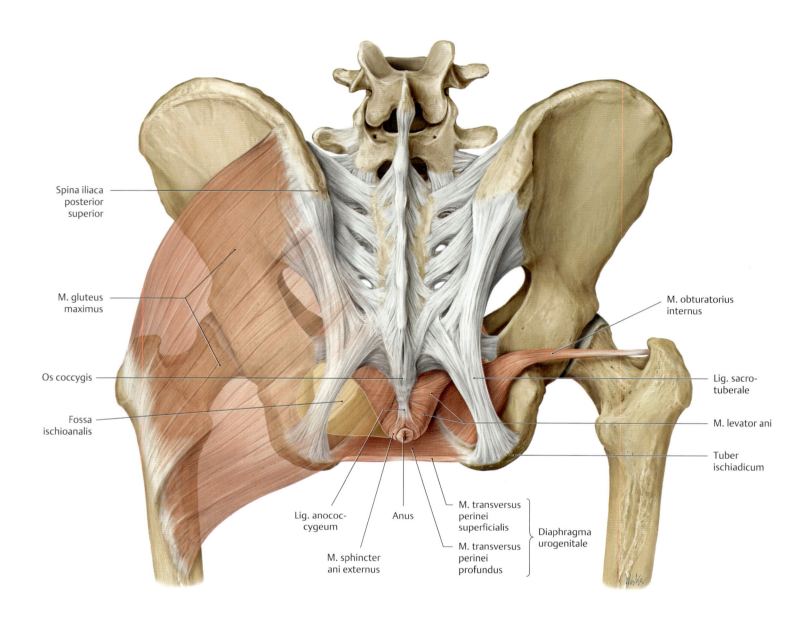

A The muscular boundaries of the fossa ischioanalis
Left and right gluteal region (regio glutealis), posterior view. The fossa ischioanalis is a pyramid-shaped space located lateral to the m. levator ani on each side. The tip of the three-sided pyramid points toward the symphysis, and the base of the pyramid faces posteriorly. The fossa ischioanalis is bounded by the following muscles:

- Superomedially by the m. levator ani
- Laterally by the m. obturatorius internus
- Inferiorly by the m. transversus perinei profundus
- The entrance to the fossa ischioanalis is bounded posteriorly by the m. gluteus maximus and lig. sacrotuberale.

The fatty tissue that occupies most of the fossa ischioanalis (corpus adiposum fossae ischioanalis) functions as a mobile pad that can slide downward and backward, for example, during bowel evacuation or during labor. It is traversed by the branches of the vasa pudenda interna and n. pudendus (see **B**), whose trunks run in the canalis pudendalis, or Alcock's, canal; see **A**, p. 558.

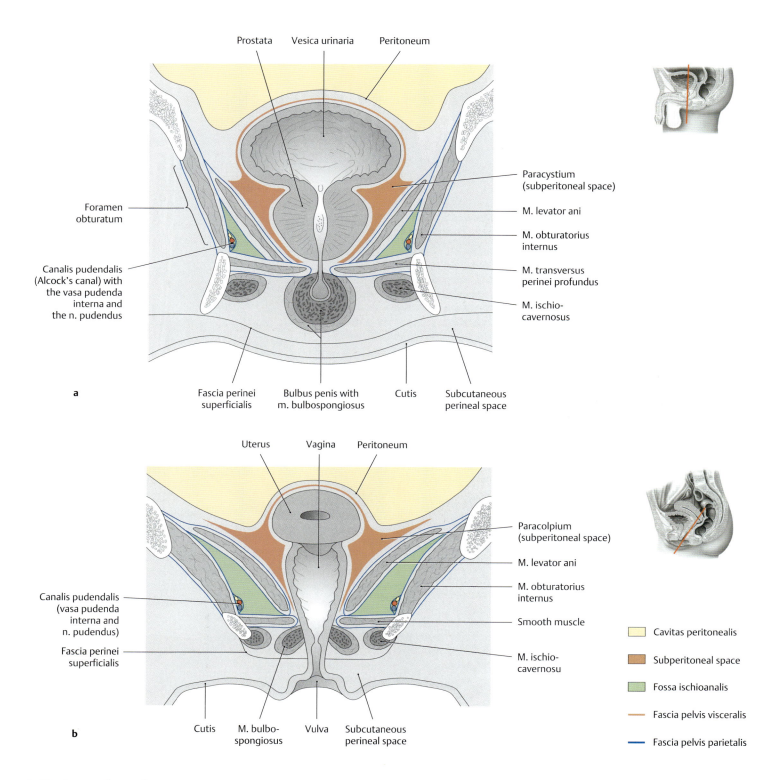

B The fossa ischioanalis
a Coronal section through the male pelvis at the level of the prostata.
b Oblique coronal section through the female pelvis at the level of the vagina.

The pelvic organs make varying contributions to the shape of the cavitas peritonealis and subperitoneal space, but they are not represented in the fossa ischioanalis. While the cavitas peritonealis is lined by the peritoneum parietale and viscerale (on intraperitoneal organs such as the ovarium), the subperitoneal space is lined by the pelvic fasciae (fascia pelvis parietalis and fascia pelvis visceralis, see p. 183).

23.8 The Canalis pudendalis and Regio perinealis (Regio urogenitalis and Regio analis)

A Location of the canalis pudendalis (Alcock's canal) and the neurovascular structures it contains

Right half of the pelvis, medial view. All of the muscles have been removed except for the mm. psoas major, piriformis, and obturatorius internus. For clarity, the individual veins are not shown. The canalis pudendalis is formed by the m. obturatorius internus fascia. It begins just below the spina ischiadica and courses in the lateral wall of the fossa ischioanalis below the arcus tendineus m. levatoris ani, passing toward the symphysis pubica and the posterior border of the diaphragma urogenitale (see p. 162). The neurovascular structures that are transmitted by the canal (the vasa pudenda interna, of which only the artery is shown, and the n. pudendus, see B) exit the lesser pelvis through the foramen ischiadicum majus and enter the canalis pudendalis through the foramen ischiadicum minus. They pass through the canal toward the symphysis pubica and the posterior border of the diaphragma urogenitale.

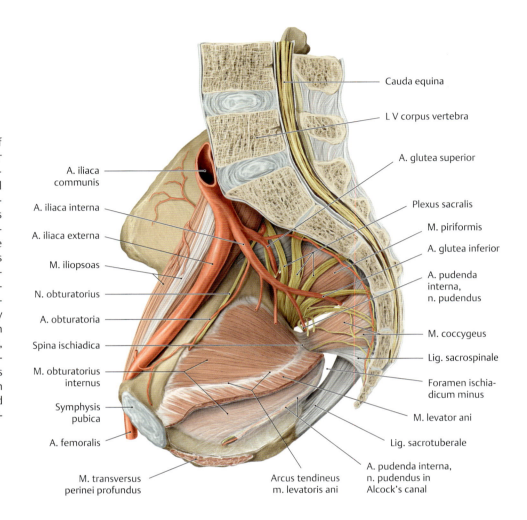

B Distribution of the n. pudendus and vasa pudenda interna to the anus, perineum, and external genitalia

Regio glutealis and fossa ischioanalis on the right side, posterior view. The m. gluteus maximus and lig. sacrotuberale have been partially removed, and all fatty tissue has been removed from the fossa ischioanalis to demonstrate the course of the n. pudendus and the vasa pudenda interna. On their way through the canalis pudendalis (not shown here in order to display the course of the nerve and vessels below the lig. sacrotuberale), the various neural and vascular branches are successively distributed in a fan-shaped pattern to the anus, perineum, and external genitalia. It is very common in obstetrics to perform a pudendal nerve block by anesthetizing the n. pudendus at the level of the spina ischiadica (i.e., before it branches into the nn. rectales inferiores, perineales, dorsalis clitoridis, and labiales posteriores; see p. 542).

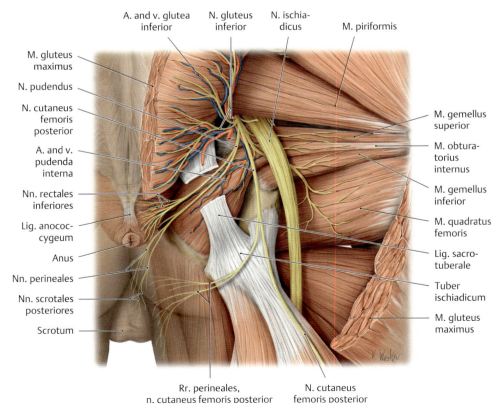

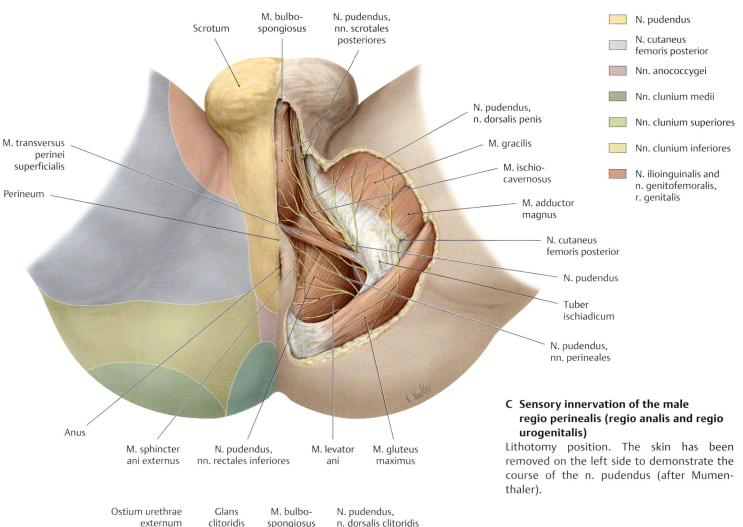

C Sensory innervation of the male regio perinealis (regio analis and regio urogenitalis)
Lithotomy position. The skin has been removed on the left side to demonstrate the course of the n. pudendus (after Mumenthaler).

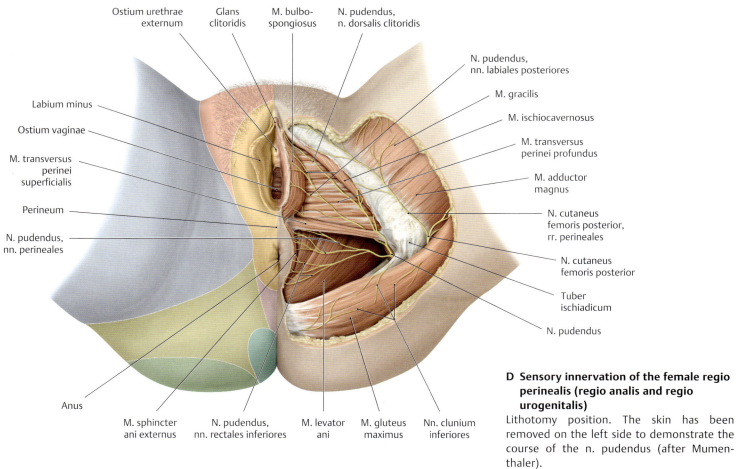

D Sensory innervation of the female regio perinealis (regio analis and regio urogenitalis)
Lithotomy position. The skin has been removed on the left side to demonstrate the course of the n. pudendus (after Mumenthaler).

23.9 The Posterior Thigh Region (Regio femoris posterior) and Popliteal Region (Regio genus posterior)

A The vessels and nerves of the posterior thigh

Right thigh, posterior view. To demonstrate the vessels and nerves in their course from the regio glutealis down the back of the thigh to the fossa poplitea (see C), the skin and muscle fasciae have been removed, and the following muscles have been partially removed: mm. glutei maximus, medius, and biceps femoris. The m. semimembranosus has been retracted slightly medially to display the hiatus adductorius (transmits the a. and v. femoralis) (deeper neurovascular structures in the fossa poplitea are shown in F). The regio femoris posterior receives most of its blood supply from branches of the a. profunda femoris (aa. perforantes I–III) and the r. profundus (not shown) of the a. circumflexa femoris medialis. The proximal part of the n. ischiadicus is supplied by the a. comitans n. ischiadici, which is a branch of the a. glutea inferior, and its distal part is supplied by branches of the aa. perforantes I–III.

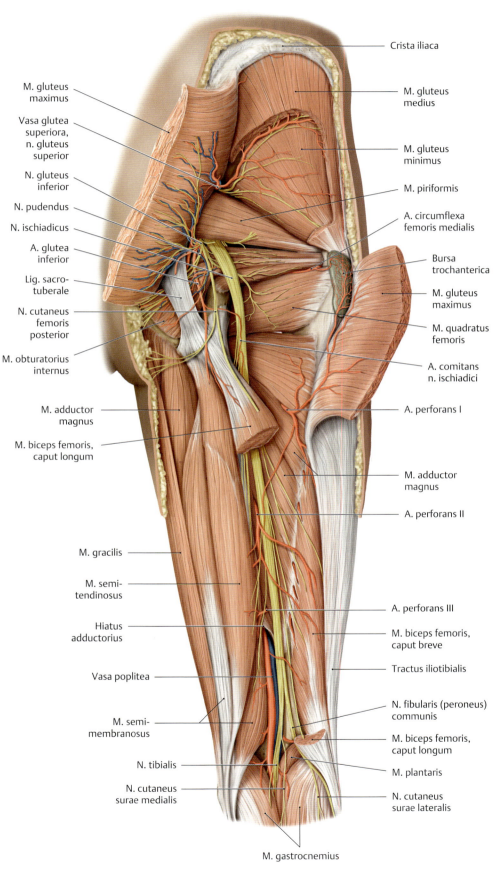

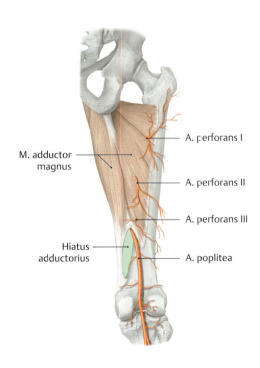

B Sites of emergence of the aa. perforantes on the back of the thigh

Right thigh, posterior view. All of the muscles have been removed except for the m. adductor magnus.
Note: The a. femoralis enters the fossa poplitea through the hiatus adductorius, thereafter becoming the a. poplitea.

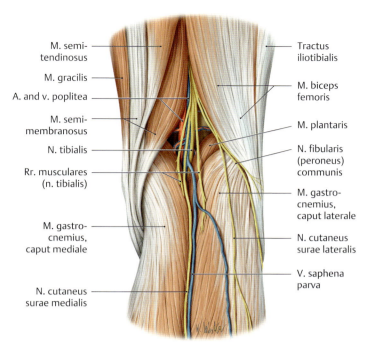

C The muscular boundaries of the fossa poplitea
Right fossa poplitea, posterior view. For clarity the skin, fasciae, and fat pads have been removed.

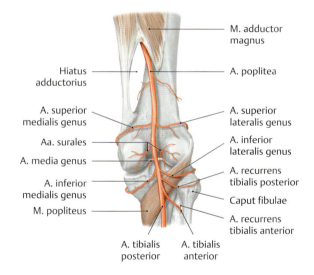

D Branches of the a. poplitea that course in the fossa poplitea
Right art. genus, posterior view. The a. poplitea begins at the outlet of the canalis adductorius and ends at the level of the m. popliteus, where it divides into the anterior tibial a. tibialis anterior and a. tibialis posterior.

E Palpation of the a. poplitea in the fossa poplitea

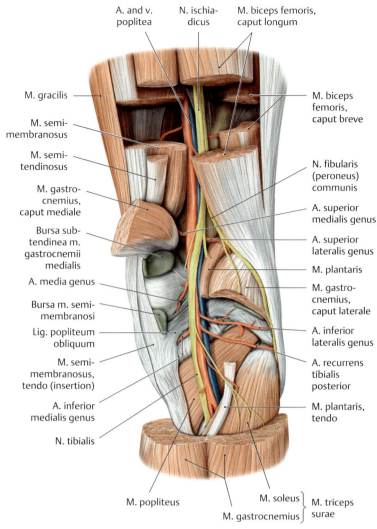

F Deep neurovascular structures in the fossa poplitea
Right art. genus, posterior view. Portions of both heads of the m.gastrocnemius and portions of the hamstrings have been removed to demonstrate the course of the deep neurovascular structures in the fossa poplitea. Five vessels, some paired, branch from the middle portion of the a. poplitea to supply the art. genus (see **D**):

- Aa. superiores lateralis and medialis genus
- A. media genus
- Aa. inferiores lateralis and medialis genus

One of these vessels, the a. media genus, pierces the capusla articularis of the knee in the area of the lig. popliteum obliquum and supplies the ligg. cruciata. The other vessels run forward on the medial and lateral sides to form the arterial network (rete articulare genus) of the knee. The aa. recurrentes tibialis posterior and anterior contribute to the rete articularis genus. The paired aa. surales supply the two heads of the m. gastrocnemius (see **D**, removed in **F**).
Note on the medial side the bursa subtendinea m. gastrocnemii, which consistently communicates with the joint cavity of the knee, and the bursa m. semimembranosi, which occasionally communicates with the bursa of the m. gastrocnemius head (this forms an extensive recess in the joint cavity of the knee, which may become abnormally enlarged to form a Baker cyst; see p. 442).

23.10 The Posterior Leg Region (Regio cruris posterior) and the Tarsal Tunnel (Canalis tarsalis)

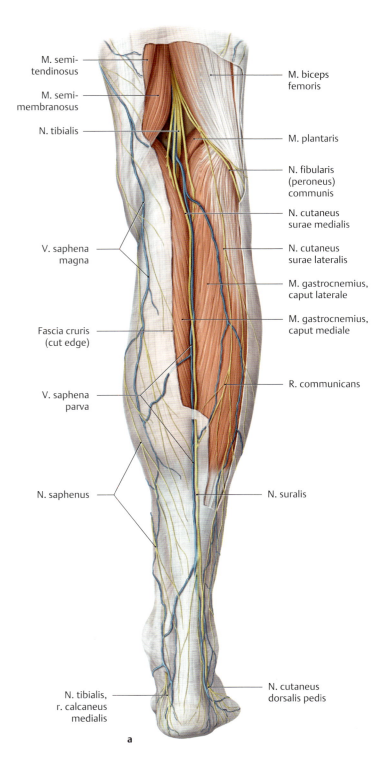
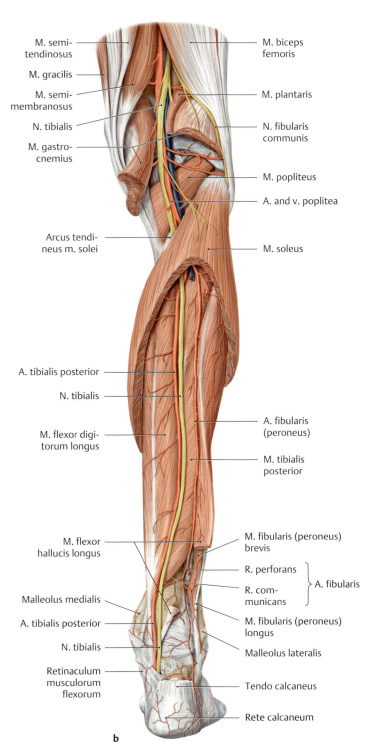

A The neurovascular structures in the superficial and deep posterior compartments
Right leg, posterior view.

a Neurovascular structures in the *superficial posterior compartment*: The superficial layer of the fascia cruris ensheaths the m. triceps surae and has been partially removed proximally.
b Neurovascular structures in the *deep posterior compartment* after partial removal of the m. triceps surae and the deep layer of the fascia cruris. The a. poplitea divides into the aa. tibiales anterior and posterior at the distal border of the m. popliteus. The a. tibialis anterior pierces the membrana interossea cruris (not shown here, see **B**) and passes to the anterior side of the leg, entering the anterior compartment. The a. tibialis posterior, accompanied by the n. tibialis, passes below the arcus tendineus m. solei into the deep posterior compartment, almost immediately gives off the a. fibularis (peronea), and then continues distally behind the malleolus medialis to the plantar side of the foot. The deep posterior compartment is one of four poorly distensible muscle compartments in the leg ("fibroosseous canals"), which are potential sites for the development of a compartment syndrome following a vascular injury (see p. 567).

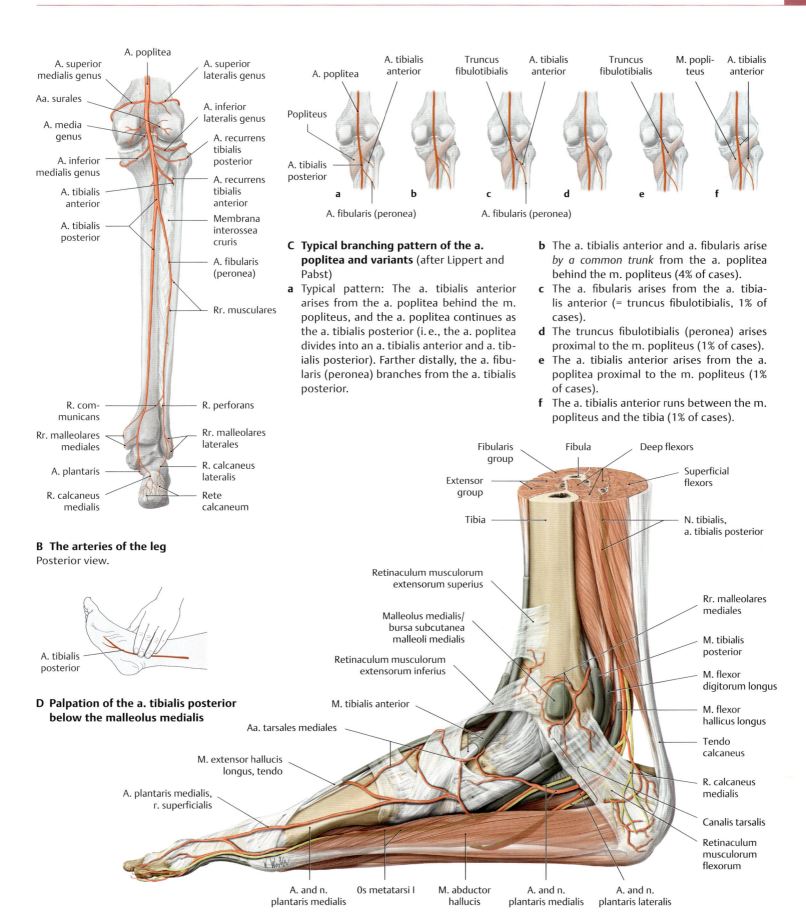

B The arteries of the leg
Posterior view.

C Typical branching pattern of the a. poplitea and variants (after Lippert and Pabst)

a Typical pattern: The a. tibialis anterior arises from the a. poplitea behind the m. popliteus, and the a. poplitea continues as the a. tibialis posterior (i.e., the a. poplitea divides into an a. tibialis anterior and a. tibialis posterior). Farther distally, the a. fibularis (peronea) branches from the a. tibialis posterior.

b The a. tibialis anterior and a. fibularis arise by a common trunk from the a. poplitea behind the m. popliteus (4% of cases).

c The a. fibularis arises from the a. tibialis anterior (= truncus fibulotibialis, 1% of cases).

d The truncus fibulotibialis (peronea) arises proximal to the m. popliteus (1% of cases).

e The a. tibialis anterior arises from the a. poplitea proximal to the m. popliteus (1% of cases).

f The a. tibialis anterior runs between the m. popliteus and the tibia (1% of cases).

D Palpation of the a. tibialis posterior below the malleolus medialis

E The neurovascular structures of the medial malleolar region
Right foot, medial view. The neurovascular structures pass from the deep flexor compartment to the plantar side of the foot through the tarsal tunnel (canalis tarsalis), between the retinaculum musculorum flexorum and malleolus medialis. They are accompanied by the tendons of insertion of the long flexors (mm. tibialis posterior, flexor digitorum longus, flexor hallucis longus) in their synovial sheaths.

Note the division of the n. tibialis posterior into the nn. plantares medialis and lateralis and the division of the a. tibialis posterior into the aa. plantares medialis and lateralis within the malleolar canal. Compression of the nerves at this site can cause a medial or posterior tarsal tunnel syndrome (see p. 541).

563

23.11 The Sole of the Foot (Planta pedis)

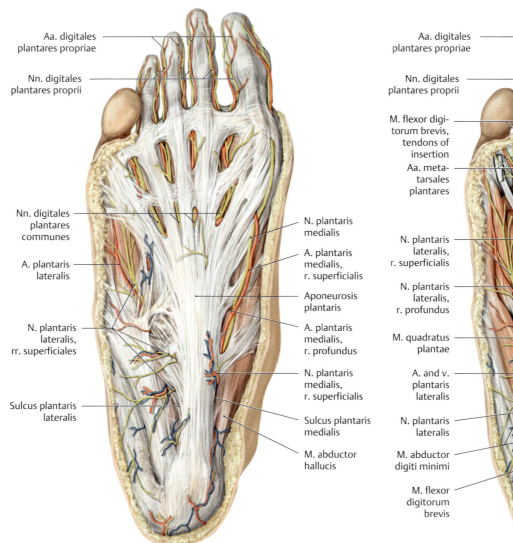

A The arteries and nerves of the sole of the foot (superficial layer)
Right foot, plantar view. The skin and subcutaneous tissue have been removed to demonstrate the aponeurosis plantaris and superficial neurovascular structures.

B The arteries and nerves of the sole of the foot (middle layer)
Right foot, plantar view, with the aponeurosis plantaris and m. flexor digitorum brevis.

C The arteries of the sole of the foot: possible variants
Right foot, plantar view. Any of four basic anatomic variants may be seen (after Lippert and Pabst):

a The arcus plantaris profundus and the aa. metatarsales plantares arising from it are supplied entirely by the r. plantaris profundus of the a. dorsalis pedis (53% of cases).
b The aa. metatarsales plantares I–III are supplied by the r. plantaris profundus of the a. dorsalis pedis, the a. metatarsalis plantaris IV by the r. profundus of the a. plantaris lateralis (19% of cases).
c The aa. metatarsales plantares I and II are supplied by the r. plantaris profundus of the a. dorsalis pedis, the aa. metatarsales III and IV by the r. profundus of the a. plantaris lateralis (13% of cases).
d The arcus plantaris profundus and the aa. metatarsales plantares I – IV are supplied entirely by the r. profundus of the a. plantaris lateralis (7% of cases).

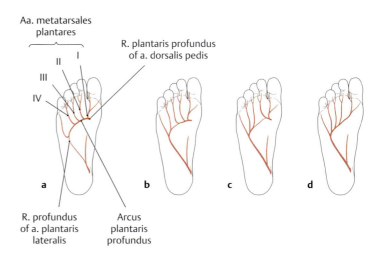

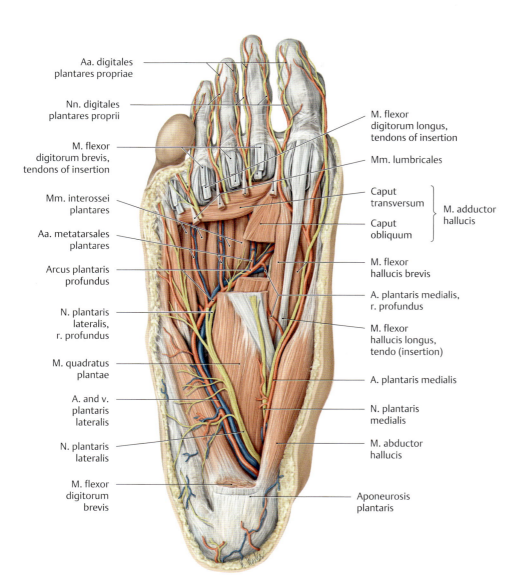

D The arteries and nerves of the sole of the foot (deep layer)

Right foot, plantar view. The aponeurosis plantaris, m. flexor digitorum brevis, the tendons of m. flexor digitorum longus, and the caput obliquum of m. adductor hallucis have been removed to demonstrate the arcus plantaris profundus and the r. profundus n. plantaris lateralis.

Note: The r. superficialis of the n. plantaris lateralis and the a. plantaris lateralis course in the *sulcus plantaris lateralis,* while the rr. superficiales of the n. plantaris medialis and the a. plantaris medialis course in the *sulcus plan- taris medialis* (see **A**). The rr. superficiales of the aa. plantares medialis and lateralis help to supply the critically important pressure-chamber system in the sole of the foot (see p. 472).

E Overview of the arteries of the sole of the foot

Right foot, plantar view.

The *arcus plantaris profundus* is an arterial arcade in the sole of the foot that is formed by the r. plantaris profundus of the *a. dorsalis pedis* and also by the r. profundus of the *a. plantaris lateralis*. Often these two arteries that supply the arcus plantaris profundus differ in size and thus make different contributions to the aa. metatarsales plantares I–IV that consistently branch from the arcus plantaris profundus (see **C**).

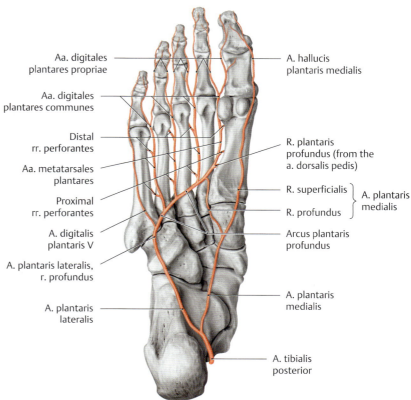

565

23.12 The Anterior Leg Region and Dorsum of the Foot (Regio cruris anterior and Dorsum pedis): Cutaneous Innervation

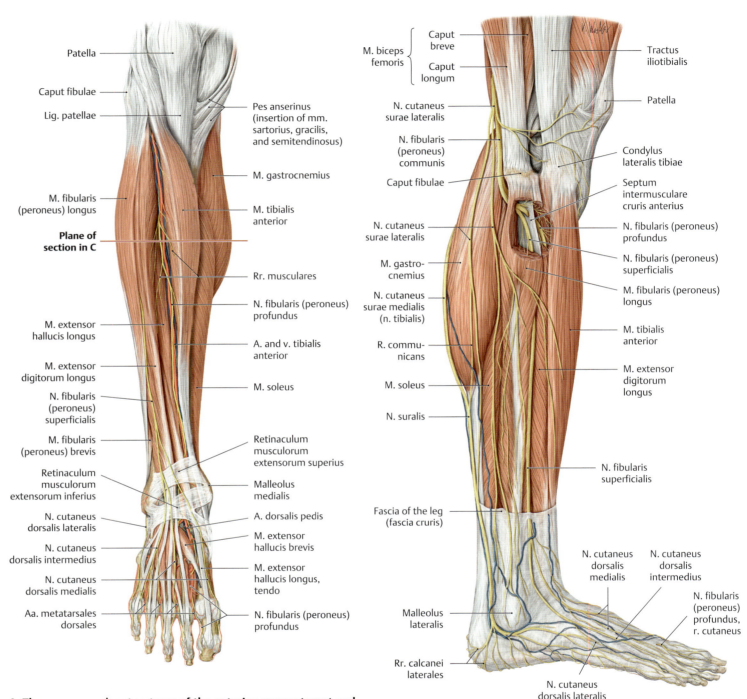

A The neurovascular structures of the anterior compartment and dorsum of the foot

Right leg with the foot in plantar flexion, anterior view. The skin, subcutaneous tissue, and fasciae have been removed and the mm. tibialis anterior and extensor hallucis longus have been retracted to demonstrate the vasa tibialia anteriora (= a. and v. tibialis anterior). The *a. tibialis anterior* crosses beneath the tendon of the m. extensor hallucis longus at the junction of the leg with the dorsum of the foot. Below the retinaculum musculorum extensorum inferius it becomes the *a. dorsalis pedis*, which runs lateral to the tendon of m. hallucis longus on the dorsum of the foot, accompanied by the terminal branch of the *n. fibularis profundus* (the site for taking the pedal pulse is shown in **E**). The *n. fibularis profundus* may be compressed in its passage beneath the retinaculum musculorum extensorum inferius (with sensory disturbances affecting the first and second toes).

B Division of the n. fibularis communis into the nn. fibulares profundus and superficialis

Right leg, lateral view. The origins of the mm. fibularis longus and extensor digitorum longus have been excised below the caput fibulae and the condylus lateralis tibiae. After the *n. fibularis communis* bifurcates in the proximal part of the lateral compartment, the *n. fibularis superficialis* remains in the lateral compartment. The *n. fibularis profundus* pierces the septum intermusculare cruris anterius and descends with the vasa tibialia anteriora in the extensor compartment (**C** gives a sectional view of the compartments in the leg).

Lower Limb — 23. Neurovascular Systems: Topographical Anatomy

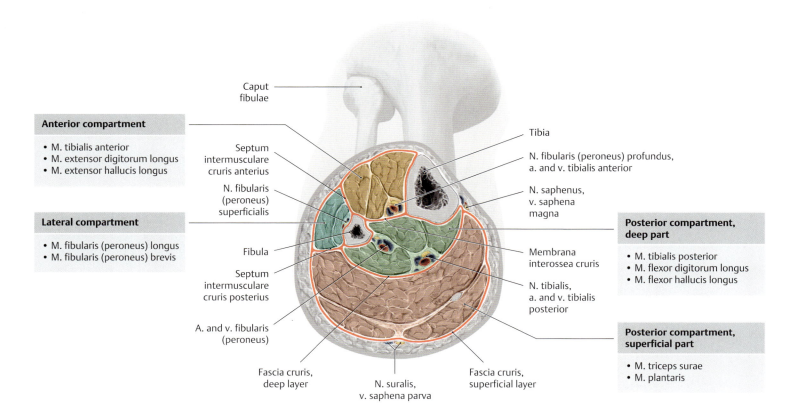

Anterior compartment
- M. tibialis anterior
- M. extensor digitorum longus
- M. extensor hallucis longus

Lateral compartment
- M. fibularis (peroneus) longus
- M. fibularis (peroneus) brevis

Posterior compartment, deep part
- M. tibialis posterior
- M. flexor digitorum longus
- M. flexor hallucis longus

Posterior compartment, superficial part
- M. triceps surae
- M. plantaris

C The compartments and neurovascular structures in the leg

Cross section through a right leg one handwidth below the collum fibulae, distal view (the level of the section is shown in **A**). The septa intermuscularia cruris and membrana interossea, together with the superficial and deep layers of the fascia cruris, define the boundaries of four distinct, poorly distensible fibro-osseous compartments in which the neurovascular structures descend through the leg. A rise in tissue pressure, which may result from conditions such as muscular edema or a fracture hematoma, can lead to neurovascular compression, inducing a local ischemia that can cause irreversible neuromuscular damage within a few hours (compartment syndromes such as the tibialis anterior syndrome). At greatest risk are the neurovascular structures of the deep posterior compartment (the a. and vv. tibiales posteriores and the n. tibialis) and the anterior compartment (the a. and vv. tibiales anteriores and the n. fibularis (peroneus) profundus). Tibialis anterior syndrome is characterized in its acute stage by severe pain and an inability to dorsiflex the toes due to the unopposed action of the plantar flexors. This causes the toes to "claw up." Generally, the only effective treatment option at this stage is emergency incision of the fascia cruris. This immediately decompresses the compartment and relieves the pressure on the vessels that supply the muscles.

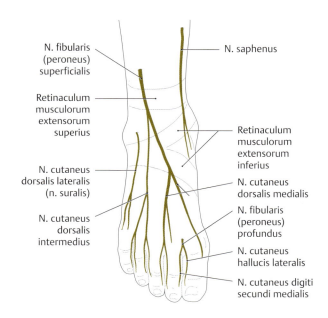

D The cutaneous nerves on the dorsum of the foot
Right foot, dorsal view.

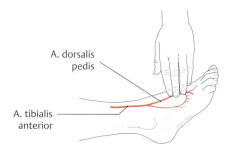

E Palpation of the pedal pulse
The a. dorsalis pedis is palpable at the level of the tuberositas ossis navicularis, just lateral to the m. extensor hallucis longus tendon. In addition to determining regional skin temperature, checking the pedal pulse is an important step in the examination of patients with suspected lower limb arterial disease (one foot is markedly colder or paler than the other due to diminished blood flow). It is generally best to begin by palpating the a. femoralis at the groin crease, then proceed distally to the fossa poplitea (a. poplitea), the malleolus medialis (a. tibialis posterior), and finally to the dorsum of the foot (a. dorsalis pedis, which is the terminal branch of the a. tibialis anterior). The palpable pulses should always be compared between the right and left sides. It should be noted that the pedal pulses may be difficult or impossible to palpate when peripheral edema is present, so it is best to examine the supine patient.

23.13 The Arteries of the Dorsum of the Foot (Dorsum pedis)

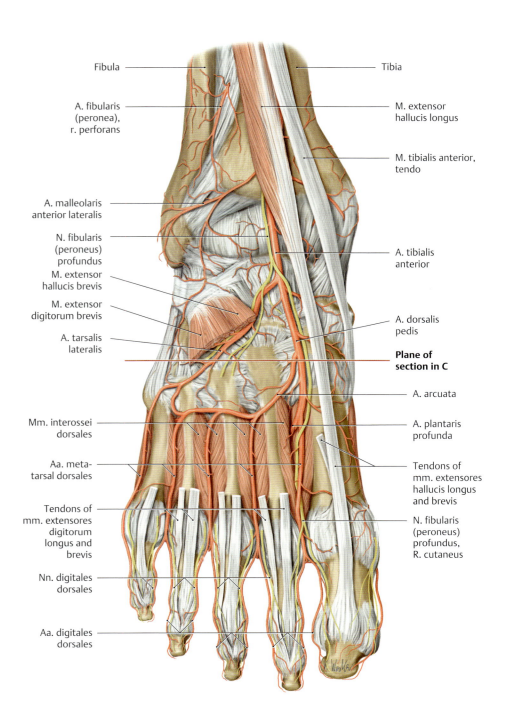
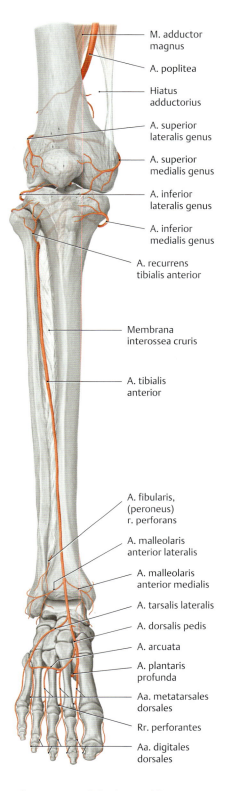

A The dorsal arteries and nerves of the foot
Right foot in plantar flexion, dorsal view. The skin, subcutaneous tissue, and superficial and deep layers of the fascia dorsalis pedis have been removed for clarity, along with the m. extensor digitorum longus tendons and the mm. extensor digitorum brevis and extensor hallucis brevis. Possible variants of the arteries are shown in **D**.

B The arteries of the leg and foot
Right lower limb, anterior view.
Note: The dorsum of the foot is supplied mainly by branches of the a. tibialis anterior.

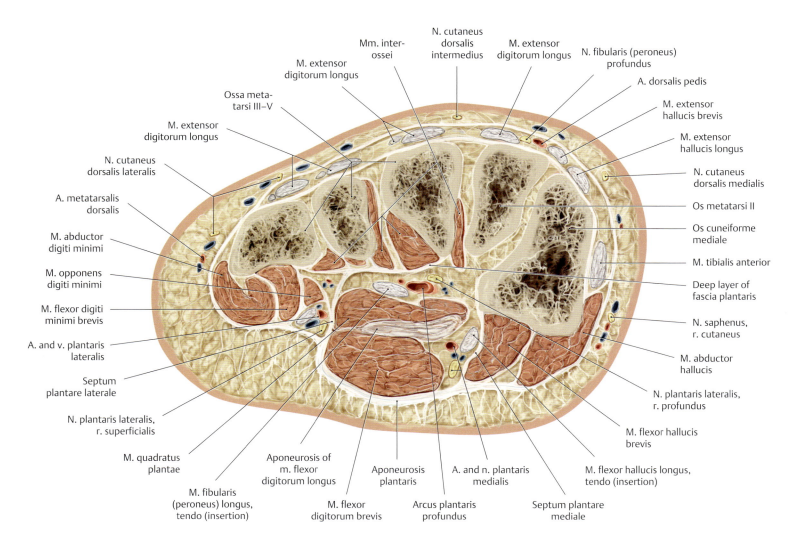

C The neurovascular structures in the sole of the foot
Cross section through the right foot at the level of the os cuneiforme mediale (the location of the section is shown in **A**), distal view (after Rauber and Kopsch).

Note the deep layer of the fascia plantaris, in which the deep neurovascular structures of the planta pedis (the arcus plantaris profundus and r. profundus n. plantaris lateralis) are embedded in connective tissue, which cushions and protects them (for the arrangement of the pedal compartments, see p. 523).

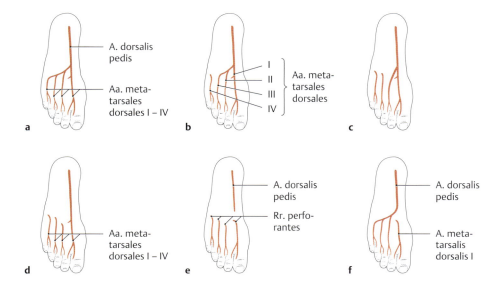

D Variants in the dorsal arterial supply of the foot (after Lippert and Pabst)
a All of the aa. metatarsales dorsales pedis arise from the a. dorsalis pedis (20% of cases).
b The a. metatarsalis dorsalis pedis IV is supplied by a r. perforans from the plantar side of the foot (6% of cases).
c The aa. metatarsales dorsales pedis III and IV are supplied by r. perforantes from the aa. metatarsales plantares (5% of cases).
d The a. metatarsalis dorsalis I is the only branch of the a. dorsalis pedis (40% of cases).
e All of the aa. metatarsales dorsales are supplied by rr. perforantes from the aa. metatarsales plantares (10%).
f Only the a. metatarsalis dorsalis I is supplied by a r. perforans (5%).

References 573
Index 575

References

Agur AMR. Grants Anatomie. Lehrbuch und Atlas. Stuttgart: Enke; 1999

Bähr M, Frotscher M. Duus' Neurologisch-topische Diagnostik. 8th ed. Stuttgart: Thieme; 2003

Baumgartl F. Das Kniegelenk. Berlin: Springer; 1969

Bohndorf K, Imhof H, Fischer W. Radiologische Diagnostik der Knochen und Gelenke. 2nd ed. Stuttgart: Thieme; 2006

Chassard J, Lapiné C. Étude radiologique de l'arcade pubienne chez la femme enceinte. J Radiol Electrol 7 (1923), 113

Christ B, Wachtler F. Medizinische Embryologie. Wiesbaden: Ullstein Medical; 1998

Dauber W. Feneis' Bild-Lexikon der Anatomie. 8th ed. Stuttgart: Thieme; 1999

Debrunner AM. Orthopädie. Die Störungen des Bewegungsapparates in Klinik und Praxis. 2nd ed. Stuttgart: Hans Huber; 1985

Debrunner HU. Gelenkmessung (Neutral-0-Methode), Längenmessung, Umfangmessung. Bern: AO-Bulletin; 1971

Drews U. Taschenatlas der Embryologie. Stuttgart: Thieme; 1993

Echtermeyer V, Bartsch S. Praxisbuch Schulter. 2nd ed. Stuttgart: Thieme; 2004

Faller A. Anatomie in Stichworten. Stuttgart: Enke; 1980

Ficat P. Pathologie Fémoro-Patellaire. Paris: Masson; 1970

Földi M, Kubik S. Lehrbuch der Lymphologie. 3rd ed. Stuttgart: Gustav Fischer; 1993

Frick H, Leonhardt H, Starck D. Allgemeine und spezielle Anatomie. Taschenlehrbuch der gesamten Anatomie, Bd. 1 u. 2. 4th ed. Stuttgart: Thieme; 1992

Fritsch H, Kühnel W. Taschenatlas der Anatomie. Bd. 2. 7th ed. Stuttgart: Thieme; 2001

Gert SD, Liebman M. Basiswissen Neuroanatomie. 4th ed. Stuttgart: Thieme; 2003

Goerke K. Taschenatlas der Geburtshilfe. Stuttgart: Thieme; 2002

Graumann W, Sasse D. CompactLehrbuch Anatomie. Bd. 4. Stuttgart: Schattauer; 2005

Hansen K, Schliack K. Segmentale Innervation. 2nd ed. Stuttgart: Thieme; 1962

Hees H. Grundriss und Atlas der Mikroskopischen Anatomie des Menschen. Bd. 1. Zytologie und Allgemeine Histologie. 12th ed. Stuttgart: Gustav Fischer; 1996

Henne-Bruns D, Dürig M, Kremer B. Chirurgie. 2nd ed. Stuttgart: Thieme; 2003

Hepp WR. Radiologie des Femoro-Patellargelenkes. Bücherei des Orthopäden. Bd. 37. Stuttgart: Enke; 1983

Hilgenreiner H. Zur Frühdiagnose der angeborenen Hüftgelenksverrenkung. Med Klein 21 1925. Stuttgart: Hippokrates; 1981

Hochschild J. Strukturen und Funktionen begreifen. Bd. 1 u. 2. Stuttgart: Thieme; 1998 u. 2002

von Hochstetter A, von Rechenberg HK, Schmidt R. Die intragluteale Injektion. Stuttgart: Thieme; 1958

Hüter-Becker A, Schewe H, Heipertz W. Physiotherapie. Bd. 1. Biomechanik, Arbeitsmedizin, Ergonomie. Stuttgart: Thieme; 1999

Junghanns, H. Die funktionelle Pathologie der Zwischenwirbelscheibe als Grundlage für klinische Betrachtungen. Langenbecks. Arch Klin Chir 1951; 267: 393–417

Kahle W, Frotscher M. Taschenatlas der Anatomie. Bd. 1. Stuttgart: Thieme; 2001

Kapandji IA. Funktionelle Anatomie der Gelenke. Bd. 1–3. 2nd ed. Stuttgart: Enke; 1992

Kaufmann P. Reife Plazenta. In: Becker V, Schiebler TH, Kubli F, Hrsg. Die Plazenta des Menschen. Stuttgart: Thieme; 1981

Kilka HG, Geiger P, Mehrkens HH. Die vertikale infraklavikuläre Blockade des Plexus brachialis. Anästhesist 1995; 44: 339–344

Klinke R, Silbernagl S. Lehrbuch der Physiologie. 3rd ed. Stuttgart: Thieme; 2001

Koebke, J. Anatomie des Handgelenkes und der Handwurzel. Unfallchirurgie 1988; 14: 74–79

Konermann W, Gruber G. Ultraschalldiagnostik der Bewegungsorgane. Kursbuch nach den Richtlinien der DEGUM und der DGOOC, Buch und DVD. 2nd ed. Stuttgart: Thieme; 2006

Kristic, RV. General Histology of the Mammals. Berlin: Springer; 1985

Kubik S. Lymphsystem der oberen Extremität. In: Földi M, Kubik S, Hrsg. Lehrbuch der Lymphologie für Mediziner und Physiotherapeuten. Stuttgart: Gustav Fischer; 1989

Kummer, B. Biomechanik der Wirbelgelenke. In: Meinicke, FW. Die Wirbelbogengelenke. Stuttgart: Hippokrates; 1983

von Lanz T, Wachsmuth W. Praktische Anatomie. Bd. I/3 Arm. 2nd ed. Berlin: Springer; 1959

von Lanz T, Wachsmuth W. Praktische Anatomie. Bd. I/4 Bein und Statik. Berlin: Springer; 1972

Lehnert G. Dopplersonographische Diagnostik der erektilen Dysfunktion unter Anwendung des Papaverintests [Dissertation] Kiel: Universität Kiel, Medizinische Fakultät; 1995

Lelièvre J. Pathologie du Pied. 2nd ed. Paris: Masson; 1961

Lippert H, Pabst R. Arterial Variations in Man. München: Bergmann; 1985

Loeweneck H. Diagnostische Anatomie. Berlin: Springer; 1981

Lüllmann-Rauch R. Histologie. 2nd ed. Stuttgart: Thieme; 2006

Lundborg G, Myrhage R, Rydevik B. The vascularization of human flexor tendons within the digital synovial sheath region – structural and functional aspects. J Hand Surg 1977; 2: 417–427

Luschka, H. Die Halbgelenke des menschlichen Körpers. Berlin: Reiner; 1858

Masuhr KF, Neumann M. Neurologie. Duale Reihe. 4th ed. Stuttgart: Hippokrates; 1998

Matzen P. Praktische Orthopädie. 3rd ed. Stuttgart: J. A. Barth Verlag im Thieme Verlag; 2002

Meier G, Bauereis C, Maurer H, Meier Th. Interscalenäre Plexusblockade. Anästhesist 2001; 50: 333–341

Merk H, Jerosch J, Hrsg. Arthroskopie des Schultergelenks. Stuttgart: Thieme, 2000

Möller TB, Reif E. Taschenatlas der Röntgenanatomie. 2nd ed. Stuttgart: Thieme; 1998

Möller TB, Reif E. Taschenatlas der Schnittbildanatomie. Bd. 3: Extremitäten, Gelenke, Wirbelsäule. Stuttgart: Thieme; 2007

Mow VC, Hou JS, Owens JM. Biphasic and quasilinear viscoelastic theories for hydrated soft tissue. In: Mow JC, Ratcliffe A, Woo SL. Biomechanics of Diarthrodial Joints. Springer: New York; 1990. Vol. I: 215–260

References

Mubarak SJ, Hargens AR. Compartment Syndromes and Volkamn's Contracture. Philadelphia: W. B. Saunders; 1981

Mumenthaler M, Stöhr M, Müller-Vahl H. Läsion peripherer Nerven und radikuläre Syndrome. 8th ed. Stuttgart: Thieme; 2003

Netter FH. Farbatlanten der Medizin. Stuttgart: Thieme; 2000

Niethard FU, Pfeil J. Orthopädie. Duale Reihe. 5th ed. Stuttgart: Thieme; 2005

Niethard FU. Kinderorthopädie. Stuttgart: Thieme; 1997

Noback CR, Strominger NL, Demarest RJ. The Human Nervous System. 4th ed. Philadelphia: Lea & Febiger; 1991

O'Rahilly, Müller RF. Developmental Stages in Human Embryos. Carnegie Institution of Washington: Publication 637; 1987

Pauwels F. Eine neue Theorie über den Einfluss mechanischer Reize auf die Differenzierung der Stützgewebe (X). Beitrag zur funktionellen Anatomie und kausalen Morphologie des Stützapparates. Z Anat Entwickl Gesch 1968; 121: 478–515

Pauwels F. Atlas zur Biomechanik der gesunden und kranken Hüfte: Prinzipien, Technik und Resultate einer kausalen Therapie. Heidelberg: Springer; 1973

Petersen W, Tillmann B. Structure and vascularization of the cruciate ligaments of the human knee joint. Z Orthop 1999; 137: 31–37

Pette D. Das adaptive Potential des Skelettmuskels. Dtsch Z Sportmed 1999; 50: 262–271

Pette D, Staron RS. Transitions of muscle fiber phenotypic profiles. Histochem Cell Biol 2001; 115 (5): 359–379

Platzer W. Taschenatlas der Anatomie. Bd. 1. Stuttgart: Thieme; 1999

Platzer W. Atlas der topographischen Anatomie. Stuttgart: Thieme; 1982

Rauber A, Kopsch F. Anatomie des Menschen. Bd. 1–4. Stuttgart: Thieme; Bd. 1. 2nd ed. 1997; Bde. 2 u. 3 1987; Bd. 4 1988

Reiser M, Kuhn FP, Debus J. Radiologie. Duale Reihe. 2nd ed. Stuttgart: Thieme; 2006

Rockwood CA. Subluxations and dislocations about the shoulder. In: Rockwood CA, Green DP (eds). Fractures. Philadelphia: Lippincott; 1984: 722–985

Rohen, JW. Topographische Anatomie. 10th ed. Stuttgart: Schattauer; 2000

Rohen JW, Yokochi C, Lütjen-Drecoll E. Anatomie des Menschen. 4th ed. Stuttgart: Schattauer; 2000

Romer AS, Parson TS. Vergleichende Anatomie der Wirbeltiere. 5th ed. Hamburg und Berlin: Paul Parey; 1983

Rudigier J. Kurzgefasste Handchirurgie. 5th ed. Stuttgart: Thieme; 2006

Sadler, TW. Medizinische Embryologie. 10th ed. Stuttgart: Thieme; 2003

Scheldrup, EW. Tendon sheath patterns in the hand. Surg Gynec-Obestetr 1951; 93: 16–22

Schmidt HM, Lanz U. Chirurgische Anatomie der Hand. 2nd ed. Stuttgart: Thieme; 2003

Schumpelick V. Hernien. 4th ed. Stuttgart: Thieme; 2000

Schünke M. Funktionelle Anatomie–Topographie und Funktion des Bewegungssystems. Stuttgart: Thieme; 2000

Silbernagl S, Despopoulos A. Taschenatlas der Physiologie. 6th ed. Stuttgart: Thieme; 2003

Sökeland J, Schulze H, Rübben H. Urologie. 12th ed. Stuttgart: Thieme; 2002

Starck D. Embryologie. 3rd ed. Stuttgart: Thieme; 1975

Streater GL. Developmental horizons in human embryos: age group XI, 13–20 somites and age group XII, 21–29 somites. Contrib Embryol 1942; 30: 211–245

Tossy JD, Newton CM, Sigmond HM. Acromioclavicular separations: useful and practical classification for treatment. Clin Orthop 1963; 28: 111–119

Uhthoff HK. The Embryology of the Human Locomotor System. Berlin: Springer; 1990

Vahlensieck M, Reiser M. MRT des Bewegungsapparates. 3rd ed. Stuttgart: Thieme; 2006

Weber U, Greulich M, Sparmann M. Orthopädische Mikrochirurgie. Stuttgart: Thieme; 1993

Wiberg G. Roentgenographic and anatomic studies on the femoropatellar joint. With special reference to chondromalacia patellae. Acta Orthop Scand 1941; 12: 319–410

Wiberg G. Studies on dysplastic acetabulum and congenital subluxation of the hip joint with special reference of the complication of osteoarthritis. Acta Chir Scand 1939; 83 (Suppl. 58)

Wolpert L, Beddington R, Brockes J et al. Entwicklungsbiologie. Weinheim: Spektrum; 1999

Index

Note: Imaging, regional anatomy, sectional anatomy, surface anatomy, and clinical applications are found under these main headings, broken out by region. *Italicized* page numbers in the main body of the index represent clinical applications.

A

Abbreviations, 26
Abdomen, 24. *See also specific organ*
　intersectiones tendineae, 31
　regional anatomy (regions) of, 36, 199
　skin of, 217
Abdominal press, 156, 160
Abdominal wall
　abdominal organ projection onto, 199
　hernias of, 220
　internal surface anatomy of, anterior, 212
　layers of, 179
　　comparison of, 217
　　schematic representation of, 217
　structure of, 178–179
　vasa lymphatica superficialia of, 195
Acetabular balcony, 433
Acetabular teardrop, 433
Acetabulum, 138–139, 143, 180, 184–185, 187, 412–413, 415–416, 427, 429, 432–434
　articular surface, 431
　development of, 15
　dysplasia, *439*
　facies lunata, 412, 430–431
　fractures of, *412*
　imaging of, 413
　inferolateral tilt of, 427
　labrum of, 42, 415–416, 430–432, 435
　limbus of, 412–413, 415, 426
　　in infant, 438
　　inferior, 427
　　superior, 439
　roof of, 427, 431, 433, 482–483
Acetylcholine, 59, 95
Acetylcholinesterase, 59
Achilles' tendon, 31, 55, 466, 471, 488–489, 510, 512, 515, 522, 546, 562–563
　rupture of, *488*
Achillodynia, *471*
Achondroplasia, 14
Acini, 208
Acraniata, 2
Acromion, 32–33, 38, 55, 61, 165–166, 238–243, 250, 258–269, 271, 273, 299, 301, 303, 305, 307, 309, 312–313, 331, 362, 374, 376, 384–385, 389–390

facies articularis clavicularis, 265, 267
　imaging of, 28
Acropodium, 20
Acrosome, 5
ACTH. *See* Adrenocorticotropic hormone
Actin, 55, 59
Actin filament(s), 59
　sliding, in muscle contraction, 59
Action potential(s), 78
ADH. *See* Antidiuretic hormone
Adhesive capsulitis, of shoulder/ art. humeri (glenohumeralis), *270*
Adipose tissue. *See* Fat
Adrenal medulla, 74
Adrenocorticotropic hormone (corticotropin, ACTH), 71
Adrenogenital syndrome, 223
Aggrecan, 45
Ala
　ossis ilii, 139, 142, 413
　ossis sacri, 114–115
Alar folds, of knee, 450
Albinism, 75
Allantois, 10
Alzheimer disease, *78*
Amelia, 14
Amnion, 6, 8
Amphiarthrosis/Amphiarthroses, 40
　of sacroiliac joint, 143
Amphibians
　arcus pharyngei in, 10
　embryonic development of, 2
　limb positions of, 21
Ampulla
　ductus deferentis, 188
　recti, 188
Anastomosis/Anastomoses
　arterial
　　at cubitus/art. cubiti, 393
　　in manus, 401
　arteriovenous, 67
　cavocaval, 192
　venous
　　in cardinal venous system, 13
　　of vitelline veins, 13
Anatomical bottlenecks, 96
Anatomical position, 26
　normal, in relation to line of gravity, 411
Anatomic snuffbox, 376
　boundaries of, 397
Androgen(s), 71
Anesthesia, *545, 547*
　axillary brachial plexus, 385
　lumbar spinal, *128*
　peripheral conduction, principle of, 384
Angiotensinogen, 71
Angle
　acetabular, of Hilgenreiner, 439
　of acetabular inlet plane (in adult)
　　sagittal, 427
　　transverse, 427

anorectal, 186
anteversion, 417
　of cavitas glenoidalis, with sagittal plane, 245
CCD (centrum-collum-diaphysis), 415, 437
center-edge, of Wiberg, 427, 439
collum femoris, 415
　articulatio coxae load and, 437
　decreased, 415, 437
　increased, 415, 437
　normal, 437
　trabecular patterns and, 415
cubital, 282
dorsopalmar inclination, 256
femorotibial, 410
first metatarsophalangeal, in hallux valgus, 469
intermetatarsal, in hallux valgus, 469
lumbosacral, 103
patellar facet, 419
pelvic inclination, 103
pennation, 57
radioulnar inclination, 256
sacral, 103
torsion, of humerus, 247
Angulus
　acromii, 243
　costae, 132, 135
　infrasternalis, 136
　mandibulae, 32–33
　scapulae
　　with clavicula, 241
　　with coronal plane, 241
　　lateralis, 30, 33, 35, 101, 165, 238–240, 243, 250, 301, 303, 307
　　　lateral rotation of, 274
　　superior, 33, 165, 239–240, 242–243, 245, 259, 301, 303, 362
　sterni, 34, 132, 134
　subpubicus, 138–139, 141
　venosus
　　dexter, 194
　　sinister, 194
Ankle mortise, 408, 410, 420, 455, 458
　distal view of, 421
　facies articularis, 458
　　inferior, 458
　stability of, 458
　syndesmotic ligamentum of, 461
Ankylosis, 40
　of articulatio genus, *40, 50*
　of shoulder/art. humeri (glenohumeralis), *275*
Anlage
　corpus vertebrae, 7, 104
　ganglion sympatheticum, 74
　glandula suprarenalis, 74
　glandula thyroidea, 10
　liver, 13
　medulla spinalis, 7
　　cornu anterius, 81, 88, 104
　　cornu posterius, 81, 104

dorsal, 7
embryology of, 104
ventral, 7
pituitary, 72
plexus myentericus, 74
tracheal, 10
ANS. *See* Autonomic nervous system
Antebrachium, 238, 280–281
　anterior, 37, 394–395
　arteriae of, 354–355
　　development of, 395
　　normal anatomy of, 395
　　rr. perforantes, 355
　　variants, 395
　cross sections, 342–343
　deep layer, anterior view, 394
　extensor side, arteriae of, 396
　length of, 239
　musculi of, 296, 314–317
　　anterior, 296, 338–339
　　cross sectional anatomy of, 342–343
　　extensors
　　　deep, 296, 318–319
　　　superficial, 296, 318–319
　　flexors
　　　deep, 296, 314–315
　　　superficial, 296, 314–315
　　posterior, 296, 340–341
　　radial, 296, 316–317
　ossa, 38, 238, 247, 250
　posterior, 37, 396–397
　pronation, with elbow flexed, 310
　superficial layer, anterior view, 394
　supination, with elbow flexed, 310
　"windowed" dissection of, 343
Antecurvation, 27
Antetarsus, 422
Antidiuretic hormone (vasopressin, ADH), 71
Anulus femoralis, 213
Anulus fibrosus, 116–118, 126, 130–131, 137
　crossing fiber systems of, 116
　embryology of, 104
　functions of, 117
　inner zone of, 116
　outer zone of, 116, 126
Anulus inguinalis
　profundus, 177, 211–214, 216–217
　superficialis, 153, 176, 210–212, 214–217, 225, 228, 532–533, 545, 549
　　crus laterale, 210, 549
　　crus mediale, 210, 549
　　fiberae intercrurales, 210
Anulus umbilicalis, 153
Anus, 24, 180, 188, 223, 230, 528, 556, 558–559
　neurovascular structures of, 558
　sensory innervation of, 543

575

A Aorta

Aorta, 63, 175
　abdominalis, 105, 160, 167, 190–191, 204, 227, 524–525, 550
　ascendens, 63, 105, 190
　　development of, 12
　descendens, 63, 190
　　development of, 12
　dorsalis (paired), 6–7, 10, 12
　embryology of, 7, 104
　luminal radius of, 64
　thoracica, 190–191, 205
　ventralis (paired), 10, 12
　wall of
　　structure of, 64
　　thickness of, 64
Aortic arch. See Arcus, aortae
Apertura
　pelvis
　　in female, 139
　　inferior, AP diameter of, 114, 141–142
　　superior, 141–142
　thoracis
　　inferior, 132
　　superior, 132
Apex
　dentis, 121
　ossis sacri, 114
　patellae, 418
Apocytosis, 70
Aponeurosis, 57
　bicipitalis, 311, 336, 338, 343, 356, 386, 392, 394
　dorsal, of toes, 492, 494, 515
　lateral abdominal wall muscles, 155
　of m. flexor digitorum longus, 569
　of m. latissimus dorsi, 55, 166, 168
　of m. obliquus externus abdominis, 153, 155, 176, 179, 210–211, 215, 216–217, 219, 549
　of m. obliquus internus abdominis, 153, 176, 179
　of m. transversus abdominis, 153, 155, 177, 179
　of m. triceps brachii, 376
　palmaris, 54, 315, 343, 348, 375, 394, 398, 404–405
　　fasciculi longitudinales, 348
　　fasciculi transversi, 348
　plantaris, 458, 464, 466–467, 471–472, 492, 494, 516–520, 523, 541, 564–565, 569
　　exostosis of, 471
Apophyseal plate(s)
　of membrum inferius, 19
　of membrum superius, 18
Apophysis, 39
Apoptosis
　in formation of joint cavity, 15
　interdigital, 14
Appendix
　epididymidis, 226
　testis, 226
　vermiformis, 68
Arcade(s)
　psoas, 160, 174–175
　quadratus, 204
　quadratus lumborum, 160, 174–175

Arch(es)
　dorsal, 3
　of manus
　　carpal, 256
　　longitudinal, 256
　　metacarpal, 256
　ventral, 3
Arcus
　aortae, 190–191
　　arteries derived from, development of, 12
　　duplicated, 12
　　embryology of, 10
　　right-sided, 12
　atlantis
　　anterior, 109, 121, 123
　　posterior, 108–109, 121–123
　costalis, 31, 35, 132–133, 160, 175, 199, 335
　of forefoot, stabilizers of, 465
　hemalis, 3
　iliopectineus, 210, 213, 477, 549
　metatarsal, stabilizers of, 465
　neuralis, 3
　palmaris, 395
　　profundus, 354–355, 395, 401–402, 405
　　rami, 401
　　relationship to carpal and digital tendon sheaths, 402
　　superficialis, 354–355, 395, 398, 400–403, 405
　　　normal anatomy of, 400
　　　rami, 400
　　　variants, 400
　pedis
　　anterior, stabilizers of, 465
　　longitudinalis, 458–459, 464, 466–467, 490, 492, 494, 516
　　　constraining force (H) needed to maintain, 467
　　　stabilizers of, 466
　　　support of, 467
　　transversus, 464, 490, 492
　　　stabilizers of, 465
　pharyngei, 2, 14
　　derivatives of, 11
　　development of, 10–11
　　hyoid, 11
　　mandibular, 11
　　musculature, 10
　　nerve, 10–11
　　skeletal element of, 10
　　structure of, 10
　plantaris, 464
　　profundus, 524–525, 564–565, 569
　pubis, 139, 141
　tarsal, stabilizers of, 465
　tendineus
　　of aorta abdominalis, 160
　　m. levatoris ani, 163, 185–187, 558
　　musculi solei, 488–489, 541, 562
　　of psoas arcade, 160
　　of quadratus lumborum arcade, 160
　venosus dorsalis pedis, 526
　venosus palmaris
　　profundus, 357
　　superficialis, 357
　venosus plantaris, 526

vertebrae, 3, 7, 106–109, 111, 113, 116–119, 121–122, 129–131, 167
zygomaticus, 32
Area intercondylaris, tibial
　anterior, 421, 451
　posterior, 421
Areola mammae, 30, 199, 208
Arm. See Brachium
Arm span, 22
Arteriae (in general, including artery types), 62
　elastic type, 62, 64
　large, 64
　luminal radius of, 64
　muscular type, 62, 64
　　wall structure of, 64
　small, 64
　　luminal radius of, 64
　　wall of, thickness of, 64
　structure of, 64
　of systemic circulation, overview of, 62
　wall of
　　structure of, 64–65
　　thickness of, 64
Arteriae (named)
　antebrachiales superficiales, 395
　arcuata, 524–525, 568
　arcus aortae, 10
　arcus pharyngei, 10
　axillaris, 63, 190, 207, 209, 261, 273, 297, 354–355, 359–361, 363, 366–367, 369, 371, 373, 379, 381, 384–387, 389, 391
　　course of, 354
　　damage to, 262
　　development of, 395
　　origin of, 382
　　rami, 354, 382
　acromialis, 382
　　clavicularis, 382
　　deltoideus, 382
　　normal anatomy of, 383
　　pectoralis, 382
　　variants, 383
　　relationship with plexus brachialis fasciculi, 360, 382
　brachialis, 63, 342, 354–355, 359, 382–383, 386–387, 391–392, 394–395
　　course of, 354, 387
　　development of, 395
　　normal anatomy of, 387, 391
　　profunda, 63, 354–355
　　rami, 354, 391
　　　acromialis, 387
　　　deltoideus, 387
　　　pectorales, 387
　　　variants, 391
　　superficialis, 387, 395
　　variants, 387, 391
　bulbi penis, 230
　bulbi vestibuli, 234
　carotis
　　communis, 63, 167, 190–191, 355, 361, 379, 382, 389
　　dextra, development of, 12
　　sinistra, arising from truncus brachiocephalicus, 12
　　externa, 63
　　　development of, 12
　　interna, 63
　　　development of, 12
　　　embryology of, 10, 12

carpalis dorsalis, 401
cervicalis
　ascendens, 379
　profunda, 191, 203, 354, 379
　superficialis, 379
circumflexa humeri
　anterior, 246, 354–355, 381–383, 387, 391
　communis, 383
　posterior, 203, 246, 273, 354–355, 361, 381–383, 387, 390–391
circumflexa ilium
　profunda, 190, 212, 524–525, 548, 550
　superficialis, 190, 206, 524–525, 548, 550
　　in groin skin flaps for plastic surgery, 207
circumflex scapulae, 203, 354–355, 379, 381–383, 387, 389–391
collateralis
　media, 354–355, 387
　radialis, 354–355, 387, 396
　ulnaris
　　inferior, 354–355, 386–387, 392–394
　　superior, 354–355, 386–387, 392–394
comitans n. ischiadici, 29, 553, 560
corona mortis, ramus obturatorius, 212–213
cremasterica, 225, 227, 524
digitales dorsales
　of manus, 354–355, 396, 398–399, 401
　of pes, 568
digitales palmares, 398–401, 403
　communes, 354, 398, 400–401, 403
　propriae, 354–355, 398, 401
digitales plantares, 565
　communes, 524–525, 565
　propriae, 525, 564–565
dorsalis clitoridis, 234
dorsalis pedis, 63, 524–525, 565–566, 568–569
　rami, 524–525
　　plantaris profundus, 564–565
dorsalis penis, 228–231
dorsalis scapulae, 379, 389–391
ductus deferentis, 212–213, 225, 227
epigastrica
　inferior, 190–191, 206, 211–217, 524–525, 548, 550
　　in arterial supply, 207
　　ramus pubica, 213
　superficialis, 190, 206–207, 524–525, 548, 550
　superior, 190–191, 206–207
femoralis, 29, 63, 189–191, 207, 210–215, 225, 227–228, 434, 522, 524–525, 528, 545, 548–551, 558, 560
　in hiatus saphenus, 215
　rami, 524, 550
　　variants of, 550
　superficial, 550

576

femoris
  circumflexa
    lateralis, 431, 524–525, 550–551
    medialis, 431, 524–525, 550–551, 555, 560
  profunda, 63
fibularis, 63, 524–525, 562, 567
  rami calcanei, 524–525
    lateralis, 563
    medialis, 563
  rami malleolares laterales, 524–525, 563
  rami musculares, 563
  ramus communicans, 524–525, 562–563
  ramus perforans, 524–525, 562–563, 568
gastric, sinistra, 63
genus
  descendens, 524, 548, 550–551
  inferior
    lateralis, 524–525, 550, 561, 563, 568
    medialis, 524–525, 550, 561, 563, 568
  media, 524–525, 561, 563
  superior
    lateralis, 524–525, 550, 561, 563, 568
    medialis, 524–525, 550, 561, 563, 568
gluteae
  inferior, 525, 550–551, 553–555, 558, 560
  superior, 525, 550–551, 553–555, 558, 560
hallucis plantaris medialis, 565
helicinae, 231
hepatica communis, 63
iliaca
  communis, 63, 204, 524–525, 550, 558
  externa, 63, 182, 190–191, 204, 207, 524–525, 550–551, 558
    rami, 524–525, 548
    ramus pubicus, 524, 548, 550
  interna, 63, 190, 204, 230, 524–525, 548, 550, 558
    rami, 550
iliolumbalis, 190
intercostales, 173, 206–208
  anteriores, 190–191
    rami of, 190
  course of, 191
  posteriores, 190–191, 205
  rami, 191
  suprema, 191, 354, 379
interossea
  anterior, 354–355, 387, 394–395, 401
    r. posterior, 355
  communis, 354–355, 387, 394, 401
    development of, 395
  development of, 395
  posterior, 354–355, 387, 394, 396, 401
  recurrens, 354, 396, 401
ligamenti teretis uteri, 524
lumbales, 190–191
malleolares anterior
  lateralis, 524–525, 568
  medialis, 524–525, 568
mammaria interna, 354
mediagenus, 446, 524
mediana, 395
  persistent, 395
  variant, 400
mesenterica
  inferior, 63
  superior, 63
metacarpales
  dorsales, 354–355, 396, 401
  palmares, 354–355, 401
metatarsales
  dorsales, 524–525, 566, 568–569
  plantares, 524–525, 564–565
    ramus perforans, 569
musculophrenica, 190–191, 205
obturatoria, 212–213, 431, 550, 558
occipitalis, 203
ovarica, 63
palmaris superficialis, 402
pericardiacophrenica, 205
perinealis, 230
phrenica superior dextra, 205
plantaris, 563
  lateralis, 524–525, 563–565, 569
    rami profundus, 564–565
  medialis, 524–525, 563, 565, 569
    ramus profundus, 524–525, 564–565
    ramus superficialis, 524, 563–565
  middle layer, 564
  overview of, 565
  profundus, 565, 568
  variants, 564
poplitea, 63, 441, 451, 453, 524–525, 529, 550, 552, 560–563, 568
  palpation of, in fossa poplitea, 561
  rami, 446, 524–525, 563
    in fossa poplitea, 561
    variants of, 563
princeps pollicis, 354–355
profunda brachii, 354–355, 382, 387, 390–391
  development of, 395
profunda clitoridis, 234
profunda femoris, 29, 431, 524–525, 548, 550–551, 560
  blood supply to thigh, 551
  course of, 550
  perforantes, 550–551, 560
profunda penis, 228–231
pudendae
  externae, 228, 230, 525, 548, 550
    profunda, 524
    superficialis, 524–525
  interna, 189, 229–230, 234, 554–555, 557–558
    distribution of, 558
pulmonalis, 62
  development of, 12
radialis, 63, 342, 354–355, 387, 392–398, 400–405
  course of, 354
  development of, 395
  indicis, 354
  perforantes, 354–355, 401
  rami, 354
    carpalis dorsalis, 354–355, 396–397
    carpalis palmaris, 354–355
    palmaris superficialis, 354–355, 398, 400–401, 403–405
radicularis
  anterior, 191
  posterior, 191
rectalis
  inferior, 230, 234
  media, 230
recurrens radialis, 354–355, 387, 392–393
recurrens tibialis
  anterior, 524–525, 561, 563, 568
  posterior, 524–525, 561, 563
recurrens ulnaris, 354–355, 387
renalis, 63
sacralis
  lateralis, 190, 204, 550
  mediana, 190
scrotalis anterior, 228
spirales uteri, 8–9
splenica, 63
subclavia, 190–191, 207, 209, 354–355, 359, 361, 363, 379, 381–384, 386, 389, 391
  course of, 354
  development of, 12
  dextra, 63
    course of, 379
    development of, 12
    origin of, 379
    rami, 379
  rami, 354, 379
    normal anatomy of, 379
    variants, 379
  sinistra, 63
    development of, 12
subcostalis, 190–191
subscapularis, 203, 273, 354–355, 379, 382–383, 387, 388–391, 389, 391
surales, 524, 561, 563
tarsalis
  lateralis, 524, 568
  medialis, 524, 563
testicularis, 212–213, 225–227
thoracica
  interna, 63, 190–191, 205–207, 209, 354–355, 379, 382
    in arterial supply, 207
  lateralis, 190, 206–207, 209, 354–355, 379, 381–383, 387
  superior, 190, 207, 354–355, 379, 381–383
thoracoacromialis, 190, 354–355, 379–383, 387
  acromialis, 354–355
  clavicularis, 354
  deltoideus, 354–355
  pectoralis, 354–355
thoracodorsalis, 190, 207, 354–355, 379, 381–383, 387, 389, 391
thyroidea inferior, 354, 379
tibialis
  anterior, 63, 522, 524–525, 550, 561, 563, 566–568
    rami, 524–525
  posterior, 63, 471, 522, 524–525, 561–563, 565, 567
    palpation of, 563
    rami, 524–525
    rami calcanei, 524
    rami calcaneus medialis, 563
    rami malleolares mediales, 524–525, 563
transversa colli (cervicis), 202–203, 378–379, 389, 391
  rami
    profundus, 203, 389
    superficialis, 203
ulnaris, 63, 342, 354–355, 387, 392–395, 398, 400–405
  carpalis dorsalis, 354–355, 396, 401
  carpalis palmaris, 354
  course of, 354
    in ulnar tunnel and deep palm, 405
  development of, 395
  palmaris profundus, 354
  profundus, 400–401, 403–405
  rami, 354
  superficialis, 356, 395, 405
umbilicalis, 9, 12
  obliterated, 217
urethralis, 228–231
vertebralis, 108, 126–127, 190–191, 203, 354–355, 379, 382, 389, 391
  in foramen transversarium, 126
  injury to, 384
  topographical relationship to processi uncinatus, 126
vitellina, 12
Arteriae (of region or organ)
  of antebrachium
    development of, 395
    extensor side, 396
    "high origin" of, 387
    normal anatomy of, 395
    variants, 395
  of brachium, 354–355
    overview of, 354
  of crus, 525, 568
  of digiti, extensor side, 396
  of fossa poplitea, 525
  of manus, 354–355
    dorsum, 396
    palma, 398
  of membrum inferius, 524–525
    overview of, 524
    segments of, 524
  of membrum superius, 354–355
    development of, 395
  of pes, 568
    dorsum, 568
    planta, 525, 564–565
  of scrotum, 230
  segmental trunk, development of, 12
  of shoulder/art. humeri (glenohumeralis), 354–355
    overview of, 354
  of testis, 227, 230
  of thigh, 550–551
  of trunk wall, 190–191
Arteriola, 64
  blood flow in, 67
  luminal radius of, 64

577

A Arteriola (cont.)

precapillary, 66
wall of, thickness of, 64
Arthrodesis, 40, 46
  for pes equinovarus, 40
  triple, 40
Arthrography, 40
Arthroscopy, 40
  of shoulder/art. humeri (glenohumeralis), 270–271
  patient positioning for, 270
  portals for, 270
Arthrosis
  latent, 47
  uncovertebral, 126–127
    in columna vertebralis pars cervicalis, 127
Articulatio coxae, 38, 408, 411
  abduction of, 436
  adduction of, 436
  anterior view of, 426
  axes of motion of, 436
    longitudinal, 436
    sagittal, 436
    transverse, 436
  biomechanics of, 436–437
  capsula of, 430–431, 435
    weak spots in, 429
  with centered caput femoris, 416
  coronal section through, 432
  cross-sectional anatomy of, 432, 434–435
  degenerative joint disease, 46–47
  development of, 15, 438–439
    radiographic, 438
  dislocation of
    congenital, 439
    after hip replacement, 429
    traumatic, 429
  dysplasia, congenital, 439
  effusions in, 435, 483
  extension of, 436
    normal, 436
    range of, determination of, 436
  external rotation of, 416, 436
  flexion of, 436
  imaging of, 432–435
  of infant
    radiographic evaluation of, 439
    sonographic classification of, 438
    sonographic evaluation of, 438
  instability of, 439
  internal rotation of, 416, 436
  ligamentum of, 43, 428–431
    actions of, as function of joint position, 429
    capsularia, twisting mechanism of, 429
  load on, 51, 437
  movements of, 49, 436, 496–497
  musculi of, 474–482
    abductors, 496–497
    adductor group, 474, 480–481, 496–497
      anterior view, 481
    anterior view, 477, 481, 503–505
    extensors, 496–497
    external rotators, 496–497
    flexors, 476, 496–497
    functional testing of, 497

functions of, 496–497
inner, 474, 476–477
  anterior view, 477
innervation of, 530
internal rotators, 496–497
lateral view, 506
medial view, 502
origins and insertions, 504–505, 508–509
outer, 474, 478–479
  deep, 479
  horizontally oriented, 478
  superficial, 479
  vertically oriented, 478
posterior view, 479, 507–509
shortening, symptoms of, 497
weakness, symptoms of, 497
ossa, 138, 408, 411–413
  articulating, 426–427
  before closure of growth plates, 41, 438
  column principle of, 412
  fusion of, 41
  growth plates of, 41, 438
osteoarthritis of, 51
  alleviation of stress on affected side in, 437
pediatric, imaging of, 438–439
posterior view of, 426
range of motion of, 436
sonographic examination of, 435
Thomas maneuver for, 436
total endoprosthesis of, 46
transverse section through, 434
Articulationes (in general, including joint types)
  abduction of, 48
  cartilaginous, 40–41
  degrees of freedom of, 48
  ellipsoidal, 40
    movements of, 49
  false, 40
  fibrous, 40–41
  flexion of, 48
  form of, and movements, 49
  as functional units, 48
  gliding of, 48–49
  hinge, movements of, 49
  imaging of, MRI, 452
  interzone of, 15
  load on, size of force-transmitting surface and, 51
  mechanics of, 50–51
  motion of, 48–49
    constraints on, 50
    convex-concave rule for, 49
  pivot, movements of, 49
  plane, 259
  range of motion
    bony constraints on, 50
    factors affecting, 50
    ligamentous constraints on, 50, 428
    measurement of, neutral-zero method, 50
    muscular constraints on, 50
    soft-tissue constraints on, 50
  rolling of, 48
  rotation of, 48–49
  saddle, 40
    movements of, 49
  sellar, 293
  sliding of, 48–49
  spheroidal, 40
    movements of, 48–49

stabilization of
  active, 50
  passive, 50
stiff, 40
synovial, 42
translation of, 48–49
true, 40, 42–45
  extra-articular structures of, 43
  intra-articular structures of, 42
  structure of, 42
Articulationes (named)
  acromioclavicularis, 31, 38, 238, 240–242, 258–259, 273
    imaging of, 28
    ligamenta of, 259–260
  atlantoaxialis, 120, 122–123
    lateralis, 120, 122–123, 126–127
    ligaments of, 122–123
    mediana, 120, 123
      ligaments of, 123
    range of motion, 125
    rotation, 125
  atlantooccipitalis, 120, 122–123
    extension, 125
    flexion, 125
    lateral flexion, 125
    ligaments of, 122–123
    range of motion, 125
  calcaneocuboidea, 454–457
  carpometacarpales, 38, 238, 249, 294
    pollicis, 238, 249, 253, 292–294
      articulating surfaces, 292
      axes of motion, 292–293
      movements of, 49, 293
      rotation-induced incongruity, 293
  clavicular
    lateral, 259
    medial, 259
  costovertebral, 106, 136–137
  craniovertebral, muscles of, 144
  cuneocuboidea, 454
  cuneonavicularis, 454–457
  elbow. See Cubitus/Art. cubiti
  femoropatellar, 38, 419, 440–441, 445
    joint pressure in, 51
    load on, 51
    movements of, 49
    net joint force and resultant force vector in, 51
  femorotibialis, 440
    coronal section through, 447
  genus, 38, 408, 411
    ankylosis of, 40, 50
    arterial anastomotic network around, 524–525, 548
    arterial network of, 524
    ballottable patella sign in, 451
    capsule of, 61, 443, 450–451, 482
    cavitas articularis, 450–451
    cross-sectional anatomy of, 452–453
    degenerative joint disease, 46
    effusion in, 451
    extension of, 448–449
    external rotation of, 449
    flexion of, 448–449
      recessus suprapatellaris unfolding during, 451
    imaging of, 441, 452–453

anteroposterior view, 441
lateral view, 441
tangential (sunrise) view, 441
internal rotation of, 449
ligamentum of, 43, 50, 442–443, 482
  collaterale, 441–445
  cruciata, 43, 441, 443–444
  extrinsic, 443
  intrinsic, 443
medial joint space of, 409
mid-sagittal section through, 451
movements of, 448–449, 498–499
musculi of, 480, 482, 484–485, 488
  dysbalances of, 499
  extensors, 498–499
  external rotators, 498–499
  flexors, 498–499
  functional testing of, 499
  functions of, 498–499
  internal rotators, 498–499
  shortening, symptoms of, 499
  weakness, symptoms of, 499
ossa, 440–441
osteoarthritis of, 410
range of motion of, 449
stabilization of, 50, 444–445, 448
  active, 50
  deficient, 482
  passive, 50
unhappy triad of, 448
valgus deformity of, 27
varus deformity of, 27
hip. See Articulatio coxae
humeri (glenohumeralis). See Shoulder/Art. humeri (glenohumeralis)
humeroradialis, 238, 249–250, 276–277
  range of motion, 283
humeroulnaris, 238, 249, 276–277
  range of motion, 283
iliosacral, development of, 15
intercuneiformes, 454–455
intermetatarsales, 454–455
interphalangeae
  manus, 239
  pollicis, 253, 294
  proximal, 494
interphalangeae distales, manus, 238, 249, 253, 290–291, 294, 347, 376
  degenerative joint disease, 46
  extension, 295
  flexion, 295
  ligg. collateralia, 284
interphalangeae distales, pedis, 454–455, 486–487, 490, 492–493
interphalangeae proximales, 238, 249, 253, 290, 294, 347, 376, 454–455
  degenerative joint disease, 46
  flexion, 295
  ligg. collateralia, 284–285
mediocarpalis, 253, 287, 294
  articular surfaces of, 255
  movements of, 295

578

metacarpophalangeales, 32, 238–239, 249, 253, 284–285, 290, 294, 347, 376, 398
  abduction, 295
  adduction, 295
  capsula articularis of, 291
  extension, 295
  flexion, 295
  ligamenta of, 291
  pollicis, 253, 294
metatarsophalangeae, 32, 454–457, 486, 490, 492–494
  capsula, 460
    of phalanx I, 461
  hallux, 455, 465
  subluxation, in hallux valgus, 469
radiocarpalis, 38, 238, 249, 253, 256, 286–287, 294
  articular surfaces of, 255
  movements of, 49, 295
  radial compartment, 255
  ulnar compartment, 255
radioulnaris, 256
  cross section through, in pronation, 281
  distalis, 38, 238, 248–249, 251, 253, 280, 285–287, 294
    cross section through, 345
    pronation, ligamenta and axes for, 280
    supination, ligamenta and axes for, 280
  movements of, 282–283
  proximalis, 38, 238, 248–249, 251, 276–277, 279–281
    movements of, 49
    pronation, ligamenta and axes for, 280
    supination, ligamenta and axes for, 280
sacrococcygea, 114
sacroiliaca, 38, 138–139, 141, 143, 186, 408, 413, 476
  articular surfaces of, 142
  functional impairment, *497*, 499
  hypermobility of, *143*
  ligaments of, 143
  locking of, *143*
  nutation of, 142
  pain in, *143*
scapulothoracic, 258, 274
  anterolateral part, 261
  dorsomedial part, 261
  location of, 261
shoulder. *See* Shoulder/Art. humeri (glenohumeralis)
sternoclavicularis, 32, 38, 238, 240–241, 258–259
  development of, 15
  ligamenta of, 259–260
  movements of, 274
  range of motion of, 274
sternocostalis, 259
subtalaris, 38, 411, 454–455, 464, 470, 486, 488–490
  anterior compartment of, 459, 466
  axis of motion, 462
  coronal section through, 455
  facies articular of, 459
  movements of, 500–501
  posterior compartment of, 459, 466
  range of motion of, 463
talocalcaneonavicularis, 454, 459, 466
talocalcaneum, 454–455, 459, 466
talocruralis, 38, 408, 410–411, 454–455, 466, 470, 486, 488–490
  articulating skeletal elements of, 458
  axis of motion, 462
  coronal section through, 455
  imaging of, 470
  ligamenta of, 461
  movements of, 500–501
  muscles of, 500–501
    dorsal extensors, 500–501
    functional testing of, *501*
    plantar flexors, 500–501
    pronators, 500–501
    shortening, symptoms of, *501*
    supinators, 500–501
    weakness, symptoms of, *501*
  partal body weight on, transfer of, 464
  range of motion of, 463
  sprains, *460*
  supination trauma, *460, 501*
talonavicularis, 454–457, 470
tarsi transversa, 454–455
  axis of motion, 462
  movements of, 500–501
  range of pronation/supination of, 463
tarsometatarsales, 454–457
  axis of motion, 462
  range of pronation/supination of, 463
temporomandibular, 38
  development of, 15
thumb saddle, 286
tibiofibularis, 38, 420, 440, 446
Articulationes (of region or structure)
  of back. *See* Columna vertebralis
  capitis costae, 137
  costotransversaria, 132, 137
  costovertebral, ligamentum of, 137
  craniovertebral, 120
    ligaments of, 122–123
  development of, 15
    through appositional growth, 15
    through segmentation, 15
    vascularization and, 15
  digiti (manus)
    classification of, 294
    range of motion of, 295
  facet, 110, 121, 124. *See also* Articulationes (of region or structure), zygapophysialis
    osteoarthritis of, *127*
  growth
    appositional, 15
    interstitial, 15
  of hallux, range of motion of, 463
  interphalangeal
    manus, 32
    pedis, 32
  intervertebral, 124, 476
  of manus, interphalangeal, 32
  of pes, 454–455, 458–459
    facies articularis, 456–457
    interphalangeal, 32
  sternocostales, 134
  uncovertebral, 126
  vertebral, 130
    degenerative joint disease, *46*
    movements of, 49
  zygapophysialis, 108–110, 112, 120, 122, 124, 127, 129
Astrocyte(s)
  fibrillary, 79
  functions of, 79
  protoplasmic, 79
Atlas (C I), 38, 102, 107–109, 120–123, 126–127, 149–151, 159, 165, 172, 241, 260, 301, 362, 367
  arcus posterior of, 171
  ligamentum cruciforme, 122
  ligamentum transversum, 121–123
  massa lateralis, 109, 122–123
  processi transversus, 33, 151, 170–171, 203, 362
  tuberculum posterius, 151
Atrial natriuretic peptide, 71
Atrium/Atria, cardiac
  dextrum, 62
  hormone secretion by, 71
  sinistrum, 62
Autonomic nervous system (ANS), 94–95
  circuit diagram of, 95
  and endocrine system, 71
  hormones and hormonelike substance secretion by, 71
  parasympathetic division, 94–95
  sympathetic division, 94–95
Autopodium, 20–21
Axilla
  anterior wall, 380–381
  lateral wall, 380
  medial wall, 380
  posterior wall, 380, 382–383
  quadrangular space of, 390
  regional anatomy (regions) of, 37
  transverse section through, 381
  triangular space of, 390
Axillary block, *384*
Axillary fold
  anterior, 380, 385
  posterior, 380
Axillary nerve palsy, *367*
Axis (C II), 38, 102, 107–109, 120, 122–123, 126, 149–151, 159, 165, 172–173, 301
  corpus, 121–123, 127
  dens of. *See* Dens axis (C II)
  processi spinosus, 123, 151, 170–171, 203
Axis/Axes
  abduction/adduction, of humerus, 304
  anatomic, of femur, 410, 415
  of articulatio coxae
    abduction/adduction, 479
    flexion/extension, 480
  of articulatio subtalaris, 500
  of articulatio talocruralis, 500
  brainstem, 77
  calcaneal, 462–463
  cardinal, in human body, 27
  collum femoris, 415, 417
  condylar, 417
  of costovertebral joint movement, 136
  deviations, 27
  dorsopalmar, 295
  of embryonic differentiation, 14
  epicondylar (of cubitus), 247
  of hindfoot, 462
  horizontal, of CNS, 77
  of human body, 6
  humeral head, 247
  instantaneous, 48
  of knee flexion/extension, 449
  long, of scapula, 240
  longitudinal (vertical), 27
  of manus movements, 295
  mechanical, of leg, 410
    longitudinal, 410
    normal, age-related changes in, 411
  of motion, 48–49
    of articulatio carpometacarpalis pollicis, 292–293
    of articulatio coxae, 436
    of articulatio genus, 61, 449
      transverse, 449, 482
    articulatio sternoclavicularis, 274
    articulatio subtalaris, 462
    articulatio talocruralis, 462
    articulatio tarsi transversa, 462
    articulatio tarsometatarsales, 462
    forefoot, 462
    in pes, 462
    of shoulder/art. humeri (glenohumeralis), 275
  of nutational movement, in sacroiliac joint, 142
  of pronation/supination
    of antebrachium, 280–281
    of manus, 283
  of rib movement, 136
  sagittal, 27
  tibial, 462
  of tibial rotation, 449
  transverse (horizontal), 27
  transverse, of articulatio genus, 449, 482
Axon hillock, 78
Axon(s), 75, 78–79
  extrapyramidal, 91
  of motor nerve cell, 59
  myelinated, 75, 92–93
  unmyelinated, 75, 92–93

B

Back
  articulationes of. *See* Columna vertebralis
  musculi of, 164, 166
    and erect posture, 50
    intrinsic, 31, 56, 144, 146–147, 155, 166–171, 174
      embryology of, 7, 145
      intertransverse system, 144, 146–147
      lateral tract, 144, 146–147, 168
      medial tract, 144, 148–149, 169
      and posture, 157
      sacrospinal system, 144, 146–147

B Back (cont.)

spinal system, 144, 148–149
spinotransverse system, 144, 146–147
transversospinal system, 144, 148–149
intrinsic and nonintrinsic, fascia thoracolumbalis as partition between, 166
ossa. See Columna vertebralis; Costae; Vertebra/Vertebrae
regional anatomy (regions) of, 36
Backpack paralysis, 363
BALT. See Bronchus-associated lymphatic tissue
Bankert lesion, 262, 270
Bartholinitis, 235
Basal lamina, in blood vessel wall, 64, 67
Basipodium, 20
Biogenetic law, 2
Bird(s), embryonic development of, 2
Blastema, limb bud, 14
Blastocyst, implantation of, 5
Bleeding. See Hematoma
Blood
  circulation of, 62
  transport function of, 66
Blood-brain barrier, 93
Blood pressure, 62
  arterial, in change from recumbent to standing position, 65
  in terminal vascular bed, 66
  venous, in change from recumbent to standing position, 65
Blood vessel(s), 62. See also Arteriae; Capillary(ies); Terminal vascular bed; Vasa; Venae
  afferent, of skeletal muscle, 58
  cranial, innervation of, 94
  diameter of, by type, 66
  embryonic, 8
  flow velocity in, by type, 66
  number of, by type, 66
  organization of, 64
  percentage of total resistance, by type, 66
  of peripheral nerve, 75
  placental
    fetal, 9
    maternal, 9
  of planta, 472
  pressure in. See also Blood pressure
    by type, 66
  in systemic circulation, structure of, 64
  in tendons, 60
  tibialia posteriora, 455
  total cross-sectional area, by type, 66
  wall of
    layers of, 64
    structure of, 64
B-lymphocytes, 68–69
BMI. See Body mass index
Bochdalek's triangle, 174
Body (human)
  cardinal planes and axes in, 27
  functional subdivisions, by organ system, 24
  internal organs of, location of, 24
  planes of section through, 25
  regional subdivisions of, 24
  structural design of, 24–25
  terms of location and direction for, 26
  upper, terms of location and direction for, 26
Body(ies)
  nerve cell. See Substantia grisea
  Pacini, 44
  Ruffini, 44
  vertebral. See Vertebra/Vertebrae
Body mass index (BMI), 23
Body measurement(s), for standing and sitting human being, 22
Body proportion(s)
  change during growth, 22
  normal, 22
Body surface anatomy
  of female, 30
  of male, 31
Body surface area
  age-related changes in, consequences of, 23
  distribution
    in adults, 23
    in children, 23
    in infants, 23
  hand area rule for, 23
  rule of nines for, 23
Bone bridges, formation of, 130
Bone marrow, 68
  fat, 39
  red (hematopoietic), 39
  spaces, 17
Bone matrix, uncalcified, 17. See also Osteoid
Bone model
  compressive stress in, 415
  tensile stress in, 415
Bone(s). See Ossa
Bone-to-bone connections. See also Articulationes
  types of, 40
Bouton en passage, 79
Bowel loops, 213
Bowlegs, 27. See also Genu varum
Brachialgia paraesthetica nocturna, 372
Brachial plexus. See Plexus (neural), brachial
Brachial plexus block, 384–385
  for axillary brachial plexus anesthesia, 385
  clinically important pathways for, 385
  interscalene approach (according to Meier), 385
  vertical infraclavicular, 384, 385
Brachium, 238. See also Antebrachium
  abduction, 275
  adduction, 275
  arteriae of, 354–355
    overview of, 354
  cross sections, 342–343, 386
  development of, 14
  extension, 275
  external rotation, 275
  flexion, 275
  internal rotation, 275
  length of, 239
  musculi of, 164, 296, 310–313
    anterior, 296, 336–337
  cross sectional anatomy of, 342–343
  extensors, caput commune of, 337, 339, 341
  flexors, caput commune of, 336–341, 343
  posterior, 296, 332–333
  neurovascular tract of, 386
  ossa, 38
  ossa of, 238, 247, 250
  "windowed" dissection of, 343
Bradykinin, 71
Brain. See Cerebrum
Branchiostoma lanceolatum. See Lancelet
Breast. See Mamma
Breast cancer, nodal metastases, 359
Breathing
  abdominal, 136
  chest, 136
  costal, 136
  costodiaphragmatic, 136
  diaphragmatic, 160
  sternocostal, 136
  thoracic, 160
Bronchus-associated lymphatic tissue (BALT), 68
Brood capsule(s), 47
Bulbus
  olfactorius, 72
  penis, 183, 189, 229, 557
  vestibuli, 182–183, 232, 234–235
Bundle(s)
  in striated skeletal muscle
    primary, 58
    secondary, 58
  in tendons, primary, 60
Büngner bands, 97
Burn(s), area of
  hand area rule for, 23
  rule of nines for, 23
Bursa/Bursae, 24
  gastrocnemio-semimembranosa, 442
  iliopectinea, 434–435, 549
  infrapatellaris, 450–451
  olecrani, 277
  prepatellaris, 441, 451
  semimembranosi, 442, 453, 561
  subacromialis, 61, 258, 266–269, 273, 389
    imaging of, 28
  subcoracoidea, 263
  subcutanea acromialis, 61, 268
  subcutanea malleoli medialis, 563
  subdeltoidea, 61, 258, 261, 266–269, 389
  subtendinea
    m. gastrocnemii
      lateralis, 442
      medialis, 442, 561
    m. infraspinatus, 266
    m. subscapularis, 261, 263, 266, 268
  suprapatellaris, 451. See also Recessus, suprapatellaris
  synoviales, 42
    in joint formation, 15
    in shoulder region, 61
  of tendinis calcanei, 466, 471
  trochanterica, 432, 434, 555, 560
Bursitis, 61
  iliopectineal, 549
  infrapatellaris, 451
  prepatellaris, 451
  trochanteric, 435
Buttocks
  regional anatomy (regions) of, 36
  sensory innervation of, 537

C

Calcaneal (Achilles') tendon reflex/triceps tendon reflex, lumbar root compression syndromes and, 131
Calcaneus, 30, 33, 38, 408, 411, 422–423, 454–455, 457–461, 464–466, 468, 470–472, 487, 489–493, 510, 512, 515, 520
  apophysitis, 471
  corpus, 422, 470
  facies articularis
    cuboidea, 422–425, 457
    talaris anterior, 422, 424
    talaris media, 422, 424–425
    talaris posterior, 422, 424–425
  ossification of, 19
Calcitonin, 71
Callus
  fibrocartilaginous, 53
  fixation, 53
  fracture, 53
  osseous, 53
Cambium layer, periosteal, 39
Canalis
  adductorius, 524–526, 550
    boundaries of, 551
    contents of, 551
    location of, 551
  carpi, 254, 288–289, 351–352, 402–403
    boundaries of, 289, 402
    cross section through, 289
    imaging of, 288
    tendines in, 402
    vagina tendinis in, 402
  cervicis uteri, 8
    mucus plug in, 8
  digestorius, 6–7, 104
  gastrointestinales, 62
    hormone secretion by, 71
    innervation of, 94–95
  inguinalis, 210–211, 224, 227, 549
    funiculus spermaticus in, 211
    location of, 210, 214
    musculus obliquus abdominis contribution to, 211
    openings and wall structures of, 211
    sagittal section through, 210
    schematic representation of, 217
    testis retention in, 224
  intestinales, 13
  malleolaris, nervus tibialis in, 541
  neuralis, 3
  n. hypoglossi, 121
  obturatorius, 186, 434
  pudendalis, 182–183, 542–543, 555–557
    location of, 558
    neurovascular contents of, 558

sacralis, 106, 114–115, 128, 138–139, 143
supracondylaris, 393
tarsalis, 541, 563
tarsi, 424, 459
vertebralis, 7, 105–106
Canal(s)
Alcock's, 182–183, 542–543, 555–557
location of, 558
neurovascular contents of, 558
cartilage, 15
external auditory, 103
fibro-osseous, of leg, 562
haversian, 17, 39
hemal, 3
resorption, in osteon development, 17
supinator, 369
Volkmann, 39
Capacitance vessels, 62
Capillary bed(s), 66
blood flow in, 67
Capillary exchange, 66–67
Capillary(ies), 62, 64
in chorionic mesoderm, 8
endothelial cells of, 67
fetal, 9
fluid exchange in, mechanism of, 67
hormone release into, 71
lymphatic, 68–69
pulmonary, 62
in striated skeletal muscle, 58
structure of, 66
Capitulum, humeri, 244–246, 249, 276–277, 279
Capsula
fibrosa, 167
joint. See Capsula articularis
Capsula articularis, 42, 121
of articularis atlantoaxialis lateralis, 120
of articularis zygapophysialis, 120, 122
of articulatio genus, 61, 482
posterior, 482
atlantooccipital, 122
laterale, 123
development of, 15
of facet joint, 124
femorotibial, 447
of humeri (glenohumeralis), 260
of interphalangealis distalis, 285
of interphalangealis proximalis, 285
ligaments in, 43
of shoulder, imaging of, 28
structure of, 44
Caput, 24. See also Cerebrum; Oculus; Orbit
organs of. See specific organ
palpable bony prominences
anterior view, 32
posterior view, 33
regional anatomy (regions) of, 36
sensory innervation of, 86
surface contours
anterior view, 33
posterior view, 33
terms of location and direction for, 26
Caput medusae, 192
Carcinoid(s), 75
Cardiovascular system, 24

overview of, 62–63
Carpal tunnel syndrome, 96, 288, 372, 403
Carpus
compartments within, 286
imaging of, 256, 286
Cartilago
articular, 16, 39, 42
biphasic viscoelastic model of, 45
of knee, 453
arytenoidea, embryology of, 11
corniculata, embryology of, 11
costalis, 41, 132–135, 159, 172, 176, 240, 259, 335, 363
cricoidea, 32, 384
embryology of, 11
cuneiformis, embryology of, 11
fibrocartilaginous, in fracture healing, 53
hyaline, 14, 42, 44, 104
in bone formation, 17
development of, 15
structure of, 44
of vertebral body endplate, 116–117
zones of, 44
Meckel's, embryology of, 11
mineralized, in long bone development, 16
of sacroiliac joint, 142
thyroidea, 199
embryology of, 11
triradiate, 413, 438–439
Cat, skeleton of, 21
Cauda equina, 105, 128–129, 131, 537, 558
Cavea thoracis, 100, 132
muscles of, 144, 158–161
respiratory movement of, 160
Cavitas
abdominis, 24–25
standard transverse planes through, 35
testis retention in, 224
amniotica, 6, 8
articularis, 42
blastocystica, 5
chorionica, 8
cranii, 25
glenoidalis, 38, 165, 240–241, 243, 245, 259, 261–262, 264–266, 269, 271–273, 301, 304
medullary, 39
primary, 16
pelvis, 24–25
pericardialis, 24–25
peritonealis, 224, 557
abdominis, 24
of lesser pelvis, 182
pelvis, 24
pleuralis, 24–25
synovial, 24
of manus, 402
thoracic, 24–25, 173
Cavity
body, 24–25
embryonic, 7
resorption
in osteoarthritis, 47
in osteon development, 17
serous, 24
Cavum
epidurale, of spine, 128
peritoneale scroti, 226

scroti, 217, 224
uteri, 188
Cell(s). See also Crista, neuralis, cell(s)
C, of thyroid gland, 75
hormone secretion by, 71
chromaffin, 74
defense, in synovial fluid, 43
endothelial, of capillaries, 67
epithelial, 70
glial, 75
functions of, 79
goblet, 70
Hofbauer, 9
Langerhans, 71
Langhans, 9
milieu of, 66
muscle. See Muscle cell(s)
nerve. See Neuron(s)
of nervous system, 78–79
neuroglial, 79
protozoan, milieu of, 66
Purkinje, 78
pyramidal, 78
satellite, 58, 75, 79
Schwann, 59, 75, 79, 91–92
synoviocytes
A-, 44
B-, 44
Cement line, of osteon, 39
Central nervous system (CNS), 24, 76
blood-brain barrier in, 93
development of, 72–73
hormones and hormonelike substance secretion by, 71
myelination in, 92
terms of location and direction in, 77
Centrum perinei of corpus perineale, 180
Centrum tendineum, 205
Cephalic flexure. See Flexura mesencephalica
Cephalochordata, 2
Cerebellum, 105
development of, 72
Cerebrum, 24, 76. See also Telencephalon
development of, 72
nuclear columns in, 73
Cervical rib syndrome, 96, 106, 363
Cervix uteri, 232
cGMP. See Cyclic guanosine monophosphate
Chassaignac's paralysis, 283
Chest. See Thorax
Chest injury(ies), 132
Chest tube placement, 207
Chiasm
crurale, 490, 513
plantare, 490, 513
Cholecystokinin, 71
Chondroblast(s), 14
Chondroclast(s), 16
Chondrocyte cluster(s), 47
Chondrocyte(s), 16
of hyaline cartilage, 44
Chondrogenic layer(s), 15
Chondroitin, 45
Chondroitin-6-sulfate, 45
Chondronectin, 45
Chondrosis, in lumbar spine, 130
Chorda dorsalis, 3, 6–7, 74, 104
segments of, 7, 104

Chorda obliqua, of antebrachium, 280
Chorda sheath, 104
Chordata, 2–3
characteristics of, 3
Chorda umbilicalis, 9
Chordoma(s), 3
Chorion frondosum, 8
Chorion laeve, 8
Chromosome number
diploid, 5
haploid, 5
Cingulum membri inferioris, 20
Cingulum membri superioris, 20, 24, 238
integration into skeleton of trunk, 240–241
movements of, 274–275
musculi of, 164, 296, 298–309
anterior, 334–335
posterior, 330–331
deep layer, 331
superficial layer, 330
nerve compression (bottleneck) syndromes in, 96
ossa of, 238, 242–243
in relation to trunk, 240
and pelvic girdle, comparison of, 241
Cingulum pelvicum, 20, 24, 138, 408, 413
integration of spinal column into, 103
nerve compression (bottleneck) syndromes in, 96
ossa of, 412–413
and shoulder girdle, comparison of, 241
Circulation
extraembryonic vitelline, 12
intraembryonic systemic, 12
placental, 12
portal, 62
pulmonary, 62
systemic, 62
blood vessels of, structure of, 64
Circulatory system, 62
embryonic, 12
Cisterna
cerebellomedullaris, 121
chyli, 68, 194
lumbalis, 128
Claudicatio spinalis, 130
Clavicula, 30, 32, 38, 55, 61, 161, 166, 199, 238–241, 250, 258–264, 268, 299, 301, 305, 307, 309, 331, 334–335, 359, 362–363, 365, 367, 374, 378–381, 383–385, 389–390
corpus of, 242
extremitas acromialis, 33, 242, 259
extremitas sternalis, 242, 259, 389
facies articularis acromialis, 242
facies articularis sternalis, 242
fractures of, 242
movements of, 274
ossification of, 18, 242
position of, 242
range of motion of, 274
shape of, 242
Claw hand, 370
Claw toes, 469

Cleft
  synaptic, 59
  uncovertebral, 126. *See also*
    Articulationes (of region or
    structure), uncovertebral
Clergyman's knee, *451*
Clinical applications
  a. axillaris damage, 262
  abdominal muscles and spinal
    column stability, 156
  accessory ossicles of pes, 425,
    471
  acetabular dysplasia, 439
  acetabular fractures, 412
  achillodynia, 471
  albinism, 75
  Alzheimer disease, 78
  anesthesia, 545, 547
    lumbar spinal, 128
  ankle sprains, 460
  ankle supination trauma, 460,
    501
  ankylosis
    of articulatio genus, 40, 50
    of shoulder/art. humeri
      (glenohumeralis), 275
  anterior cruciate ligament
    rupture, 448
  aponeurosis plantaris exostosis,
    471
  apophysitis calcanei, 471
  arcus aortae duplication, 12
  arteria carotis communis,
    sinistra, arising from truncus
    brachiocephalicus, 12
  arteria poplitea palpation, in
    fossa poplitea, 561
  arteria tibialis posterior palpa-
    tion, 563
  arthrodesis, 40, 46
    for pes equinovarus, 40
    of shoulder/art. humeri
      (glenohumeralis), 275
    triple, 40
  arthrography, 40
  arthroscopy, 40
    of shoulder/art. humeri
      (glenohumeralis), 270–271
  arthrosis
    latent, 47
    uncovertebral, 126–127
      in cervical spine, 127
  articulatio coxae dislocation
    congenital, 439
    after hip replacement, 429
    traumatic, 429
  articulatio coxae instability, 439
  a. vertebralis injury, 384
  axillary block, 384
  axillary lymphadenectomy, 362
  backpack paralysis, 363
  ballottable patella sign, 451
  Bankert lesion, 262, 270
  bartholinitis, 235
  bone bridge formation, 130
  bottle sign, 372
  brachialgia paraesthetica
    nocturna, 372
  breast cancer metastases to nodi
    lymphoidei axillares, 359
  bridging spondylophytes, 130
  brood capsule, 47
  burn area
    hand area rule for, 23
    rule of nines for, 23
  bursitis, 61

  iliopectineal, 549
  infrapatellaris, 451
  prepatellaris, 451
  trochanterica, 435
calcaneal (Achilles') tendon
  rupture, 488
caput femoris dislocation, 429
caput femoris fractures, 433
carcinoids, 75
carpal fractures, 257
carpal tunnel syndrome, 96, 288,
  372, 403
cervical rib syndrome, 96, 106,
  363
cervical spine disorders, 127
Chassaignac's paralysis, 283
chest injury, 132
chondrosis in lumbar spine, 130
chordoma, 3
claudicatio spinalis, 130
clavicula fractures, 242
claw hand, 370
claw toes, 469
clergyman's knee, 451
colpopexy, 163
compartment syndromes, 540,
  562, 567
  in pes, 523
costae fractures, 132
costoclavicular syndrome, 96,
  363
crepitus with fractures, 52
crista neuralis derivatives,
  diseases of, 75
cubital tunnel syndrome, 96, 370
cysts
  Baker, 442, 561
  branchial, 10
  lateral cervical, 10
  subchondral, 47
  synovial popliteal, 442
degenerative joint disease,
  46–47
  common locations of, 46
  in hip, 46–47
descensus uteri, 163
discus ulnocarpalis degeneration, 286
disk degeneration, 130
disk herniation, 86, 545
  lumbar, 130–131
disk prolapse, 131
disk protrusion, 131
distal radial compression
  syndrome, 96
distal radius fractures, 257
drawer sign, 448
drop hand, 368
Duchenne limp, 437, 536
Duchenne sign, 497
Dupuytren's contracture, 348
dysesthesia
  at interior lower leg, 96
  n. medianus lesions causing,
    372
dysostosis cleidocranialis, 242
dysostosis craniofacialis, 242
edema, 67
effusions
  in articulatio coxae, 435, 483
  chest tube placement for, 207
  in knee, 451
epicondylitis lateralis, 329
epicondylitis medialis, 329
episiotomy, 233, 542
exostosis, marginal, 47

femoral fractures, 432–433
femoral neck fractures, 432–433
fibular tunnel syndrome, 96
finger mal-alignment, 256
first arch syndromes, 10
foot drop, 540
fractures, 52–53
  acetabular, 412
  carpal, 257
  clavicula, 242
  Colles', 257
  costae, 132
  distal radius, 257
  femoral, 432–433
  malleolus medialis, 541
  patellar, 419
  pelvic, 412
  proximal humeral, 246
  scaphoid, 257
  Smith, 257
  of sustentaculum tali, 424
  tibial, 541
  treatment of, 53
  vertebral, 104, 130
  Weber, 461
functional muscle testing
  in articulatio talocruralis, 501
  in shoulder muscles, 325
gait disturbances, 540
ganglion stellatum injury, 384
golfer's elbow, 329
Guyon–Logen syndrome, 96
Haglund's exostosis, 471
hallux valgus, 469
hammer toes, 469
hand of benediction, 372
heart failure, right-sided, 378
heel pain, 471
heel spurs, 471
hemarthrosis, 44
hemiplegia, 473
hernia
  diaphragmatic, 175
  femoral, 549
  hiatal, 175
  sliding, 175
Hill-Sachs lesion, 262, 270
hip dysplasia, congenital, 439
hip muscle functional testing,
  497
Hirschsprung disease, 74–75
Hoffman–Tinel sign, 97
Howship–Romberg syndrome,
  96
hyperabduction syndrome, 363
hypermobility of sacroiliac joint,
  143
hypoesthesia, 545, 547
iliopsoas paralysis, 476
iliopsoas shortening, 476
iliopsoas weakness, 476
immunological diseases, 91
impotence, male, 543
incisura scapulae syndrome, 96,
  243, 362, 388
incontinence
  fecal, 163, 543
  stress, 163
  urinary, 163, 543
infraspinatus weakness/atrophy,
  243, 362
inguinal ligament syndrome, 96
injection
  intra-arterial, 356
  intra-articular, 40
  intragluteal, 555

  paravascular, 356
  ventrogluteal, 555
interdigital Morton neuralgia, 96
interscalene block, 384
ischiocrural muscle insufficiency,
  484
joint contracture, 43
joint endoprosthesis, 40, 46
joint puncture, 40
joint replacement surgery, 40,
  46
knee instability, 482
Lachman test, 448
lateral axillary hiatus syndrome,
  96
leg length discrepancy, 497
leg lengthening, functional, 497
leg pain, 96
leg shortening, 439
  functional, 481, 497
Lichtenstein technique, *219*
ligament overstretching, 43
ligament shortening, 43
ligament tears, 43, 448–449
ligamentum cruciatum posterius
  rupture, 448
ligg. acromioclaviculares injuries
  Rockwood classification, 260
  Tossy classification, 260
lordosis lumbalis, hip extension
  and, 436
lower motor neuron lesions, 90
lumbar hyperlordosis, 499
lumbar puncture, 128
lumbar root compression
  syndromes, 131
lumbar spinal anesthesia, 128
lung percussion, 160
lymphangitis in upper limb, 358
m. abductor pollicis longus
  palsy, 368
malleolus lateralis fractures, 461
malleolus medialis fractures, 541
mandibulofacial deformity, 10
measurement of joint range
  of motion, neutral-zero
  method, 50
medullary thyroid carcinoma, 75
Medusa's head, 192
melanoma, 75
meniscal degenerative changes,
  447
meniscal injury, 445, 447, 448
meniscal tears, 447–448
meralgia paresthetica, 96
m. extensor carpi ulnaris palsy,
  368
m. extensor digitorum palsy, 368
m. extensor indicis palsy, 368
m. extensor pollicis brevis palsy,
  368
m. extensor pollicis longus palsy,
  368
m. gluteus maximus paralysis,
  537
m. gluteus medius weakness,
  536
m. gluteus minimus weakness,
  536
m. supraspinatus, tendo
  damage, 269
multiple sclerosis, 92
muscle mass, age-related
  changes in, 56
muscle shortening
  in articulatio talocruralis, 501

in cubitus, 327
in hip, 497
in knee, 499
in shoulder, 325
muscular dysbalance, 499
n. axillaris injury/damage, 96, 262, 367
n. axillaris palsy, 367
n. dorsalis scapulae paralysis, 362
nerve block
  axillary, 384
  for digiti, 399
  interscalene, 384
  n. pudendus, 542, 558
  Oberst, 399
  vertical infraclavicular, 384
nerve compression (bottleneck) syndromes in, 96
nerve root compression, 131
nerve root injury, 375
nervi femoralis injury, 96
nervi plantares injury, 96
nervi suprascapularis injury, 96
nervi ulnaris entrapment, 96
nervi ulnaris injury/lesions, 96
nervus fibularis injury, 96, 540
nervus fibularis profundus compression, 566
nervus obturatorius injury, 96, 534
nervus radialis injury/lesions, 96
nervus saphenus injury, 96
nervus tibialis compression, 541
nervus tibialis injury, 96
neuroblastoma, 75
neurofibromatosis, 75
neurological disorder diagnosis, 90
n. ischiadicus injury, 96, 538, 540
n. laryngeus recurrens injury, 384
n. medianus compression syndromes, 372
n. medianus injury/lesions, 96, 372
n. pudendus injury, 543
n. radialis compression, 368, 393
n. radialis entrapment, 393
n. radialis injury/lesions, 246, 368, 393
n. suprascapularis compression, 362, 388
n. suprascapularis injury, 362
n. thoracicus longus injury, 362
n. ulnaris compression syndromes, 370
n. ulnaris injury/lesions, 370
n. ulnaris palsy, 370
nursemaid's elbow, 283
Ortolani test, 439
osteoarthritis, 46
  activated, 47
  of art. carpometacarpalis pollicis, 293
  of articulatio coxae, 51
    alleviation of stress on affected side in, 437
  compensatory mechanisms in, 47
  etiology of, 46
  of facet articularis, 127
  of knee, 410
  pathogenesis of, 46
  primary, 46
  secondary, 46
  stages of, 47
  symptoms of, 47
osteoarthrosis, 46
osteochondrosis, in lumbar spine, 130
osteochondrosis dissecans, 40
osteolytic metastases, 130
osteophytes
  in cervical spine, 126–127
  marginal, 47
osteoporosis, 130
  and femoral fractures, 432
osteosynthesis, surgical, 53
os trigonum, 471
overloading of caput of second os metatarsale, 544
pain, in thigh, 533
palpation of pedal pulse, 567
paradoxical respiration, 132
paresthesia, 96
  n. medianus lesions causing, 372
paresthesias, in thigh, 533
park bench paralysis, 96, 368
patellar fracture, 419
patellar instability, 419
pedal pulse palpation, 567
pelvic floor muscle damage/insufficiency, 163
pelvic floor repair, 163
pelvic fractures, 412
Pendelluft effect, 132
peridentinitis of calcaneal (Achilles') tendon, 471
perineal protection, 233
peripheral nerve injury, 96–97, 375, 545
pes deformities, 467
pes equinovalgus, 501
pes equinovarus, 40, 501
pes planus, 459, 464
  pain associated with, location of, 467
pheochromocytoma, 75
piano-key phenomenon, 260
piriformis syndrome, 96, 553
plantar fascitis, 471
pleural dome injury, 384
plexus brachialis block, 384
plexus brachialis compression syndromes, 363
plexus brachialis injury, 96
pregnancy testing, 9
prolaps uteri, 163
pronatio dolorosa, 283
pronator teres syndrome, 96, 372
proximal humeral fractures, 246
pseudarthrosis, 53
pseudo-osteoarthritis, scaphoid fracture and, 257
pudendal nerve blockade, 558
quadriceps femoris paralysis, 482
quadriceps femoris weakness, 482
reactive synovitis, 47
rectopexy, 163
rectus sign, 483
reflex testing, 89
resorption cavities in osteoarthritis, 47
right-sided arcus aortae, 12
rolling veins, 356
rotator cuff damage, 262
rotator cuff rupture, 269–270
sacroiliac joint impairment, 497
sacroiliac joint locking, 143
sacroiliac joint pain, 143
sarcopenia, 56
scalene syndrome, 96, 363
scalenus syndrome, 106
scaphoid fractures, 257
scapula alata, 362
scoliosis, 133, 409
sexual dysfunction, 543
shoulder/art. humeri (glenohumeralis) issues
  adhesive capsulitis, 270
  degeneration, 270
  disease, 275
  dislocation, 262–263
  inflammation, 270
  instability, 270
  synovialitis, 270
shoulder repositioning, 263
Shouldice technique, *219*
sonographic examination, of hip, 435
spider veins, 527
spina bifida, 104
spinal stenosis, 127, 130
splayfoot, 464, 467, 469
spondyloarthritis in lumbar spine, 130
spondyloarthrosis, 130
spondylodesis, 40
spondylolisthesis, 104
spondylolysis, 104
spondylophytes in columna vertebralis pars cervicalis, 127
"steppage" gait, 540
stork legs, 501
subacromial impingement syndrome, 269–270
subchondral sclerosis, 47
sulcus n. ulnaris syndrome, 370
sulcus ulnaris syndrome, 96
supination trauma of ankle, 501
supinator syndrome, 96, 368
supraspinatus (impingement) syndrome, 268
supraspinatus weakness/atrophy, 243, 362
sustentaculum tali fractures, 424
synovectomy, 40, 44
talipes calcaneus, 501
talipes valgus, 501
tarsal tunnel syndrome, 96, 541, 563
tennis elbow, 329
Thomas maneuver, 436
thoracic outlet syndrome, 96
thrombophlebitis, 527
tibial fractures, 541
tibialis anterior syndrome, 567
tibiofibular syndesmosis tears, 461
toe deformities, 469
toeing-in gait, 417
toeing-out gait, 417
total endoprosthesis of hip, 46
totally extraperitoneal repair, *219*
transabdominal preperitoneal repair, *219*
Trendelenburg sign, 479, 497, 536
truncus brachiocephalicus variants, 12
tumors derived from crista neuralis cells, 74
ulnar tunnel syndrome, 96, 370
ulnocarpal complex injury, 287
unhappy triad, 448
upper motor neuron lesions, 90
varicose veins, 527, 545
vena-saphena varicosity, 527
venous thrombosis, 527
vertebral assimilation disorders, 103
vertebral endplate sclerosis, 130
vertebral fractures, 104, 130
vertical infraclavicular block, 384
visceral descent, 163
Wartenberg syndrome, 96
weakness
  in ankle, 501
  in cubitus, 327
  in hip, 497
  iliopsoas, 476
  infraspinatus, 243, 362
  in knee, 499
  quadriceps femoris, 482
  in shoulder, 325
  small gluteal, 536
  supraspinatus, 243, 362
  in upper leg, 96
Weber fractures, 461
wrist drop, 368
Clitoris, 188, 232, 543
  in adrenogenital syndrome, 223
  crus of, 182–183
  development of, 222
  erectile tissues of, 235
  glans of, 180, 183–185, 559
  lymphatic drainage of, 234
  preputium of, 180
Club foot. *See* Pes equinovarus
CNS. *See* Central nervous system
Coelom, intra-embryonic, 6
Collagen
  fibers, of tendon, 60
  fibrils, in hyaline cartilage, 44–45
  meniscal, 446
Colliculus/Colliculi
  of brain, 72
  seminalis, 183
Colloid osmotic pressure, 67
Collum, 24. *See also* Columna vertebralis; Larynx; Vertebra/Vertebrae
  nervi of
    deep, 378
    superficial, 378
  organs of. *See specific organ*
  palpable bony prominences
    anterior view, 32
    posterior view, 33
  regional anatomy (regions) of, 36
  surface contours
    anterior view, 32
    posterior view, 33
  terms of location and direction for, 26
  venae of, superficial, 378
Colon, 24
  innervation of, 94
Colpopexy, 163
Columna vertebralis, 2, 38, 100, 408. *See also* Vertebra/Vertebrae
  bony, 102–103
  development of, 104–105
  embryology of, 14, 104

C Columna vertebralis (cont.)

evolution of, 3
ligaments of, 118–119
motion segments
  formation of, 104
  loading of, 124
  position of discus intervertebralis, 117
  structure of, 124
  nutrient vessels of, 104
os sacrum. *See* Ossa (named), sacrum
pars cervicalis, 24, 100–101, 105
  anterior view, 109
  curvature of, 103, 105
  degenerative changes in, *127*
  extension, 125
  flexion, 125
  lateral flexion, 125
  lateral view, 108
  ligaments of, 120–121
  MRI, 121
  radiograph, 121
  range of motion, 125
  rotation, 125
  superior view, 109
  uncovertebral arthrosis of, *127*
  uncovertebral joints of, 126–127
pars lumbalis, 24, 100–101, 105, 112–113, 476, 502
  anterior view of, 113
  cross-sectional anatomy of, 128–129
  curvature of, 103, 105
  disk herniation, *130–131*
    MRI, 131
  extension, 125
  flexion, 125
    measurement of (method of Schober and Ott), 125
  hyperlordosis, *499*
  lateral flexion, 125
  lateral view of, 112
  MRI, 129–130
  range of motion, 125
  rotation, 125
  sagittal section of, 129
  stenosis, imaging of, *130*
  superior view of, 113
  transverse section of, 129
pars thoracica, 101, 105, 110–111, 476
  anterior view of, 111
  curvature of, 103, 105
  extension, 125
  flexion, 125
    measurement of (method of Schober and Ott), 125
  lateral flexion, 125
  lateral view of, 110
  range of motion, 125
  rotation, 125
  superior view of, 111
  straightening of, in development, 105
Commissura labiorum
  anterior, 232
  posterior, 232
Compartment syndrome(s), 540, *562*, *567*
  of pes, *523*
Compressive stress, in bone model, 415
Computed tomography (CT). *See* Imaging

Condylus
  femoris, 446
    lateralis, 38, 408, 414, 416–419, 440–441, 444–445, 447–448, 450, 453, 484
    medialis, 38, 408, 414, 416–417, 419, 440–441, 444–445, 447–448, 450
  humeri, 244–245, 247, 260, 267
  occipitalis, 122
  tibiae, 446
    lateralis, 32–33, 409, 418, 420–421, 440, 486–487, 510, 566
    medialis, 32–33, 409, 420–421, 440, 449, 484–485, 487, 489, 491
    superior margin of, 409
Conjugata, pelvic
  diagonalis, 141
  externa, 141
  obstetric, 141
  vera, 141
Connective tissue
  of blood vessel wall, 64
  interlobular, 208
  meniscal, 446
  of muscle fasciae, 61
  perivascular, 65
  in striated skeletal muscle, 58
  in traction tendon, 60
Connective-tissue spaces, 24
Conus medullaris, 105, 128–129
Convection, of interstitial fluid, 45
Cor, 62, 105
  embryology of, 12
  innervation of, 94–95
  primitive, 6
  pumping action of, 62
  suction effect of, 65
  venous drainage of, development of, 13
  venous return to, 65
Cornua
  coccygea, 114
  sacralia, 114
Corona glandis, 228–230
Corona radiata, 5
Corpora cavernosa, 188
  clitoridis, 235
  penis, 228–231
    development of, 222
    erect state, 231
    flaccid state, 231
Corpus adiposum
  of cubitus/art. cubiti, 277
  fossae ischioanalis, 556
  Hoffa's, 452
  infrapatellar, 61, 450–451
Corpus callosum, 105
Corpuscle(s)
  Meissner tactile, 374
  Vater-Pacini, 374
Corpus clitoridis, 235
Corpus luteum, 5, 9
Corpus ossis hyoidei, 32
  lower part, embryology of, 11
  upper part, embryology of, 11
Corpus perineale, 180–181, 184
Corpus spongiosum, 183, 188
  penis, 228–231
    development of, 222
Cortex, of nodi lymphatici, 69
Corticotropin. *See* Adrenocorticotropic hormone

Costae, 32–33, 41, 61, 100–101, 106, 132, 147, 153, 155, 161, 165, 174, 207, 258, 261, 297, 301, 362, 365, 381
  accessory, 106
  articulations with vertebral bodies, 110
  caput, 107, 135–137
    joint of, 137
  cervicalis, 106, 363
  collum, 107, 135–137
  corpus, 107, 135
  "counting," for anatomical orientation in thorax, 34
  false, 133
  first, 34, 151, 159, 165, 172–173, 191, 240–241, 258–260, 268, 334, 361–363, 365, 367, 379
    ligamentous attachment to, 363
  floating, 133
  fractures of, *132*
  lumbalis, 106
  movement during breathing, 136, 158–159
  second, 34
  segments of, 135
    thoracic, structure of, 135
  shape of, 135
  size of, 135
  true, 133
  twelfth, 34, 101, 133, 160–161, 169, 172–173, 240, 477
Costoclavicular syndrome, *96*, *363*
Cotyledon(s), placental, 9
Coxa anteverta, 417
Coxa retroverta, 417
Coxa valga, 415, 437
Coxa vara, 415, 437
Cranium, 24–25
  imaging planes for, 28
Crena
  analis, 201
  ani, 30, 181
Crepitus, with fractures, *52*
Cristae
  capitis costae, 137
  colli costae, 135
  humeri
    tuberculi majoris, 165, 244, 303, 307, 309, 311
    tuberculi minoris, 244, 303, 307, 311
  iliacae, 31–33, 35, 55, 101, 138–140, 142, 147, 155, 161, 164, 166, 168–169, 175, 186–187, 199, 201–202, 307, 330, 365, 408–409, 411–413, 426–428, 477, 479, 481, 502–503, 506–507, 522, 532, 546, 554–555, 560
    labium externum of, 138–139, 152
    labium internum of, 138–139, 152
    linea intermedia of, 138–139, 152
  intertrochanterica, 414, 426, 428, 478–479, 485
  musculi supinatoris, of ulna, 250
  neuralis, 6–7
    cell(s)
      development of, 74

differentiation of, 74
  migratory pathways, 74
  tumors derived from, *74*
cross section through, 73
derivatives of, 74–75
  diseases of, *75*
  in head and neck, 75
occipitalis
  externa, 122
  interna, 120
sacralis
  lateralis, 114–115
  medialis, 114
  mediana, 106, 114–115, 138–139
supracondylaris (humerus)
  lateralis, 244, 246, 276, 278, 317
  medialis, 244, 246, 276
Cross-sectional anatomy. *See* Section anatomy
Crown-rump length (CRL), 4
Crural sling, 175
Crus (Leg). *See also* Articulationes (named), genus; Articulationes (named), talocruralis; Membrum inferius; Pes; Thigh
  anterior compartment, 486–487, 567
    musculi of, 474, 486–487
    neurovascular structures of, 566–567
  arteriae of, 525, 568
  cross-sectional anatomy of, 522
  extensor compartment, 566
  fascia of, 545
  interior lower, dysesthesia at, *96*
  lateral compartment, 486–487, 566–567
    musculi of, 474, 486–487
  lengthening of, functional, *497*, *501*
  mechanical axis of, 410
  mediale, 211
    sensory innervation of, 535
  movements of, during gait cycle, 473
  musculi of, 474, 486–491
    anterior compartment, 474, 486–487
    anterior view, 510–511
    extensor group, 563
    fibularis group, 563
    flexors
      deep, 563
      superficial, 563
    lateral compartment, 474, 486–487
    lateral view, 510
    origins and insertions, 511, 513
    posterior compartment, 474, 488–491
      deep, 474, 490–491, 541
      superficial, 474, 488–489
    posterior view, 512–513
    superficial flexor group, 488–489
  nervi, 530–531
  neurovascular structures in, 566–567
  ossa of, 38, 408, 411
    overview of, 408
  posterior compartment
    deep, 567

584

neurovascular structures in, 562
musculi of, 474, 488–491, 541
superficial, 567
neurovascular structures in, 562
shortening of, 439
functional, 409, 481, 497
true, 409
upper front, pain/weakness in, 96
Crus/Crura
clitoridis, 182–183, 232, 234–235
of diaphragma, pars lumbalis, 160, 174–175
penis, 183
Cubital tunnel syndrome, 96, 370
Cubitus/Art. cubiti, 38, 238, 249
anterior view, 276
arteriae of, 354–355, 393, 396
arterial anastomoses about, 393
articulating skeletal elements of, 276
capsula articularis of, 278–279
in extension, 279
development of, 15
extension, 326
flexion, 326
lateral view, 276
ligamenta of, 278–279
in extension, 279
medial view, 276
movements of, 49, 282–283, 326
musculi of
extensors, 326
functional testing of, 327
shortening/weakness, symptoms of, 327
flexors, 326
functional testing of, 327
shortening/weakness, symptoms of, 327
functional testing of, 327
innervation of, 326
nerve segments associated with, 326
overview of, 326–327
pronators, 326
functional testing of, 327
shortening/weakness, symptoms of, 327
shortening, symptoms of, 327
supinators, 326
functional testing of, 327
shortening/weakness, symptoms of, 327
weakness, symptoms of, 327
nervi of, cutaneous, 356
neurovascular systems of, 392–393
normal valgus of, 282
ossa of, 276–277
overview of, 276–277
posterior view, 276
pronation, 326
range of motion, 326
skeletal elements of, 277
soft-tissue elements of, 277
supination, 326
venae of, cutaneous, 356
Cubitus valgus, 282
Cupula optica, 72
Cutis, 24
of abdominal wall, 179, 213

innervation of, 95
Cyclic guanosine monophosphate (cGMP), 231
Cyst(s)
Baker, 442, 561
branchial, 10
lateral cervical, 10
subchondral, 47
synovial popliteal, 442
Cytokines, 71
Cytotrophoblast, 8–9

D

Decidua basalis, 8–9
Decidua capsularis, 8
Decidua parietalis, 8
Defecation, 156, 187
Degenerative joint disease, 46–47
common locations of, 46
Dendrite(s), 78
Dens axis (C II), 102–103, 105, 108–109, 121, 123, 126–127, 241, 260
apex, 121
articular surface, posterior, 122
ligamenta apicis, 121–123
Dermatoglyphs, 374
Dermatome(s), 82
course of sensory fibers from posterior root to, 84
derivatives of, 7
development of, 82
distribution of, 83
embryonic, 7
learning, simplified scheme for, 83
lumbar root compression syndromes and, 131
malfunction of, 82
of membrum inferius, 545
of membrum superius
anterior view, 375
posterior view, 377
phylogenetic development of, 84
in spinal development, 104
of trunk wall
anterior, 198
posterior, 200
Dermis, 7, 374
Descensus testis, 224
Descensus uteri, 163
Development, antenatal
Carnegie stages of, 4
early, 4
embryonic period of, 4
fetal period of, 4
first week of, 5
gastrulation in, 6
neurulation in, 6
Diameter(s), pelvic
obliqua dextra, 141
obliqua sinistra, 141
transversa, 141
transversa apertura pelvis superior, 141
Diaphragm, in capillary endothelium, 67
Diaphragma, 105, 156, 160–161, 173–175, 178, 205
anterior view, 175
apertures in
aortic, 175
oesophageal, 174–175
vena caval, 174–175
domes of, 161

foramen vena cavae, 160–161, 174
inferior view, 174
overview of, 160
pars costalis, 144, 160–161, 174–175, 178
pars lumbalis, 144, 160–161, 204
crura of, 160, 174–175
dextrum and sinistrum crura, 160
lateral parts, 160
medial parts, 160
pars sternalis, 144, 160–161, 174
position of, 175
respiratory movement of, 160
shape of, 175
superior view, 174
urogenitale, 230
Diaphragma pelvis, 144, 162–163, 182, 542
in female, 182
function of, 162
Diaphragma urogenitale, 556
Diaphysis/Diaphyses, 16, 39, 41
primary ossification centers in, 14
Diarthrosis/Diarthroses, 40, 42
Diencephalon, 72
hormone secretion by, 71
Digestive system, 24
Digiti (manus), 238. See also Ossa digitorum, of manus
anularis, 374
articulationes of
classification of, 294
range of motion of, 295
development of, 14
distal interphalangeal joint crease, 374
extensor side, arteriae of, 396
flexor tendons of, blood supply to, 399
index, 374
lymphatic drainage of, 358
ligamenta of, 290–291
mal-alignment of, 256
medius, 374
arteriae of, 398
nervi of, 398
minimus, 374
nerve block for, 399
ossa, 238. See also Ossa (named), metacarpi
ossification of, 18
and pollicis, relationship of, in neutral position, 293
proximal interphalangeal joint crease, 374
Direction, terms of, 26
Disc
embryonic, 6
bilaminar, 6
intra-articular, 15
Discus
articularis, 42, 259
intervertebralis, 3, 41, 102, 118–122, 124, 126–131, 133, 137, 143
degeneration of, 130
embryology of, 104
function of, 117
herniation of, 86, 545
ligamentous reinforcement of, 119

load-dependent fluid shifts in, 117
lumbar, herniation, 130–131
position in motion segment, 117
prolapse, 131
protrusion, 131
structure of, 116
ulnocarpalis, 255–256, 281, 286–287, 294
degenerative changes in, 286
Distal radial compression syndrome, 96
Distance(s)
intercondylar, 410–411
intermalleolar, 410–411
Distantia
intercristalis, 141
interspinosa, 141
Drop hand, 368
Duchenne limp, 437, 536
Ductuli efferentes, 226
Ductus
arteriosus, 12
deferens, 212–213, 224–226, 231
ejaculatorius, 188
epididymidis, 226
excretorius, 70
lactifer, 208
colligens, 208
lymphaticus dexter, 68, 194–195, 209
thoracicus, 68, 194–195
venosus, 13
Duodenum, development of, 13
Dupuytren's contracture, 348
Dural bags, 129
Dural sac, 128–131
Dura mater spinalis, 128, 131
Dysesthesia
at interior lower leg, 96
n. medianus lesions causing, 372
Dysostosis cleidocranialis, 242
Dysostosis craniofacialis, 242

E

Ectoderm, 104
derivatives of, 7
dorsal, cross section through, 73
embryonic, 6
surface, 6–7
Edema, 67
Egg, amniotic, 2
Eicosanoids, 71
Ejaculation center, 231
Elastic fiber(s), of tendon, 60
Elastic recoil, 62
Elbow/Elbow joint. See Cubitus/Art. cubiti
Electrical nerve stimulation
brachial plexus sheath and, 384
motor responses to, 385
of nervi mediani, 385
of nervi musculocutaneus, 385
of nervi radialis, 385
of nervi ulnaris, 385
Electron microscopy
of neurons, 78
of synapses in central nervous system, 79
transmission (TEM), of collagen fibrils, 44
Embryoblast, 5
Embryonic genital anlage, undifferentiated, 222

E Embryonic period

Embryonic period
  development in, 4
  sensitivity to teratogens in, 4
Embryo(s)
  human, 8
    Carnegie stages of, 4
    circulatory system of, 12
  vertebrata, 2
Eminentia
  carpi radialis, 289
  carpi ulnaris, 289
  hypothenaris, 374, 402, 404
  iliopubica, 138, 549
  intercondylaris, 420–421, 440, 447
  plantaris
    lateralis, 523
    medialis, 523
  thenaris, 374, 402
Encephalon, 76
Endocrine system, 24. *See also* Hormone(s)
  and autonomic nervous system, 71
  functions of, 71
  glandulae of, 70–71
  and immune system, 71
Endoderm
  derivatives of, 7
  embryonic, 6
Endometrium, 5, 8
Endomysium, 58
Endoneurium, 75, 93
Endosteum, 17
Endothelium
  capillary, types of, 67
  vascular, 64
Endplate(s), vertebral, 116–117
  sclerotic, *130*
Enteric nervous system, 75
Epiblast, 6
Epicondylitis lateralis, *329*
Epicondylitis medialis, *329*
Epicondylus
  femoris
    lateralis, 32–33, 414, 440, 445, 487–489
    medialis, 32–33, 414, 440, 445, 487–489
  humeri, 373
    lateralis, 32–33, 239, 244–246, 249, 260, 267, 276–279, 282, 311–313, 317, 319, 335, 339, 341, 393
    medialis, 32–33, 239, 244–247, 249, 267, 276–279, 282, 311–313, 315, 317, 319, 334–335, 338–341, 343, 356, 369, 371, 387, 391–395
Epidermis, 7, 374
Epididymis, 216–217, 224–226, 231
  caput, 226
  cauda, 226
  corpus, 226
  examination of, 227
  structure of, 226
  surface anatomy of, 226
  vasa lymphatica, 227
Epididymitis, 227
Epigastrium, 36, 199
Epimere, 7, 145
Epimysium, 58
Epinephrine, 71

Epineurium, 75
Epiphyseal plate(s), 41
  blood supply of, 16
  cellular processes in, 16
  before closure, 41
  of membrum inferius, 19
  of membrum superius, 18
  ossified, 41
  structure of, 16
  zones of, 16
Epiphysis anularis, corpus vertebrae, 111, 113, 116–117
Epiphysis/Epiphyses, 16, 41
  distalis, 16, 39
  proximal, 39
Episiotomy, *233*, *542*
  advantages and disadvantages of, 233
  types and techniques, 233
Epispadias, 223
Epithelial ridge, 14
Epithelio-mesenchymal transition, 74
Erectile dysfunction, 231
Erectile tissue(s)
  in female, 235
  mechanism of, 231
  penile, 228–229, *543*
Erection center, 231
Erythropoietin, 71
Estrogen(s), 71
  placental production of, 9
Evolute, of knee, 449
Excavatio rectouterina, 188
Exocytosis, 70
Exostosis, marginal, *47*
Expiration
  muscles of, 158–159
  position of diaphragma and costae in, 160
Extinction phenomenon, 384
Extracellular fluid, 66
Extracellular matrix (ECM), of hyaline cartilage, 44
  function of, 45
  structure of, 45
Eye. *See* Oculus

F

Face
  muscles of
    embryology of, 2
    mimetic, embryology of, 11, 145
  palpable bony prominences, 32
  surface contours, 32
Facies
  articularis
    calcanea
      anterior, 423
      media, 423
      posterior, 423, 425
    capitis costae, 137
    malleoli lateralis, 421, 455, 457–458
    malleoli medialis, 421, 455, 457–458
    of vertebrae, 124
      anterior, 108–109
      inferior, 108–112, 119, 126, 130
      posterior, 108–109
      superior, 106, 108–113, 119, 123, 126–127
  malleolaris
    lateralis, 422, 425

    medialis, 422, 424
  patellaris
    lateral, 419, 441
    medial, 419, 441
  trochlea tali with, superior, 422, 424, 457–458
  urethralis, 229
  of vertebrae
    articular, 124
      anterior, 108–109
      inferior, 108–112, 119, 126, 130
      posterior, 108–109
      superior, 108–113, 118–119, 123, 126–127
    inferior, 118
    superior, 118
Falx cerebri, 105
Fascia/Fasciae
  abdominis superficialis, 155, 179, 208, 210, 216–217, 380–381
  antebrachii, 348, 398
  axillaris, 380–381
  biceps brachii, 392
  brachii, 61, 380–381
  cervicalis
    lamina pretrachealis, 167
    lamina prevertebralis, 167
    lamina superficialis, 167
    space between middle and deep layers of, 24
  clavipectoralis, 380–381
  Colles', 181, 183
  cremasterica, 176, 215
    musculus cremaster and, 216–217, 225–226
  cruris, 545, 562, 566
    deep, 567
    superficial, 486, 567
  Denonvilliers, 188
  diaphragmatis pelvis
    inferior, 181–183
    in female, 182
    superior, 182
    in female, 182
  diaphragmatis urogenitalis inferior, 229
  endothoracica, 173–174, 178, 205, 207
  glutea, *552*
  iliac, 477
  infraspinata, 388
  lata, 206, 210, 215–216, 219, 545, 548, 552
  of thigh, 533
  muscle, 58, 61
  nuchae
    lamina profundus, 55, 166–168
    lamina superficialis, 167
  obturatoria, 181, 186
    internus, 187, *543*
  pectoralis, 208
  of pelvic floor, superficial, 181
  pelvis
    parietalis, 557
    visceralis, 557
  penis, 228
    profunda, 228, 230
    superficialis, 228
  perinei, superficialis, 181–183, 557
  phrenicopleuralis, 173
  plantaris profunda, 523, 569
  popliteus, 484

  renalis
    lamina anterior, 167
    lamina posterior, 167
  Sibson's, 173
  spermatica
    externa, 210, 216–217, 225–226, 228
    interna, 210, 214, 216–217, 225–226
  superficial, 392
  thoracica
    deep, location of, 381
    superficial, location of, 381
  thoracolumbalis, 164, 166–167, 202–203, 307, 330–331, 365, 478–479
    course of, 168
    lamina profunda, 155, 167, 169
    lamina superficialis, 55, 155, 166–168
    m. erector spinae, 203
    transverse section
      at C VI level, 167
      at L III level, 167
  transversalis, 155, 167, 177–179, 206, 210–216
    imbricated, 219
  visceralis, 167
Fasciculi
  longitudinales, ligamenta cruciforme atlantis, 122–123
  of plexus brachialis, 261
    lateralis, 85, 297, 360–361, 365–366, 373, 381–384
      relationship with a. axillaris, 382
      spinal segments associated with, 85, 361
    medialis, 85, 297, 360–361, 365, 371, 373, 381–384
      relationship with a. axillaris, 382
      spinal segments associated with, 85, 361
    posterior, 85, 297, 360–361, 365, 367, 369, 381–384
      rami, 382
      relationship with a. axillaris, 382
      spinal segments associated with, 85, 361
    spinal segments associated with, 85, 361
    subdivision into main branches, 360
  transversi, 516
Fat
  epidurale, 128–129, 131
  intraforaminal, in spine, 129
Fat pad. *See* Corpus adiposum
Femoral ring, 549
Fenestration(s), in capillary endothelium, 67
Fertilization, 5
Fetal period
  development in, 4
  sensitivity to teratogens in, 4
Fetus, human, 8
  crown-rump length (CRL), 4
  female, 222
  greatest length (GL), 4
  longitudinal growth of, 4
  male, 222
  weight, by age in weeks, 4

Fiber(s). *See also* Nervus/Nervi (of region or organ), craniales
　elastic, in venous wall, 65
　muscle. *See* Muscle fiber(s)
　nerve. *See* Axon(s)
　prerectal, 163, 185–186
　Sharpey, 39
Fibrae intercrurales, 210–211
Fibroblast(s), subsynovial, 44
Fibrocartilage
　meniscal, 446
　of pressure tendons, 60
Fibula, 20, 38, 41, 43, 411, 418, 420–421, 440, 444–445, 448–450, 454–455, 458, 460–461, 468, 470–471, 479, 481, 483–485, 488–491, 510, 515, 522, 563, 567–568
　anterior view, 420
　caput, 31–33, 38, 408, 411, 418, 420–421, 440, 443–448, 484–487, 489, 491, 503, 506, 510–511, 540, 544, 550, 561, 566–567
　collum, 420, 440, 540
　corpus, 420, 440
　facies lateralis, 420–421
　facies medialis, 420–421
　facies posterior, 420–421, 491
　ossification of, 19
　posterior view, 420
Fibular tunnel syndrome, 96
Filtration, capillary, 67
Filum terminale, 128
Finger ray(s), 14
Finger(s). *See* Digiti (manus)
Finger-to-floor distance (FFD), measurement of, 125
First arch syndromes, 10
Fish
　arcus pharyngei in, 10
　embryonic development of, 2
Fistula(s), in first arch syndromes, 10
Flatfoot. *See* Pes planus
Flexura cervicalis, 72
Flexura mesencephalica, 72
Follicle(s)
　glandula endocrina with, 70
　Graafian, 5
Follicle-stimulating hormone (follitropin, FSH), 71
Follitropin. *See* Follicle-stimulating hormone
Fontanelle(s), 41
Fonticulus
　anterior, 41
　posterior, 41
Foot. *See* Pes
Foot drop, *540*
Footprint, 464, 467, 546
Foramen magnum, 122
Foramina/Foramen
　infraorbitale, 32
　infrapiriforme, 553–554
　intervertebralia, 76, 102, 110–112, 117–119, 121–122, 124, 127, 129–131, 137
　　nerve root in, 130
　ischiadicum, 554
　　majus, 140, 143, 543, 553
　　　boundaries of, 554
　　　location of, 554
　　　transmitted structures, 554
　　minus, 140, 143, 543, 558

boundaries of, 554
location of, 554
transmitted structures, 554
mentale, 32
nutricia, 119
obturatum, 138–139, 142, 187, 229, 412–413, 433, 557
sacralia, 41
　anteriora, 102, 114–115, 138, 143, 537
　posteriora, 102, 114–115, 138, 537
scapulae, 243, 388
suprapiriforme, 553–554
supratrochleare, of humerus, 244
transversarium, 106, 108–109, 120, 122–123, 126–127
　arteria vertebralis in, 126
　vena cavae, 160–161, 174
　vertebrale, 106–107, 113, 116, 123, 126, 130, 135, 137
Force(s), on spinal motion segment
　counterforce (S'), 124
　longitudinal (L), 124
　normal (Sn), 124
　resultant *(R)*, 124
　shear *(S)*, 124
　tangential (St), 124
Forearm. *See* Antebrachium
Forebrain, development of, 72
Forefoot, 423
　arch stabilizers in, 465
　load on, 464
　range of motion of, 463
　variants of, 544
Fornix humeri, 259, 263, 266–268
Fossa/Fossae
　acetabuli, 412, 427, 430–432, 435, 549
　axillary, 37
　coronoidea (humerus), 244, 276–277, 279
　cubitalis, 30, 356, 386
　　deep dissection of, 392
　iliaca, 138–139, 155, 413
　infraclavicularis, 36–37, 378, 380, 384–385
　infraspinata, 238, 243, 262, 389
　inguinalis
　　lateralis, 212–213, 217
　　medialis, 211–213, 215, 217
　intercondylaris, 416, 419, 440, 444, 447
　ischioanalis, 181–183, 189, 542–543, 556–557
　　in female, 557
　　in male, 557
　　muscular boundaries of, 556
　　nervi of, 555
　　vasa of, 555
　jugularis, 30, 36, 134, 199, 384–385
　lumbalis, 204
　malleoli, lateralis, 420–421
　Mohrenheim's, 384–385
　navicularis, 229
　olecrani, 244–246, 276–278
　ovalis, 212
　paravesicalis, 183
　poplitea, 30, 55, 442, 507, 541, 546
　　arteriae of, 525
　　arteria poplitea branches in, 561

　　arteria poplitea in, palpation of, *561*
　　capsule of, 442
　　deep neurovascular structures in, 561
　　ligaments of, 442
　　muscular boundaries of, 561
　　periarticular bursae of, 442
　radialis (humerus), 244, 246, 276, 279
　retromandibularis, 36
　subscapularis, 243
　supraclavicularis, 31
　　major, 36
　　minor, 36
　supraspinata, 238, 241–243, 245, 260, 267, 389
　supravesicalis, 212–213
　trochanterica, 414, 480
Fovea
　capitis femoris, 414, 416–417, 430, 433
　costalis
　　inferior, 102, 106, 110–111, 124, 137
　　processi transversi, 102, 106–107, 110–111, 124, 137
　　superior, 102, 106–107, 110–111, 124, 137
　　of twelfth thoracic vertebra, 110–111
　dentis, 109
Fracture(s), *52–53*
　acetabular, *412*
　avulsion, 52
　bending, 52
　carpal, *257*
　classification of, 52
　clavicula, *242*
　closed, 52
　Colles', *257*
　comminuted, 52
　conservative (cast) treatment of, *53*
　costae, *132*
　dislocations, types of, 52
　distal radius, *257*
　　extra-articular, *257*
　　imaging of, *257*
　　intra-articular, *257*
　fatigue, 52
　femoral, 432–433
　greenstick, 52
　healing of, 53
　humeral shaft, *246*
　humerus, 246–247
　intraperiosteal, 52
　of long bones
　　distal, 52
　　proximal, 52
　of malleolus lateralis, *461*
　of malleolus medialis, *541*
　march, 52
　mechanism of injury and, 52
　metatarsal, 52
　morphology of, 52
　nontraumatic, 52
　oblique, 52
　open, 52
　patellar, *419*
　pathologic, 52
　pelvic, *412*
　proximal femoral, *432*
　proximal humerus, *246*
　　comminuted, *246*

　　extra-articular, *246*
　　intra-articular, *246*
　scaphoid, *257*
　　imaging of, *257*
　shearing, 52
　Smith, *257*
　and soft-tissue damage, 52
　spontaneous, 52
　surgical osteosynthesis for, 53
　of sustentaculum tali, *424*
　tibial, *541*
　torsion (spiral), 52
　transverse, 52
　traumatic, 52
　treatment of, *53*
　types of, 52
　vertebral, *104*, *130*
　Weber, *461*
Frenulum preputii, 228
FSH. *See* Follicle-stimulating hormone
Fundus, uteri, 182
Funiculus, spermatica, 54, 176–177, 189, 210, 214, 216–217, 219, 225, 230, 434, 532–533, 548
　contents of, 225
　coverings of, 217
　fascia spermatica externa, 210–211
　fascia spermatica interna, 210–211
　with m. cremaster, 211, 214–215

G

Gait, 472–473
　disturbances of, *540*
　stability in, 473
　"steppage," *540*
　toeing-in, *417*
　toeing-out, *417*
Gait cycle, leg movements during, 473
Galea aponeurotica, 54
GALT. *See* Gut-associated lymphatic tissue
Gamete(s), formation of, 5
Ganglioblasts, 74
Ganglion/Ganglia
　cervicale
　　inferius, 94
　　medium, 94
　　superius, 94
　coeliaca, 94
　mesenteric
　　inferius, 94
　　superius, 94
　parasympathetic
　　cranial, 95
　　near organs, 95
　　visceral, 75
　paravertebral sympathetic, 74, 94–95
　prevertebral sympathetic, 94–95
　spinal, 76, 78, 81, 88–89
　spinal (sensory), 129
　stellatum, 94
　　injury to, *384*
　trunci sympathici, 81, 94–95
Ganglion sensorium, nervi spinalis
　dural sleeve with, 131
　of radix anterior, 75, 104
　of radix posterior, 7, 74–75, 104, 126

587

Gaster, 105
  innervation of, 94
Gastrin, 71
Gastrulation, 6
Gemmae membrorum, 2, 10, 14, 82
Genitalia. *See also specific organ*
  external
    development of, 222–223
    examination of, 227
    intersexuality, 223
    neurovascular structures of, 230–231, 234–235, 558
    undifferentiated embryonic genital anlage, 222
  female
    adrenogenital syndrome, 223
    erectile tissues and muscles, 235
    lymphatic drainage of, 234
    neurovascular structures of, 234–235
    organs of, 232
    vestibule and vestibular glands, 235
    innervation of, 94
  internal
    female, 232
    male, 224–225
  male
    descensus testis, 224
    fasciae and erectile tissues, 228–229
    funiculus spermaticus, 224–225
    hypospadias, 223
    innervation of, 95
    neurovascular structures of, 230–231
    organs of, 224
      external, 224–227
      internal, 224–225
    testis and epididymis, 226–227
Genu recurvatum, 499
Genu valgum, 27, 447
  mechanical axis in, 410
  normal, 411
Genu varum, 27, 447
  mechanical axis in, 410
  normal, 411
Germ layers. *See also* Ectoderm; Endoderm; Mesoderm
  differentiation of, 7
Gestation, length of, 4
Glandulae, 70–71. *See also specific glandulae*
  areolares, 208
  Bartholin's, 232, 235
  bulbourethrales, 183, 188, 224, 230
  classification of, 70
  Cowper, 188
  development of, 70
  endocrine, 70
    overview of, 71
  exocrine, 70
    secretory mechanisms of, 70
  lacrimal, innervation of, 94
  mammariae lobi, 208
  multicellular intraepithelial, 70
  parathyroideae, 71
    hormone secretion by, 71
  pineale, 71
    hormone secretion by, 71
  pituitaria, 71
    hormone secretion by, 71
  prostata, 105, 183, 188–189, 543, 557
  salivary, 70
    innervation of, 94–95
  seminal, 231
  sudorifera, 70
    of manus, 374
  suprarenales, 71
    hormone secretion by, 71
  thyroidea, 71, 167
    C cells, hormone secretion by, 71
    hormone secretion by, 71
  vesiculosa, 189, 224, 434
  vestibulares
    majores, 232, 235
    minores, 232, 235
Glans clitoridis, 180, 183, 232, 235
  development of, 222
Glans penis, 31, 180, 188, 223–224, 228–231, 543
  development of, 222
Glucagon, 71
Glucocorticoid(s), 71
Glucuronic acid (GlcUA), 45
Glycoprotein(s), in extracellular matrix, 45
Glycosaminoglycans, 45
Golfer's elbow, *329*
Golgi apparatus, in exocytosis, 70
Gomphosis, 41
Gravity. *See also* Line(s), of gravity
  partial-body center of, 415, 437
  whole-body center of, 27, 103, 411
Greatest length (GL), of fetus, 4
Grip, types of, 292
Groin skin flaps, arteria circumflexa ilium superficialis in, 207
Growth plate(s)
  fractures and, 52
  of hip bones, 41, 438
Grynfeltt's hernia, 202, 221
Gut, pharyngeal, 10
  embryology of, 2
Gut-associated lymphatic tissue (GALT), 68
Guyon–Logen syndrome, *96*
Gyri, of brain, 72

H

Haeckel, Ernst, 2
Haglund's exostosis, *471*
Hallux valgus, *469*
  etiology of, 469
  pathogenesis of, 469
Hammer toes, *469*
Hamulus ossis hamati, 32, 239, 252, 255, 285, 288–289, 293, 315, 321, 323, 405
Hand. *See* Manus
Hand area rule, for estimating burn area, *23*
Hand lines, 374
Hand of benediction, *372*
Haversian vessel(s), 17, 39
HCG. *See* Human chorionic gonadotropin
Head height, as fraction of body length, change during growth, 22
Heart failure, right-sided, *378*
Heel pad, 472
Heel pain, *471*
Heel spur
  plantar, *471*
  posterior, *471*
Height, adult, prediction of, 19
Hemarthrosis, *44*
Hematoma, fracture, 53
Hemiarthroses, laterales, 126
Hemiplegia, *473*
Hemolymphatic system, 24
Hepar, 24, 62, 105
  hormone secretion by, 71
  innervation of, 94
  venous drainage of, development of, 13
Hermaphroditismus verus, 223
Hernia
  abdominal
    definition, occurrence and structure of, 213
    internal and external openings for, 212
    internal surface anatomy for, 212
  axial hiatal, *175*
  diaphragmatic, *175*
  epigastrica, 220
  femoralis, 212, 214–215, *549*
    course, 214
    examination of, 218
    external openings, 214
    internal openings, 213–214
  Grynfeltt's, 202, 221
  hiatal, *175*
  incarceration of, 218
  incisional, 220
  infrapiriformis, 221
  inguinalis
    directa, 212, 214–215
      comparison of, 217
      condition after, 219
      course, 214
      examination of, 218
      external openings, 214
      internal openings, 213–214
      palpation of, 218
    indirecta, 212, 214
      comparison of, 217
      course, 214
      examination of, 218
      external openings, 214
      internal openings, 213–214
      palpation of, 218
    repair of, 219
    scrotal swelling, 227
    topographic anatomy of, 216–217
  ischiadica, 221
  ischiorectalis, 221
  lumbalis, 202
    inferior, 221
    superior, 221
  obturatoria, 221
  omphalocele, 220
  perinealis
    anterior, 221
    posterior, 221
  Petit's, 202, 221
  rare external, 220–221
  rectus diastasis, 220
  reduction of, 218
  repair of, 218
  sac of, 213, 215
    peritoneum of, 214–216
  sliding, *175*
  Spieghelian, 221
  spinotuberosa, 221
  suprapiriformis, 221
  supravesicalis, 212
  umbilicalis, 220
Hiatus
  adductorius, 481, 504–505, 508, 524–526, 550, 560–561, 568
  aorticus, 160–161, 174
  basilicus, 357, 375, 386
  levatorius, 185–186
  oesophageus, 160, 174
  sacralis, 114, 138–139, 143, 185, 187
  saphenus, 212, 216, 528, 545
    vasa femoralia in, 215
  triceps, 390–391
  of ulnar tunnel
    distalis, 405
    proximalis, 405
  urogenitalis, 163
Hill-Sachs lesion, *262, 270*
Hindbrain. *See also* Rhombencephalon
  structure of, 73
Hindfoot, 423
  axis of, 462
  load on, 464
  range of motion of, 463
  sensory innervation of, 535
Hirschsprung disease, *74–75*
Histamine, 71
Histogenesis, 4
Hofbauer cell, 9
Holocytosis, 70
Hormone(s). *See also specific hormone*
  actions of, 71
  amino acid derivatives, 71
  anti-müllerian, 222
  classification of, 71
  fatty acid derivatives, 71
  formation of, 70
    principal sites of, 71
  inhibitory, 71
  peptide, 71
  placental, 9
  release of, 70
  releasing, 71
  secretion of, 71
  steroid, 71
  thymic, 71
  tissue, 71
Howship–Romberg syndrome, *96*
Hueter triangle, 278
Human chorionic gonadotropin (HCG)
  placental production of, 9, 71
  in pregnancy testing, *9*
Humeroscapular rhythm, 275
Humerus, 20, 38, 61, 164–165, 238, 244–245, 249–250, 258–260, 262, 266–267, 269, 276–279, 282, 307, 309, 317, 342–343, 393
  caput of, 38, 244–247, 258–259, 261–262, 264, 269, 271–273, 304, 307, 312, 381
  collum anatomicum, 244–246, 262, 265
    fractures of, 246
  collum chirurgicum, 244, 246
    fractures of, 246
  corpus of, 244, 246, 303, 305, 311–313
  development of, 14

distal, inferior view, 245
facies anterolateralis, 244, 246
facies anteromedialis, 244, 246
facies posterior, 244
fractures of, 246–247
margo lateralis, 244, 246, 276
margo med, 244
margo medialis, 246
nervi closely related to, 391
ossification of, 18
processus supracondylaris of, 244, 393
proximal
  fractures of, *246*
  superior view, 245
torsion of, 247
Hyaluronate, 43, 45
Hydrostatic indifference level, 65
Hydrostatic pressure, 65, 67
Hymen, 232
Hyperabduction syndrome, *363*
Hypoblast, 6
Hypoesthesia, *545, 547*
Hypogastrium, 199
Hypomere, 7, 145
Hypospadias, 223
  glandis, 223
  penis, 223
  perinealis, 223
  scrotalis, 223

I
Imaging
  of acetabulum, radiography, 413
  of acromioclavicular joint, 28
  of acromion, 28
  of ankle joint, 470
  of articulatio coxae
    in children, 438–439
    during development, 438–439
    MRI, 432–434
    radiography, 433, 438
    ultrasound, 435, 438
  of articulatio femoropatellaris, 441
  of articulatio genus
    MRI, 452–453
    radiography, 441
  of Baker cyst, *442*
  of biceps femoris tendon, 28
  of bursa subacromialis, 28
  of canalis carpi, MRI, 288
  of carpus, 286
    radiography, 256
  of cervical spine
    MRI, 121
    radiograph, 121
  of collagen fibrils, TEM, 44
  of collum femoris angle, radiography, 415
  computed tomography (CT), sectional planes for, 28
  of distal radius fractures, 257
  of femoral fractures
    CT, 433
    MRI, 433
    radiography, 433
  of gastrocnemio-semimembranosa bursa, 442
  of glenohumeral ligament, 28
  of joints, MRI, 452
  of labrum glenoidale, 28
  of ligamentum coracoacromiale, 28
  of lig. glenohumerale, 271
  of lumbar disk herniation, MRI, 131
  of lumbar spinal stenosis, *130*
  of lumbar spine, MRI, 129
  of m. infraspinatus, 28
  MRI
    of joints, 452
    sectional planes for, 28–29
  of m. supraspinatus, 28
  pelvis survey radiograph, 433
  of pes
    MRI, 471
    radiography, 470
  planes for, 28–29
  of proteoglycans, light microscopy, 44
  of rotator cuff muscles, 28
  of scaphoid fractures, 257
  of shoulder/art. humeri (glenohumeralis)
    arthroscopy, 271
    MRI, 273
      standard planes in, 28, 273
    radiography, 272
    ultrasound, 272
  of shoulder joint capsule, 28
  of skull, planes for, 28
  of subacromial space, 28
  of thigh, axial MRI, 29
  of thorax, axial CT, 29
Immune response
  cell-mediated, 68
  humoral, 68
  specific, 68
Immune system. *See also* Lymphatic organs
  and endocrine system, 71
  hormones and hormonelike substance secretion by, 71
  specific, 68
Immunological disease(s), *91*
Implantation, of blastocyst, 5
Impotence, male, *543*
Incisura
  acetabuli, 412
  clavicularis, of sternum, 132, 134
  costalis, of sternum, 134
  frontalis, 32
  ischiadica
    major, 139, 412, 554
    minor, 139, 189, 412, 554
  jugularis, of sternum, 132, 134
  radialis, of ulna, 248, 250–251, 277, 279
  scapulae, 240–241, 243, 245, 259–260, 262–263, 267, 303, 362, 382, 388, 390
  supraorbitalis, 32
  thyroidea, 384
    superior, 32
  trochlearis, 277, 279–280
    of ulna, 248, 251
  vertebralis
    inferior, 110–112
    superior, 106–107, 110–113, 116, 127
Incisura scapulae syndrome, *96, 243, 362, 388*
Incontinence
  fecal, *163, 543*
  stress, *163*
  urinary, *163, 543*
Incus, embryology of, 11
Inguinal ligament syndrome, *96*
Injection(s)
  intra-arterial, *356*
  intra-articular, *40*
  paravascular, *356*
  ventrogluteal, *555*
Inner cell mass, of embryoblast, 5
Innominate bones, 408
Inspiration
  muscles of, 158–159
  position of diaphragm and ribs in, 160
Insula, 72
Insulin, 71
Interbrain, development of, 72
Intercondylar distance, 410–411
Interdigital Morton neuralgia, *96*
Intermalleolar distance, 410–411
Internal organs, location of, in body, 24
Interneuron(s), 75, 88, 90–91
Interscalene block, *384*
Intersectio tendineae, 31, 199
Intersexuality, 223
Interstitial fluid, 66, 68
  flow of, in viscoelastic model of joint cartilage, 45–46
Interstitium, 66
Intervertebral disk(s). *See* Discus, intervertebralis
Intestine
  innervation of, 94
  large. *See* Appendix, vermiform; Colon
  small. *See* Duodenum
Intestinum tenue, 24
Intima, synovial, 43–44, 450
Intracellular fluid, 66
Intraperitoneal organs, 204
Islets of Langerhans, 71
  hormone secretion by, 71

J
Joint capsule. *See* Capsula articularis
Joint contracture, *43*
Joint endoprosthesis, *40, 46*
Joint pressure, 51
Joint puncture, *40*
Joint replacement surgery, *40, 46*
Joint(s). *See* Articulationes
Joint space(s), 42
  art. interphalangealea distalis (DIP), 376
  art. interphalangealea proximalis (PIP), 376
  art. metacarpophalangea (MCP), 376
  of intervertebral facet joints, 124
  of ribs, 134
Junction
  cervicothoracic, 103
  craniocervical, 103
  lumbosacral, 103, 476
  thoracoabdominal, 174–175
  thoracolumbar, 103
    ligaments of, 118–119

K
Keratan sulfate, 45
Knock knees, 27. *See also* Genu valgum
Kyphosis
  neonatal, 105
  sacralis, 103, 105
  thoracica, 103, 105

L
Labia majora
  in adrenogenital syndrome, 223
  development of, 222
  pudendi, 180
Labia minora, 235
  in adrenogenital syndrome, 223
  development of, 222
  pudendi, 180, 232, 559
Labioscrotal swelling, 224
Labrum/Labra
  acetabuli, 42, 415–416, 430–432, 435
  articularia, 42–43
  glenoidale, 42, 261–262, 266, 269, 273
  imaging of, 28
Lacerta viridis, limb positions of, 21
Lachman test, *448*
Lacuna/Lacunae, Howship, 17
Lacuna musculorum, 533, 535
Lacuna vasorum, 524, 533, 549
Lamella/Lamellae, bony, 17, 39
Lamina arcus vertebrae, 106, 109, 111, 113, 118–119, 126, 137
Lamina chorionica, 8–9
Lamina neuralis, 6, 72
Lamina parietalis, 24
Lamina visceralis, 24
Lancelet, 2–3
Landmarks, 32–33, 35
  in deep nuchal region, 170
  lines. *See* Line(s); Reference line(s)
  in membrum inferius, 409
  palpable. *See* Palpable bony prominences
  of plexus brachialis, 384
  processi spinosi as, 35, 101. *See also* Processus/Processi, spinosi vertebrales
Langhans cell, 9
Larrey's cleft, 174
Larynx, 31–32, 105
Lateral axillary hiatus syndrome, *96*
Leg. *See* Crus
Leg length
  discrepancy, 409, *497*
  measurement of, in standing position, 409
Levator raphe, 185–186
Lever(s)
  force arm of, 51
  load arm of, 51
  one-arm, 51
  two-arm, 51
Leydig cells, 222
LH. *See* Luteinizing hormone
Liberins, 71
Lichtenstein technique, 219
Ligamentum flavum/Ligamenta flava, 41, 117–122, 124, 128–130
Ligamentum/ligamenta (in general, including ligament types)
  embryology of, 11
  and erect posture, 50
  extracapsularia, 42–43
  guiding, 43
  intra-articular, 43
  intracapsularia, 42–43
  overstretching of, *43*
  restraining, 43
  shortening of, *43*
  tears of, *43*

L Ligamentum/ligamenta

Ligamentum/ligamenta (named)
  acromioclaviculare, 260, 263, 268, 389
    injuries to
      Rockwood classification, 260
      Tossy classification, 260
  alaria, 122–123
  anococcygeal, 163, 181, 184, 187, 543, 556, 558
  anulare radii, 277–282
    course of, in proximal radioulnar joint, 279
  anularia
    of fingers, 290–291, 344, 347–348
    of toes, 517
  arcuatum
    and articulatio genus stabilization, 50
    laterale, 160, 174–175, 204
    mediale, 160, 174–175
    pubis, 185, 230
  arteriosum, development of, 12
  bifurcatum, 459–461
  calcaneocuboid
    dorsale, 459–460
    plantare, 492
  calcaneofibulare, 460–461
  calcaneonaviculare plantare (spring), 424, 458, 460, 466–467
    course of, 459
  capitatohamatum, 286
  capitis costae radiatum, 137
  of capitis fibulae
    anterius, 444–445
    posterius, 444–445
  of caput femoris, 43, 428, 430–432, 434
  cardinale (transverse cervical), 182
  carpi
    classification of, 284–285
    extrinsic, 284
    intrinsic, 284
    palmare, 289, 400, 402, 404–405
    transversum, 288–289, 398, 402, 404. See also Retinaculum/Retinacula, musculorum flexorum
  carpometacarpalia
    dorsalia, 284
    palmaria, 285
  collaterale fibulare, 43, 443, 445
  collaterale radiale, 277–280, 282
    carpi, 284–286, 294
  collaterale tibiale, 43, 443, 445
  collaterale ulnare, 277–280, 282
    carpi, 255, 284, 286–287, 294
    pars anterior, 278
    pars posterior, 278
    pars transversa, 278
  conoideum, 259
  Cooper's, 208, 549
  coracoacromiale, 259–260, 263, 265–268, 389
    imaging of, 28
  coracoclaviculare, 258–260, 263, 268, 388–389
  coracohumerale, 263–265
  costoclaviculare, 258–259
  costotransversaria, 137
    laterale, 137, 172
    superius, 137
  costoxiphoideus, 134

cruciata
  of fingers, 290, 344, 348
  of toes, 517
cruciata (of knee), 43, 441, 444, 447, 450
  anterius, 443–444, 446, 448–453
    rupture of, 448
    in extension, 448
    in flexion, 448
    in flexion and internal rotation, 448
  posterius, 443–444, 446, 448–449, 452–453
    rupture of, 448
  sites of attachment of, 446
  tears of, 449
deltoideum, 424, 461
  pars tibiocalcanea, 460–461
  pars tibionavicularis, 460–461
  pars tibiotalaris
    anterior, 460–461
    posterior, 460–461
fundiforme penis, 176
glenohumerale
  imaging of, 28, 271
  inferius, 263–265, 271
    anterior band, 264
    posterior band, 264
  medium, 263–265, 271
  superius, 263–265
iliofemorale, 43, 429, 497, 504
  and erect posture, 50
  and joint motion, 50
iliolumbale, 140, 143, 428
inguinale, 140, 153, 155, 176–177, 198–199, 206, 210–211, 214–215, 219, 428, 430, 477, 503, 524, 526, 528–530, 532–533, 535, 544–545, 548–549, 551
intercarpalia
  dorsalia, 282, 284
  palmaria, 285
interclaviculare, 259
interfoveolare, 211, 213
interosseous, of carpus, 287
  degenerative changes in, 286
intertransversaria, 172
ischiofemorale, 43, 428–429
lacunare, 210, 215, 219, 549
longitudinale anterius, 117–121, 128–130, 140, 160, 428, 503
longitudinale posterius, 117–119, 121–122, 128–130
lunotriquetrum, 286
mallei, anterius, embryology of, 11
meniscofemorale posterius, 443–444, 446
metacarpalia
  dorsalia, 284, 294
  palmaria, 285
  transversa profunda, 285, 290–291, 347–349
metatarsea
  dorsalia, 461
  transversum
    profundum, 465, 468, 492
    superficiale, 348, 516
nuchae, 118, 120–123, 165, 172, 299
palmaria, 285, 291, 352

patellae, 43, 54, 61, 441, 443–447, 450–452, 482–483, 502–504, 506, 510, 566
  and femoropatellar joint load, 51
pectinale, 210, 219
pectineum, 549
phalangoglenoidale, 290–291
pisohamatum, 405
plantare longum, 519–520
plantaria, 465, 520
  longum, 458, 460, 466–467, 492–493
popliteum
  arcuatum, 442–443
  and articulatio genus stabilization, 50
  obliquum, 442–443, 484, 561
pubofemorale, 43, 428–429
radiocarpale
  dorsale, 282, 284
  palmare, 282, 285
radiotriquetrum, 287
radioulnare
  dorsale, 280–281, 284, 287
  palmare, 280–281, 285, 287
reflexum, 210–211, 549
sacrococcygeum, anterius, 143
sacroiliaca
  anteriora, 140, 143, 187, 428
  interossea, 140, 143
  posteriora, 140, 143, 428
sacrospinale, 140, 143, 163, 175, 184, 187, 428, 430, 477, 543, 554, 558
sacrotuberale, 140, 143, 163, 184, 187, 221, 428, 430, 477–479, 484–485, 507–508, 537, 541, 553–556, 558, 560
scapholunatum, 286
sphenomandibulare, embryology of, 11
sternoclaviculare
  anterius, 259–260
  posterius, 260
sternocostalia radiata, 134, 172, 259
Struthers', 372, 393
stylohyoideum, embryology of, 11
supraspinale, 172
suspensorium
  mammaria, 208
  ovarii, 182
  penis, 228
talocalcaneum interosseum, 455, 458–460, 466
talofibulare
  anterius, 460–461
  posterius, 460–461, 471
talonaviculare, dorsale, 460–461
tarsi, dorsalia, 460–461
teres hepatis, 212
teres uteri, 182, 188, 215, 533
tibiofibulare
  anterius, 460–461
  posterius, 460–461
transversum
  genus, 443–444, 446
  humeri, 263, 266
  perinei, 230
scapulae
  inferior, 389–390

    superius, 259–260, 263, 267–268, 362, 388–390
  transversum acetabuli, 431
  trapezoideum, 259
  trapezoideumcapitatum, 286
  ulnocarpale
    dorsale, 282
    palmare, 282, 285
  ulnolunatum, 287
  ulnotriquetrum, 287
  zona orbicularis (anular ligament), of hip joint, 428, 431
Ligamentum/ligamenta (of structure)
  of apicis dentis, 121, 123
  of articulatio coxae, 43, 428–431
  of articulatio genus, 43, 50, 442–443, 482
    collaterale, 441–445
      in extension, 448
      fibulare, 441–450, 453
      in flexion, 448
      in flexion and internal rotation, 448
      tears of, 448
      tibiale, 441–449
    cruciata, 43, 441, 443–444
    extracapsulare, 450
    intracapsularia, 450
  of articulatio talocruralis, 461
    laterale, 461
    mediale, 461
  of atlantoaxial joint, 122–123
  atlantooccipitale laterale, 120
  of atlantooccipital joint, 122–123
  capitis costae intraarticulare, 137
  capsular, and articulatio genus stabilization, 50
  of cervical spine, 120–121
    upper, 122–123
  of columna vertebralis, 118–119
  of costovertebral articularis, 137
  of craniovertebral joints, 122–123
  cruciforme atlantis, 122
  of fingers, 290–291, 294
  of interphalangealis distalis, collateralia, 284
  of interphalangealis proximalis, collateralia, 284–285
  interspinalia, 40, 118, 121, 128
  intertransversaria, 118–119, 123
  intracapsularia, 43–44
  of joints, 43
    and erect posture, 50
  of manus, collateralia, 347
  of metacarpophalangealis, collateralia, 284–285
  pelvic, 140
  of pes, 460–461
  of sacroiliac joint, 143
  supraspinale, 40, 118, 121, 128
  of thoracolumbar junction, 118–119
  transversum atlantis, 121–123
  vertebral arch, 118–119
  vertebral body, 118–119
Limb girdle, 20
Limb(s)
  five-ray (pentadactyl) tetrapod
    bony constituents of, 20
    skeletal structure of, 20
  human. See also Membrum inferius; Membrum superius

basic skeletal structure of, 20
development of, 14
ossification of, 18–19
terms of location and direction for, 26
positions of, in primitive terrestrial tetrapod, 21
rotation of, in mammalian evolution, 21
Limbus acetabuli, 138–139
Linea alba, 31, 153–155, 157, 176–179, 199, 210, 220
Linea aspera, 38, 414, 417, 426, 480, 482, 484–485
labium laterale, 414
labium mediale, 414
Lineae
arcuata, 138–139, 141–143, 153, 155, 177–178, 187, 206, 210, 212–213, 221, 412–413
axillaris
anterior, 34
media, 34
posterior, 34
epiphysialis (bony), 39
gluteae
anterior, 139, 412, 478
inferior, 412
posterior, 139, 412, 478–479
intercondylaris, 414
intermedia, of crista iliaca, 138–139
intertrochanterica, 414, 426, 428, 430, 477, 482–483
mediana
anterior, 34
posterior, 34
medioclavicularis, 34
midclavicularis, 199
musculi solei, 420, 440, 488–489, 491
nuchalis
inferior, 120, 149, 151, 171
superior, 120, 122, 147, 149, 151, 165, 169–171, 299, 331
parasternalis, 34
paravertebralis, 34
pectinea, 413–414, 417, 426, 480
primitiva, 6
scapularis, 34
semilunaris, 31, 177, 199, 221
sternalis, 30, 34
supracondylaris
lateralis, 414
medialis, 414
transversae, of os sacrum, 41, 114
Linea terminalis, 141, 182, 534
Line(s)
acetabular roof, 439
boundary (tide mark), in hyaline cartilage, 44
cement, of osteon, 39
clavicular, 125
epiphyseal, 432, 439
femoral, 415
of gravity, 27, 103, 411, 482
and articulatio genus stability, 50
and stability in bipedal stance, 421
hand, 374
Hilgenreiner, 439

Hueter, 278
ilioischial, 433
iliopectineal, 433
Lisfranc joint, 455
Ménard-Shenton, 439
Mikulicz, 410
Perkins-Ombredanne, 439
Roser-Nélaton, 429
Sölder, 86
spina-trochanter, 554
spina-tuber, 554
tuber-trochanter, 554
Lingua, 105
Liquor cerebrospinalis sampling, lumbar puncture for, 128
Lithotomy position, 35, 180–181, 542
Lizard, limb positions of, 21
Lobi glandulae mammariae, 208
Lobuli testis, 226
Location, terms of, 26
Locomotor system, 24
Lordosis
cervicis, 103, 105
lumbalis, 103, 105, 157, 476
hip extension and, 436
Lower limb. See Membrum inferius
Lower motor neuron lesion(s), 90
L-system, in skeletal muscle, 58
Lubricin, 43
Lumbar puncture, 128
Lumbar spinal anesthesia, 128
Lung bud(s), 10
Luteinizing hormone (lutropin, LH), 71
Lutropin. See Luteinizing hormone
Lymph, 68
flow, direction of, 69
transport of, 69
Lymphadenectomy, axillary, 362
Lymphangitis, in upper limb, 358
Lymphatic organs, 68
functions of, 68
primary, 68
secondary, 68
Lymphatic vascular system, 68
compartments of, 69
deep, 69
organization of, 69
organ-specific, 69
perforator vessels, 69
regions of, 69
superficial, 69
Lymph node(s). See Nodi lymphoidei
Lymphokines, 71
Lymph vessels. See Vasa lymphatica

M

Macrophage(s), in epiphyseal plate, 16
Magnetic resonance imaging (MRI). See Imaging, MRI
Malleolus
lateralis, 30, 32–33, 38, 408, 411, 420–421, 454–455, 458, 460–461, 468, 470–471, 487, 489–491, 510, 512, 514–515, 524, 526, 540–541, 544, 546, 562, 566
fractures of, 461

medialis, 30, 32–33, 38, 408–409, 420–421, 454–455, 458, 460–461, 465–466, 468, 470, 487, 489–491, 510, 512, 514–515, 523–524, 544, 546, 562–563, 566
fractures of, 541
Malleus, embryology of, 11
MALT. See Mucosa-associated lymphatic tissue
Mamilla, 30
Mamma, 30, 199
blood supply to, 209
gross and microscopic anatomy of, 208
lymphatic drainage of, 209
malignant tumor distribution in, 209
nerves, blood vessels, and lymphatics in, 208–209
nerve supply to, 209
ridges of, 208
shape and appearance of, 208
Mammal(s), 2
characteristics of, 2–3
embryology of, 2
evolution of, rotation of limbs in, 21
Mandibula, 38, 41, 105, 151
embryology of, 11
Mandibulofacial deformity(ies), 10
Manubrium, of sternum. See Sternum
Manus, 238. See also Digiti (manus); Pollex
arteriae of, 354–355
rete carpale dorsale, 354–355
rete carpale palmare, 354–355
arterial anastomoses in, 401
articulationes of, 238
column model of, 254, 256
connexus intertendinei, 319, 340, 345–347
distal interphalangeal joint crease, 374
distal transverse crease, 374
dorsal digital expansion, 291, 319, 340, 345–347
interosseous slip, 347
lumbrical slip, 347
tractus intermedius, 347
tractus lateralis, 347
dorsum, 37, 252–253
arteriae of, 396
connexus intertendinei, 340
extension creases of, 376
distal, 376
proximal, 376
lymphatic drainage of, 358
nerve supply to, 397
nervi of, cutaneous, 397
sensory innervation of, 397
skin of, 376
superficial venae of, 357
extensor tendons of, 346
dorsal tendon compartments for, 345
functional position of, 295
interphalangeal joint crease, 374
length of, 239
ligamenta of
intrinsic, 286–287
overview of, 284–285
metacarpophalangeal joint crease, 374

middle crease, 374
movements of, 256, 282, 295
musculi of, 296
dorsal, origins and insertions of, 346
hypothenar, 296, 320–321, 394, 404–405
intrinsic, 320–323
deep layer, 352–353
middle layer, 350–351
superficial layer, 348–349
palmar, origins and insertions of, 353
thenar, 296, 320–321, 343, 373, 394, 404
ossa of, 38, 238, 247, 249–250, 252–253
dorsal view, 253
palmar view, 252
palma, 37, 252, 374
arteriae of, superficial, 398
epifascial nervi of, 398–399
innervation of, normal and variant, 399
nervi of, 398–399
sensory innervation of, exclusive and overlapping areas, 399
skin of, structure of, 374
vasa of, 398–401
vascular supply to, 400–401
pronation, 282
range and axis of, 283
proximal interphalangeal joint crease, 374
proximal transverse crease, 374
rete carpale dorsale, 396, 401
rete carpale palmare, 401
rete venosum dorsale, 357, 377
supination, 282
range and axis of, 283
terms of anatomic orientation for, 252
thenar crease (linea vitalis), 374
transverse crease, 374
vaginae tendinum of, 344–345
Marsupialia, 2
Maxilla, 32, 38, 121
Medial malleolar region, neurovascular structures of, 563
Mediastinum, 24
testis, 226
Medulla, of nodi lymphatici, 69
Medulla oblongata, 105
development of, 72
Medullary thyroid carcinoma, 75
Medulla spinalis, 76, 105, 121, 126, 128, 167, 205
embryology of, 7
growth of, 82
in male sexual reflexes, 231
rami of, 197
Medusa's head, 192
Meiosis, in gametogenesis, 5
Melanoblasts, 74
Melanocyte(s), 74–75
Melanocyte-stimulating hormone (melanotropin, MSH), 71
Melanoma, 75
Melanotropin. See Melanocyte-stimulating hormone
Melatonin, 71
Membrana
atlantooccipitalis
anterior, 120–121
posterior, 120–123, 171

M Membrana (cont.)

elastica externa, in blood vessel wall, 64–65
elastica interna, in blood vessel wall, 64–65
fibrosa
  of joint capsule, 43–44
  of tendon sheath, 61
intercostalis externa, 172
interossea, 41
  antebrachii, 280, 282, 294, 342, 351, 355, 395–396, 401
  cruris, 420, 443–444, 461, 479, 487, 513, 522, 524–525, 563, 567–568
  cruris, 421, 486, 490
  of membrum superius, 251, 315, 317, 341
obturatoria, 140, 143, 189, 430, 478, 480
perinei, 181–183
  in female, 182
  in male, 229
sterni, 134
suprapleuralis, 173
synovialis
  of joint capsule, 43–44
    adipose, 44
    areolar, 44
    fibrous, 44
  of tendon sheath, 61
tectoria, 121–123
Membrane(s)
  acrosomal, 5
  basal, of skeletal muscle, 58–59
  of femoropatellar joint
    fibrosa, 441, 446, 450
    synovialis, 441, 446, 450
  fetal, development of, 8
  of hip joint
    fibrosa, 430–431, 434
    synovialis, 430–431
  oropharyngeal, 6
  postsynaptic, 59
  vastoadductoria, 535, 550–551
Membrum inferius, 24. See also Articulationes (named), genus; Crus (Leg); Pes; Thigh; Toe(s)
  arteries of, 524–525
  basic skeletal structure of, 20
  cross-sectional anatomy of, 522
  dermatomes of, 545
  directional terms for, 477
  lymphatic system of
    profundi, 528
    superficiales, 528
  medial malleolar region, neurovascular structures of, 563
  musculi of, 408, 474–475
    arrangement of, 21
    hip and gluteal, 474
    leg, 474
    motor innervation of, 475
    short muscles of pes, 474
    thigh, 474
  nerve compression (bottleneck) syndromes in, 96
  nerves of, 545
  nervi, 530–531
  nodi lymphatici of, 528–529
  ossa of, 38, 408, 411
    overview of, 408
    ossification of, 18–19
  palpable bony prominences, 409
    anterior view, 32
  posterior view, 33
  peripheral sensory innervation of, 545
  regional anatomy (regions) of, 37
  surface anatomy of
    anterior view, 544
    posterior view, 546
  surface contours
    anterior view, 32
    posterior view, 33
  vasa lymphatica of, 528
  venae of, 526–527
    cutaneous, superficial, 545
    overview of, 526
    perforantes, 526–527
    profunda, 526
    superficiales, 526, 545
      varicosities, 527, 545
Membrum superius, 24. See also Antebrachium; Axilla; Brachium; Cubitus/Art. cubiti; Digiti (manus); Manus; Shoulder/Art. humeri (glenohumeralis); Wrist
  arteriae of, 354–355
    development of, 395
  basic skeletal structure of, 20
  cutaneous innervation of
    peripheral sensory, 375, 377
    radicular (segmental), 375, 377
  dermatomes of
    anterior view, 375
    posterior view, 377
  musculi of
    arrangement of, 21
    classification of
      functional-topographical, 296
      by innervation, 296–297
      principles of, 296
      functional groups, 296–297
  nerve compression (bottleneck) syndromes in, 96
  nervi of, cutaneous, 375, 377
  nodi lymphoidei of, 358–359
  ossa of, 238, 244–255
  ossification of, 18–19
  palpable bony prominences, 239
    anterior view, 32
    posterior view, 33
  regional anatomy (regions) of, 37
  segment lengths, 239
  skeleton of, 238
  surface anatomy of
    anterior view, 374
    posterior view, 376
  surface contours
    anterior view, 32
    posterior view, 33
  vasa lymphatica of, 358–359
  venae of, 356–357
    cutaneous, 356
    deep, 357
    perforantes, 357, 375
    subcutaneous, 356, 377
    superficial, 357
    superficial cutaneous, 375, 377
Meninges, spinal, 128
Meniscoid synovial folds, in joint capsule, 124
Meniscus/Menisci
  articularis, 42
  of articulatio genus, 61, 443, 446–447
    blood supply to, 446
    degenerative changes in, 447
    external, 452–453
    injury of, 445, 447
    internal, 453
    lateralis, 43, 444–450
      shape of, 446
      sites of attachment, 446
    medialis, 43, 444–450
      injury of, 447
      shape of, 446
      sites of attachment, 446
      tearing of, 448
    movements of, during flexion, 447
    during rotational movements, 449
    structure of, 446
    tears, 447–448
      bucket-handle, 447
      longitudinal (flap), of anterior horn, 447
      peripheral, 447
      radial, of posterior horn, 447
Meniscus ulnocarpalis homologue, 255, 286–287
Meralgia paresthetica, 96
Meromelia, 14
Mesencephalon, 72
Mesenchyme, 10, 14
  in bone formation, 17
  derivatives of, 16
Mesenterii, radix, 179
Mesoderm
  axial, 7
  chorionic, 8–9
  derivatives of, 7, 14
  embryonic, 6
  extraembryonic, 6
  intermediate, 6–7
  laminae lateralis, 6–7
  paraxial, 6–7
  prechordal, 7
Mesogastrium, 199
Mesotendineum, 61
Metabolic rate, relationship to body surface area and body volume, 23
Metacarpus, architecture of, 256
Metameric arrangement, 2
Metaphysis, 16
Metapodium, 20
Metarteriole, 67
Metencephalon, 72
Michaelis rhomboid, 554
Microcirculation, 64. See also Terminal vascular bed
Microglia, 79
  functions of, 79
Microscopy
  electron. See Electron microscopy
  light, of proteoglycans, 44
Micturition, 156
Midbrain. See also Mesencephalon
  development of, 72
Midfoot, 423
Mineralization zone, in hyaline cartilage, 44
Mineralocorticoid(s), 71
Mitochondria, in skeletal muscle, 58
Mitosis, in gametogenesis, 5
M-line, 59
Monotremata, 2
Mons pubis, 30, 180, 199, 232
Morphogenesis, 4
Morula stage, 5
Motion segments, spinal. See Columna vertebralis, motion segments
Motoneuron, a-, 59
Motor end plate(s), 58–59
Motor innervation, 88–91
  neural circuit for, 91
Motor neuron, 59
Motor unit(s), 59
  muscle fibers of, 56
MSH. See Melanocyte-stimulating hormone
Mucosa-associated lymphatic tissue (MALT), 68
Mucus plug, in canalis cervicis uteri, 8
Multiple sclerosis (MS), 92
Muscle cell(s), 58–59
Muscle fiber(s)
  length of, 57
  of muscles of movement, 56
  pennation angle of, 57
  of postural muscles, 56
  structure of, 58
  thickness of, 57
  tonic, 56
  twitch, 56
  type I (slow-twitch; ST), 56, 476, 499
    histochemistry of, 56
  type II (fast-twitch; FT), 56
    histochemistry of, 56
Muscle mass, age-related changes in, 56
Muscular dysbalance, 499
Musculoskeletal system, 24
Musculus/Musculi (in general, including muscle types)
  adductores, 56
  agonist/antagonist, at joint, 42
  bellies of, 57
  bicipital (two heads), 57
  bipennate, 57
  branchiogenic (branchial arch), 145
  contraction of, actin-myosin interaction in, 59
  cross section
    anatomical, 57
    physiological, 57
  digastric (two bellies), 57
  embryology of, 11
  epaxial, 7, 145
  and erect posture, 50
  extensors, embryology of, 145
  extrafusal, 54
  extraocular, 145
  flat, 57
  flexors, embryology of, 145
  force development in, 57
  force production in, 57
  fusiform, 57
  hypaxial, 7, 145, 164
  innervation of, 95
    monosegmental, 88
    polysegmental, 88
  insertions of, 57
  lifting force of, 57
  lifting height of, 57
  morphological formation of, 57
  multigastric (multiple bellies), 57

Musculus/Musculi **M**

multipennate, 57
non-pennate (parallel-fibered), 57
origins of, 57
pennate, 57
phasic (locomotive; muscles of movement; white muscles), 54, 499
   characteristics of, 56
   examples of, 56
postural (anti-gravity; red muscles), 50, 54
   characteristics of, 56
   examples of, 56
"power output" of, 57
quadricipital (four heads), 57
radial, 57
segment-indicating, 88
skeletal
   anterior view of, 54
   histochemistry of, 56
   non-pennate (parallel-fibered), 57
   overview of, 54–55
   pennate, 57
   posterior view of, 55
   structure of, 58
somatic, 145
sphincter, 57. *See also* Sphincter(s)
tricipital (three heads), 57
unipennate, 57
Musculus/Musculi (named)
  abductor digiti minimi, 55, 294, 296, 320–321, 344–346, 348–353, 398, 400–401, 403, 405, 455, 466, 471, 474–475, 492–493, 514–521, 523, 541, 564, 569
    actions of, 320
    innervation of, 297, 320, 370, 538
    origin and insertion of, 320
  abductor hallucis, 455, 466, 471, 474–475, 492–493, 516–521, 523, 525, 541, 563–565, 569
    innervation of, 538
  abductor pollicis brevis, 54, 296, 320–321, 344, 348–353, 398, 400–401, 403, 405
    actions of, 320
    innervation of, 297, 320, 372
    origin and insertion of, 320
  abductor pollicis longus, 54–55, 296, 318–319, 338–343, 346, 349, 368–369, 394, 396, 404
    actions of, 318, 328
    innervation of, 297, 318, 328, 368
    origin and insertion of, 318
    palsy, *368*
  adductor brevis, 474–475, 480–481, 496, 504–505, 509, 522, 534, 550–551
    innervation of, 532, 534
  adductores (of thigh), 183, 189, 432, 544
  adductor hallucis, 475, 492–493, 518–521, 523, 541
    caput transversum and obliquum, 465, 467–468, 474, 565
    innervation of, 538

  adductor longus, 29, 54, 221, 474–475, 480–481, 496, 502–505, 509, 522, 534, 548, 550–551
    innervation of, 532, 534
  adductor magnus, 29, 54–55, 474, 480–481, 496, 502–505, 507–509, 522, 525–526, 534, 548, 550–553, 555, 559–561, 568
    deep part, 475
    innervation of, 532, 534, 538, 541
    medial part, 541
      innervation of, 538, 541
    superficial part, 475
    tendinous part, 481
  adductor minimus, 474–475, 480–481, 505
    innervation of, 534
  adductor pollicis, 54, 296, 320–321, 346, 353, 398, 400, 403
    actions of, 320
    caput obliquum, 320–321, 344, 348–352, 401
    caput transversum, 320–321, 344, 348–352, 401
    innervation of, 297, 320, 370
    origin and insertion of, 320
  anconeus, 55, 277, 296, 312, 330, 332–333, 340–341
    actions of, 312, 326
    innervation of, 297, 312, 326, 368
    origin and insertion of, 312
    posterior (dorsal) view, 312–313
  articularis genus, 474, 482–483, 504–505
  auricularis, embryology of, 11
  biceps brachii, 56–57, 61, 296, 310, 337–339, 343, 356, 359, 366, 374, 380–384, 386, 392, 394
    actions of, 310, 324, 326
    anterior (ventral) view, 311
    caput breve, 54, 266, 268, 273, 310–311, 324, 334–337, 342–343, 366, 381, 389
    caput longum, 54, 266, 268, 273, 310–311, 324, 334–337, 342–343, 366, 381
    innervation of, 297, 310, 326
    origin and insertion of, 310
    pronating action, with elbow flexed, 310
    supinating action, with elbow flexed, 310
  biceps femoris, 410, 474, 484–485, 504–505, 508–509, 512–513, 529, 546, 561–562
    arterial supply to, 550–551
    caput breve, 29, 475, 484–485, 498, 508–510, 522, 540–541, 560–561, 566
    innervation of, 538

    caput longum, 29, 55, 475, 484–485, 496, 498, 506–508, 510, 522, 540–541, 552–553, 555, 560–561, 566
      innervation of, 538, 541
    innervation of, 538, 541
  brachialis, 54, 61, 296, 310, 333–339, 342–343, 366, 368–369, 386, 392, 394
    action of, 310, 326
    anterior (ventral) view, 311
    innervation of, 297, 310, 326, 368
    origin and insertion of, 310
  brachioradialis, 54–55, 277, 296, 316–317, 330, 332–333, 337–343, 355, 368–369, 374, 392, 394–396, 400, 403–405
    actions of, 316, 326
    innervation of, 297, 316, 326, 368
    origin and insertion of, 316
  buccinator, 54
  bulbospongiosus, 144, 162–163, 181–184, 188, 229–230, 233–235, 543, 557, 559
    innervation of, 542
  coccygeus, 144, 162, 175, 184–187, 558
    innervation of, 542
  compressor urethrae, 144
  coracobrachialis, 54, 165, 261, 268, 296, 334–337, 343, 366, 380–382, 384–386, 389
    actions of, 308, 324
    anterior view, 309
    innervation of, 297, 308
    origin and insertion of, 308
  corrugator supercilii, 54
  cremaster, 54, 176–177, 210
    fascia cremasterica and, 216–217, 225–226
    funiculus spermaticus and, 211, 214–215
    in males, innervation of, 532–533
  deltoideus, 30, 54–55, 61, 166, 199, 201–203, 261, 268–269, 272–273, 296, 304, 330–337, 343, 367, 374, 376, 379–381, 386, 388–391
    actions of, 304, 324
    anterior view, 305
    embryology of, 145
    innervation of, 297, 304
    lateral view, 305
    origin and insertion of, 304
    pars acromialis, 273, 304–305, 324, 332–333, 335
      actions of, 304
    pars clavicularis, 273, 304–305, 324, 332–333, 335
      actions of, 304
    pars spinalis, 273, 304–305, 324, 332–333
      actions of, 304
    posterior view, 305
  depressor anguli oris, 54
  depressor labii inferioris, 54
  depressor supercilii, 54
  diaphragma urogenitale, 556

  digastricus, 57
    embryology of, 11
  erector spinae, 31, 56, 144, 146–147, 155, 166–171, 174, 201, 205, 331
    embryology of, 7, 145
    intertransverse system, 144, 146–147
    lateral tract, 144, 146–147, 168
    medial tract, 144, 148–149, 169
    and posture, 157
    sacrospinal system, 144, 146–147
    spinal system, 144, 148–149
    spinotransverse system, 144, 146–147
    transversospinal system, 144, 148–149
  extensor carpi radialis, 376
  extensor carpi radialis brevis, 55, 296, 316–317, 330, 332–333, 337–343, 346, 394, 396
    actions of, 316, 328
    innervation of, 297, 316, 328, 368
    origin and insertion of, 316
  extensor carpi radialis longus, 55, 277, 296, 316–317, 330, 332–333, 337–339, 341–343, 346, 392, 394, 396–397
    actions of, 316, 328
    innervation of, 297, 316, 328, 368
    origin and insertion of, 316
  extensor carpi ulnaris, 55, 281, 296, 318–319, 330, 332, 340–342, 346, 353, 396
    actions of, 318, 328
    innervation of, 297, 318, 328, 368
    origin and insertion of, 318
    palsy, *368*
  extensor digiti minimi, 55, 296, 318–319, 340–342, 346
    actions of, 318, 328
    innervation of, 297, 318, 328, 368
    origin and insertion of, 318
  extensor digitorum, 55, 296, 318–319, 330, 332, 340–342, 346, 368–369, 376, 396, 455
    actions of, 318, 328
    innervation of, 297, 318, 328, 368
    origin and insertion of, 318
    palsy, *368*
  extensor digitorum brevis, 54, 474–475, 492–493, 510–511, 514–515, 523, 567–568
    innervation of, 538
  extensor digitorum longus, 54, 471, 474–475, 486–487, 500, 510–511, 514–515, 522–523, 540, 566, 569
    innervation of, 538
  extensor hallucis, 455
  extensor hallucis brevis, 54, 474–475, 492–493, 510–511, 514, 523, 566, 568–569

593

M Musculus/Musculi (cont.)

innervation of, 538
extensor hallucis longus, 54, 471, 474–475, 486–487, 500, 510–511, 514–515, 522–523, 540, 566–569
　innervation of, 538
extensor indicis, 55, 296, 318–319, 340–341, 346, 396
　actions of, 318, 328
　innervation of, 297, 318, 328, 368
　origin and insertion of, 318
　palsy, 368
extensor pollicis brevis, 55, 296, 318–319, 340–342, 346, 350–352, 368–369, 396
　actions of, 318, 328
　innervation of, 297, 318, 328, 368
　origin and insertion of, 318
　palsy, 368
extensor pollicis longus, 55, 296, 318–319, 340–342, 346, 368–369, 376, 396
　actions of, 318, 328
　innervation of, 297, 318, 328, 368
　origin and insertion of, 318
　palsy, 368
faciei, embryology of, 11, 145
fibularis brevis, 55, 455, 474–475, 486–487, 500, 510–515, 520–522, 540, 562, 566–567
　innervation of, 538
fibularis longus, 54–55, 455, 465, 467, 474–475, 486–487, 500, 506, 510–518, 520–522, 540, 546, 562, 566–567
　innervation of, 538
fibularis tertius, 54, 474–475, 486–487, 510–511, 514–515
　innervation of, 538
flexor carpi radialis, 54, 296, 314–315, 338, 342–344, 348–349, 353, 372–373, 392, 394, 400, 403–405
　actions of, 314, 328
　innervation of, 297, 314, 328, 372
　origin and insertion of, 314
flexor carpi ulnaris, 54–55, 296, 314–315, 330, 332, 338–344, 348–349, 353, 355, 370–371, 392, 394–395, 400–401, 403–405
　actions of, 314, 328
　innervation of, 297, 314, 328, 370
　origin and insertion of, 314
flexor digiti minimi, 296, 351–352, 405
　innervation of, 297, 370
flexor digiti minimi brevis, 54, 320–321, 344, 348–353, 398, 400–401, 403, 405, 474–475, 492–493, 516–521, 523, 569
　action of, 320
　innervation of, 320, 538
　origin and insertion of, 320

flexor digitorum brevis, 455, 466, 471, 474–475, 494–495, 513, 517–521, 523, 541, 564–565, 569
　innervation of, 538
flexor digitorum longus, 455, 466, 474–475, 490–491, 495, 500, 508–509, 512–513, 515–518, 521–523, 562–563, 567
　innervation of, 538
flexor digitorum profundus, 55, 296, 314–315, 322, 332, 337, 339–342, 353, 370–373, 394
　actions of, 314, 328
　innervation of, 297, 314, 328, 370, 372
　origin and insertion of, 314
flexor digitorum superficialis, 54, 296, 314–315, 338, 342–344, 348–349, 353, 372–373, 394, 400, 403–405
　actions of, 314, 328
　caput humeroulnare, 394
　caput radiale, 314, 339, 394
　caput ulnare, 314, 339
　innervation of, 297, 314, 328, 372
　origin and insertion of, 314
flexor hallucis brevis, 466, 468, 492–493, 516–521, 523, 565, 569
　caput mediale and laterale, 468, 474–475
　innervation of, 538
flexor hallucis longus, 55, 455, 466–468, 471, 474–475, 490–491, 500, 512–513, 515–518, 521–522, 562–563, 567
　innervation of, 538
flexor pollicis brevis, 296, 320–321, 353, 401
　actions of, 320
　caput profundum, 320, 351–352
　caput superficiale, 320–321, 344, 348–352, 398, 400, 403, 405
　innervation of, 297, 320, 370, 372
　origin and insertion of, 320
flexor pollicis longus, 54, 296, 314–315, 338–339, 342–344, 348–351, 353, 372–373, 394, 400, 403–405
　actions of, 314, 328
　innervation of, 297, 314, 328, 372
　origin and insertion of, 314
gastrocnemius, 31, 55–56, 441, 488–489, 502, 506, 522, 529, 541, 544, 546, 560–562, 566
　caput mediale and laterale, 54–55, 442, 452, 474, 488–489, 498, 507–510, 512–513, 561–562
gemellus/gemelli, 474–475
　inferior, 55, 478–479, 507–509, 522, 553, 555, 558
　innervation of, 537

innervation of, 536–537
　superior, 55, 478–479, 507–509, 522, 553, 555, 558
　innervation of, 537
gluteus maximus, 55–56, 157, 166, 168–169, 175, 181, 189, 221, 230, 233, 410, 432, 434, 474–475, 478–479, 496, 499, 502, 506–509, 522, 542, 546, 552–553, 555–556, 558–560
　innervation of, 536–537
　paralysis, 537
gluteus medius, 55, 166, 432, 434–435, 474–475, 478–479, 496, 504, 506–509, 522, 536, 546, 553, 555, 560
　and erect posture, 50
　innervation of, 536
　weakness, 536
gluteus minimus, 55, 183, 432, 474–475, 478–479, 496, 504–505, 507–509, 522, 536, 555, 560
　and erect posture, 50
　innervation of, 536
　weakness, 536
gracilis, 29, 54–55, 410, 474–475, 480–481, 496, 498, 502–505, 507–508, 510, 512, 522, 534, 548, 551–553, 555, 559–561
　innervation of, 532, 534
hamstrings, 56, 157, 474, 485
　arterial supply to, 550–551
　and joint motion, 50
hypothenar, 296, 320–321, 394, 404–405
iliacus, 154–155, 175, 178, 182, 204, 212, 213, 432, 474–477, 496, 502–505, 532–533, 535, 549
　innervation of, 532
iliococcygeus, 144, 162–163, 185–187
iliocostalis, 144, 146–147, 168
　actions of, 146
　innervation of, 146
　origins and insertions of, 146
iliocostalis cervicis, 146–147, 169
iliocostalis lumborum, 146–147, 169
iliocostalis thoracis, 146–147, 169
iliopsoas, 54, 56, 154–155, 157, 175, 189, 210, 213, 434–435, 474, 476–477, 496, 499, 503–505, 509, 535, 548–549, 551, 558
　innervation of, 532, 535
　paralysis of, 476
　shortening of, 476
　weakness of, 476
infrahyoidei, 167
　embryology of, 11
infraspinatus, 55, 166, 203, 261, 265–266, 273, 296, 302, 331–333, 362, 388, 390–391
　action of, 302, 324
　imaging of, 28
　innervation of, 297, 302

lateral view, 303
　origin and insertion of, 302
　posterior view, 303
　weakness/atrophy, 243, 362
interossei/interossei, 54, 370–371, 455, 510, 514, 569
　dorsales, 55, 294, 296, 322–323, 344–350, 352–353, 396–397, 400, 403, 474–475, 494–495, 517–521, 523, 568
　actions of, 322
　innervation of, 297, 322, 370, 538
　origin and insertion of, 322
　palmares, 296, 322–323, 346, 351–353
　actions of, 322
　innervation of, 297, 322, 370
　origin and insertion of, 322
　plantares, 474–475, 494–495, 516–521, 523, 565
　innervation of, 538
interspinales, 144, 148–149
　actions of, 148
　innervation of, 148
　origins and insertions of, 148
interspinales cervicis, 148–149, 169, 171
　action of, 148
　innervation of, 148
　origin and insertion of, 148, 171
interspinales lumborum, 148–149, 169
　action of, 148
　innervation of, 148
　origin and insertion of, 148
intertransversarii, 144, 146–147
　actions of, 146
　innervation of, 146
　origins and insertions of, 146
intertransversarii cervicis, 171
　anteriores, 146
　posteriores, 146–147
intertransversarii lumborum
　innervation of, 532
　laterales, 146–147, 169
　mediales, 146–147, 169
ischiocavernosus, 144, 162–163, 181–184, 229, 233–235, 557, 559
latissimus dorsi, 54–55, 144, 164, 166–167, 174, 202, 205, 269, 273, 296, 306, 330–331, 334–337, 359, 365, 376, 380–383, 386
　actions of, 306, 324
　embryology of, 145
　innervation of, 297, 306
　origin and insertion of, 306–307
　pars costalis, 306
　pars iliaca, 164, 306–307
　pars scapularis, 164, 306–307, 332–333
　pars vertebralis, 164, 306–307
　posterior view, 307
levator anguli oris, 54
levator ani, 144, 162–163, 181–189, 221, 233–234, 542–543, 556–559
　funnel shape of, 187

594

Musculus/Musculi (cont.)

gender-related differences in, 187
hiatus ani in, 163
innervation of, 542
levatores costarum, 144, 146–147, 169
　actions of, 146
　innervation of, 146
　origin and insertion of, 146
levatores costarum breves, 146–147, 169
levatores costarum longi, 146–147, 169
levator labii superioris, 54
levator labii superioris alaeque nasi, 54
levator scapulae, 55, 144, 164–167, 203, 296, 300, 331–333, 362, 389
　actions of, 300
　innervation of, 297, 300
　origin and insertion of, 300
　posterior view, 301
longissimus, 144, 146–147, 168
　actions of, 146
　innervation of, 146
　origins and insertions of, 146
longissimus capitis, 146–147, 169–171, 203
　insertion of, 171
longissimus cervicis, 146–147
longissimus thoracis, 146–147, 168–169
longus capitis, 144, 150–151
longus colli (cervicis), 144, 150–151, 167
　pars obliqua inferior, 150–151
　pars obliqua superior, 150–151
　pars recta, 150–151
lumbricales, 54, 296, 322–323, 344, 347–352, 373, 400–401, 474–475, 494–495, 517–520, 538, 541, 565
　actions of, 322
　innervation of, 297, 322, 370, 372, 538
　origin and insertion of, 322
masseter, 54
　embryology of, 11
masticatorii, 56
　embryology of, 145
mentalis, 54
multifidus, 144, 148–149, 169
　actions of, 148
　innervation of, 148
　origin and insertion of, 148
mylohyoideus, embryology of, 11
nasalis, 54
obliquus capitis inferior, 144, 150–151, 169–171, 203
　origin and insertion of, 171
obliquus capitis superior, 144, 150–151, 170–171, 203
　origin and insertion of, 171
obliquus externus abdominis, 7, 54–55, 57, 144, 152–153, 155, 157, 166, 168, 174, 176–179, 199, 202, 205–206, 210–211, 216–217, 219, 330–331, 334, 380–381, 532, 549

obliquus internus abdominis, 7, 54–55, 144, 152–153, 155, 157, 166, 168–169, 174, 176–179, 202, 206, 210–211, 216–217, 219, 330, 532
　canalis inguinalis contribution of, 211
　innervation of, 532
obturatorius externus, 183, 189, 221, 432, 474–475, 480–481, 496, 504–505, 509, 534
　innervation of, 532
obturatorius internus, 55, 163, 175, 181–186, 189, 233, 434, 474–475, 478–479, 496, 502, 507–509, 522, 543, 553–558, 560
　innervation of, 536–537
occipitalis, embryology of, 11, 145
occipitofrontalis, venter frontalis, 54
omohyoideus, 296, 298, 379, 384, 388–389
　actions of, 298
　anterior view, 299
　innervation of, 297–298
　lateral view, 299
　origin and insertion of, 298
　venter inferior, 359
opponens digiti minimi, 296, 320–321, 344, 346, 349–353, 401, 474–475, 492–493, 519–521, 523, 569
　action of, 320
　innervation of, 297, 320, 370, 538
　origin and insertion of, 320
opponens pollicis, 54, 294, 296, 320–321, 344, 348–353, 400–401, 403, 405
　actions of, 320
　innervation of, 297, 320, 372
　origin and insertion of, 320
orbicularis oculi, 54
orbicularis oris, 54
palmaris brevis, 54, 296, 320, 343, 348, 398, 405
　action of, 320
　innervation of, 297, 320, 370
　origin and insertion of, 320
palmaris longus, 54, 296, 314–315, 338, 342–343, 372–373, 392, 394, 398, 400, 404–405
　actions of, 314, 328
　innervation of, 297, 314, 328, 372
　origin and insertion of, 314
pectineus, 54, 189, 210, 219, 221, 434, 474–475, 480–481, 496, 503–505, 534–535, 548, 551
　fascia lata, 215
　innervation of, 532, 534–535
pectoralis major, 31, 54, 61, 144, 164–165, 199, 207–208, 261, 273, 296, 308, 335–337, 343, 359, 365, 374, 378–383, 386
　actions of, 308, 324
　anterior view, 309
　innervation of, 297, 308

origin and insertion of, 308
pars abdominalis, 165, 176, 308–309, 334
pars clavicularis, 165, 308–309, 324, 334–335, 380
pars sternocostalis, 165, 176, 308–309, 324, 334–335, 380
pectoralis minor, 54, 144, 164–165, 208–209, 261, 273, 296, 300, 335–337, 355, 359, 363, 365, 379, 381–384, 386, 389
　actions of, 300
　anterior view, 301
　innervation of, 297, 300
　origin and insertion of, 300
piriformis, 175, 184–187, 221, 474–475, 478–479, 496, 502–505, 508–509, 522, 525–526, 543, 550, 553–555, 558, 560
　innervation of, 536–537
plantaris, 55, 474–475, 488–489, 498, 500, 507–509, 512–513, 522, 529, 560–562, 567
　innervation of, 538
platysma, 54
popliteus, 55, 442, 474–475, 484–485, 498, 508–509, 512–513, 561–563
　innervation of, 538
procerus, 54
pronator quadratus, 54, 296, 314–315, 338–339, 344, 349, 372–373, 394, 400–401, 404–405
　actions of, 314, 326
　innervation of, 297, 314, 326, 372
　origin and insertion of, 314
pronator teres, 54, 296, 314–315, 336–339, 341–343, 392, 394–395
　actions of, 314, 326
　caput humerale, 314, 339, 373, 392–394
　caput ulnare, 314, 339, 373, 392–394
　innervation of, 297, 314, 326, 372
　n. medianus and, 393
　origin and insertion of, 314
psoas major, 144, 154–155, 157, 160, 167, 174–175, 204, 212, 213, 474–477, 496, 499, 502, 504–505, 530, 532–533, 535, 549
　deep layer, 154
　innervation of, 532
　superficial layer, 154
psoas minor, 175, 204, 474–477, 502
pterygoideus
　lateralis, embryology of, 11
　medialis, embryology of, 11
pubococcygeus, 144, 162–163, 185–187
puborectalis, 144, 162–163, 185–187
pyramidalis, 144, 154–155, 177, 216

quadratus femoris, 55, 183, 474–475, 478–479, 496, 505, 507–509, 522, 553, 555, 558, 560
　innervation of, 536–537
quadratus lumborum, 144, 154–155, 160, 167, 169, 174–175, 204, 532–533
　innervation of, 532
quadratus plantae, 455, 466, 471, 474–475, 494–495, 518–521, 541, 564–565, 569
　innervation of, 538
quadriceps femoris, 31, 57, 61, 157, 419, 445, 474–475, 482–483, 498–499, 504–505, 522, 535, 548
　and articulatio genus stabilization, 50
　and erect posture, 50
　and femoropatellar joint load, 51
　innervation of, 532, 535
　paralysis of, *482*
　weakness of, *482*
radialis group, 369
rectus abdominis, 7, 54, 57, 144, 154–155, 157, 174, 176–179, 199, 206, 210–213, 216–217, 220, 434, 533
　embryology of, 145
rectus capitis
　anterior, 144, 150–151
　lateralis, 144, 150–151
　posterior
　　major, 144, 150–151, 169–171, 203
　　minor, 144, 150–151, 169–171, 203
rectus femoris, 29, 54, 56, 189, 434, 474, 476, 482–483, 496, 498–499, 502–506, 508–510, 522, 535, 544, 551
rhomboideus major, 55, 144, 164–166, 168, 202–203, 261, 296, 300, 331–333, 362
　actions of, 300
　innervation of, 297, 300
　origin and insertion of, 300
　posterior view, 301
rhomboideus minor, 55, 144, 164–166, 168, 203, 296, 300, 331–333, 362
　actions of, 300
　innervation of, 297, 300
　origin and insertion of, 300
　posterior view, 301
risorius, 54
rotator cuff. *See* Rotator cuff
rotatores breves, 144, 148–149
　actions of, 148
　innervation of, 148
　origin and insertion of, 148
rotatores longi, 144, 148–149
　actions of, 148
　innervation of, 148
　origin and insertion of, 148
rotatores thoracis breves, 169
rotatores thoracis longi, 169

595

M *Musculus/Musculi (cont.)*

sartorius, 29, 54, 189, 199, 410, 434, 474–476, 482–483, 496, 498, 502–506, 510, 522, 535, 548, 551
  fascia lata, 215
  innervation of, 532, 535
scaleni, 144, 158–159, 167, 363, 382
  anterior, 158–159, 172–173, 355, 361, 363, 366–367, 369, 373, 379, 384
  medius, 158–159, 172–173, 355, 361, 363, 367, 379, 384
  posterior, 158–159, 172–173, 379
semimembranosus, 29, 31, 55, 442, 474–475, 484–485, 496, 498, 502, 504, 507–509, 512, 522, 529, 541, 546, 552–553, 555, 560–562
  arterial supply to, 550–551
  innervation of, 538, 541
semispinalis, 144, 148–149, 203
  actions of, 148
  innervation of, 148
semispinalis capitis, 55, 148–149, 168–171, 203, 330–331
  origin and insertion of, 148, 171
semispinalis cervicis, 148–149, 170, 203
semispinalis thoracis, 148–149
semitendinosus, 29, 54–55, 410, 474–475, 484–485, 496, 498, 502, 507–509, 512, 522, 541, 546, 552–553, 555, 560–562
  arterial supply to, 550–551
  innervation of, 538, 541
serratus anterior, 54–56, 144, 164–166, 176, 199, 205, 261, 273, 296, 300, 331, 334, 336–337, 362, 380–383, 386
  actions of, 300
  embryology of, 145
  innervation of, 297, 300
  lateral view, 301
  origin and insertion of, 300
serratus posterior
  inferior, 55, 144, 164, 166–168, 202, 331
  superior, 144, 164, 168
soleus, 31, 54–56, 471, 474, 488–489, 508–510, 512–513, 522, 541, 561–562, 566
sphincter ani externus, 57, 144, 162–163, 181–182, 184, 188, 230, 233, 235, 543, 556, 559
  innervation of, 542
  motor innervation of, 543
sphincter ani internus, 188
sphincter urethrae, 144
spinalis, 144, 148–149, 168–169
  actions of, 148
  innervation of, 148
  origins and insertions of, 148
spinalis cervicis, 148–149, 169
spinalis thoracis, 148–149, 169
splenius, 144, 146–147

splenius capitis, 55, 146–147, 168–171, 202–203, 330–331
splenius cervicis, 146–147, 168–170, 331
stapedius, embryology of, 11
sternocleidomastoideus, 31, 54–55, 166–167, 170, 199, 203, 296, 298, 330–331, 334–335, 378, 384–385
  actions of, 298
  caput claviculare, 298–299
  caput sternale, 298–299
  embryology of, 145
  innervation of, 297–298
  lateral view, 299
  origin and insertion of, 171, 298
  relationship to neck veins, 378
stylohyoideus, embryology of, 11
stylopharyngeus, embryology of, 11, 145
subclavius, 54, 144, 164–165, 273, 296, 300, 335–337, 362–363, 379, 381, 383
  action of, 300
  anterior view, 301
  innervation of, 297, 300
  origin and insertion of, 300
subscapularis, 54, 61, 261, 265–266, 268–269, 273, 296, 302, 335–337, 365, 381–383, 386, 389
  action of, 302, 324
  anterior view, 303
  innervation of, 297, 302
  lateral view, 303
  origin and insertion of, 302
supinator, 54–55, 277, 296, 318–319, 332, 337–341, 392, 396
  action of, 318, 326
  caput humerale, 341
  innervation of, 297, 318, 326, 368
  n. radialis and, 393
  origin and insertion of, 318
  pars profunda, 393
  pars superficialis, 393
supraspinatus, 55, 166, 265–267, 269, 273, 296, 302, 331–333, 335–337, 362, 382, 388–391
  action of, 302, 324
  anterior view, 303
  imaging of, 28
  innervation of, 297, 302
  lateral view, 303
  origin and insertion of, 302
  posterior view, 303
  starter function of, 362
  weakness/atrophy, *243*, *362*
temporalis, embryology of, 11
tensor fasciae latae, 54–55, 189, 434, 474–476, 478–479, 496, 503–509, 522, 536, 544, 548, 551, 555
tensor tympani, embryology of, 11
tensor veli palatini, embryology of, 11

teres major, 54–55, 166, 201, 203, 268–269, 273, 296, 306, 330–337, 343, 365, 376, 382–383, 386–388, 390–391
  action of, 306, 324
  innervation of, 297, 306
  origin and insertion of, 306–307
  posterior view, 307
teres minor, 55, 203, 266, 273, 296, 302, 331–333, 367, 390–391
  action of, 302, 324
  innervation of, 297, 302
  lateral view, 303
  origin and insertion of, 302
  posterior view, 303
thenar, 296, 320–321, 343, 373, 394, 404
tibialis anterior, 54, 56, 455, 474–475, 486–487, 500, 502, 506, 510–511, 513–515, 521–522, 540, 544, 563, 566–567, 569
  and erect posture, 50
  innervation of, 538
tibialis posterior, 55, 455, 465, 467, 474–475, 490–491, 500, 508–509, 512–513, 515–518, 521–522, 562–563, 567
  innervation of, 538
transversus abdominis, 7, 54, 144, 152–153, 155, 157, 169, 174–175, 177–179, 204, 206, 210–211, 213, 216–217, 219–220, 532
transversus perinei
  profundus, 144, 162–163, 182–184, 187–188, 221, 556–559
    in females, 163, 233
    innervation of, 542
    in males, 229
  superficialis, 144, 162–163, 181, 183–184, 233, 235, 556, 559
    innervation of, 542
transversus thoracis, 144, 158–159, 172–173, 175, 178
trapezius, 30–31, 54, 56, 144, 164–165, 167–168, 170–171, 202–203, 268–269, 273, 296, 298, 331–337, 376, 378–379, 384, 397
  actions of, 298
  embryology of, 145
  innervation of, 297–298
  origin and insertion of, 171, 298
  pars ascendens, 55, 165–166, 298–299, 330, 388
  pars descendens, 55, 165–166, 298–299, 330, 388
  pars transversa, 55, 165–166, 298–299, 330, 388
  posterior view, 299
triceps brachii, 54–55, 61, 166, 277, 296, 312, 331, 338, 340–341, 368–369, 380–381, 383, 392
  actions of, 312, 324, 326

  caput laterale, 312–313, 330, 332–333, 342, 374, 376, 388, 390–391, 396
  caput longum, 265, 312–313, 324, 330, 332–333, 342, 355, 376, 382, 388, 390–391
  caput mediale, 312–313, 332–333, 342, 376, 382, 391
  innervation of, 297, 312, 326, 368
  origin and insertion of, 312
  posterior (dorsal) view, 312–313
triceps surae, 31, 57, 467, 474–475, 488–489, 500–501, 510, 513–515, 522, 538, 561, 567
  and erect posture, 50
urethrae externus, 144, 162–163, 182, 184, 188
  innervation of, 542
vastus intermedius, 29, 54, 474, 482–483, 498, 503–505, 508–509, 522, 535, 551
vastus lateralis, 29, 54, 56, 189, 432, 443, 474, 482–483, 498, 503–506, 508–510, 522, 535, 544, 546, 551
vastus medialis, 29, 54, 56, 443, 474, 482–483, 498, 502–505, 509–510, 522, 535, 544, 551
zygomaticus major, 54
zygomaticus minor, 54
Musculus/Musculi (of region or organ)
  abdominal, embryology of, 7, 145
  of abdominal wall, 144, 152–157
    actions of, 156–157
    anterior (straight), 144, 154–155, 177
    arrangement of, 157
    course of, 157
    embryology of, 7
    functions of, 157
    in male, 177
    and trunk movement, 157
    embryology of, 145
    lateral (oblique), 144, 152–153, 167, 176
      arrangement of, 157
      course of, 157
      embryology of, 7, 145
      functions of, 157
      in male, 176
      and trunk movement, 157
    posterior (deep), 144, 154–155, 176
    strap, anterior, 154–155
  of antebrachium, 296, 314–317
    anterior, 296, 338–339
    arrangement of, 21
    cross sectional anatomy of, 342–343
    extensors
      deep, 296, 318–319
      superficial, 296, 318–319
    flexors, 277
      deep, 296, 314–315
      superficial, 296, 314–315
    posterior, 296, 340–341
    radial, 296, 316–317

Musculus/Musculi (cont.)

of articulatio coxae, 474–482.
  See also Musculus/Musculi
    (of region or organ), of
    thigh
  abductors, 496–497
  adductor group, 474,
    480–481, 496–497
    anterior view, 481
  anterior view, 477, 481,
    503–505
  extensors, 496–497
  external rotators, 496–497
  flexors, 476, 496–497
  functional testing of, 497
  functions of, 496–497
  inner, 474, 476–477
    anterior view, 477
  innervation of, 530
  internal rotators, 496–497
  lateral view, 506
  medial view, 502
  origins and insertions,
    504–505, 508–509
  outer, 474, 478–479
    deep, 479
    horizontally oriented, 478
    superficial, 479
    vertically oriented, 478
  posterior view, 479, 507–509
  shortening, symptoms of, 497
  weakness, symptoms of, 497
of articulatio genus, 480, 482,
  484–485, 488
  dysbalances of, 499
  extensors, 498–499
  external rotators, 498–499
  flexors, 498–499
  functional testing of, 499
  functions of, 498–499
  internal rotators, 498–499
  shortening, symptoms of, 499
  weakness, symptoms of, 499
of articulatio talocruralis,
  500–501
  dorsal extensors, 500–501
  functional testing of, 501
  plantar flexors, 500–501
  pronators, 500–501
  shortening, symptoms of, 501
  supinators, 500–501
  weakness, symptoms of, 501
of back, 164, 166
  and erect posture, 50
  intrinsic and nonintrinsic,
    fascia thoracolumbalis as
    partition between, 166
of brachium, 164, 296, 310–313
  anterior, 296, 336–337
  arrangement of, 21
  cross sectional anatomy of,
    342–343
  extensors, caput commune of,
    337, 339, 341
  flexors, caput commune of,
    336–341, 343
  posterior, 296, 332–333
of chest wall, 144, 172
of craniovertebral joint, 144
of crus, 474, 486–491
  anterior compartment, 474,
    486–487
  anterior view, 510–511
  arrangement of, 21
  and erect posture, 50
  extensor group, 563
  fibularis group, 563

flexors
  deep, 563
  superficial, 563
lateral compartment, 474,
  486–487
lateral view, 510
origins and insertions, 511,
  513
posterior compartment, 474,
  488–491
  deep, 474, 490–491, 541
  superficial, 474, 488–489
posterior view, 512–513
superficial flexor group,
  488–489
of cubitus/art. cubiti
  extensors, 326
    functional testing of, 327
    shortening/weakness,
      symptoms of, 327
  flexors, 326
    functional testing of, 327
    shortening/weakness,
      symptoms of, 327
  functional testing of, 327
  innervation of, 326
  nerve segments associated
    with, 326
  overview of, 326–327
  pronators, 326
    functional testing of, 327
    shortening/weakness,
      symptoms of, 327
  shortening, symptoms of, 327
  supinators, 326
    functional testing of, 327
    shortening/weakness,
      symptoms of, 327
  weakness, symptoms of, 327
epaxial, 104
erectile, 162–163, 182. See
  also Musculus/Musculi
    (named), bulbospongiosus;
    Musculus/Musculi (named),
    ischiocavernosus
facial, embryology of, 2, 11, 145
gluteal region, 474–481,
  552–553
  anterior view, 503–505
  embryology of, 145
  and erect posture, 50
  lateral view, 506
  medial view, 502
  origins and insertions,
    504–505, 508–509
  posterior view, 507–509,
    552–553
  and posture, 157
hypaxial, 104
hypothenar, 296, 320–321, 394,
  404–405
intercostales, 56, 132, 144,
  158–159, 174, 208, 273
  actions of, 158–159
  externi, 54, 158–159,
    168–169, 172–173, 176,
    205, 207
  actions of, 158–159
  interni, 54, 158–159,
    172–173, 175–176, 205,
    207
  actions of, 158–159
  intimi, 158, 173, 205, 207
ischiocrural, 484, 499
  insufficiency, 484
laryngis, embryology of, 11

of manus, 296
  dorsal, origins and insertions
    of, 346
  hypothenar, 296, 320–321,
    394, 404–405
  intrinsic, 320–323
    deep layer, 352–353
    middle layer, 350–351
    superficial layer, 348–349
  palmar, origins and insertions
    of, 353
  thenar, 296, 320–321, 343,
    373, 394, 404
of membrum inferius, 408,
  474–475
  arrangement of, 21
  crus, 474
  hip and gluteal, 474
  motor innervation of, 475
  short muscles of pes, 474
  thigh, 474
of membrum superius
  arrangement of, 21
  classification of
    functional-topographical,
      296
    by innervation, 296–297
    principles of, 296
  functional groups, 296–297
metacarpal, 296
nuchal
  prevertebral, 144, 150–151
  short (deep), 144, 150–151,
    170–171
    actions of, 170
    course of, 170–171
pedal, 486–487
  deep layer, plantar view, 520
  dorsal, 474
  middle layer, plantar view,
    518–519
  origins and insertions, 521
  plantar, 474, 516–521
  short, 466, 474, 492–495
    of central compartment,
      plantar surface, 494–495
    dorsal, 492–493
    of medial and lateral com-
      partments, plantar surface,
      492–493
    plantar view, 517–520
    superficial layer, plantar view,
      516–517
pelvic floor, 144, 156, 162–163.
  See also Sphincter(s)
  damage/insufficiency, 163
  in female, 163, 184–185
  functions of, 163
  innervation of, 542
pelvic wall, in female, 185
pharyngis, embryology of, 11
plantar
  central compartment, 474
  lateral compartment, 474
  medial compartment, 474
of shoulder/art. humeri (glenohumeralis), 296
  abductors, 324
    functional testing of, 325
    weakness/shortening,
      symptoms of, 325
  adductors, 324
    functional testing of, 325
    weakness/shortening,
      symptoms of, 325
  anterior, 296, 334–337

extensors, 324
  functional testing of, 325
  weakness/shortening,
    symptoms of, 325
external rotators, 324
  functional testing of, 325
  weakness/shortening,
    symptoms of, 325
flexors, 324
  functional testing of, 325
  weakness/shortening,
    symptoms of, 325
functional testing of, 325
functions of, 324
innervation of, 324
internal rotators, 324
  functional testing of, 325
  weakness/shortening,
    symptoms of, 325
nerve segments associated
  with, 324
posterior, 296, 330–333, 388
  deep layer, 331
  superficial layer, 330
shortening of, symptoms of,
  325
weakening of, symptoms of,
  325
of shoulder girdle, 164, 296,
  298–309
  anterior, 334–335
  posterior, 330–331
    deep layer, 331
    superficial layer, 330
small gluteal, weakness, 536
spinocostal, 144, 164–165
spinohumeral, 144, 164–165
subcostales, 144, 159, 173
suboccipitales, 144, 150–151,
  170–171
thenar, 296, 320–321, 343, 373,
  394, 404
of thigh, 474, 482–485
  anterior, 474, 482–483,
    503–505
  arrangement of, 21
  extensor group, 482–483
  flexor group, 484–485
  lateral view, 506
  medial view, 502
  origins and insertions,
    504–505, 508–509
  posterior, 474, 484–485,
    507–509
of thoracic cage, 144, 158–161
thoracohumeral, 144, 164–165
of trunk, 164, 296
  anterior, 296
  posterior, 296
of trunk wall
  in broad sense, 144
  development of, 145
  embryology of, 145
  origin of, 145
  overview of, 144–145
  secondarily incorporated, 144,
    164–165
  in strict sense, 144
urogenital, deep, 182
of wrist
  dorsal extensors, 328–329
  innervation of, 328
  nerve segments associated
    with, 328
  overview of, 328–329
  palmar flexors, 328–329

597

M Musculus/Musculi (cont.)

radial abductors, 328–329
shortening of, symptoms of, 329
ulnar abductors, 328–329
weakness/shortening, symptoms of, 329
Myelin sheath, 59, 91–93
Myoblasts, 58
Myofibril(s), 58–59
Myomere(s), 3
Myometrium, 5
Myosin, 55, 59
Myosin filament(s)
head of, 59
neck of, 59
structure of, 59
tail of, 59
Myotome, 7, 82, 104, 145
derivatives of, 7

N

N-Acetylgalactosamine (GalNAc), 45
Neck region
nerves of, superficial, 378
venae of, superficial, 378
Nerve block
axillary, 384
for digiti, 399
interscalene, 384
n. pudendus, 542, 558
Oberst, 399
vertical infraclavicular, 384, 385
Nerve conduction velocity, 97
Nerve fiber(s)
afferent, 77, 80–81, 89, 231
efferent, 77, 80–81, 89, 231
myelinated, 75, 80, 96
unmyelinated, 75, 80, 96
Nerve root(s), injury to, 375
Nervous system, 24
autonomic. See Autonomic nervous system
cells of, 78–79. See also Neuron(s)
central. See Central nervous system
information flow in, 77
peripheral. See Peripheral nervous system
topography of, 76
tubular, of chordates, 3
Nervus/Nervi (in general, including nerve types)
cutaneous
exclusive area of, 545, 547
maximum area of, 545, 547
embryology of, 11
peripheral
compression, 96
damage to
by cutting, 96
and loss of sensitivity, 97
by pressure, 96
development of, 75
injury to, 96–97, 375, 545
regeneration of, 97
structure of, 75
territories of, 545
"problematic," 385
Nervus/Nervi (named)
accessorius, 202, 378
musculi innervated by, 297
anococcygei, 201, 542–543, 559
ansa cervicalis, musculi innervated by, 297

auricularis magnus, 87, 378
axillaris, 85, 87, 203, 273, 297, 360–361, 364, 366–367, 369, 371, 373, 375, 377, 382–383, 388, 390–391
course of, 367
injury/damage to, 96, 262, 367
musculi innervated by, 297, 324, 364, 367
palsy, 367
rami
cutaneous, 364
motor, 367
musculares, 367
sensory, 367
spinal segments associated with, 360, 364, 367
clunium
inferiores, 87, 196, 200–202, 531, 536, 547, 552, 559
medii, 87, 196–197, 200–202, 537, 547, 552, 559
spinal segments associated with, 196
superiores, 87, 196–197, 200–202, 537, 547, 552, 559
spinal segments associated with, 196
coccygeus, 530–531, 542–543
course of
in female, 543
in male, 543
muscles innervated by, 542
rami cutanei, 542
rami dorsales, 542
rami ventrales, 542
spinal segments associated with, 542
cutaneus antebrachii
lateralis, 356, 364, 366, 375, 377, 392
medialis, 87, 356, 361, 366–367, 369, 371, 373, 375, 377, 383, 386, 392
sensory distribution of, 364
spinal segments associated with, 364
posterior, 364, 368–369, 377
cutaneus brachii
lateralis
inferior, 364, 368–369, 375, 377, 388
sensory distribution of, 367
superior, 200, 364, 367, 375, 377, 388
medialis, 87, 198, 361, 366–367, 369, 371, 373, 375, 377, 386
sensory distribution of, 364
spinal segment associated with, 364
posterior, 364, 368–369, 377, 381–383, 386, 388
cutaneus digiti secundi medialis, 538–539, 567
cutaneus dorsalis (of membrum inferius)
intermedius, 538–540, 545, 566–567, 569
lateralis, 538–539, 547, 566–567, 569
medialis, 538–540, 545, 566–567, 569
pedis, 562

cutaneus femoris
lateralis, 87, 198, 200, 204, 213, 530–531, 533, 545, 547–549, 551
course of, 533
injury to, 96
muscles innervated by, 532
rami cutanei, 532
spinal segment associated with, 533
posterior, 87, 200–201, 530–531, 547, 552–555, 558–560
muscles innervated by, 536
rami cutanei, 536
rami perineales, 536–537, 542, 553, 558–559
sensory distribution of, 537
spinal segments associated with, 536
cutaneus hallucis lateralis, 538–539, 567
cutaneus surae
lateralis, 531, 538–539, 545, 547, 552, 560–562, 566
ramus communicans, 566
medialis, 538–539, 552, 560–562, 566
digitales
dorsales
of manus, 364, 368–371, 398–399
of pes, 568
palmares, 398–400, 403
communes, 364, 370–373, 375
proprii, 364, 370–373, 375, 377, 398–399
plantares
communes, 541, 564
injury to, 96
proprii, 538–539, 541, 564–565
dorsalis clitoridis, 234–235, 542–543, 559
dorsalis penis, 228–231, 543, 559
dorsalis scapulae, 297, 361–362, 390
musculi innervated by, 297, 362
paralysis, 362
spinal segments associated with, 362
facialis (VII), 94
embryology of, 11
femoralis, 87, 189, 196, 204, 210, 213, 434, 475–476, 480, 482, 496, 498, 530–531, 534–535, 545, 548–549, 551
course of, 535
injury to, 96
muscles innervated by, 532, 535
rami
cutanei, 198, 204, 530, 532–533, 535
musculares, 530, 535
spinal segment associated with, 532, 535
fibularis, 475
communis, 87, 475, 484, 498, 530–531, 538, 540, 545, 547, 552–553, 560–562, 566

course of, 540
injury to, 96, 540
rami cutanei, 538
ramus communicans fibularis, 538–539, 562
spinal segments associated with, 538
profundus, 87, 475, 486, 492, 500, 522, 531, 538–540, 545, 566–569
compression of, 566
course of, 540
injury to, 96, 540
muscles innervated by, 538, 540
rami cutanei, 538–539, 566, 568
rami, 566
superficialis, 87, 475, 486, 500, 531, 538–540, 545, 566–567
course of, 540
injury to, 540
muscles innervated by, 538, 540
rami cutanei, 538
genitofemoralis, 87, 196, 204, 530–531, 533
course of, 533
in females, 533
in males, 533
muscles innervated by, 532
rami
cutaneus, 532
femoralis, 204, 213, 532–533, 545, 549
genitalis, 201, 204, 210–211, 213, 225, 532–533, 545, 559
spinal segment associated with, 532
glossopharyngeus (IX), 94
embryology of, 11
gluteus
inferior, 475, 478, 496, 530–531, 537, 553–555, 558, 560
motor distribution of, 537
muscles innervated by, 536
spinal segments associated with, 536
superior, 475, 478, 496, 530–531, 553–555, 560
gluteal injections and, 555
location of, 555
motor distribution of, 536
muscles innervated by, 536
spinal segments associated with, 536
hypogastrici, 231
iliohypogastricus, 87, 196–198, 204, 219, 530–531, 537
course of, 532
muscles innervated by, 532
rami cutaneus, 532
anterior, 204, 211, 532–533, 545
lateralis, 198, 200, 204, 532–533, 545, 547, 552
spinal segment associated with, 532
ilioinguinalis, 87, 196–198, 204, 210–211, 216, 219, 228, 530–531, 533, 545, 559
dermatomes and, 201
course of, 532–533

muscles innervated by, 532
spinal segment associated with, 532
rami
cutaneous, 532
scrotales, anteriores, 533
intercostales, 173, 196–197, 202, 206, 208, 365
course of, 197
rami cutanei
anteriores, 87, 197–198, 205, 364, 375
collateralis, 205
laterales, 87, 197–198, 200, 205, 375
mediales, 197
rami mammarii
laterales, 198, 209
mediales, 198, 209
rami ventrales, 205
spinal segments associated with, 196
intercostobrachiales, 197–198, 297, 361, 364–365, 375, 377
anastomosis with n. cutaneus brachii medialis, 365
cutaneous distribution of, 364–365
origin of, 365
r. cutaneus anterior, 365
rr. cutanei laterales, 364–365
spinal segments associated with, 364
interosseus
antebrachtii anterior, 297, 342, 373, 401
posterior, 368–369
ischiadicus, 29, 189, 196, 221, 475, 522, 530–531, 534, 537, 540–541, 553–555, 558, 560–561
course of, 538, 540–541, 554
in relation to piriformis, variability, 553
fibular part (Fib), 538, 540
gluteal injections and, 555
high division, 475, 553
injury to, 96, 538, 540
location of, 555
low division, 475
motor distribution of, 540–541
muscles innervated by, 538, 541
overview of, 538
rami, 530, 538
cutaneous, 538
musculares, 538, 540–541
sensory distribution of, 539
spinal segments associated with, 538
tibial part (Tib), 538, 541
laryngeus
embryology of, 11
recurrens
embryology of, 11
injury to, 384
superior, embryology of, 11
mandibularis, 87
embryology of, 11
maxillaris, 87

medianus, 85, 87, 288, 297, 342, 360–361, 364, 366–367, 369, 371–373, 377, 379, 381–383, 386–387, 392–394, 400, 402–405
compression syndromes, 372
course of, 373
electrical nerve stimulation of, 385
injury/lesions, 96
distal, 372
proximal, 372
traumatic, 372
innervation of dorsum of manus, 397
musculi innervated by, 297, 326, 328, 364, 372–373
pronator teres relationship to, 393
radix lateralis, 360–361, 373
radix medialis, 360–361, 373
rami
articulares, 372–373
communicans ulnaris, 372, 399
cutaneous, 364
motor, 372–373, 403, 405
palmaris, 364, 372–373, 375, 403, 405
sensory, 372–373
sensory distribution of, 373
sensory innervation of palma, 399
spinal segments associated with, 360, 364, 372
union of radices, 360
musculi obturatorii interni, 536
musculi piriformis, 536
musculi quadrati femoris, 536
musculocutaneus, 85, 87, 297, 342, 360–361, 364, 366–367, 369, 371, 373, 375, 377, 379, 381–383, 386–387, 392
course of, 366
electrical nerve stimulation of, 385
musculi innervated by, 297, 324, 326, 364, 366
rami
articulares, 366
cutaneous, 364
motor, 366
musculares, 366
sensory, 366
spinal segments associated with, 360, 364, 366
obturatorius, 87, 196, 204, 212–213, 475, 480, 496, 498, 530–531, 551, 558
course of, 534
injury to, 96, 534
muscles innervated by, 532
rami
anterior, 530, 532, 534
cutaneous, 532, 534, 545, 547, 551
posterior, 530, 532, 534
spinal segment associated with, 532, 534
occipitalis
major, 87, 196, 200, 203
spinal segments associated with, 196
minor, 87, 200, 203
tertius, 196, 200, 202–203

spinal segments associated with, 196
oculomotorius, 94
ophthalmicus, 87
pectorales
laterales, 297, 361, 364–365, 379–381
musculi innervated by, 297, 324, 364
spinal segments associated with, 364
mediales, 297, 361, 364–365, 379–381
musculi innervated by, 297, 324, 364
spinal segments associated with, 364
perinealis, 230–231, 542, 558–559
phrenicus, 205, 297, 361, 367, 369, 379, 384
plantaris
injury to, 96
lateralis, 87, 475, 492, 494, 531, 538, 541, 563–565
muscles innervated by, 538
rami cutanei, 538, 547
ramus profundus, 564–565, 569
ramus superficialis, 541, 564, 569
medialis, 87, 475, 492, 494, 531, 538, 541, 563–565, 569
muscles innervated by, 538
rami cutanei, 538, 547
rami superficialis, 564
middle layer, 564
profundus, 565
superficial layer, 564
pudendus, 183, 189, 201, 230–231, 234–235, 530–531, 542–543, 554–555, 557–560
blockade, 542, 558
course of
in female, 543
in male, 543
distribution of, 558
injury to, 543
muscles innervated by, 542
rami, 543
cutaneous, 542
perineales, 553
sensory distribution in female, 542
spinal segments associated with, 542
radialis, 85, 87, 297, 342, 360–361, 364, 366–369, 371, 373, 375, 377, 381–383, 388, 390, 393
compression syndromes, 368, 393
course of, 369
in sulcus n. radialis, 369, 391
in supinator canal, 369
electrical nerve stimulation of, 385
entrapment, 393
injury/lesions, 96, 246, 393
distal, 368
midlevel, 368
proximal, 368
traumatic, 368

innervation of dorsum of manus, 397
musculi innervated by, 297, 324, 326, 328, 364, 368
rami
articulares, 368
communicans ulnaris, 368
cutaneous, 364
motor, 368–369, 382
musculares, 368–369, 391–392
profundus, 297, 368–369, 392–393
sensory, 368–369
superficialis, 96, 297, 342, 364, 368–369, 375, 377, 392–394, 402
relationship to supinator, 369
sensory distribution of, 369
sensory innervation of palma, 399
spinal segments associated with, 360, 364, 368
in sulcus n. radialis, 369, 391
supinator relationship to, 393
rectales inferiores, 230, 234–235, 542–543, 558–559
sacrales, 115
emerging, 537
radix dorsalis, 537
radix ventralis, 537
ramus dorsalis, 537
ramus ventralis, 537
saphenus, 29, 87, 530–532, 535, 545, 547, 551, 567
injury to, 96
rami
cutanei cruris mediales, 535
cutaneus, 569
infrapatellaris, 535, 545
scrotalis, posteriores, 230, 558–559
splanchnici pelvici, 231
subclavius, 297, 361–362
musculi innervated by, 297, 362
spinal segments associated with, 362
subcostalis, 196–197, 204, 530–531
spinal segments associated with, 196
suboccipitalis, 196, 203
spinal segments associated with, 196
subscapularis, 273, 297, 361, 364–365, 381–383
musculi innervated by, 297, 324, 364
spinal segments associated with, 364
supraclaviculares, 87, 198, 200, 209, 364, 366–367, 369, 371, 373, 375, 377, 388
suprascapularis, 203, 243, 273, 297, 361, 382, 388–390
compression, 362, 388
in incisura scapulae, 362
injury to, 96, 362
musculi innervated by, 297, 324, 362
spinal segments associated with, 362
suralis, 87, 531, 538–539, 545, 547, 562, 566–567

599

N Nervus/Nervi (cont.)

thoracicus longus, 197, 297, 361–362, 379, 381, 383
  injury to, *362*
  musculi innervated by, 297, 362
  spinal segments associated with, 362
thoracodorsalis, 297, 361, 364–365, 381–383
  musculi innervated by, 297, 324, 364
  spinal segments associated with, 364
tibialis, 87, 475, 484, 488, 490, 496, 498, 500, 522, 530–531, 538, 540–541, 547, 552, 560–563, 566–567
  compression of, *541*
  course of, *541*
  injury to, *96*
  motor distribution of, *541*
  muscles innervated by, 538
  posterior, 471
  rami
    calcanei, 538–539, 547, 562–563
    cutaneous, 538
    musculares, 561
  spinal segments associated with, 538
transversa cervicis, 87, 378
trigeminus (V), 11, 86–87
  divisions of, 87
ulnaris, 85, 87, 297, 342, 360–361, 364, 366–367, 369–371, 373, 379, 381–383, 386, 391–394, 398, 400, 402–405
  compression syndromes, *370*
  course of, *371*
    in ulnar tunnel and deep palm, 405
  electrical nerve stimulation of, 385
  entrapment, *96*
  injury/lesions, *96*, *370*
    distal, *370*
    midlevel, *370*
    proximal, *370*
    traumatic, *370*
  innervation of dorsum of manus, 397
  musculi innervated by, 297, 328, 364, 370–371
  palsy, *370*
  rami
    articulares, 370–371
    communicans medianus, 399
    cutaneous, 364
    dorsalis, 364, 370–371, 377
    motor, 370–371
    musculares, 370–371
    palmaris, 364, 370–371, 375, 403, 405
    profundus, 371, 400–401, 403–405
    sensory, 370–371
    superficialis, 403, 405
  sensory distribution of, *371*
  sensory innervation of palma, 399
  spinal segments associated with, 360, 364, 370
vagus (CN X), 11, 94, 167

Nervus/Nervi (of region or organ)
  anococcygei, 530
  craniales, 77. *See also specific nerve under* Nerve(s) (named)
    embryology of, 11
  of crus, 530–531
  of cubitus/art. cubiti, 356
  of digiti medius, 398
  of female external genitalia, 235
  of fossa ischioanalis, 555
  labiales (of labia majora, in females)
    anterior, 532
    posteriores, 542–543, 559
  of mamma, 209
  of manus, palma, 398–399
  of membrum inferius, 530–531, 545
  of membrum superius, 374–377
  of neck
    deep, 378
    superficial, 378
  of penis, 230
  perineales, 230, 234, 235, 543
  of pes, 530–531, 568
    planta, 564
  regio glutealis, 552–555
  scrotales, 230
    anterior, 532–533
    posteriores, 230, 542–543
  of shoulder/art. humeri (glenohumeralis). *See also* Plexus (neural), brachial
    anterior view, 378
    posterior view, 388
  spinales, 77, 88, 124, 128, 361. *See also* Ramus/Rami, of nervi spinales
    embryology of, 7
    rami, 80, 196, 197
      communicans albus, 80–81, 89, 95, 197
      communicans griseus, 80–81, 89, 95, 197
      dorsalis, 80, 196–197, 200, 202
      functions of, 80
      meningeus, 80–81, 197
      sensory territory of, 197
      ventralis, 80, 196–197
    segmental, development of, 104
    topographical relationship to processi uncinatus, 126
  of thigh, 530–531
  of toes
    lateralis, 538–539
    medialis, 538–539
  of trunk wall, 196–197
    course of, 197
    superficial cutaneous
      anterior, 198–199
      posterior, 200–201
Neurite(s), 75
Neuroblastoma, *75*
Neuroectoderm, 6, 74
Neurofibril(s), 78
Neurofibromatosis (NF), *75*
Neuroglia, 79
Neurohemal regions, 71
Neurohormone(s), 71
Neurohypophysis, development of, 72
Neurological disorders, diagnosis of, *90*

Neuroma, scar, *97*
Neuromodulator(s), 71
Neuromuscular junction, 90
Neuromuscular synapse, 59
Neuron(s)
  afferent (sensory), 7, 73, 75, 78, 81, 89, 91
  associative, 91
  bipolar, 78
  efferent (motor), 7, 73, 75, 78, 80–81, 89
  electron microscopy of, 78
  embryonic migratory movements of, 73
  functions of, 78–79
  lower motor, 90–91
  multipolar
    with long axon, 78
    with short axon, 78
  postganglionic autonomic, 95
  preganglionic autonomic, 73, 94–95
  pseudounipolar, 78
  in sensory cortex, 91
  structure of, 78
  upper motor, 90–91
Neurosecretion, 71
Neurotransmitter(s), 71, 79, 95
Neurulation, 6
Nissl substance, 78
Nitrous oxide (NO), 231
Node of Ranvier, 93
Nodi lymphoidei (in general), 62, 68
  B-lymphocyte region, 69
  collecting, 69
  regional, 69
  structure of, 69
  T-lymphocyte region, 69
Nodi lymphoidei (named)
  axillares, 68, 194–195, 358–359
    apicales, 209, 359
    breast cancer metastases to, *359*
    centrales, 209, 359
    classification of, by level, 359
    interpectorales, 209, 359
    laterales, 209
    level I (lower group), 359
    level II (middle group), 359
    level III (upper, infraclavicular group), 359
    pectorales, 209, 359
    subscapulares, 209
  brachiales, 359
  cervicales, 68, 194–195, 359
  cubitales, 358–359
  iliaci
    communes, 194
      drainage pathways, 529
    externi, 194, 528
      drainage pathways, 529
    interni, 194, 234
      drainage pathways, 529
    profundi, drainage pathways, 529
  inguinales, 68
    inferiores, 528–529
    profundi, 194, 215, 234, 528–529
    superficiales, 194–195, 227, 234, 528
      drainage pathways, 529
    superolaterales, 528–529
    superomediales, 528–529
  lumbales, 194, 227

  paramammarii, 209, 359
  parasternales, 194–195, 209
  popliteal
    profundi, 528–529
      drainage pathways, 529
    superficiales, 528–529
      drainage pathways, 529
  Rosenmüller, 215, 528, 549
  supraclaviculares, 359
  supratrochleares, 359
Nodi lymphoidei (of region or organ)
  of axilla, 358–359
  intestinal, 68
  lumbales, 529
  of membrum inferius, 528–529
    drainage pathways, 529
    groups, 529
  of membrum superius, 358–359
  pelvis, 528
  of trunk wall, 194–195
Nodus primitivus, 6
Norepinephrine, 71, 95
Nucleus/Nuclei
  cranial nerve, 73
  of muscle cell, 58
Nucleus pulposus, 3, 116–118, 126, 129–131, 137
  embryology of, 104
  functions of, 117
Nursemaid's elbow, *283*
Nutation, of sacroiliac joint, 142

O

Obersteiner-Redlich zone, 91–92
Occlusal plane, 125
Oculus
  development of, 72
  innervation of, 94–95
Oesophagus, 105, 167, 175, 205
  embryology of, 10
Olecranon, 31, 33, 38, 166, 238–239, 247–248, 250–251, 276–280, 312–313, 317, 319, 330, 332, 340, 376, 396
  and joint motion, 50
Oligodendrocyte(s), 79, 92
  functions of, 79
Ontogeny
  definition of, 4
  human, 4–9
Ontogeny recapitulates phylogeny, 2
Oocyte(s)
  primary, 5
  secondary, 5
Oogenesis, 5
Oogonium/Oogonia, 5
Opening zone, epiphyseal, 16
Orbita, 38
Organogenesis, 4
Ortolani test, *439*
Os cuneiforme, 38, 423, 464, 466, 472
  facies articularis
    distal, 457
    proximal, 456
  intermedium, 422–423, 425, 455, 465, 470, 490, 492–493
    facies articularis
      distal, 457
      proximal, 456
    ossification of, 19

laterale, 422–423, 425,
454–455, 465, 470–471,
490, 492
  facies articularis
    distal, 457
    proximal, 456
  ossification of, 19
mediale, 422–425, 454–455,
458–460, 464–466, 470,
486–487, 490, 492–493,
495, 569
  facies articularis
    distal, 457
    proximal, 456
  ossification of, 19
Os femoris, 20, 29, 38, 61, 189,
408, 411, 418, 428, 440,
442–443, 445, 447,
450–453, 478, 480–482,
484–485, 487, 491, 511,
522
  anterior view of, 414
  caput, 38, 414–417, 426–427,
    430, 434–435
    blood supply to, 430–431
    development of, 15
    fractures of, 433
    ligamentous stabilization of,
      428–429
    luxatio, 429
      iliaca, 429
      ischiadica, 429
      obturatoria, 429
      pubica, 429
    nutrition of, 430–431
  collum, 38, 408, 414–417,
    426–427, 430, 432,
    434–435
    fractures of, 432
      lateral, 432
      medial, 432–433
    rotational deformities of, 417
  corpus, 414–415, 432
  development of, 14
  distal view of, 416
  facies patellaris, 414, 416, 419,
    441, 444–445, 450. See also
    Trochlea, femoral
  facies poplitea, 414
  fractures of, 432
    lateral, 433
    medial, 432–433
      classification of, 433
      peritrochanteric, 432
      proximal, 432–433
      subtrochanteric, 432
    greater trochanter of. See
      Trochanter, major
    lesser trochanter of. See Trochan-
      ter, minor
    medial view of, 417
    ossification of, 19, 438
    posterior view of, 414
    proximal view of, 416
    structure of, 39
    torsion of, 417
Ossa (named)
  accessory, 38, 425, 471
  anomalous, 38
  brevia, 38
  capitatum, 33, 38, 239,
    252–256, 286–289,
    292–294, 321, 402
    ossification of, 18–19
  carpi, 20, 38, 238, 249, 252,
    254–255

development of, 14
distal row, 254
  proximal view, 255
fractures, 257
lunate column, 254, 256
ossification of, 18–19
proximal row, 254
  distal view, 255
  proximal view, 255
scaphoid column, 254, 256
triquetral column, 254, 256
coccygis, 38, 100–102, 105,
  114–115, 128, 138–143,
  163, 180–181, 184–187,
  556
compacta (cortical), 17, 39
  of corpus vertebrae, 117
costale, 135
coxae, 138, 408
cuboideum, 38, 411, 422–425,
  454–455, 459–461,
  464–465, 470–471, 487,
  490, 492
  facies articularis
    distal, 457
    proximal, 456
  ossification of, 19
development of, 16
epiphyseal, 16
formation of. See also Osteogenesis
  direct, 16–17
  indirect, 16–17
frontale, 32, 38
  desmal osteogenesis of, 14
growth
  appositional, 17
  interstitial longitudinal, 17
hamatum, 38, 239, 252–255,
  284, 286–289, 293–294,
  321, 402
  ossification of, 18
hip. See Articulatio coxae, ossa;
  specific bone
hyoideum, 299, 384
  cornu majus, embryology of,
    11
  cornu minus, embryology
    of, 11
ilium, 41, 138, 143, 164, 307,
  415, 432, 478
  ala (wing) of, 139, 142, 413
  corpus of, 412–413
  facies auricularis, 41, 142,
    412–413, 427
  facies glutea, 139–140, 412,
    479
  facies symphysialis, 141–143,
    412–413
  ossification of, 19
innominate, 408
intermetatarseum, 425
irregularia, 38
ischii, 38, 41, 138, 434, 484–485
  corpus, 412–413, 433
  ossification of, 19
lamellar (mature), 16–17, 39
  in fracture healing, 53
  remodeling of, 17
longa, 38
  development of, 16
  fractures of, 52
lunatum, 38, 252–255,
  286–287, 289, 292, 294,
  321
  ossification of, 18

metacarpi, 20, 38, 238–239,
  247, 249–250, 252,
  254–255, 282, 284–287,
  289–292, 294, 315, 317,
  321, 323, 347, 376, 397,
  399
  basis, 238, 252, 292, 315,
    317, 319, 321
  caput, 238, 252, 290, 292,
    317
  corpus, 238, 252, 292, 317,
    321
  cross section through, 291
  development of, 14
  ossification of, 18
metatarsi, 20, 38, 408, 422–423,
  454–455, 457, 466,
  470–472, 569
  fifth (V), 422–423, 425,
    455, 459–460, 486–487,
    490–491, 494, 523
    basis, 456, 465, 541
    caput, 464
  first (I), 38, 423–425,
    454–455, 458, 460–461,
    464–465, 468, 486–487,
    489–491, 493–495, 515,
    523, 563
    basis, 422, 424, 456–457,
      465
    caput, 422, 424, 457, 464,
      468
    corpus, 422, 424, 457
    deviation, in hallux valgus,
      469
  fourth (IV), 490–494
  fractures of, 52
  ossification of, 19
  second (II), 490–494, 523
    caput of, overloading of,
      544
  third (III), 490–494
nasale, 32
naviculare, 38, 422–425,
  454–455, 458–461, 464,
  466–467, 470–472, 489
  facies articularis
    distal, 457
    proximal, 456
  ossification of, 19
occipitale, 33, 38, 120, 122, 170
  pars basilaris, 120–121, 151
  osteolytic metastases to, 130
parietale, 33, 38, 170
peroneum, 425
pisiforme, 32, 38, 239, 252,
  254–255, 285, 288–289,
  293–294, 315, 321, 323,
  402, 404–405
  ossification of, 18–19
plana, 38
pneumatica, 38
pubis, 38, 41, 138–139, 434
  corpus, 413, 433
  facies symphysialis, 41
  medial portion of, 412
  ossification of, 19
remodeling of, 16–17
sacrum, 33, 38, 41, 100–102,
  105, 114–115, 128, 131,
  138–140, 143, 147, 149,
  164, 180, 186–188,
  307, 365, 408, 413, 428,
  478–479, 485, 502, 554
  ala, 114–115
  anterior (pelvic) view of, 114

basis, 106, 114–115, 138
curvature of, 103, 105
facies articularis superior, 106
facies auricularis, 102, 114,
  142
facies dorsalis, 115
facies pelvica, 106, 114–115,
  138, 142
gender differences in, 139
lateral view of, 115
pars lateralis, 106, 114,
  138–139
posterior (dorsal) view of, 114
rotation of
  anterior, 142
  posterior, 142
structure of, 106–107
superior view of, 115
transverse section through,
  115
scaphoideum, 38, 252–256,
  286–287, 289, 292, 294,
  321, 397, 402
  fractures of, 257
  ossification of, 18
sesamoidea, 38, 419. See also
  Patella
  functional significance of, 61
  ossification of, 18–19
spongiosa (cancellous, trabecular), 39
  of corpus vertebrae, 117
  primary, 16
  remodeling of, 16–17
  secondary, 16
subchondral, 44
supernumerary, 38
supranaviculare, 425
tarsi, 422
temporale, 33, 122
  desmal osteogenesis of, 14
  embryology of, 11
tibiale externum, 425
trapezium, 38, 252–255, 286,
  288–289, 292–294, 321,
  402
  ossification of, 18
trapezoideum, 38, 252–255,
  286–289, 292–294, 323
  ossification of, 18
trigonum, 425
triquetrum, 33, 38, 239,
  252–255, 284, 286–287,
  289, 293–294, 321, 402
  ossification of, 18
tubular, 38
  structure of, 39
vesalianum, 425
woven (immature) (fiber bone),
  16–17
  in fracture healing, 53
Ossa (of region or structure). See
  also specific os
  of antebrachium, 38, 238, 247,
    250
  of articulatio coxae, 138, 408,
    411–413
  articulating, 426–427
  before closure of growth
    plates, 41, 438
  column principle of, 412
  fusion of, 41
  growth plates of, 41, 438
  of articulatio genus, 440–441
  of back. See Columna vertebralis;
    Costae; Vertebra/Vertebrae

O Ossa (cont.)

of brachium, 38, 238, 247, 250
of cingulum pelvicum, 412–413
cranii, 41
of crus, 38, 408, 411
    overview of, 408
of cubitus/art. cubiti, 276–277
of digiti, 238. *See also* Ossa
    (named), metacarpi; Ossa
    digitorum, of manus
of manus, 38, 238, 247, 250,
    252–253
    dorsal view, 253
    palmar view, 252
of membrum inferius, 38, 408,
    411
    overview of, 408
of membrum superius, 238,
    244–255
of pelvis, fractures of, *412*
of pes, 38, 408, 422–425,
    454–455
    articulating, 454–455
    dorsal view, 422
    lateral view of, 425
    medial view of, 424
    plantar view, 423
of shoulder girdle, 238, 242–243
    in relation to trunk, 240
of thigh, 38, 408
Ossa digitorum
    of hallux
        distalis, 422–424, 455,
            460–461, 468, 470
        proximalis, 422–424, 455,
            460–461, 465, 468, 470
        basis, 422, 424
        caput, 422, 424
        corpus, 422, 424
    of manus, 20, 33, 38, 239, 247,
        252
        basis, 252, 292
        caput, 252, 292
        corpus, 252, 292
        distalis, 238, 249–250, 252,
            282, 284–285, 290–292,
            315, 321, 347, 376, 399
            basis, 319, 323
            caput, 323
            corpus, 323
        media, 238, 249, 252, 282,
            284–285, 290–292, 315,
            323, 376, 399
        proximalis, 238, 249–250,
            252, 282, 284–285,
            290–292, 323, 376, 399
            basis, 319, 321
            caput, 321
    of pes, 20, 38, 408, 422–423,
        454–455, 486–487, 490,
        492–494. *See also* Toe(s)
        distalis, 38, 422
        facies articularis
            distal, 457
            proximal, 456
        media, 38, 422
        proximalis, 38, 422
            basis, 465
    of phalanx V
        distalis, 423, 425, 470
        media, 423, 425, 470
        proximalis, 423, 425, 470
Ossa sesamoidea
    of hallux, 468
        articular surfaces of, 468
        laterale, 468, 492–493
            in hallux valgus, 469
        mediale, 468, 492–493
    manus, 252, 254
    ossification of, 18–19
    pes, 423, 454, 457, 468, 470,
        492–493
    radial, 320
    ulnar, 320
Ossification
    angiogenic, in fracture healing,
        53
    apophyseal, 18
    definition of, 16
    diaphyseal, 18
    endochondral, 14
        in fracture healing, 53
    epiphyseal, 18
    membranous, 242
    perichondral, 14
Ossification center(s), 14
    apophyseal
        of membrum inferius, 19
        of membrum superius, 18
    of caput femoris, 438–439
    epiphyseal, 16
        of membrum inferius, 19
        of membrum superius, 18
    of membrum inferius, 19
    of membrum superius, 18
    primary, 14, 16, 18
    secondary, 15–16, 18
    in spinal column development,
        104
    of trochanter major, 438–439
Osteoarthritis, 46
    activated, 47
    of art. carpometacarpalis
        pollicis, *293*
    of art. coxae, *51*
        alleviation of stress on affected
            side in, *437*
        compensatory mechanisms in,
            47
    etiology of, *46*
    of knee, *410*
    pathogenesis of, *46*
    primary, *46*
    secondary, *46*
    stages of, *47*
    symptoms of, *47*
Osteoarthrosis, 46
Osteoblast(s), 14, 16–17, 39
    in fracture healing, 53
Osteochondrosis, in lumbar spine,
    130
Osteochondrosis dissecans, *40*
Osteoclast(s), 17, 39
    in fracture healing, 53
Osteocyte(s), 17, 39, 44
Osteogenesis, 17
    definition of, 16
    desmal, 14
    endochondral, 14, 16–17
    membranous, 14, 16–17
Osteogenic zone, in osteon
    development, 17
Osteoid, 16–17
Osteon, 39
    development of, 17
    in fracture healing, 53
Osteophyte(s)
    in cervical spine, *126–127*
    marginal, *47*
Osteoporosis, *130*
    and femoral fractures, *432*
Osteoprogenitor cell(s), 17
    in fracture healing, 53
Osteosynthesis, 53
    external fixation procedure for,
        53
    intramedullary nail/rod/fixation
        for, 53
    plate, 53
    principles of, 53
    screw, 53
    surgical, *53*
    tension-band wiring for, 53
Ostium
    ureteris, 183
    urethrae
        externum, 180, 184, 223,
            228–229, 232, 235, 559
        internum, 183, 229
    uteri, 8
    vaginae, 180, 232, 235, 559
Os trigonum, *471*
Ovarium, 71, 182, 188, 232, 543
    hormone secretion by, 71
Ovum/Ova
    formation of, 5
    postovulatory, 5
Oxytocin, 71

P

Pain
    heel, *471*
    in lateral thigh, *533*
    lumbar root compression
        syndromes and, *131*
    of pes planus, location of, *467*
    sacroiliac joint, *143*
    of splayfoot, location of, *467*
    in upper front leg, *96*
Palpable bony prominences
    of head and neck
        anterior view, 32
        posterior view, 33
    of membrum inferius, 32–33,
        409
    of membrum superius, 32–33,
        239
    of trunk and limbs
        anterior view, 32
        in female, 32
        in male, 33
        posterior view, 33
Pancreas, 71
    innervation of, 94–95
Pancreatic polypeptide, 71
Papilla mammae, 30, 199, 208
Paracolpium, 182
Paracortex, of nodi lymphatici, 69
Paralysis
    flaccid, 90, 96
    spastic, 90
Paraphimosis, 228
Paratenon, 60
Parathyroid hormone, 71
Paresthesia, 96, *96*
    n. medianus lesions causing, *372*
Park bench paralysis, *96*, *368*
Patella, 20, 30, 32, 38, 54, 61, 408,
    411, 416, 419, 440–441,
    443–445, 447, 450, 452,
    481, 483, 487, 502–503,
    506, 510–511, 544, 566
    anterior view of, 418
    basis of, 418
    bipartita, 419
    distal view of, 418
    dysplasia of, 419
    facies articularis, 418–419, 450
    fractures of, *419*
    hunter's hat configuration of,
        419
    instability of, *419*
    location of, 418
    medial hypoplasia of, 419
    ossification of, 19
    posterior view of, 418
    rete, 551
    shape of, evaluation, 419
    stabilization of, 443
Patellar tendon reflex, lumbar root
    compression syndromes
    and, *131*
Pecten ossis pubis, 138–139,
    142–143, 480
Pedal pulse, palpation of, *567*
Pediculus arcus vertebrae, 106,
    109, 111, 113, 118–119,
    129
Pedunculus connectans, 6, 8, 10,
    14
Pelvic floor. *See also* Musculus/
    Musculi (of region or organ),
    pelvic floor
    fascia of, superficial, 181
    lower level, 182–183
    middle level, 182–183
    repair, *163*
    structure of, 182–183
    upper level, 182–183
Pelvic measurement(s)
    external, 141
        in men, 141
    internal, 141
        in women, 141
Pelvic ring, 138, 408
Pelvic tilt, 409
    excessive, 157
    normal, 157
Pelvic torsion, 476
Pelvimetry, 141
Pelvis, 24, 408. *See also specific
    organ*
    bones of, fractures of, *412*
    bony, 138–139
    female, 139, 141
        cross section through, 189
        midsagittal section through,
            188, 232
    gender-specific features of, 139,
        182–183
    greater, 141, 182
    lesser, 141, 182
    ligamentous stabilization of, 428
    ligamentum, 140
    lower level, 182
    male, 138–139, 141, 183
        cross section through, 189
        ligamentum, 140
        midsagittal section through,
            188
    middle level, 182
    movements of, 156–157
    nerves of. *See specific nerve*
    nodi lymphatici of, 194
    organs of. *See specific organ*
    ossification of, 19
    subdivisions of, 182
    survey radiograph of, 433
    upper level, 182
Pendelluft effect, *132*
Penis, 31, 105, 180, 224–225, 528,
    543
    arterial supply to, 230
    bulbus, 183, 189, 224, 557
    corpus, 229

crus, 229
distal, 229
dorsum, 229
   vessels and nerves of, 230
fasciae, 228
nerve supply of, 230
radix, 229
skin of, 228
urethra emergence at, 223
venous drainage of, 230
Pericardium, 205
Perichondrial sleeve, 16
Perichondrium, 16, 44
Pericytes, 67
Peridentinitis, of calcaneal (Achilles') tendon, 471
Perimysium, 58
Perineal protection, 233
Perineum, 180, 559
   neurovascular structures of, 558
   regional anatomy of, 37
   sensory innervation of, 543
Perineurium, 75
Periosteum, 16, 39, 61
Peripheral nervous system (PNS), 24, 76–77
   development of, 73–75
   myelination in, 92
   node of Ranvier in, 93
   perineural sheath in, 93
Peritendineum, 60
   externum, 60
Peritoneum, 182, 211, 213, 557
   of hernial sac, 214–216
   parietale, 155, 167, 178–179, 210, 213, 216, 219, 557
   viscerale, 557
Pes
   arcus venosus dorsalis of, 526–527
   arcus venosus plantaris of, 526
   arteriae of, 568
      dorsum, 568
   arterial supply of, dorsal, variants, 569
   articulationes of, 454–455, 458–459
      facies articularis, 456–457
   axes of motion in, 462
   compartments of
      central, 523
      interosseous, 523
      lateral, 523
      medial, 523
      muscular contents of, 492–495, 523
   compartment syndromes of, 523
   coronal section through, 455
   cross-sectional anatomy of, 523
   deformities of, 467
      static, 467
   dorsum of, 31, 37, 544, 566–567
      cutaneous nerves on, 567
   functional position of, 462
   imaging of, 470–471
   inversion/eversion, 462–463
   ligamenta of, 460–461
   ligamentum of, 465
   load on, 464–465
   movements of, 462–463, 500–501
   musculi of, 486–487
      deep layer, plantar view, 520
      dorsal, 474
      middle layer, plantar view, 518–519

      origins and insertions, 521
      plantar, 474
      plantar view, 516–521
      short, 474, 492–495
         of central compartment, plantar surface, 494–495
         dorsal, 492–493
         of medial and lateral compartments, plantar surface, 492–493
      superficial layer, plantar view, 516–517
   nervi, 530–531, 568
   oblique transverse section through, 455
   ossa of, 38, 408, 422–425, 454–455
      articulating, 454–455
      dorsal view, 422
      lateral view of, 425
      medial view of, 424
      plantar view, 423
   planta, 37, 546
      arteriae of, 525, 564–565
      course of fibularis longus tendon on, 487
      nervi, 564
      neurovascular structures of, 569
      pressure chamber system in, 472, 546
   planta pedis, 546
   plantar architecture of, 464
   plantar flexion/dorsiflexion, 462–463
   plantar vault of, 464
   plantigrade position of, 462
   pronation/supination, 462–463
   rete venosum dorsale, 526–527, 544–545
   rete venosum plantare, 526
   retinacula of, 514–515
   second ray of, 466
   shape of
      "Egyptian" type, 544
      "Greek" type, 544
      square type, 544
      variants of, 544
   vaginae tendineum of, 514–515
   weight-bearing
      dynamics of, 546. See also Footprint
      transfer of compressive stresses in, 464
   X-ray anatomy of, 470
Pes anserinus, 54, 483–485, 502–504, 507, 510, 566
Pes cavus, 467
Pes equinovalgus, 501
Pes equinovarus, 501
   treatment of, 40
Pes planus, 459, 464
   pain associated with, location of, 467
Pes rectus, 462, 467
Pes transversoplanus. See Splay foot
Pes valgus, 462
Pes varus, 462
Petit's hernia, 202, 221
Peyer patches, 68
Phalanx distalis, manus, 238, 249–250, 252, 282, 284–285, 290–292, 315, 321, 347, 376, 399
   basis, 319, 323

   caput, 323
   corpus, 323
Phalanx media, manus, 238, 249, 252, 282, 284–285, 290–292, 315, 323, 376, 399
Phalanx proximalis, manus, 238, 249–250, 252, 282, 284–285, 290–292, 323, 376, 399
   basis, 319, 321
   caput, 321
Phasic muscle, 499
Pheochromocytoma, 75
Phimosis, 228
Phylogeny, human, 2–3
Phylum Chordata, 2–3
Piano-key phenomenon, 260
Piriformis syndrome, 96, 553
Placenta
   characteristics of, 9
   development of, 8–9
   fetal part, 8
   function of, 9
   hormone secretion by, 71
   lamina chorionica, 12
   maternal part, 8
      blood circulation of, 9
   mature, 9
   postpartum, 9
   size of, 9
   total volume of, 9
   transport functions of, 9
   villous surface area of, 9
   weight of, 9
Placental barrier, structure of, 9
Placentalia, 2
Placodes, ectodermal, 7
Plane(s)
   acetabular inlet, 427
      sagittal angle of, in adult, 427
      transverse angle of, in adult, 427
   anatomic, 27–28
   canthomeatal, 28
   cardinal, in human body, 27
   coronal (frontal)
      anatomic, 27–28
         axis deviations in, 27
      for imaging, 28
   Frankfurt horizontal, 28
   for imaging, 28–29
   midsagittal (median), 27
   radiographic, 28
      for skull, 28
   sagittal
      anatomic, 27–28
         axis deviations in, 27
      for imaging, 28
   transverse (axial)
      anatomic, 27–28
         through cavitas abdominis, 35
      for imaging, 28
Plantar fascitis, 471
Plantar vault, 464
   lateral rays of, 464
   medial rays of, 464
Planum
   interspinale, 35
   intertuberculare, 35
   scapulare, 241
   subcostale, 35
   supracristale, 35
   transpyloricum, 35

Pleura
   parietalis, 207
      pars costalis, 173, 178, 205
      pars diaphragmatica, 173, 178
   visceralis (pulmonalis), 173, 207
Pleural dome, injury to, 384
Pleural effusion, chest tube placement for, 207
Plexus (neural), 76
   brachialis, 76, 167, 196, 273, 363, 365, 367, 379
      anatomic landmarks, 384
      compression syndromes, 363
      course of, 361
      divisiones of, 384
         anteriores, 297, 360
         posteriores, 297, 360
      fasciculi of. See Fasciculi, of plexus brachialis
      formation of, 85
      injury to, 96
      motor branches, 297
      nervi of, 360–361
      pars infraclavicularis, 364–373
         long branches, 364
         nervi of, 360–361
         overview of, 364–365
         short branches, 364–365
         spinal segments associated with, 360–361
      pars supraclavicularis, 362–363
         musculi innervated by, 362
         nervi of, 360–363
         spinal segments associated with, 360–362
      radices of, 360
      spinal segments associated with, 196, 360–361, 384
      structure of, 360–361
      topography, 384
      trunci of, 360, 384
         inferior, 85, 360–361, 384
         medius, 85, 360–361, 384
         spinal segments associated with, 85, 360–361
         superior, 85, 360–361, 384
   cervicalis, 196, 388
      spinal segments associated with, 196
   coccygeus, 196, 530–531, 542–543
      spinal segments associated with, 196
   definition of, 85
   formation of, 85, 88
      principles of, 85, 88
   in gastrointestinal tract, 74
   hypogastricus inferior, 94
   ischiadicus, 530
   lumbalis, 196, 475, 530, 532–535
      muscles innervated by, 532
      nervi of, 475, 532–535
      overview of, 532
      rami cutanei, 532
      sensory innervation from, 537
      short, direct (muscular) branches, 475–476, 496, 530, 532
      spinal segments associated with, 196, 530, 532
      structure of, 531
   lumbosacralis, 76, 530–531
      nervi, 530
      rami, 530

P *Plexus (neural) (cont.)*

structure of, 531
topography of, 531
pudendus, 530, 542
sacralis, 196, 230, 235, 475, 530, 543, 551, 558
  direct (muscular) branches, 475, 478, 496, 530, 536–537
  muscles innervated by, 536
  nervi of, 475, 536–543
  rami cutanei, 536
  spinal segments associated with, 196, 530, 536
  structure of, 531
testicularis, 225
Plexus (vascular)
  lumbalis, 204
  pampiniformis, 225–227
  venosus
    areolaris, 192
    pelvic, 183
    prostaticus, 230
    in spatium epidurale, 128
    vertebralis externus
      anterior, 193
      posterior, 193
    vertebralis internus
      anterior, 128, 193
      posterior, 128, 193
    vesicalis, 230
  in vesicula umbilicalis, 12
Plexus lymphaticus axillaris, 359
Plicae
  alares, 42, 450
  analis, 222
  cloacalis, 222
  neurales, 6, 72
  synoviales, 42
  umbilicales
    laterales, 212, 217
    medialis, 212–213, 217
    mediana, 212
PNS. *See* Peripheral nervous system
Podogram, 464, 467, 546
Polar body(ies), 5
Pollex, 374
  art. carpometacarpalis, 238, 249, 253, 292–294
    articulating surfaces, 292
    axis of motion in, 292–293
    movements of, 49, 293
    rotation-induced incongruity, 293
  articulationes of, 294
  art. interphalangea pollicis, 253, 294
  art. metacarpophalangea, 253, 294
  and digiti, relationship of, in neutral position, 293
  interphalangeal joint crease, 374
  lymphatic drainage of, 358
  metacarpophalangeal joint crease, 374
  nn. digitales palmares, 398, 400, 403
  opposition of, axis for, 293
Polydactyly, 20
Pons, 105
  development of, 72
Portio cervicis
  supravaginalis, 8
  vaginalis, 8
Portio uteri, 182, 188
Porus acusticus externus, 103

Posture (of body)
  active
    normal, 157
    rigid, 157
  erect, 50
  passive slumped, 157
Pregnancy
  hormones in, 9
  testing for, 9
Preperitoneal fat tissue, 216
Preputium
  clitoridis, 180, 232, 235
  penis, 228–229, 543
    constriction of, 228
Pressure, blood. *See* Blood pressure
Pressure chamber(s)
  of pes, 472, 546
    structure of, 472
  plantar, 472
Primate(s), 2
  spine of, 408
PRL. *See* Prolactin
Processus/Processi
  accessorius, 106, 113
  articularis vertebrales, 138
    inferior, 102, 106, 108–113, 116–117, 119, 124, 127
    superior, 102, 106–114, 116–119, 124, 127, 130–131, 137, 139
  chordalis, 6
  coracoideus, 32, 38, 61, 165, 238–243, 245, 250, 258–268, 272, 301, 303, 305, 307, 309, 311, 313, 335, 363, 365–366, 382–383, 386, 389
  coronoideus, 248–251, 278–280, 315
  costales vertebrales, 102, 106, 112–113, 116–119, 130–131, 147, 149, 161, 167, 169
  mamillaris, 106, 113, 147
  mastoideus, 33, 120, 122, 147, 151, 170–171, 203
  posterior tali, 422–425, 470
  spinosi vertebrales, 7, 33, 41, 101–102, 106–113, 116–124, 127, 130–133, 135–137, 140, 143, 149, 165, 167, 201, 240, 301, 365, 554
    as posterior landmarks, 35
  styloideus
    embryology of, 11
    radii, 30, 32–33, 239, 247–255, 280–282, 284–286, 289, 292, 317, 376
    of temporal bone, 120, 122
    ulnae, 31–33, 38, 239, 248, 280–282, 284–287, 289, 292, 376
  supracondylaris, of humerus, 244, 393
  transversi vertebrales, 7, 41, 102, 106–111, 118–120, 122–124, 126, 132, 135–137, 149, 165, 240, 301
  tuberis calcanei
    lateralis, 423, 425
    medialis, 423–425
  uncinatus, 106, 108–109, 126–127

vaginalis peritonei, obliterated, 217
vaginalis testis, 224–225
  obliterated, 224
xiphoid. *See* Sternum, processi xiphoideus
xiphoideus, 32, 38, 132, 134, 153, 155, 159, 161, 172, 176, 199, 335
Progesterone, placental production of, 9, 71
Progestin(s), 71
Prolactin (PRL), 71
Prolaps uteri, *163*
Prominentia
  cardiaca, 2, 10, 14
  laryngea, 32
  liver, 10
  maxillaris, 10
Promontorium, 102–103, 105, 114–115, 128, 138–140, 142–143, 187, 428, 477, 481, 502–503
  in male, 139
Pronatio dolorosa, *283*
Pronator teres syndrome, *96*, *372*
Pronucleus
  female, 5
  male, 5
Proprioceptors, 44
Prosencephalon, 10, 72
Prostaglandin(s), 71
Prostata, 105, 183, 188–189, 212, 223, 224, 229, 231, 543, 557
Protein(s)
  compound, in hyaline cartilage, 45
  nuclear, in hyaline cartilage, 45
Proteoglycans, 44–45, 64
Protuberantia mentalis, 32
Protuberantia occipitalis
  externa, 33, 120–122, 165, 170–172, 299
  interna, 120
Pseudarthrosis, 40, *53*
  with femoral fractures, 433
Pseudohermaphroditismus, 223
  femininus, 223
  masculinus, 223
Pseudo-osteoarthritis, scaphoid fracture and, *257*
Pudendum femininum, 232
Pulmo, 24, 207
  innervation of, 94–95
  percussion, *160*
Punctum fixum
  of mm. intercostales externi, 159
  of mm. intercostales interni, 159
Punctum mobile
  of mm. intercostales externi, 159
  of mm. intercostales interni, 159
Pyramidal decussation, 90
Pyramid(s), in brainstem, 90

R

Radial tunnel, 369, 392
Radial zone, in hyaline cartilage, 44
Radiography. *See* Imaging
Radius, 20, 38, 41, 238, 247–249, 252–254, 276–282, 284–289, 292, 294, 312, 315, 317, 319, 323, 342–343, 345, 350–353, 366, 393, 395
  articular surfaces of, 250–251

distal, 251
proximal, 251
caput of, 33, 38, 238–239, 247, 249–251, 276–277, 279, 281, 393
  circumferentia articularis, 248, 251, 276–277
  lunula obliqua, 279
collum of, 248, 250, 276, 278, 280, 282
corpus of, 248, 251
development of, 14
distal
  articulating surfaces, inclination angles of, 256
  fractures of, *257*
  length of, relative to ulna, 256
facies anterior, 248, 250–251
facies articularis carpalis, 248, 251, 255, 281
facies lateralis, 248, 250–251, 280
facies posterior, 248, 251, 280
fovea articularis, 248, 251, 277, 280
margo anterior, 248, 251, 280
margo interosseus, 248, 251, 280
margo posterior, 248, 280
ossification of, 18
proc. styloideus of, 30, 32–33, 239, 247–255
in pronation, 249
rotation, during pronation and supination, 281
in supination, 249
Radix
  anterior, 7
  posterior, 7
Ramus communicans
  albus, 80–81, 89, 95, 197
  griseus, 80–81, 89, 95, 197
Ramus/Rami
  collateralis, 191
  cutaneus
    anterior, 197–198
    lateralis, 190–191, 197–198, 200, 202
    medialis, 190–191, 200, 202
  dorsalis, 190–191, 196
    spinal segments associated with, 196
  iliacus, 190
  intercostales anteriores, 190
  labiales
    anteriores, 234
    posteriores, 234
  lumbalis, 190
    posterior, 235
  mammarii, 209
    laterales, 190–191, 198, 209
    mediales, 190, 198, 209
  musculares, 230
  of nervi spinales, 89
    dorsales, 7, 80–81, 87–88, 115, 145, 196–197, 360, 365, 537
      cutanei mediales and laterales, 80, 388
    ventrales, 7, 80–81, 88, 115, 126, 145, 196–197, 360, 362, 537
  obturatorius, 212–213
  ossis ischii, 139, 180, 184, 187, 413, 480

ossis pubis
  angle between, gender differences in, 139
  inferior, 139, 163, 180, 182–183, 185, 189, 229, 412–413, 480
  superior, 139, 180, 184–185, 210, 219, 229, 412–413, 480–481
  perforantes, 190–191, 205, 209
  pubica, 213
  scrotales posteriores, 230
  spinalis, 190–191
  sternales, 191
Raphe
  iliococcygeal, 163
  musculi iliococcygei, 185–186
  penis, development of, 222
  perinei, 180, 232
    development of, 222
Rathke's pouch, 10
Reabsorption, capillary, 67
Recess(es). See also Ankle mortise
  articular, 42
Recessus
  axillaris, 263–264, 266, 269, 270–271
  costodiaphragmaticus, 173
  sacciformis, 277–279
  subpopliteus, 442, 450
  suprapatellaris, 450–451, 482
    unfolding during knee flexion, 451
Rectopexy, 163
Rectum, 105, 182, 189, 229, 232, 434, 543
  ampulla recti, 188
  innervation of, 94
Rectus diastasis, 220
Recurvation, 27
Reference line(s), 34
  in regio glutealis, for locating neurovascular structures, 554
  vertical, on trunk, 34
Reflex arc, 90
Reflex(es)
  biceps, 89
  calcaneal (Achilles') tendon, 89
  clinically important, 89
  lumbar root compression syndromes and, 131
  monosynaptic proprioceptive, 88
  patellar tendon, 89
  polysynaptic exteroceptive, 88
  quadriceps, 89
  testing, 89
  triceps, 89
Regio abdominalis lateralis, 36
Regio analis, 36–37, 180, 558–559
  sensory innervation of, 542
    in female, 559
    in male, 559
Regio antebrachialis
  anterior, 37, 394–395
  posterior, 37, 396–397
Regio axillaris, 36–37. See also Axilla
Regio brachialis
  anterior, 37, 386–387
  posterior, 37, 390–391
Regio buccalis, 36
Regio calcanea, 37
Regio carpalis
  anterior, 37, 404
  posterior, 37

Regio cervicalis
  anterior, 36
  lateralis, 36
  posterior, 36
Regio cruris
  anterior, 37, 566–567
  posterior, 37, 562–563
Regio cubitalis, 392–393
  anterior, 37
  posterior, 37
Regio deltoidea, 36–37
Regio epigastrica, 36
Regio femoralis
  anterior, 548–549
  internal hernial openings, 213
Regio femoris
  anterior, 37
  posterior, 37
    neurovascular structures of, 560
Regio frontalis, 36
Regio genus
  anterior, 37
  posterior, 37
Regio glutealis, 30, 36–37
  deep, vessels and nerves of, 553
  fasciae of, 552
  musculi of, 474, 552–553
    anterior view, 503–505
    lateral view, 506
    medial view, 502
    origins and insertions, 504–505, 508–509
    posterior view, 507–509, 552–553
  nervi of, 552–555
  posterior view, 556
  sensory innervation of, 537
  superficial, fasciae and cutaneous nerves of, 552
  vasa of, 553–555
Regio hypochondriaca, 36
Regio inframammaria, 36
Regio infraorbitalis, 36
Regio infrascapularis, 36
Regio infratemporalis, 36
Regio inguinalis, 36, 549
  internal hernial openings, 213
  lacuna musculorum, 549
  lacuna vasorum, 549
  male, sensory innervation of, 533
  nodi lymphatici, 194
    profundi, 528
  sagittal section through, 219
Regio interscapularis, 36
Regio mentalis, 36
Regional anatomy, 36–37
  of abdomen, 36
  axillary, 37
  of back, 36
  of buttocks, 36
  of head and neck, 36
  of membrum inferius, 37
  of membrum superius, 37
  perineal, 37
  of thorax, 36
Regio nasalis, 36
Regio occipitalis, 36
Regio oralis, 36
Regio orbitalis, 36
Regio parietalis, 36
Regio parotideomasseterica, 36
Regio pectoralis, 36
  lateralis, 36
Regio pectoralis lateralis, 36

Regio perinealis, 37, 558–559
  female, 180
    neurovascular structures of, 234–235
  innervation of, 537
  male, 180
    neurovascular structures of, 230–231
  sensory innervation of
    in female, 559
    in male, 559
Regio presternalis, 36
Regio pubica, 36
Regio retromalleolaris lateralis, 37
Regio sacralis, 36
Regio scapularis, 36–37
  arterial supply to, 391
  deep, 203
Regio sternocleidomastoidea, 36
Regio suprascapularis, 36
Regio temporalis, 36
Regio umbilicalis, 36
Regio urogenitalis, 37, 180, 558–559
  female, sensory innervation of, 542
  sensory innervation of
    in female, 559
    in male, 559
Regio vertebralis, 36
  with spinal furrow, 30
Regio zygomatica, 36
Ren, 24, 167
  hormone secretion by, 71
  innervation of, 94
  venous drainage of, development of, 13
Renin, 71
Reproductive system, 24
Reptile(s)
  embryonic development of, 2
  limb positions of, 21
Reserve zone, epiphyseal, 16
Respiration. See also Expiration; Inspiration
  abdominal wall muscles in, 156–157
  diaphragma in, 160–161
  paradoxical, 132
  rib movement during, 136
  thoracic muscles in, 158–159
Respiratory system, 24
Rete
  calcaneum, 562–563
  patellare, 551
  testis, 226
  venosum
    dorsale, 526–527, 544–545
    plantare, 526
Retinaculum/Retinacula
  musculorum extensorum
    of manus, 55, 345–346, 376, 396–397
    of pes
      inferius, 514–515, 563, 566–567
      superius, 514–515, 563, 566–567
  musculorum fibularium
    inferius, 515
    superius, 515
  musculorum flexorum, 371, 373, 394, 398, 400, 402–405
    of manus, 288–289, 343–344, 348–352
    of pes, 515, 562–563

patellae
  laterale, 443, 482–483
  mediale, 443, 482–483
  transversale
    laterale, 443
    mediale, 443
  of pes, 514–515
Retroperitoneal organs, 204
Rhombencephalon, 72–73
Rhomboid of Michaelis, 30, 201
Ribs. See Costae
Rima ani, 409
Ring
  hernial, 213
  pelvic, 413
  pharyngeal lymphatic (Waldeyer's), 68
Root(s), spinal nerve. See Spinal radix nervi
Rotator cuff, 258, 302–303
  damage to, 262
  degenerative changes, 269
  imaging of, 28
  rupture, 269–270
Rotator interval, 265
Rough endoplasmic reticulum (RER), 78
Rough endoplasmic reticulum, in exocytosis, 70
"Rule of nines," 23

S

Sacci pharyngei, 10, 12
  endodermal, 10
Sacral triangle, 31, 201
Sacrum. See Ossa (named), sacrum
Saltatory nerve conduction, 93
Sarcolemma, 58
Sarcomere, 58
  structure of, 59
Sarcopenia, 56
Sarcoplasm, 59
Sarcoplasmic reticulum, 58
Scalene syndrome, 96, 363
Scalenus syndrome, 106
Scapula, 24, 38, 161, 164, 238, 240, 250, 261, 267, 269, 273, 299, 301, 304–305, 307, 381
  acromion of. See Acromion
  alata, 362
  angulus inferior, 201
  anterior view, 243
  collum, 243, 245, 263
  development of, 14
  facies anterior, 311
  facies articularis clavicularis, 245
  facies costalis, 240, 242–243, 258–259, 263, 267, 305, 307
  facies posterior, 165, 240, 243, 301, 305, 312
  lateral view, 243
  margo lateralis, 164, 240, 243, 262–263, 266, 303, 307, 312
  margo medialis, 31, 33, 55, 165–166, 201, 239–240, 243, 259, 301, 303, 331, 362, 389–391
  margo superior, 242–243, 245, 267, 299, 303
  movements of, 274
  neurovascular tracts associated with, 390
  ossification of, 18

S Scapula (cont.)

posterior view, 243
spina scapulae, 201
Scapular arcade, 389
Sclerosis, subchondral, 47
Sclerotome, 7, 104
  derivatives of, 7
Scoliosis, 133, 409
Scrotum, 31, 105, 180, 188, 223–225, 528, 543, 558
  arterial supply to, 230
  development of, 222
  hernia palpation from, 218
  nerve supply of, 230
  sensory innervation of, 559
  skin of, 217, 226
  swelling of, 227
  urethra emergence at, 223
  vasa lymphatica, 227
  venous drainage of, 230
Secretin, 71
Secretion. See also Exocytosis; Neurosecretion
  apocrine, 70
  autocrine, 71
  eccrine, 70
  endocrine, 71
  holocrine, 70
  merocrine, 70
  neurocrine, 71
  paracrine, 71
Section anatomy
  of abdominal wall, 178–179
    at L III, 155
  of acetabular inlet plane, 427
  of antebrachium, 342–343, 345
  of art. digitorum, 294
  of articulatio coxae, 415, 432, 434–435
  of articulatio femorotibialis, 446–447
  of articulatio genus, 441, 446–447, 451
  of articulatio humeri (glenohumeralis), 261
  of articulatio subtalaris, 455
  of articulatio talocruralis, 455
  of art. mediocarpalis, 294
  of art. radiocarpalis, 294
  of axilla, 381
  of brachium, 342–343, 386
    at tuberositas radii, 310
  of canalis carpi, 289, 402
  of canalis pudendalis, 558
  of caput femoris, 427, 431, 434–435
  of carpus, 286
  of cauda equina, 128
  of cavitas abdominis, 23, 156
  of cavitas pelvis, 23, 182, 188, 543, 557
  of columna vertebralis
    pars cervicalis, 127
    pars lumbalis, 128
  of cranial cavity, 25
  of crus, 421, 522, 567
  of cubitus/art. cubiti, 277
  of diaphragma, 175
  of digiti, 291, 345
  of femoropatellar joint, 419, 441
  of head, 105
  of manus, 294
  of membrum inferius, 29
  of metatarsi, 468, 523
  of ossa metacarpi, 291
  of pelvic floor muscles
    female, 182

male, 183
of pelvis, 558
  bony, 143
  female, 182, 188–189, 543, 557
  male, 183, 188–189, 543, 557
of pes, 455, 466, 471–472, 523, 569
of radioulnaris
  distalis, 345
  proximalis, 281
of recessus costodiaphragmaticus, 173
of rectus sheath, 178–179
of regio calcanea, 471
of sacrum, 537
of shoulder/art. humeri (glenohumeralis), 269, 272–273, 304
of thigh, 522
of thoracic cavity, 25
of thoracic wall, 173
of thoracoabdominal junction, 174
of thoracolumbar fascia
  at C VI, 167
  at L III, 167
of trunk, 105, 175
Sella turcica, 121
Sensory innervation
  neural circuit for, 91
  nuclear, in head, 86
  overview of, 82–83
  peripheral, 87
    of head and neck, 87
  radicular (segmental), 86
Sensory organs, 24
Septula testis, 226
Septum/Septa
  decidual, 9
  femorale, 549
  intermusculare
    of brachium
      laterale, 342, 391
      mediale, 342, 386
    cruris, 486, 522
      anterius, 566–567
      posterius, 567
    of membrum superius
      laterale, 61
      mediale, 61
  longitudinal, in epiphyseal plate, 16
  nasale, 105
  plantare
    laterale, 516, 523, 569
    mediale, 516, 523, 569
  of plantar pressure chamber, 472
  transverse, in epiphyseal plate, 16
Serosa, 24
Sex cell(s). See Gamete(s)
Sexual dysfunction, 543
Sexual reflexes, male, 231
Sheath
  brachial plexus, 384
  connective tissue, in striated skeletal muscle, 58
  perineural, 93
  rectus. See Vagina musculi recti abdominis
  synovial. See Vagina synovialis
  of tendon. See Vagina tendinis
Shoulder/Art. humeri (glenohumeralis), 38, 238
  abduction, 275, 324

adduction, 275, 324
angled coronal plane of
  anatomic structures visible in, 28
  position of, 28
angled sagittal plane of
  anatomic structures visible in, 28
  position of, 28
arteriae of, 354–355
  overview of, 354
arthroscopic anatomy of, 271
arthroscopy, 270–271
articular capsule of, 262–263
articular cavity of, 262–263
articular surfaces of, 262–263
articulationes of, 258–267
axial (transverse) plane of
  anatomic structures visible in, 28
  position of, 28
bursae in, 268–269
bursae synoviales in, 61
capsula articularis of, 262–263, 266–268, 390
  ligamenta reinforcing, 264–265
cavitas of, 262–263
coronal section through, 269
degenerative joint disease, 46, 270
development of, 15
disease of, 275
dislocation, 262
  recurrent, 262
  repositioning of, 263
extension, 275, 324
external rotation, 324
flexion, 275, 324
functional articulations, 258
imaging of, 271–273
  MRI, standard planes in, 28
inflammatory changes, 270
instability, 270
internal rotation, 324
ligamenta of, 258–259, 263
mobility of, 258
movements of, 49, 261, 324
  horizontal, 275
  rotational, 275
  vertical, 275
musculi of, 296
  abductors, 324
    functional testing of, 325
    weakness/shortening, symptoms of, 325
  adductors, 324
    functional testing of, 325
    weakness/shortening, symptoms of, 325
  anterior, 296, 334–337
  extensors, 324
    functional testing of, 325
    weakness/shortening, symptoms of, 325
  external rotators, 324
    functional testing of, 325
    weakness/shortening, symptoms of, 325
  flexors, 324
    functional testing of, 325
    weakness/shortening, symptoms of, 325
  functional testing of, 325
  functions of, 324
  innervation of, 324

internal rotators, 324
  functional testing of, 325
  weakness/shortening, symptoms of, 325
  nerve segments associated with, 324
posterior, 296, 330–333
  deep layer, 331
  superficial layer, 330
  posterior view, 388
  shortening of, symptoms of, 325
  weakening of, symptoms of, 325
nervi of. See also Plexus (neural), brachial
  posterior view, 388
  superficial, 378
posterior view, 388
range of motion, 324
range of motion of, 262
regio brachialis
  anterior, 386–387
  posterior, 390–391
stability of, 262
superior view, 389
suprascapular region
  posterior view, 388
  superior view, 389
transverse section through, 261
venae of, superficial, 378
Shoulder girdle. See Cingulum membri superioris
Shoulder region
  anterior view, 378–379
  nervi of, superficial, 378
  venae of, superficial, 378
Shouldice technique, 219
Sign
  ballottable patella, 451
  bottle, 372
  drawer, anterior or posterior, 448
  Duchenne, 497
  Froment, 370
  Hoffman–Tinel, 97
  rectus, positive, 483
  Trendelenburg, 536
    positive, 479, 497, 536
Sinus(es)
  coronarius, development of, 13
  epididymidis, 226
  lactifer, 208
  portal, development of, 13
  sphenoidalis, 121
  tarsi, 422–423, 459
  urogenitalis, 222
    in adrenogenital syndrome, 223
Sinusoid(s), hepatic, 63
  development of, 13
Sinus venosus, 13
Skeletal age, 19
Skeletal muscle, 24
Skeleton (cat), 21
Skeleton (human), 24
  anterior view, 38
  embryology of, 11, 14
  laryngeal, embryology of, 11
  of membrum superius, 238
  overview of, 38
  pedal. See also Pes, ossa of
    anatomical subdivisions of, 422
    functional subdivisions of, 423
  posterior view, 38
  prenatal development of, 14

606

primordial, 14
remodeling of, 16
thoracic, 132–133
of trunk
anterior view, 100
posterior view, 101
visceral, embryology of, 2
Skin. *See* Cutis
Skin appendages, 24
Soft tissue, damage, with fracture, 52
Somatomedins, 71
Somatopleura, 6–7
Somatostatin, 71
Somatotopic organization, 86
Somatotropic hormone (somatotropin, STH), 71
Somatotropin. *See* Somatotropic hormone
Somite(s), 2, 6, 82, 104, 145
derivatives of, 7
Space
Colles'
boundaries of, 183
contents of, 183
costoclavicular, narrowing of, *363*
infralevator, 182–183
interscalene, 361, 379
joint. *See* Joint space(s)
pleural, 173
preperitoneal, 179, 216, 219
quadrangular, axilla, 390
rectoprostatic, 188
rectovaginal, 188
subacromial, 258, 266–267
imaging of, 28
subarachnoid, 121, 128, 384
supralevator, 182–183
triangular, axilla, 390
Spatium
extraperitoneale, 24–25
intervillosum, 8–9
volume of, 9
perinei
profundum, 163, 182–183, 229, 230
boundaries of, 183
contents of, 183
subcutaneous, 182–183, 557
superficialis, 182–183, 230, 557
boundaries of, 183
contents of, 183
retroperitoneale, 24–25
retropubicum, 188
subperitoneale, 24, 182–183, 557
Spermatid(s), 5
Spermatocele, 227
Spermatocyte(s)
primary, 5
secondary, 5
Spermatogenesis, 5
Spermatogonium/Spermatogonia, 5
Spermatohistogenesis, 5
Spermatozoa, 5
formation of, 5
Sphincter(s), 162–163, 182
ani externus, 188
ani internus, 188
postcapillary, 67
precapillary, 67

Sphincter urethrae. *See* Musculus/Musculi (named), sphincter urethrae
Sphincter urethrae externus. *See* Musculus/Musculi (named), sphincter urethrae
Spider veins, *527*
Spina bifida, *104*
Spinal column. *See also* Columna vertebralis
abdominal muscles and, *156*
curvatures of, 103, 119
development of, 105
lateral (scoliosis), *133*
neonatal, 105
evolution of, 3
growth of, 82
integration into pelvic girdle, 103
regions of, 103
stabilization of, 156
Spinal cord segment(s)
cross section of, 81
embryology of, 81
functional anatomy of, 81, 89
location, in relation to vertebral canal, 82
structure of, 80–81
topography of, 81, 89
Spinal furrow, 201
Spinal nerve(s). *See* Nervus/Nervi (of region or organ), spinales
Spinal radix nervi, 129. *See also* Cauda equina
compression of, *131*
dorsalis, 7, 75–76, 80–81, 88, 90, 126, 197, 360
dural sleeve with, 131
in foramen intervertebrale, 130
lumbar, compression of, *131*
sensory fibers from, course to dermatome, 84
ventralis, 76, 80–81, 88, 90, 126, 197, 360
dural sleeve with, 131
Spinal stenosis, *127*
lumbar, *130*
Spina/Spinae
iliaca
anterior inferior, 138–140, 142, 157, 180, 184, 186–187, 412–413, 428, 481–483, 485
anterior superior, 30, 32, 35, 103, 138–140, 142–143, 153, 157, 176, 180, 184, 186–187, 199, 408–409, 411–413, 426, 428–429, 433, 477–479, 481–483, 485, 502–503, 506–507, 533, 536, 540, 544, 548–549, 551, 554
hernia palpation from, 218
posterior inferior, 138–139, 412, 426, 485
posterior superior, 33, 35, 103, 138–139, 142–143, 161, 186–187, 201, 408–409, 412, 426, 428, 506, 554–556
ischiadica, 138–143, 163, 180, 184–187, 411–413, 426–428, 433, 477–479, 485, 542–543, 549, 554, 558

scapulae, 31, 33, 35, 38, 55, 101, 165–166, 238–240, 242–243, 245, 260, 262, 265, 267, 299, 301, 305, 312, 330–331, 362, 388–391
Spine (of body). *See also* Columna vertebralis
presacral, 103
Splanchnopleura, 6–7
Splayfoot, *464*, *469*
pain associated with, location of, *467*
Spleen, 68
Spondyloarthritis, in lumbar spine, *130*
Spondyloarthrosis, *130*
Spondylodesis, *40*
Spondylolisthesis, *104*
Spondylolysis, *104*
acquired, *104*
congenital, *104*
Spondylophyte(s), 131
bridging, *130*
in columna vertebralis
lumbalis, 130
pars cervicalis, *127*
Stance, stability in, 473
Stapes, embryology of, 11
Statins, 71
Step length, 473
Step width, 473
Sternum, 41, 100, 132–133, 135, 157, 159, 161, 165, 174, 177, 193, 205, 260, 299, 309, 363, 365, 378
corpus, 32, 38, 132, 134, 153, 155, 172–173, 175–176, 178, 240, 334–335
manubrium, 32, 38, 132, 134, 159, 172–173, 175, 178, 240–241, 258–259, 334–335
superior border, 35
processi xiphoideus, 32, 38, 132, 134, 153, 155, 159, 161, 172, 176, 335
STH. *See* Somatotropic hormone
Stork legs, *501*
Stratum basale, 374
Stratum corneum, 374
Stratum fibrosum, 39
Stratum granulosum, 374
Stratum lucidum, 374
Stratum spinosum, 374
Stretch protection, principle of, 60
Stylopodium, 20–21
Subacromial impingement syndrome, *269–270*
Subcutaneous tissue(s), 374
of abdominal wall, 179
of regio cubitalis, 392
Subcutis, 7, 213, 374
Subintima, synovial, 43–44, 450
Subneural apparatus, 59
Suboccipital region, muscles of, 144, 150–151, 170–171
origin and insertion of, 171
Subphylum Invertebrata, 2
Subphylum Vertebrata, 2–3
Substantia alba, of spinal cord, 81
Substantia grisea, of spinal cord, 81
Succinate dehydrogenase (SDH), 56
Sulcus/Sulci
arteriae subclaviae, 135

arteriae vertebralis, 108, 123
bicipitalis medialis, 386
of brain, 72
calcanei, 422, 459
capitulotrochlearis, 245, 276–277, 279
carpi, 289
costae, 135, 173, 205, 207
cutis, 374
deltoideopectoralis, 199, 357, 378, 380, 384
glutealis, 30, 201, 546, 552
innervation of, 537
inguinalis, 30
intertubercularis, 165, 244–247, 259, 262–265, 267, 303, 307, 309, 311, 337, 366, 391
malleolus, 420
for m. flexor digitorum longus, in sustentaculum tali, 424
for musculi subclavii, in clavicula, 242
neuralis, 6, 72
n. radialis, 244, 246, 313, 333, 369, 391
n. spinalis, 108–109, 120, 126–127
n. ulnaris, 244, 276, 278, 297, 319, 371, 386, 391
pharyngei, 10
ectodermal, 10
plantar
lateralis, 564
medialis, 564
primitivus, 6
supraacetabularis, 412
tali, 423, 459
telodiencephalicus, 72
tendinis musculi flexoris hallucis longi
in calcaneus, 423
in talus, 422–423
tendinis musculi peronei, in cuboid, 423
venae subclaviae, 135
Sulcus ulnaris syndrome, *96*, *370*
Supination trauma, of ankle, *501*
Supinator syndrome, *96*, *368*
Supraspinatus (impingement) syndrome, *268*
Surface anatomy. *See also* Landmarks; Reference line(s)
of abdominal wall, anterior, 212
of epididymis, 226
of female, 30
of male, 31
of manus, 374, 376, 397
of membrum inferius
anterior view, 544
posterior view, 546
of membrum superius
anterior view, 374
posterior view, 376
of pes, 544, 546
of regio perinealis
in female, 180
in male, 180
of testis, 226
of trunk wall
anterior, 199
posterior, 201
of wrist, 404
Surface contours
of head and neck
anterior view, 32

607

S Surface contours (cont.)

posterior view, 33
of trunk and limbs
  anterior view, 32
    in female, 32
    in male, 33
  posterior view, 33
Sustentaculum tali, 422–424, 454, 457–460, 465–466, 468, 470, 495
  fractures of, 424
Sutura
  coronalis, 28
  lambdoidea, 33
  sagittalis, 28, 33
Symphysis
  definition of, 40
  pubica, 32, 38, 41, 103, 105, 138–141, 143, 153, 155, 157, 163, 175, 184–189, 224, 229, 232, 408, 413, 428, 477, 480, 502–503, 543, 548–549, 558
Synapse(s), 78
  axoaxonal, 79
  axodendritic, 79
  axosomatic, 79
  in central nervous system, electron microscopy of, 79
  neuromuscular, 59
  spine, 79
  in sympathetic nervous system, 95
Synarthrosis/Synarthroses, 40, 42
Synchondrosis/Synchondroses, 40–41
  of costae, 134
    manubriosternalis, 134
    xiphosternalis, 134
Syncytiotrophoblast, 8–9
Syndactyly, 14, 20
Syndesmosis/Syndesmoses, 40–41
  tibiofibularis, 38, 41, 420, 455, 460
    tears of, 461
    and Weber fractures, 461
Synostosis/Synostoses, 18–19, 40–41
  pelvic, 19
Synovectomy, 40, 44
Synovial fluid, 15, 42, 44, 61, 450
  composition of, 43
  function of, 43
  volume of, 43
Synoviocyte(s)
  type A, 44
  type B, 44
Synovitis
  reactive, 47
  of shoulder/art. humeri (glenohumeralis), 270

T

Tactile cutaneous stimuli, 231
Talar dome, 455
Talipes calcaneus, 501
Talipes valgus, 501
Talus, 38, 41, 422–423, 455, 458–461, 464–468, 470–471, 489–490, 493, 520
  caput, 422, 424–425, 458
  collum, 422–425, 458
  corpus, 422–425
  facies articularis, 459
    distal, 457
    navicularis, 422–425, 458

ossification of, 19
processi posterior, 495
trochlea, facies superior, 425
Tangential fiber zone, in hyaline cartilage, 44
TAPP repair. See Transabdominal preperitoneal repair
Tarsal tunnel syndrome, 541, 563
  anterior, 96
  posterior, 96
Tarsus, 20, 38, 408, 422
  accessory, 425
  development of, 14
  ossification of, 19
TDLU. See Terminal duct lobular unit
Tela subserosa, 210, 219
Telencephalon, 72
Tendinocyte, 60
Tendo/Tendines (in general, including tendon types), 58
  chondral apophyseal
    function of, 60
    structure of, 60
  function of, 60
  of insertion, 57
  intermediate, 57
  of origin, 57
  periosteal diaphyseal
    function of, 60
    structure of, 60
  pressure, 60
  structure of, 60
  traction, 60
Tendo/Tendines (named)
  abductor hallucis, in hallux valgus, 469
  adductor hallucis, in hallux valgus, 469
  adductor magnus, 505
  biceps femoris, imaging of, 28
  calcaneal (Achilles'), 31, 55, 466, 471, 488–489, 510, 512, 515, 522, 546, 562–563
    rupture of, 488
  centrum, of diaphragma, 160–161, 174–175, 178
  extensor digitorum brevis, 493, 568
  extensor digitorum longus, 487, 500, 568
  extensor hallucis brevis, 493, 568
  extensor hallucis longus, 487, 500, 544, 563, 566, 568
    in hallux valgus, 469
  fibularis brevis, 471, 487, 500, 510, 519
  fibularis longus, 471, 487, 493, 495, 500, 510, 518–520, 569
  fibularis tertius, 510
  flexor digitorum brevis, 517–519, 564–565
  flexor digitorum longus, 466, 471, 491, 494–495, 500, 519, 523, 541, 564–565
  flexor hallucis longus, 424, 466, 468, 471, 491, 500, 519, 523, 541, 564–565, 569
    in hallux valgus, 469
  gracilis, 481, 484, 503–505, 510
  iliopsoas, 431
  intermedius, 298–299
  lumbricales, 523

of m. abductor pollicis longus, 345–346, 350–352, 397, 402
of m. biceps brachii, 261, 263–266, 271–273, 277, 311, 336, 338, 343, 381–383, 386, 392, 394
of m. brachialis, 311
of m. brachioradialis, 317, 346
of m. extensor carpi radialis brevis, 340, 345–346, 396–397, 402
of m. extensor carpi radialis longus, 340, 345–346, 396, 402
of m. extensor carpi ulnaris, 281, 287, 345, 396, 402
of m. extensor digiti minimi, 345, 396, 402
of m. extensor digitorum, 291, 345, 347, 376, 396–397, 402
  dorsal digital expansion, 291, 340
of m. extensor indicis, 345–346, 397, 402
of m. extensor pollicis brevis, 345, 396–397, 402
of m. extensor pollicis longus, 340, 345–346, 396–397, 402
of m. flexor carpi radialis, 350–352, 402, 404
of m. flexor carpi ulnaris, 285, 350–352, 405
of m. flexor digitorum profundus, 290, 323, 338–339, 347, 349–351, 399, 402
of m. flexor digitorum superficialis, 290–291, 338, 347, 349–351, 394, 399, 402, 405
of m. flexor pollicis longus, 338–339, 349–351, 402
of m. palmaris longus, 348, 374, 404–405
of m. pectoralis major, 308
m. quadricipitis femoris, 61
of m. subscapularis, 265, 271
of m. supraspinatus, 265, 269, 273
  damage to, 269
of m. triceps brachii, 313
obturator internus, 478
plantaris, 489, 512, 522, 561
quadratus plantae, 523
quadriceps femoris, 419, 441, 445, 450–451, 483
rectus femoris, 443
sartorius, 480, 484, 503–505, 510
semimembranosus, 485, 561
semitendinosus, 480, 503–505, 510
tibialis anterior, 471, 500, 520, 544, 568
tibialis posterior, 420, 471, 490–491, 500, 519–520
triceps surae, 500, 515
vastus intermedius, 443
Tendo/Tendines (of structure)
  of digiti, flexor, 399
  of manus, extensor, 346
    dorsal tendon compartments for, 345
Tennis elbow, 329

Tensile stress, in bone model, 415
Tension band, principle of, 415
TEP repair. See Totally extraperitoneal repair
Teratogen(s), stages sensitive to, 4
Terminal duct lobular unit (TDLU), 208
Terminal vascular bed, 66–67. See also Arteriola; Capillary bed(s); Capillary(ies); Venula
  afferent arterial limb of, 66
  characteristics of, 66
  efferent venous limb of, 66
  luminal radii of, 64
  wall thickness in, 64
Terminology, 26, 252
Testis/Testes, 24, 71, 216–217, 224, 226, 231, 543
  abdominal, 224
  blood supply of, 227
  coverings of, 217, 224
    vasa lymphatica, 227
  descensus, 224
  ectopic, 224
  examination of, 227
  feminization of, 223
  gubernaculum, 224
  hormone secretion by, 71
  hydrocele, 227
  inguinal, 224
  schematic representation of, 217
  structure of, 226
  surface anatomy of, 226
  tumor, 227
  vasa lymphatica, 227
Testosterone, 71
Thigh, 552
  arterial supply to, 550–551
  cross-sectional anatomy of, 522
  imaging of, axial MRI, 29
  lateral, pain in, 533
  musculi of, 474
    anterior, 474, 482–483, 503–505
    extensor group, 482–483
    flexor group, 484–485
    lateral view, 506
    medial view, 502
    origins and insertions, 504–505, 508–509
    posterior, 474, 484–485, 507–509
  nervi of, 530–531
  ossa of, 38, 408
  paresthesias in, 533
  sensory innervation of, 533, 535, 537
  soft-tissue envelope of, and joint motion, 50
  transverse (axial) plane of, 29
    anatomic structures in, 29
    imaging in, 29
Thomas maneuver, 436, 476
Thoracic diameter
  sagittal, breathing and, 136
  transverse, breathing and, 136
Thoracic inlet, 132
Thoracic outlet, 132
Thoracic outlet syndrome, 96
Thorax, 24, 132
  anatomical orientation in, rib counting for, 34
  imaging of, axial CT, 29
  organs of. See specific organ
  regional anatomy (regions) of, 36

transverse (axial) plane of, 29
Thrombophlebitis, *527*
Thumb. *See* Pollex
Thymus, 68
Thyroid-stimulating hormone (thyrotropin, TSH), 71
Thyrotropin. *See* Thyroid-stimulating hormone
Thyroxine (T4), 71
Tibia, 20, 30, 38, 41, 61, 408, 411, 418, 420, 440, 444, 447–455, 458, 460–461, 464, 466, 468, 470–472, 479, 481, 483, 486–491, 502, 510, 514–515, 522, 544, 563, 567–568
   anterior view, 420
   caput of, 38, 420
   corpus of, 420
      fractures of, *541*
   development of, 14
   facies articularis inferior, 421
   facies lateralis, 420
   facies medialis, 32, 420–421, 445
   facies posterior, 420–421, 491
   margo anterior, 420–421
   normal orientation of, 421
   ossification of, 19
   posterior view, 420
   rotational movements of, with knee flexed, 449
Tibialis anterior syndrome, *567*
Tibialis posterior reflex, lumbar root compression syndromes and, *131*
Tibial plateau, 38, 408, 420, 440, 446–447
   proximal view of, 421
Tight junction(s), in capillary endothelium, 67
Titin, 59
T-lymphocytes, 68–69
Toe(s), 544. *See also* Ossa digitorum, of pes
   cutaneous nerves of
      lateralis, 538–539
      medialis, 538–539
   deformities of, *469*
   hallux, 544, 546
   ossification of, 19
   variants of, 544
Tonic muscle, 499
Tonsilla
   lingualis, 68
   palatina, 68
   pharyngea, 68
Torque, 51
Total endoprosthesis (TEP), of hip joint, *46*
Totally extraperitoneal (TEP) repair, 219
Trabecula/Trabeculae
   bony, 16
   cancellous, 17, 39
   of pes, 464
   compression, 415
   tension, 415
Trachea, 105, 167
Tracts, pyramidal, 90
Tractus
   corticospinalis, 90
   iliopubicus, 210, 213

iliotibialis, 54–55, 410, 478–479, 503–507, 510, 512, 522, 536, 548, 551, 555, 560–561
   as tension band, 415
Transabdominal preperitoneal (TAPP) repair, 219
Transformation zone, in osteon development, 17
Transitional zone, in hyaline cartilage, 44
Triangle
   Hesselbach's, 211–213, 215, 217, 219
   Hueter, 278
   sacral, 31, 201
   von Hochstetter, 555
Triangular fibrocartilage complex, 287
Trigonum
   arteriae vertebralis, 203
   Bochdalek's, 174
   caroticum, 36
   clavipectorale, 36–37, 357, 380
   colli laterale, 362, 378–379
   femorale, 548
   femoris, 36–37
   lumbale, 36, 166
      fibrosum, 202
   lumbocostale, 161, 174
   musculare (omotracheale), 36
   Petiti, 202
   sternocostale, 174
   submandibulare, 36
   submentale, 36
   suboccipitale, 203
Triiodothyronine (T3), 71
Trochanter
   major, 32–33, 38, 408–409, 411, 414–417, 426–428, 430, 432–434, 477–479, 481, 483, 485, 507, 554–555
   minor, 38, 155, 408, 411, 414, 417, 426–428, 430–431, 433, 477, 479, 481, 483, 485
Trochlea
   femoral, 416, 419
      dysplasia of, 419
      morphology of, 419
   humeri, 244–245, 249, 276–277, 279
   tali, 422, 424, 457–458, 470, 493
Trophoblast, 5, 8
   syncytial, 8
Tropocollagens, 45
Tropomyosin, 55, 59
Troponin, 55, 59
Truncus
   brachiocephalicus, 191, 354–355, 361, 363, 382, 389
      variants of, *12*
   bronchomediastinalis
      dexter, 194
      sinister, 194
   coeliacus, 63
   costocervicalis, 191, 354–355, 379
   fibulotibialis, 563
   intestinales, 194
   jugularis, 194
      dexter, 194
      sinister, 194

lumbalis
   dexter, 194
   sinister, 194
mediastinalis, 194
subclavius, 194
   dexter, 194
   sinister, 194
sympathicus, 204
thyrocervicalis, 191, 354–355, 379, 382, 389, 391
Truncus sympatheticus, 94–95
Trunk (of body), 24
   arteriae of, 190–191, 207
   canalis inguinalis of, 210–211
   cutaneous innervation of peripheral sensory
      anterior, 198
      posterior, 200
      segmental
         anterior, 198
         posterior, 200
   flexion, 157
   movements of, 156–157
   musculi of, 164, 296
      anterior, 296
      posterior, 296
   nervi of, 196–197
      course of, 197
   neurovascular structures of
      anterior, 204–207
      posterior, 202–203
   palpable bony prominences
      anterior view, 32
      posterior view, 33
   posterior
      anterior view, 204–205
      posterior view, 202–203
   rotation, 157
   skeleton of, 100–101
   superficial vessels and nerves of
      anterior, 198–199
      posterior, 200–201
   surface anatomy of
      anterior, 198–199
      posterior, 200–201
   surface contours
      anterior view, 32
      posterior view, 33
   terms of location and direction for, 26
   venae of, 192–193
   vertical reference lines on, 34
Trunk(s) (lymphatic), 68–69
Trunk(s) (venous), 64
   cardinal, 13
TSH. *See* Thyroid-stimulating hormone
T-system, in skeletal muscle, 58
Tuba, uterina, 5, 182, 188, 232, 543
Tuber
   calcanei, 422–425, 454, 464, 488–493, 495, 515–519, 546
      processi lateralis, 492–493
      processi medialis, 492–493
   ischiadicum, 32–33, 38, 138–140, 142–143, 180–181, 184–185, 187, 189, 230, 234, 411–413, 426–429, 478–480, 484–485, 507, 522, 542, 549, 553–554, 556, 558–559
Tuberculum
   adductorium, 414, 505

anterius atlantis, 123
of arcus posterior atlantis posterius, 121, 123
conoideum, 242
costae, 107, 132, 135–137
dorsale (Lister's), 248, 250–251, 254–255, 280–282, 284, 287, 319, 340–341, 345
genitale, 222
humeri
   majus, 38, 165, 239, 244–247, 259–260, 262, 264–267, 271–273, 303, 307, 309, 311–313, 335, 362, 391
   minus, 38, 165, 239, 244–247, 259–260, 262, 264, 267, 272–273, 278, 303, 307, 309, 311, 391
iliaca, 35
infraglenoidale, 243, 262, 312–313
labioscrotale, 222
majus, 32–33
mamillaris, 72
minus, 32
m. scaleni anterioris, 135
ossis scaphoidei, 32, 239, 252, 255, 289
ossis trapezii, 32, 239, 252, 254–255, 285, 288–289, 292, 315
pubicum, 32, 138–141, 155, 186–187, 210, 411–412, 426, 428, 477, 548
supraglenoidale, 243, 262, 271, 311
talar
   laterale, 422–424, 470
   mediale, 422–424, 466, 468, 470
vertebral
   anterius, 106, 108–109, 120, 126, 151
   posterius, 106, 108–109, 120, 126
Tuberositas
   deltoidea, 305
      of humerus, 244
   glutea, 414, 417, 426, 478–479, 485
   iliaca, 139, 142–143, 412–413
   m. serrati anterioris, 135
   ossis cuboidei, 423, 490
   ossis metatarsi V, 32, 422–423, 425, 454, 456, 465, 470, 486–487, 490–493, 514–516, 519–520
   ossis navicularis, 32, 457, 490
   ossis sacri, 114–115, 142–143
   phalangis distalis, 252, 290–292
   radii, 248–249, 251, 276, 278–280, 282, 311, 315, 337
   tibiae, 31–32, 38, 54, 61, 411, 418, 420–421, 440, 443, 445, 448, 480–483, 487, 506, 510–511, 544
   ulnae, 248–249, 251, 276, 279–280, 282, 311, 315
Tubule(s)
   longitudinale, in skeletal muscle, 58
   transversale, in skeletal muscle, 58
Tubuli seminiferi contorti, 226

Tubus neuralis, 3, 7, 10, 72
  closure of, 6
  cross section through, 73
  differentiation of, in spinal cord region, during development, 73
Tunica
  albuginea, 226, 228, 230–231
  dartos, 217, 225–226
  externa, of blood vessel wall, 64–65
  intima, of blood vessel wall, 64–65
  media, of blood vessel wall, 64–65
    arterial vs venous, 65
  vaginalis
    lamina parietalis, 217, 225–226
    lamina visceralis, 217, 225–226
    testis, 226
Tunicata, 2

U

Ulna, 20, 38, 41, 238, 247–249, 252–254, 276–282, 284–289, 292, 294, 312, 317, 319, 323, 342–343, 345, 350–353, 366, 376, 393, 395
  articular surfaces of, 250–251
    distal, 251
    proximal, 251
  caput of, 238, 248, 250–252, 280–281, 289
    circumferentia articularis, 248
  carpal articulating surface, 256
  cartilage-free strip of, 251, 279
  corpus of, 33, 239, 248, 251, 280
  development of, 14
  facies anterior, 248, 251, 281
  facies medialis, 248, 251, 281
  facies posterior, 239, 248, 251, 281
  margo interosseus, 248, 250–251
  margo posterior, 248, 281, 319
  ossification of, 18
  proc. styloideus of, 31–33, 38, 239, 248, 250–255
  in pronation, 249
  rotation, during pronation and supination, 281
  in supination, 249
Ulnar tunnel, 289, 370–371, 398, 402
  apertures of, 405
  bony landmarks in, 405
  walls of, 405
Ulnar tunnel syndrome, 96, 370
Ulnocarpal complex, 287
  injury to, 287
Ultrasound. See Imaging
Umbilicus, 30–31, 176–178, 199, 206, 212, 220
Uncus corporis vertebrae. See Processus/Processi, uncinatus
Upper limb. See Membrum superius
Upper motor neuron lesion(s), 90
Ureter, 24, 188
  female, 543
  male, 224

Urethrae
  female, 188, 232, 543
  male, 230–231
    anomalies in, 223
    course of, 229
    pars membranacea, 183, 188, 229
    pars prostatica, 188–189, 229
    pars spongiosa, 188, 228–229
    development of, 222
Urinary bladder. See Vesica urinaria
Urinary system, 24
Uterus, 182, 188, 232, 557
  blastocyst implantation in, 5
  corpus, 8
  pregnant, 8

V

Vagina, 8, 180, 182, 188–189, 232, 543, 557
  ostium of, 180, 559
  vestibulum of, 182, 184, 188
Vagina carotica, 167
Vagina musculi recti abdominis, 155, 176, 334
  lamina anterior, 153, 155, 176–177, 179, 210, 216–217
  lamina posterior, 155, 175, 177–179, 206, 212, 216
  lateral wall, 221
  structure of, 178–179
Vagina synovialis
  intertubercularis, 263
  structure of, 61
Vagina tendinis
  in canalis carpi, 402
  carpal, palmar arches and, 402
  digital, 290
    ligamenta reinforcing, 290
    palmar arches and, 402
  of extensor carpi ulnaris, 287
  of fibularis longus, plantar, 492–493
  of flexor digitorum profundus, 402
  of flexor digitorum superficialis, 402
  of flexor pollicis longus, 402
  intertubercularis, 266, 268
  of manus, 344–345
    carpales, 344
      communication with digital, 344
      dorsales, 345
      palmares, 344
      communis tendinum musculorum flexorum, 344
    digital, 344
      communication with carpal, 344
  of pes, 514–515
  structure of, 61
Valgus, 27
Valvula, venosa, 64–65
Varices, of superficial leg veins, 527, 545
  primary, 527
  reticular, 527
  secondary, 527
Varicocele, 227
Varicose veins, 527, 545
Varus, 27
Vasa. See also Arteriae; Capillary(ies); Terminal vascular bed; Venae

in bone marrow, 44
  of vertebrae, 117
of caput femoris, 431
diaphyseal, 16
epiphyseal, 16
iliolumbalia, 204
intercostalia posteriora, rami cutanei laterales, 202
intrachondral, 15
intraforaminal, in spine, 129
and joint development, 15
lumbalia, 204
of mamma, 209
metaphyseal, 16
nutricia, of substantia compacta, 39
Vasa lymphatica, 62, 67–69
  afferent peripheral, 68
  collectors, 69
  of female genitalia, 234
  of funiculus spermaticus, 225
  initial, 69
  intercostalia, 194
  of male genitalia, 227
  of mamma, 209
  of membrum inferius, 528
    profundi, 528
    superficiales, 528
      anteromedial bundle of, 528
      posterolateral bundle of, 528
  of membrum superius, 358–359
    anastomoses of, 358
    deep, 358
    dorsal descending, 358
    dorsolateral arm territory, 358
    dorsomedial arm territory, 358
    middle arm territory, 358
    middle forearm territory, 358
    radial bundle territory, 358
    radial group, 358
    superficial (epifascial), 358
    ulnar group, 358
  precollectors, 69
  superficialia, 195
  of trunk wall, 194–195
    areas drained by, 195
    left and right venous angles, 194
Vasa nutricia, 104
Vasa vasorum, 64
Vascular system, 62
Vas deferens, 224
Vasoconstriction, 62
Vasodilation, 62
Vasopressin. See Antidiuretic hormone
Vena cava/Venae cavae
  inferior, 62–63, 167, 175, 192–194, 204, 227, 529
    development of, 13
    tributaries of, 192
    wall of, 64
    luminal radius of, 64
  superior, 62–63, 192–194, 378
    development of, 13
    tributaries of, 192
  wall thickness of, 64–65
Venae (in general, including vein types), 62
  blood volume in, 62
  deep, 63
  luminal radius of, 64
  medium-size, 64
  of membrum superius, deep, 357

perforantes, 62, 526
pulmonary, 62
reservoir function of, 62
rolling, 356
small, 64
structure of, 64
superficial, 63
of systemic circulation, overview of, 62
varicose, 527
wall of
  structure of, 65
  thickness of, 64–65
Venae (named)
  antebrachii, mediana, 356–357, 375, 392
    variable, 357
  arcuata cruris posterior, 526–527
  axillaris, 63, 192, 194, 209, 273, 357, 363, 378–379, 381, 384
  azygos, 63, 192–193, 205
    development of, 13
    tributaries of, 192
  basilica, 63, 209, 356–357, 359, 374–375, 377, 383, 392
    mediana, 356–357
  basivertebrales, 129
  brachialis, 63, 342, 357, 359, 381, 392
  brachiocephalica, 194, 378
    dextra, 192–193, 378
    sinistra, 63, 192, 378
      development of, 13
  bulbi vestibuli, 234
  cardinalis caudis
    anterior, 12–13
    communis, 13
    development of, 13
    posterior, 12–13
  cephalica, 63, 192, 198, 206, 356–357, 374–375, 377–381, 384, 392
    accessoria, 356–357, 377
    mediana, 356–357
  circumflexa, 231
  circumflexa femoris
    laterales, 526
    mediales, 526
  circumflexa humeri posterior, 273
  circumflexa ilium
    profunda, 192, 212
    superficialis, 192, 198, 206–207, 526–528, 545
  columnae vertebrales, 192
  cremasterica, 227, 524
  cubiti, mediana, 356–357, 374–375
    profunda, 356–357
  cutanea anterior femoris, 527–528, 545
  digitales
    dorsales, 357, 377
    palmares, 357
  dorsalis clitoridis profunda, 234
  dorsalis penis
    profunda, 228–231
    superficiales, 228, 230
  ductus deferentis, 212–213, 225, 227
  emissariae, 231
  epigastrica
    inferiora, 192–193, 206, 211–217

superficialis, 192, 198, 206, 526–528, 545
superior, 193, 206
facialis, 378
femoralis, 29, 63, 189, 192–193, 198, 210–215, 225, 228, 434, 522, 526–529, 545, 548–549, 551
in hiatus saphenus, 215
femoropoplitea, 526–527
fibulares, 526, 567
geniculares, 526
glutea
inferior, 553–555, 558
superior, 553–554, 560
hemiazygos, 192
accessoria, 192
development of, 13
hepatica, 63
development of, 13
iliaca
communis, 63, 529
development of, 13
externa, 63, 182, 192–193, 526–529, 551
interna, 63, 192, 230, 529
iliolumbalis, 192
intercapitulares, 357, 377
intercostales, 173, 206–208
anteriores, 192–193
posteriores, 192–193, 205
superiores, 192
suprema, 192
interosseae (membrum superius), 357
jugularis. See also Venous junction(s), jugulo-subclavian
externa, 63, 192, 198, 378–379
interna, 63, 68, 167, 192–194, 359, 378–379, 385
labiales posteriores, 234
lumbales, 192–193
ascendens, 192
mesenterica
inferior, 63
development of, 13
superior, 63
development of, 13
metacarpales palmares, 357
metatarsales
dorsales, 526
plantares, 526
musculophrenica, 193
obturatoria, 192, 212–213
occipitalis, 378
ovarica, 63
sinistra, development of, 13
palmaris superficialis, 402
paraumbilicales, 212
pericardiacophrenica, 205
perineales, 234
periumbilicales, 192, 198
plantaris
lateralis, 526, 564, 569
medialis, 526
poplitea, 63, 441, 451, 453, 526–527, 529, 545, 552, 560–562
portae hepatis, development of, 13
portal, 63
profunda femoris, 29, 526

pudenda
externae, 192, 198, 228, 230, 526–528, 545
interna, 189, 230, 234, 554–555, 557
distribution of, 558
radiales, 63, 357
rectalia inferiora, 230, 234
renalis sinistra, 63, 227
development of, 13
sacralis
laterales, 192
mediana, 192
sacrocardinalis, 13
saphena
accessoria, 526–527, 545
magna, 29, 63, 192, 198, 206, 215, 526–529, 545, 562, 567
varicosity, *527*
parva, 471, 526–529, 545, 547, 561–562, 567
varicosity, *527*
scrotalis
anterior, 228
posterior, 230
splenica, 63
development of, 13
subcardinalis, 13
subclavia, 63, 68, 192–194, 209, 357, 363, 378–379, 381, 383–386. See also Venous junction(s), jugulo-subclavian
subcostalis, 192–193
subscapularis, 273
supracardinalis, 13
suprascapular, 273
surales, 526
temporalis superficialis, 378
testicularis, 212–213, 225–227
development of, 13
sinistra, 227
thoracica
interna, 192–193, 206, 209
lateralis, 192, 194, 206, 209, 359
thoracodorsalis, 357
thoracoepigastrica, 192, 198, 206, 357
tibialis
anterior, 63, 522, 526, 566–567
posterior, 63, 471, 522, 526–527, 567
ulnares, 63, 357
umbilicalis, 9, 12–13
development of, 13
vitellina, 12–13
development of, 13
Venae (of region or organ)
of cubitus/art. cubiti, 356
intercostal, 13
intracranial, 63
intraforaminal, in spine, 129
of manus, dorsal, superficial, 357
of membrum inferius, 526–527
cutaneous, superficial, 545
overview of, 526
perforantes, 526–527
Boyd group, 526–527
Cockett group, 526–527
Dodd group, 526–527
profunda, 526
superficialis, 526, 545
varicosities, *527*, 545

of membrum superius, 356–357
cutaneous, 356
deep, 357
perforantes, 357, 375
subcutaneous, 356, 377
superficial, 357
superficial cutaneous, 375, 377
of neck, 378
omphalomesentericae, 13
ovarica sinistra, development of, 13
profundae clitoridis, 234
profundae penis, 230
of shoulder/art. humeri (gleno-humeralis), 378
testicularis, development of, 13
of testis, 227
of trunk wall, 192–193
superficial cutaneous anterior, 198–199
posterior, 200–201
Venous junction(s), jugulo-subclavian, 68
Venous system(s)
azygos
development of, 13
tributaries of, 192
cardinal, development of, 13
Venous thrombosis, *527*
Venous trunk. See Trunk(s) (venous)
Ventricle(s), cardiac, 62
Venula, 62, 64
blood flow in, 67
luminal radius of, 64
postcapillary, 66–67
wall of, thickness of, 64
Vermiform appendix. See Appendix
Vertebral column. See Columna vertebralis
Vertebra prominens (C VII), 31, 33, 36, 101–102, 105–106, 108–109, 120–121, 149, 159, 165, 171, 201, 240, 299, 301, 334, 361–362, 365, 367
body of, 126–127
spinous process of, 35
Vertebrata, 2
basic anatomy of, 3
characteristics of, 2–3
embryology of, 2
primitive, model of, 2
terrestrial, 2
embryology of, 10
limbs, basic skeletal structure of, 20
Vertebra/Vertebrae
assimilation disorders, *103*
cervicales, 102, 108–109, 126, 151, 167, 301, 361, 367
C I (first). See Atlas (C I)
C II (second). See Axis (C II)
C IV (fourth), 108–109
C VII (seventh). See Vertebra prominens (C VII)
morphology of, 108–109
structure of, 106–109
coccygeal, 114
corpus, 41, 106–113, 116–119, 121–122, 124, 126–129, 133, 135–137, 174
cancellous bone of, 117
embryology of, 7, 104

epiphysis anularis, 111, 113, 116–117
fractures of, *130*
hyaline cartilage endplates of, 116–117
intervertebralis, 111, 113, 117
embryology of, 104
fractures of, *104*
lumbales, 102, 105–106, 128–129, 140, 147, 155, 160–161, 167, 175, 187–188, 240, 428, 476–477, 502, 506–507, 530, 534–535, 554, 558
fourth, processi spinosus, as landmark, 35
sacralization of, 103
structure of, 106–107
sacrales, 102, 114, 530
lumbarization of, 103
pars lateralis, 106, 114, 138–139
second, processi spinosus, as landmark, 35
stabilization, 156
structure of, 106–107
thoracicae, 102, 110–111, 124, 128, 137, 151, 158, 164–165, 175, 240, 301, 335, 361, 365, 476, 530
seventh, processi spinosus, as landmark, 35
structure of, 106–107
third, processi spinosus, as landmark, 35
twelfth, processi spinosus, as landmark, 35
transitional, 103
Vertical infraclavicular block, *384*, 385
Vesica urinaria, 24, 105, 224, 229, 232, 434
in female, 188–189, 543
innervation of, 94
in male, 183, 188, 212, 543, 557
Vesicle(s)
brain, 72
derivatives of, 72
cytoplasmic, in capillaries, 67
secretory, 70
synaptic, 59
Vesicula
optica, 2, 10, 14
umbilicalis, 6, 8, 13
Vestibulum vaginae, 182, 184, 188, 232, 235
development of, 222
Villus/Villi
chorionic, 8–9
mainstem, 8
placental, 9
synoviales, 42
Vincula brevia, 61, 347, 399
Vinculum longum, 61, 347, 399
Viscera, 24
innervation of, 89, 94–95
Visceral descent, *163*
Vomiting, 156
von Hochstetter triangle, location of, 555
von Luschka, Hubert, 126
Vulva, 182, 232, 557

W

Waller degeneration, 97
Wartenberg syndrome, *96*
Watershed, 195
Wing, iliac. *See* Ossa, ilium, ala (wing) of
Wrist
  cross section through, 402
  distal
    functional muscle testing of, 329
    movements of, 328
    musculi of, weakness/shortening, symptoms of, 329
  dorsal extension, 328–329
  movements of, 328–329
  musculi of
    dorsal extensors, 328–329
    innervation of, 328
    nerve segments associated with, 328
    overview of, 328–329
    palmar flexors, 328–329
    radial abductors, 328–329
    shortening of, symptoms of, 329
    ulnar abductors, 328–329
    weakness of, symptoms of, 329
  palmar flexion, 328–329
  proximal
    functional muscle testing of, 329
    movements of, 328
    musculi of, weakness/shortening, symptoms of, 329
  radial abduction, 328–329
  range of motion, 328
  surface anatomy of, 404
  ulnar abduction, 328–329
Wrist crease
  distal, 374
  proximal, 374
Wrist drop, *368*

X

X chromosome, 5
Xiphoid process. *See* Sternum, processi xiphoideus

Y

Y chromosome, 5

Z

Z-disk, 59
Zona pellucida, 5
Zone of calcification, epiphyseal, 16
Zone of maturation and hypertrophy, epiphyseal, 16
Zone of ossification, epiphyseal, 16
Zone of proliferation, epiphyseal, 16
Zone of vesicular cartilage, epiphyseal, 16
Zygopodium, 20–21
Zygote, formation of, 5